CONTROLS ON CARBONATE PLATFORM AND REEF DEVELOPMENT

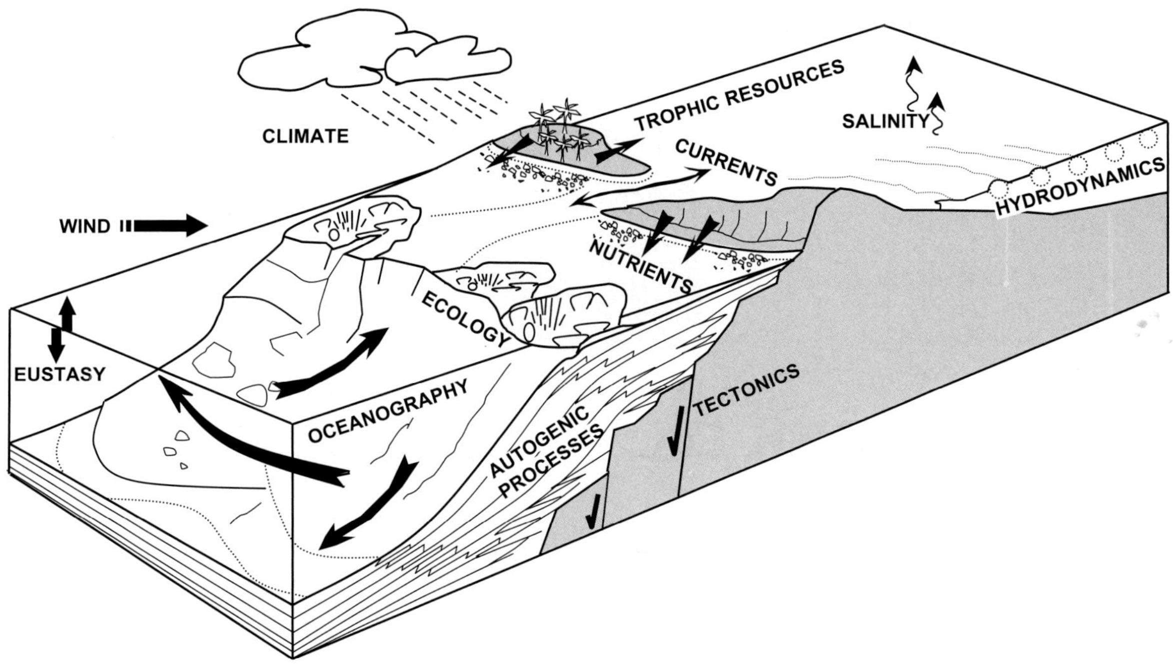

Edited by:

JEFF LUKASIK
StatoilHydro, Oslo N-0246, Norway
e-mail: jeluk@statoilhydro.com

AND

J.A. (TONI) SIMO
ExxonMobil Upstream Research Company, GW3 908B,
P.O. Box 2189, Houston, Texas, 77252, U.S.A.

Donald F. McNeill and Gary J. Nichols, SEPM Special Publications Editors
SEPM Special Publication 89

Tulsa, Oklahoma, U.S.A. *April, 2008*

SEPM and the authors are grateful to the following
for their generous contribution to the cost of publishing

Controls on Carbonate Platform and Reef Development

ConocoPhillips ExxonMobil

Chevron

MAERSK OIL StatoilHydro

Contributions were applied to the cost of production, which reduced the purchase price, making the volume available to a wide audience

---

SEPM (Society for Sedimentary Geology) is an international not-for-profit Society based in Tulsa, Oklahoma. Through its network of international members, the Society is dedicated to the dissemination of scientific information on sedimentology, stratigraphy, paleontology, environmental sciences, marine geology, hydrogeology, and many additional related specialties.

The Society supports members in their professional objectives by publication of two major scientific journals, the Journal of Sedimentary Research (JSR) and PALAIOS, in addition to producing technical conferences, short courses, and Special Publications. Through SEPM's Continuing Education, Publications, Meetings, and other programs, members can both gain and exchange information pertinent to their geologic specialties.

For more information about SEPM, please visit **www.sepm.org.**

ISBN 978-1-56576-130-8

SEPM (Society for Sedimentary Geology)
6128 E. 38th Street, Suite 308
Tulsa, Oklahoma 74135-5814, U.S.A.

Printed in the United States of America

Controls on Carbonate Platform and Reef Development

JEFF LUKASIK AND J.A. (TONI) SIMO, Editors

CONTENTS

Improved Characterization of Carbonate Platforms and Reefs

CONTROLS ON CARBONATE PLATFORM AND REEF DEVELOPMENT

ACKNOWLEDGMENTS

This volume represents ideas and concepts initially presented in the SEPM Special Session "Controls on Carbonate Platform and Reef Development: A Tribute in Honor of Dr. Wolfgang Schlager" at the AAPG-SEPM Conference in Calgary, May 2005.

We sincerely thank all contributing authors for their eagerness to participate and their patience with the review and editing process. Your efforts and ideas do make a difference. We would also like to express our gratitude to each of the reviewers for all of their time and effort spent in assessing the manuscripts, helping to ensure the authors' observations and interpretations were clearly represented. We could not have done this without your help.

Steve Bachtel
Eric Blanc
Dan Bosence
Brian Coffey
Mario Coniglio
Peter Droz
Bob Erlich
Evan Franseen
Kate Giles
Bob Goldstein
Mark Harris
Mitch Harris
Heiko Hillgartner
Tina Hughes
Dan Lehrmann
Bill Martindale
Maria Mutti
Bill Morgan
Jeff Pietras
Peir Pufahl
Amy Ruf
Art Saller (twice)
Rick Sarg
Beverly Saylor
Wolfgang Schlager
Ian Sharp
Taury Smith
Kelly Steffen
Finn Surlyk
Conxita Taberner
Elizabeth Turner
Georg Warrlich
Greg Wahlman
Alun Williams
Brian Witzke

Thanks also to our industry sponsors StatoilHydro, Chevron, ExxonMobil, Maersk, and ConocoPhillips for their generous donations and encouragement of their technical staff to present their research in a public forum.

Many thanks to the SEPM, especially Howard Harper and Theresa Scott at the SEPM office and Special Publications editor Don McNeill, for their enthusiastic responses to our book proposal. Last but not least, our sincere gratitude goes to John Southard and Bob Clarke, whose keen editing and late night work sessions ensured a publication of the highest quality.

Controls on Carbonate Platform and Reef Development
SEPM Special Publication No. 89, Copyright © 2008
SEPM (Society for Sedimentary Geology), ISBN 978-1-56576-130-8, p. 1–1.

Introduction and Synthesis

CONTROLS ON DEVELOPMENT OF PHANEROZOIC CARBONATE PLATFORMS AND REEFS—INTRODUCTION AND SYNTHESIS

JEFF LUKASIK
StatoilHydro, Oslo NO-0246, Norway
e-mail: jeluk@statoilhydro.com
AND
J.A. (TONI) SIMO
ExxonMobil Upstream Research Company, GW3 908B, P.O. Box 2189, Houston, Texas, 77252, U.S.A.
e-mail: toni.t.simo@exxonmobil.com

DEPOSITIONAL CONTROLS AND SEDIMENTARY RESPONSES

Carbonate platforms and reefs emerge, grow, and die in response to intrinsic and extrinsic mechanisms forced primarily by tectonics, oceanography, climate, ecology, and eustasy. These mechanisms, or controls, create the physical, biological, and chemical signals accountable for the myriad of carbonate depositional responses that, together, form the complex depositional systems present in modern and ancient settings. If we are to fully comprehend these systems, it is critical to ascertain which controls ultimately govern the "life cycle" of carbonate platforms and reefs and understand how these signals are recorded and preserved. Deciphering which signals produce a dominant sedimentological response from the plethora of physical and biological information generated from superimposed regional to global-scale controls is critical to achieving this goal. With this understanding, it may be possible to extract common time- and space-independent depositional responses to specific mechanisms that may, ultimately, be used in a predictive sense.

Extensive research on a wide variety of carbonate platform and reefal systems in the past few decades has provided the foundation and understanding necessary to take carbonate research to a new level. With assistance from rapidly advancing computer software and an increasing use of cross-disciplinary integration, carbonate research is shifting from description and morphological analysis towards a science that is more focused on the assessment of process and genetic relationships. The aim of this special publication is to present a cross section of recent research that shows this evolution from a variety of perspectives and scales using examples distributed throughout the Phanerozoic. As such, this article pauses to reflect on our present knowledge of the controls on development of carbonate platforms and reefs and where it seems to be heading in the future.

DEVELOPMENT OF CARBONATE PLATFORMS AND REEFS—FROM THE MORPHOLOGICAL TO THE GENETIC

Many strides have been made in the past three decades to study, classify, and understand the complexity of carbonate platform systems. First came the differentiation of several carbonate platform types (Ahr, 1973, Ginsburg and James, 1974), followed by Wilson (1974, 1975), who provided a comprehensive overview of development of carbonate platforms and carbonate platform margins through time as a response to the interactions between subsidence, hydrodynamics (currents, wave energy), and climate (aridity, sediment supply, salinity). The classification of carbonate platforms into ramps (homoclinal and distally steepened), rimmed shelves, isolated platforms, and drowned platforms (Read, 1982, 1985) provided a common terminology and morphological framework into which most platform types could be placed. Growing awareness of eustasy and the application of its principles (e.g., Vail et al., 1977) turned these static models into dynamic examples of how various platforms and reefs can respond to relative sea-level change (Kendall and Schlager, 1981) and how their stratigraphic components fit into a sequence stratigraphic framework (e.g., Loucks and Sarg, 1993).

Although subsequent classification attempts have focused on refining these models as influenced by environmental factors (e.g., Burchette and Wright, 1992, Lukasik et al., 2000), distinguishing carbonate platforms based primarily on morphological criteria remains the most common way of describing the depositional disposition of the carbonate system regardless of setting, scale, or age. Morphological classification of platforms is limited, however, in its static approach to defining physiological and facies relationships as temporal "snapshots" and by indirectly addressing questions regarding the mechanisms responsible for their initiation, growth, and demise. As such, the classification of platforms can become equivocal due to the lack of genetic processes tied to their definition, development, and character through time.

The bulk of research on carbonate platforms and reefs over the past two decades has focused on understanding the large-scale controls responsible for the morphological variability among different platform and platform-margin types, and on detailing the temporal genetic factors associated with the morphological aspects of these systems (see Scholle et al., 1983, Crevello et al., 1989, Tucker et al., 1990, Simo et al., 1993, Wright and Burchette, 1998, Homewood and Eberli, 2000, Insalaco et al., 2000, and Pedley and Carannante, 2006, for specific examples). Our need and desire to distinguish and comprehend genetic processes has pushed our knowledge of carbonate systems ahead in a series of significant leaps. Together, each of these advances has helped to increase our understanding of carbonate platform and reef systems, allowing us to ask more accurate and specific questions on the controls responsible for their initiation, growth, and demise. Some of the most noteworthy were:

1. Linking the influence of global-scale climate-related changes in ocean chemistry to fluctuations in the dominant mineralogical and taxonomic composition of biota through time,

Controls on Carbonate Platform and Reef Development
SEPM Special Publication No. 89, Copyright © 2008
SEPM (Society for Sedimentary Geology), ISBN 978-1-56576-130-8, p. 5–12.

resulting in measurable differences of carbonate-platform composition between icehouse and greenhouse periods (e.g., Wilkinson, 1979, Sandberg, 1983, Read, 1995, Kiessling et al., 2003).

2. Recognizing and predicting internal facies and stacking patterns within every type of carbonate platform system through the application of sequence stratigraphic concepts (e.g., Sarg, 1988, Goldhammer et al., 1990, Handford and Loucks, 1993).

3. Recognizing the interaction between seawater temperatures and trophic resources on the development of heterozoan carbonate sediments (e.g., Hallock, 1987, James, 1997), thus allowing tropical and nontropical carbonate systems to be unchained from their presumed latitudinal constraints.

4. Increasing the focus on the role of tectonics in localizing carbonate platform domains, thus creating a potentially new spectrum of platform geometries with distinctive internal complexity (e.g., Bosence, 2005).

5. Invention of new and improved technology in subsurface seismic fidelity and earth surface satellite imaging, making it possible to detail ancient and modern carbonate systems at a higher resolution than ever before and to establish three-dimensional relationships between sedimentary bodies and genetic oceanographic processes (Eberli et al., 2004).

CONTROLS ON CARBONATE SYSTEMS

The complexity of carbonate platform and reef systems reflects the multifarious nature of their inherent physical, biological, and geochemical foundations. At the simplest level, carbonate-secreting organisms grow in almost all conditions, from the tropics to the poles, clastics-rich to clastics-poor systems, high- and low-energy areas, etc., but they form platforms and reefs only when the sum of the feedback between certain depositional processes and their controls allow the initiation and growth of carbonate nuclei. Under conditions conducive to carbonate production, the size, shape, duration, and internal architecture of these nucleated platforms and reefs are ultimately driven by sediment accommodation and accumulation rates as defined here:

Carbonate Platform and Reef Development =

Conditions Conducive to Carbonate Production	+	[Accommodation	+	Accumulation (Production	x	Distribution)]
- decreased fine-grained clastic input		- eustasy - tectonics - ecology		- ecology - temperature - trophic resources - nutrients - turbidity		- oceanography - hydrodynamics - salinity

Although simply stated, this formula represents a complex and interrelated network of factors responsible for the development of carbonate platforms and reefs. Multiple controls (tectonics, eustasy, climate, oceanography, trophic resources, nutrients, and water temperature among many) directly and indirectly affect, via feedback, all stages of evolution of carbonate platforms and reefs (Fig. 1). Their relevancies differ, however, depending on the geographic location (e.g., most reefs today are on the western sides of oceans) and temporal character (i.e., short-lived, early and late stages of initiation and demise vs. relatively long-lived growth stages) of the carbonate system. As such, each of them may play major or minor roles at different stages of platform and reef development (Table 1) and occur at all scales (Table 2). Changes in accommodation created via tectonics (including subsidence) and relative sea level (which may or may not be created solely through eustatic rise) can, for example, have a direct impact on oceanographic, climatic, and hydrodynamic factors, which, in turn, may directly influence the accumulation rate, type, and distribution of sedimentary facies generated across a platform or reef complex.

These convoluted interactions form a nested hierarchy of scale (incorporating observation + interpretation) in the overall evolution of platforms and reefs, regardless of their age (Table 2). The existence of carbonate-producing conditions, rates of accommodation change, and magnitude of accumulation rates can be strongly influenced on a global scale by eustasy, plate tectonic movements, secular variations of mineralogy and biota, and associated ecological trends characterizing greenhouse and icehouse periods. On a regional (hemisphere to continental) to basin (single depocenter) scale, the mechanisms which control the conditions conducive to carbonate production may also include climatic shifts and a reorganization of large-scale oceanographic patterns created by hinterland uplift and the tectonically derived opening or closing of oceanic gateways.

While eustasy can be distinguished at every scale thanks to the stratigraphic responses created by different orders of eustatic change, climate and oceanography can have a profound effect on the ecology and type of carbonate factory developed and its stratigraphic expression on the regional to platform scale through their controls on turbidity (input of clastic sediment), temperature, and salinity of the water column. Climate can also modify the development and stratal character of platforms and reefs at these scales in a more subtle fashion through the alteration of carbonate sediment production rates via changes in climate-induced trophic resource levels (e.g., Mallinson et al., 2003, Lukasik and James, 2006). On a more detailed platform level, environmental controls, including water temperature, salinity, and trophic resources, become major influences on the generation and character of smaller-scale depositional and sedimentological features within the succession during the growth phase of individual platforms and reef complexes.

This hierarchy of controls and their geographic scales of response are unidirectional, in that tectonics and eustasy impart a large influence on the development of climatic, oceanographic, and other environmental factors affecting the depositional system but not vice versa. From a temporal standpoint, tectonics, eustasy, and climate are controls mostly relevant at long-term scales reflecting their generation from global mechanisms, while environmental changes are significant at shorter-term scales caused by regional to local perturbations in the oceanic system.

To reflect these hierarchical relationships, the seventeen papers of this volume are arranged into four distinct sections that follow this hierarchical scale. These include:

a) Global Perspectives on Development of Carbonate Platforms and Reefs;

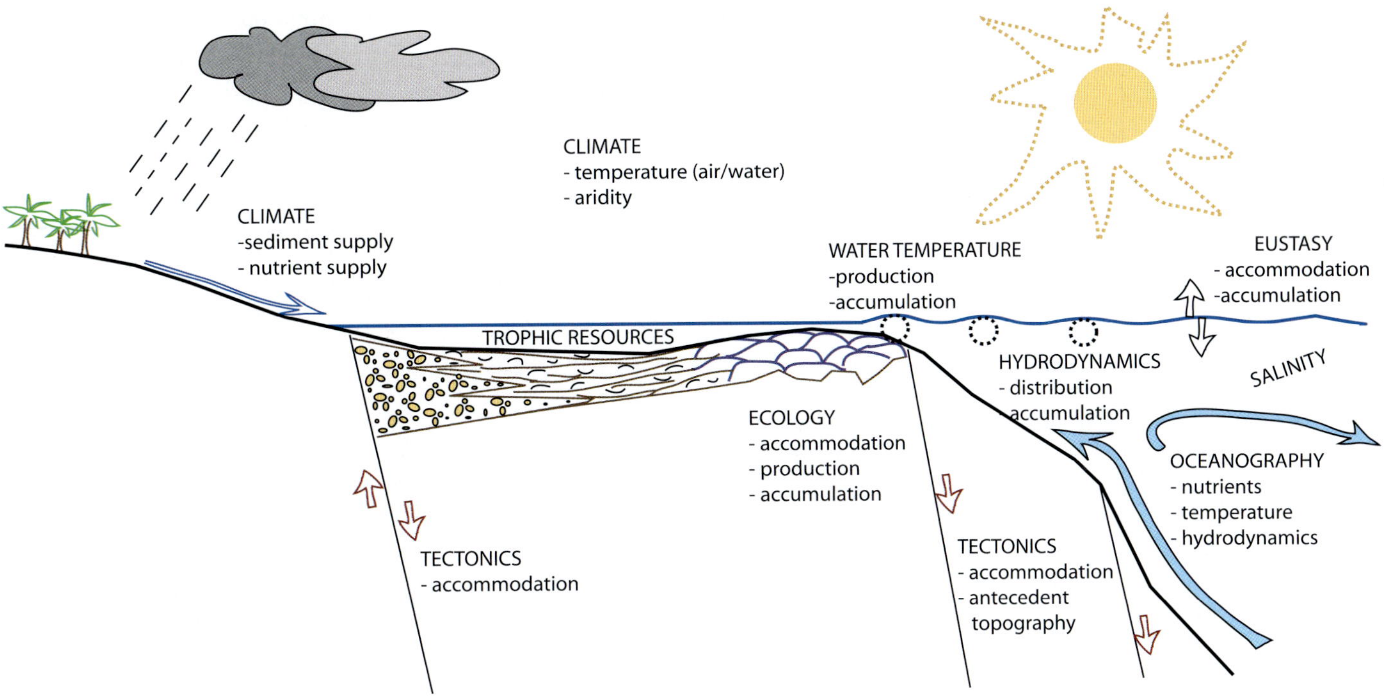

FIG. 1.—Controls on development of carbonate platforms and reefs.

b) Tectonic Controls;

c) Environmental Controls (Oceanography, Climate, and Trophic Resources); and

d) Improved Characterization of Carbonate Platforms and Reefs.

GLOBAL PERSPECTIVES ON DEVELOPMENT OF CARBONATE PLATFORMS AND REEFS

Many processes and factors have contributed to the generation of carbonate platforms and reefs through time. Although it is important to delve into the details of carbonate systems to resolve the magnitude of these various influences, stepping back from them allows carbonates to be viewed at the holistic scale, where influences of plate tectonics and eustasy set up climatic and oceanographic patterns which impart a direct control on platforms and reefs from deposition to diagenetic modification. Evaluating these systems at the global level forces investigation on a scale that transcends the particulars and details typically associated with local variations. This approach can provide a fresh and exciting perspective on carbonate platform and reef systems, resulting in thought-provoking ideas and concepts.

The work of **Markello, Koepnick, Waite, and Collins** takes such a broad-scale approach. Stemming from their research at the Mobil Exploration and Production Center in the 1990s, the authors summarize previous work on the carbonate processes (biotic evolution + mineralogic variation) and earth processes (tectonics, climate, eustasy, ocean chemistry, ocean circulation) from Phanerozoic carbonate platforms, including information from over 8300 carbonate-hosted hydrocarbon reservoirs (database included here) and place them into 29 time intervals to create predictive, age-specific models for carbonate systems. These carbonate analogues through time (CATT) can be used to predict occurrence, composition, and stratal attributes of Phanerozoic carbonate platforms and their reservoirs around the world. In creating these maps, they have performed a systematic, comparative analysis of all known carbonate and earth processes responsible for carbonate sedimentation through time. Results of this study are presented here as a series of predictive paleogeographic maps and Phanerozoic field posters. These, together with an updated Phanerozoic carbonate trends chart, will serve as valuable reference material for those working on carbonate-platform patterns and trends throughout the Phanerozoic.

Not all of the interpreted global controls on carbonate-platform development are tectonically or eustatically derived.

TABLE 1.—Relative controls on the life cycle of carbonate platforms and reefs.

	Eustasy	Tectonics	Climate	Oceanography **(currents, upwelling)**	Trophic Resources	Temperature
Initiation	●	●	○	○	○	○
Growth	●	●	●	●	●	●
Demise	●	●	●	●	●	○

● Major ○ Minor

TABLE 2.—Controls on development of carbonate platforms and reefs, and their relative geographic scale of dominance.

	Eustasy	Tectonics	Climate	Oceanography (currents, upwelling)	Trophic Resources	Temperature
Global Scale	●	●	○	○	○	○
Regional Scale	●	●	●	●	○	○
Basin Scale	●	●	●	●	●	●
Platform Scale	●	●	●	●	●	●

● Major ○ Minor

The contribution of **Wright and Cherns** provides a thought-provoking discussion on syndepositional to shallow-burial aragonite dissolution, and its implications for the preservation potential and, thus, depositional interpretation of platformal carbonate systems. Early aragonite dissolution is a relatively new concept that carries potentially large implications for our understanding of carbonate depositional processes (see Walter and Burton, 1990, Cherns and Wright, 2000, and James et al., 2005 for more information) because it can selectively alter the primary sedimentary record leading to the preservation of an unrepresentative skeletal and mineralogical record. Because the sediment record is the only one we have to distinguish depositional conditions and develop interpretations on facies patterns and depositional geometries, this early diagenetic process may directly impact our paleo-environmental assessment of nutrients and trophic resources on the carbonate factory, faunal trophic levels, carbonate production rates (and, subsequently, forward modeling of carbonate systems), sedimentation rates, diagenetic potentials, taphonomic fidelity, and diagenetic bedding.

TECTONIC CONTROLS

Deciphering the relative dominance of tectonic vs. nontectonic controls preserved in the rock record is not always straightforward. Although observational bias and depth of investigation can play critical roles in distinguishing the relative impact they have had on the development of carbonate platforms and reefs, another possible explanation lies in the scale of their expression. Together, tectonics and tectonic setting impart a dominant signal that determines the three-dimensional morphology, large-scale stratigraphic architecture, and depositional sequences of carbonate platforms, while environmental mechanisms including climate, oceanography, and hydrodynamics mostly determine biotic and depositional responses comprising the type and distribution of grain types (facies) across the platform. Environmental mechanisms are often considered to overpower tectonic signals and are commonly invoked to explain the character of small-scale stratigraphic packages, especially in passive-margin settings (e.g., Wilson, 1975, Read, 1985). Although this relationship between scale and mechanism seems reasonable, it is important to keep an open mind when interpreting appropriate mechanisms for stratigraphic responses, even at the smallest scale of cyclicity (e.g., De Benedictus et al., 2007).

If tectonics imparts a primary control on gross stratigraphic architecture and platform geometry during platform initiation and growth, how and when do these responses become blurred and obscured by those generated from environmental and eustatic factors? Because environmental signatures are recorded mostly in the sedimentological details and small-scale cyclical stratal patterns superimposed on larger-scale tectonic expressions, recognizing this change of signal dominance requires an examination of the platform and reef system at all scales. As such, it is important to recognize that our interpretations of signal dominance may be significantly influenced by the scale at which we examine the strata in hand.

Focusing on tectonic controls as the dominant force in development of carbonate platforms, **Dorobek** provides a comprehensive review of syn-rift carbonate platforms, including their stratigraphic packaging and facies distributions generated in response to syn-rift deformation. Stratal patterns are determined largely by fault geometry, spacing, length, and displacement history and moderated by hydrodynamic energy, oceanographic, and climatic factors. Recurring patterns of internal stratigraphic complexity emphasize the need for carbonate facies analysis to be done in conjunction with a detailed analysis of local fault systems and regional tectonic history to fully understand the genetic history of platform development within these settings.

Also focusing on syn-rift to post-rift carbonate platform development, **Cross and Bosence** illustrate how rift-basin, tropical carbonate platform geometries are formed in response to fault-block rotation and demonstrate how recurring tectonosedimentary signatures related to the timing of fault movement can allow the differentiation of syn-rift and post-rift platforms in the rock record. Stratigraphic character within syn-rift carbonate platforms is formed primarily by tectonic mechanisms, while stratal patterns within post-rift platforms reflect a passive exploitation of preexisting rift topography and accommodation space. Transfer zones generate complex stratigraphic and platform geometries related to episodic fault activity.

Giles, Druke, Mercer, and Hunnicutt-Malk examine isolated carbonate platforms of Maastrichtian age developed on seafloor highs generated by rising salt diapirs. Initiation and growth of these platforms were influenced by a variety of short-term, near-diapir conditions, and longer-term regional factors including tectonism, sea-level change, and sediment supply. Through the complex interaction of these controlling factors affecting sedimentation, a recognizable and predictable stratigraphic architecture emerges. What makes these platforms even more distinctive is their predominantly heterozoan carbonate composition, despite having accumulated in warm subtropical to tropical waters. This is possibly one example where the role of tectonics has a direct influence on sediment composition through oceanic nutrification related to methane seeps surrounding the rising diapirs.

The influence of basement structuring on carbonate-platform development and hydrocarbon distribution and production is highlighted in the work of **Watney, Franseen, Byrnes, and Nissen**. Using well data, 3-D seismic, and rock properties measured from Carboniferous hydrocarbon fields spanning the US midcontinent, they found that syn-depositional and post-depositional structuring exerted an episodic but continual influence on accommodation, depositional pattern, paleotopography, diagenesis, and hydrocarbon emplacement. The reactivation of Precambrian structures resulted in a complex distribution of mostly structural–stratigraphic reservoirs and traps formed from the segmentation of the broad depositional platform into a mosaic of kilometer-scale blocks, each containing internally variable stratigraphic growth patterns modified only by local environmental conditions.

Batt, Pope, Isaacson, Montañez, and Abplanalp integrate basin chronostratigraphy with analysis of meter-scale stacking patterns to distinguish regional tectonic from global eustatic controls on deposition from a late Mississippian carbonate ramp in the Antler Foreland Basin, western USA. Tectonics exerted a major influence on platform generation, resulting in a tripartite subdivision of the broad shelf into distinct regions. Each region generated its own distinctive facies associations and stratigraphic packaging as determined by variable subsidence rates and bathymetry caused by differential tectonic loading and activation–reactivation of faults. Strong eustatic signatures, recorded as small-scale depositional sequences, dominated only during periods of tectonic stabilization.

Through correlation of shelf and foreslope to basinal strata within the Oligocene Costa Blanca succession, **Stoklosa and Simo** determine that faulting and increased subsidence associated with a larger-scale extensional setting governed overall platform morphology by allowing an accumulation of a thick stratigraphic succession which overprinted the effects of sea-level fall. Changes in the carbonate factory from photozoan-dominated (coralgal) to heterozoan-dominated (rhodalgal) assemblages were also tectonically controlled through the opening of the Valencia trough, which brought deeper, cooler, and possibly more nutrient-rich waters onto the shelf.

ENVIRONMENTAL CONTROLS (OCEANOGRAPHY, CLIMATE, TROPHIC RESOURCES, TEMPERATURE)

Biota and carbonate cement form the fundamental building blocks of carbonate platforms and reefs and are the expression of an array of environmental conditions affecting deposition, including, but not limited to, water temperature, salinity, nutrification, oxygenation, hydrodynamic energy, and substrate rigidity. It is how these sediments form stratigraphic units and how these units stack together that gives carbonate platforms and reefs their character and complexity. This character may often develop independently of the tectonic regime, especially during periods of tectonic quiescence where biota and carbonate sediments preserve such a strong environmental signature that their deposition, stratal stacking patterns, and, ultimately, large-scale platform geometry seem to be dictated primarily by environmental mechanisms. It is during these times where we can start to resolve relative environmental controls on carbonate sedimentation at many scales of detail.

Distinguishing environmental controls on carbonate sedimentation is critical in any assessment of carbonate platforms. Knowing which factors can drive sedimentation at the smallest scale will allow a thorough discrimination of a wide variety of signals that are ultimately preserved in the stratigraphic record. Building on the concepts established in Pomar (2001a, 2001b), **Pomar and Kendall** advocate the study of carbonate platforms through analytical, iterative, and successive backstripping of sediment accumulation from youngest to oldest. These patterns are then reassembled to determine the genetic character and individual geometries of the sequence stratigraphic components (sequences, parasequences, etc.) as products of changing accommodation. Accommodation, they contend, is strongly controlled by ecological factors (ecological accommodation) as determined by a variety of environmental controls working in combination with physical processes. In their interpretation, oceanographic change can affect the ecological development of biota so much that it can have a larger impact on stratal patterns and facies architecture than sea-level change alone. As such, it is possible that the fluctuation in ecological and physical accommodation through time accounts for the generation of highly diverse and complex stacking patterns ultimately responsible for the development of different carbonate-platform types.

Accommodation plays a very important role in the generation and preservation of carbonate stacking patterns from the kilometer to decimeter scale. **Peterhänsel and Egenoff** focus on accommodation to explain the complex facies architecture and lateral variations of high-frequency stacking patterns found in shallow-water facies from the Triassic Latemàr platform in the Italian Dolomites. They outline an interesting conundrum where, opposite to the notion that platformal regions with relatively higher levels of accommodation should record a greater number of cycles, topographically elevated regions with lesser amounts of accommodation seem to record a high frequency of low-amplitude sea-level oscillations that go unnoticed in the "deeper-water" shallow lagoon. This research describes a delicate balance between accommodation and sediment accumulation at very high orders of sea-level fluctuation identified at a very detailed scale of observation.

Overprinting global eustatic and climatic signals with regional to local environmental controls can have a significant impact on the sediment composition, generation of discrete surfaces, and stratigraphic interpretation within platformal successions, as described by **James and Bone** in their comparison of two (Eocene and Oligocene) mixed heterozoan carbonate and siliceous complexes in southern Australia. Although the complexes formed in similar extensional basin settings under estuarine conditions, they have pronounced sedimentological differences. These differences are interpreted to be not of eustatic origin as previously thought but rather the result of changing local depositional controls, especially terrestrial climate, hydrodynamic energy, and trophic resource levels possibly related to the ongoing northward shift of the southern Australian margin.

Similar controls on carbonate platform development are found in **Carannante, Cherchi, Graziano, Ruberti, and Simone**'s study of post-Turonian, peri-Tethyan rudist-bearing limestone. They conclude that the sedimentological and morphological differences between central and northern Tethyan carbonate platforms (rudist-rich foramol vs. rhodalgal respectively) were generated by a clear latitudinal partitioning of local paleoceanographic and paleoclimatic conditions and not as a direct result of the larger-scale pattern of post-Turonian climate change.

Nutrient availability and trophic resource levels within the water column are important and often underestimated factors in the development of carbonate platforms and reefs (e.g., Hallock and Schlager, 1986). **MacNeil and Jones** provide a late Devonian example from the Alexandra Reef Complex in northern Canada that outlines how an increase in levels of nutrient and trophic resources in the water column led to a switch from oligotrophic stromatoporoid reefs to more mesotrophic reefs rich in stromatoporoids and microbes. Mechanisms responsible for nutrification included changes in platform geometry and oceanography generated during times of lowered sea level. In this situation, changes in sea level imparted a large control on the delivery of nutrients to the water column, corresponding to an ecological shift of reefal biofacies.

New approaches to the dating and correlation of carbonate strata can help elucidate the type and scale of the controls responsible for carbonate deposition and carbonate-platform development. **Whalen and Day** use a stratigraphic approach comprising several high-resolution techniques, including magnetic susceptibility, sequence stratigraphy, and high-frequency biostratigraphy, to document and improve upon the timing and determination of the relative effect local versus eustatic sea-level fluctuation, climate, and biotic change had upon an Upper Devonian isolated platform-to-basin carbonate system. They found that the magnetic-susceptibility technique may record a global tectonic and climatic signature useful for regional correlation where other techniques fail.

IMPROVED CHARACTERIZATION OF CARBONATE PLATFORMS AND REEFS

As technology advances in conjunction with our understanding of carbonate depositional systems and processes, so does our ability and desire to quantify and model the natural geological realm. Development and use of detailed geological and forward stratigraphic models are increasing our knowledge of specific carbonate depositional systems, prediction of facies distributions, and determination of reservoir-body geometries in the subsurface. Predictability of facies trends and their stacking patterns in response to a variety of controlling factors is especially critical in the oil and gas industry, where the accurate delineation of reservoir, source, and seal components can lead to exploration success. As such, it comes as no surprise that industry is driving the need and impetus for modeling of carbonate systems and is working with academic research institutions in a group effort to develop and improve this technology.

The use of high-resolution satellite imagery, increasing awareness and functionality of reservoir modeling, and further development of software for predictive modeling of carbonate platforms is changing and enhancing the way geoscientists study carbonate systems and their attributes in the subsurface. Computer-assisted geological modeling tools such as CARB3D+ (Smart et al., 2005) and Dionisos (Granjeon and Joseph, 1999), for example, are useful in validating existing depositional models, generating alternative models for use in exploration and reservoir prediction, and improving our understanding of geological processes and stratigraphic responses related to variable depositional controls. As with any model, however, accurate input data are critical to decrease the uncertainty in geologic models and to enhance their overall utility. The papers presented here provide some current examples of the scientific interaction between industry and academic geoscientists in meeting these goals of reservoir and platform characterization.

Building on the satellite image database presented in Harris and Kowalik (1994), **Rankey and Harris** use recent satellite imaging technology to develop a remote-sensing image database of modern carbonate environments representing a wide spectrum of geomorphological patterns generated in coastal and shallow marine systems. Each analogue is imaged and described in terms of depositional setting (reefs, shoals, and tidal flats), interpreted facies distribution, and, where possible, environmental factors associated with their development. These images provide (1) examples of the types of heterogeneity that might be present in ancient analogues, (2) actualistic samples of the scales at which they occur, and (3) a useful database of images that can be used in a dynamic and interactive teaching environment.

In a quest to develop realistic geological models, **Harris and Vlaswinkel** measure multiple facies attributes from Landsat images of modern isolated platforms (size, shape, orientation, and distribution) to assess the quantitative relationships between them. Through their analysis, they identify several trends in geomorphological and reefal facies attributes that may be of particular value to subsurface geologists involved in the integration of well and seismic data to generate realistic models of isolated platform reservoirs.

Taking two-dimensional geological modeling to the next level, forward stratigraphic modeling is used to analyze and spatially predict facies distributions through time. **Bassant and Harris** use Dionisos, software developed by Institut Français du Pétrole (IFP) with assistance from an industry consortium, to model grainstone distribution on an isolated platform exposed to different eustatic amplitudes and wave-energy fluctuations in icehouse and greenhouse modes. The models show a relationship between gross platform morphology and net reservoir volume consistent with analogous hydrocarbon reservoirs and illustrate the utility of forward stratigraphic models in providing a way to test our assumptions on which factors ultimately control and characterize the different aspects of development of carbonate platforms and reefs.

WHERE DO WE GO FROM HERE? THE IMPORTANCE OF SCALE AND EVALUATION OF GENETIC RELATIONSHIPS

On the large scale, tectonic systems vary, biotic evolution occurs, and global sea level fluctuates. Each of these systems generates numerous structural, climatic, and oceanographic signals that are ultimately responsible for carbonate sedimentation and deposition at all scales. As shown by the research highlighted in this volume, carbonate systems respond to each of these controls in varied ways, resulting in a wide variety of platform geometries with diverse character not directly linked to one dominant factor at any time in geological history. How we define these controls via their physical responses hinges on our knowledge of sedimentological processes and the stratal character generated from them.

Scale is the one theme common to all of the research presented here. Externally and internally generated mechanisms forced upon a depositional system generate widely variable sedimentary responses. Accommodation influenced by the principal controls of tectonics, eustasy, and biotic evolution affect the geometry and character of the platform or reef complex at all scales—from the global to the local. Climate and changes in oceanography, which may or may not be influenced by mechanisms involved with changing accommodation, strongly influence platform development on the regional to local scale (epicontinental, basinal, and local) and modified by trophic resources and hydrodynamics which can significantly alter depositional patterns on the platform. Of course, there is a range of frequencies involved with each of these controls that are expressed differently in the sedimentology and stratigraphy of the platform and reef system, meaning that each control may affect sedimentological responses on the bed, sequence, supersequence, and platform scales.

Global patterns of carbonate and reef development are mostly perceived as being influenced by truly global mechanisms (e.g., **Markello et al.**), while smaller-scale stratal character is more likely to be explained by the influence of regional to local controls on sedimentation (e.g., **James and Bone**). Occasionally, however, when carbonate platforms and reefs are investigated in great detail at the finest level, it is possible to explain the generation of sedimentological processes and their stratigraphic expressions through the interaction of many controls acting at all scales (e.g.,

Peterhänsel and Egenoff). As such, it is likely that the scale of investigation may significantly influence or perhaps even bias our interpretations on which controls are considered to play a dominant role in the different stages of development of carbonate platforms and reefs. The more detailed the investigation, the more specific and local the controls seem to be.

An increasing amount of research is devoted to evaluating carbonate platform and reef systems from a holistic point of view in an attempt to understand the processes involved and the genetic relationships responsible for initiation, growth geometry, stratal character, and demise of platforms or reefs . One example of this approach is our evolving view of carbonate platforms and their classification. This is the classic conundrum, where, although platforms have been generated in all kinds of tectonic settings during the Phanerozoic through a varied history of eustatic, biologic, climatic, and oceanographic evolution, we classify them equally according to their morphological profile. Is it our desire to sort natural systems into defined boxes that prevents us from acknowledging the integrated and genetic relationships responsible for the generation of carbonate platforms and reefs, or is the multi-scale network of mechanisms responsible for their development too complex to measure and too integrated to separate? Then again, perhaps it is the history of how our science developed that provides the limiting factor.

Carbonate platforms were first studied in terms of facies types, distribution of facies belts, and depositional profile from the modern to the ancient in order to assess the morphological patterns generated across carbonate systems. As such, our knowledge and classification of them adopted the physiographic focus we use today. Though useful in describing the shape of the platform profile and characterizing platform types in a general sense, differentiating platform types as ramps, open shelves, rimmed shelves, and isolated platforms (and all of their varieties) limits our assessment to one static morphological style in time and provides little information regarding the nature or scale of the processes involved. As carbonate research continues to grow into increasingly detailed, systematic studies of all carbonate systems, so does our understanding of the relationships that exist between extrinsic and intrinsic controls and their sedimentological and stratigraphic expressions.

The science surrounding development of carbonate platforms and reefs is at a turning point. As our scale of assessment becomes finer and more detailed, so too does our need to accurately clarify and quantify these natural systems from an increasingly holistic perspective with the hope of developing realistic and universal models that span the boundaries of time and space. In light of this, today we cannot help but examine platforms and reefs from a genetic perspective, thus enabling us to develop the detailed interpretations necessary to explain temporal trends in platform development and to generate accurate, predictive models for exploration in the subsurface. How these models are generated, tested, and applied could form the next significant advancement in the science of carbonate systems.

ACKNOWLEDGMENTS

This volume represents ideas and concepts initially presented at the SEPM Special Session "Controls on Carbonate Platform and Reef Development: A Tribute in Honor of Dr. Wolfgang Schlager" at the AAPG-SEPM Conference in Calgary, May 2005. We would also like to express our gratitude to each of the reviewers for all of their time and energy spent in assessing the manuscripts, helping to ensure the authors' observations and interpretations were clearly represented. Many thanks to Noel James for reviewing the introduction and summary.

REFERENCES

AHR, W.M., 1973, The carbonate ramp: an alternative to the shelf model: Gulf Coast Association of Geological Societies, Transactions, v. 23, p. 221–225.

BOSENCE, D., 2005, A genetic classification of carbonate platforms based on their basinal and tectonic settings in the Cenozoic: Sedimentary Geology, v. 175, p. 49–72.

BURCHETTE, T.P., AND WRIGHT, V.P., 1992, Carbonate ramp depositional systems: Sedimentary Geology, v. 79, p. 3–57.

CHERNS, L., AND WRIGHT, V.P., 2000, Missing molluscs as evidence of large-scale, early aragonite dissolution in a Silurian sea: Geology, v. 28, p. 791–794.

CREVELLO, P.D., WILSON, J.L., SARG, J.F., AND READ, J.F., 1989, Controls on Carbonate Platform and Basin Development: SEPM, Special Publication 44, 415 p.

DE BENEDICTUS, D., BOSENCE, D., AND WALTHAM, D., 2007, Tectonic control on peritidal carbonate parasequence formation: an investigation using forward tectono-stratigraphic modelling: Sedimentology: v. 54, p. 508–605.

EBERLI, G., MASAFERRO, J.L., AND SARG, J.F., 2004, Seismic Imaging of Carbonate Reservoirs and Systems: American Association of Petroleum Geologists, Memoir 81, 376 p.

GINSBURG, R.N., AND JAMES, N.P., 1974, Holocene carbonate sediments of continental shelves, *in* Burke, C.A., and Drake, C.L., eds., The Geology of Continental Margins: New York, Springer-Verlag, p. 137–155.

GOLDHAMMER, R.K., DUNN, P.A., AND HARDIE, L.A., 1990, Depositional cycles, composite sea-level changes, cycle stacking patterns, and the hierarchy of stratigraphic forcing: Examples from Alpine Triassic platform carbonates: Geological Society of America, Bulletin, v. 102, p. 535–562.

GRANJEON, D., AND JOSEPH, P., 1999, Concepts and applications of a 3-D multiple lithology, diffusive model in stratigraphic modelling, *in* Harbaugh, J.W., Watney, L., Rankey, E.C., Slingerland, R., Goldstein, R.H., and Franseen, E., eds., Numerical Experiments in Stratigraphy: Recent Advances in Stratigraphic and Sedimentologic Computer Simulations: SEPM, Special Publication 62, p. 197–210.

HALLOCK, P., 1987, Fluctuations in the trophic resource continuum: A factor in global diversity cycles?: Paleoceanography, v. 2, p. 457–471.

HALLOCK, P., AND SCHLAGER, W., 1986, Nutrient excess and the demise of coral reefs and carbonate platforms: Palaios, v. 1, p. 389–398.

HANDFORD, C.R., AND LOUCKS, R.G., 1993, Carbonate depositional sequences and systems tracts—responses of carbonate platforms to relative sea-level changes, *in* Loucks, R.G., and Sarg, J.F., eds., Carbonate Sequence Stratigraphy: American Association of Petroleum Geologists, Memoir 57, p. 3–42.

HARRIS, P.M., AND KOWALIK, W.S., 1994, Satellite Images of Carbonate Depositional Settings: American Association of Petroleum Geologists, Methods in Exploration Series, 10, 147 p.

HOMEWOOD, P.W., AND EBERLI, G.P., 2000, Genetic Stratigraphy on the Exploration and Production Scales—Case Studies from the Pennsylvanian of the Paradox Basin and the Upper Devonian of Alberta: Bulletin Centre Recherche, Elf Exploration and Production, Memoir 24, 290 p.

INSALACO, E., SKELTON, P.W., AND PALMER, T.J., 2000, Carbonate Platform Systems: Components and Interactions: Geological Society of London, Special Publication 178, 231 p.

JAMES, N.P., 1997, The cool-water carbonate depositional realm, *in* James N.P., and Clarke, J.A.D., eds., Cool-Water Carbonates: SEPM, Special Publication 56, p. 1–22.

JAMES, N.P., BONE, Y., AND KYSER, T.K., 2005, Where has all the aragonite gone? Mineralogy of Holocene neritic cool-water carbonates, southern Australia: Journal of Sedimentary Research, v. 75, p. 454–463.

KENDALL, C.G.ST.C., AND SCHLAGER, W., 1981, Carbonates and relative changes in sea level: Marine Geology, v. 44, p. 181–212.

KIESSLING, W., FLÜGEL, E., AND GOLONKA, J., 2003, Patterns of Phanerozoic carbonate platform sedimentation: Lethaia, v. 36, p. 195–226.

LOUCKS, R.G., AND SARG, J.F., 1993, Carbonate Sequence Stratigraphy: Recent Developments and Applications: American Association of Petroleum Geologists, Memoir 57, 545 p.

LUKASIK, J., AND JAMES, N.P., 2006, Carbonate sedimentation, climate change, and stratigraphic completeness on a Miocene cool-water epeiric ramp, Murray Basin, South Australia, *in* Pedley, H.M., and Carannante, G., eds., Cool-Water Carbonates: Depositional Systems and Palaeoenvironmental Controls: Geological Society of London, Special Publication 255, p. 217–244.

LUKASIK, J.J., JAMES, N.P., MCGOWRAN, B., AND BONE, Y., 2000, An epeiric ramp: low-energy, cool-water carbonate facies in a Tertiary inland sea, Murray Basin, South Australia: Sedimentology, v. 47, p. 851–881.

MALLINSON, D.J., HINE, A.C., HALLOCK, P., LOCKER, S.D., SHINN, E.A., NAAR, D.F., DONAHUE, B.T., AND WEAVER, D., 2003, Development of small carbonate banks on the South Florida platform margins: response to sea level and climate change: Marine Geology, v. 199, p. 45–63.

PEDLEY, H.M., AND CARANNANTE, G., 2006, Cool-Water Carbonates: Depositional Systems and Palaeoenvironmental Controls: Geological Society of London, Special Publication 255, 373 p.

POMAR, L., 2001a, Ecological control of sedimentary accommodation: evolution from a carbonate ramp to rimmed shelf, Upper Miocene, Balearic Islands: Palaeogeography, Palaeoclimatology, Palaeoecology, v. 175, p. 249–272.

POMAR, L., 2001b, Types of carbonate platforms, a genetic approach: Basin Research, v. 13, p. 313–334.

READ, J.F., 1982, Carbonate platforms of passive (extensional) continental margins: types, characteristics and evolution: Tectonophysics, v. 81, p. 195–212.

READ, J.F, 1985, Carbonate platform facies models: American Association of Petroleum Geologists, Bulletin, v. 69, p. 1–21.

READ, J.F., 1995, Overview of carbonate platform sequences, cycle stratigraphy and reservoirs in greenhouse and icehouse worlds, *in* Read, J.F., Kerans, C., and Weber, L.J., eds., Milankovitch Sea Level Changes, Cycles and Reservoirs on Carbonate Platforms in Greenhouse and Ice-House Worlds: SEPM, Short Course 35, p. 1–102.

SANDBERG, P.A., 1983, An oscillating trend in Phanerozoic non-skeletal carbonate mineralogy: Nature, v. 305, p. 19–22.

SARG, J.F., 1988, Carbonate sequence stratigraphy, *in* Wilgus, C.K., Hastings, B.S., Posamentier, H., Van Wagoner, J., and Kendall, C.G.St. C., eds., Sea-Level Changes: An Integrated Approach: SEPM, Special Publication 42, p. 155–182.

SCHOLLE, P.A., BEBOUT, D.G., AND MOORE, R.C., 1983, Carbonate Depositional Environments: American Association of Petroleum Geologists, Memoir 33, 708 p.

SIMO, J.A.T., SCOTT, R.W., AND MASSE, J.-P., 1993, Cretaceous Carbonate Platforms: American Association of Petroleum Geologists, Memoir 56, 479 p.

SMART. P.L., WALTHAM, D., FELCE, D., AND WHITAKER, F.F., 2005, CARB3D+: a new forward simulation model for sedimentary architecture and near-surface diagenesis in isolated carbonate platforms (abstract): American Association of Petroleum Geologists International Meeting, Abstracts, Paris.

TUCKER, M.E., WILSON, J.L., CREVELLO, P.D., SARG, J.F., AND READ, J.F., 1990, Carbonate Platforms: Facies, Sequences and Evolution: International Association of Sedimentologists, Special Publication 9, 328 p.

VAIL, P.R., MITCHUM, R.M., JR., TODD, R.G., WIDMIER, J.M., THOMPSON, S., III, SANGREE, J.B., BUBB, J.N., AND HATLELID, W.G., 1977, Seismic stratigraphy and global changes of sea level, *in* Payton, C.E., ed., Seismic Stratigraphy—Applications to Hydrocarbon Exploration: American Association of Petroleum Geologists, Memoir 26, p. 49–212.

WALTER, L.M., AND BURTON, E.A., 1990, Dissolution of platform carbonate sediments in marine pore fluids: American Journal of Science, v. 290, p. 601–643.

WILKINSON, B.H., 1979, Biomineralization, paleoceanography, and the evolution of calcareous marine organisms: Geology, v. 7, p. 524–527.

WILSON, J.L., 1974, characteristics of carbonate-platform margins: American Association of Petroleum Geologists, Bulletin, v. 58, p. 810–824.

WILSON, J.L., 1975, Carbonate Facies in Geologic History: New York, Springer-Verlag, 470 p.

WRIGHT, V.P., AND BURCHETTE, T.P., 1998, Carbonate Ramps: Geological Society of London, Special Publication 149, 165 p

Global Perspectives on Carbonate Platform Development

THE CARBONATE ANALOGS THROUGH TIME (CATT) HYPOTHESIS AND THE GLOBAL ATLAS OF CARBONATE FIELDS—A SYSTEMATIC AND PREDICTIVE LOOK AT PHANEROZOIC CARBONATE SYSTEMS

JAMES R. MARKELLO
ExxonMobil Upstream Research Company, P.O. Box 2189, Houston, Texas, 77252, U.S.A.
e-mail: jim.markello@exxonmobil.com
RICHARD B. KOEPNICK
Koepnick Geological Consulting, Inc., 600 Cambridge Rd., Blacksburg, Virginia 24060, U.S.A.
e-mail: rbkoepnick@hotmail.com
LOWELL E. WAITE
Pioneer Natural Resources USA, Inc., 5205 North O'Connor Blvd., Irving, Texas, 75039, U.S.A.
e-mail: lowell.waite@pxd.com
AND
JOEL F. COLLINS
ExxonMobil Development Company, 12450 Greenspoint Dr., Houston, Texas, 77060, U.S.A.
e-mail: joel.f.collins@exxonmobil.com

Abstract: The *Carbonate Analogs Through Time* hypothesis states that "high-confidence, age-specific predictive models for carbonate systems and for reservoir occurrence, composition, stratal attributes, and reservoir properties can be developed by summing the ambient conditions of the carbonate processes and Earth processes at any geologic age." We term these models age-sensitive patterns or themes. The hypothesis is built upon the cumulative body of knowledge that demonstrates carbonate processes (biotic evolution and mineralogic variation) and Earth processes (tectonics, climate, eustasy, ocean chemistry, and ocean circulation) have varied differentially throughout Phanerozoic time. The concepts of the stratigraphic hierarchy and the Sloss sequences are employed to temporally define and anchor age-specific predictive models. Age-based themes correspond to second-order supersequence scale of stratal packages, and these are fixed in geologic time (Ma). Two products developed are the *Phanerozoic Carbonate Trends Chart* and the *Global Atlas of Carbonate Fields*.

The *Carbonate Analogs Through Time* hypothesis defines an approach for developing systematic evaluations and predictive models of Phanerozoic carbonate systems and reservoirs for use in exploration, development, and production businesses. Exploration geoscientists employ a host of concepts, tools, and data to develop predictive models for occurrence and quality of fields and reservoirs. However, as exploration successes decrease, alternative approaches are needed to refresh the exploration mindset. The *CATT* hypothesis and approach represent a basis for developing an alternative mindset for exploration for carbonate reservoirs. It provides a thought-provoking perspective on known occurrences of carbonate reservoirs, and offers a different way of thinking about predicting where undiscovered carbonate reservoirs may exist.

Reservoir engineers require detailed geology-based reservoir parameters for simulations of reservoir/field performance. Such simulations form the bases for field development/depletion plans that carry with them implications of huge capital and operating expense. Thus, it is imperative to provide the best possible input to simulation so that investments are optimal. Typically, the input, if not derived directly from data collected within a field under development, is obtained from "analog" fields. Thus, choosing the most appropriate analog is a critical task. We contend that this approach provides the conceptual basis for choosing the most appropriate analogs for development and production-based reservoir characterization. Further, this approach provides a framework or context for the systematic organization and evaluation of the concepts, facts, and carbonate-reservoir case studies one encounters throughout a career.

INTRODUCTION

The *Carbonate Analogs Through Time* Project was conducted from 1991 to 1999 at Mobil Exploration and Production Technology Center. The results include (1) the overarching idea that is articulated as a hypothesis, (2) an approach, based on the idea, to develop numerous applications including analog selection, systematic comparative analyses, and predictive concepts, and (3) a summative database of geologic data, facts, and concepts integrated into specific products such as the *Phanerozoic Carbonate Trends Chart* and *Global Atlas of Carbonate Fields*. The major objective of this work has been to build upon known fundamentals of earth history, tectonics, and sedimentary geology to develop carbonate-reservoir-specific concepts and principles that both explain and predict characteristics of carbonate reservoirs. These characteristics include spatial occurrences, internal stratal architecture, facies distribution and geometry, diagenetic modifications, and reservoir-rock properties.

Appropriate or optimal carbonate field and reservoir analogs are often lacking in play, prospect, and reservoir assessment. Our current understanding of carbonate systems indicates that analogs must be chosen to reflect suitable age, stratigraphic, and facies characteristics, because carbonate systems are diverse with respect to age, tectonic setting, paleogeography, stratigraphic architecture, stratal geometry, facies, biota, and diagenesis. Predictive concepts of carbonate systems are centered on "age-sensitive patterns and characteristics" of carbonate systems that are the net result from the interactions of "earth processes" and "carbonate processes." Earth processes include plate tectonics, climate, sea level, and oceanography,

Controls on Carbonate Platform and Reef Development
SEPM Special Publication No. 89, Copyright © 2008
SEPM (Society for Sedimentary Geology), ISBN 978-1-56576-130-8, p. 15–45.

whereas carbonate processes include platform type (isolated vs. attached), depositional profile (ramp vs. shelf margin), biotic composition and evolution (including extinctions and radiations), mineralogy, CO_3 chemistry, and diagenesis. It has been long recognized that these earth and carbonate-specific processes vary secularly throughout geologic time (Wilson, 1975; Tucker and Wright, 1990; Veevers, 1990; Schlager, 2005). Some vary independently, whereas others vary dependently. Regardless, an age-sensitive pattern or theme is the net result of the sum of the states of each of these processes at any given geologic age. Thus, our objective is to select age ranges or time intervals within which predictions are constrained, to identify the key characteristics of the carbonate systems for each specific time interval, and to understand the causality such that these predictive concepts, principles, and/or themes can be applied effectively.

CARBONATE ANALOGS THROUGH TIME

Statement of the Hypothesis

The *CATT* hypothesis states: "high-confidence, age-specific predictive models and concepts for (1) ancient carbonate systems and (2) carbonate reservoirs in terms of occurrence, composition, stratal attributes, and reservoir properties can be developed by summing the **ambient conditions** of the carbonate processes and Earth processes at any geologic age." The summations are termed **age-sensitive patterns or themes**. Graphically, the hypothesis can be expressed as:

Geologic Age	+	Carbonate Processes	+	Earth Processes	=	Age-Sensitive Patterns (themes)
Time-Slice Definition		Biotic evolution		Tectonics		Occurrence and Location
		Mineralogy		Climate		Reefs vs Grainstones vs Mudstones
				Eustasy		Platform types/profiles
				Ocean Circulation		Stratal architecture
				Ocean Chemistry		Diagenetic potential
						Analogs

The hypothesis is built upon the cumulative body of knowledge that demonstrates that carbonate and Earth processes have varied differentially throughout Phanerozoic time. These carbonate and Earth processes include: (1) ecologic, oceanographic, and sedimentologic process-based controls on development of a carbonate factory; (2) stratigraphic and accommodation process-based controls on carbonate stratal architecture; (3) secular trends of evolution, grain mineralogy, tectonics, climate, eustasy, ocean circulation, and ocean chemistry; (4) the stratigraphic hierarchy and the constraint that first- and second-order Phanerozoic stratigraphic successions (Sloss sequences; Sloss, 1963) are age-fixed in geologic time (Ma).

The *Carbonate Analogs Through Time* hypothesis provides the underpinnings for an approach to develop numerous applications that include formulating predictive concepts of carbonate systems and reservoirs. The distinguishing characteristic of such an approach to making predictions about carbonate systems is that the predictions are age-specific. Two contentions are embedded in the approach. First, we reject the idea that a single comprehensive model of carbonate systems or reservoir models can be built to describe and predict all features of all Phanerozoic carbonate systems, and second, we reject the idea that every Phanerozoic carbonate system is unique, requiring an infinite number of models to predict all features. Rather, we propose that there are a limited and manageable number of models that can be constructed to encompass the range of variability recognized within Phanerozoic carbonate systems. The first widely available summary of age-sensitive themes for carbonate systems is contained in Wilson (1975). Other work that synthesizes carbonate systems into predictive models is based on (1) depositional geometric criteria, e.g., ramps vs. rimmed shelves; attached vs. detached platforms (Ahr, 1973; Wilson, 1974; Read, 1985), (2) combined geometric and sequence stratigraphic criteria (Sarg, 1988; Handford and Loucks, 1993), (3) tectonic criteria (Read, 1982; Dorobek, 1995; Pomar, 2001; Bosence, 2005), (4) global climatic criteria, e.g., greenhouse vs. icehouse systems (Wright, 1992; Read, 1995; 1998), or (5) biotic communities and local to regional climates (Kiessling et al., 1999; Kiessling et al., 2002).

The *Carbonate Analogs Through Time* models were developed to capture and predict geologic age–based variability within carbonate systems. The hypothesis provides an approach that integrates the geometric, tectonic, climatic, and other significant controlling factors for development of carbonate system characteristics and combines them into predictive models for specific geologic ages. Geologic age provides constraints and context, and thus limits the number of models.

The Global Atlas of Carbonate Reservoirs—Key Development Elements and Products

The Phanerozoic Carbonate Trends Chart.—

A major product that captures the secular variability of Earth and carbonate processes is the *Phanerozoic Carbonate Trends Chart*, compiled by author Joel Collins (Fig. 1. Posters 1a, the 1998 original version, and Poster 1b, the 2005 update, are on the enclosed CD). This chart brings together a myriad of secularly varying data from multiple sources and synchronizes all of the curves to the geologic time scale that serves as the backbone of the poster (Table 1). New and original data were not collected for this product. It can be viewed as an integrated summative database of existing data and described as a derivative data product. Data points and reference annotations are included on the posters.

→

FIG. 1 (opposite page).—*Phanerozoic Carbonate Trends Chart*. The figure was compiled by author Joel Collins (1996 to 1998). Column headings and data references are given in Table 1. Large-scale poster formats for the original 1998 and updated 2005 versions are available for plotting from electronic files on the enclosed CD. This chart combines many different types of data from a variety of sources. Every attempt has been made to faithfully register data points to the time scale as accurately as possible; however, some inaccuracies may exist. Many transitions and variations are depicted as gradational rather than abrupt in columns with sparse interpolated data or columns of continuously changing characteristics (e.g., sea level, global climate, frame-builder reef times, biotic evolution). This may or may not be the true nature of these characteristics in the rock record, but we have chosen to stylistically represent them in this manner. Discrete events and intervals and abruptly changing phenomena are depicted as such (e.g., seafloor magnetic striping, extrusive events, first occurrences and extinctions). Full appreciation of this chart requires systematic and persistent study.

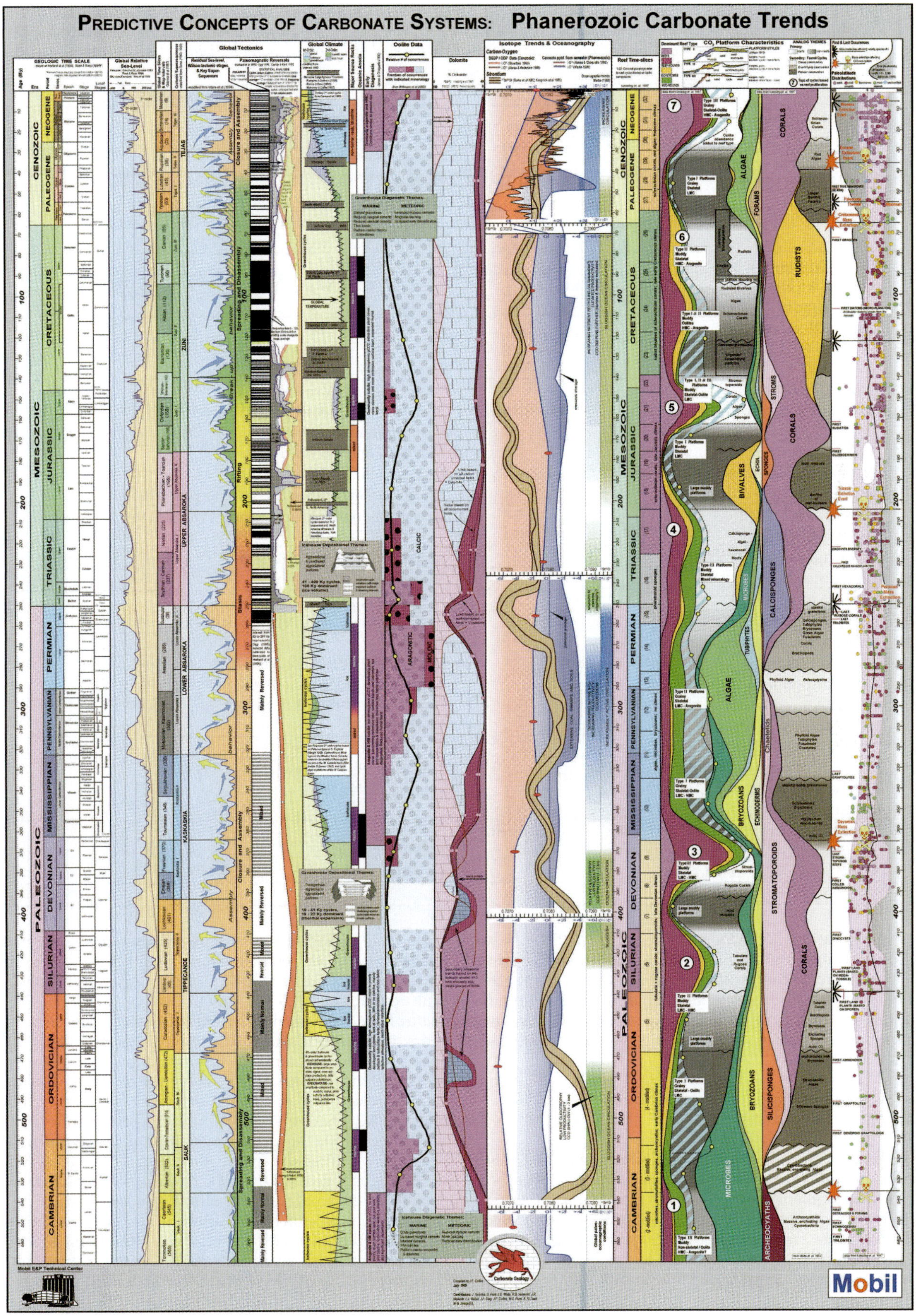
Predictive Concepts of Carbonate Systems: Phanerozoic Carbonate Trends
CENOZOIC
MESOZOIC
PALEOZOIC
NEOGENE
PALEOGENE
CRETACEOUS
JURASSIC
TRIASSIC
PERMIAN
PENNSYLVANIAN
MISSISSIPPIAN
DEVONIAN
SILURIAN
ORDOVICIAN
CAMBRIAN
Mobil E&P Technical Center

TABLE 1.—Components of the *Phanerozoic Carbonate Trends Chart* (text Figure 1). Each column heading on the chart is listed along with the corresponding source references for the data presented in the columns. This chart is a major component of the *Global Atlas of Carbonate Fields.*

Column Heading	Authors for Data Displayed
Geologic time scale	Harland et al., 1990 Ross and Ross, 1994
Global relative sea level	Greenlee and Lehmann, 1993 Ross and Ross, 1988 Haq et al., 1987
Global time slices and paleogeographic map intervals	Heritage-Mobil Global Themes Project, 1993 Golonka et al., 1997
Cratonic supersequences	Sloss, 1963, 1988
Residual sea level, Wilson tectonic stages, and key supersequences	Modified from Waite et al., 1994; Wilson, 1988; Haq et al., 1987
Paleomagnetic reversals	Harland et al., 1990 Ogg, 1995 Cande and Kent, 1992
Global climate Mesozoic large igneous provinces Tertiary 2nd-order cycles Late Paleozoic 2nd-order cycles	Modified from Veevers, 1990 Rampino and Stothers, 1988 Renne and Basu, 1991 Mahoney and Coffin, 1997 Davies et al., 1989 Wright, 1988; Martindale and Boreen, 1997
Major source rock times Oceanic anoxia	Heritage-Mobil Global Themes Project, 1993
Early diagenesis	James and Choquette, 1990b
Oolite occurrences	Wilkinson et al., 1985
Dolomite trends (reefs) Dolomite trends (fields)	Kiessling et al., 1999 In-house Mobil and PetroConsultants databases
Isotope trends and oceanography: Carbon-oxygen Marine cement data Strontium Oceanographic trends	 Mountain, 1991; Abreu and Anderson, 1998 James and Choquette, 1990a; Martin, 1996 Burke et al., 1982 ; Koepnick et al., 1985 Martin, 1996
Geologic time scale and reef time slices	Kiessling et al., 1999
Biotic evolutionary trends Carbonate platform characteristics Analog themes	James, 1983; Kiessling et al., 1999 Wilson, 1974; 1975
Paleontology: first and last occurrences, extinction events Total biodiversity Paleolatitude distribution of reefs	 Jablonski, 1999 Kiessling et al., 1999

The principal utility of the *Phanerozoic Carbonate Trends Chart* is to facilitate solving the *CATT* equation. Because each of the trends is registered to the geologic time scale, it is straightforward to read across the poster at any geologic age and sum the ambient states of the Earth and carbonate processes for that age. Such a summation is the basis for the "age-sensitive patterns or themes."

Time-Slice Definition for Development of Global Paleogeographic Maps.—

The cornerstone product for the *Carbonate Analogs Through Time* project is the *Global Atlas of Carbonate Fields.* The atlas contains 29 global paleogeographic maps that span the Phanerozoic. Each map encompasses an interval of geologic time, designated as a time slice, and the paleogeography depicted on each map is an interpretation or judgment of the generalized and representative paleogeography for the time increment or time slice. Time increments for maps are not equal. Determining the increment of time for the maps was a key decision. Two basic criteria were used: (1) stratigraphic validity and character of the time increments and (2) practicality of number of maps. For example, if the time increment were too large, e.g., 50 million years, then only 10 maps would have been needed to cover the entire Phanerozoic, and the environments of deposition plotted on each of the 10 maps would have little relevance for any specific smaller-scale cycle of deposition within a 50-million-year time interval. Alternatively, if the time increment for one map were too small, e.g., 100,000 years, over 5000 maps would have been required to cover the entire Phanerozoic. Completing global-scale paleogeography for 5000 maps was untenable within the project time frame, resource allocation, and budget.

For the global paleogeographic maps, the second-order supersequence (Haq et al., 1988; Vail et al., 1991; Sarg, Markello et al., 1999) stratigraphic scale was chosen as the framework for time increments (Fig. 2). There were three basic reasons for this deci-

ORDER	DURATION (My) RANGE	DURATION (My) MODE	STRATIGRAPHIC NOMENCLATURE GREENHOUSE	STRATIGRAPHIC NOMENCLATURE ICEHOUSE
1st	350–500	450	Megasequence Set	
1st	50–100 +	80	Megasequence	
2nd	20–50 +	30	Supersequence Set	
2nd	5–20 +	10	Supersequence	
3rd	0.5–5	1	Sequence	Composite Sequence
4th	0.05–0.5	0.1; 0.45	Parasequence Set Parasequence	Sequence
5th	0.01–0.05	0.02; 0.04	Parasequence Bed Sets	Parasequence Set Parasequence
> 6th	< 0.01		Beds Laminae	Bed Sets Beds Laminae

FIG. 2.—The Stratigraphy Hierarchy Chart (modified from Sarg et al., 1999). The concepts of the stratigraphic hierarchy (Haq et al., 1987; Tucker and Wright, 1990; Vail et al., 1991; Sarg et al., 1999) were employed to determine the validity and most practical time increment for the global paleogeographic maps. The basic concept that underpins the stratigraphic hierarchy is that the Phanerozoic sedimentary rock record consists of well organized, recognizable, and superimposed scales of depositional cycles. The different scales have been previously defined (Haq et al., 1987; Tucker and Wright, 1990; Vail et al., 1991), and the definitions are based on time, not on depositional or stratal thickness. Also, each depositional cycle, regardless of scale, is bounded below and above by unconformities or sequence boundaries. They range from large-scale tectonically driven first-order basin fill (Wilson, 1966, 1988) and cratonic-flooding–draining successions (Sloss, 1963) to climate-driven, small-scale fifth-order Milankovitch-type parasequences (Goldhammer et al., 1987; Vail et al., 1991; Drummond and Wilkinson, 1993). Importantly, within each scale of the hierarchy the defined subdivisions of strata are contiguous. The top unconformity or sequence boundary of a stratal package is the basal unconformity or sequence boundary of the next overlying stratal package. There are no gaps between and no overlaps of strata constituting cycles within a given hierarchical scale.

sion. First, most of the second-order supersequences were already defined for the Phanerozoic with sequence-bounding or unconformity surfaces tied to the geologic time scale (Haq et al., 1988; Ross and Ross, 1988; Greenlee and Lehmann, 1993). At the time of decision for the project—1993—it was estimated the Phanerozoic consists of approximately 41 to 45 second-order supersequences. The variability in the estimate depends on number of second-order subdivisions defined in the Ordovician and Cambrian (Table 2). Second, it was recognized that although the second-order supersequence scale encompasses significant geologic time, global-scale paleogeography could be mapped that is representative of the large-scale and long-lived depositional systems for regions and areas within basins. Building this number of global-scale paleogeographic maps (between 25 and 35) was deemed achievable for the project. Third, significant changes in the Earth and carbonate processes that impact the age-based characteristics of carbonate systems and reservoirs, such as plate motions, plate positions, basin subsidence, paleoclimate, ocean chemistry, mineralogy, and major biota, occur approximately at the second-order supersequence time scale (Fig. 1). Ultimately, the final numbers of maps was 29 (Table 2).

Subsequent to completion of this work, refinements have been made to the geologic time scale. Golonka and Kiessling (2002) summarize these modifications and present an updated geologic time scale to use in defining the time-slice ages and for naming the paleogeographic maps (in Kiessling et al., 2002).

Paleogeography: A Cautionary Note.—

It is understood that paleogeographic maps should depict the physiography and depositional systems of the Earth at some instant in geologic time. However, since it is impossible to identify sufficient rocks and / or strata of exactly the same age (to within some instant in time) across the globe, we are forced to make paleogeographic maps that represent Earth physiography and depositional systems throughout some time interval. Obviously, the shorter the time duration of the interval, the more accurate the paleogeographic map will be. As an example, during global icehouse times, when high-amplitude, high-frequency sequences are formed, the paleogeography of the sequence components, lowstand, highstand, and transgressive systems tracts, can be dramatically different. Accurate paleogeography mapping requires maps for each of the systems tracts, rather than trying to capture on one map all of the variability of an icehouse high-frequency sequence. We admit that the best approach would be mapping at the sub-sequence scale. However, because of project constraints (resources and timeline) we made the decision to lengthen our paleogeographic mapping time intervals to the second-order supersequence scale of stratigraphy, and attempted to represent the global physiography and depositional systems within that scale as best we could.

Development of Upgraded Global Paleogeographic Base Maps.—

As stated above, the cornerstone product for the project was the *Global Atlas of Carbonate Fields*. As a data-rich foundation for development of age-based predictive concepts for carbonate systems, upgraded global paleogeographic maps were made to provide the framework, both temporal and spatial, to empirically map known carbonate systems and known carbonate fields/reservoirs. The methodology employed to make the upgraded global paleogeographic base maps was multi-stepped and labor-intensive, requiring on the order of 20–30 person work-years to accomplish. Steps included:

1. Define geologic ages (in Ma) for each time slice to be mapped (Table 2).

2. Name each time slice (usually the geologic stage name encompassed by ages; Table 2).

3. Determine the characteristic or representative "type" stratal sections for the time slice from one or more basins from multiple continents.

4. Choose the time (geologic age) of maximum flooding of continental shelves and cratonic interiors for the time slice.

TABLE 2.—Name and age definitions for each of the time intervals and paleogeographic maps constituting the *Global Atlas of Carbonate Fields* map series (Appendices 2–5). Also, these are the definitions for time intervals used to partition fields in the carbonate fields database (Appendix 1). In 1993, the initial time scale used for the Global Themes mapping project had the Permian–Triassic boundary dated at 245 Ma. The time scale used in 1999 for naming the maps in the *Global Atlas of Carbonate Fields* had the Permian–Triassic boundary dated at 250 Ma. Also, the names of the supersequences and paleogeographic maps are the names of the geologic stages that best represent the time interval. As a consequence, some strata from adjacent underlying and overlying geologic stages are included within the named time intervals and thus maps.

1993 Global Themes Project 2nd-Order Supersequences and Paleogeographic Map Names:	1993 Global Themes Project 2nd-Order Supersequences and Time-Slice Definitions:	1999 Global Atlas of Carbonate Fields 2nd-Order Supersequences and Paleogeographic Map Names:	1999 Global Atlas of Carbonate Fields 2nd-Order Supersequences and Time-Slice Definitions:
Messinian	0 to 10 Ma	Messinian (6 Ma)	0 to 10 Ma
Serravallian	10 to 20 Ma	Serravallian (14 Ma)	10 to 20 Ma
Aquitanian	20 to 30 Ma	Aquitanian (22 Ma)	20 to 30 Ma
Rupelian	30 to 39 Ma	Rupelian (36 Ma)	30 to 40 Ma
Lutetian	39 to 49 Ma	Lutetian (45 Ma)	40 to 50 Ma
Ypresian	49 to 58 Ma	Ypresian (53 Ma)	50 to 59 Ma
Danian	58 to 66 Ma	Danian–Campanian (65 Ma)	59 to 85 Ma
Campanian	66 to 85 Ma		
Turonian	85 to 94 Ma	Turonian (90 Ma)	85 to 94 Ma
Albian	94 to 112 Ma	Albian-upper Aptian (112 Ma)	94 to 118 Ma
Aptian	112 to 122 Ma		
Barremian–Hauterivian	122 to 140 Ma	Barremian (130 Ma)	118 to 137 Ma
Berriasian	140 to 146 Ma		
Kimmeridgian–Oxfordian–Callovian	146 to 163 Ma	Callovian–Oxfordian (155 Ma) and Kimmeridigian-Berriasian (140 Ma)	137 to 163 Ma
Bathonian–Bajocian	163 to 176 Ma	Bathonian–Bajocian (166 Ma)	163 to 176 Ma
Toarcian	176 to 187 Ma	Toarcian–Pliensbachian (195 Ma)	176 to 205 Ma
Pliensbachian	187 to 205 Ma		
Norian	205 to 224 Ma	Norian (225 Ma)	205 to 225 Ma
Carnian	224 to 235 Ma		
Ladinian–Anisian–Scythian	235 to 245 Ma	Carnian–Scythian (237 Ma)	225 to 250 Ma
Guadalupian	245 to 256 Ma	Guadalupian (255 Ma)	250 to 259 Ma
Artinskian	256 to 272 Ma		
Sakmarian	272 to 281 Ma		
Asselian	281 to 293 Ma	Asselian (285 Ma) includes Sakmarian and Artinskian	259 to 293 Ma
Kasimovian	293 to 311 Ma	Moscovian–Kasimovian (302 Ma) includes Bashkirian	293 to 323 Ma
Bashkirian	311 to 323 Ma		
Serpukhovian	323 to 334 Ma		
Visean	334 to 343 Ma		
Tournaisian	343 to 364 Ma	Mississippian (348 Ma) includes Serpukhovian, Visean, and Tournaisian	323 to 364 Ma
Frasnian	364 to 380 Ma	Frasnian (370 Ma)	364 to 380 Ma
Emsian	380 to 393 Ma	Emsian ((388 Ma)	380 to 393 Ma
Lochkovian	393 to 409 Ma	Lochkovian (401 Ma)	393 to 409 Ma
Ludlovian	409 to 430 Ma		
Llandoverian	430 to 439 Ma	Llandoverian (435 Ma) includes Ludlovian	409 to 439 Ma
Caradocian	439 to 466 Ma	Caradocian (452 Ma)	439 to 466 Ma
Llandeilian	466 to 476 Ma	Llandeilian–Arenigian (472 Ma)	466 to 493 Ma
Arenigian	476 to 493 Ma		
Tremodocian	493 to 510 Ma		
Croixian	510 to 517 Ma	Croixian–Tremodocian (514 Ma)	493 to 517 Ma
Albertan	517 to 548 Ma	Albertan (522 Ma)	517 to 535 Ma
Caerfaian	548 to 570 Ma	Caerfaian (545 Ma)	535 to 550 Ma
		Tommotian (565 Ma)	550 to 570 Ma

5. Start with the basic plate reconstructions of C.R. Scotese modified by J. Golonka (from Mobil global geology company files) depicting (a) deep ocean, (b) shallow shelf, (c) land, and (d) mountains. [NOTE: updated versions of these basic plate reconstructions can be accessed from websites of Christopher R. Scotese @www.scotese.com and Jan Golonka @ www.dinodata.net/Golonka/Golonka.htm].

6. Assign time slices and maps to an individual or small group for detailed analyses and paleogeographic mapping.

7. For each paleogeographic map, complete:
   a. Quality control check on starting plate position and orientation; modifications made as required.
   b. Add unmapped exotic terranes, if applicable
   c. Investigate the literature, worldwide by basin and by geologic age, and accumulate and document the key stratigraphic sections, depositional environmental interpretations, and paleogeography for local areas, regions, basins, and continental margins.
   d. Transfer upgraded and detailed paleogeography to the Scotese base maps and redraft.
   e. Develop a comprehensive map legend for use with all maps.

8. Contract with Robertson Research (now Fugro Robertson, Ltd.) to complete several time slices and paleogeographic maps.

In all, 29 upgraded global-scale paleogeographic maps were completed (Middle Pennsylvanian global paleogeography is given as an example in Figure 3). It is recognized that each map encompasses much stratigraphic section, and that at the local scale the paleogeography will not reflect all of the variability present. The purpose of these maps is to provide the global, second-order-scale perspective of paleogeography, and the goal for each map is to depict representative paleogeography for carbonate systems at times of maximum continental flooding during each second-order time interval, inasmuch as that would represent time of maximum extent of carbonate-producing environments. As local areas, regions, and basins are further investigated by the geoscience community, we anticipate further upgrades, modifications, and corrections to these maps. Also, it is recognized that depicting general paleogeography at the 5–20 million year scale can never capture and accurately portray the rapid change of paleogeography that can occur at the 100,000 year scale. Our intent was not to undertake making global-scale paleoenvironmental reconstructions at that scale of time interval. As

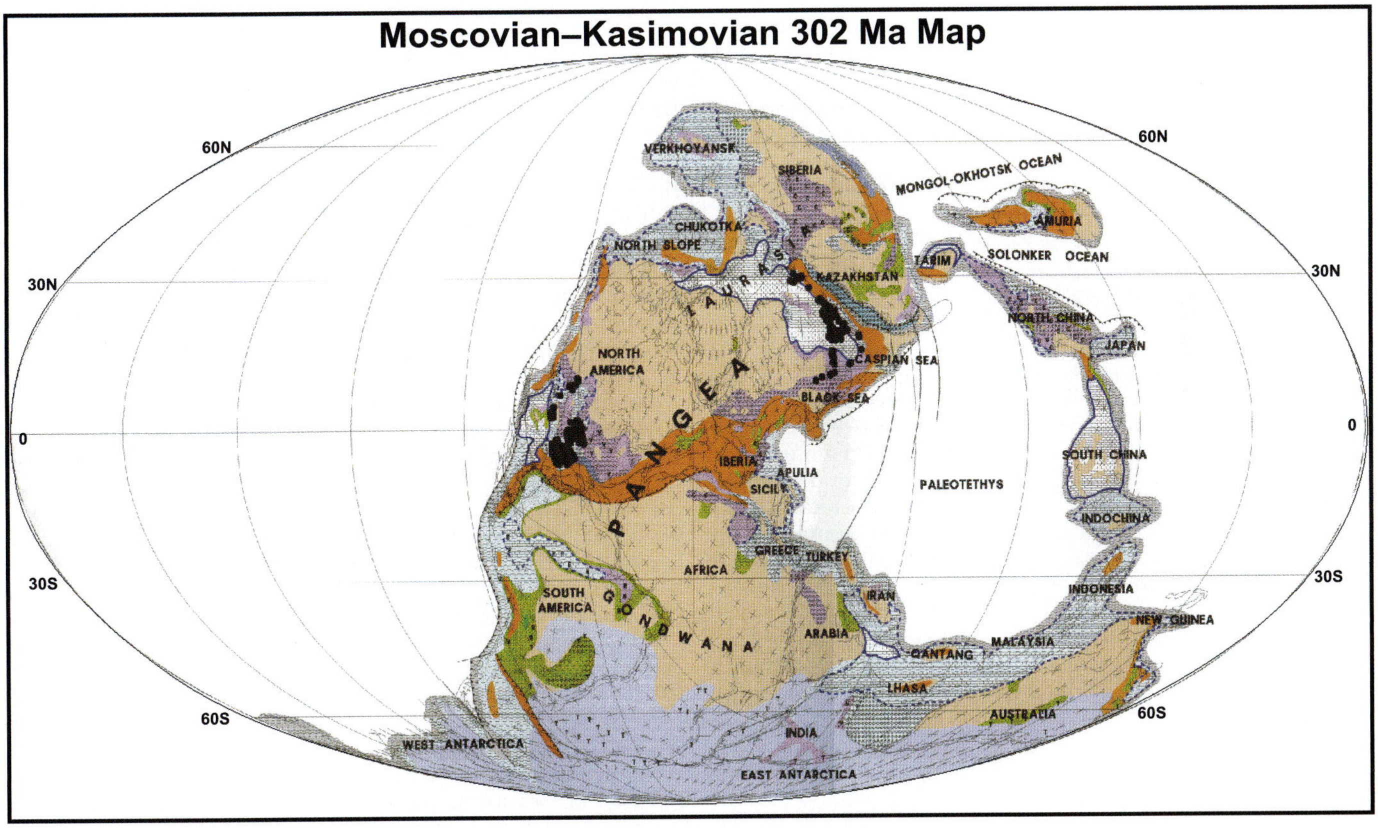

FIG. 3.—Example of an upgraded paleogeographic map (Moscovian–Kasimovian 302 Ma Map—Pennsylvanian or Late Carboniferous). Starting paleogeographic maps were from Scotese (PALEOMAP Project @www.scotese.com), and showed only (1) deep ocean basin—white color, (2) shallow marine shelf—light blue color, (3) land—tan color, and (4) mountains—red color. The Mobil Global Themes Project provided significant detail to these basic maps by adding (1) 11 tectonic elements (symbols), (2) 22 differentiated environments of deposition (colors), (3) 18 depositional qualifiers and sediment types (symbols), and (4) 19 lithologies (black and white patterns). The differentiated environments are grouped into four depositional categories: continental erosional, continental depositional, marginal marine, and marine. Lithologies are grouped into six categories: siliciclastics, carbonates, evaporites, mixed systems, volcanics, and organic-rich. Refer to second page of Appendices 2 to 5 for the complete pictorial map legend.

with the *Phanerozoic Carbonate Trends Chart*, these upgraded paleogeographic maps can be considered as summative database products. There are no new and original data on the maps from Mobil-funded field studies. Rather, these products are integrated summaries of extant data and interpretations. Excerpts and cut-out windows of some of the maps have been published by Ford and Golonka (2003) and Golonka et al. (2003).

Development of the Carbonate Analog Through Time *Maps (Appendices 1 and 2).—*

A principal goal of the project was to empirically map known Phanerozoic carbonate platforms and carbonate fields/reservoirs in their proper temporal and spatial context on a series of global-scale paleogeographic base maps. It was anticipated that such maps would provide the basis for development of concepts that could be used to predict characteristics of carbonate systems and carbonate reservoirs. To this end, a world-wide, Phanerozoic carbonate fields database was compiled (Appendix 1). These carbonate fields were merged with the upgraded paleogeographic maps to make the "*CATT* Maps" (Appendix 2).

Preparation of the world-wide, Phanerozoic carbonate fields database was accomplished by:

1. Accessing both in-house proprietary and commercially available databases from Petroconsultants S.A./IHS Energy, Inc.

2. Extracting information on carbonate fields from both data sources. Key mapping data that were extracted included:
   a. Field name
   b. Location; country, state/province, and present-day X, Y coordinates
   c. Basin name
   d. Geologic age
   e. Reservoir name and main lithology
   f. Various field-size and reservoir-quality parameters

3. Initial extractions resulted in over 14,000 entries. Quality control checking of extracted fields was undertaken to ensure field name, age, basin, and reservoir lithology to be carbonate. This first round of quality checking reduced the number of potential fields for mapping to 8570. The next round of quality control occurred during the mapping phase. The final number of Phanerozoic carbonate fields used in the summary frequency histogram of field by age was 8369. Total number of fields on the 29 maps was 8325.

4. Assigning each field to its appropriate geologic stage if the field had not been previously assigned within the database, and then assigning each field to a specific paleogeographic map.

5. Partitioning the database into individual files corresponding to the paleogeographic maps.

The database of carbonate fields used for the project consists of data from 1999 and earlier and is given in Appendix 1. Approval to publish this list has been granted by Petroconsultants S.A./IHS Energy, Inc. These data are copyrighted with all rights reserved by them, and may not be republished without further authorization. Appendix 1 is an Excel workbook file with two worksheets. The first worksheet is the carbonate fields list, and the second worksheet is the summary frequency histogram of fields by time interval. For the fields list, columns of data include: (1) map age, (2) field name, (3) country, (4) basin name, 5) present-day latitude, (6) present-day longitude, (7) paleo-latitude, (8) paleo-longitude, (9) reservoir name, (10) reservoir lithology, and (11) comments. This list contains the 8570 carbonate fields that survived the initial quality control screening. Color annotation has been employed along with comments that explain the exclusion of another 201 fields to arrive at 8369 as the final number of fields used in the summary histogram. 44 fields were not mapped because of lack of lat/long data, but were included in the histogram counts because of field age, basin, formation name, and reservoir lithology verification. Thus, 8325 total fields were mapped on the 29 paleogeographic maps. Reasons for elimination of a field from mapping include: (1) duplicate sample, (2) errors in present-day location coordinates, (3) errors in paleo-rotation to yield inaccurate paleo-coordinates, (4) clastic reservoir lithology (missed during initial screening), (5) assigned to wrong map or age, and (6) Russian fields data are sparse; many fields were eliminated for lack of verified formation and lithology information. Although every effort was made to screen and quality check fields in the database, verification of every field in the database was impossible, and thus, erroneous entries and/or omissions in the database may exist.

Upon completion of the database, two summary diagrams were produced: (1) the present-day global distribution of carbonate fields by size (Fig. 4), and (2) a histogram of field abundance versus age (Fig. 5). Note that we did not gain approval to include reservoir parameters such as volumes and rock properties in the fields database (Appendix 1). Figure 4 is based on proprietary data originally gathered for the project. The combination of these two figures reveals temporal and spatial trends of carbonate fields and demonstrates that carbonate fields are neither uniformly nor randomly distributed in time and space. Rather, these diagrams suggest that carbonate fields and reservoirs are causally distributed. This conclusion is the rationale and underpinning for development of predictive concepts. As above, the carbonate fields database and these two figures represent an integrated summative database of extant data.

Following compilation of the carbonate fields database, the *CATT* maps were made using both the field files and the Phanerozoic paleogeographic base maps. Methods employed for making each map were:

1. Using the modern-day plate-tectonic map of the world, each field within a time-slice data file was assigned to a specific tectonic plate or plate fragment. Plate reconstructions of Scotese have naming conventions for each plate or fragment, and these were adopted.

2. Using early versions of the plate reconstruction rotation programs of Scotese (***Plate Tracker*** and ***Point Tracker*** from PALEOMAP Project), each carbonate field was rotated from its present-day latitude and longitude location to its paleogeographic location at time of deposition. The paleo-latitude and paleo-longitude values were recorded in the database files.

3. Each carbonate field was plotted as a black dot onto the paleogeographic base map, using the rotation-derived paleo-latitude and paleo-longitude coordinates.

4. Again the rotated fields and their corresponding paleo-map positions were quality controlled.
   a. If the paleo-map position of a field was correct, no action was taken.

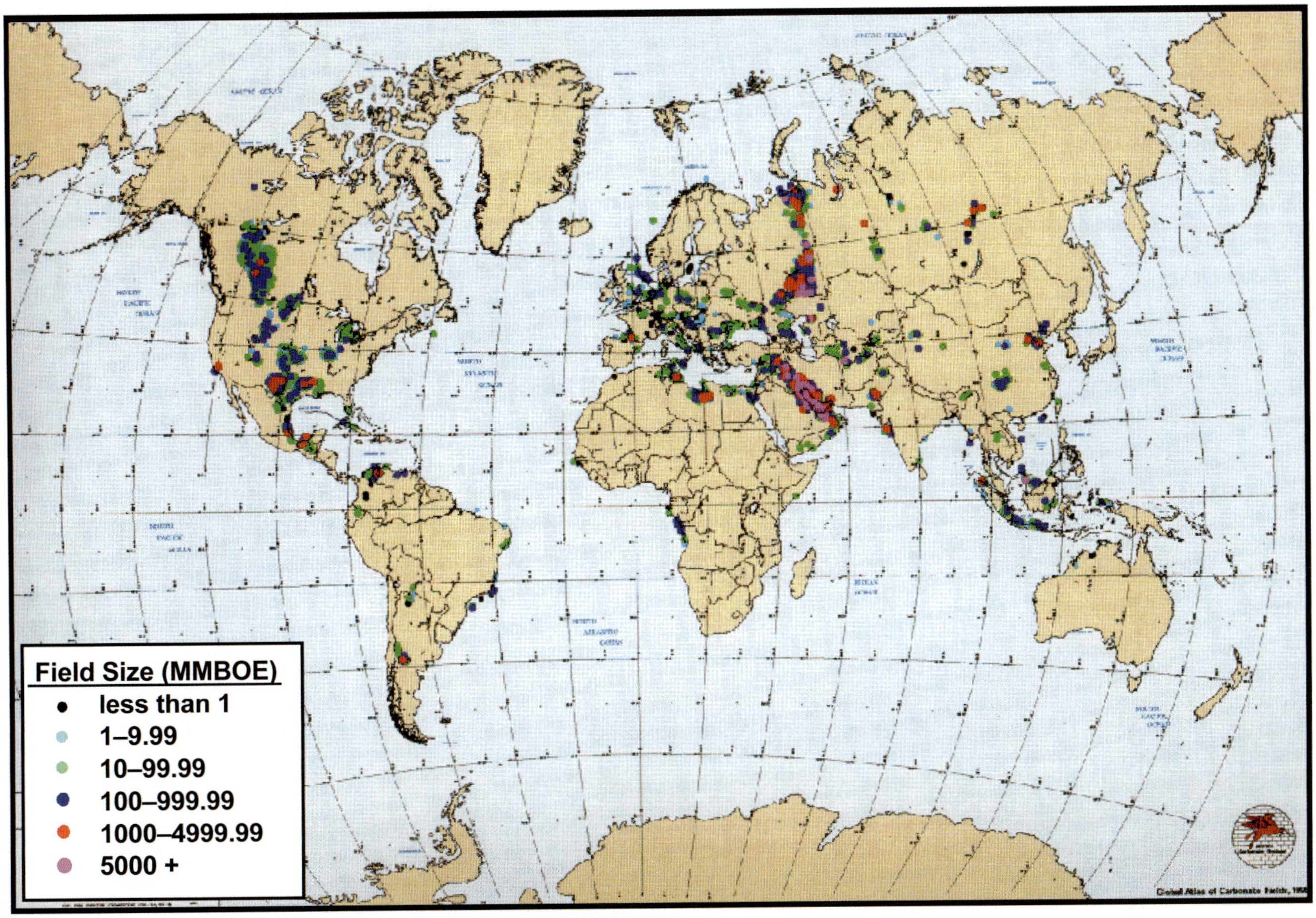

FIG. 4.—World map with Phanerozoic carbonate fields plotted in present-day locations. Size and color of dot for each field represents field size measured in hydrocarbon volume (MMBOE = millions of barrels of oil equivalent). Many of the famous hydrocarbon-rich basins with large and / or numerous carbonate fields / reservoirs are easily recognized on this map. They are the Arabian basin in the Middle East, Volga–Ural and Timan–Pechora basins in Russia, Sirte basin in Libya, Bohai and Sichuan basins in China, Alberta basin in western Canada, the cratonic basins in the U.S.A., coastal basins of Mexico, Maracaibo basin in Venezuela, Neuquén basin in Argentina, and the small complex tectonic Neogene basins of Southeast Asia. Based on this map, it is obvious that carbonate fields and reservoirs are clustered to specific basins, and that they are neither uniformly nor randomly distributed across the present-day world.

b. If the paleo-map position of a field was obviously incorrect, e.g., plotted at a pole, at the origin, in the middle of an ocean, or at the top of a mountain belt, the rotations were evaluated and corrected if possible (see comments Appendix 1).
c. If a field plotted within a siliciclastics-dominated craton-interior region, the geology of the field was investigated, and if it was found to be a sandstone reservoir, it was removed from the database.

5. Once the carbonate field points were plotted on a map, the map was finalized by outlining and accentuating known carbonate platforms. Platforms both with and without known fields were outlined.

Thus a for a given time interval, the completed *CATT* map consists of the upgraded paleogeography as the base map, an overlay of the carbonate fields / reservoirs (black dots) identified for that time slice, and emphasized and annotated carbonate platforms for the time slice whether a platform has discovered fields or not (e.g., Late Jurassic map, Figure 6). All 29 maps for the Phanerozoic and the symbol explanations are included on the enclosed CD as Appendix 2.

Regrettably, we can publish only scanned versions of the original maps (Appendix 2). Although these maps are small and have poor readability as text figures, readability is improved when they are examined on the enclosed CD using 200% or more magnification. In total, the map series attempts to capture, on a time-interval and paleogeographic basis, the total sum of carbonate platforms, deposition systems, and reservoirs that were recognized at the time of the project. It is anticipated that these maps will be updated, modified, and corrected as further investigations are completed and as new fields are discovered.

Summary of Products.—

The *Carbonate Analogs Through Time* project resulted in development of two key products: (1) the *Phanerozoic Carbonate Trends*

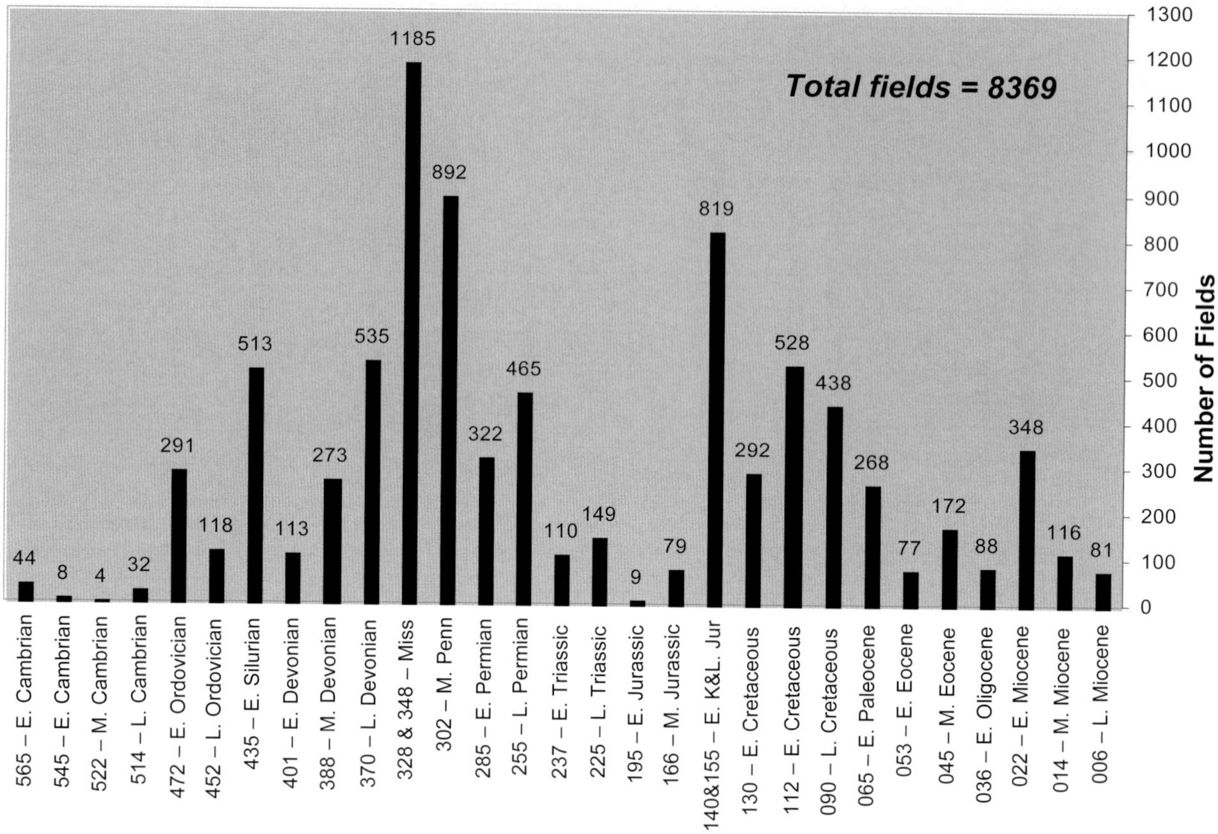

FIG. 5.—Frequency distribution by age of carbonate fields for the Phanerozoic. This histogram was built from the carbonate fields database (Appendix 1). Each histogram bar is labeled with geologic age (years Ma and geologic period name; E. = early, M. = middle, L. = late) on X axis and with number of fields above bar. It is clear from this plot that carbonate reservoirs are not uniformly distributed across all geologic time periods. Major abundance spikes in carbonate reservoirs occur during the Early Ordovician, Silurian, Late Devonian, Carboniferous, Late Permian, Late Jurassic, middle to Late Cretaceous, and Miocene. Conversely, times when few carbonate reservoirs are known include Cambrian, Early Devonian, Triassic to Early Jurassic, and Paleogene. It is important to remember that this plot does not imply that carbonate platforms did not develop during geologic times with few known carbonate reservoirs. In fact the *CATT* maps outline the known carbonate platforms of these ages. These observations provide the bases for proposing two research questions. (1) Why did Cambrian, Early Devonian, Late Triassic, Early Jurassic and Paleogene carbonate platforms not yield or not become more abundant carbonate fields/reservoirs? (2) Have we missed some exploration opportunities?

Chart, and (2) the *Global Atlas of Carbonate Fields*, which includes the worldwide Phanerozoic carbonate fields/reservoirs database, and the 29 *CATT* maps. Combined, these products represent an integrated summative database for carbonates. Although new data were not generated for this project, extant data were gathered and integrated into these new products using an approach defined by a hypothesis. We refer to these products as our pictorial integrated carbonate database.

Development of Nonquantitative Paleo-Oceanographic Maps.—

To increase versatility and go another step beyond the Phanerozoic paleogeography maps with superimposed carbonate fields, oceanographic characteristics were integrated with the maps to generate three new map series (Appendices 3, 4, and 5). Paleo-oceanography is included in the *CATT* equation as an Earth process, and clearly, oceanographic processes today are and in the geologic past have been important and critical controls on location, size, and vitality of carbonate depositional systems (e.g., James, 1978; Hallock and Schlager, 1986; Scott, 1988; Handford and Loucks, 1993; Wood, 1995; Kuznetsov, 1996; Flügel and Kiessling, 2002; Kiessling et al., 2003). It is recognized that significant efforts have been undertaken by various workers and institutes to develop quantitative parametric modeling algorithms for determining local to global-scale paleoclimates and paleo-oceanography (e.g., Parrish, 1982; Parrish and Curtis, 1982). However, it has never been within the scope of the *Carbonate Analogs Through Time* endeavors to complete comprehensive and systematic quantitative parametric models for the 29 paleogeographic maps. Instead, a nonquantitative, analog, and conceptual approach was undertaken to develop global-scale paleo-oceanographic characterizations for the 29 maps.

This nonquantitative, analog approach was based on modern-day oceanographic phenomena and corresponding global-scale maps from Earle (2001) and Mehler et al. (2002). Oceanographic phenomena evaluated for this phase of the study included:

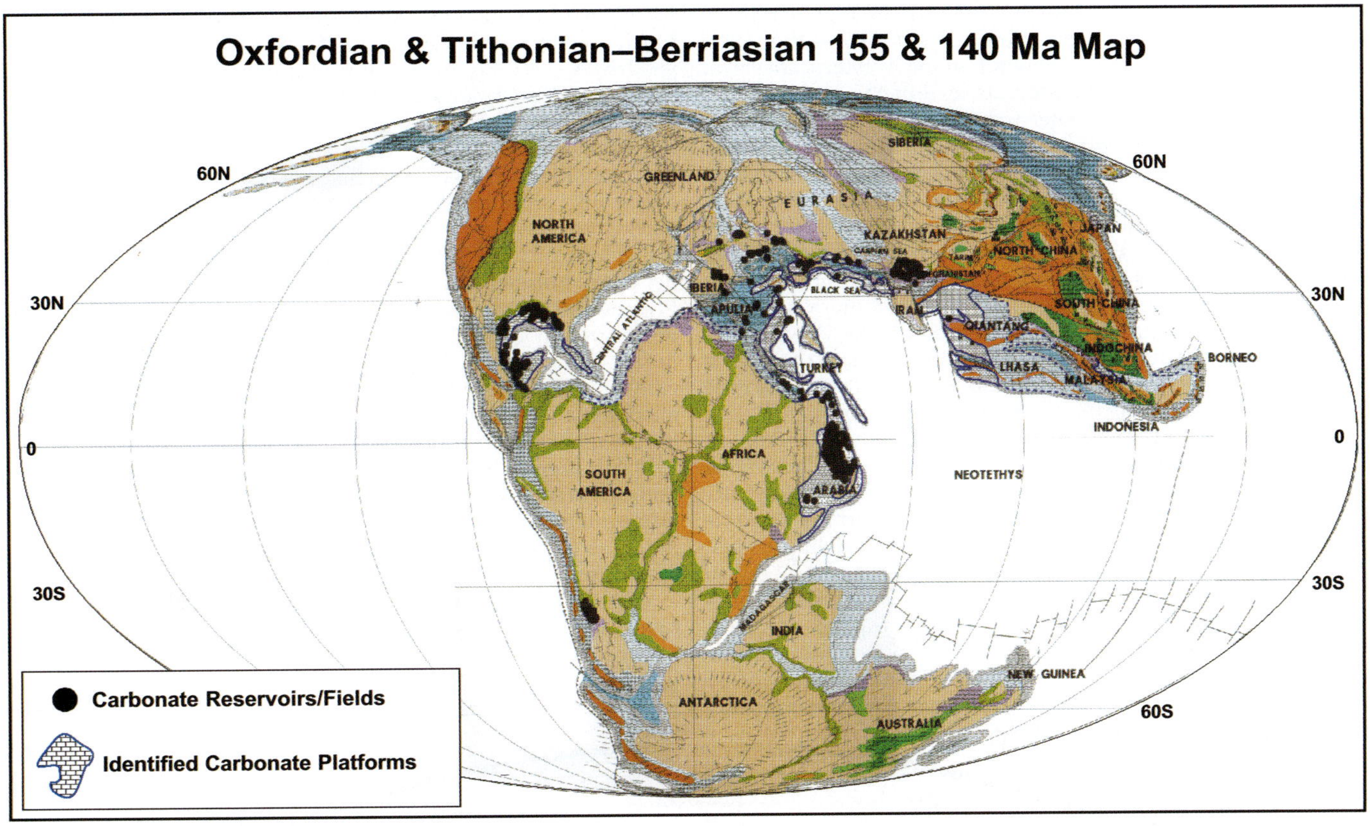

FIG. 6.—Example of a *"Carbonate Analogs Through Time (CATT)* Map". This is the combined map for the Late Jurassic to Early Cretaceous time interval (Oxfordian—155 Ma and Tithonian to Berriasian—140 Ma map). Note at this time there are four significant clusters of carbonate fields: (1) Arabian basin in the Middle East at the paleo-equator at the western edge of the Neotethys–Panthalassic "super-ocean"; (2) coastal basins around the Gulf of Mexico, US and Mexico, between 15° N and 25° N latitude in the newly formed Gulf and adjacent to the Central Atlantic; (3) Kopeh Dagh basin of Turkmenistan at 32° to 35° N latitude sheltered behind the Iranian continental block along the northern edge of the Neotethys Ocean; and (4) Neuquén basin, southern Argentina, at 33° to 35° S latitude along the southeastern edge of the Panthalassic Ocean. Other solitary fields and small groups of fields are scattered among the complex tectonic settings of the northwest Neotethys Ocean. Other carbonate platforms are outlined within the Neotethys ocean, and note that some of the platforms lack discovered carbonate fields. All 29 *CATT* maps are presented in Appendix 2 and are on the enclosed CD.

1. World climate zones (Mehler et al., 2002, p. 34–35)
2. Prevailing global wind patterns (Earle, 2001, p. 33)
3. Global hurricane tracts (Mehler et al., 2002, p. 37)
4. Global ocean currents and circulation (Mehler et al., 2002, p. 34–35)
5. World ocean surface water temperature (Earle, 2001, p. 50–51, 98-99, 108–109; Mehler et al., 2002, p. 12–13)
6. World ocean vertical temperature profiles (Earle, 2001, p. 26–27)
7. World ocean tidal ranges (Mehler et al., 2002, p. 37)

Of these, world ocean surface water temperature, world ocean tidal ranges, and global hurricane tracts have been applied to the 29 paleogeography maps.

Global Paleo-Ocean Surface Water Temperature CATT Maps (Appendix 3).—

In general, the present-day global circulation pattern of ocean surface currents is driven by a combination of prevailing surface wind directions and geostrophic currents generated by temperature and salinity gradients (Mehler et al., 2002; Flügel and Kiessling, 2002). Open ocean circulation is typically clockwise in the northern hemisphere between 40° N latitude and the equator, and counterclockwise in the southern hemisphere between 40° S latitude and the equator. This is because of west-to-east flow between 30° and 40° north or south, and east-to-west flow just north and south of the equator. In high latitudes, typically at and above 60°, flow is east to west. A major consequence of this global ocean circulation pattern is that it sets up the spatial distribution or surface geometry for ocean surface water temperature.

Typically cool to colder, mid-latitude ocean water travels west to east until it encounters a western continental shelf and/or coast. If the current is at about the 30° latitude range, the continental coast will/may deflect the east-moving current toward the equator. If the east-moving current is at a higher latitude range, the coast will/may deflect the current toward the poles. Water deflected toward the equator along western coasts of continents is of cool to cold temperature; thus the surface water temperature isotherms of cooler waters will be closer together north and south of the equator along such western coasts. As north-flowing and/or south-flowing ocean surface currents converge on the equator, they are turned westward. Since the equatorial belt receives the highest and most sustained solar energy flux (Mehler et al., 2002), the ocean surface waters warm as they flow east to west. Thus, because of this circulation and warming process, the ocean surface water temperature isotherms diverge from the equator in an east-to-west expanding wedge. Consequently, western equatorial regions of oceans and of eastern coasts of continents typically have the warmest waters and the greatest north-to-south expanse of those warm waters.

These present-day open ocean circulation patterns and ocean surface water temperature trends are interpreted to have been similar for past ocean-basin and continental-plate reconstructions. It is beyond the scope of this work to undertake rigorous quantitative parametric paleo-oceanographic modeling to test these interpretations. To move forward, these concepts for ocean surface water temperature were systematically applied by analogy to each of the 29 maps (e.g., Fig. 7). Also applied was the *a priori* historical geologic knowledge of specific times of global Earth greenhouse and icehouse climates (see *Phanerozoic Carbonate Trends chart*, Figure 1). During greenhouse times, the ocean surface water temperatures were characterized as being warmer at higher latitudes, and during global icehouse times, warmer ocean surface waters were restricted, on a conceptual, nonquantitatively calculated basis to between 15° and 25° north and south latitudes. It is understood that these maps cannot be substitutes for rigorous quantitative parametric modeling. The intent is to provide a thought process and qualitative guidelines for expectations of surface water temperatures throughout the Phanerozoic. Such intent assists with comparative analyses and development of predictive concepts. It is anticipated that these maps will be superseded in the future by quantitative parametrically modeled oceanography. The 29 maps with overlays of qualitatively mapped ocean water circulation and surface water temperature are included on the enclosed CD as Appendix 3.

With regard to present-day ocean currents and circulation patterns, it is understood that in regions of complex tectonics with numerous, closely spaced small continental blocks and fragments separated by deep seaways, and proximal to one or more major continents with submerged, shallow-water shelves, for example southeast Asia, ocean currents flowing along and between the continents and smaller fragments result in very com-

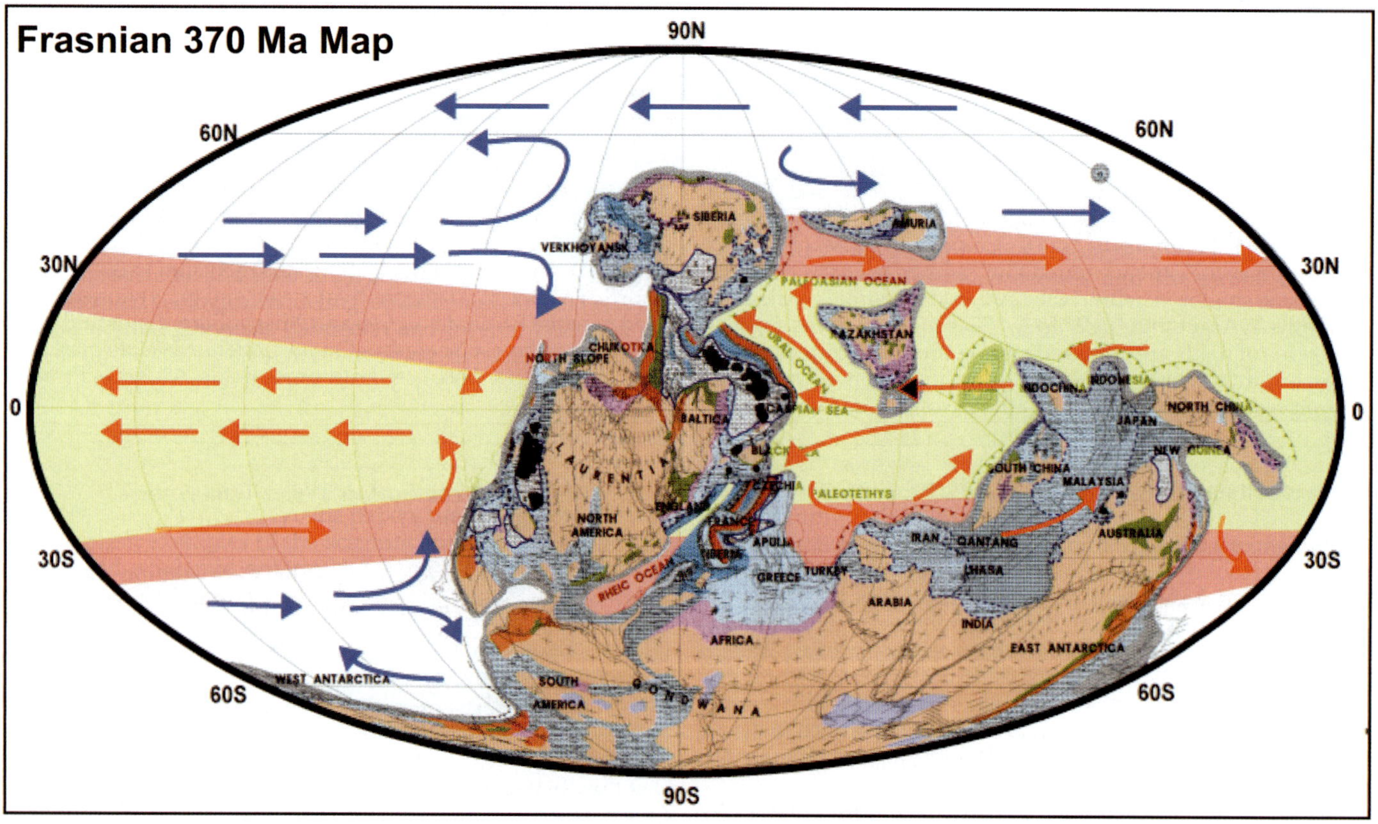

Fig. 7.—Example of paleo-oceanographic map displaying interpreted paleo–ocean surface water temperature (Late Devonian, Frasnian—370 Ma map). This is a nonquantitative, analog depiction of surface ocean water temperature with superimposed generalized global ocean circulation patterns. Interpreted global ocean surface water temperature is depicted by color (yellow represents temperatures from 35° C to 25° C; red represents 25° C to 15° C; white represents less than 15° C). Blue arrows represent cool-water and cold-water surface currents; red arrows represent warm-water surface currents. See text discussion for explanation of polygon wedges that depict interpreted geometries of surface-water isotherms. All 29 interpreted paleo–ocean surface water temperature maps are presented in Appendix 3 and are on the enclosed CD.

plicated circulation patterns. Although paleo-plate reconstructions can show locally to regionally complex continent and continental-fragment orientations, no attempts were made to interpret complex ocean current circulations or local variations or anomalies in ocean surface water temperature within complex tectonic regions displayed on the maps (Appendix 3).

Global Paleo–Tidal Ranges CATT Maps (Appendix 4).—

Mehler et al. (2002, p. 37) present a global map of modern-day tidal ranges. Global tides and tidal flow are complex oceanographic phenomena that have been related to many factors, including lunar gravitational pull, ocean current circulation patterns, continental position, orientation, and coastline geometry, just to list a few. Consequently, development of quantitative parametric modeling algorithms for calculating tidal range is a complicated task, the complexity of which increases as uncertainty increases with the input variables. Thus, parametric modeling of paleo–tidal ranges would be a highly uncertain, if not an intractable, endeavor, given high uncertainties of paleo–plate positions and orientations that continually increase with increasing age toward the Early Cambrian. However, the map from Mehler et al. (2002, p. 37) displays a straightforward, first-order, nonquantitative, and observable pattern for present-day tidal range.

Combining ocean surface current flow patterns and tidal ranges, it is evident that there are three observable situations where large tidal ranges are found:

1. where surface currents impinge upon or intersect a continental coastline (e.g., west coast of Canada or the west coast of U.K.);
2. where ocean currents are forced to flow through narrow and shallow constrictions between continental masses (e.g., between Africa and Madagascar);
3. where surface currents flow along a shallow shelf and are forced into coastal "cul de sac" type estuaries (e.g., Bay of Fundy and the Gulf Stream current of the eastern coast of North America).

Further, tidal ranges will be complex and highly variable where the tectonics are complex and continents and continental blocks are numerous and closely spaced. These observations provide a basis or set of concepts with which to make qualitative, straightforward interpretations of paleo–tidal ranges on the 29 paleogeographic maps (e.g., Fig. 8). In the absence of rigorous parametric models, these interpretations, as for the ocean surface water temperature maps, are presented to provide first approximations and / or guidelines to paleo-tides, and as thought-provoking maps for predictive concept development. We expect these maps to be upgraded, modified, and corrected with continued investigations and more quantitative analyses.

Global Paleo-Climate CATT Maps (Appendix 5).—

Quantitative modeling of global-scale paleoclimates in the early 1980s by Parrish (1982) and Parrish and Curtis (1982) focused on the purpose of predicting conditions conducive to coastal upwelling, with the goal of predicting areas of source-rock deposition. Golonka et al. (1994) furthered global paleoclimate modeling as a joint effort between Mobil and the University of Texas at Arlington. As a result of these efforts, a series of paleoclimate maps was published for 29 Phanerozoic time slices (Golonka et al., 1994). Because these maps are company documents, they are included with this publication as Appendix 5 to complement the other Phanerozoic paleogeographic map sets of Appendices 2, 3, and 4.

The paleoclimate maps (e.g., Fig. 9) display the originally modeled climate regions with (1) weather cells of high and low atmospheric pressure cells, (2) calculated wind vectors, (3) regions of predicted high atmospheric humidity (blue regions on maps), and (4) regions of predicted coastal upwelling, (Golonka et al., 1994). Interpreted paleo-hurricane storm tracks based on analogy with modern-day storm tracks after Mehler et al. (2002) have been added to the maps.

Summary of Paleo-Oceanographic and Paleo-Climate Maps (Appendices 3, 4, and 5).—

The three paleo-oceanographic (paleo–ocean surface water temperature; paleo-tides) and paleo-climate) map series are post–year 2000 additions to the *Global Atlas of Carbonate Fields*. The intent in building these map series was to take the original *CATT* maps into new directions that would provide new perspectives to understand temporal and spatial causes for carbonate platform and field / reservoir occurrences and characteristics and to facilitate within-time-slice and between-time-slice comparative analyses. It is fully recognized that these are nonquantitatively derived maps. They represent an attempt to apply modern-day physical observations, principles, and / or concepts, in an analog manner, to the Phanerozoic maps. We recognize that, by the very nature of the *Carbonate Analogs Through Time* idea, the modern is not the key to the past for many geologic characteristics. However, these maps can be viewed as a test of the idea that some physical oceanographic and climatic processes, observed and understood in present-day settings, can be appropriately applied to Phanerozoic paleogeographic reconstructions in qualitative ways. In the absence of support and time for rigorous quantitative parametric modeling of all 29 Phanerozoic maps, we provide these analog maps as an interpretive summative database of ideas for paleo-oceanographic conditions.

Applications of the CATT *Hypothesis and Key CATT Elements and Products: Introduction.—*

To date, many categories of applications have been developed using the hypothesis, approach, and derivative key elements and products. Five applications have been approved for publication:

1. Age-sensitive patterns and time-slice carbonate themes.
2. Synthesized field posters of *CATT* maps and time-slice themes.
3. Intra-time-slice and inter-time-slice comparative analysis.
4. Methodology for selection of analogs for carbonate fields.
5. Predictive concepts for carbonate systems and reservoirs.

It is fully anticipated that new applications will be developed. One of our goals for the hypothesis and approach is to deliver an overall concept design and associated products with sufficient flexibility and breadth that ideas and materials can be combined and customized to formulate new research directions or can be used to solve reservoir problems at hand. The following sections describe the ideas and methods for these applications. Typically, only one example per application is presented here.

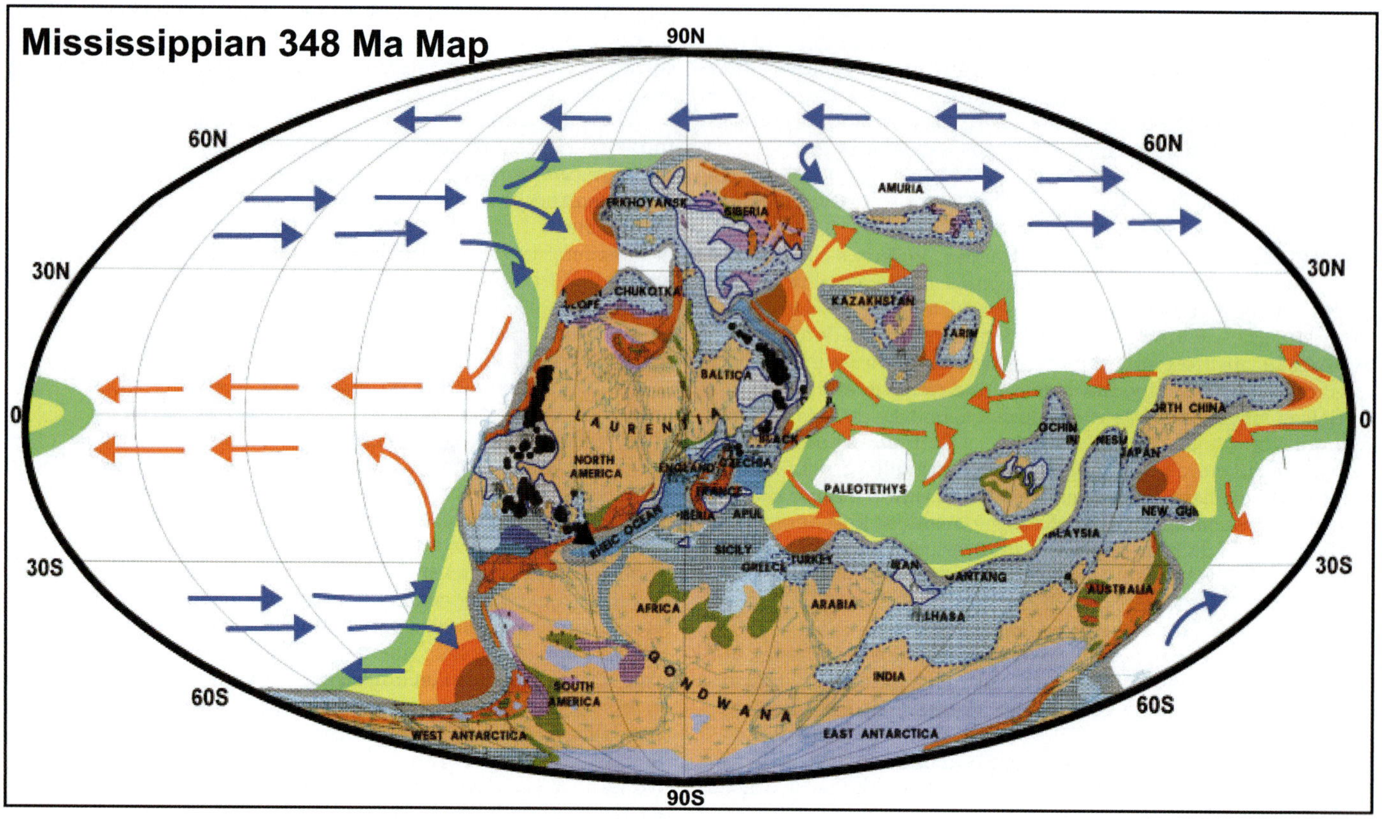

FIG. 8.—Example of paleo-oceanographic map displaying interpreted paleo–tidal levels (Mississippian—328 Ma map). This is a non-quantitative, analog depiction of interpreted paleo–tidal ranges with superimposed generalized global ocean circulation patterns. Tidal range is represented by color. White and green are microtidal levels (0–1 m). Yellow and light brown are mesotidal ranges (1–3 m). Red and dark brown colors are macrotidal ranges (> 3 m). Blue arrows represent cool-water and cold-water currents; red arrows represent warm-water currents. See text discussion for explanations of interpreted locations of high vs. low tidal regime. All 29 interpreted paleo–tidal regime maps are presented in Appendix 4 and are on the enclosed CD.

Age-Sensitive Patterns and Time-Slice Carbonate Themes (Appendix 6).—

The *CATT* Hypothesis contends that for any given geologic age or time interval (defined as a time slice), the summation of the ambient states of carbonate and Earth processes defines a unique set of characteristics or a sedimentary signature for the strata, stratigraphic sequences, successions, and carbonate reservoirs of that geologic time slice. This is a testable hypothesis in that analyses and characterizations of age-equivalent systems across basins and continents will reveal similarities and differences between systems. Definition of the word "unique," as it applies here, requires tolerating some variability within the idea of the time-slice "sedimentary signature."

Time-slice carbonate themes were prepared for nine geologic time periods (Table 3). All of the age-based themes, including text and figures, are presented in Appendix 6 (see enclosed CD). None of the themes is recapped in this text; only the structure and style of the themes are discussed. As can readily be seen in Table 3, the span in geologic age for the time intervals with carbonate themes is variable and in some cases quite large (from 9 million years for the Guadalupian theme to more than 60 million years for the Sauk theme).

Time-slice carbonate themes are organized into a standardized format that facilitates intra-time-slice and inter-time-slice comparative analysis. The outline for a time-slice theme with its key characterization elements is:

Tectonics and Stratigraphic Architecture
I. Global Tectonics
II. Sequence Architecture
III. Key Tectonic and Stratigraphic Themes

Carbonate Themes
I. Global Climate
II. Age-Based Themes
III. Reservoir Types and Diagenesis
IV. Geographic Platform Development and Distribution of Carbonate Fields.
V. Key Carbonate Themes

Embedded within each time-slice theme are characterizations of the carbonate reservoirs of that age (Appendix 6). While the statements may appear to be generalized, they attempt to capture highlights, key characteristics, and through-going global themes. Certainly, specific examples may have local and individualistic qualities that are not covered by the large-scale themes. In these cases, investigation of important local factors such as diagenesis or unusual environmental factors may be warranted to explain causes of discrepancies between global themes and local variabil-

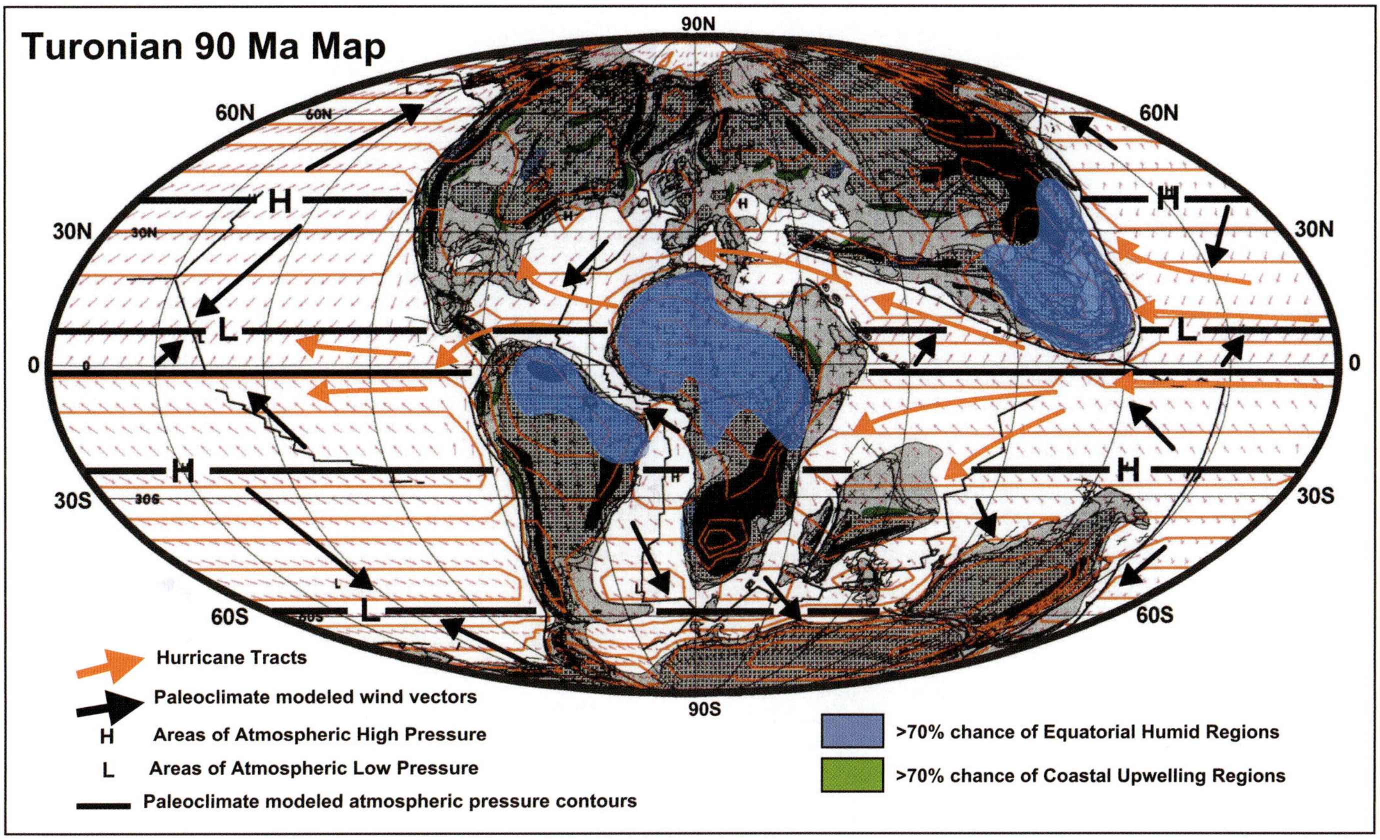

FIG. 9.—Example of a paleoclimate-modeled map from the map series of Golonka et al. (1994) (Late Cretaceous, Turonian—90 Ma). This map was first developed using 1990s computer models for paleoclimate reconstruction. Atmospheric attributes and climate phenomena probabilities were calculated from parametric modeling. Hurricane or large oceanic storm tracks were nonquantitatively determined, and were included on the maps as analog interpretations for this research project. See text discussion for explanation of paleo-storm tract interpretations. All interpreted paleoclimate maps with superimposed storm tracks are presented in Appendix 5 and are on the enclosed CD.

ity. Further, it is recognized that the characterizations for each theme are not exhaustive; they do not contain all that is known about the time interval. Our intent was not to recap all of Phanerozoic geologic history, but rather provide highlights. The theme structure is flexible enough to include additional information as needed or desired.

An age-based theme is developed by convolving data from the *Phanerozoic Carbonate Trends Chart* (Fig. 1), observations from the paleogeographic maps, and results from literature and case studies (Fig. 10). Articulation of an age-based theme for carbonate reservoirs takes the form of a series of statements or points that capture the key characteristics of stratigraphic architecture, reservoir layering, facies, grain types, lithology, mineralogy, pore types, and fracture enhancements. The purpose of an age-based theme is to set expectations for the characteristics of a carbonate system in advance of acquiring platform-, field-, or reservoir-specific data.

Development of Integrated CATT-*Map-Based Field Posters (Appendix 7).—*

Another application is formulation of nine field posters that span the Phanerozoic, and which are organized according to the Sloss (1963) first-order and second-order scales of stratigraphy. The names of the posters are Upper Tejas, Lower Tejas, Upper Zuni, Lower Zuni, Upper Absaroka, Lower Absaroka, Kaskaskia, Tippecanoe, and Sauk. Each poster is a separate file in Appendix 7. Each field poster is a matrix with three to five rows representing geologic age or time intervals and six columns containing the key element maps and themes from Appendices 2 through 6. The first column on the left establishes the geologic ages and defines the time intervals for the poster. Columns 2 through 5 contain the various global-scale maps with fields and interpreted paleo-oceanography and paleo-climates. The final column 6 is a recap of the key Earth and carbonate processes that comprise the time-interval themes. These posters have been used on field trips and schools to set large-scale (global, basinal, regional) context for the details and focused discussions of local-scale geology.

Intra-Time-Slice Comparative Analysis.—

Although the *CATT* Hypothesis is primarily designed to develop holistic themes for any given geologic age or time slice, the framework of the hypothesis also provides an approach for systematic comparative analysis of coeval carbonate systems. The qualitative summing of Earth and carbonate processes to produce a theme can be rewritten such that for any given carbon-

TABLE 3.—Definitions of the geologic time intervals for which age-sensitive patterns or time-slice themes were written and included as a synthesis section in the *Global Atlas of Carbonate Fields*. The complete text and figures for these themes are included as Appendix 7 on the enclosed CD.

Geologic Time Interval for Theme development	Sloss (1963) Nomenclature	CATT Maps included in Time Interval	Ages of CATT Maps
Tertiary	Tejas	Messinian Serravallian Aquitanian Rupelian Lutetian Ypresian	006 Ma 014 Ma 022 Ma 036 Ma 045 Ma 053 Ma
Early Tertiary to Early Cretaceous	Zuni III and Zuni II	Danian Turonian Albian Barremian	065 Ma 090 Ma 112 Ma 130 Ma
Early Cretaceous to Middle Jurassic	Zuni I	Berriasian & Oxfordian Bajocian–Bathonian	140 Ma & 155 Ma 166 Ma
Early Jurassic to Triassic	Upper Absaroka	Pliensbachian–Toarcian Norian Scythian–Carnian	195 Ma 225 Ma 237 Ma
Late Permian	Uppermost of Lower Absaroka II	Guadalupian	255Ma
Early Permian to Pennsylvanian	Lower Absaroka I and II	Asselian Moscovian–Kasimovian	285 Ma 302 Ma
Mississippian	Kaskaskia II	Mississippian	348 Ma
Late Devonian to Late Middle Ordovician	Kaskaskia I and Tippecanoe	Frasnian Emsian Lochkovian Landoverian Caradocian	370 Ma 388 Ma 401 Ma 435 Ma 452 Ma
Middle Ordovician Early Cambrian	Sauk	Arenigian–Llandeilian Croixian Albertan Caerfaian Tommotian	472 Ma 514 Ma 522 Ma 545 Ma 565 Ma

ate system or systems, the Earth and carbonate processes are characterized and contrasted, character for character. This systematic approach facilitates understanding of the sometimes large and contrasting differences between age-equivalent carbonate systems. For example, the Late Jurassic time period is one during which many significant carbonate systems were deposited that later became important reservoirs. Two key reservoir systems of this time period are the Smackover Formation in the Gulf of Mexico, U.S.A., and the Arab formations in the Arabian Basin, Saudi Arabia.

These two systems constitute an example for comparative analysis. It is well known that field size and reservoir quality are dramatically different between them. For example, Smackover fields in Louisiana and Arkansas typically have estimated in-place oil accumulations measured in tens to a few hundreds of millions of barrels, whereas reservoirs within the Arab formations in Saudi Arabia have estimated in-place oil volumes measured in the units to tens of billions of barrels. Further, reservoir quality as measured by daily production or well rates demonstrate that Smackover carbonates have poorer reservoir quality than carbonates of the Arab formations. Typical Smackover reservoir wells produce on the order of tens to hundreds of barrels of oil per day, compared to wells of the Arab formations, which can produce on the order of tens of thousands of barrels of oil per day. Using the comparative analysis approach, an understanding of why these differences exist is possible.

The initial step in the analysis is to assemble the time-slice *CATT* maps (Appendices 2–5) and the *Phanerozoic Carbonate Trends Chart* (Fig. 1) and then begin with the generalized time-slice theme (using methodology of Figure 10; see Late Jurassic theme in Appendix 6 and Zuni I field poster in Appendix 7). The synthesis steps are to construct a table that lists the Earth and carbonate processes, record and compare observations and interpretations of processes for each system in the table, and then supplement observations and interpretations from the literature and personal knowledge. For the Late Jurassic example, key comparative maps are the paleo-oceanography and paleo-climate maps (Fig. 11). Table 4 is the final comparative list of characteristics contrasting the Smackover Formation reservoirs with those of the Arab formations.

The systematic, comparative list of characteristics (Table 4) provides the basis for understanding the key differences in reservoir performance between the Jurassic carbonate systems of the Smackover versus the Arab formations. First and foremost, there are two key differences in reservoir facies and pore types. The Arab formations are dominantly grainstone facies of

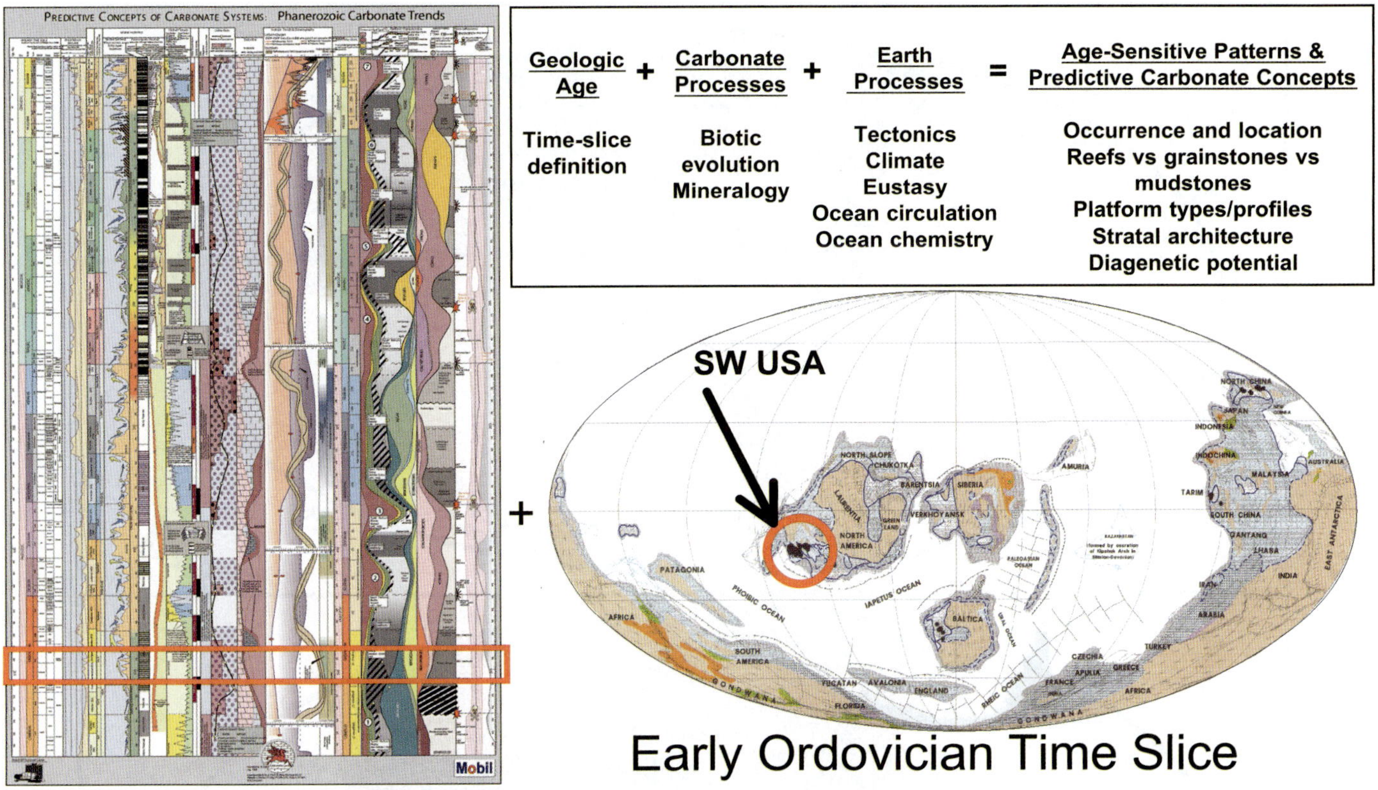

FIG. 10.—An age-based pattern or time-slice theme is developed by solving the CATT equation and convolving the Phanerozoic Carbonate Trends Chart with the age-corresponding paleogeographic map. The time-slice theme example is for Early Ordovician reservoirs, and is developed from characteristics of numerous fields containing Ellenburger Formation reservoirs in the Permian basin of west Texas. The thematic statements for these types of carbonate reservoirs are: (1) Reservoir strata are beds immediately beneath the top-Sauk unconformity. Strata formed during long-term sea-level fall and continental draining that culminated in the inter-continental Middle Ordovician (Sauk) unconformity. (2) Reservoir layers within fields are typically meter-scale peritidal mud-dominated cycles. (3) Layers are thin-bedded with heterogeneous peritidal facies within layers; grain types are intraclasts, peloids, and minimal skeletal fragments, and rare ooids. (4) Carbonate buildups are small (< 1 m to a few tens of meters) and are composed of stromatolites, thrombolites, and / or microbialites. (5) Reservoir quality is moderate and is commonly present only if the strata are dolomitized. Early Ordovician and Cambrian limestones typically lack good depositional primary porosity and permeability. (6) Many Ellenburger reservoirs in west Texas consist of cave-associated facies (e.g., roof-collapse breccia, crackle breccia, etc.) and reservoir quality related to dolomitization of those facies. (7) Typically Early Ordovician reservoirs of the Sauk Megasequence have fracture-enhanced permeability. This is because many times, rocks of this early Phanerozoic age have experienced one or more episodes of postdepositional tectonics. This method of characterizing carbonate reservoirs on a time-slice basis was accomplished for all other time periods (Appendix 6; included on field posters in Appendix 7).

calcitic mineralogy with preserved primary intergranular pores and little to no early marine cements; some interpretations of high-Mg calcite have been made, and minor occurrences of dolomite are observed (e.g., Wilson, 1985; Saner and Abdulghani, 1995; Alsharhan and Whittle, 1995; Saner and Sahin, 1999). Contrastingly, the Smackover Formation contains interbedded grainstone, mud-rich facies, and local microbial or stromatolitic-type buildups of mixed mineralogy (abundant dolomite and calcite converted from aragonite) with both moldic and dolomite intercrystalline pore types (e.g., Baria et al., 1982; Feazel, 1985; Druckman and Moore, 1985; Humphrey et al., 1986; Petta and Rapp, 1990; Chimene, 1991; Heydari and Moore, 1994; Kopaska-Merkel et al., 1994; Bliefnick and Kaldi, 1996; Heydari, 2000). Intergranular pore systems (moderate to high porosity, high permeability) typically perform much better than moldic pore systems (high porosity, low permeability).

Although these differences in the pore systems may explain the relative production performance differences between the Smackover and Arab formations, questions remain concerning the origins of the differences in reservoir facies, mineralogies, and pore-system geometries. For example, why do Smackover reservoirs have thin interbeds of mudstones and local microbial mounds (Baria et al., 1982) interbedded with grainstones while most Arab formation reservoir facies are dominantly skeletal grainstones with rare fine-grained interbeds? Why do Smackover reservoir rocks have mixed calcite, aragonite, and dolomite mineralogies, whereas Arab formation reservoir rocks are almost exclusively calcitic? Dolomite is minor in the Arab reservoirs. Why do Smackover reservoirs have moldic and intercrystalline dolomite pore systems, when Arab reservoirs have overwhelmingly preserved primary pore systems? It is suggested that the maps and *Phanerozoic Carbonate Trends Chart* coupled with established car-

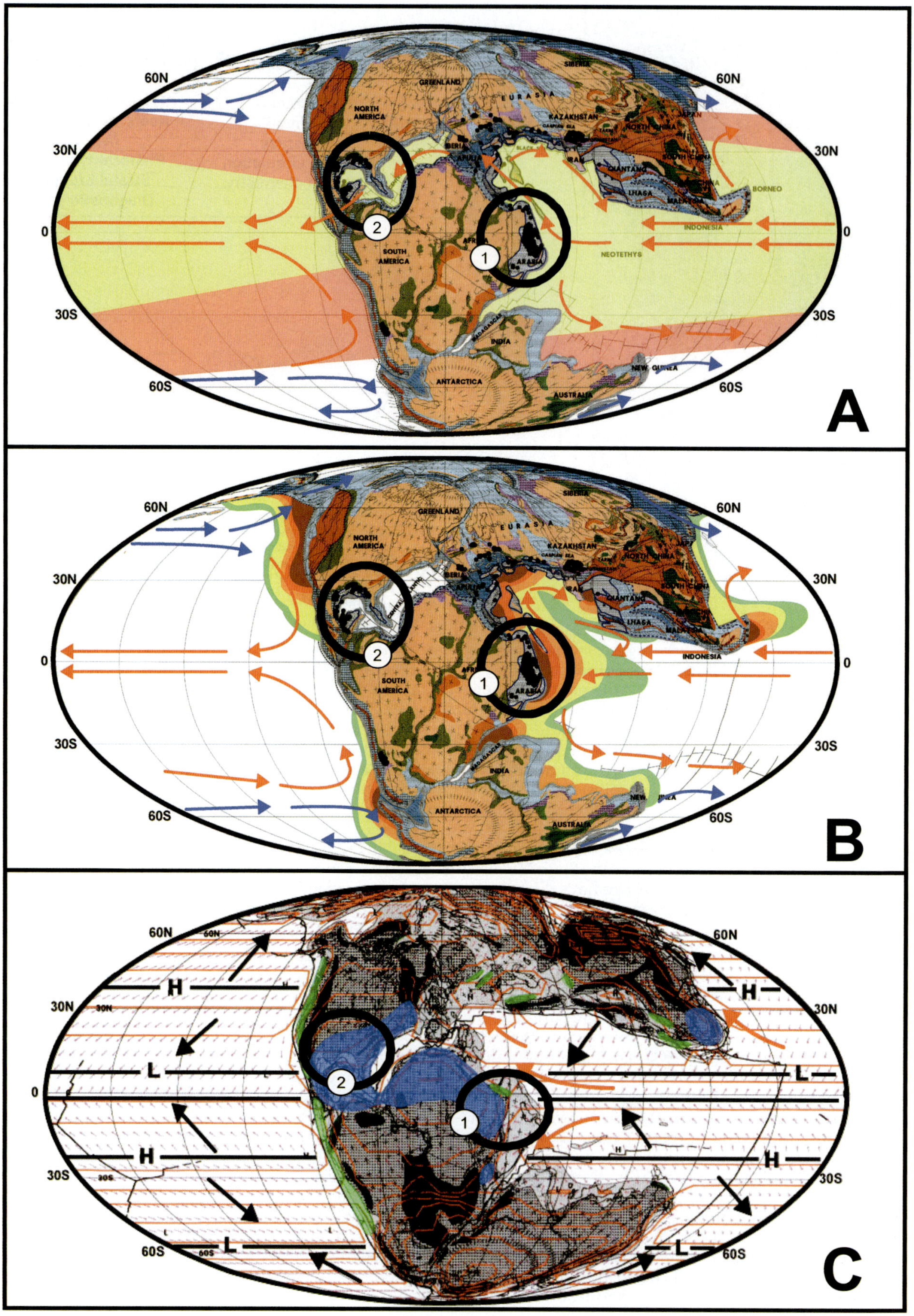
60N
30N
0
30S
60S
GREENLAND
NORTH AMERICA
SOUTH AMERICA
EURASIA
SIBERIA
KAZAKHSTAN
NORTH CHINA
SOUTH CHINA
QIANTANG
LHASA
INDONESIA
BORNEO
NEOTETHYS
ARABIA
INDIA
MADAGASCAR
ANTARCTICA
AUSTRALIA
NEW GUINEA
1
2
A
B
H
L
C

TABLE 4.—Comparative-analysis table for coeval Late Jurassic carbonate reservoirs of the Arab formations, Saudi Arabia, and the Smackover Formation, Gulf of Mexico, U.S.A. Characteristics used for comparison are from the variables contained with the *CATT* equation. Facts and interpretations about the two reservoir systems under comparison are based on observations and inferences from the *Phanerozoic Carbonate Trends Chart* (Fig. 1), the global maps (Fig. 11), literature case studies, and personal experience and knowledge.

CHARACTERISTICS	ARABIA	GULF OF MEXICO
Tectonics at deposition	passive margin	passive margin
Tectonic history	passive to collisional	still passive
Ocean circulation	open	restricted
Global climate	greenhouse	greenhouse
Paleo-latitude	equatorial	25° N
Regional air temp.	constantly hot	seasonally variable
Regional climate	mostly humid	semi-arid (dolo & evaps)
Relative humidity	low	very low
Winds	onshore; west to east	offshore; northeast to southwest
Wind strength	low (equatorial doldrums)	moderate?
Ocean temperature	very high (35°C avg.)	not as high (30°C avg.)
Tide level	macro	micro
Storms	abundant	rare
Overall ocean energy	high	low
Key reservoir facies	grainstones	grainstones/dolomudstones
reefs	stromatoporoid/coral framestones	microbial mounds
Depositional mineralogy	calcite	aragonite
Marine cements	rare/none	early/abundant
Early diagenesis	minor cementation minor dolomitization	significant dissolution significant dolomitization
Dominant pore type	primary	moldic/dolomite
Fractures	yes (foreland history)	maybe (passive margin history)

bonate principles may provide various perspectives and key insights that can address these questions. The following is a brief summary of this analysis.

Analysis explaining the key differences of reservoir facies and pore types is based on two principal factors: (1) overall ocean energy, and (2) global climate with local variability. The Late Jurassic was a time of continued Pangea rifting, continental drift divergence, global first-order sea-level high, and global greenhouse climate. The Panthalassic and Neotethys oceans merged to form an equatorial super-ocean. The Mid-Atlantic and Gulf of Mexico were newly forming ocean basins. From the *Phanerozoic Carbonate Trends Chart*, expectations of two key carbonate system characteristics are:

1. Grain types are mostly skeletal with some frame builders and rare to absent ooids, so reservoir rocks should be dominated by skeletal grainstones.

2. Abiotic mineralogy is calcitic, so aragonitic ooids and marine cement should be rare.

Observations different from these expectations require further explanation.

The Smackover carbonates formed in the Gulf of Mexico under passive-margin conditions. At that time, the Gulf was a relatively small oceanic water body, and circulation was probably restricted to small gyres within the Gulf. Currents bringing warm equatorial waters into the Gulf would have been from the northwest Neotethys Ocean, which passed through the Middle Atlantic (Fig. 11A). Most likely, tidal levels were low in the Gulf, and the probability of hurricanes influencing the northern Gulf seafloor was minimal due to small areal extent of the young Middle Atlantic Ocean (Figs. 11B, C). Together, these conditions suggest that the overall oceanic energy in the Gulf of Mexico during the Late Jurassic was low. Thus, ramp-profile environments of deposition for the Smackover facies included low-energy, shallow-water settings (tidal-flat muds; microbial

←

FIG. 11 (opposite page).—Paleo-oceanographic and paleo-climate *CATT* maps used for comparative analysis of late Jurassic carbonate reservoir systems in the Arab formations, Saudi Arabia (inset circle 1) and the Smackover Formation of the Gulf of Mexico, USA (inset circle 2). A) Paleo-oceanography map with interpreted paleo-temperatures of ocean surface water and interpreted paleocurrent directions (warm currents = red arrows; cold currents = blue arrows). B) Paleo-oceanography map with interpreted paleo–tidal levels and interpreted paleocurrent directions (warm currents = red arrows; cold currents = blue arrows). C) Paleo-climate map with interpreted humid zones (blue), upwelling regions (green), wind vectors (black arrows), and interpreted storm tracks (red arrows). Refer to maps in Appendices 2, 3, 4, and 5 for further explanations of symbols.

mounds) that accumulated mud along with the well documented Smackover ooid shoals (Baria, Stoudt, Harris, and Crevello, 1982).

Contrastingly, the Arabian basin straddled the equator during the Late Jurassic. The ocean surface water that circulated into the basin is interpreted, based on the maps, to have been the highest-temperature ocean surface water ever known to have formed during the Phanerozoic (Fig. 11A). This interpretation is based on the total distance the equatorial water traveled from the western coasts of Central and South America to the eastern coast of the Arabian Plate. As stated previously, ocean-surface water temperature is directly related to the travel time at the equator. During the Late Jurassic, the global plate reconstruction shows that there were no intervening continental masses at the equator between the west coast of South America and the east coast of the Arabian Plate. Thus, in the Late Jurassic, the ocean surface water traveled in excess of 24,000 kilometers at the equator, and when it arrived at the Arabian basin it must have been extreme in temperature.

Further, it is interpreted that the tidal range along the eastern margin of the Arabian plate was large, because of the enormous size and force of the oceanic circulation gyres that are interpreted to have been present in the Late Jurassic because of the global plate configuration (Fig. 11B). Lastly, it is suspected that the Late Jurassic may have been a time when significant high-energy hurricanes were both strong and abundant in the western Neotethys Ocean and were likely to have affected sedimentation styles in the Arabian basin (Fig. 11C). These three phenomena combined to impart an overall high oceanic energy to the carbonate system that formed the Arab Formation carbonates, and it is interpreted that this explains why these reservoir carbonates are virtually all grainstone. Thus, a straightforward analysis of overall ocean energy delivered to the different carbonate provinces, low for the Gulf of Mexico and very high for the Arabian Basin, provides a rationale for understanding the differences in development of reservoir facies.

Overall global climate for the Late Jurassic was greenhouse; however, there must have been significant local variation between the settings for Smackover vs. Arab sediment mineralogy development. The paleogeographic maps (Fig. 11) show that the Arabian Basin was equatorial and at the western edge of the Panthalassic–Neotethys super-ocean. It is interpreted that this setting was continuously hot and humid, with open ocean circulation, and was conducive for producing calcitic mineralogy. Contrastingly, the maps show that the northern Gulf of Mexico was at about 25° N latitude in the hot but more semiarid and possibly arid belt, with restricted ocean circulation, and that was more conducive to aragonite precipitation, evaporite deposition, and syndepositional to early postdepositional dolomitization. Whereas the climate conditions of the Late Jurassic in Arabian Basin promoted more stable calcitic mineralogies and preservation of primary pore systems, the coeval local climate conditions in the Gulf of Mexico appear to have resulted in more metastable aragonite and evaporite mineralogy, and were optimal for dolomitization. Thus, it is suggested here that local climate variations (very hot and continuously humid vs. warm to hot and semiarid) within overall global greenhouse climate conditions can be drivers for differences in formation of carbonate minerals between age-equivalent depositional systems. Such differences in mineralogy profoundly affect diagenesis, porosity evolution, and thus reservoir quality.

This style of intra-time-slice comparative analysis is a process for explaining regional variability and key differences between age-equivalent carbonate systems using global-scale perspectives, by taking advantage of established and well documented principles of Earth and carbonate processes that control development and characteristics of carbonate systems. Satisfactory explanation of the differences can yield new insights about carbonate systems, and these insights can become the bases for predictive concepts.

Inter-Time-Slice Comparative Analysis.—

As with comparative analysis of carbonate systems within a time slice, the *CATT* hypothesis can be used as an approach for systematic comparative analysis of carbonate systems of different time slices within one basin or from different basins, to derive insights that would explain the differences and/or similarities. As an example, the carbonate systems of the Late Jurassic and the Late Permian–Early Triassic of the Arabian Basin are compared (Table 5; Fig. 12).

Systematic comparative analysis between the Late Jurassic and Late Permian carbonate systems in the Arabian basin reveals that there are significant differences between the two systems, and the key differences are manifested in the rock properties of the reservoir facies (Table 5). Published key characteristics of the Khuff Formation for this comparison are from Al-Jallal (1995). The interpretations for the characteristics of the Late Jurassic units are explained in the previous section. It is proposed that the characteristics of the Late Permian also can be understood in the context and perspective of the paleo-oceanographic and paleoclimate maps (Fig. 12).

It is interpreted that the most important control on the differences between the Late Jurassic and Late Permian carbonate systems in the Arabian Basin is position of tectonic plates for each time period. In the Late Jurassic, the Kimmerian continental blocks had docked with southern Asia, the Paleotethys Ocean had been completely closed, and the Neotethys Ocean was merged with the Panthalassic Ocean to form an enormous "super-ocean". Circulation of this ocean imparted very high overall thermal, tidal, and storm energy to the equatorially located, humid-climate Arabian basin, and ocean chemistry resulted in calcitic mineralogy.

Contrastingly, during the Late Permian, the Kimmerian continental blocks were still in close proximity after just having rifted away from the Arabian plate in the Early Permian. This plate configuration shows the expanse of the Paleotethys Ocean and the small, narrow infant geometry of the Neotethys Ocean. Also, at this time, the Arabian basin was at about 25° S latitude in the southern hemisphere desert belt. It is interpreted that the Kimmerian continental fragments severely damped if not completely dissipated the Panthalassic and Paleotethys oceanic current, tidal, and storm energy, prior to entering the Neotethys Ocean. The combination of desert latitude, climatic aridity, restricted ocean circulation, and diminished ocean energy resulted in Late Permian aragonitic and dolomitic mineralogies, and complex mixed stratification of mudstones, grainstones, and microbial mounds.

Also, the Late Permian and Early Triassic carbonate systems lacked framestone reefs because frame-building organisms were not extant at the time. Following the Late Devonian extinction, robust frame builders did not return to the ecologic "reef" niche until the Late Triassic and Late Jurassic (see *Phanerozoic Carbonate Trends Chart*, Figure 1).

There are many more specific contrasts that can be made between these two carbonate systems in the Arabian basin. The point is that the *CATT* hypothesis provides the approach and methodology for conducting a systematic comparative analysis, whether the analysis is for coeval carbonate systems or for carbonate systems of different ages in the same basin or in different basins. Obviously, one would expect much more vari-

TABLE 5.—Comparative-analysis table for carbonate reservoirs of different ages from the Late Jurassic Arab formations and the Late Permian Khuff Formation, both being within the Arabian Basin. As with the comparative-analysis method for coeval carbonate reservoirs, characteristics used for comparison are from the variables contained with the *CATT* equation. Facts and interpretations about the two reservoir systems under comparison are based on observations and inferences from the *Phanerozoic Carbonate Trends Chart* (Fig. 1), the global maps (Fig. 12), literature case studies, and personal experience and knowledge.

CHARACTERISTICS	ARABIA (LATE JURASSIC)	ARABIA (LATE PERMIAN)
Tectonics at deposition	passive margin	passive margin
Tectonic history	passive to collisional	passive to collisional
Ocean circulation	open	very restricted
Global climate	greenhouse	transitional from icehouse
Paleo-latitude	equatorial	25–30° S; desert belt
Regional air temp	constantly hot	seasonal; desert-style variability
Regional climate	mostly humid	arid
Winds	onshore; west to east	mostly offshore; SE to NW
Wind strength	low (equatorial doldrums)	moderate?
Ocean temperature	very high (35°C avg.)	warm (25–30°C avg.)
Tide level	macro	micro
Storms	abundant	rare
Overall energy levels	high	low
Key reservoir facies	grainstones	grainstones & mudstones
Reefs/organic mounds	stromatoporoid/coral framestones	thrombolite/stromatolite microbial mounds
Depositional mineralogy	calcite	aragonite & dolomite
Marine cements	rare/none	early/abundant
Early diagenesis	minor cementation minor dolomitization	significant dissolution significant dolomitization
Dominant porosity type	primary	molds & vugs/dolomite intercrystal
Fractures	yes (foreland history)	yes, and faults (foreland history)

ability and many more differences between carbonate systems of different ages and from different basins. As will be discussed in the next section, this perspective provides the conceptual and practical basis for selection of appropriate field and reservoir analogs for geologic model building and reservoir simulation.

Selection of Appropriate Carbonate Field Analogs.—

Field performance predictions made from reservoir simulations are fundamental to the business decisions during the life of a field or reservoir. Reservoir simulations are built from constrained geologic models that include reservoir layering, facies, and rock properties. Commonly, in exploration or field development activities, such as post-discovery volumes assessment or depletion planning, the geologic modeler must build the model for a newly discovered or incompletely developed field with minimal and / or sparse data. Under these circumstances, input data for the geologic model must be selected from other similar or "analog" fields and reservoirs. The hypothesis provides a conceptual basis and methodological approach for identifying the most appropriate analog(s) for use in geologic model building and reservoir simulation.

When a geologic model of a field or reservoir is built, it is important to ensure that selected analog fields or reservoirs have characteristics that are most likely to be representative of the field being modeled. Important characteristics include reservoir layering, facies, and rock-properties distributions. Reservoir layering has two components: lateral continuity and vertical connectivity. Reservoir facies reflect geometries of depositional and diagenetic environments, which usually combine to produce distinct pore systems and rock properties. Rock properties of most concern to the modeler are porosity and permeability. The following example concerns Tengiz Field, and serves to illustrate how the *Carbonate Analogs Through Time* approach can be used to make appropriate analog selections.

Tengiz field is a large isolated carbonate platform in the North Caspian basin (Weber et al., 2003). The carbonates constituting the platform are approximately 3000 m thick and range in age from Middle Devonian at the base to Early Pennsylvanian (Bashkirian) at the top. The Tengiz platform is therefore long lived, on the order of 65 million years, and the carbonates span and record significant global tectonic, climatic, eustatic, mineralogic, and biotic changes (see the *Phanerozoic Carbonate Trends Chart*, Figure 1). The Tengiz platform formed after an Early Devonian rifting event, and terminated during initial Pennsylvanian collisions leading up to Pangea development. The following reservoir characterization summaries are from Collins et al. (2006) and Kenter et al. (2006).

To date, the deepest wells in Tengiz penetrate the Upper Famennian, at the top of the Devonian section. These carbonates formed during global greenhouse climate, and they consist of meter-scale reservoir layers with lateral facies segregation from platform margin to interior, driven by low-amplitude, high-frequency greenhouse-style eustatic sea-level fluctuations. The rocks are mostly calcitic, and have low porosity due to muddy platform-interior facies and porosity-reducing cements in grain-rich facies. These facies postdate the Frasnian–Famennian extinction of reef-building stromatoporoids and corals. Well data suggest that the Famennian margin was formed by a combination of

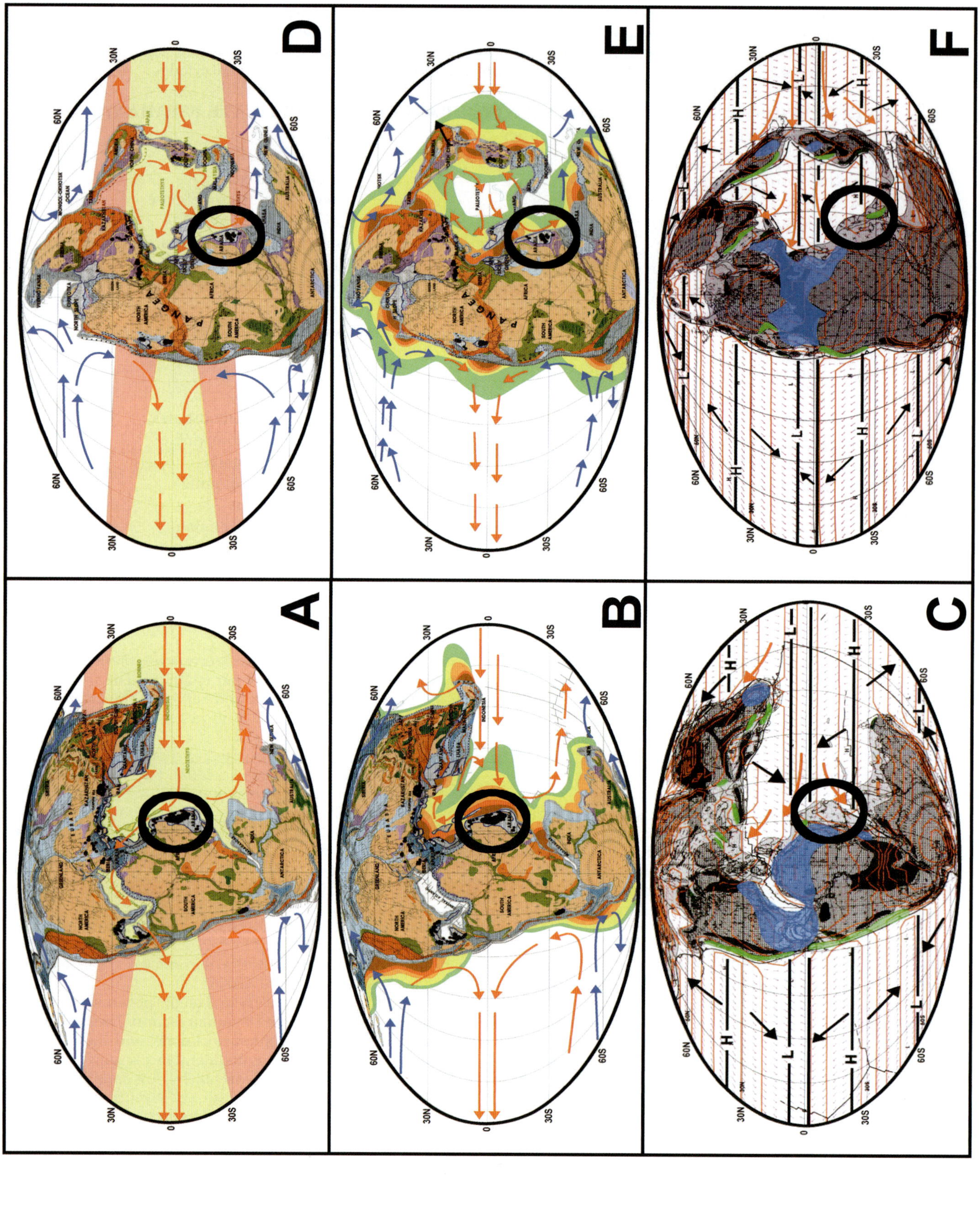
D
E
F
A
B
C

grainstone shoals and microbial boundstones. In fact, Tengiz is interesting because much of the reservoir section exists as an isolated platform during a time when reef builders were poorly represented or absent (Late Devonian–Famennian to Early Pennsylvanian–Bashkirian, Fig. 1). Although not penetrated by wells, pre-extinction Devonian carbonates are likely to be present below the Famennian section. Well known Middle and Late Devonian (Frasnian) analogs can be used to predict the nature of the underlying section. Analogs from the western Canada Alberta and Australian Canning basins, for example, would predict (1) stromatoporoid-dominated reef margins with muddy platform interiors, (2) formation of reef-rimmed isolated platforms, and (3) possible presence of dolomitized reservoir facies. The propensity for pre-extinction stromatoporoid-rich carbonates to form isolated buildups may in fact be the reason for Tengiz's existence as a post-extinction isolated platform.

Overlying Mississippian carbonates formed during global transitional climates that changed from greenhouse to icehouse by the Pennsylvanian. Lower Mississippian (Tournaisian and Early Visean) depositional facies are consistent with global themes (Fig. 1; Appendix 5), and include echinoderm-rich skeletal sediments and abundant organic-rich mudstones. Transitional climate-driven, moderate-amplitude, high-frequency eustatic sea-level fluctuations resulted in stratigraphic architecture consisting of less laterally continuous and less vertically connected reservoir layers compared to Devonian layers. Echinoderm-rich grainstones characteristically are cemented with syntaxial overgrowths that significantly reduce porosity and permeability.

The Upper Mississippian (Late Visean and Serpukhovian) consists of high-porosity fourth-order cycles (10–30 m thick) that are separated by tight or low-porosity facies at sequence boundaries. Cycles are grainstone-dominated at top and packstone to grainstone at base. Grainstone porosity is characterized by dissolution-modified interparticle porosity, while packstones are characterized by small matrix vugs and microporosity. Despite differences in pore types, these facies have nearly equal porosity ranges. Low-porosity facies at sequence boundaries are mudstones, volcanic ash-tuffs, and cemented peloidal grainstones. Upper Mississippian facies exhibit increased skeletal diversity, including colonial corals, brachiopods, and a diverse algal assemblage. The Serpukhovian margin is dominated by microbial boundstone facies (Collins et al., 2006).

The youngest reservoir zone in Tengiz Field is of Early Pennsylvanian Bashkirian age. These carbonates formed in true global icehouse conditions, and are dominated by high-frequency, high-amplitude cycles (a few meters to tens of meters thick). Where preserved, cycle tops consist of subaerially exposed grainstones and bases consist of skeletal grainstones and packstones. Fine-grained algal-rich grainstones and coarse ooid-rich grainstones are also present. Bashkirian facies show even greater skeletal diversity than the Serpukhovian. Patch reefs and mounds formed by *Chaetetes* and other sponges occur locally in the platform interior and along the platform margin. Reservoir quality is highly variable. Ooid grainstones may exhibit oomoldic porosity or solution-enhanced interparticle porosity, or may be tightly cemented. Algal grainstones tend to have less dissolution and more cementation. Skeletal packstones have attributes similar to those of the Visean–Serpukhovian section. Facies at sequence boundaries tend to be of low porosity because of the presence of volcanic ash-tuffs and tight peloidal mudstones.

The above abbreviated descriptions of the carbonates of Tengiz Field have been provided to establish a reference for analog comparison. These described characteristics have been verified after extensive reservoir-characterization efforts by the operating company TengizChevroil (TCO) (Weber et al., 2003) and partner companies (Collins et al., 2006; Kenter et al., 2006). The following analog comparisons would not be valid if these facts were unknown about Tengiz. However, it is proposed the Tengiz example demonstrates that the *CATT* hypothesis, concepts, and methodology provide a basis for analog selection in the absence of such descriptions.

The original question was: what carbonate fields and/or reservoirs would be good analogs for Tengiz field, if the previously described facies and reservoir characteristics were unknown from Tengiz? One approach would be to use the equation, and evaluate and compare potential analogs with the target field / reservoir on a variable-by-variable basis, starting with age, followed by Earth processes (tectonics, climate, eustasy, ocean circulation, and ocean chemistry), and lastly the two carbonate processes (mineralogy and biota).

In this example, several carbonate fields are proposed to evaluate appropriateness as Tengiz analogs (Fig. 13). These include Arun (Jordan and Abdullah, 1992), Golden Lane (Viniegra and Castillo-Tejero, 1970), Walker Creek (Chimene, 1991; Bliefnick and Kaldi, 1996), Reeves (Chuber and Pusey, 1985; as example of Late Permian San Andres Formation fields similar to Means and Slaughter), Horseshoe Atoll (Vest, 1970, including Kelly-Snyder, Cogdell, and Salt Creek), Swan Hills (Hemphill et al., 1970), and Rainbow (Barrs et al., 1970) (Fig. 13).

As recommended above, comparable geologic age between analog and field is the first variable for analog selection. Of the fields listed in Figure 13, Salt Creek (Pennsylvanian) and Swan Hills and Rainbow (Late Devonian) are partly age equivalent to Tengiz. Arun (Miocene), Golden Lane (Cretaceous), Walker Creek (Late Jurassic), and Slaughter (Permian) are not age equivalent. Other comparative characteristics are best illustrated in table format (Table 6). Such a format allows immediate visual inspection to find similarities and differences between the subject field and the possible analogs. From Table 6 it is clear that none of the fields is a perfect analog to Tengiz, but by characteristics, the Horseshoe Atoll, Swan Hills, and Rainbow fields may be the best overall combination analogs for Tengiz.

As a related topic, there is an important distinction between field analogs and reservoir analogs, particularly for isolated carbonate platforms. A field that is an entire, long-lived isolated carbonate platform, such as Tengiz, typically contains many reservoir zones, reservoir layers, and reservoir geobodies that have geometries and reservoir quality controlled by stratigraphic, depositional, and diagenetic processes. The best field analogs are those that have similar distributions of zones, layers, and geobodies. These analogs are much more likely to be of the same age as the modeled field. On the other hand, a field model could be composited from a series of reservoir analogs, as defined by intra-layer continuity, inter-layer connectivity, pore types, and/or grain types. This allows much greater flexibility for analog selection within the *Carbonate Analogs Through Time* framework (Fig. 1).

←

FIG. 12 (opposite page).—Paleo-oceanographic and paleo-climate *CATT* maps for the comparative analysis between Late Jurassic and Late Permian carbonate systems of the Arabian Basin. Maps of pale–ocean surface water temperature for **A)** Late Jurassic and **D)** Late Permian. Paleo–tidal range maps for **B)** Late Jurassic and **E)** Late Permian. Paleoclimate and storm-track maps for **C)** Late Jurassic and **F)** Late Permian. Inset black circles identify the location of the Arabian basin. Refer to maps in Appendices 2, 3, 4, and 5 for further explanations of symbols.

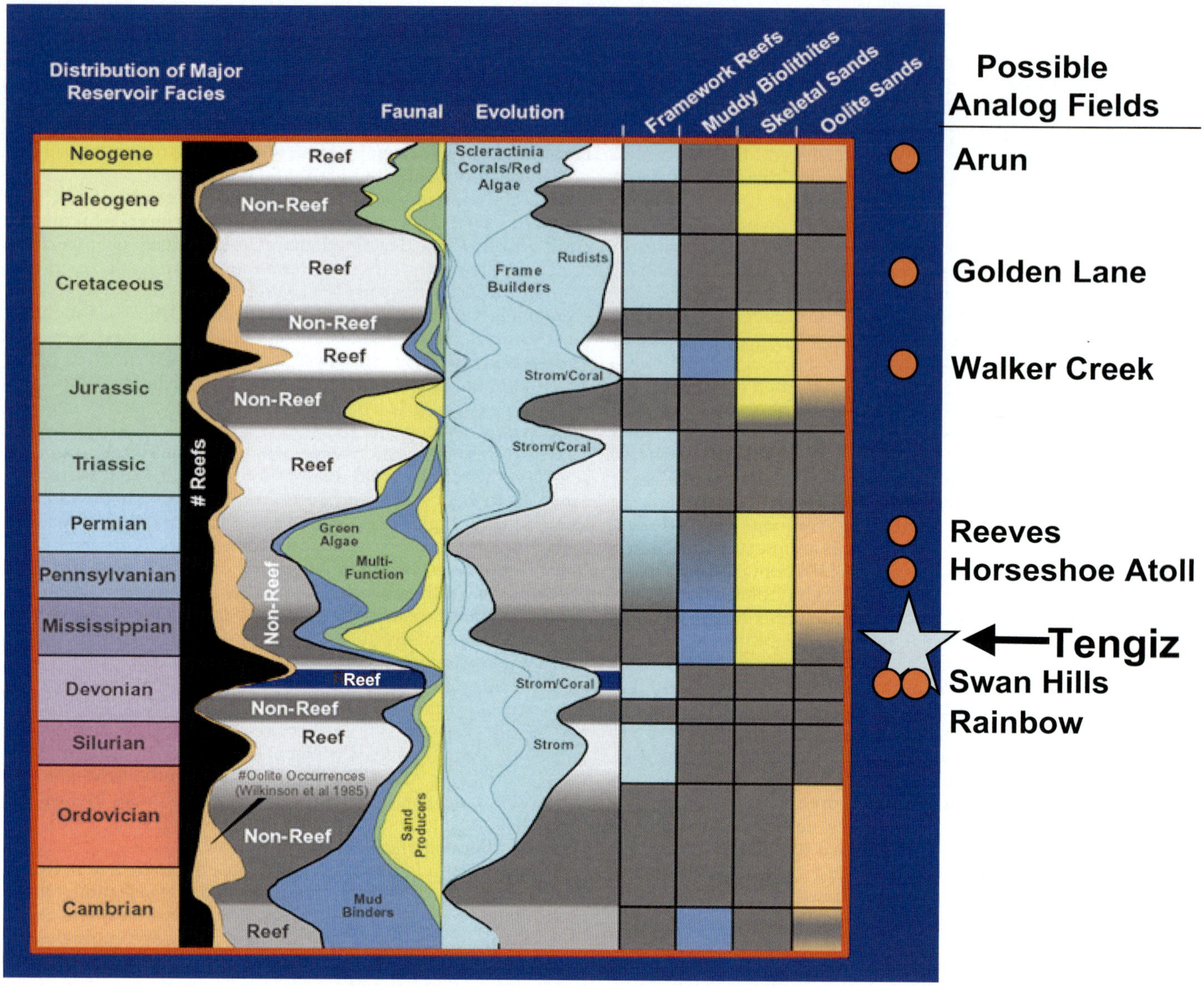

FIG. 13.—CATT chart displaying Phanerozoic age, biotic evolution, and carbonate reservoir types as historical context and perspective for evaluating and selecting analog(s) to Tengiz field. Color codes for reservoir types are: light blue = reefs, dark blue = mud mounds, yellow = skeletal grainstones, and orange = ooid grainstones. Red dots represent ages of fields (names to right of dots) that were suggested as possible analogs. The Tengiz isolated platform was a long-lived carbonate system that spanned the geologic time period from Middle Devonian to Early Pennsylvanian. For age-equivalent analogs, the Late Devonian Rainbow and Swan Hills systems are synchronous with the lower reservoir zones of Tengiz, and the Middle Pennsylvanian Horseshoe Atoll is slightly younger than the uppermost Bashkirian reservoir zones of Tengiz.

In summary, analogs are valuable sources of input data to geologic models in the absence of first-hand data. However, the quality of results is predicated on the quality of the inputs. If the inputs are from an inappropriate analog, the results will not be satisfactory.

Development of Predictive Concepts.—

In the absence of possessing all facts about all geologic systems and/or reservoirs, a major goal of geologic investigation is to examine one or more sedimentary systems, synthesize collected data and observations, and then develop predictions of characteristics for data-poor sedimentary systems. Throughout the past hundred years, geologists studying carbonates have developed numerous predictive concepts for carbonate sedimentary systems and reservoirs, and workers in the petroleum industry have used those concepts for significant exploration, appraisal, and production success. Interestingly, most predictive concepts have been restricted to "single-variable" ideas. For example, depositional facies can be predicted by knowing the depositional profile type: ramp, rimmed shelf, or isolated platform (Ahr, 1973; Wilson, 1975; Read, 1985). Handford and Loucks (1993) upgraded this concept by adding in the concepts of sequence stratigraphy. Another example is predicting pore type

TABLE 6.—Comparative-analysis table showing systematic and ordered comparisons of reservoir characteristics for seven possible analog fields to the supergiant carbonate Tengiz Field, Kazakhstan. The first column lists the analog factors for comparison, and the subsequent columns are for facts and interpretations about each possible analog field. Entries of characteristics for each field analog were found in published literature and case studies. Graphical comparisons for analog selection can be made using diagrams such as Figure 13, and these complement any tabulated comparisons.

Analog Characteristics	Tengiz Field	Arun Field	Golden Lane Fields	Walker Creek Field	Reeves Field	Horseshoe Atoll	Swan Hills and Rainbow Fields
Geologic age	Devonian-Carboniferous	Miocene	Mid Cretaceous	Late Jurassic	Late Permian	Mid Pennsylvanian	Late Devonian
Present location	N. Caspian, Kazakhstan	Indonesia	Mexico GOM	Arkansas, USA	West Texas, USA	West Texas, USA	Alberta, Canada
Platform type	open oceanic isolated	open oceanic isolated	shelf-adjacent isolated	ramp	ramp to rim-shelf	intra-basin isolated	intra-basin isolated
Tectonic history	passive to collisional	passive to collisional	passive	passive	passive to collisional	passive to collisional	passive to collisional
Global climate	greenhouse to icehouse	transitional to icehouse	greenhouse	greenhouse	transitional	icehouse	greenhouse
Eustasy	low to high order	high order	low order	low order	transitional	high order	low order
Ocean circulation	open	open	restricted	restricted	restricted	restricted	open
Ocean chemistry/ Mineralogy	calcite to aragonite	aragonite	calcite	mixed	aragonite	aragonite	calcite
Reef/mound builders	stromatoporoid/coral	scleractinian coral	rudist mounds	microbial	sponge/algal	phylloid algal mounds	stromatoporoid/coral
Reservoir layering	continuous (Devonian) to discontinuous (Penn)	discontinuous	continuous	continuous	mixed	discontinuous	continuous
Vertical connectivity	good to poor	poor	good	good	mixed	poor	good
Reservoir facies	skeletal/ooid grsts & fractured bdsts	skeletal grsts/mdsts coral bdsts	rudist bdsts/skel grsts	ooid grsts	ooid/pellet grsts dolomudsts	ooid grsts phylloid algal mounds	skel grsts & bdsts
Reservoir rock Mineralogy	calcite	aragonite/calcite	calcite	dolomite	dolomite	calcite	calcite/dolomite
Dominant ϕ type	molds, vugs	molds, vugs, micro ϕ	vugs, caverns	molds, intercrystal	intercrystal	molds	molds, vugs, intercrystal
Fractures	abundant	moderate	moderate	rare	moderate	moderate	abundant

from processes of diagenesis (Choquette and Pray, 1970; Morse and Mackenzie, 1990).

It is rare to encounter predictive concepts that encompass "multiple-variable" ideas. An example of a multiple-variable concept would relate times of global climate character to ocean chemistry and abiotic grain development. During greenhouse climate, carbonate systems produce abiotic grains that are dominantly calcitic, versus during icehouse climate, when carbonate systems produce abiotic grains that are dominantly aragonitic (Schlager, 2005). Subsequently, these conclusions can be convolved with diagenetic concepts to make predictions of pore-system geometry. A simplistic example, well known from the literature, would be: during icehouse times, deposited ooids have a high probability of having aragonitic mineralogy, and if these are subjected to meteoric diagenesis, the metastable aragonitic ooids dissolve. Overall, this process results in development of oomoldic grainstones. Moldic grainstones are high-porosity rocks that provide excellent storage capacity in hydrocarbon reservoirs.

Although many high-quality predictive concepts for carbonate systems are in use today, it is proposed that the *Carbonate Analogs Through Time* hypothesis can be used as the framework and conceptual basis to develop "multiple-variable" predictive concepts that are beyond what are available now. The previously stated conclusion that carbonate platforms, fields, and reservoirs are causally distributed in time and space (Figs. 4, 5), provides the underpinnings for this application of predictive concept development. Invoking the *CATT* equation, the *Phanerozoic Carbonate Trends Chart*, the carbonate fields database, and the global maps, predictive concepts can be developed that are practical and that range from single-variable concepts to multiple-variable concepts. To illustrate the methodology for developing predictive concepts, the following discussion concerns predicting reservoir layering characteristics and pore geometry.

Reservoir layering has two components: lateral continuity and vertical connectedness. These components are fundamentally controlled by the stratigraphic architecture and depositional profile of the system. Stratigraphic architecture can be related to the depositional history of a system through the history of relative-sea-level changes (Fig. 14). Much work has been completed concerning the stratigraphic hierarchy (Haq et al., 1988; Goldhammer et al., 1990; Tucker and Wright, 1990; Vail et al., 1991; Sarg et al., 1999) that demonstrates that global climate is a major driver for high-order fourth- and fifth-order-scale depositional cycles. Global climate is also a significant component that combines with tectonics to control development of third-order-scale depositional cycles (Fig. 14).

The key reason for defining greenhouse versus icehouse cycles and establishing the mechanisms responsible for the frequency and amplitude of sea-level variation is that these facts and theory provide the basis for understanding the relative velocity (both speed and direction) of migration of depositional environments on any given depositional profile during relative-sea-level change. Quantitative forward modeling is based on this theory. Invoking sequence stratigraphic principles, and knowing the depositional profile and velocity of migration of depositional environments, it is straightforward to anticipate or predict stratigraphic architecture. This leads directly to predicting reservoir layering.

To illustrate this line of argument, four variations on the reservoir layering of an isolated platform are analyzed: greenhouse versus icehouse climates and reef-rimmed versus grainstone shoal-rimmed (Fig. 15). In the case of global icehouse climate, the stratigraphic signature is controlled by the fourth-order high-amplitude, high-frequency sea-level variation. If, for example, an isolated platform bank top is flooded with only 10–20 m of water, yet fourth-order amplitude changes are in excess of 100 m, then the bank top is repetitively drained (exposed) and flooded (submerged) at the 100 ky frequency. Under these conditions, depositional environments are forced to migrate at very high rates, forming complex facies mosaics

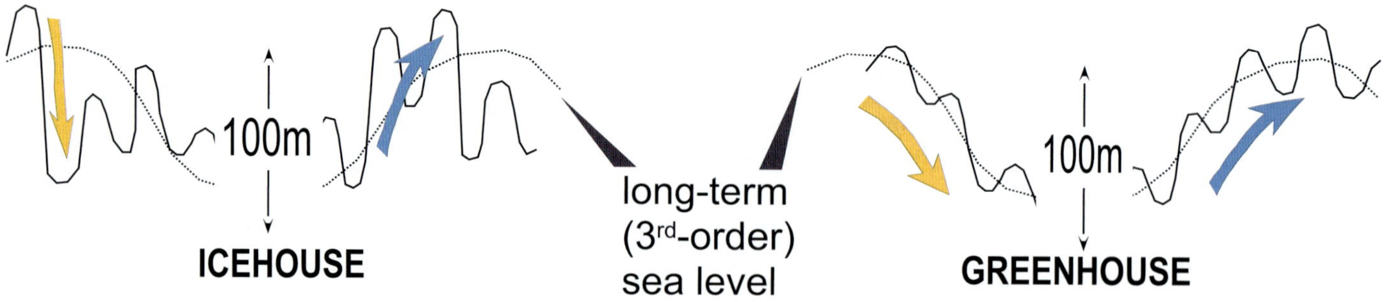

FIG. 14.—Schematic representations of third- and fourth-order sea-level fluctuations and amplitude related to icehouse and greenhouse climates. During icehouse global climate, glacial advance and retreat are the two mechanisms attributed to be the controls governing the 100 meter amplitude variations of sea level responsible for deposition of fourth-order high-frequency, high-amplitude sedimentary cycles. This fourth-order high-amplitude signal overwhelms the longer-term third-order sea-level variation, and thus controls the stratigraphic architecture of the system. Contrastingly, during times of global greenhouse climate, ice-volume fluctuation is minimal, so the fourth-order high-frequency, high-amplitude sea-level variation is muted. Instead, the long-term third-order sea-level variation related to both tectonic subsidence rates and other mechanisms of water-volume change (continental-aquifer volume change and ocean water temperature–driven volume change) is the principal control on the depositional signature of a sedimentary succession. For discussion purposes, the term icehouse cycle refers to depositional systems formed during times of global icehouse climates, and conversely, greenhouse cycle refers to depositional systems formed during times of global greenhouse climates.

and juxtapositions. This results in extreme lateral facies heterogeneity in the platform interior, and thus lateral discontinuity of reservoir layers. Further, bank-top facies and strata constituting cycles are vertically separated by numerous exposure surfaces with associated subaerial deposits and diagenetic overprints and flooding surfaces with deeper-water deposits that form during lag times prior to restart of the shallow-water carbonate factory. Such conditions lead to vertical unconnectedness of reservoir layers. Certainly, if the isolated platform formed during a geologic time when reef frame-builder biota was extant (Fig. 1), then the marginal reef framestones also will be both lateral and vertically separated (Fig. 15).

This same line of thinking can be applied to platform deposits formed during times when reef-building organisms were absent and high-energy platform-margin facies were ooid and skeletal grainstone shoals. As in the first case, it is interpreted that distributions of platform-interior facies will be complex, resulting in lateral layer heterogeneity and discontinuity. Also, cycles will be vertically unconnected. This applies to marginal grainstone bodies (Fig. 15).

The two cases that illustrate the reservoir-layering characteristics due to global greenhouse climate are significantly different from the icehouse cases (Fig. 15). During global greenhouse time, variation in sea-level amplitude at the fourth order is on the order of meters to tens of meters. Longer-term third-order variation in sea-level amplitude is on the order of many tens of meters to about 100 meters. Thus, deposition on the bank top of an isolated platform is relatively stable. Depositional environments are not forced to migrate rapidly or far, because sea level does not change as drastically as during icehouse times. These conditions result in much more uniform and/or orderly deposition of facies. Thus, facies are laterally well connected, and bank-top layers are laterally continuous. Further, because the bank top is not frequently and repetitively exposed and submerged, the depositional cycles tend to stack, both in the platform interior and at the platform margin. Obviously, the consequence is that reservoir layers tend to be vertically connected. During reef times, framestones do not dramatically shift, and favor stacking vertically at the margin. During non-reef times, marginal grainstone bodies exhibit similar behavior and tend also to stack vertically (Fig. 15).

These four cases can be used as the starting points to develop a series or family of single-variable to multiple-variable predictive concepts. Examples of single-variable predictive concepts can be articulated as declarative statements or propositions:

1. Icehouse climates result in high-amplitude, high-frequency sea-level change, and such sea-level change behavior results in significant complexity in reservoir layering: poor lateral continuity and vertical unconnectedness.

2. Greenhouse climates result in low-amplitude, low-frequency sea-level change, and such sea-level change behavior results in simplicity in reservoir layering: good lateral continuity and vertical connectedness.

Development of multiple-variable predictive concepts requires convolving these predictions of reservoir-layering characteristics resulting from climate-driven controls with facts and observations from the *Phanerozoic Carbonate Trends Chart* (Fig. 1) and with the *CATT* equation. The following is a progression of predictive concepts that increases in complexity by addition of variables, facts, and ideas.

1. Icehouse climates are positively correlated with ocean chemistry that tends to produce aragonitic abiotic grains (Fig. 1; Mackenzie and Morse, 1992; Schlager, 2005). Thus, carbonate facies deposited during times of icehouse climates are expected to form strata which include ooid grainstones where ooids originally had aragonite mineralogy.

2. Icehouse platforms that consist of depositional cycles or reservoir layers which are vertically separated by exposure surfaces and which contain aragonitic ooid marginal shoals are predicted to have a high probability of secondary-dissolution oomoldic pore systems that formed during times of bank-top exposure.

3. From the *Phanerozoic Carbonate Trends Chart* (Fig. 1), there are two time periods in the Phanerozoic when global icehouse climates occurred: the Pennsylvanian and the Neogene. Dur-

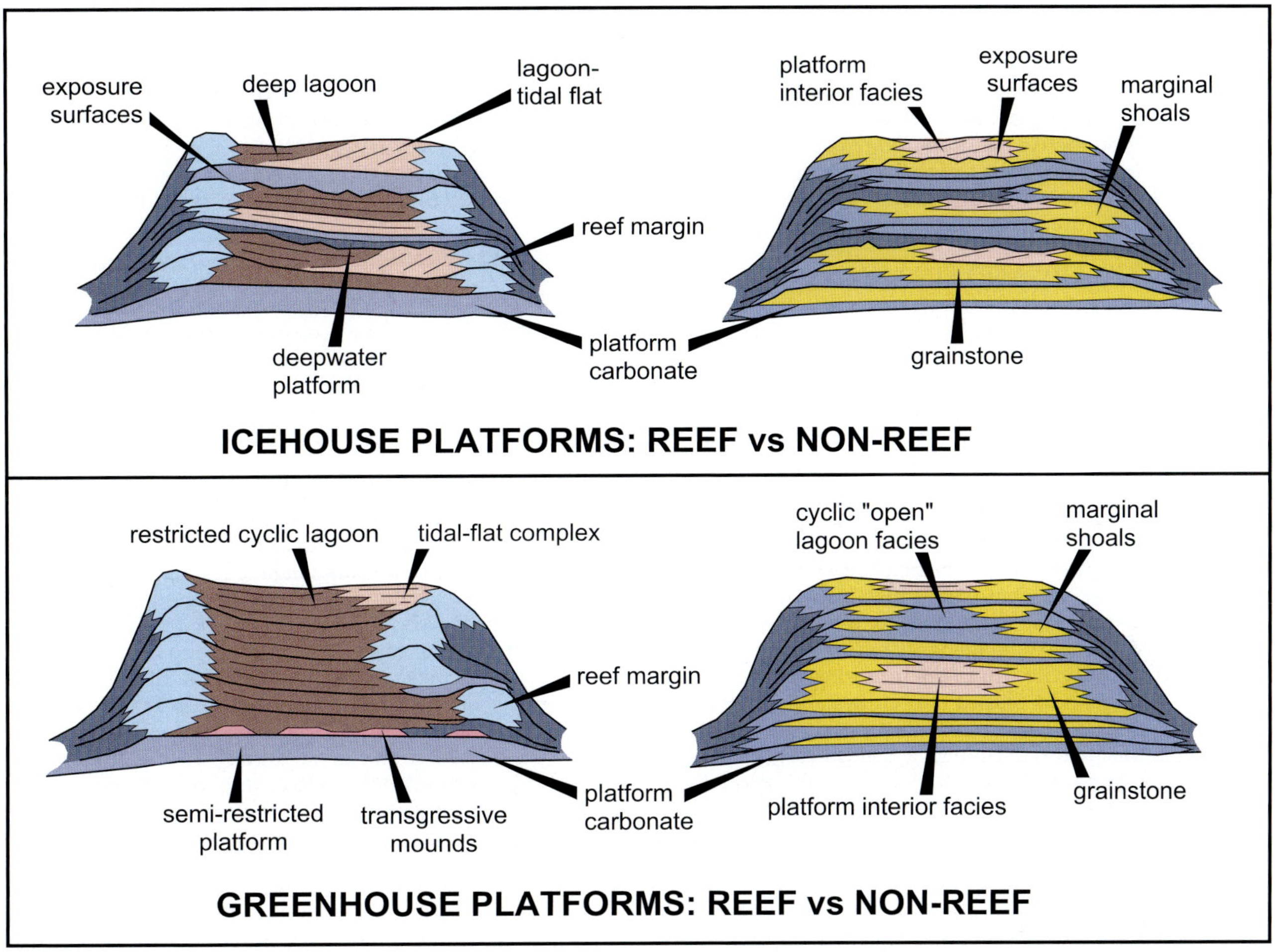

FIG. 15.—Four schematic representations of predicted facies, sequence stratigraphic architecture, and reservoir layering for a generic isolated platform. The generalized platform has two starting margin types: (1) framestone or boundstone reefal margins vs. (2) grainstone shoal margins. It is recognized that isolated platforms rarely develop significant relief in the absence of boundstone margins. The two depictions of platforms with grainstone shoal margins may not be realistic, but they are included to facilitate and schematically illustrate the discussion points. The upper panel shows facies, architecture, and layering resulting from icehouse-climate-driven eustasy, and the lower panel shows these same attributes from greenhouse-climate-driven eustasy. See text discussion for explanations of the similarities and differences between these four cases. This analysis of predicting stratigraphic architecture and reservoir layering characteristics for an isolated carbonate platform based on global climate conditions is an example of a predictive concept developed using the *Carbonate Analogs Through Time* hypothesis and approach. This is only one of many types of predictive concepts that can be developed using the components of the *CATT* equation. Further, such predictive concepts can range from single-variable concepts to multiple-variable concepts.

ing these times, the dominant abiotic mineralogy was aragonite. The Pennsylvanian was a time when large reef frame-building biotas were not common, but the Neogene is a time when scleractinian corals became prolific reef builders. Thus, age-specific predictive concepts are proposed. Expectations of Pennsylvanian carbonate reservoirs are that they should have complex reservoir layering (lateral discontinuities and vertical unconnectedness) and may contain layers with oomoldic grainstone reservoir facies with locally occurring vuggy phylloid algal mounds, whereas Neogene (Miocene) isolated platforms that are reservoirs should also have complex layering but could have both reefal and ooid–skeletal grainstone shoal margins, also with secondary dissolution porosity.

4. The time of Pennsylvanian global icehouse climate correlates positively with development and assembly of the Paleozoic Pangean supercontinent (Fig. 1; Veevers, 1990). Thus, it is possible to make an *a priori* predictive concept that Pennsylvanian isolated-platform carbonate reservoirs most likely will have complex reservoir layering, secondary enhanced dissolution porosity, and significant fracturing because of the overall compressional tectonic settings required to form Pangean supercontinents (see maps in Appendix 1).

This progression of four increasingly complex or multiple-variable predictions for carbonate reservoirs, starting with the global icehouse climate variable, provides the flavor for the types and styles of predictive concepts that can be developed using the *Carbonate Analogs Through Time* approach. Although predictive concepts based on global climate have been proposed previously, our contention here is that this approach and equation provide the framework and conceptual basis for systematic development of broadly ranging predictive concepts. One can now imagine the numerous styles and families of predictive concepts that can be derived or developed by using the equation variables singly and/or in varying combinations.

SUMMARY AND FINAL NOTE

The main goal of the *Carbonate Analogs Through Time* research effort was to propose a testable hypothesis and to develop an approach with sound conceptual underpinning that can be used to systematically evaluate, compare, and predict characteristics of carbonate systems and reservoirs. The hypothesis and approach are represented in the form of an equation:

Geologic Age	+	Carbonate Processes	+	Earth Processes	=	Time-Slice Themes and Predictive Concepts

Along with the approach, the principal deliverables of the project were the various global-scale products that include the *Phanerozoic Carbonate Trends Chart* (Fig. 1), the carbonate fields database (Appendix 1), the paleogeographic map series (Appendices 2, 3, 4, and 5), the write-ups of the time-slice themes or age-specific patterns (Appendix 6), and the synthesized Phanerozoic field posters (Appendix 7). It is contended that the approach is a large-scale idea that has wide-ranging usefulness. It does not replace previous work, facts, ideas, concepts, interpretations, and/or syntheses. Rather, the design of the approach is to take full advantage of and build upon all of the past work on carbonate systems and reservoirs.

Convolving the *Phanerozoic Carbonate Trends Chart* with the global paleogeographic maps produces a four-dimensional, temporal and spatial framework for systematically compiling, organizing, storing, retrieving, and integrating facts, ideas, concepts, principles, case studies, etc... that one encounters throughout a career. Industry and academic geologists have careers that range up to fifty years, maybe more. During those years, each geologist will undertake and complete numerous projects, attend many conventions, meetings, and conferences, and continuously read the literature. There is a lot of information that is encountered over a lifetime, and commonly some, if not much, of that information is lost to an individual because of the lack of a structured mental container within which to organize and store that information. Perhaps the *CATT* hypothesis, approach, and products can be used as that structured mental container. The *Phanerozoic Carbonate Trends Chart* provides the time context, and the global maps provide the spatial buckets. As a perspective, this time–spatial framework encompasses the entire Phanerozoic sedimentary rock record worldwide. Most probably, one individual cannot fill the entire framework during one lifetime. In the long term, however, filling the entire framework is the goal of our worldwide geoscience community.

Whether one considers this Phanerozoic global framework to be complete or not, it is proposed that the *CATT* approach leverages all of the materials within this global framework, and becomes the facilitator for three key outcomes: (1) conduct systematic comparative analyses to gain new and better understandings of carbonate systems in general, (2) develop more accurate predictive concepts and models, and (3) make better business decisions concerning carbonate reservoirs.

This approach is predicated on the foundation that geologic time is an overriding variable or control on the signatures (fabric and character) of carbonate systems. Using this time-based approach facilitates synthesis of carbonate themes within a temporal slice. Further, the approach provides a rigorous basis for systematic comparative analysis of different carbonate systems within one time slice, and, importantly, for systematic comparative analysis of carbonate systems from different time slices. Advances in understanding the characteristics of carbonate systems and predicting those characteristics are made by such analyses.

As a result of rigorous and systematic comparative analysis of carbonate systems and reservoirs, and by using the hypothesis and approach, we suggest that more accurate predictive concepts for carbonate systems and reservoirs can be developed. These predictive concepts can be single- variable, based on one element in the *CATT* equation, or they can be multiple-variable, based on combinations of elements in the equation. Again, the approach design is flexible enough to allow each user to customize as desired or required by the project or application. The intent of the hypothesis and approach in predictive concept development is to be unconstrained and to leverage all or as much as is needed from past work on carbonate sedimentary systems.

Lastly, as the *Carbonate Analogs Through Time* hypothesis name implies, we contend that invoking the hypothesis provides an approach to identifying and selecting better field analogs for use as input data sources to geologic models and reservoir simulators. Appropriate selection of field analogs is a difficult task, because commonly a field under evaluation by reservoir simulation is lacking significant and critical data needed for performance calculations. However, if the geologic age and basin name (given present-day location) of the field or reservoir under evaluation are known, this hypothesis provides the conceptual justification and the global products for identifying and selecting appropriate age-equivalent analogs. It is suggested that the approach to analog selection can lead to better business decision-making.

It is our hope that the CATT hypothesis and approach and its derivative products will provide the stimulation for future work such as refining current plate reconstructions at both global and regional scales, and testing the nonquantitative paleo-oceanography maps against quantitative parametric climate-modeling results. Several types of applications have been proposed in this paper, and it is known that many more types of applications have been and can be developed.

ACKNOWLEDGMENTS

The *Carbonate Analogs Through Time* research program was accomplished with Mobil Research management support. It was first proposed in late 1991 at the Mobil Dallas Research Laboratory by the authors, and was completed with the help of many contributors who worked with the authors in the Mobil carbonate research group. They are: Aus Al-Tawil, Julia Caldaro-Baird, Jan Golonka, Mike Pope, Jim Weber, Bill Zempolich, and geotech Dave Guido. Julia Turner and Roxanne Clary were instrumental in extracting the carbonate fields data from the Petroconsultants S.A./IHS fields databases.

Petroconsultants S.A./IHS has approved release of the carbonate fields database (Appendix 1) for this publication, and retains all copyrights. Carbonate fields data included in this listing are from 1999 and earlier only.

In 1992, the Mobil Global Geology Team was formed with the principal objective of bringing into the company state-of-the-art global paleogeographic maps. The Mobil Global Themes Team collaborated with the carbonate team and was responsible for defining the ages of time slices and preparing the new paleogeographic maps. The Global Themes project members were: Jeff

Brown, Wayne Cross, Mary Edrich, Dave Ford, Ulrich Franz, George Gail, Jan Golonka, Carla Lacerda, Bob Pauken, Richard Teiss, and Bill Werner. In 1995 the Mobil carbonate research team initiated work on development of *Predictive Concepts of Carbonate Systems.* Completion of the research occurred in 1999 immediately prior to the Exxon–Mobil merger in 2000.

Special acknowledgments go to Aus Al-Tawil and Jan Golonka, inasmuch as they were the team members who made the *CATT* maps using the paleo-rotation programs. The project objectives were successfully completed on time due to their efforts. Special thanks are extended to Laura Parnell, who electronically drafted all of the paleo-oceanographic maps as supplements to the original *CATT* maps, and who helped with preparation of the field posters. Further, she assisted with conversions of the original *Global Atlas of Carbonate Fields* text and figures to Microsoft PowerPoint files so they could be included as Appendix 6.

The manuscript was significantly enhanced by ExxonMobil internal reviewers James H. Anderson, Sean Guidry, Gareth Jones, and Lisa Roehl. The authors thank Jeff Lukasik, Robert Goldstein, and Taury Smith for also providing outside critical reviews and suggestions for clarifying concepts, focusing key points, and overall improving the manuscript. We have incorporated Goldstein's suggestion for characterizing the *CATT* products as an integrated summative database for carbonates, recognizing that no new data were generated for the project. The authors gratefully acknowledge ExxonMobil management for support and approval to publish these research results.

REFERENCES

Abreu, V.S., and Anderson, J.B., 1998, Glacial eustasy during the Cenozoic: Sequence stratigraphic implications: American Association of Petroleum Geologists, Bulletin, v. 82, p. 1385–1400.

Ahr, W.M., 1973, The carbonate ramp: an alternative to the shelf model: Gulf Coast Association of Geological Societies, Transactions, v. 23, p. 221–225.

Al-Jallal, I.A., 1995, The Khuff Formation: its regional reservoir potential in Saudi Arabia and other Gulf countries; depositional and stratigraphic approach, *in* Al-Husseini, M.I., ed., Middle East Petroleum Geosciences: GEO'94, Gulf PetroLink, Bahrain, v. 1, p. 103–119.

Alsharhan, A.S., and Whittle, G.L., 1995, Carbonate–evaporite sequences of the Late Jurassic, southern and southwestern Arabian Gulf: American Association of Petroleum Geologists, Bulletin, v. 79, p. 1608–1630.

Baria, L.R., Stoudt, D.L., Harris, P.M., and Crevello, P.D., 1982, Upper Jurassic reefs of Smackover Formation, United States Gulf Coast: American Association of Petroleum Geologists. Bulletin, v. 66, p. 1449–1482.

Barrs, D.L, Copland, A.B., and Ritchie, W.D., 1970, Geology of Middle Devonian Reefs, Rainbow Area, Alberta, Canada, *in* Halbouty, M.T., ed., Geology of Giant Petroleum Fields: American Association of Petroleum Geologists, Memoir 14, p. 19–49.

Bliefnick, D.M., and Kaldi, J.G., 1996, Pore geometry: control on reservoir properties, Walker Creek Field, Columbia and Lafayette Counties, Arkansas: American Association of Petroleum Geologists, Bulletin, v. 80, p. 1027–1044.

Bosence, D., 2005, A genetic classification of carbonate platforms based on their basinal and tectonic settings in the Cenozoic: Sedimentary Geology, v. 175, p. 49–72.

Burke, W.H., Denison, R.E., Hetherington, E.A., Koepnick, R.B., Nelson, H.F., and Otto, J.B., 1982, Variation of seawater $^{87}Sr/^{86}Sr$ throughout Phanerozoic time: Geology, v. 10, p. 516–519.

Cande, S.C., and Kent, D.V., 1992, A new geomagnetic polarity time-scale for the late Cretaceous and Cenozoic: Journal of Geophysical Research, v. 97, p. 13,917–13,951.

Chimene, C.A., 1991, Walker Creek Field—USA Gulf of Mexico Basin, Arkansas, *in* Foster, N.H., and Beaumont, E.A., eds., Stratigraphic Traps II: American Association of Petroleum Geologists, Treatise of Petroleum Geology—Atlas of Oil and Gas Fields, p. 55–116.

Choquette, P.W., and Pray, L.C., 1970, Geologic nomenclature and classification of porosity in sedimentary carbonates: American Association of Petroleum Geologists, Bulletin, v. 54, p. 207–250.

Chuber, S., and Pusey, W.C., 1985, Productive Permian carbonate cycles, San Andres Formation, Reeves Field, West Texas, *in* Roehl, P.O., and Choquette, P.W., eds., Carbonate Petroleum Reservoirs: New York, Springer-Verlag, p. 289–308.

Collins, J.F., Kenter, J.A.M., Harris, P.M., Kuanysheva, G., Fischer, D.J., and Steffen, K.L., 2006, Facies and reservoir-quality variations in the late Visean to Bashkirian outer platform, rim, and flank of the Tengiz buildup, Precaspian Basin, Kazakhstan, *in* Harris, P.M., and Weber, L.J., eds., Giant Hydrocarbon Reservoirs of the World: From Rocks to Reservoir Modeling: American Association of Petroleum Geologists, Memoir 88, p. 1–41.

Davies, P.J., Symonds, P.A., Feary, D.A., and Pigram, C.J., 1989, The evolution of the carbonate platforms of northeast Australia, *in* Crevello, P.D., Wilson, J.L., Sarg, J.F., and Read, J.F., eds., Controls on Carbonate Platform and Basin Development: SEPM, Special Publication 44, p. 233–258.

Dorobek, S.L., 1995, Synorogenic carbonate platforms and reefs in foreland basin: controls on stratigraphic evolution and platform/reef morphology, *in* Dorobek, S.L., and Ross, G.M., eds., Stratigraphic Evolution of Foreland Basins: SEPM, Special Publication 52, p. 127–147.

Drummond, C.N., and Wilkinson, B.H., 1993, Carbonate cycle stacking patterns and hierarchies of orbitally forced eustatic sea-level change: Journal of Sedimentary Petrology, v. 63, p. 369–377.

Druckman, Y., and Moore, C.H., 1985, Late subsurface secondary porosity in the Jurassic grainstone reservoir, Smackover Formation, Mt. Vernon Field, Southern Arkansas, *in* Roehl, P.O., and Choquette, P.W., eds., Carbonate Petroleum Reservoirs: New York, Springer-Verlag, p. 369–383.

Earle, S.A., 2001, Atlas of the Ocean: Washington, D.C., National Geographic Society, 192 p.

Feazel, C.T., 1985, Diagenesis of Jurassic grainstone reservoirs in the Smackover Formation, Chatom Field, Alabama, *in* Roehl, P.O., and Choquette, P.W., eds., Carbonate Petroleum Reservoirs: New York, Springer-Verlag, p. 357–367.

Flügel, E., and Kiessling, W., 2002, A new look at ancient reefs, in Kiessling, W., Flügel, E., and Golonka, J., eds., Phanerozoic Reef Patterns: SEPM, Special Publication 72, p. 3–10.

Ford, D., and Golonka, J., 2003, Phanerozoic paleogeography, paleoenvironment, and lithofacies maps of the circum-Atlantic margins: Marine and Petroleum Geology, v. 20, p. 249–285.

Goldhammer, R.K., Dunn, P.A., and Hardie, L.A., 1987, High frequency glacio-eustatic sea level oscillations with Milankovitch characteristics recorded in Middle Triassic platform carbonates in Northern Italy: American Journal of Science, v. 287, p. 853–892.

Goldhammer, R.K., Dunn, P.A., and Hardie, L.A., 1990, Depositional cycles, composite sea-level changes, cycle staking patterns, and the hierarchy of stratigraphic forcing: examples from Alpine Triassic platform carbonates: Geological Society of America, Bulletin, v. 102, p. 535–562.

Golonka, J., Ross, M.I., and Scotese, C.R., 1994, Phanerozoic paleogeographic and paleoclimatic modeling maps, *in* Embry, A.F., Beauchamp, B., and Glass, D.J., eds., Pangea; Global Environments and Resources: Canadian Society of Petroleum Geologists, Memoir 17, p. 1–47.

Golonka, J., Ford, D., and Bednarczyk, J., 1997, Sauk and Tippecanoe (Cambrian – Silurian) paleoenvironments and Lithofacies (abstract); and Kaskaskia (Devonian–Mississippian) paleoenvironments and lithofacies (abstract); and Absaroka (Pennsylvanian–Triassic) paleoenvironments and lithofacies (abstract); and Zuni (Jurassic–Cretaceous) paleoenvironments and Lithofacies (abstract), *in* Canadian Society of Petroleum Geologists–SEPM Joint Convention, Calgary, Programs with Abstracts, June 1–6, 1997, B. Beauchamp, ed., p. 109–111.

GOLONKA, J., AND KIESSLING, W., 2002, Phanerozoic time scale and definition of time slices, *in* Kiessling, W., Flügel, E., and Golonka, J., eds., Phanerozoic Reef Patterns: SEPM, Special Publication 72, p. 11–20.

GOLONKA, J., BOCHAROVA, N.Y., FORD, D., EDRICH, M.E., BEDNARCZYK, J., AND WILDHARBER, J., 2003, Paleogeographic reconstructions and basins development of the Arctic: Marine and Petroleum Geology, v. 20, p. 211–248.

GREENLEE, S.M., AND LEHMANN, P.J., 1993, Stratigraphic framework of productive carbonate buildups, *in* Loucks, R.G.. and Sarg, J.F., eds., Carbonate Sequence Stratigraphy: Recent Developments and Applications: American Association of Petroleum Geologists, Memoir 57, p. 43–62.

HALLOCK, P., AND SCHLAGER, W., 1986, Nutrient excess and the demise of coral reefs and carbonate platforms: Palaios, v. 1, p. 389–398.

HANDFORD, C.R., AND LOUCKS, R.G., 1993, Carbonate depositional sequences and systems tracts—responses of carbonate platforms to relative sea-level changes, *in* Loucks, R.G., and Sarg, J.F., eds., Carbonate Sequence Stratigraphy: American Association of Petroleum Geologists, Memoir 57, p. 3–42.

HARLAND, W.B., ARMSTRONG, R.L., COX, A.V., CRAIG, L.E., SMITH, A.G., AND SMITH, D.G., 1990, A Geologic Time Scale 1989: Cambridge, U.K., Cambridge University Press, 263 p.

HAQ, B.U., HARDENBOL, J., AND VAIL, P.R., 1987, Chronology of fluctuating sea levels since the Triassic: Science, v. 235, p. 1156–1166.

HAQ, B.U., HARDENBOL, J., AND VAIL, P.R., 1988, Mesozoic and Cenozoic chronostratigraphy and cycles of sea level change, *in* Wilgus, C.K., Posamentier, H., VanWagoner, J.C., Ross, C.A., and Kendall, C.G.St.C., eds., Sea Level Changes: An Integrated Approach: SEPM, Special Publication 42, p. 71–108.

HEMPHILL, C.R., SMITH, R.I., AND SZABO, F., 1970, Geology of Beaverhill Lake Reefs, Swan Hills Area, Alberta, *in* Halbouty, M.T., ed., Geology of Giant Petroleum Fields:, American Association of Petroleum Geologists, Memoir 14, p. 50–90.

HEYDARI, E., 2000, Porosity loss, fluid flow, and mass transfer in limestone reservoirs: application to the Upper Jurassic Smackover Formation, Mississippi: American Association of Petroleum Geologists, Bulletin, v. 84, p. 100–118.

HEYDARI, E., AND MOORE, C.H., 1994, Paleoceanographic and paleoclimatic controls on ooid mineralogy of the Smackover Formation, Mississippi Salt Basin: implications for Late Jurassic seawater composition: Journal of Sedimentary Research, v. A64, p. 101–114.

HUMPHREY, J.D., RANSOM, K.L., AND MATTHEWS, R.K., 1986, Early meteoric diagenetic control of Upper Smackover Production, Oaks Field, Louisiana: American Association of Petroleum Geologists, Bulletin, v. 70, p. 70–85.

JABLONSKI, D., 1999, The future of the fossil record: Science, v. 284, p. 2114–2116.

JAMES, N.P., 1978, Facies Models 10. Reefs: Geoscience Canada, v. 5, p. 16–26.

JAMES, N.P., 1983, Reef environment (Chapter 8), *in* Scholle, P.A., Bebout, D.G., and Moore, C.L., eds., Carbonate Depositional Environments: American Association of Petroleum Geologists, Memoir 33, p. 345–462.

JAMES, N.P., AND CHOQUETTE, P.W., 1990a, Limestones—the sea-floor diagenetic environment, *in* McIlreath, I.A., and Morrow, D.W., eds., Geoscience Canada, Reprint Series 4, p. 13–34.

James, N.P., and Choquette, P.W., 1990b, Limestones—the meteoric diagenetic environment, *in* McIlreath, I.A., and Morrow, D.W., eds., Geoscience Canada, Reprint Series 4, p. 35–74.

JORDAN, C.F., AND ABDULLAH, M., 1992, Arun Field—Indonesia, North Sumatra Basin, Sumatra, *in* Foster, N.H., and Beaumont, E.A., eds., Stratigraphic Traps III: American Association of Petroleum Geologists, Treatise of Petroleum Geology and Atlas of Oil and Gas Fields, p. 1–39.

KENTER, J.A.M., HARRIS, P.M., WEBER, L.J., COLLINS, J.F., KUANYSHEVA, G., AND FISCHER, D.J., 2006, Late Visean to Bashkirian platform cyclicity in the central Tengiz buildup, Precaspian Basin, Kazakhstan: Depositional evolution and reservoir development, *in* Harris, P.M., and Weber, L.J., eds., Giant Hydrocarbon Reservoirs of the World: From Rocks to Reservoir Modeling: American Association of Petroleum Geologists, Memoir 88, p. 45–70.

KIESSLING, W., FLÜGEL, E., AND GOLONKA, J., 1999, Paleo reef maps: Evaluation of a comprehensive database on Phanerozoic reefs: American Association of Petroleum Geologists, Bulletin, v. 83, p. 1552–1587.

KIESSLING, W., FLÜGEL, E., AND GOLONKA, J., eds., 2002, Phanerozoic Reef Patterns: SEPM, Special Publication 72, 775 p.

KIESSLING, W., FLÜGEL, E., AND GOLONKA, J., 2003, Patterns of Phanerozoic carbonate platform sedimentation: Lethaia, v. 36, p. 195–226.

KOEPNICK, R.B., BURKE, W.H., DENISON, R.E., HETHERINGTON, E.A., NELSON, H.F., OTTO, J.B., AND WAITE, L.E., 1985, Construction of the seawater $^{87}Sr/^{86}Sr$ curve for the Cenozoic and Cretaceous: supporting data: Chemical Geology, v. 58, p. 55–81.

KOPASKA-MERKEL, D.C., MANN, S.D., AND SCHMOKER, J.W., 1994, Controls on reservoir development in a shelf carbonate: Upper Jurassic Smackover Formation of Alabama: American Association of Petroleum Geologists, Bulletin, v. 78, p. 938–959.

KUZNETSOV, V.G., 1996, Evolution and cyclicity of reef formation in Russia and adjacent countries: Lithology and Mineral Resources, v. 31, no. 2, p. 101–111.

MARTIN, R.E., 1996, Cyclic and secular variation in microfossil biomineralization and preservation: clues to calcium carbonate saturation and nutrient levels of Phanerozoic oceans (abstract), *in* Mutti, M., Simo, T., Wiessert, H., and Baker, P., eds., Carbonates and Global Climate Change, abstracts, SEPM/IAS Research Conference, Wildhaus, Switzerland, June 22–27, 1996.

MARTINDALE, W., AND BOREEN, T.D., 1997, Temperature-stratified Mississippian carbonates as hydrocarbon reservoirs—examples from the foothills of the Canadian Rockies, *in* James, N.P., and Clarke, J.A.D, eds., Cool-Water Carbonates: SEPM, Special Publication 56, p. 391–410.

MACKENZIE, F.T., AND MORSE, J.W., 1992, Sedimentary carbonates through Phanerozoic time: Geochimica et Cosmochimica Acta, v. 56, p. 3281–3295.

MAHONEY, J.J., AND COFFIN, M.F., eds., 1997, Large Igneous Provinces; Continental, Oceanic, and Planetary Flood Volcanism: Washington, D.C., American Geophysical Union, Geophysical Monograph 100, 438 p.

MEHLER, C. (project editor), et al., 2002, Family Reference Atlas of the World: Washington, D.C., National Geographic Society, 352 p.

MORSE, J.W., AND MACKENZIE, F.T., 1990, Geochemistry of Sedimentary Carbonates: Amsterdam, Elsevier, Developments in Sedimentology 48, 707 p.

MOUNTAIN, G., 1991, Oligocene–Upper Miocene high resolution seismic stratigraphy and the mid-Atlantic transect, *in* Lamont-Doherty Geological Observatory Industrial Associates Symposium: New Developments in Sequence Stratigraphy and Sea Level Change, September 12–13, 1991.

OGG, J.G., 1995, Magnetic polarity time scale of the Phanerozoic, *in* Ahrens, T.J., ed., Global Earth Physics: A Handbook of Physical Constants: Washington, D.C., American Geophysical Union, p. 240–270.

PARRISH, J.T., 1982, Upwelling and petroleum source beds with reference to Paleozoic: American Association of Petroleum Geologists, Bulletin, v. 66, p. 750–774.

PARRISH, J.T., AND CURTIS, R.C., 1982, Atmospheric circulation, upwelling and organic-rich rocks in the Mesozoic and Cenozoic Eras:

Palaeogeography, Palaeoclimatology, Palaeoecology, v. 40, p. 31–66.

PETTA, T.J., AND RAPP, S.D., 1990, Appleton Field—USA Gulf of Mexico Basin, Alabama, *in* Beaumont, E.A., and Foster, N.H, eds., Structural Traps IV, Tectonic and Nontectonic Fold Traps: American Association of Petroleum Geologists, Treatise of Petroleum Geology—Atlas of Oil and Gas Fields, p. 299–318.

POMAR, L.W.C., 2001, Types of carbonate platforms—a genetic approach: Basin Research, v. 13, p. 313–334.

RAMPINO, M.R., AND STOTHERS, R.B., 1988, Flood basalt volcanism during the past 250 million years: Science, v. 241, p. 663–668.

READ, J.F., 1982, Carbonate platforms of passive (extensional) continental margins: types, characteristics, and evolution: Tectonophysics, v. 81, p. 195–212.

READ, J.F., 1985, Carbonate platform facies models: American Association of Petroleum Geologists, Bulletin, v. 69, p. 1–21.

READ, J.F., 1995, Overview of carbonate platform sequences, cycle stratigraphy and reservoirs in greenhouse and icehouse worlds, *in* Read, J.F., Kerans, C., and Weber, L.J., eds., Milankovitch Sea Level Changes, Cycles and Reservoirs on Carbonate Platforms in Greenhouse and Ice-House Worlds: SEPM, Short Course 35, p. 1–102.

READ, J.F., 1998, Phanerozoic carbonate ramps from greenhouse, transitional, and ice-house worlds: clues from field and modeling studies, *in* Wright, V.P., and Burchette, T.P., eds., Carbonate Ramps: Geological Society of London, Special Publication 149, p. 107–135.

RENNE, P.R., AND BASU, A.R., 1991, Rapid eruption of the Siberian Traps flood basalts at the Permo-Triassic boundary, Science, v. 253, p. 176–179.

ROSS, C.A., AND ROSS, J.R.P., 1988, Late Paleozoic transgressive–regressive deposition, *in* Wilgus, C.K., Posamentier, H., Van Wagoner, J.C., Ross, C.A., and Kendall, C.G.St.C., eds., Sea Level Changes: An Integrated Approach: SEPM, Special Publication 42, p. 227–247.

ROSS, C.A., AND ROSS, J.R.P., 1994, Permian sequence stratigraphy and fossil zonation, *in* Embry, A.F., Beauchamp, B., and Glass, D.J., eds., Pangea: Global Environments and Resources: Canadian Society of Petroleum Geologists, Memoir 17, p. 219–231.

SANER, S., AND ABDULGHANI, W.M., 1995, Lithostratigraphy and depositional environments of the Upper Jurassic Arab-C carbonate and associated evaporites in Abqaiq Field, Eastern Saudi Arabia: American Association of Petroleum Geologists, Bulletin, v. 79, p. 394–409.

SANER, S., AND SAHIN, A., 1999, Lithological and zonal porosity–permeability distributions in the Arab-D reservoir, Uthmaniyah Field, Saudi Arabia: American Association of Petroleum Geologists, Bulletin, v. 83, p. 230–243.

SARG, J.F., 1988, Carbonate sequence stratigraphy, *in* Wilgus, C.K., Hastings, B.S., Kendall, C.G.S.C., Posamentier, H.W., Ross, C.A., and Van Wagoner, J.C., eds., Sea Level Changes: An Integrated Approach: SEPM, Special Publication 42, p. 155–182.

SARG, J.F., MARKELLO, J.R., AND WEBER, L.J., 1999, The second-order cycle, carbonate-platform growth, and reservoir, source, and trap prediction, *in* Harris, P.M., Simo, J.A., and Saller, A.H, eds., Advances in Carbonate Sequence Stratigraphy: Applications to Reservoirs, Outcrops, and Models: SEPM, Special Publication 63, p. 1–24.

SCHLAGER, W., 2005, Carbonate Sedimentology and Sequence Stratigraphy: SEPM, Concepts in Sedimentology and Paleontology no. 8, 200 p.

SCOTT, R.W., 1988, Evolution of Late Jurassic and Early Cretaceous reef biotas: Palaios, v. 3, p. 184–193.

SLOSS, L.L., 1963, Sequences in the cratonic interior of North America: Geological Society of America, Bulletin, v. 74, p. 93–114.

SLOSS, L.L., ed., 1988, Tectonic evaluation of the craton in Phanerozoic time: Geological Society of America, The Geology of North America, vol. D-2, Sedimentary Cover—North American Craton, p. 25–51.

TUCKER, M.E., AND WRIGHT, V.P., 1990, Carbonate Sedimentology: Oxford, U.K., Blackwell, 496 p.

VAIL, P.R., AUDEMARD, F., BOWMAN, S.A., EISNER, P.N., AND PEREZ-CRUZ, C., 1991, The stratigraphic signatures of tectonics, eustasy, and sedimentology—an overview, *in* Einsele, G., Ricken, W., and Seilacher, A., eds., Cycles and Events in Stratigraphy: Berlin, Springer-Verlag, p. 617–659.

VEEVERS, J.J., 1990, Tectonic–climatic supercycles in the billion-year plate tectonic eon: Permian Pangean icehouse alternates with Cretaceous dispersed-continents greenhouse: Sedimentary Geology, v. 68, p. 1–16.

VEST, E.L., 1970, Oil fields of Pennsylvanian–Permian Horseshoe Atoll, West Texas, *in* Halbouty, M.T., ed., Geology of Giant Petroleum Fields: American Association of Petroleum Geologists, Memoir 14, p. 185–203.

VINIEGRA, F., AND CASTILLO-TEJERO, C., 1970, Golden Lane Fields, Veracruz, Mexico, *in* Halbouty, M.T., ed., Geology of Giant Petroleum Fields: American Association of Petroleum Geologists, Memoir 14, p. 309–325.

WAITE, L.E., MARKELLO, J.R., ZEMPOLICH, W.G., KOEPNICK, R.B., SARG, J.F., WEBER, L.J., AND EDRICH, M.E., 1994, Stratigraphic hierarchy: Part III. First- and second-order depositional sequences: A framework for stratigraphic analyses: Internal Memorandum No. 64 (MM 941202-1), Mobil Exploration and Technology Company, Dallas, Texas, 49 p.

WEBER, L.J., FRANCIS, B.P., HARRIS, P.M., AND CLARK, M., 2003, Stratigraphy, lithofacies, and reservoir distribution, Tengiz field, Kazakhstan, *in* Ahr, W.M., Harris, P.M., Morgan, W.A., and Somerville, I.D., eds., Permo-Carboniferous Carbonate Platforms and Reefs: SEPM, Special Publication 78 and American Association of Petroleum Geologists, Memoir 83, p. 351–394.

WILKINSON, B.H., OWEN, R.M., AND CARROLL, A.R., 1985, Submarine hydrothermal weather, global eustasy, and carbonate polymorphism in Phanerozoic marine oolites: Journal of Sedimentary Petrology, v. 55, p. 171-183.

WILSON, A.O., 1985, Depositional and diagenetic facies in the Jurassic Arab-C and –D reservoirs, Qatif field, Saudi Arabia, *in* Roehl, P.O., and Choquette, P.W., eds., Carbonate Petroleum Reservoirs: New York, Springer-Verlag, p. 319–340.

WILSON, J.L., 1974, Characteristics of carbonate-platform margins: American Association of Petroleum Geologists, Bulletin, v. 58, p. 810–824.

WILSON, J.L., 1975, Carbonate Facies in Geologic History: New York, Springer-Verlag, 471 p.

WILSON, J.T., 1966, Did the Atlantic close and then reopen?: Nature, v. 207, p. 676–681.

WILSON, J.T., 1988, Convection tectonics: some possible effects upon the Earth's surface of flow from the deep mantle: Canadian Journal of Earth Sciencex, v. 25, p. 1199–1208.

WOOD, R., 1995, The changing biology of reef-building: Palaios, v. 10, p. 517–529.

WRIGHT, V.P., AND ROBINSON, D., 1988, Early Carboniferous flood plain deposits from South Wales: A case study of the controls on paleosol development: Geological Society of London, Journal, v. 145, no. 5, p. 847–857.

WRIGHT, V.P., 1992, Speculations on the controls on cyclic peritidal carbonates: ice-house versus greenhouse eustatic controls: Sedimentary Geology, v. 76, p. 1–5.

THE SUBTLE THIEF: SELECTIVE DISSOLUTION OF ARAGONITE DURING SHALLOW BURIAL AND THE IMPLICATIONS FOR CARBONATE SEDIMENTOLOGY

V. PAUL WRIGHT AND LESLEY CHERNS

School of Earth, Ocean and Planetary Sciences, Cardiff University, Cardiff, Wales, CF10 3YE, UK

e-mail wrightvp@Cardiff.ac.uk; cherns@Cardiff.ac.uk

ABSTRACT: Syndepositional aragonite dissolution at very shallow depths, above the lysocline, is a major process that affects carbonate deposition and skews the composition of carbonate sediments. Such dissolution is capable of altering sediment composition in many settings, and during microfacies analysis it is critical to be aware of this early, selective filtering by identifying taphonomic signatures. These effects are also capable of distorting the trophic composition of fossil biotas, potentially restricting the ability to identify nutrient levels and other controls. The evidence from widespread diagenetic limestones in shale (marl)–limestone rhythms supports the dissolution model, but the source of the precursor aragonite is unresolved; allochthonous aragonite mud is one possible source, but, especially during calcite-sea intervals, another possibility is from the autochthonous aragonitic fauna. Forward models for carbonate sedimentation will need to compensate for aragonite dissolution if realistic models are to be developed, but our knowledge of the environmental distribution and magnitude of aragonite dissolution is still woefully incomplete. Another major consequence of early aragonite loss is that the diagenetic potentials of many carbonate sediments have been changed, drastically reducing secondary porosity potentials long before they are affected by meteoric or burial processes.

INTRODUCTION

There is a growing body of evidence that early, effectively syndepositional dissolution of aragonite is a major process affecting carbonate sediments well above the lysocline (that depth in the oceans at which there is a marked increase in carbonate dissolution) and even in very shallow-water settings. These effects are documented from modern tropical platform interiors (Walter and Burton, 1990; Ku et al., 1999), from a range of ancient tropical, shallow-platform to mid-outer-ramp settings (Cherns and Wright, 2000; Sanders, 2001; Wright et al., 2003), and from modern (James et al., 2005) and ancient (Kyser et al., 1998; Nelson and James, 2000; Knoerich and Mutti, 2006) cool-water systems. Aragonite loss from deeper-water settings, but related to undersaturation in the sediment, not water, column, has been documented from the Quaternary (Brachert and Dullo, 2000; Schwarz and Rendle-Buhring, 2005). The importance of relatively shallow-burial marine fluids in causing dissolution and mimicking meteoric diagenesis has been discussed by Melim et al. (2002).

The aim of this paper is to expand and update an earlier review (Wright and Cherns, 2004) highlighting and exploring the implications of the effects of early aragonite dissolution on a range of aspects of carbonate sedimentology, and to emphasize the need for a more sophisticated approach to understanding carbonate accumulation. If we are guilty of overemphasizing certain problems it is caused by our experience with taphonomic windows in Silurian and Jurassic successions (Cherns and Wright, 2000; Wright et al., 2003) and with as yet unpublished examples, which has led us to question the completeness, even at the bed scale, of the fossil record of marine skeletal invertebrates; the fossilization potential of such groups directly controls limestone genesis. A subtle thief has been at work selectively altering the sedimentary record in ways that we are only just beginning to appreciate; the implications are wide-reaching and will require a reassessment of many basic tenets in carbonate sedimentology. We have to adapt to a level of uncertainty in reconstructing carbonate systems, and to appreciate that there may be major limitations on what we can know as a result of destructive processes leaving little direct trace.

EARLY ARAGONITE DISSOLUTION

Even during very shallow, effectively syndepositional burial of as little as a few centimeters, in shallow tropical seas at supralysoclinal depths, and without the ingress of undersaturated meteoric waters, aragonite can dissolve. The carbonate sedimentological community was slow to react to the implications of suggestions by Walter and Burton (1990) and Ku et al. (1999) that an estimated 50% of the annual carbonate produced in some modern platform interiors is lost by dissolution. James et al. (2005) estimated that 40–90% of the carbonate produced on the Southern Australian Shelf may never enter the geological record because of early aragonite loss. They based this on a comparison of Holocene shelf carbonate sediments from that area and Cenozoic limestones (James et al., 2005, p. 461). Evidence from detailed taphonomic analyses (Cherns and Wright, 2000; Sanders, 2001, 2003; Wright et al., 2003) also supports the case for early aragonite loss. Cherns and Wright (2000) estimated, by comparing time-averaged assemblages with silicified life assemblages dominated by former aragonitic molluscs, that the aragonitic molluscan component of the Silurian benthos of Gotland, Sweden, had been depleted at least a hundred-fold by early dissolution. Such processes must have skewed the fossil record, as confirmed recently by Bush and Bambach (2004) in their estimate of a 29% loss in diversity of Paleozoic marine faunas as a consequence of taphonomic processes controlled by original mineralogy.

If these estimates of aragonite loss are more widely representative, they predict significant effects on carbonate sedimentary systems, affecting sediment accumulation rates and profiles, sediment budgets, diagenetic potentials, and porosity evolution.

An excellent review of the processes responsible for early aragonite dissolution is provided by Sanders (2003). Major processes producing undersaturation with respect to aragonite in oxic sediment include aerobic oxidation of organic matter, and the oxidation of by-products of anaerobic decay of organic matter (oxidation of H_2S). In the anaerobic zone, sulfate reduction and anaerobic methane oxidation also contribute to aragonite dissolution. Factors that control the nature of dissolution include composition of the aragonite grains (especially their organic

Controls on Carbonate Platform and Reef Development
SEPM Special Publication No. 89, Copyright © 2008
SEPM (Society for Sedimentary Geology), ISBN 978-1-56576-130-8, p. 47–54.

content), the grain size of the sediment, the contents of organic matter, oxygen, iron, and clay, the degree of irrigation by burrowers, and the deposition rate (affecting the residence time of the grains in the taphonomically active zone). The saturation state for aragonite of the ambient sea water is another variable, but early, shallow aragonite dissolution is not limited to calcite seas, and appears to be largely due to processes in the shallowest levels of the sediment column (Sanders, 2003).

These multiple factors result in aragonite dissolution being spatially very variable. Low-energy settings where organic matter can accumulate seem to be especially prone to aragonite dissolution, the organic matter providing the drive for microbial oxidation and related processes. Clean, winnowed carbonate sands seem less prone, and so there are gradients in aragonite preservation related to depth and energy. For example, in the Lower Jurassic successions of South Wales, Wright et al. (2003) note that inner-ramp grainstones contain abundant open and occluded molds of former aragonitic molluscs, but muddy coeval mid-ramp to outer-ramp facies had almost all the former molluscan aragonite removed during very early, shallow burial. Sanders and Krainer (2005) provide a detailed analysis of how early aragonite dissolution relates to facies variability from Permian platform-interior settings.

Differential preservation of former aragonitic fossils in shallower waters is influenced by another taphonomic effect: micrite envelopes. In many limestones former aragonitic grains are almost invariably defined by micrite envelopes; the preservation of these envelopes is evidence that they were not aragonite. Micrite envelopes are most common on bioclasts in very shallow-water settings, preserving shell outlines, and they seem to be an additional factor why former aragonitic grains are found preferentially in shallow-water carbonates. Betzler et al. (1997) have stressed the absence of micrite envelopes as a factor explaining the loss of evidence of former aragonitic shells in cool-water carbonates.

What is the fate of all the carbonate released from aragonite dissolution? As has been shown by Munnecke and co-workers (Munnecke and Samtleben, 1996; Munnecke et al., 1997; Munnecke and Westphal, 2004), the very common diagenetic bedding in shale (marl)–limestone rhythms is one consequence of the redistribution of aragonite. Melim et al. (2003) have shown that diagenesis in marine waters can also redistribute carbonate from aragonitic grains to low-Mg calcite cements in late Cenozoic carbonates, and Knoerich and Mutti (2006) have shown this effect in Oligo-Miocene heterozoan carbonates from the central Mediterranean. What other effects are there waiting to be discovered? Perhaps even some early marine dolomites may relate to carbonate released by aragonite dissolution (Hendry et al., 2000)?

High Mg calcite is not removed (Munnecke and Westphal, 2005); the relative solubility depends on the Mg content (Burton and Walter, 1987). In the case of Cenozoic cool-water carbonates from South Australia, high-Mg calcite stabilizes to low-Mg calcite in marine pore waters (Kyser et al., 1998; James et al., 2005).

PALEOENVIRONMENTAL AND PALEOECOLOGICAL ANALYSES

Microfacies analysis is the main tool for interpreting depositional environments in ancient carbonates, largely based on the identification of environmental indices (Flügel, 2004). There is an appreciation of the selective skewing of sediment composition caused by many taphonomic processes, and that many (most?) microfacies are what should be seen as microtaphofacies (Brachert et al., 1996; Dullo, 1990; Perry, 2000; Wright and Burgess, 2005). However, the effects of early aragonite dissolution are not routinely considered. Put simply, is the absence in a thin section of a grain type originally composed of aragonite, such as the depth-sensitive dasycladacean algae, due to an environmental or taphonomic effect? There is a need to look for subtle evidence in limestones of the former presence of aragonitic components (Sanders, 2003; Sanders and Krainer, 2005); we are unwise to take at face value the compositions of many microfacies, and we need to "interrogate" microfacies much more carefully than we have in the past.

A major influence on carbonate systems is nutrient availability (e.g., Mutti and Bernoulli, 2003, Pomar et al., 2004; Wilson and Vecsei, 2005; Wood, 1993). Lukasik et al. (2000) and Lukasik and James (2005) provide an important example in relating carbonate production to trophic resources for the Oligo-Miocene of the Murray Basin, Australia, compiling, from the literature, the faunal and sedimentological attributes of a trophic continuum. Evaluating the influences of trophic resources may prove an important interpretive tool for understanding ancient carbonates, but what needs to be determined is whether the criteria used, such as trophic structure and diversity, are reliable when aragonite dissolution can so dramatically skew faunas (Cherns and Wright, 2000; Wright et al., 2003). What must be considered when using such criteria is whether the trends or changes seen in trophic structure and diversity are due to true trophic or to taphonomic effects. For example, Lukasik and James (2003) note a shift from eutrophic facies with infaunal components to oligotrophic facies with epifaunal forms. This shift from infaunal to epifaunal dominance is a common taphonomic effect of aragonite dissolution (Cherns and Wright, 2000; Wright et al., 2003), which indicates the need for caution when applying faunal criteria where some of the fauna (or flora) could have been selectively removed syndepositionally.

PROVENANCE OF CARBONATE MUDROCKS

Carbonate muds in the past, as today, are polygenetic. In settings accumulating such muds several sources are possible (Fig. 1), but one source may dominate. Many deeper-water ramp settings are characterized by shale (marl)–limestone rhythms. Extensive studies by Munnecke and co-workers (Munnecke and Samtleben, 1996; Munnecke et al., 1997; Munnecke and Westphal, 2004) have demonstrated that many limestones in such successions are the products of the dissolution of aragonite and its precipitation as micritic or microsparitic calcite in nodules and beds. This is not to imply that all deeper-water muds were originally aragonite and have been diagenetically remobilized, but the huge volume of diagenetic bedding in the stratigraphic record is clear evidence of the importance of these processes. If such mud-grade carbonates, including microspars, in ancient shale (marl)–limestone rhythms simply constitute replaced former aragonite mud in a closed diagenetic system, as proposed by Munnecke and co-workers (see review in Munnecke and Westphal, 2005), there is little need to rethink our conceptual understanding of carbonate deposition and sediment budgets. We could continue to envisage such former aragonitic muds as having been largely derived from shallow platform areas, with some input from aragonite precipitated in the water column. However, as these authors acknowledge, this could not have been the case in calcite-sea intervals inasmuch as current wisdom implies that during such intervals primary aragonite muds could not have been derived by precipitation from the water column, even if this is likely to be a significant process during aragonite-sea intervals (e.g., Dix et al., 2005).

At present many aragonite muds in platform interiors appear to be largely derived from the breakdown of aragonitic biota (Gischler and Zingeler, 2002, but see also Yates and Robbins,

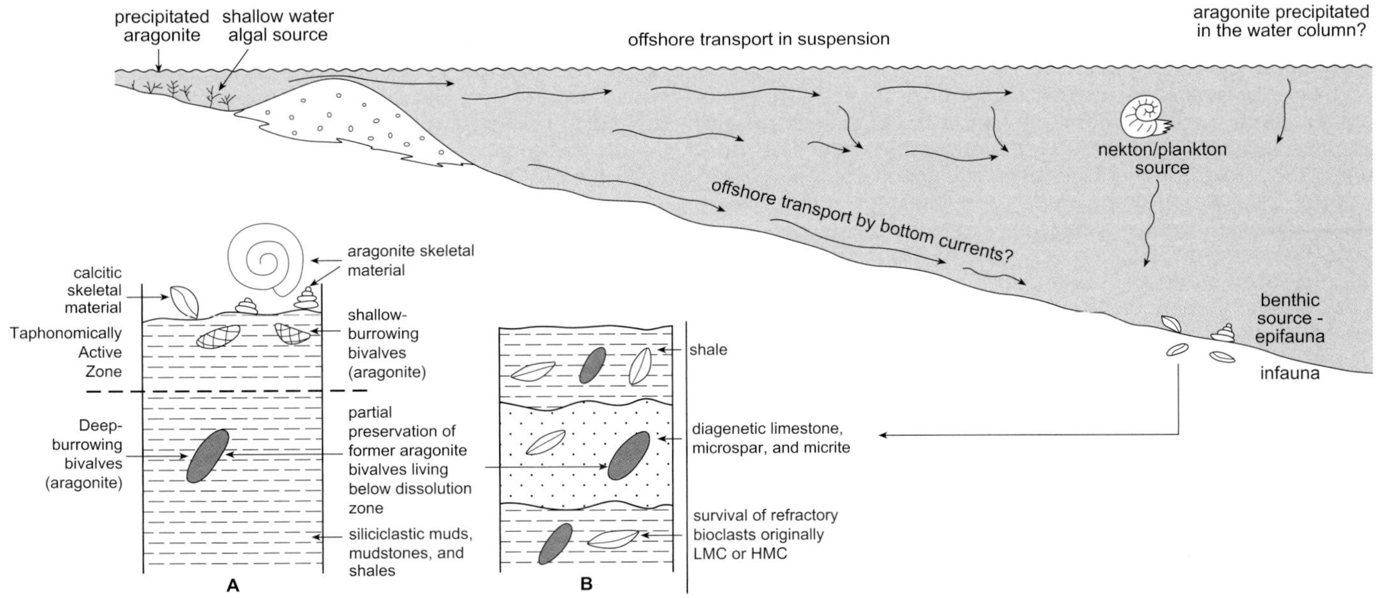

FIG. 1.—Sources of aragonite in offshore mid–outer-ramp settings. During calcite seas (most of the Phanerozoic) the potential for the supply of aragonite from abiogenic precipitation and probably also from green-algal sources was reduced, and skeletal aragonite from the benthic and nektic fauna was a likely major but underappreciated source of aragonite. The aragonitic benthos (epifaunal and infaunal) composed of gastropods and shallow burrowing bivalves **A)** was capable of supplying large volumes of carbonate. In many offshore settings, such as the Lower Jurassic of South Wales (Wright et al., 2003), as evidenced by shale (marl)–limestone rhythms **B)**, calcitic fauna represents the refractory remains of the original biota. In such successions former aragonitic deep-burrowing bivalves are preserved as molds, indicating that the aragonite-dissolution zone was less deep in the sediment than the depths inhabited by these bivalves. Their molds were preserved because they lived below the main depth where bioturbation destroyed the molds produced after the dissolution of shallow burrowers and epifaunal molluscs.

1998), and derivation from a shallow-water algal source seems a more likely explanation for limestones in shale (marl)–limestone rhythms during calcite-sea intervals. However, there are some complications here. Firstly, the contribution by such algae in calcite seas was likely reduced compared to that of aragonite-sea intervals (Ries, 2005, 2006; Stanley and Hardie, 1998, 1999). Secondly, the remains of calcareous green algae are not common in many ancient successions; for example in the Jurassic of southern Britain, in a calcite sea, shale (marl)–limestone rhythms are widespread in offshore facies but calcareous green algae are conspicuously rare in shallow-water facies (Elliott, 1982). Dodd and Nelson (1998, p. 121) have commented on the lack of abundance in general of calcareous green algae in Paleozoic successions. It may be unreasonable to extrapolate the importance of green algae as aragonite producers in the late Cainozoic to earlier times. Of course the calcareous green algae are amongst the most important depth indicators in microfacies analysis and are prone to dissolution. With regard to this important group we need to ask whether they were less common in the past, were less calcified, or simply did not get preserved.

Another potential source for diagenetic carbonate in shale (marl)–limestone rhythms could be benthic molluscan aragonite (and potentially other aragonite benthos such as some bryozoans; James et al., 2005). Wheeley et al. (2008) have calculated that aragonite from bivalves can provide a more than sufficient source of carbonate for the generation of diagenetic bedding. This is proposed as at least a contributory source, to the likely pelagic input. In some shale (marl)–limestone successions the thinly laminated nature of the diagenetic carbonate would seem to preclude former benthic sources (Munnecke and Westphal, 2005). There are other sources of skeletal aragonite besides the benthos; many nektonic molluscs are / were also aragonitic, and in the past the nautiloid and ammonoid cephalopods were likely sources, as were pteropods and scaphopods. We are struck by how rare many former aragonitic fossils are in many mid–outer-ramp successions, not only benthos such as infaunal bivalves and gastropods but also cephalopods; where ammonoids are found, they lack replaced shells, implying that the aragonite was dissolved relatively early during burial. Taphonomy is just as important for the sedimentologist as is isotope geochemistry or time-series analysis, and that early dissolution is a critical process has been evident for some years (e.g., Jeans, 1980).

One possible test to establish whether diagenetic bedding is sourced from *in situ* benthic production or from allochthonous aragonite (muds or aragonitic nekton, etc.) would be to examine successions lacking a skeletal benthos, such as black shale deposits indicating anoxic bottom waters; if diagenetic limestones are less common in such deposits, it would lend support to the role of aragonitic benthos as a major source of carbonate.

PRODUCTION RATES AND MODELING

Does it really matter that some of the carbonate in some environments has been lost at an early stage so that it never contributed to what gets preserved? Is this just an issue for the academic paleontologist or paleoecologist? Among the many reasons why this does matter is the current huge interest in forward modeling of carbonate deposystems. The carbonate production rates used in many models are based on measurements over very short intervals, and are not net accumulation rates. If early-dissolution effects are as important as proposed, then modelers must factor them in; otherwise, their models are likely missing a critical component. The drawback is that as yet we know virtually nothing, quantitatively, about the rates of dissolution and their spatial distribution.

One important element of forward models of platform development is the depth–productivity profile; Pomar (2001a) has emphasized the importance of different depth–productivity profiles and their influences on platform geometry. The classic tropical, phototroph-dominated profile has high production in shallow waters, rapidly decreasing with depth below the photic zone (Bosscher and Schlager, 1993), but this is not applicable to all ancient carbonate deposystems (Aurell et al., 1998; Brandano and Corda, 2002; Pomar, 2001b, Wright and Faulkner, 1990). Most models include a function with a decrease in production rate with depth. Very high production in shallow water is one cause of strongly aggradational growth that favors the development of steep platform margins. To produce ramp profiles it appears necessary to have different levels of shallow to offshore water production compared to those of steep-sided platforms, and/or efficient offshore transport of sediment from shallow waters; the result is a less shallow-water-biased depth–productivity profile (Aurell et al., 1998; Read, 1998; Wright and Faulkner, 1990). Where production is dominated by phototrophs, reduced rates at depth are easily explained because light levels decrease, but should this apply equally to heterotroph-dominated systems? In settings such as mid–outer ramps, could the decrease reflect lower accumulation rates due to dissolution of aragonite? This is not to imply that aragonite dissolution is limited to offshore settings, which it is clearly not, but that we may have been missing a critical process when trying to model sediment dynamics in deeper water settings (Fig. 2). Could the fact that deeper oligophotic and aphotic zones appear dominated by calcitic forms (Fig. 2) simply reflect a taphonomic bias?

Models can be tuned to simulate geometries and facies compositions of outcropping successions (Aurell et al., 1998). If these models do not allow for the loss of the aragonite component by early dissolution in either shallow-water or deeper-water settings, they may potentially be flawed. Sanders (2004) has provided a preliminary attempt to test the importance of syndepositional dissolution on platform growth.

For shale (marl)–limestone rhythms the case for their diagenetic origin is now very strong, indicating that what we perceive as carbonate mudrocks are actually diagenetic micrites and microspars. When trying to understand sedimentary processes and budgets this is not a problem if depositional muds become diagenetic micrite or microspar. However, if the mud was originally skeletal aragonite, we cannot use the volumes of offshore mudrocks as an indicator of allochthonous mud-grade sediment of whatever source. Depositional packstones or wackestones with aragonitic benthos could become diagenetic mudstones with micrite or microspar. These diagenetic, not depositional, mudrocks now contain the refractory remains of the originally calcitic skeletal biotas, but the aragonitic component has been reconstituted into the diagenetic matrix. The unsettling evidence is growing, discussed above, of massive loss of skeletal aragonite during very shallow burial, especially of molluscs. Extensive mud-grade carbonates in the stratigraphic record are due to the early dissolution of aragonite and the precipitation of calcite micrite and microspar.

Perhaps we can model these systems relatively simply; if they were diagenetically closed, the volume of diagenetic carbonate gives a good indication of the total carbonate input from *in situ* and allochthonous sources. In the case of shale (marl)–limestone rhythmic successions Munnecke and co-workers (Munnecke et al., 2001; Munnecke and Westphal, 2004, 2005) have invoked a

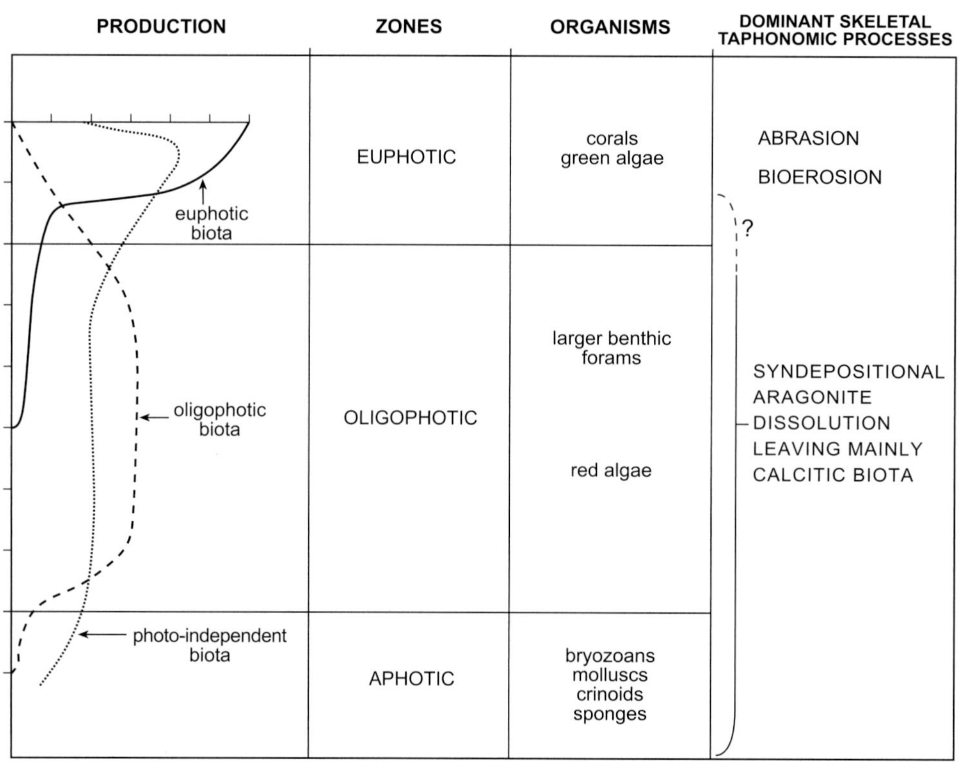

FIG. 2.—Carbonate production vs. bathymetry model of Pomar (2001a). The apparent dominance of calcite-producing biota in the deeper-water, lower-energy areas may reflect early aragonite dissolution (see text), as supported by the data in James et al. (2005).

closed diagenetic system, although they acknowledge that this is an oversimplification (e.g., Munnecke and Westphal, 2005). Geochemical studies (e.g., Ku et al., 1999) lead us to expect that dissolution of aragonite is most marked in the upper, oxic, bio-irrigated part of the sediment column, where pore-water circulation would be essential (Sanders, 2003, p. 118). A closed system seems improbable, and thus carbonate must be lost to the water column in solution. Calculating the mass balances for such systems is a challenge, but accurate forward modeling of carbonate deposystems will have little real credibility until such problems are addressed.

Being able to understand and quantify the loss of aragonite is not just an issue for low-energy, mid–outer-ramp settings. Ku et al. (1999) show that these processes also affect very shallow-water platform-interior settings at present, and Sanders and Krainer (2005) provide a Permian example. Sedimentation rates across carbonate platform tops are highly varied (Schlager, 2003; Demicco and Hardie, 2002; Yang et al., 2004). Such deposystems in large part represent facies mosaics (see discussion by Wright and Burgess, 2005), with each component habitat having its own complex and poorly understood sediment budgets (for example, see Sanders, 2003, fig. 23) An analogy might be drawn with soils (Wright and Cherns, 2004) (Fig. 3) of *in situ* and allochthonous inputs (gains), losses through the export of carbonates, abrasion, bioerosion and aragonite dissolution, and syndepositional mineral transformations and translocations. Not all these types of processes will have operated in every sediment, but perhaps we need to view carbonate sediments with a more critical approach, meaning that we need to move away from simply interpreting sediment composition in terms of energy level and grain types and to attempt to reconstruct the process domain or domains under which the sediment was produced, modified, and finally buried to a depth at which little further textural and compositional changes occurred. Compare the scientific rigor with which a paragenetic sequence is identified in a limestone or dolomite, perhaps identifying ten events after burial and the likely temperature, depth, and fluid regimes through which that carbonate passed, with the typically cursory interpretation of where that sediment was deposited: inner platform, or shoal, or patch reef. If we are good at interpreting depositional settings from microfacies, why is it that we still struggle to readily identify in limestones the two main Cenozoic inner-platform settings, sea-grass meadows and mangrove coasts (Beavington-Penney et al., 2006; Wright and Azeredo, 2006)? What we can state is that it is likely that dissolution is a major process removing aragonite from shallow platform interiors and that this affects carbonate accumulation in varying degrees that are spatially complex. Even the more advanced attempts to model shallow-water platform-interior deposystems (e.g., Burgess and Wright, 2003) cannot include a consideration of the budgets of individual environments.

Estimates of long-term rates of carbonate accumulation are much lower than rates derived from short-term estimates (Schlager, 1999). These differences can reflect scaling problems, but these can be avoided if short time intervals are studied, as in the case of the late Jurassic–early Cretaceous and Holocene shallow-water deposits discussed by Strasser and Samankassou (2003). However, these authors calculated accumulation rates four to five times less in the former than for the Holocene. Although Strasser and Samankassou did not consider aragonite dissolution as a possible contributory factor, it seems likely that it must have played some role.

DIAGENETIC POTENTIAL

The diagenetic potential of carbonate sediments is controlled principally by their initial composition. If they are composed of diagenetically relatively stable low-Mg calcite, they are likely to undergo little diagenetic alteration unless buried to depths where pressure solution takes place; chalks are an example of a limestone where a reduced primary diagenetic potential results in relatively little diagenesis at shallow burial depths. The potential for secondary porosity is also strongly influenced by sediment composition. Dolomitization and burial corrosion can produce secondary porosity in even stabilized calcitic or already dolo-

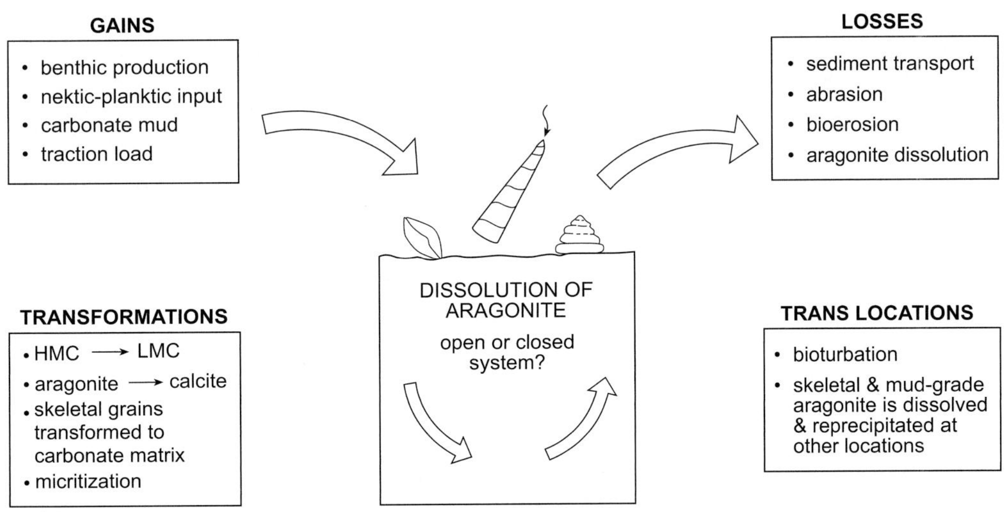

FIG. 3.—Carbonate sediments can be compared with the soils. Soil features are the product of the interplay of four main groups of processes: gains, losses, transformations, and translocations, and the near-surface levels in carbonate sediments can also be conceptualized as the product of four sets of processes. What is not resolved is the extent to which aragonite dissolution occurs in a closed or open system (see text). If the former, the main consequence is that aragonitic grains, such as shells, are transformed into matrix-grade carbonate. If the latter, much of the dissolved aragonite, of whatever source, is likely lost from the system. Translocation (movement from one position to another) is an important process in some soils (such as clay eluviation–illuviation) but can also result from aragonite dissolution, diffusion, and reprecipitation as calcite in sediments.

mitized carbonates, but dissolution of aragonite is arguably the most critical process producing secondary porosity in limestones that have not been dolomitized. Sediments with a high proportion of aragonite grains are potentially likely to produce moldic porosity, which may later become filled by cement. Carbonate sediments that originally lacked significant proportions of aragonitic grains could not produce moldic porosity as a result of aragonite dissolution. This relationship was thought to be seen clearly when comparing, for example, tropical Paleozoic to younger limestones (Fig. 4A, B; Tucker and Wright, 1990, p. 412) or tropical with cool-water carbonates (James and Bone, 1989). Molluscs, as the principal aragonitic fauna, were thought relatively unimportant in early to mid Paleozoic biotas (Sepkoski, 1984), a feature clearly supported by even cursory examination of limestones of that age. This reduced content of skeletal aragonite led some researchers (Dodd and Nelson, 1998) to propose that Cenozoic cool-water carbonates might make better diagenetic analogues for Paleozoic limestones than modern tropical ones. However, cool-water carbonates may also have had a much higher porosity potential that previously thought (James et al., 2005).

The orthodox view that tropical Paleozoic limestones were different from younger ones is shown in Figure 4A and B (modified from Tucker and Wright 1990, p. 412). The stabilization pathways and secondary-porosity potentials were very different for modern and early–mid Paleozoic carbonates. The release of aragonite during dissolution provides an important source for cementation during meteoric or burial diagenesis, and the presence of "unstable" aragonite is one of the main drivers for diagenesis. Figure 4C shows a postulated stabilization pathway for early Jurassic carbonates from South Wales based on data in Wright et al. (2003).

If aragonite dissolution at very shallow depth is such an important process, wouldn't we expect to see evidence of molds and porosity, in, for example, shale (marl)–limestone rhythms? The explanation is that dissolution takes place in unlithified sediment that is still prone to bioturbation and later compaction. It the sediment becomes indurated early in its history, to form hardgrounds, the molds can be preserved (James et al., 2005; Palmer et al., 1988). The formation of many hardgrounds may also relate to localized aragonite dissolution and calcite precipitation. However, oversized pores are potentially evidence of early aragonite dissolution and cementation (Knoerich and Mutti, 2006).

It would be instructive to determine, from the literature, whether, as seems likely, reservoirs with moldic porosity are more common in post-Paleozoic successions, bearing in mind the likely skewing of any analysis by the abundance of Cretaceous rudist-bearing reservoirs and Neogene coral reefs. Intuitively we regard it as correct that reservoirs with moldic porosity after aragonite are relatively rare in the Paleozoic, excluding the latest Paleozoic phylloid algal and *Paleoaplysina*-bearing reservoirs. This would confirm the orthodox view, but it presents a problem, inasmuch as the case has been made for the potential abundance of aragonitic molluscs even in Silurian shallow-water settings (Cherns and Wright, 2000). Why are reservoirs with moldic porosity after molluscs seemingly less common in the Paleozoic? This must be due to such forms being much less common.

SUMMARY

A more dynamic concept of carbonate accumulation is needed, in which some shallow-water and deeper-water settings, well above the lysocline or ACD, are likely to be major sites of aragonite loss ("black holes" of Wright and Cherns 2004). The evidence from diagenetic limestones in shale (marl)–limestone rhythms supports this view, although the source of the precursor aragonite is in dispute: is it all derived from allochthonous aragonite mud, or is at least some, and perhaps most, derived from autochthonous animal skeletal remains? Syndepositional dissolution has the potential to skew sediment composition in many environmental settings, and a new approach is needed to investigate more thoroughly the taphonomic signatures in sediments. Modeling carbonate systems will require a thorough understanding of the distribution and extent of dissolution and of

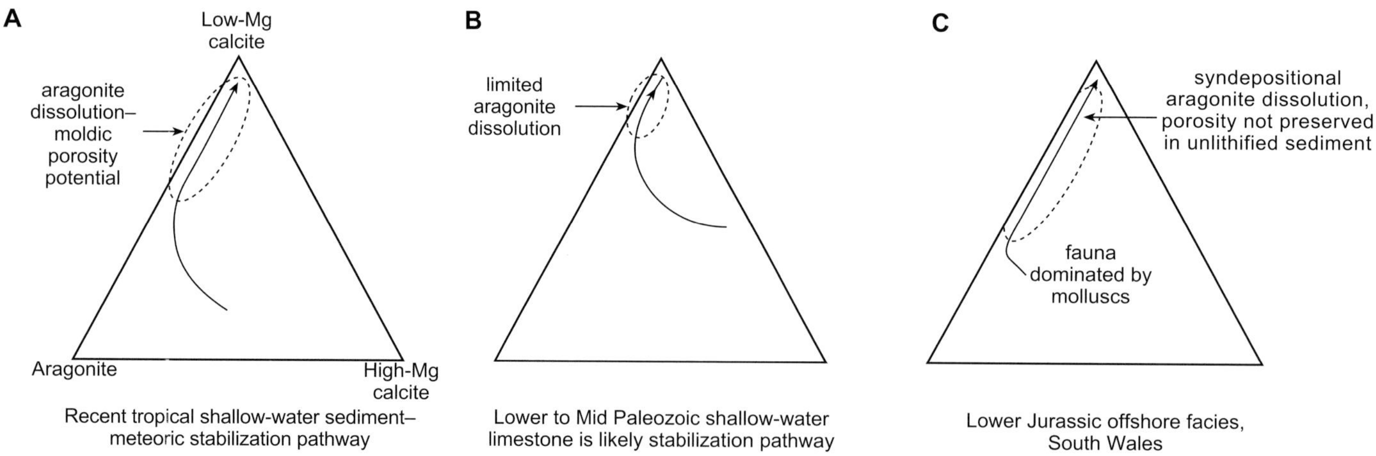

FIG. 4.—Ternary diagrams contrasting different diagenetic pathways for carbonate sediments. **A)** Diagenetic (stabilization) pathway for Recent, tropical shallow-marine sediments in which aragonite dissolution has the capacity to produce secondary porosity in the sediment. **B)** Hypothetical pathway for lower–mid Paleozoic shallow-water carbonate in which aragonitic skeletal grains were relatively unimportant and any ooids were primary calcite in composition. A and B after Tucker and Wright, 1990. **C)** Proposed trend for Lower Jurassic outer-ramp sediments based on data in Wright et al. (2003). Former high-Mg calcite grains in the successions were rare, but the fauna was originally dominated by aragonitic gastropods and infaunal bivalves. The aragonite was lost during shallow burial, and the molds were not preserved.

the "open-ness" of such diagenetic systems. Nodular and diagenetic bedding are indicators that early cementation has occurred, but what other features can be attributed to the release of large volumes of carbonate in solution? Not only has aragonite dissolution altered the grain populations and grain sizes of many sediments, but it has also reduced their diagenetic and secondary-porosity potentials.

ACKNOWLEDGMENTS

The original paper (Wright and Cherns, 2004) from which we have developed this review was published in a volume dedicated to the memory of the late and much respected Fransesc Calvet. We thank James Wheeley for many stimulating discussions on aragonite dissolution and taphonomy. We also thank the reviewers, Tina Hughes, Peir Pufahl, and an anonymous reviewer, for their constructive comments, and the editors for the invitation to publish these ideas. This paper is a contribution to the Cardiff Geobiology Initiative.

REFERENCES

AURELL, M., BADENAS, B., BOSENCE, D.W.J., AND WALTHAM, D.A., 1998, Carbonate production and offshore transport on a Late Jurassic carbonate ramp (Kimmeridgian, Iberian Basin, NE Spain): evidence from outcrops and computer modelling, *in* Wright, V.P., and Burchette, T.P., eds., Carbonate Ramps: Geological Society of London, Special Publication 149, p. 237–161.

BEAVINGTON-PENNEY, S.J., WRIGHT, V.P., AND RACEY, A., 2006, The Middle Eocene Seeb Formation of Oman: an investigation of acyclicity, stratigraphic completeness, and accumulation rates in shallow marine carbonate settings: Journal of Sedimentary Research, v. 76, p. 1137–1161.

BETZLER, C., BRACHERT, T.C., AND NEBELSICK, J., 1997, The warm-temperate carbonate province. A review of facies, zonations and delimitations: Courier Forschungsinstitut Senckenberg, v. 201, p. 83–99.

BOSSCHER, H., AND SCHLAGER, W., 1993, Accumulation rates of carbonate platforms: Journal of Geology, v. 101, p. 345–355.

BRACHERT, T.C., BETZLER, C., BRAGA, J.C., AND MARTIN, J.M., 1996, Record of climatic change in neritic carbonates: turnover in biogenic associations and depositional modes (Late Miocene, southern Spain): Geologishe Rundschau, v. 85, p. 327–337.

BRACHERT, T.C., AND DULLO, W-C., 2000, Shallow burial diagenesis of skeletal carbonates: selective loss of aragonite shell material (Miocene to Recent, Queensland Plateau and Queensland trough, NE Australia)—implications for shallow cool-water carbonates: Sedimentary Geology, v. 136, p. 169–187.

BRANDANO, M., AND CORDA, L., 2002, Nutrients, sea level and tectonics: constraints for the facies architecture of a Miocene carbonate ramp in central Italy: Terra Nova, v. 14, p. 257–262.

BURGESS, P.M., AND WRIGHT, V.P., 2003, Numerical forward modelling of carbonate platform dynamics: an evaluation of complexity and completeness in carbonate strata: Journal of Sedimentary Research, v. 73, p. 637–652.

BURTON, E.A., AND WALTER, L.M., 1987, Relative precipitation rates of aragonite and Mg calcite from sea water: temperature or carbonate ion control?: Geology, v. 15, p. 111–114.

BUSH, A.M., AND BAMBACH, R.K., 2004, Did alpha diversity increase during the Phanerozoic? Lifting the veils of taphonomic, latitudinal and environmental biases: Journal of Geology, v. 112, p. 625–642.

CHERNS, L., AND WRIGHT, V.P., 2000, Missing molluscs as evidence of large-scale, early aragonite dissolution in a Silurian sea: Geology, v. 28, p. 791–794.

DEMICCO, R.V., AND HARDIE, L.A., 2002, The "carbonate factory" revisited: a reexamination of sediment production functions used to model deposition on carbonate platforms: Journal of Sedimentary research, v. 72, p. 849–857.

DIX, G.R., JAMES, N.P., KYSER, T.K., BONE, Y., AND COLLINS, L., 2005, Genesis and dispersal of carbonate mud relative to late Quaternary sea-level change along a distally steepened carbonate ramp (Northwestern Shelf, Western Australia): Journal of Sedimentary Research, v. 75, p. 665–678.

DODD, J.R., AND NELSON, C.S., 1998, Diagenetic comparisons between non-tropical Cenozoic limestones of New Zealand and tropical Mississippian limestones from Indiana, USA: is the non-tropical model better than the tropical model?: Sedimentary Geology, v. 121, p. 1–21.

DULLO, W.-C., 1990, Facies, fossil record and age of Pleistocene reefs from the Red Sea (Saudi Arabia): Facies, v. 22, p. 1–46.

ELLIOTT, G.F., 1982, Calcareous algae and Middle Jurassic facies of southern England: Geological Magazine, v. 119, p. 309–313.

FLÜGEL, E., 2004, Microfacies of Carbonate Rocks; Analysis, Interpretation and Application: Berlin, Springer, 976 p.

GISCHLER, E., AND ZINGELER, D., 2002, The origin of carbonate mud in isolated carbonate platforms of Belize, Central America: International Journal of Earth Sciences, v. 91, p. 1054–1070.

HENDRY, J.P., WILKINSON, M., FALLICK, A.E., AND TREWIN, N.H., 2000, Disseminated 'jigsaw piece' dolomite in Upper Jurassic shelf sandstones, Central North Sea: an example of cement growth during bioturbation: Sedimentology, v. 47, p. 631–644.

JAMES, N.P., AND BONE, Y., 1989, Petrogenesis of Cenozoic, temperate water calcarenites, South Australia: a model for meteoric/shallow water calcite sediments: Journal of Sedimentary Petrology, v. 59, p. 191–203.

JAMES, N.P., BONE, Y., AND KYSER, T.K., 2005, Where has all the aragonite gone? Mineralogy of Holocene neritic cool-water carbonates, southern Australia: Journal of Sedimentary Research, v. 75, p. 454–463.

JEANS, C.V., 1980, Early submarine lithification in the Red Chalk and Lower Cahlk of eastern England: a bacterial control model and its implications: Yorkshire Geological Society, Proceedings, v. 43, p. 81–157.

KNOERICH, A.C., AND MUTTI, M., 2006, Missing aragonitic biota and the diagenetic evolution of heterozoan carbonates: a case study from the Oligo-Miocene of the central Mediterranean: Journal of Sedimentary Research, v. 76, p. 871–888.

KU, T.C.W., WALTER, L.M., COLEMAN, M.L., BLAKE, R.I, AND MARTINI, A.M., 1999, Coupling between sulphur cycling and syndepositional carbonate dissolution: evidence from oxygen and sulphur isotope composition of pore water sulphate, South Florida Platform, USA: Geochimica et Cosmochimica Acta, v. 63, p. 2529–2546.

KYSER, K.T., JAMES, N.P., AND BONE, Y., 1998, Alteration of Cenozoic coolwater carbonates to low-Mg calcite in marine waters, Gambier Embayment, South Australia: Journal of Sedimentary Research, v. 68, p. 947–955.

LUKASIK, J.J., AND JAMES, N.P., 2003, Deepening-upwards subtidal cycles, Murray Basin, South Australia: Journal of Sedimentary Research, v. 73, p. 653–671.

LUKASIK, J.J., AND JAMES, N.P, 2005, Carbonate sedimentation, climate change and stratigraphic completeness on a Miocene cool-water eperic ramp, Murray Basin, South Australia, *in* Pedley, H.M., and Carannante, G., eds., Cool-Water Carbonates: Depositional Systems and Palaeoenvironmental Controls: Geological Society of London, Special Publication 255, p. 217–244.

LUKASIK, J.J., JAMES, N.P., MCGOWRAN, B., AND BONE, Y., 2000, An epeiric ramp: low energy, cool-water carbonate facies in a Tertiary inland sea, Murray Basin, South Australia: Sedimentology, v. 47, p. 851–881.

MELIM, L.A., WESTPHAL, H., SWART, P.K., EBERLI, G., AND MUNNECKE, A., 2002, Questioning carbonate diagenetic paradigms: evidence from the Neogene of the Bahamas: Marine Geology, v. 185, p. 2753.

MUNNECKE, A., AND SAMTLEBEN, C., 1996, The formation of micritic limestones and the development of limestone–marl alternations in the Silurian of Gotland, Sweden: Facies, v. 34, p. 159–176.

MUNNECKE, A., AND WESTPHAL, H., 2004, Shallow-water aragonite recorded in bundles of limestone–marl alternations—the Upper Jurassic of SW Germany: Sedimentary Geology, v. 164, p. 191–202.

MUNNECKE, A., AND WESTPHAL, H., 2005, Variations in primary aragonite, calcite, and clay in fine-grained calcareous rhythmites of Cambrian to Jurassic age—an environmental archive?: Facies, v. 51, p. 592–607.

MUNNECKE, A., WESTPHAL, H., ELRICK, M., AND REIJMER, J.J.G., 2001, The mineralogical composition of precursor sediments of calcareous rhythmites: a new approach: International Journal of Earth Sciences, v. 90, p. 795–812.

MUNNECKE, A., WESTPHAL, H., REIJMER, J.J.G., AND SAMTLEBEN, C., 1997, Microspar development during early marine burial diagenesis: a comparison of Pliocene carbonates from the Bahamas with Silurian limestones from Gotland (Sweden): Sedimentology, v. 44, p. 977–990.

MUTTI, M., AND BERNOULLI, D., 2003, Early marine lithification and hardground development on a Miocene ramp (Mailella, Italy): key surfaces to track changes in trophic resources in nontropical carbonate settings: Journal of Sedimentary research, v. 73, p. 296–308.

NELSON, C.S., AND JAMES, N.P., 2000, Marine cements in mid-Tertiary coolwater shelf limestones of New Zealand and southern Australia: Sedimentology, v. 47, p. 609–629.

PALMER, T.J., HUDSON, J.D., AND WILSON, M.A., 1988, Palaeoecological evidence for early aragonite dissolution in ancient calcite seas: Nature, v. 335, p. 809–810.

PERRY, C.T., 2000, Factors controlling sediment preservation on a north Jamaican fringing reef: a process-based approach to microfacies analysis: Journal Sedimentary Research, v. 70, p. 633–648.

POMAR, L., 2001a, Types of carbonate platforms: a genetic approach: Basin Research, v. 13, p. 313–324.

POMAR, L., 2001b, Ecological control on sediment accommodation: evolution from a carbonate ramp to rimmed shelf, Upper Miocene, Balearic Islands: Palaeogeography, Palaeoclimatology, Palaeoecology, v. 175, p. 249–272.

POMAR, L., BRANDANO, M., AND WESTPHAL, H., 2004, Environmental factors influencing skeletal grain sediment associations: a critical review of Miocene examples from the western Mediterranean: Sedimentology, v. 51, p. 627–651.

READ , J.F., 1998, Phanerozoic carbonate ramps from greenhouse, transitional and ice-house worlds: clues from field and modelling studies, *in* Wright, V.P., and Burchette, T.P., eds., Carbonate Ramps: Geological Society of London, Special Publication 149, p. 107–135.

RIES, J.B., 2005, Aragonite production in calcite seas: effects of seawater Mg/Ca ratio on the calcification and growth of the calcareous alga *Penicillus capitatus*: Paleobiology, v. 31, p. 445–438.

RIES, J.B., 2006, Aragonitic algae in calcite seas: effect of seawater Mg/Ca on algal sediment production: Journal of Sedimentary Research, v. 76, p. 515–523.

SANDERS, D., 2001, Burrow-mediated carbonate dissolution in rudist biostromes (Aurisina, Italy): implications for taphonomy in tropical, shallow subtidal carbonate environments: Palaeogeography, Palaeoclimatology, Palaeoecology, v. 168, p. 41–76.

SANDERS, D., 2003, Syndepositional dissolution of calcium carbonate in neritic carbonate environments: geological recognition, processes, potential significance: Journal of African Earth Sciences, v. 36, p. 99–134.

SANDERS, D., 2004, Potential significance of syndepositional carbonate dissolution for platform banktop aggradation and sediment texture: a graphic modelling approach: Austrian Journal of Earth Sciences, v. 95/96, p. 71–79.

SANDERS, D., AND KRAINER, K., 2005, Taphonomy of Early Permian benthic assemblages (Carnic Alps, Austria): carbonate dissolution versus biogenic carbonate precipitation: Facies, v. 51, p. 522–540.

SCHLAGER, W., 1999, Scaling of sedimentation rates and drowning of reefs and carbonate platforms: Geology, v. 27, p. 183–186.

SCHLAGER, W., 2003, Benthic carbonate factories of the Phanerozoic: International Journal of Earth Sciences, v. 92, p. 445–464.

SCHWARZ, J., AND RENDLE-BUHRING, R., 2005, Controls on modern carbonate preservation in the southern Florida Straits: Sedimentary Geology, v. 175, p. 153–167.

SEPKOSKI, J.J., JR., 1984, A kinetic model of Phanerozoic taxonomic diversity. III. Post-Paleozoic families and mass extinctions: Paleobiology, v. 10, p. 246–267.

STANLEY, S.M., AND HARDIE, L.A., 1998, Secular oscillations in the carbonate mineralogy of reef-building and sediment-producing organisms driven by tectonically forced shifts in seawater chemistry: Palaeogeography, Palaeoclimatology, Palaeoecology, v. 144, p. 3–19.

STANLEY, S.M., AND HARDIE, L.A., 1999, Hypercalcification: paleontology links plate tectonics and geochemistry to sedimentology: GSA Today, v. 9, no. 2, p. 2–7.

STRASSER, A., AND SAMANKASSOU, E., 2003, Carbonate sedimentation rates today and in the past: Holocene of Florida Bay, Bahamas vs. Upper Jurassic and Lower Cretaceous of the Jura Mountains (Switzerland and France): Geologica Croatica, v. 56, p. 1–18.

TUCKER, M.E., AND WRIGHT, V.P., 1990, Carbonate Sedimentology: Oxford, U.K., Blackwell, 491 p.

WALTER, L.M., AND BURTON, E.A., 1990, Dissolution of platform carbonate sediments in marine pore fluids: American Journal of Science, v. 290, p. 601–643.

WHEELEY, J.R., CHERNS, L., AND WRIGHT, V.P., 2005, Missing molluscs: prolific providers of carbonate cement? (abstract): Palaeontological Association, Annual Conference, Oxford, U.K., 2005; Palaeontological Association Newsletter, v. 60, p. 59.

WHEELEY, J.R., CHERNS, L., AND WRIGHT, V.P., 2008, Provenance of microcrystalline carbonate cement in limestone–marl alternations (LMA): aragonite mud, or molluscs: Geological Society of London, Journal, v. 164, p, 395–403.

WILSON, M.E.J., AND VECSEI, A., 2005, The apparent paradox of abundant foramol facies in low latitudes: their environmental significance and effect on platform development: Earth-Science Reviews, v. 69, p. 133–168.

WOOD, R., 1993, Nutrients, predation and the history of reef building: Palaios, v. 8, p. 26–543.

WRIGHT, V.P., AND AZEREDO, A.C., 2006, How relevant is the role of macrophyte vegetation in controlling peritidal carbonate facies? Clues from the Upper Jurassic of Portugal: Sedimentary Geology, v. 186, p. 147–156.

WRIGHT, V.P., AND BURGESS, P.M., 2005, The carbonate factory continuum, facies mosaics and microfacies: an appraisal of some of the key concepts underpinning carbonate sedimentology: Facies, v. 51, p. 17–23.

WRIGHT, V.P., AND CHERNS, L., 2004, Are there "black holes" in carbonate deposystems? Fransesc Calvet Memorial Volume: Geologica Acta, v. 2, p. 285–290.

WRIGHT, V.P., CHERNS, L., AND HODGES, P., 2003, Missing molluscs: field testing taphonomic loss in the Mesozoic through early large-scale aragonite dissolution: Geology, v. 31, p. 211–214.

WRIGHT, V.P., AND FAULKNER, T.J., 1990, Sediment dynamics of early Carboniferous ramps: a proposal: Geological Journal, v. 25, p. 139–144.

YANG, W., MAZZULLO, S.J., AND TEAL, C.S., 2004, Sediments, facies tracts, and variations in sedimentation rates of Holocene platform carbonate sediments and associated deposits, northern Belize—implications for "representative" sedimentation rates: Journal of Sedimentary Research, v. 74, p. 498–512.

YATES, K.K., AND ROBBINS, L.L., 1998, Production of carbonate sediments by a unicellular green alga: American Mineralogist, v. 83, p. 1503–1509.

Tectonic Controls

TECTONIC AND DEPOSITIONAL CONTROLS ON SYN-RIFT CARBONATE PLATFORM SEDIMENTATION

STEVEN L. DOROBEK
Department of Geology & Geophysics, Texas A&M University, College Station, Texas 77843, U.S.A.
Present address: Maersk Oil and Gas, Esplanaden 50, 1263 Copenhagen K, Denmark
e-mail: SLD@maerskoil.com

ABSTRACT: All scales of tectonic deformation influence the location, sizes, shapes, and internal stratigraphy of carbonate platforms that form in active rift settings. Normal and oblique-slip faults bound the tectono-geomorphologic features that are typically found across rift settings. These fault-bounded structural elements can provide substrates for shallow-water carbonate platforms if they are submerged to shallow water depths. Thus, the incremental and long-term growth of tectonic structures, and interactions between surface processes and carbonate depositional systems that develop around or on top of these structures, determines nearly all stratigraphic aspects of syn-rift carbonate platforms.

Isolated carbonate platforms are the most common type of platforms in active rift settings because they form on fault-bounded syn-rift highs. Steeply dipping fault scarps and tilted to flat-topped depositional surfaces control where shallow-marine carbonate deposition is possible. Fault scaling laws and rules for fault growth, spacing, and linkage / interaction are important for understanding the internal stratal patterns within syn-rift carbonate platforms. Footwall highs are common nucleation sites for carbonate platforms, although paleo-wind directions and siliciclastic supply to adjacent depocenters also influence facies distributions, platform morphology, and overall stratigraphic development. Active fault displacements and related surface deformations during platform growth can control platform-margin locations, facies distributions across fault-bounded basement highs, siliciclastic–carbonate interactions (especially in updip fault-bounded depocenters), and the internal growth stratal patterns within syn-rift platforms. Wedge-like growth stratal patterns within syn-rift isolated platforms are characteristic of half-graben structural elements and are well documented in outcrop and subsurface examples. Flexural uplift of footwall margins of large, fault-bounded horsts is also documented by stratal relationships from syn-rift isolated platforms that build on horst highs. Syn-rift thermal subsidence may influence where carbonate facies are distributed across the rift system, as well as time-dependent accumulation rates for each platform.

Syn-rift carbonate platform strata can form important petroleum reservoirs within rift-basin systems. They also provide critical records for understanding the tectonic evolution and depositional history of rift systems.

INTRODUCTION

Rifting occurs where tensile stresses stretch the lithosphere, causing extensional faulting of the brittle upper crust, either brittle faulting or ductile flow in the lower crust, and probable ductile thinning of the mantle lithosphere. The mechanical behavior of the lithosphere during stretching likely depends on a variety of factors, including the magnitude, orientation, and dynamic evolution of tectonic stresses, crustal and lithospheric thickness prior to rifting, the thermal state of the lithosphere before and during rifting, strain rate, three-dimensional lithologic variations within the lithosphere, and the location and orientation of preexisting zones of weakness in the lithosphere. Deformation of the lithosphere may not be uniformly distributed in vertical section, but the entire thickness of the lithosphere must be modified by extension to fit the definition of a rift (Burke, 1977; Olsen and Morgan, 1995). Thus, local normal faults or small graben-like features created by minor, near-surface extension that does not involve the entire lithosphere (e.g., some thin-skinned pull-apart basins) are not usually classified as rifts, even though the style of deformation where shallow crustal extension occurs may produce stratigraphic successions similar to those in lithosphere-scale rift systems. This distinction between lithosphere-scale extension and "thin-skinned" extensional settings is important, however, because subsidence patterns will be different for each case, both locally (i.e., around individual faults) and regionally (i.e., basin-scale). Many previous studies have focused on the patterns of sedimentation and resulting stratigraphy that develop during active rifting, although most studies have focused on siliciclastic depositional systems in syn-rift settings (e.g., Leeder and Gawthorpe, 1987; Frostick and Reid, 1989; Gawthorpe et al., 1997; Bosence, 1998; Gawthorpe and Leeder, 2000; Gupta and Cowie, 2000; Contreras and Scholz, 2001). Carbonate platforms can also form during syn-rift stages of basin evolution and provide important records of differential subsidence and uplift across a rift zone. This paper provides a review of syn-rift carbonate platforms and attempts to integrate the growing knowledge of fault kinematics, related surface deformation, and the responses of carbonate platforms to syn-rift deformation in order to understand the stratal and facies patterns found in these platforms. All scales of syn-rift tectonic deformation are considered here, although the emphasis is on local, fault-controlled differential subsidence and uplift, which can strongly influence the stratigraphic record of syn-rift carbonate platforms that develop on fault-bounded crustal blocks. More regional tectonic controls (strain partitioning, syn-rift regional subsidence, and stretching gradients) on carbonate platform development are also considered, as are interactions between syn-rift carbonate platforms and siliciclastic depositional systems.

Tectonic models for extensional basins are first reviewed. Emphasis is on the spatial and temporal scales of differential tectonic subsidence that are thought to result from different stretching models in order to investigate how tectonic accommodation might change across extensional settings during rifting. The behavior of individual faults is then examined to see how

Controls on Carbonate Platform and Reef Development
SEPM Special Publication No. 89, Copyright © 2008
SEPM (Society for Sedimentary Geology), ISBN 978-1-56576-130-8, p. 57–81.

local, fault-controlled differential subsidence and uplift affects accommodation space. The focus then turns to carbonate-platform systems and their response to tectonic deformation. Carbonate platforms that form on extensional salt structures are not considered here because these structures do not utilize basement-involved fault systems. That is, there may be no crustal stretching while the salt layers flow and form extensional structures in overlying stratigraphic successions. Nonetheless, carbonate platforms that form on extensional salt structures can exhibit growth stratal patterns similar to those in syn-rift platforms from continental rift settings, and the models presented here may have some utility for interpreting carbonate-platform strata in extensional salt provinces.

EXTENSIONAL SETTINGS: GENERAL TECTONIC ASPECTS AND IMPLICATIONS

Understanding how extensional strain is partitioned across an active rift zone and how various wavelengths of deformation interfere provides insight into the location and evolution of carbonate depositional systems that can develop in these settings. At every stratigraphic level in an active rift setting, tectonic deformation can strongly influence the development of shallow-marine carbonate systems. Some aspects of intracontinental rifting (e.g., simple-shear vs. pure-shear stretching models; shallow, brittle deformation vs. deeper, ductile deformation) are not considered here because they tend be more important during the post-rift phase of basin evolution and have less obvious influence during the syn-rift phase. Nonetheless, for rift settings that have experienced protracted or multiple phases of rifting, long-wavelength thermal subsidence is possible during the syn-rift phase, which might influence where shallow-water carbonate platform systems can develop.

Extensional basins can be classified according to the amount of lithospheric stretching that caused them to form (Fig. 1; Falvey, 1974; Kinsman, 1975; Salveson, 1978; Allen and Allen, 1990). Where rifting occurs within continental interiors, but there is only limited stretching of the lithosphere before rifting ends (typically less than 10% extension), broad, relatively shallow, *intracontinental sag basins* are likely to form. Sag basins are characterized by slow, long-term subsidence rates (typically less than a few cm/kyr). These characteristics likely reflect either: (1) very small amounts of stretching to initiate subsidence, (2) development of a younger thermal anomaly beneath a much older failed continental rift, which causes renewed thermal subsidence as the thermal anomaly dissipates, (3) reactivation of an old failed continental rift by intra-plate stress, or (4) phase changes in the mantle part of the lithosphere that might possibly be triggered by minor amounts of stretching. In most intracontinental sag basins, major displacements on basement faults are rarely observed and deformation of the lithosphere is best described as broad-wavelength subsidence or "sagging" that is centered about the middle of these typically semicircular basins.

Intracontinental rift basins reflect greater amounts of stretching within continental lithosphere (10–80% stretching). Intracontinental rift basins may be located in interior parts of continental plates, or they may project inboard from passive or convergent continental margins (cf. Sengör, 1995). The tectonic evolution of intracontinental rift basins is often temporally linked to tectonic events along plate margins, although causal links between the deformation in both regions may be unclear. Passive margins commonly pass landward into multiple rift depocenters with intervening basement highs. Fault systems beneath these intracontinental rift depocenters may be oriented at high angles to the fault systems that underlie the passive-margin wedge and can control carbonate depositional trends long into the post-rift history of the intracontinental basins.

If continental lithosphere is stretched to less than about 20% of its original thickness, a spreading ridge typically develops where new oceanic crust forms. The rifted continental margins on either side of the spreading axis then evolve into conjugate passive margins that "drift" away from the newly formed mid-ocean ridge. The onset of seafloor spreading typically marks the "rift-to-drift transition" in the development of passive continental margins (Falvey, 1974; Klitgord et al., 1988; Withjack et al., 1998) because stretching of continental lithosphere generally stops once seafloor spreading begins. Any additional extension is then largely accommodated along the new mid-ocean spreading axis. The conjugate passive margins on either side of the spreading ridge typically experience much less tectonic deformation, other than broad-wavelength thermal subsidence, after continental breakup and the conjugate margins begin their drift or post-rift phase of evolution.

Transition into the post-rift phase marks an important time in the development of rift depocenters. Most faults become inactive, although more outboard faults near the evolving mid-ocean ridge system may remain active and remnant tectonic topography created during rifting may continue to influence carbonate-platform development long into the post-rift history of a rift depocenter.

Rifts in continental lithosphere have variable geomorphologic expression throughout the rifting process. Rift topography can vary from: (1) a well-defined central axis where extensional strains are typically highest; (2) rift networks with no obvious axis and broadly distributed strain; or (3) a series of rift axes separated by intervening basement "highs," where strain is complexly partitioned across the stretched region. Continental rifts may consist of a linked rift network that extends at variable orientations from a plate margin and appears to be dividing the plate. Alternatively, the rift zone may be isolated within a plate's interior, far from a plate margin. The rift's distance from a plate margin and the intervening topography determines if and when the stretched region is ever inundated by seawater. Rift basins that form in plate interiors can remain nonmarine settings throughout their entire history as depocenters.

The variable topography that develops across rift systems during stretching reflects the combined effects of: (1) pre-rift topography, (2) interactions between erosional and depositional processes that continuously modify surface topography during stretching, and (3) the three-dimensional distribution of extensional strain that is unique to each rift system.

Pre-rift topography may be a peneplain surface, such as existed across the Red Sea and Gulf of Aden region prior to Oligocene–Miocene rifting (Bosworth and McClay, 2001). Where pre-rift topography is minimal, regional topographic variations are created almost entirely by the rifting processes. Conversely, pre-rift topography can be very complex, such as when an orogenic belt undergoes extensional collapse (Molnar and Lyon-Caen, 1988). In these cases, significant topographic variability may exist prior to rifting and extensional structures become superimposed on this preexisting topography. Surface processes also can either create topographic complexity (e.g., alluvial incision along hanging-wall dip slopes of half-graben structural elements) or smooth and fill topography via sediment accumulation. Syn-rift carbonate depositional systems respond to any antecedent topography, whether it is created by erosion, sediment accumulation, or tectonic deformation.

Three-dimensional extensional strain can be simply distributed about a single rift axis (e.g., the Gulf of Suez and Red Sea rift system) or more complexly partitioned, resulting in a rift system

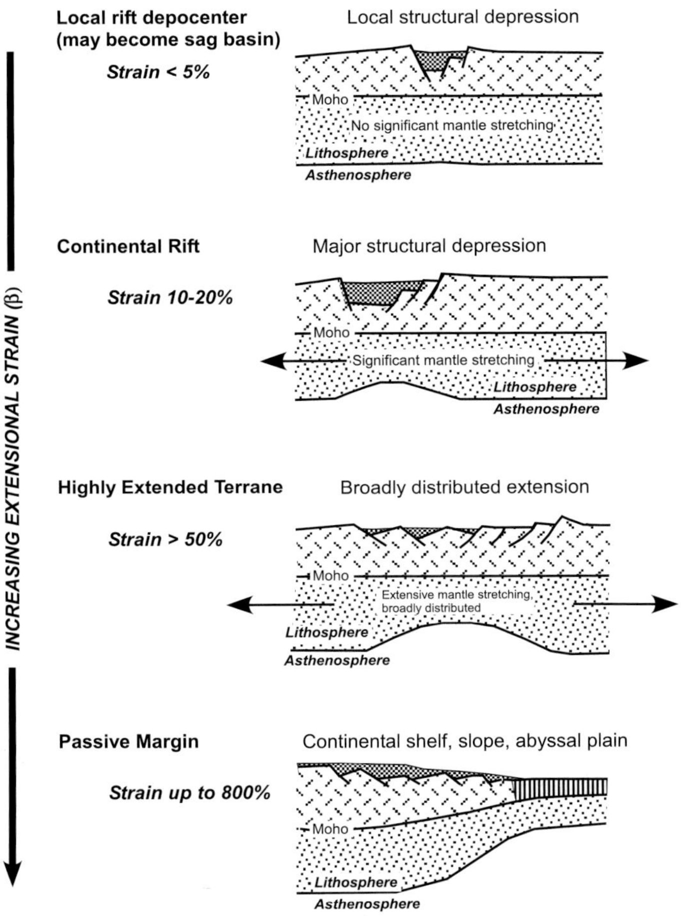

FIG. 1.—The rifting continuum within continental lithosphere. Amount of extensional strain increases from upper to lower diagrams. Details about the distribution of three-dimensional extensional strain within the lithosphere are not shown. Modified from Olsen and Morgan (1995).

characterized by elevated ribbons and platforms of fault-bounded basement blocks that separate fault-bounded depocenters (e.g., offshore Tertiary rift basins across the South China Sea region; Mesozoic rift basins beneath the Atlantic passive margin of northeastern Canada). The amount and direction of stretching can also vary between segments of more complex rift systems.

Some rift settings are characterized by abrupt, comparatively narrow transitions (50–150 km wide) between stretched and unstretched lithosphere. Unstretched lithosphere is found landward of the tectonic hinge (commonly defined as the updip limit of syn-rift faulting); the amount of lithospheric stretching progressively increases towards the most highly stretched part of the rift system. In contrast, stretching can also be broadly distributed over much wider areas, which results in some rift systems and passive margins that are underlain by > 500-km-wide zones of stretched continental lithosphere between the tectonic hinge and the area of greatest stretching.

It is difficult to estimate when shallow-water flooding of syn-rift topographic features might begin during the rift history of an intracontinental rift system. For example, Miocene carbonate systems formed on pre-rift Precambrian basement substrates very early in the rift history of the Gulf of Suez and Red Sea region. This is largely due to the very low pre-rift relief across the Middle East and northeastern Africa. In contrast, Miocene Cenozoic carbonate systems were initiated at a variety of times in the syn-rift history of the South China Sea region, depending on local time of rifting, regional pre-rift topography, the location within the complexly evolving rift network, and proximity to major siliciclastic input points (Dorobek, 1995). Significant amounts of syn-rift strata had already accumulated in many South China Sea depocenters before any significant carbonate platforms began to form. Thus, in some examples from the South China Sea, syn-rift carbonate platforms developed unconformably on faulted, previously deposited, syn-rift marine strata. In general, however, shallow-marine carbonate sedimentation typically commences during later syn-rift stages, when enough stretching and syn-rift subsidence has occurred and the evolving rift system becomes linked to an adjacent ocean basin.

These complexities emphasize that all rifted regions are unique. Generalized large-scale tectonic models that are commonly applied to continental rifts and passive margins (e.g., McKenzie, 1978; Wernicke, 1985; Lister et al., 1986) also do not adequately explain the actual topography and bathymetry that can develop across a rifted region. It is important to realize, nonetheless, that shallow-marine carbonate platform facies can provide important regional to local base-level markers at any time in the syn-rift evolution of a region (e.g., Omar and Steckler, 1995).

Stretching Gradients and Transform Segments

In plan view, the top of faulted basement across rift networks or young passive margins can be regionally characterized as (cf. Fig. 2): (1) narrow, with steep, often step-like structural and depositional profiles toward the rift axis or continent–ocean boundary, (2) wide, with more gentle, regional depositional gradients, or (3) complex, with many localized depocenters that are separated by intervening, fault-bounded, and elevated crustal blocks. Narrow, steep, and step-like tectono-geomorphology within extensional settings probably develops where stretching directions are highly oblique to the strike of rift structural elements and possibly where strain rates are high. Wide rift zones with much gentler structural and bathymetric gradients toward the rift axis are characteristic of rift systems where stretching directions are nearly orthogonal to the strike of rift structural elements and where strain rates are low. More complex tectono-geomorphologic characteristics are typical of extensional settings where stretching directions may have changed over time or where preexisting basement fabrics strongly influence the development of rift structures.

Stretching gradients describe the horizontal distance over which stretching of continental lithosphere has occurred. Stretching gradients can be expressed as β/km—β can be defined as the inverse of the amount of fractional thinning of the lithosphere in a rift zone. For example, $\beta = 2$, where continental lithosphere has been thinned (stretched) to one-half of its original thickness.—(Rihm and Henke, 1998) and are measured from the updip tectonic hinge to the zone of greatest stretching. Rihm and Henke (1998) showed that the width of a passive margin, as measured from the updip tectonic hinge to the continent–ocean boundary, and the associated stretching gradient, are a function of the stretching direction with respect to the regional strike of the passive margin. Where the stretching direction is nearly perpendicular to regional strike of the passive margin, stretching gradients are commonly an order of magnitude smaller (on order of 0.01–0.02) and the area of stretched continental crust beneath the passive margin is much wider.

Segments of rift systems and passive margins with very high stretching gradients (i.e., β/km on order of 0.1–0.2) are com-

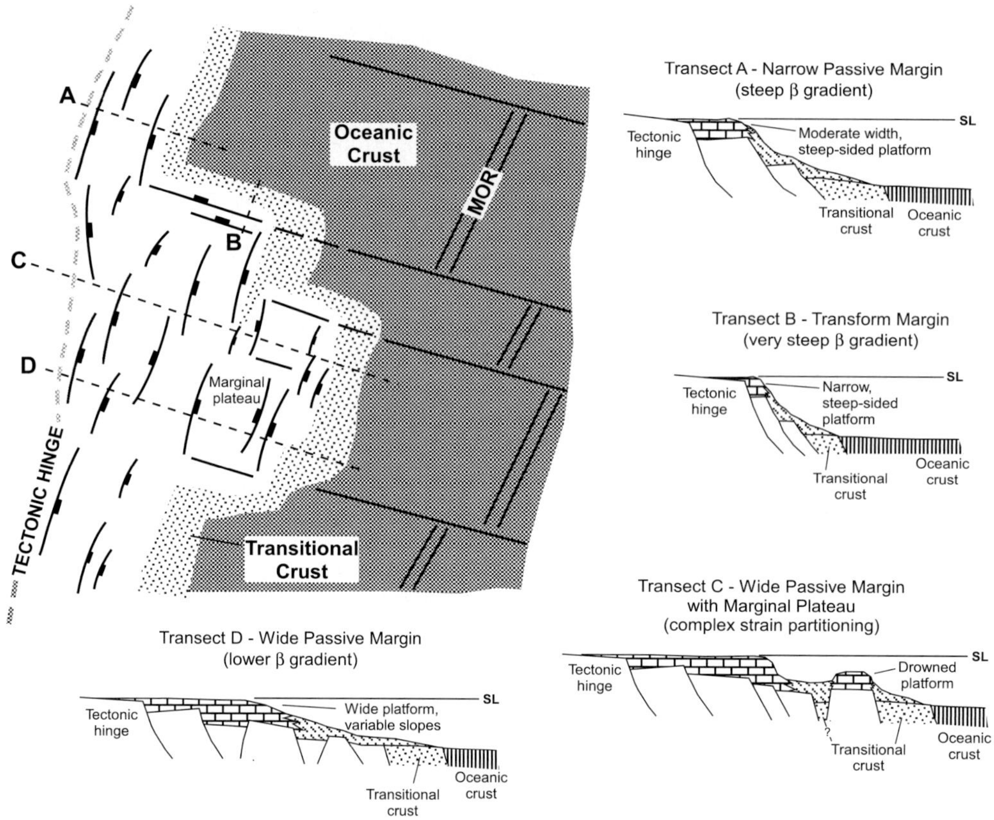

FIG. 2.—Tectonic styles and stretching gradients along passive margins and controls on associated carbonate-platform development. Note how gradient (see text for discussion) and partitioning of extensional strain influence syn-rift basement topography, long-term subsidence rates and patterns, and long-term depositional gradients, all of which in turn affect the location and growth history of carbonate platforms that develop along different segments of a passive continental margin. Although not all intracontinental rifts may progress to the passive-margin stage, spatial variations in stretching gradients within any rift system affect the location and growth history of carbonate platforms that might be initiated on fault-bounded basement highs during syn-rift and post-rift stages of basin evolution. MOR = mid-ocean ridge system.

monly found along narrow transform segments. Transform segments form where fault systems trend nearly parallel to regional (i.e., basin-scale) stretching directions. Thus, strike-slip and transtensional styles of deformation dominate transform segments. Examples of transform segments include parts of the Egyptian and Sudanese segments of the Red Sea, the conjugate transform margins of the Agulhas Shelf, offshore southern Africa and the Falklands Plateau, offshore southern Argentina (Rabinowitz and LaBreque, 1979; Ben-Avraham et al., 1997), and the conjugate passive margins of offshore French Guyana (Demerara Plateau area) and offshore Guinea, Africa (Guinea Plateau; cf. Gouyet et al., 1994; Rosendahl and Groschel, 2000). Carbonate platforms that form in these transform settings typically have very steep flanks throughout their history.

Stretching gradients influence long-term subsidence patterns and surface morphology during the syn-rift and post-rift stages of extensional basins. For very high stretching gradients, such as those found along transform-margin segments, syn-rift and post-rift subsidence rates rapidly increase seaward of the tectonic hinge. If the transform-margin segment is relatively sediment-starved, steep depositional gradients that develop during early rifting are maintained for much of the margin's history. Where sediment supply is higher, sediment forms aprons on down-faulted blocks. Large-scale, gravity-driven failure may occur repeatedly as low-angle intrastratal detachments develop within the prograding sediment wedge (e.g., Togo–Ghana Atlantic passive margin). The high syn-rift stretching gradients, however, characteristically control slope declivity long into the margin's post-rift history, even where there is high sediment supply.

Where stretching gradients are high, syn-rift carbonate platforms will likely be limited to a narrow zone, just downdip from the tectonic hinge. Where stretching gradients are low, syn-rift isolated carbonate platforms may form on numerous fault-bounded highs across the rift system, and these platforms may coalesce into larger composite platforms during the late syn-rift to early post-rift phase. Where stretching is complexly partitioned, the distribution and sizes of carbonate platforms that develop on structural highs are also highly variable.

Syn-Rift Subsidence Across Extensional Settings

Differential subsidence across extensional settings can be described according to its wavelength (i.e., local fault-controlled subsidence vs. long-wavelength, basin-scale patterns) and the tectono-evolutionary phase of the extensional basin (i.e., syn-rift or post-rift phase). All scales of syn-rift differential subsidence are examined here.

Fault-Controlled, Syn-Rift Differential Subsidence.—

Most studies of syn-rift strata focus on the differential subsidence that occurs around individual fault-bounded basement blocks or growing folds within rift systems. Local, fault-controlled patterns of subsidence are related to fault growth and the related surface deformations that develop as a result of displacement along faults. Fault displacement can occur either aseismically or during earthquakes (i.e., incremental *cosesimic* displacements) and displacement can occur either semicontinuously (i.e., creeping faults), quasi-periodically (i.e., coseismic displacements), or in a pulsed fashion (i.e., where regional tensile stresses wax and wane over geologic time scales). Accumulation of displacement along faults can ultimately result in major differential subsidence or uplift on either side of a fault zone. Tectonically created surface deformations range from development of terraced profiles, curved to ramp-like surfaces bounded by fault scarps, or fault-bounded blocks with characteristic horst–graben or half-graben cross-sectional profiles.

Long-Wavelength, Syn-Rift Differential Subsidence.—

Subtle, longer-wavelength patterns of subsidence and uplift become superimposed on local fault-controlled differential subsidence during the syn-rift stage. There have been few studies, however, that have directly investigated long-wavelength (> 100 km distance) syn-rift subsidence patterns across rift basins.

Theoretical models suggest that long-wavelength thermal subsidence should occur if rifting takes place over a protracted time period, and not "instantaneously" as in the original pure-shear model of McKenzie (1978) because part of the thermal anomaly that develops during lithospheric stretching is dissipated during the syn-rift stage (Jarvis and McKenzie, 1980; Cochran, 1983). In practical terms, some amount of thermal subsidence should occur during the syn-rift stage if rifting is a protracted process, and accordingly there will be less post-rift thermal subsidence. The total amount of tectonic subsidence (syn-rift plus post-rift), however, should be the same, regardless of the duration of the syn-rift phase, because total tectonic subsidence is solely a function of the amount of total lithospheric stretching.

Evolution of Crustal Structures in Extensional Settings

Normal faults are the dominant structural elements of rifted continental crust, although oblique slip may be more typical on many normal faults across a rift system. Local areas of either strike-slip deformation or shortening may develop along accommodation zones between major normal faults or where regional trends of normal faults change orientation. Transtensional styles of deformation may dominate where stretching directions are oblique to regional normal-fault trends or to preexisting zones of mechanical weakness in continental crust (Withjack and Jamison, 1986). If stretching directions change over long time scales, old extensional faults may be abandoned and then crosscut by younger extensional faults, or be utilized as transfer faults to accommodate the younger stretching direction.

It may be difficult to constrain the timing of deformation in many extensional basins, especially in rifts that are deeply buried. The timing and duration of deformation is typically constrained by crosscutting relationships between active faults and syn-rift strata. The rifting process can be protracted or brief, continuous or pulsed, with episodic stretching events that separate intervening periods of tectonic quiescence (e.g., Tertiary sub-basins of the East African Rift System, Morley et al., 1999a, 1999b; Nam Con Son Basin, offshore Vietnam, Matthews et al., 1997). The entire rifting history in many basins may occur over tens of millions of years, but actual pulses of extension, when active displacement occurs on fault systems, may be of much shorter duration. All of these spatial and temporal variations in the deformation histories of rift basins create additional complexities for structural and stratigraphic analyses.

Fault-bounded extensional features, such as full grabens, half-grabens, and horsts, are the basic tectono-geomorphologic building blocks of rift systems. Half-grabens are the most common element. All structural elements can be further broken by secondary synthetic and antithetic normal faults. The differential subsidence and surface geomorphology across fault-bounded features is largely controlled by displacement along their bounding faults, adjustments of ground elevation during and after fault displacement events, whether secondary faults cut the main structural element, and whether any erosion or deposition occurs along the top of the fault-bounded feature during the syn-rift phase.

Accommodation zones (also described as transfer zones, relay zones, relay ramps, and segment boundaries) separate fault-bounded depocenters and basement highs. "Accommodation zone" is often used to describe the deformation zones between normal-faulted domains where the dominant fault dip changes polarity, or dip direction, across the accommodation zone (cf. Lambiase and Bosworth, 1995). In contrast, "transfer zone" is generally used to describe the deformation zone between adjacent normal faults, even faults with the same dip, where displacement or net extension is "transferred" from one fault or fault system to the next (Larsen, 1988; Peacock and Sanderson, 1991). During active rifting, accommodation zones and transfer zones tend to have intermediate elevations that are lower than horst highs and half-graben footwall crests but are higher than adjacent full-graben and half-graben depocenters. During syn-rift phases, displacement is transferred across transfer zones by: (1) oblique slip on either a single fault or a series of faults that hard-link neighboring normal faults, with the associated surface deformation expressed as a single fault scarp or as a series of stepped, fault-bounded terraces, or (2) highly distributed shear between the faults, where surface deformation is typically expressed as a ramp-like surface (i.e., a relay ramp) with more gradual surface gradients.

Syn-Rift Fault Growth, Linkage and Interactions, and Related Surface Deformation.—

It is important to understand how faults grow and interact with neighboring faults during rifting if we are to understand how carbonate depositional systems respond to the deformation. Only a few active faults, however, have been observed through enough earthquake cycles to allow measurement of successive displacements. In most ancient rifts, the record of fault growth can be captured only through combined analysis of crosscutting relationships between active faults and *growth strata*, or the stratigraphy that forms around the fault zone during active fault displacement. Even with excellent 3D data from the subsurface, the absolute ages of syn-rift carbonate facies are rarely constrained at a resolution finer than ± 1 Myr. Thus, attributing meter-scale stratigraphic relationships to either eustatic fluctuations or syn-rift deformation will always be problematic, unless growth stratal patterns can be identified and interpreted in three-dimensions and chronostratigraphic relationships are well constrained.

Outcrop-based studies of ancient and active fault networks, analysis of regional seismic profiles, scaled analog models, and theoretical considerations have contributed greatly to our understanding of the kinematic history of fault systems within rifted continental crust (e.g., Bosworth, 1985; Watterson, 1986; McClay and Ellis, 1987a, 1987b; Vendeville et al., 1987; Ellis and McClay, 1988; Walsh and Watterson, 1987, 1988, 1989; Vendeville and Cobbold, 1988; McClay, 1990a, 1990b, 1995a; Schlische, 1991; Morley, 1995; Lambiase and Bosworth, 1995; McClay and White, 1995; Gupta and Cowie, 2000; Ackermann et al., 2001). For single, isolated normal faults, maximum displacement is typically located near the mid-point of a fault, with displacement progressively diminishing to zero at the tips of the fault (Rippon, 1985; Watterson, 1986; Barnett et al., 1987; Walsh and Watterson, 1989). Both incremental and cumulative displacement increase as faults increase in length (Walsh and Watterson, 1988; Scholz and Cowie, 1990; Marrett and Allmendinger, 1991). These rules for displacement seem to be appropriate for isolated, kilometer-scale and smaller, normal faults (Watterson, 1986; Barnett et al., 1987; Walsh and Watterson, 1987, 1988, 1989; Dawers et al., 1993; Schlische et al., 1996).

During early syn-rift stages, normal faults may initially develop as short, unconnected fault segments that are widely distributed across a rift zone (Fig. 3). Initial fault locations and orientations are typically strongly controlled by preexisting basement fabrics, stretching directions, and the thermal–mechanical properties of continental lithosphere. As stretching proceeds, short-segment faults propagate in all directions, but at the surface, where sedimentation and erosion occur, related surface deformation is expressed in complex ways (e.g., Schlische and Anders, 1996; Contreras et al., 1997; Gawthorpe et al., 1997; Ravnås and Bondevik, 1997; Gawthorpe and Leeder, 2000; Gupta and Cowie, 2000; Contreras and Scholz, 2001).

For isolated, blind normal faults that propagate upward toward the ground surface, an initial monoclinal bend is created in the unfaulted rock units or sedimentary layers above the propagating fault (Fig. 4; Gawthorpe et al., 1997). Continued dip-slip displacement on these normal faults causes the forelimb of the monocline to steepen progressively, until the fault eventually breaks the surface and a fault scarp is exposed. Thereafter, hanging-wall rotation may lower or even reverse the dips of strata that were previously deposited near the bounding fault. Similar changes in surface deformation also develop along strike, as normal faults grow at their lateral tips. Early-formed ramps along the ground surface above lateral tips of blind normal faults may evolve into fault scarps as displacement accumulates on the growing fault and the fault ultimately breaks through to the surface. Once the fault breaks to the surface, the mid-point of the fault tends to show the greatest hanging-wall subsidence and greatest footwall uplift.

Surface deformations become even more complex where neighboring normal faults begin to interact with each other as they propagate laterally. At some point in their growth history, isolated normal faults may develop either "hard" or "soft" links with nearby faults (Larsen, 1988; Davison, 1996). Hard linkage develops where fault segments propagate toward each other and directly link together to become a longer, through-going fault zone. In contrast, soft linkage is characterized by more highly distributed extensional strain in deformation zones between neighboring faults, without the development of through-going faults. Soft linkage may also be a precursor stage before faults become hard-linked. Surface deformations related to soft linkage are commonly expressed as relay ramps, which are gently dipping surfaces that can develop between subparallel but unlinked synthetic normal faults with overlapping fault tips. Surface deformations related to hard linkage are commonly expressed as terraced or step-like topography between the main extensional faults.

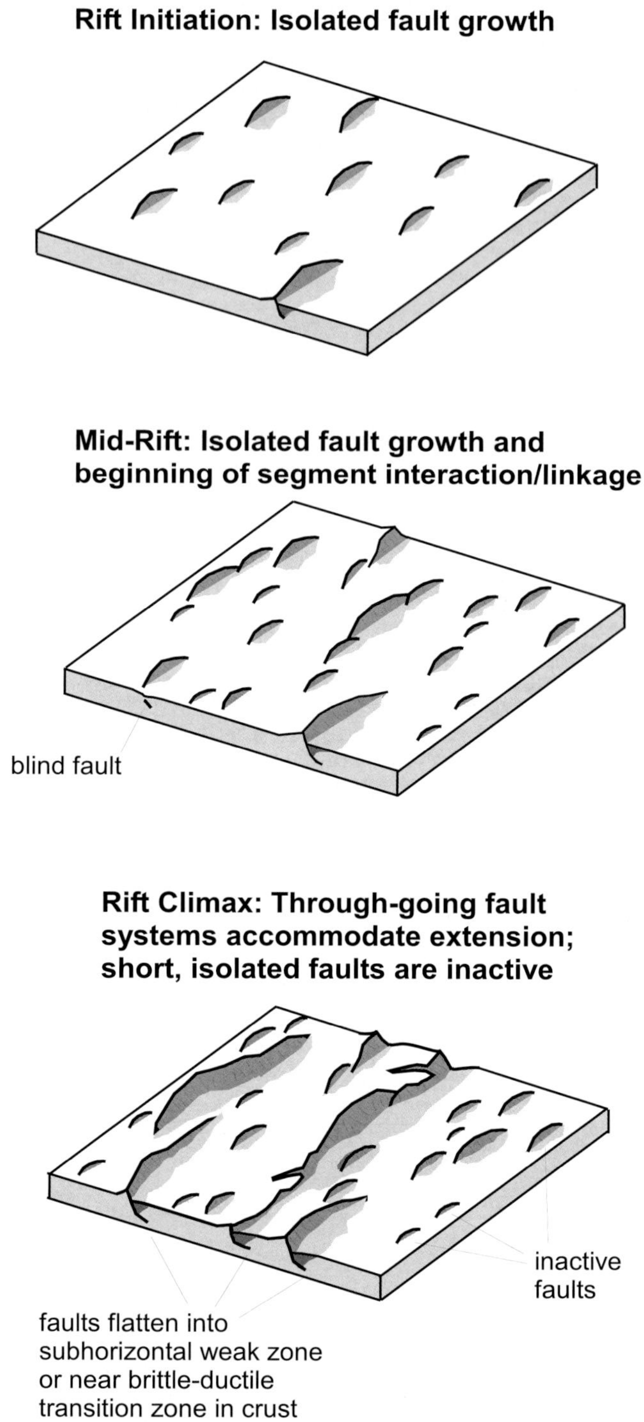

FIG. 3.—Fault growth and linkage in rift systems. Model for progressive fault linkage and depocenter evolution across a rift zone. Normal faults during early syn-rift stages are diffusely distributed, isolated, and unlinked. With time and continued stretching, fault segments begin to interact and link, with related enlargement of hanging-wall depocenters. Regional through-going fault systems develop during late stages of rifting. Smaller isolated faults progressively lock up as extension is progressively accommodated along larger, through-going fault systems. Modified from Cowie et al. (2000).

As early-formed normal faults propagate toward each other, the transfer zone between them typically becomes progressively narrower in plan view, and, as the fault tips approach, the transfer zone may be uplifted slightly. Once the fault tips link, the former uplifted transfer zone becomes part of the new "composite" hanging wall and begins to subside rapidly as displacement rapidly accumulates in the zone where faults just became linked. Thus, the topography of the transfer zone can undergo dramatic transitions from being an initial generally low-gradient ramp when fault segments were far apart, to an actively uplifting area as fault tips propagate toward each other, to a rapidly subsiding area once fault segments are hard-linked.

The evolutionary sequence of fault displacements and related surface deformations also depends on fault spacing, fault dip, and the amount of overlap at fault tips. Different models have been proposed to explain the growth stratigraphy within linked half-graben depocenters to account for these factors (cf. Anders and Schlische, 1994; Schlische and Anders, 1996).

Fault networks and syn-rift strata also exhibit regional patterns that reflect the progressive partitioning of extensional strain across a rift zone. The first normal faults to break the surface during early stages of rifting are isolated, unlinked, and broadly distributed and produce low-relief fault scarps where they cut to the surface (Anders and Schlische, 1994; Lambiase and Bosworth, 1995; Bosence, 1998; Gawthorpe and Leeder, 2000). During the earliest stages of rifting, each fault segment with enough displacement may create its own isolated half-graben basin, where the oldest stratigraphic units are restricted to the central parts of each sub-basin and thin toward the sub-basin edges. With continued stretching, however, faults that are along strike and aligned with neighboring faults begin to lengthen rapidly as formerly isolated normal faults grow and link with neighboring faults (Fig. 3). As through-going fault networks are established, these faults preferentially accommodate most of the later extension, whereas more isolated, typically shorter faults lock and become abandoned.

The response of syn-rift carbonate platform systems to ground deformations caused by interacting normal faults has not been thoroughly documented in the literature. In most rift systems this is not a serious issue, inasmuch as syn-rift carbonate systems characteristically develop late in the syn-rift phase, after most major fault systems have become linked and significant interactions between adjacent faults are not an important consideration. Regardless, surface deformation around fault tips and in overlap zones between neighboring faults should always be considered in investigations of fine-scale stratal patterns within syn-rift carbonate platforms.

Incremental and Long-Term Slip Rates on Normal Faults.—

It is important to understand displacement or slip rates on faults in rift settings because fault displacements create surface deformations that modify accommodation space and can offset facies across the fault zone. For aseismic, creeping faults, surface deformations should gradually develop and subtle growth stratal patterns should reflect the slow, steady deformation. There are few direct observations, however, of active, aseismic extensional fault zones and no reported examples of syn-rift stratigraphy from creeping, aseismic fault zones. Although not discussed in this paper, rafted carbonate platforms on extensional salt structures may exhibit stratal patterns related to aseismic extensional fault systems.

For seismically active faults, however, incremental slip events create incremental surface deformations that instantly modify accommodation space around the fault zone. Detailed geodetic surveys show that incremental coseismic displacements on normal faults are commonly on order of a few millimeters to over 10 m, although maximum hanging-wall displacements, as measured from pre-slip hanging-wall elevation, typically are tens of centimeters to less than a few meters (e.g., Jankhof, 1945; Whitten, 1957; Richter, 1958; Myers and Hamilton, 1964; Stein and Barrientos, 1985; Leeder, 1995; Cowie and Roberts, 2001). As described previously, incremental displacement is typically greatest near the mid-point of normal faults and progressively diminishes toward fault tips. Thus, the greatest instantaneous and long-term influence on syn-rift growth strata should be near the mid-points of faults.

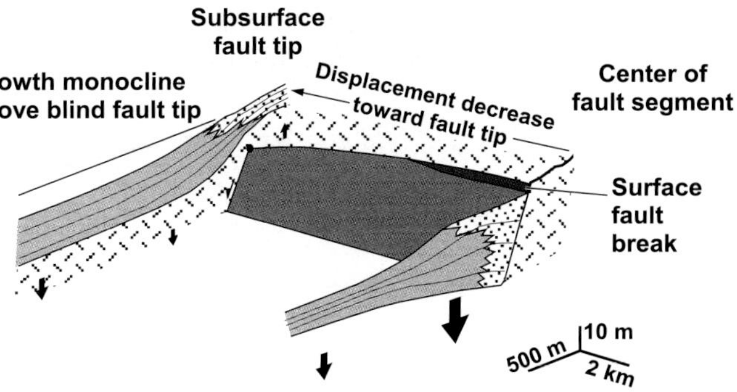

FIG. 4.—Growth stratigraphy related to growth of blind, subsurface fault that approaches and ultimately breaks to the ground surface. Modified from Gawthorpe et al. (1997).

Multiple slip events can cause significant displacements and differential subsidence to accrue. Earthquake recurrence intervals probably are on order of 10^1–10^3 yr in active rift settings. Most active normal faults, however, have not been observed through enough earthquake cycles to measure slip during multiple events. Average vertical slip rates on Neogene normal faults have been estimated from measured offset of well-dated stratigraphic surfaces or layers. These observations suggest average slip rates that range from cm / kyr to a few m / kyr (cf. Nicol et al., 1996; Contreras et al., 1997; McLeod et al., 2000; Cowie and Roberts, 2001; Morewood and Roberts, 2002).

Surface Deformation around Isolated Normal Faults During the Earthquake Cycle.—

Incremental displacement along normal faults also causes three-dimensional patterns of ground deformation around the fault zone. Carbonate depositional systems must respond to these ground deformations, which ultimately control the internal stratigraphy of syn-rift carbonate platforms.

Observational evidence suggests that surface deformation occurs during the coseismic and interseismic phases of the earthquake cycle. Coseismic slip on normal faults is accompanied by hanging-wall subsidence and footwall uplift, with footwall uplift accounting for 5–20% of the total fault throw (Jackson and McKenzie, 1983; King et al., 1988; Stein et al., 1988; Roberts and Yielding, 1994). Steeper fault dips apparently produce greater footwall uplift; vertical faults, in theory, should produce equal amounts of footwall uplift and hanging-wall subsidence. Assuming a purely elastic response to coseismic slip, a hangingwall rollover anticline and footwall syncline should characterize the ground surface on either side of the fault (Fig. 5). Following the coseismic deformation, both the footwall and the hanging-wall should experience additional interseismic deformation as the elastic strain field created during slip is relieved and isostatic

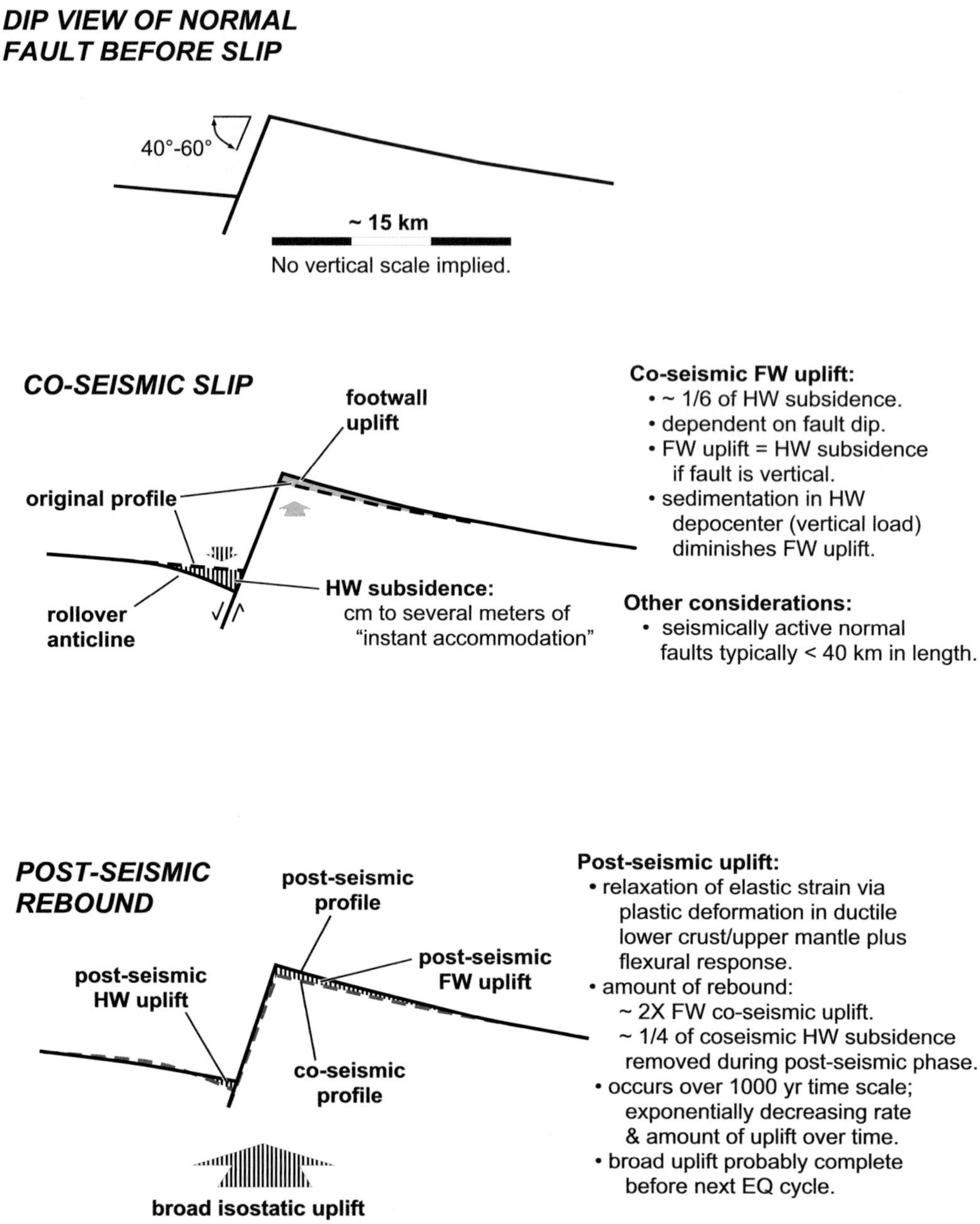

FIG. 5.—Generalized cross-sectional view of surface deformation around an isolated normal fault during a single earthquake cycle. Amount of footwall uplift is approximately 15–20% of hanging-wall displacement. Long-wavelength (~ 15 km) ground deformation is expressed as roll-over monocline in the hanging wall and broad syncline in the footwall. Later post-seismic isostatic uplift occurs as a result of footwall unloading; this flexural uplift is centered about the fault. Note how ground deformation is expressed over 15 km wavelength from the fault and progressively diminishes away from the fault.

equilibrium is restored within the lithosphere. This results in interseismic, or "post-seismic" rebound, until local isostatic equilibrium is reestablished.

Interseismic rebound has been measured across some normal faults following historical coseismic slip events. Interseismic rebound creates a broad uplift that is centered about the fault trace. The amount of interseismic rebound probably adds or removes approximately 30–100% to the original footwall uplift, or hanging-wall subsidence, that occurred during coseismic slip. Theoretical considerations, geodetic surveys of active normal faults, and an understanding of typical earthquake recurrence intervals suggest that interseismic rebound occurs at an exponentially decreasing rate over a period of about a thousand years after an episode of coseismic slip. In most cases, restoration of isostatic equilibrium is largely accomplished before the next earthquake cycle (Wallace, 1984; King et al., 1988; Roberts and Yielding, 1994).

A two-dimensional tectonic model for the surface deformations associated with coseismic slip on large, isolated normal faults might be as follows: (1) An earthquake occurs along the normal fault and coseismic slip *instantaneously* causes hanging-wall subsidence and footwall uplift. The greatest amount of coseismic displacement develops near the mid-point of the normal fault. The coseismic surface deformation is best described by longer-wavelength (i.e., tens of kilometers) folding of both the hanging wall and the footwall. (2) The rock volume around the normal fault then undergoes interseismic rebound, which increases the amount of footwall uplift and decreases the amount of hanging-wall subsidence immediately adjacent to the fault as measured relative to a pre-deformation horizontal baseline. The wavelength and amplitude of the footwall syncline gradually increases over time during interseismic rebound, although surface gradient changes caused by rebound are on the order of fractions of a degree as measured over tens of kilometers horizontal distance, even without erosion of the uplifted surface. In contrast, the surface deformation of the hanging wall tends to flatten, as gradients gently decrease toward the fault surface as rebound progresses.

Three-dimensional footwall and hanging-wall surface deformations are more complicated, depending on slip distribution along the fault, interactions between neighboring faults during coseismic displacements, and possible mass wasting during earthquakes. Satellite observation methods, such as synthetic aperture radar interferometry (InSAR) techniques, of modern active faults provide high-resolution characterization of coseismic and interseismic ground deformation, although these observations are strictly from present-day subaerial settings. Future research efforts should focus on three-dimensional characterization of carbonate stratal patterns and facies development around historically active normal fault systems.

Fault-Block Rotation.—

On strike-normal cross-sectional views, half-graben blocks appear to be rotated about horizontal axes as displacement accrues along the normal faults that bound each block (Jackson and McKenzie, 1983; Gibbs, 1984; Leeder and Gawthorpe, 1987). Pre-rift strata or unconformity surfaces are commonly used as paleo-horizontal markers to record the amount of rotation. In the southern Gulf of Suez, pre-rift strata in half-graben fault blocks were rotated as much as 20° from the onset of rifting (~ 25 Ma) to the end of Nukhul deposition (~ 21 Ma), for an average rate of rotation of about 5°/Myr (Winn et al., 2001). More detailed analyses of syn-rift strata indicate that rotation rates have varied over time in the Gulf of Suez (Bosworth et al., 1998).

These estimates for the amount and rates of block rotation in the Gulf of Suez provide reasonable constraints for stratigraphic modeling, but actual rotation histories for other syn-rift carbonate platforms depend on fault topology (i.e., fault dip, length, planar or curved, ramp-flat segments, and depth to possible detachment horizons), whether secondary antithetic or synthetic faults cut the hanging wall, and the rates of stretching.

Other Controls on the Regional Geomorphologic Evolution and Differential Subsidence Patterns Across Extensional Settings

Stretching of the lithosphere clearly causes dramatic geomorphologic changes across extensional settings throughout their pre-rift to late post-rift history. There are other factors, however, that also influence the geomorphologic evolution and subsidence history of extensional settings.

Pre-Rift Topography.—

Pre-rift topography is difficult to characterize in any extensional setting, but it may strongly influence syn-rift and post-rift topography. In modern rifts, even those that are in their earliest stages of development, pre-rift topography may be significantly modified by syn-rift uplift, erosion, and differential subsidence. In fossil rifts, pre-rift topography is obscured by erosion that occurs during syn-rift deformation, by long-term, long-wavelength, post-rift thermal subsidence, and by burial beneath syn-rift and post-rift strata. In deeply buried fossil rifts or passive margins, basement lithologies and pre-rift geologic history may also be poorly understood. It is possible, however, to infer pre-rift topography if there is some knowledge about the regional pre-rift geologic history of the extended region.

In most cratonic interiors, pre-rift topography should be subdued and have lower average elevation simply because long-term continental denudation prior to rifting would result in a regional, peneplained surface. These peneplains may experience broad, domal uplift (> 1000 km wavelengths) before or during early stages of rifting. At the other extreme are orogenic belts or high continental plateaus that undergo extension during orogenic collapse (cf. Artyushkov, 1973; Dewey, 1988; England and Houseman, 1989). Collapsing orogenic belts have much higher average elevation and short-wavelength, high-relief topographic irregularities. The rift-generated structural topography that is superimposed on the irregular pre-rift topography of orogenic belts or high continental plateaus should create a very complex regional geomorphology.

The nature of pre-rift topography may be important for carbonate-platform development because it affects the timing of initial marine flooding of rift depocenters and fault-bounded basement highs. Low-elevation (i.e., near sea level) peneplain surfaces that undergo rifting may be flooded over wide areas during the earliest stages of rifting. Local syn-rift topography in these settings may largely reflect fault-block rotation, local footwall and regional rift-flank uplift, and isostatic adjustments, with little significant influence from any inherited pre-rift topography. Local fault-controlled topography also becomes superimposed on broader-wavelength thermal doming.

In contrast, initial marine flooding may not occur across rift zones within high-elevation orogenic belts until the lithosphere is highly stretched. A network of linked rift depocenters may be needed to create a seaway through which oceanic communication can be established between the rift zone and an adjacent ocean basin. In these cases, syn-rift carbonate platforms would record the progressive flooding of the evolving rift network and might be important stratigraphic marker horizons that record progressive isostatic compensation across the collapsing orogen.

Rift-Flank Uplift.—

Topographic profiles from many modern high-strain intracontinental rifts are characterized by: (1) an axial zone, where topography is strongly fault-controlled, and (2) conspicuously elevated to low-relief rift flanks that border the axial zone, where topography diminishes with distance from the rift axis.

The syn-rift basement topography of the axial rift zone is usually rugged, but it may be smoothed significantly by syn-rift and post-rift erosion and sediment fill. Rift-flank uplifts may be symmetrical, with roughly equal amounts of elevation on both sides of the rift axis, or asymmetrical, with one side of the rift system having higher topography than the other. These differences in rift-flank elevation may reflect either pre-rift topography or the kinematics of lithospheric stretching (Wernicke, 1985; Bosence, 1998).

The elevation of rift-flank uplifts can exceed 3 km in some rifts and young passive continental margins (e.g., Gulf of Suez, Steckler and Omar, 1994; Rhine Graben, Ziegler, 1992b, 1996; Angola–Namibia–South Africa passive margin; Brown et al.,

2000), whereas some rifts and passive margins show little evidence of significant flank uplift at any stage in their evolution (Ziegler, 1990; McHargue et al., 1992). In the Gulf of Suez, rift-flank topography varies from a few meters elevation in the northern part to approximately 2–3 km over most of the remaining length of the rift, which creates a north-plunging basement surface (Schütz, 1994). If rifting has ended in the Gulf of Suez, it might be expected that erosion and post-rift thermal subsidence would cause the rift flanks in the northern part of the system to be flooded before most of the remaining length of the rift flanks are inundated by seawater.

CARBONATE PLATFORMS IN EXTENSIONAL SETTINGS

Siliciclastic sedimentologists and stratigraphers have long recognized the influence of tectonic deformation on sedimentation in extensional basins. Generalized tectono-sedimentary models for rift basins have been presented by Leeder and Gawthorpe (1987), Frostick and Reid (1989), Surlyk (1990), and Gawthorpe and Leeder (2000). These models describe general aspects of syn-rift carbonate sedimentation. Recent studies of Miocene syn-rift carbonate platforms in the Red Sea, the Gulf of Suez, and the Gulf of Aden have illustrated many details of stratigraphic development in syn-rift platforms (Burchette, 1988; James et al., 1988; Bosence, 1998; Cross et al., 1998; Perrin et al., 1998; Purser et al., 1998), but the Red Sea–Gulf of Aden region is not representative of all rift basins. Little work has investigated the development of carbonate platforms during the transition from syn-rift to post-rift phases of extensional basins.

It may be difficult to determine whether a carbonate platform formed during active rifting or on remnant fault-bounded highs that were created during the syn-rift stage. In either case, it is still the syn-rift topography that controls the earliest carbonate deposition, so understanding the tectonic evolution of the rift system during the syn-rift phase is important.

This section builds on previous tectono-stratigraphic models for extensional basins and presents models that address other aspects of carbonate-platform sedimentation in extensional settings. Additional factors include windward–leeward effects on facies development, the influence of rift-generated topography on sedimentation and platform morphology, lateral facies transitions at accommodation zones, platform development on different types of structural highs, the influence of fault propagation and linkage on platform stratigraphy, and a variety of environmental factors that may or may not be related to rifting, such as nutrient supply, upwelling, paleoclimate, and interactions with siliciclastic and evaporite depositional systems.

Outcrop examples, along with seismic data from well-imaged subsurface examples, are used to constrain the tectono-sedimentary models presented here. Some aspects of the platform-evolution models presented here, while well-founded in tectonic theory and supported by some observational evidence, remain to be tested by detailed studies of carbonate platforms in extensional basins. This section begins with general comments about methods for understanding growth strata within syn-rift carbonate platforms and then ends with a presentation of tectono-stratigraphic models for syn-rift carbonate platform systems.

Tectonic Versus Eustatic Controls on Stratigraphic Development and General Growth Stratal Relationships in Syn-Rift Carbonate Platforms

In any tectonically active setting, differential fault-controlled or long-wavelength differential subsidence interacts with eustatic fluctuations to create or remove accommodation space. High-frequency sea-level changes (i.e., those that occur with 10^3–10^6 yr periodicities) interfere significantly with the short-term changes in accommodation space created by fault displacements. These eustatic fluctuations occur at average rates (cm to 10 m / 1000 yr) that are similar to the average rate of accommodation change created by fault displacements. In contrast, longer-term eustatic changes (> 10^6 yr) will interact more significantly with long-wavelength syn-rift tectonic subsidence across rift systems because accommodation change caused by either of these mechanisms occurs at slower rates and over longer periods of time than accommodation changes related to coseismic displacements and glacio-eustatic fluctuations.

It will be difficult to isolate the eustatic versus tectonic components of accommodation change in many syn-rift carbonate platforms unless excellent age control exists and three-dimensional stratigraphic relationships can be determined. Certain large-scale stratal patterns within syn-rift carbonate platforms, however, can be explained only by differential, fault-controlled tectonic subsidence. These stratal patterns include: (1) inner-platform lithofacies that are offset or juxtaposed with non-inner-platform strata across a fault zone; (2) isolated platforms with dips on inner-platform strata that increase downsection due to fault-block rotation about a horizontal axis; (3) platform-margin and slope facies that directly overlie or are cut by syn-rift faults; and (4) regional patterns of onlap or progradation which can be related to basin-scale patterns of syn-rift tectonic subsidence (Fig. 6).

At local observational scales, we might expect growth strata from syn-rift carbonate platforms to reflect several kinds of response to active faulting: (1) gradual asesimic fault movements might produce gradual lateral and vertical facies transitions and stratigraphic stacking patterns that at least partly reflect high-frequency eustatic fluctuations, but overall growth stratal patterns (e.g., increasing dips downsection, divergent strata, and fault-truncated strata) should be similar to those produced by episodically moving faults; (2) fault zones with more episodic, coseismic displacement histories should create abrupt lateral and vertical facies changes and complex stacking patterns, especially where incremental displacements are on the order of a meter or more; (3) pulsed or waxing and waning of fault displacements should create similarly abrupt to gradual facies transitions and stratal relationships.

Timing of Platform Initiation in Extensional Basins

The time of initial marine flooding of syn-rift topography is critical for the initiation of carbonate platforms within the rift system. If seawater incursion into an extensional setting occurs during syn-rift stages, growth strata within the carbonate platforms will be key indicators of fault displacements, fault growth and linkage, and the rotation or internal deformation of fault blocks. The morphology and longer-term subsidence of remnant syn-rift topographic features and the history of long-term eustatic fluctuations (probably 1–10 Myr trends) determine which structures within the rift network become submerged to shallow, photic depths.

The exact timing of initial carbonate-platform sedimentation in extensional settings can be highly variable in time and space. During early to middle syn-rift stages of tectonic development, most intracontinental rift basins are characterized by terrestrial depositional systems, which typically include alluvial, fluvio-deltaic, and lacustrine facies (cf. Leeder, 1995). Transition into marine conditions commonly occurs during later syn-rift to early post-rift stages of basin evolution when the amount of stretching

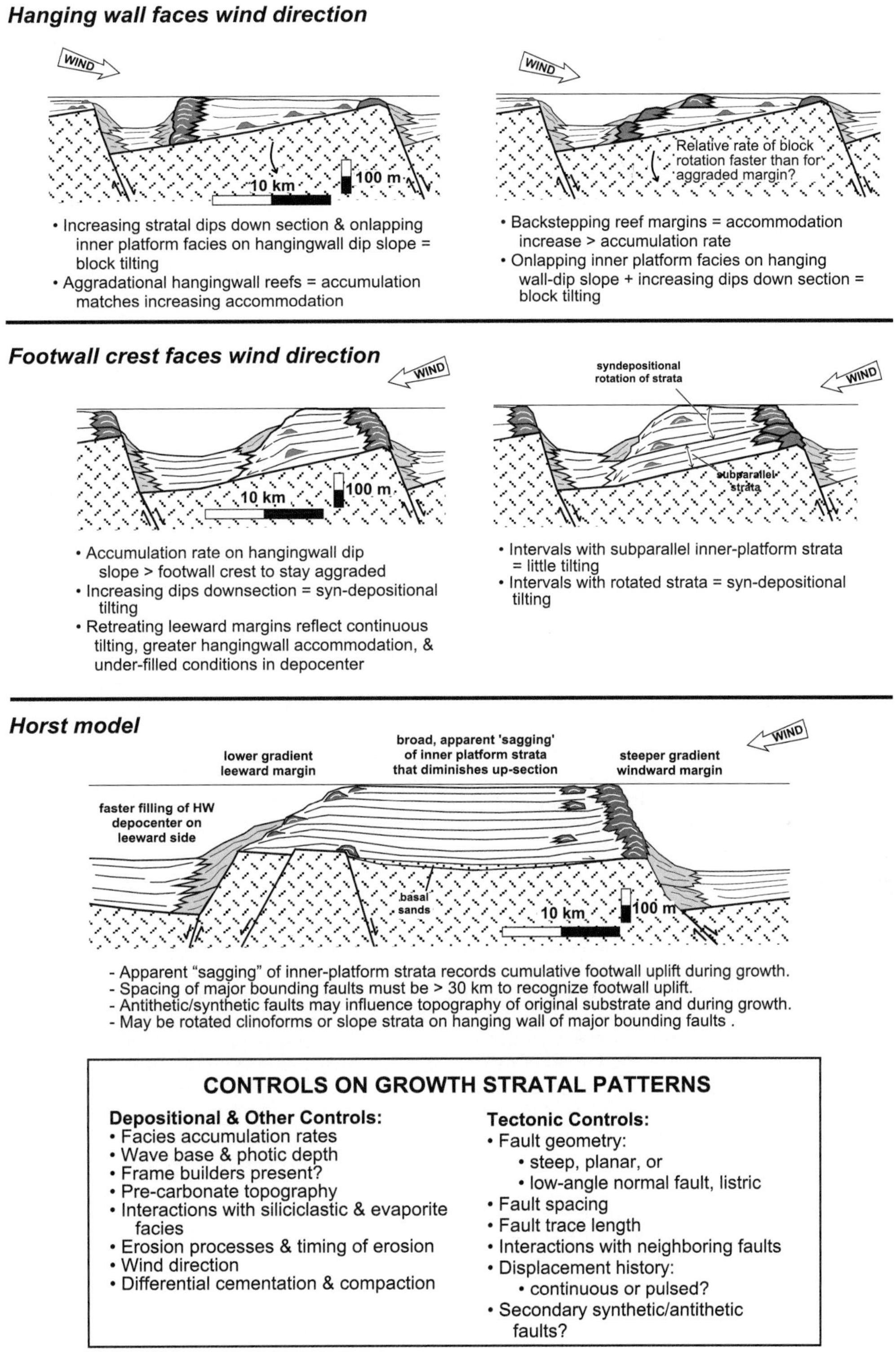

FIG. 6.—Typical growth stratal patterns within syn-rift isolated platforms. Half-graben and horst platforms are shown. Internal facies distributions are also influenced by wind direction relative to the tilt of individual fault blocks; paleo-wind direction is a major determinant of facies distributions within syn-rift isolated platforms. It is important to recognize that the growth stratigraphies shown here represent only a few possible end members. Actual syn-rift platforms might show growth strata that include elements from any of these models, as well as other patterns that reflect the unique kinematic history of individual fault-bounded structural elements and interactions with other allocyclical controls on sedimentation (e.g., climate, siliciclastic sediment flux, paleowind directions, local upwelling, biotic components, etc.). Growth stratal patterns that reflect differential, basin-scale, syn-rift tectonic subsidence are shown schematically in Figure 7.

has progressed to the point that part of the rift network has sufficiently subsided and an interconnected network of rift depocenters links to an adjacent ocean basin and marine incursion begins. The actual time and location of initial marine incursion within a rift network may depend on:

1) *whether rifting was "active" or "passive."* Active rifts begin with regional thermal doming and development of a broad topographic and structural dome. Early fluvial drainage networks flow radially away from the crest of the thermal dome, and these drainage patterns may persist some time after the initial rift has evolved into a passive margin (Cox, 1989; Frostick and Reid, 1989; Frostick and Steel, 1993). In rift basins where most of the stretching is preceded by thermal doming, the syn-rift stratigraphy may be dominated by nonmarine facies because these basins typically are not flooded by seawater until late syn-rift to early post-rift stages. In contrast, the early syn-rift stage of basins associated with passive rifts are characterized by regional subsidence with only minor associated igneous activity and a general absence of broad domal uplift during early phases of rifting. Early regional subsidence is conducive to earlier marine flooding and may allow carbonate deposition to begin earlier in passive rifts than in active rifts.

2) *whether the rift formed by pure shear, simple shear, or combined shear.* If a rift system formed by pure-shear extension (McKenzie, 1980), both sides of the rift system should be flooded at approximately the same time. If the rift system formed by simple-shear extension (Wernicke, 1985), however, the lower plate should be flooded first. If extension developed in more complex ways, with variable amounts of stretching at different levels of the lithosphere, then the distribution of syn-rift and early post-rift shallow-marine carbonate platforms might provide important information about partitioning of extensional strain.

3) *the amount of stretching* (β). Marine incursion is more likely to occur earlier in the tectonic history of extensional basins and over larger areas if β values are higher. In many continental rifts that evolve into passive margins, more outboard depocenters and fault-bounded highs are generally flooded first, simply because they are located on more highly stretched lithosphere with higher β values. Thus, syn-rift platforms with the earliest onset histories are typically located farthest offshore relative to the updip tectonic hingeline. Local variations in pre-rift topography, the patterns of stretching, and syn-rift erosion, however, influence the exact timing of initial marine flooding of local structural features.

4) *pre-rift topography.* Where regional pre-rift topography is low, broad areas may be rapidly submerged by seawater during early syn-rift stages of basin development. In contrast, topographically high areas may never be flooded during syn-rift or even late post-rift stages, unless the amount of lithospheric stretching is high.

5) *proximity of the rift to an existing ocean basin.* Marine incursion is likely to occur earlier where the rift system develops close to an existing ocean basin. Some land-locked rift basins may never become marine, even though basement may subside well below sea level (e.g., Basin and Range Province, USA). In contrast, even minor amounts of stretching might allow marine flooding if the rift system develops adjacent to a preexisting ocean basin.

6) *history of eustatic sea level during syn-rift to early post-rift stages of basin development.* Marine flooding is obviously more likely to occur sooner in the history of an extensional basin during times in Earth history when eustatic sea level was rising or at a long-term highstand. For example, rift basins that formed during Late Ordovician or Late Cretaceous time, which are long-term sea-level highstands (Sloss, 1963, 1984; Vail et al., 1977; Haq et al., 1987), might have been flooded by seawater much earlier than basins that formed during Middle Ordovician or Late Permian time, which are major lowstands.

Models of Syn-Rift Carbonate Platforms

Leeder and Gawthorpe (1987) and Bosence (1998, 2005) presented basic tectono-stratigraphic models for syn-rift carbonate platforms; these models focus on half-grabens as the main tectono-geomorphologic element in rift basins (cf. Rosendahl, 1987; Bosworth et al., 1998). Variations on these models account for differential subsidence or uplift of hanging-wall and footwall areas, erosion, or basin-scale patterns of subsidence during fault growth and linkage (e.g., Collier and Gawthorpe, 1993; Anders and Schlische, 1994; Lambiase and Bosworth, 1995; Gawthorpe and Leeder, 2000; Contreras and Scholz, 2001; Cowie et al., 2006). In important comparative studies of Miocene carbonate platforms in the Gulf of Suez and northern Red Sea, Purser et al. (1998) and Plaziat et al. (1998) showed more complex relationships between local structural features and syn-rift carbonate sedimentation.

The half-graben model, however, does not adequately describe the variety of structural features and associated carbonate platforms that can develop during syn-rift stages. Interactions between syn-rift carbonate systems and paleowind direction, siliciclastic influx, fault displacements and fault propagation, and characteristic growth stratigraphies and morphologies of syn-rift platforms are also additional considerations that are investigated here.

Tectonic Controls on the Dimensions, Morphologies, and Growth Histories of Syn-Rift Carbonate Platforms in Extensional Basins.—

Syn-rift carbonate platforms form on the complex topography that is created during rifting. Most syn-rift platforms across many extensional settings are initiated during late syn-rift stages, although the exact timing of platform initiation depends on many factors, as discussed below.

General Effects of Deformation on Growth of Syn-Rift Carbonate Platforms.—

Syn-rift deformation influences growth of carbonate platforms in many ways. At regional scales, syn-rift deformation influences platform morphology and location within the rift system (Fig. 6). At local scales, syn-rift deformation influences facies development, the location of facies tracts, the internal growth stratal patterns of individual platforms, and the degree of mixing between coeval siliciclastic and carbonate facies (Figs. 6–12).

Syn-rift isolated platforms mimic the tectonic geomorphology created during syn-rift deformation. In intracontinental extensional settings that begin as nonmarine basins, initial marine incursion first occurs in structural lows (i.e., topographically low graben and half-graben depocenters). Shallow-marine carbonate sedimentation may be initiated in these lows, but it will generally be short-lived as marine incursion progresses.

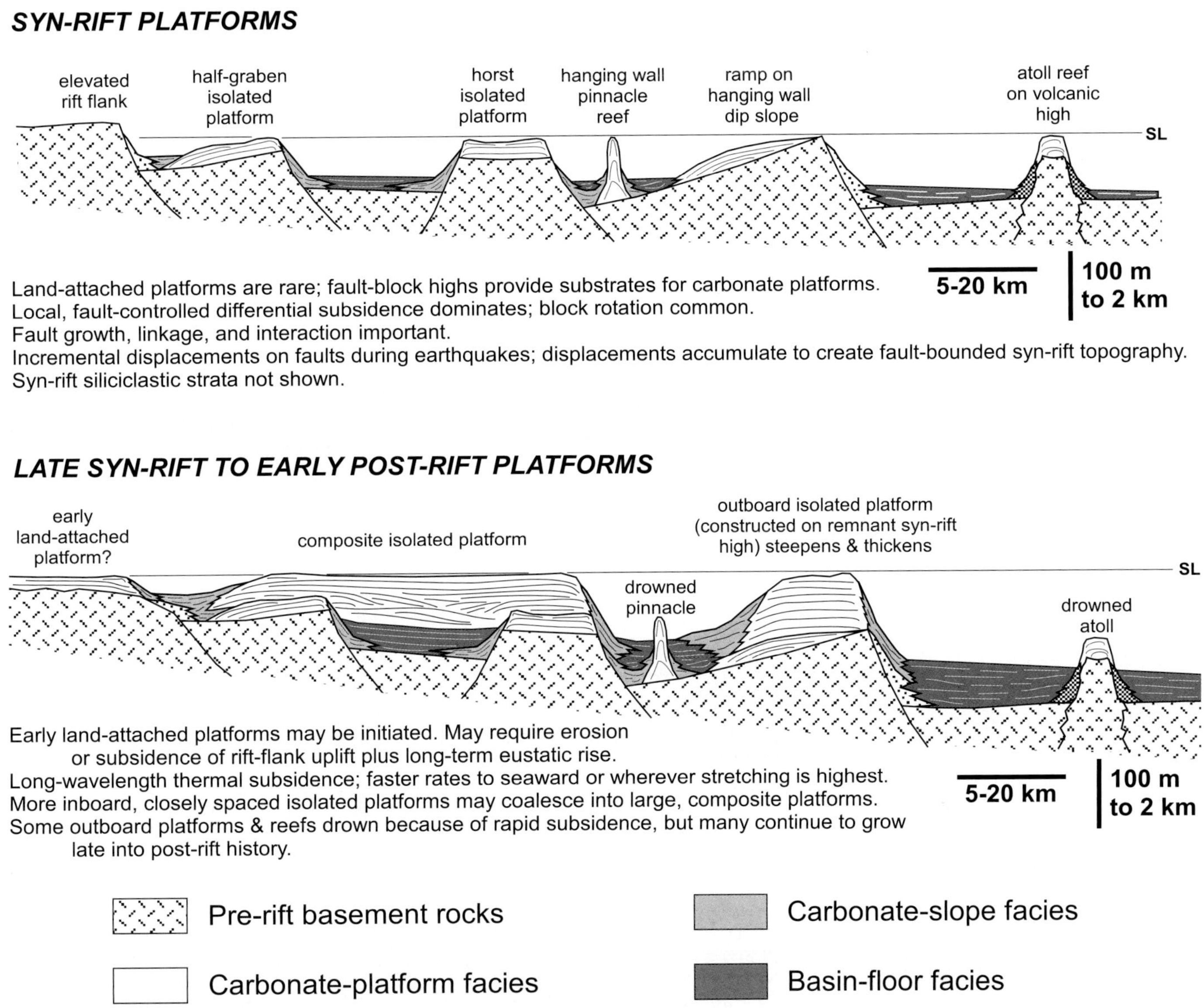

FIG. 7.—Hypothetical, basin-scale, syn-rift growth stratal patterns. Regionally, isolated syn-rift platforms should thicken towards the zones of greatest extension and thin toward the basin margin. These thickness trends reflect both long-wavelength differential tectonic subsidence during syn-rift phases of basin development and earlier marine flooding (and therefore earlier initiation of carbonate platforms) of more highly stretched zones. These regional patterns of subsidence and differential carbonate-platform growth can continue to develop well into the post-rift history of an extensional basin. There are few examples in the rock record where these types of stratal patterns have been documented, although thickness trends within Cenozoic platforms of the South China Sea and Miocene platforms of the Red Sea and Gulf of Suez indicate that long-wavelength differential syn-rift subsidence was an important determinant of regional trends in platform thickness during syn-rift to early post-rift stages.

As flooding progresses, fault-bounded structural highs are progressively submerged and can become sites of shallow-water carbonate sedimentation. The plan-view shapes of these fault-bounded highs control the initial plan-view dimensions of carbonate platforms. Flooding of tilted half-grabens typically creates oblong to crescent-shaped isolated platforms in plan view, because this is the generalized surface geomorphology that is initially flooded by seawater. Steep-sided platform margins at the mid-points of footwall crests may grade laterally into more ramp-like profiles near the tips of normal faults or at relay-ramp transfer zones. Ramp-like profiles also may be more characteristic along hanging-wall dip slopes, although the direction of wind and wave approach also influences facies development and resultant platform morphology (see below).

The plan-view dimensions of isolated platforms on horsts are strongly controlled by the trends of extensional bounding fault networks. Isolated platforms that form on horsts are likely steep-sided for most of their history unless thick stratigraphic successions, typically dominated by siliciclastic facies, fill adjacent fault-bounded depocenters and allow carbonate platform margins to prograde across the filled depocenter.

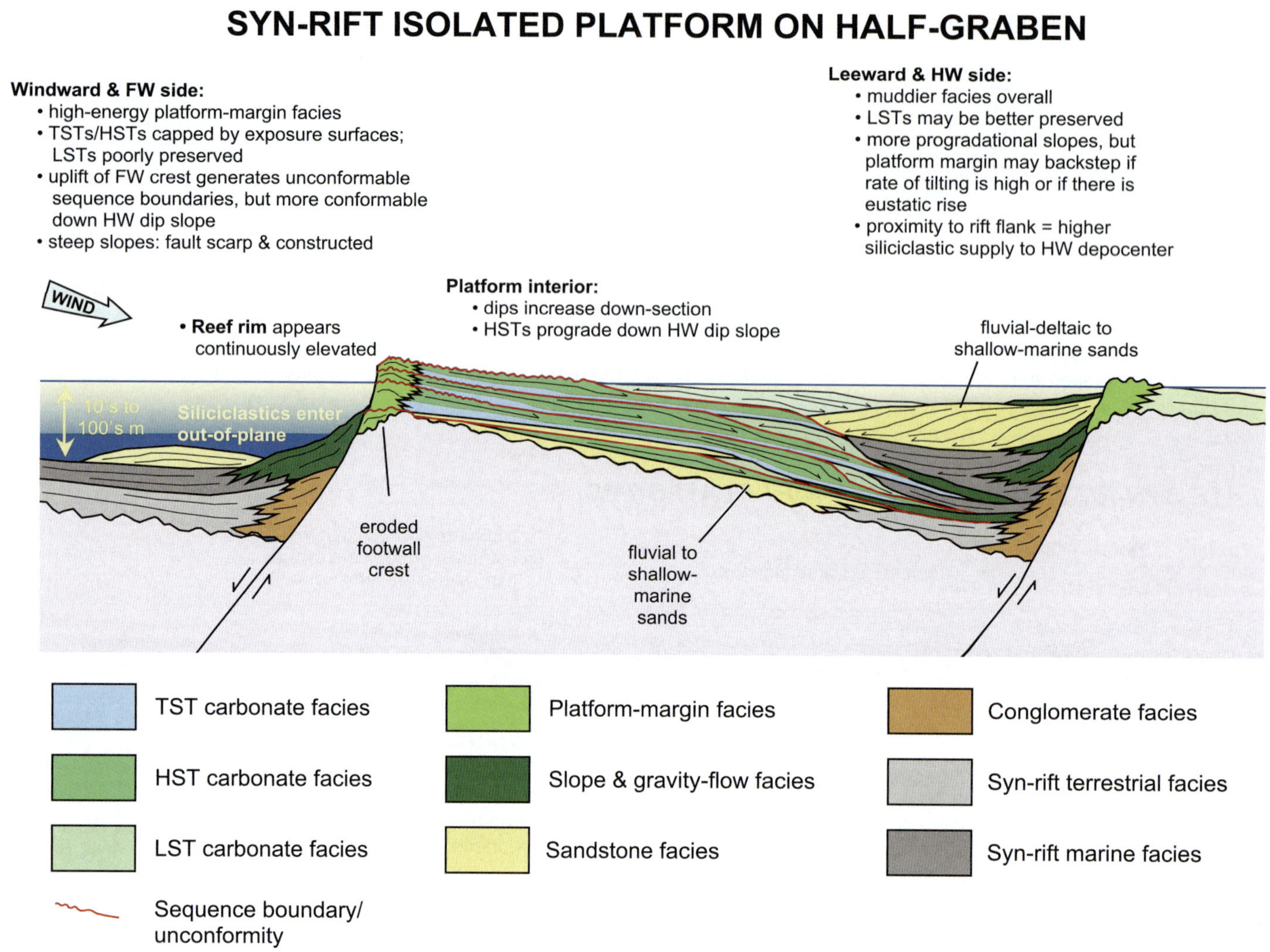

FIG. 8.—Tectono-stratigraphic model for syn-rift isolated platform on half-graben structural element. Note general wedge-like growth stratal units with superimposed sequence stratigraphic systems tracts (hypothetical). Actual facies patterns and growth strata relationships within half-graben syn-rift platforms depend on paleo-wind direction relative to fault-controlled topography (cf. Figure 6), eustatic fluctuations during syn-rift deformation, interactions between different subfacies within a carbonate-platform system, nature of carbonate factory (presence or absence of framebuilders; mud- or grain-dominated inner platform; diagenetic systems, etc.), and supply of siliciclastic sediment. Regardless, this conceptual diagram illustrates several key stratal relationships that are indicative of (but not entirely exclusive to) hanging-wall block rotation during active extension: inner-platform strata that show progressive dip increase down-section; greater abundance and possible amalgamation of disconformity surfaces toward footwall crest; wedge-like stratigraphic units within platform interior; general onlap toward footwall crest and very low-angle downlap across hanging-wall dip slope.

Morphologies of Syn-Rift Carbonate Platforms.—Nearly all syn-rift carbonate platforms are isolated platforms, which are constructed on the fault-block topography that forms during rifting (Figs. 6–11). Cumulative displacement on bounding faults creates sufficient bathymetric relief (> 100 m) between the tops of high-standing blocks and adjacent down-faulted depocenters so that shallow-water carbonate sedimentation is precluded in the lows. Syn-rift siliciclastic deposits also generally inhibit or mix with carbonate sediments in down-faulted depocenters.

Although rare, other carbonate platform morphologies or buildups besides isolated platforms may form in syn-rift settings. These morphologies include:

1) *Pinnacle reefs and mounds.* Small pinnacle reefs and mounds can form on hanging-wall dip slopes and reflect local, highly aggradational buildups attempting to keep pace with syn-rift subsidence (Fig. 12). Examples of these types of buildups can be found in hanging-wall depocenters in some Neogene sub-basins of the Nam Con Son Basin, offshore Vietnam.

2) *Fringing ramps and thin platform successions on hanging-wall dip slopes during syn-rift stages.* Thin ramp successions typically form after syn-rift fault-bounded highs have been initially inundated by seawater and if siliciclastic influx is limited (Fig. 7). Carbonate strata in these thin intervals may be complexly interstratified with siliciclastic facies that are

locally derived from pre-rift basement of the hanging wall. The updip limit of initial syn-rift ramp successions commonly reflects where sea level intersected the hanging-wall dip slope during progressive flooding. The thickness of these onlapping and backstepping ramp packages is typically < 100 m, which indicates they either are short-lived systems that are terminated soon after initiation by siliciclastic influx or relative sea-level rise, or they evolve into isolated platforms as the fault block is progressively submerged.

3) *Small isolated platforms and pinnacles on volcanic substrates.* Dormant submarine volcanic edifices may provide suitable substrates for carbonate-platform sedimentation in some rift settings (Fig. 7). Examples of these types of platforms are found in the South China Sea near the boundary between continental and oceanic crust (e.g., Pratas Island, offshore China and Sea Horse Bank, offshore Palawan) or on local volcanic highs within oceanic crust. Although it may be difficult to determine when submarine volcanism occurred

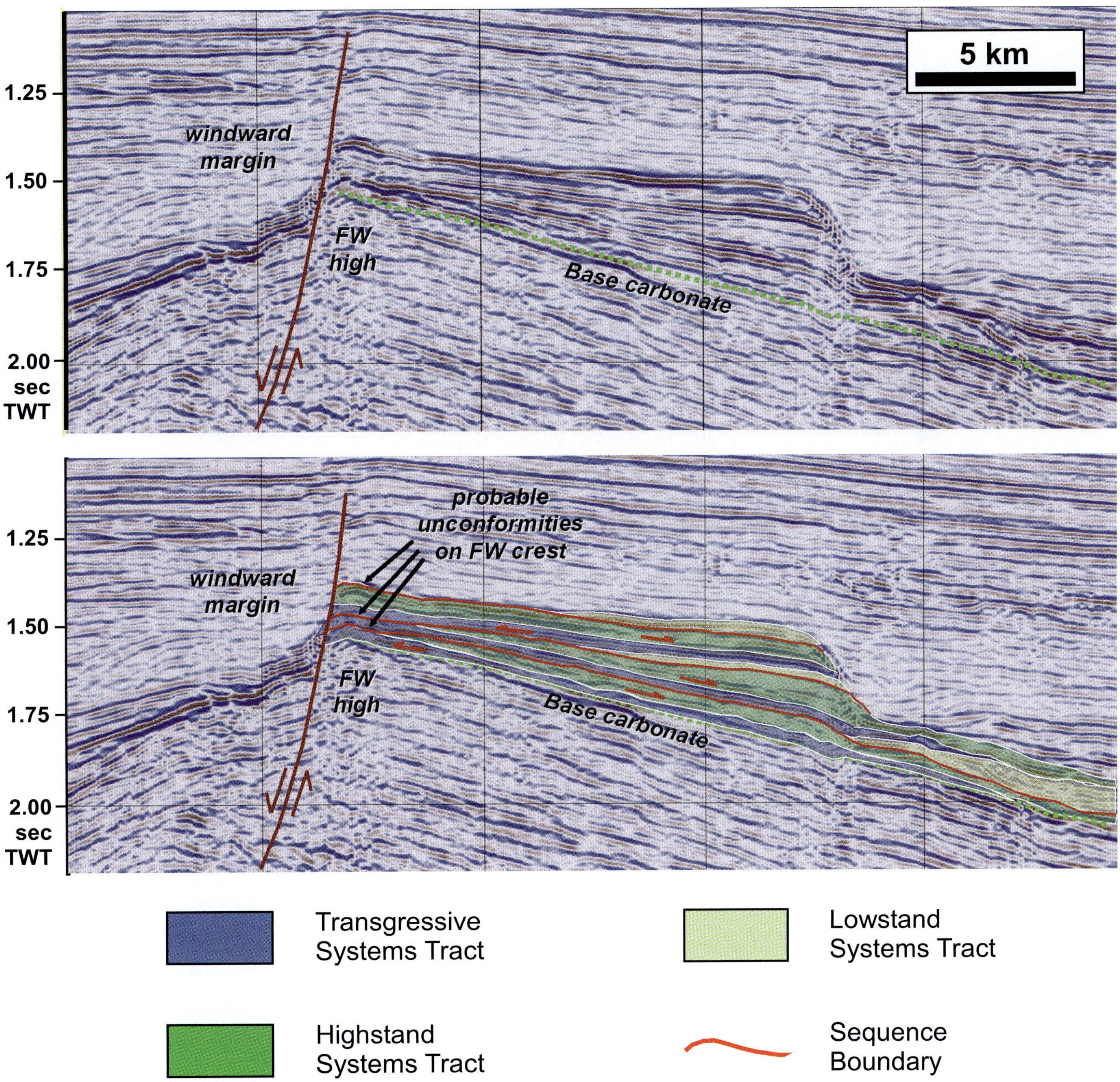

FIG. 9.—Syn-rift growth strata within an isolated platform constructed on a half-graben high; Miocene example from the subsurface of offshore Vietnam. Interpreted sequence stratigraphic relationships are based entirely on interpretation of stratal geometries observed on this seismic section. Seismic data provided by TGS-NOPEC.

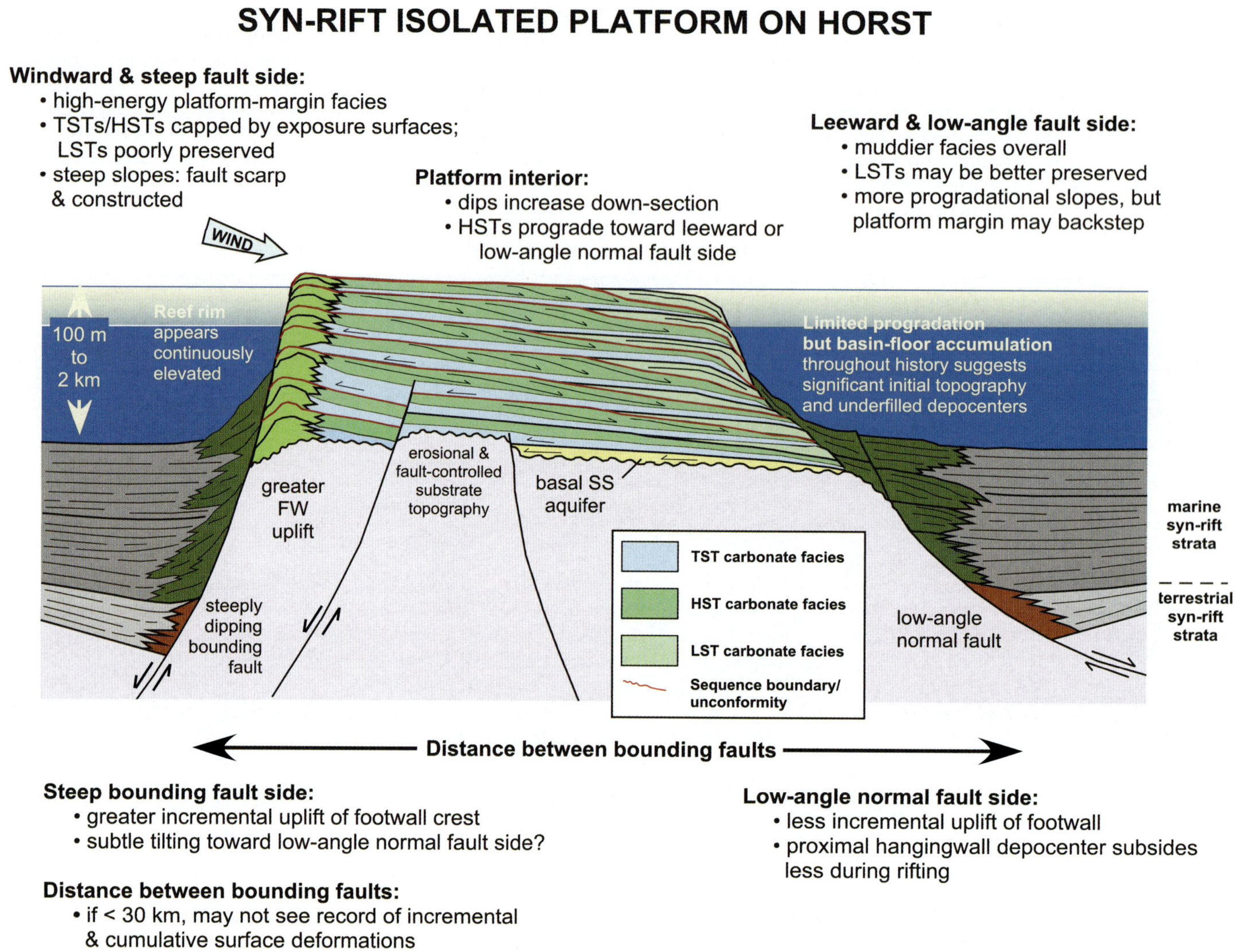

FIG. 10.—Tectono-stratigraphic model for syn-rift isolated platform on horst high. Note windward–leeward facies differentiation across platform top and in flanking slope facies. Internal growth stratal patterns are controlled by topology and spacing of the horst's main bounding faults. Apparent "uplifted" margins and broad sagging within platform interior are not recognizable within syn-rift horst platforms unless the spacing between the bounding faults is at least 30 km. Secondary antithetic or synthetic faults within the horst interior may also affect growth stratal patterns. Although not shown in detail on this diagram, lowstand systems tracts may be best expressed along leeward margins of syn-rift horst platforms.

relative to the basin's overall tectonic history, successive submarine eruptions from local vents can build enough seafloor topography to reach the photic zone where shallow-water carbonate platforms and reefs can be initiated during the syn-rift stage.

Shallow-water carbonate platforms and buildups in full-graben depocenters also are rare, probably because: (1) full grabens are more highly subsided than half-graben depocenters and therefore are rapidly submerged below photic depths at the time of initial marine incursion within a rift network, and (2) full grabens are more likely to be bounded by fault networks that divert fluvial, deltaic, or deeper-water siliciclastic dispersal systems from updip areas into the full-graben depocenters, thus inhibiting shallow-water carbonate sedimentation there.

Broad, land-attached carbonate platforms also seem to be rare in syn-rift settings, most likely because rift-flank uplifts are effective topographic highs that prevent sweeping transgression of coastal settings and establishment of wide carbonate platforms along the rift flank. Seaward of the rift-flank uplifts, fault-partitioned syn-rift topography must be eroded or filled with sediment in order for wide, shelf-like substrates to extend any significant distance from the rift-flank uplifts (Fig. 7). Wide, land-attached carbonate platforms cannot form because a wide, flat-topped depositional surface is not available along the edge of the rift system during much of the syn-rift phase. In almost every case, land-attached carbonate platforms do not form until the beginning of post-rift thermal subsidence across the rift system. Even where broad carbonate platform systems already existed prior to the onset of rifting, extensional deformation typically segments these platforms into smaller-scale, fault-controlled, isolated platforms.

Once syn-rift platforms are initiated, continued syn-rift deformation may control the dimensions and morphology of the initial

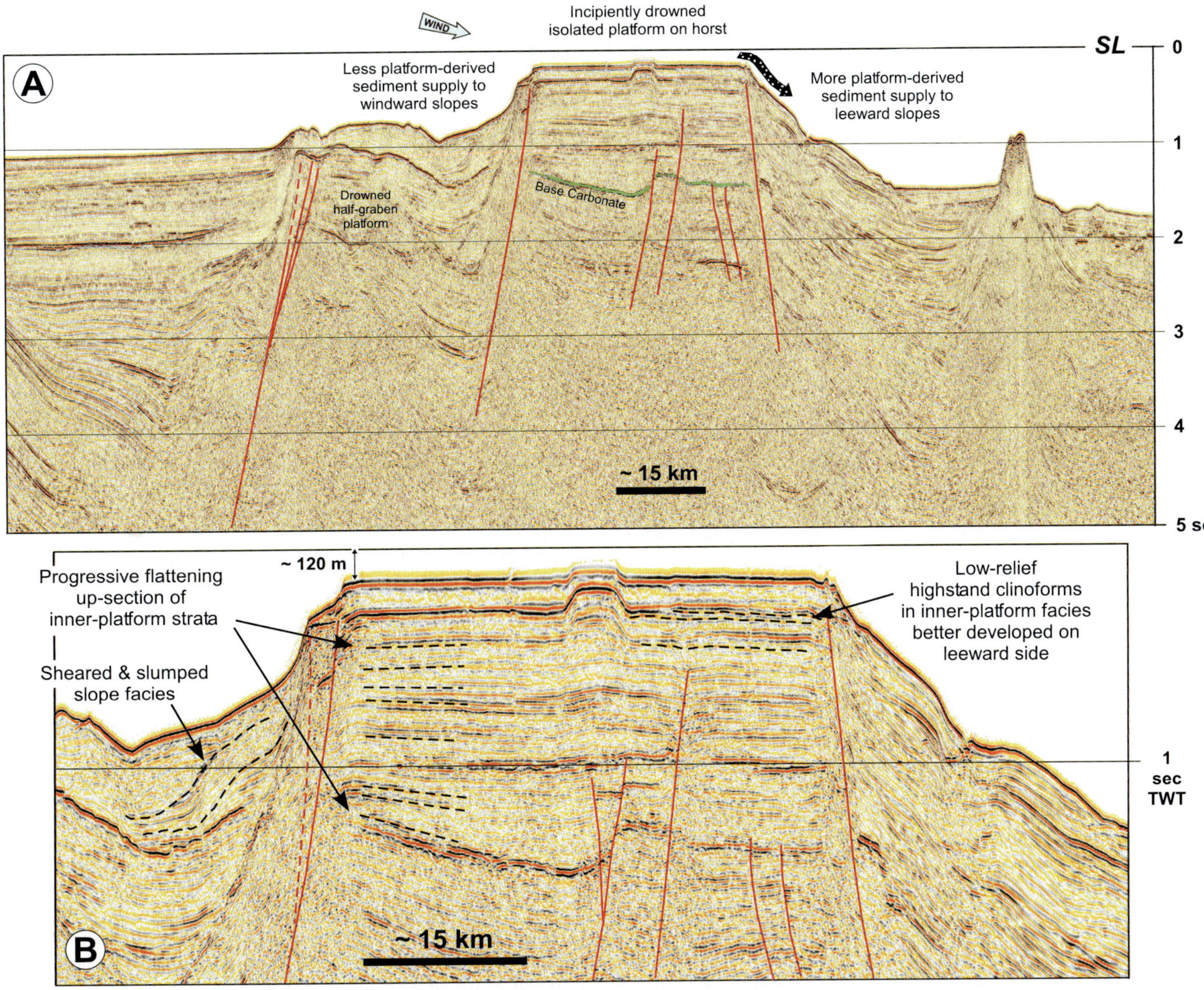

FIG. 11.—Example of syn-rift isolated platform on horst high; Miocene–Recent, offshore Vietnam. **A)** Regional seismic section showing several isolated carbonate platforms on fault-bounded highs. Major fault systems are shown, but many smaller displacement faults are not shown. The isolated platform on the left formed on a half-graben structure and apparently was drowned while isolated platform on the horst high to the right continued to grow. Note windward–leeward facies and stratal differences, with relatively sediment-starved windward slopes and greater sediment supply to leeward slopes. **B)** Close-up view of internal stratal geometries within isolated platform on a horst high. Note how the main bounding faults on both sides of platform appear to cut very close to the present-day seafloor, which indicates very recent extensional deformation. Inner-platform strata show progressive flattening, moving progressively up-section. Apparent "sagging" of the inner-platform setting is attributed here to progressive uplift of the footwall crest during rifting, although differential compaction may explain some of the apparent sag. Very low-relief and low-angle clinoforms are apparent within younger inner-platform facies along the leeward side of the platform. These are probably highstand inner-platform strata that are preferentially preserved on the leeward side of the platform. Seismic data provided by TGS-NOPEC.

isolated platforms and prevent smaller platforms from coalescing into larger composite platforms. Syn-rift fault networks also can tectonically segment larger syn-rift platforms into smaller platforms and buildups, although tectonic segmentation of preexisting platforms may be less common than platform coalescence. The general tendency for isolated syn-rift platforms to coalesce into larger platform complexes probably reflects the progressively diminishing displacements on faults during later syn-rift stages, when most syn-rift platforms develop (Fig. 7).

Location of Carbonate Facies Tracts on Individual Structures During Syn-Rift Deformation.—Understanding the likely fault-controlled surface deformations that develop during syn-rift phases of basin evolution is useful for predicting facies patterns within syn-rift carbonate platforms. Normal faults that break to the surface typically create surface scarps that dip tens of degrees, which is steep enough to prevent significant accumulation of shallow-water carbonate facies. Reef-building organisms and microbial communities, however, can construct organic frame-

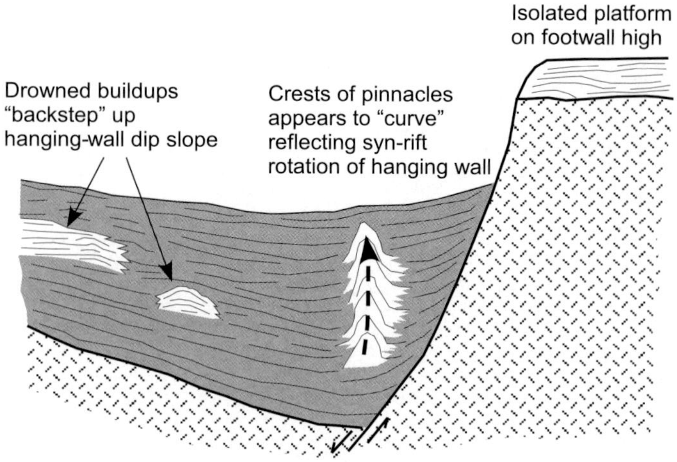

FIG. 12.—Schematic model for syn-rift pinnacle reef and backstepping buildups on rotated hanging-wall dip slope. Some highly aggradational pinnacle buildups that are initiated on the hanging-wall block close to the bounding fault may have a curved appearance in cross section, which reflects aggradation during rotation of the hanging-wall dip slope.

works along faulted margins, including directly on fault-scarp surfaces. Reefal frameworks form only on parts of the fault scarp that are within photic depths, whereas non-phototrophic microbial biofilms can encrust any exposed part of a submarine fault scarp (e.g., stromatolites on steep fault scarps and depositional slopes along Miocene isolated platforms of the Red Sea; Purser and Bosence, 1998). In general, however, local reefs and microbial encrustations that form on normal-fault scarps are volumetrically minor and tectonically generated relief and depositional gradients along fault scarps are nearly always great enough to limit significant shallow-water facies development and progradation of platform-margin and slope facies.

Sea level may intersect fault-block topography during syn-rift stages without completely submerging the entire surface of a fault block. In these cases, carbonate platforms are more limited in areal extent and the locations of facies tracts may be more strongly controlled by erosional topography that developed across the top of the fault block prior to marine flooding, rather than by fault-controlled topography.

Other factors that influence facies distributions and stratigraphic development within syn-rift carbonate platforms that form on partially or completely submerged basement highs include:

1) *Wave approach with respect to the flooded basement topography.* Windward-facing, emergent footwall highs may dissipate oncoming wave energy and create low-energy shadows on hanging-wall dip slopes (Figs. 6–11). Low-energy carbonate or mixed carbonate–siliciclastic ramp facies typically develop on leeward-facing slopes of partially submerged, fault-bounded highs (Fig. 6).

2) *Depositional gradients across the tops of fault-bounded basement blocks.* The depositional gradients across fault-bounded basement blocks influence the initial width of syn-rift carbonate platforms and the location and lateral extent of facies tracts. Platform widths generally decrease and lateral facies transitions should be more abrupt and more closely spaced as depositional gradients increase across the tops of syn-rift highs.

3) *Location and density of secondary faults between major block-bounding fault systems.* Secondary faults can create local tectonic topography within block interiors or they may offset strata within syn-rift carbonate platforms. These secondary synthetic or antithetic faults may be subparallel to or strike at high angles to major block-bounding faults. If the faults are active prior to marine flooding, they create topography that can influence early facies development and stratal architecture. The relative effects of this fault-generated topography on facies development depend on the total surface displacements that have accumulated on the faults at the time of initial marine flooding, as long as this topography has not been erosionally beveled at the time of initial flooding. Topography created by displacement along secondary faults during carbonate platform growth also influences facies development, although the incremental displacements on these secondary faults and the corresponding surface deformation may not be great enough to force major facies reorganizations across syn-rift platforms.

4) *Rates of sedimentation and amount of basin-fill in adjacent depocenters.* Depocenters next to syn-rift carbonate platforms can be filled with carbonate detritus derived from platforms or with siliciclastic sediment. Siliciclastic sediment can be supplied by local footwall highs or regional rift-flank highs and continental hinterlands that are located some distance from the rift zone. Tectonically generated bathymetric relief between syn-rift platform margins and adjacent depocenters can be filled rapidly where there is abundant supply of siliciclastic sediment. Carbonate-platform margins can then prograde from basement highs and across adjacent, partially filled, syn-rift depocenters. Even where sediment accumulation rates in depocenters are relatively high, continued syn-rift displacement on platform-margin fault networks may renew relief and steep depositional gradients along fault scarps and limit the progradation of platform-margin and slope facies. Thus, syn-rift platform margins that prograde significant distances from fault-bounded basement highs and into adjacent depocenters appear to be rare. The margins of some Miocene isolated platforms from more onshore subbasins in the South China Sea were able to prograde during latest syn-rift stages because fault displacement rates were probably decreasing and depocenters were simultaneously being filled by siliciclastic sediment eroded from nearby continental areas (unpublished data). The combined effects of waning fault displacements and basin filling allowed some carbonate platforms to prograde across former syn-rift depocenters.

Internal growth stratal patterns within syn-rift platforms.—Syn-rift carbonate platforms may contain a variety of growth stratal patterns that reflect the response of carbonate depositional systems to changes in surface gradients and differential subsidence across fault-bounded highs (Figs. 6–12). Factors that should be considered when interpreting these growth strata include: (1) whether basement topography is fully submerged during platform growth, (2) the unique spacing and kinematic history of fault networks that delineate the platforms, (3) longer wavelength patterns of differential syn-rift subsidence that might modify local, fault-controlled subsidence patterns, (4) wind and oceanic circulation patterns during platform growth, (5) the amplitude and periodicities of high-frequency eustatic fluctuations that interact with fault-controlled differential subsidence, (6) the growth potential of various carbonate facies tracts that construct syn-rift platforms, and (7) the volume, grain size, and dispersal patterns of siliciclastic sediment that is delivered to adjacent rift-related depocenters during platform growth.

Half-grabens and horsts, the typical fault-bounded highs that serve as substrates for carbonate platforms, respond in different ways to syn-rift deformation. For half-grabens, incremental syn-rift deformation results in rotation of the hanging wall, such that there are instantaneous changes in surface gradients across the hanging-wall dip slope during coseismic events. Many coseismic displacements result in progressive rotation of hanging-wall dip slopes within half-graben structural elements. Deposition across the rotated hanging wall results in the divergent stratal patterns that are characteristic of syn-rift growth strata in half-graben depocenters (Figs. 6, 8, 9).

In addition to hanging-wall rotation in half-grabens, incremental uplift occurs at the footwall crest of the half-graben due to internal deformation of the footwall near the fault scarp and isostatic adjustments after each displacement (Stein and Barrientos, 1985; King et al., 1988; Stein et al., 1988). The incremental footwall uplift that develops during a single earthquake cycle can be characterized as a long-wavelength (> 20 km) uplift that diminishes away from the bounding fault. At the crest of the footwall, vertical uplift is typically on the order of centimeters to several meters. The effect of incremental uplift on syn-rift carbonate-platform sedimentation may be almost inconsequential except in the immediate vicinity (i.e., within about 10–20 km horizontal distance) of the normal fault. The incremental uplift and surface-gradient changes that develop during a single earthquake cycle progressively diminish with horizontal distance from the fault. Many earthquake cycles, however, can cause significant uplift to accumulate at the crest of the footwall.

For horsts, block rotation is not likely to occur, but there should be similar patterns of footwall uplift along the normal-faulted sides of a horst as in the uplifted footwalls of half-graben structures (Figs. 6, 10, 11). Horst width probably determines whether footwall uplift is recognizable within a platform's growth stratigraphy. Subtle growth stratal patterns may be recognizable within syn-rift isolated platforms on horsts only if the horst is at least 30 km wide. This is due to the horizontal wavelength of footwall uplift, which is on the order of 15–20 km (Stein and Barrientos, 1985; Cisne, 1986; King et al., 1988; Stein et al., 1988). For horsts bounded by normal faults that are < 25 km apart, footwall uplift from both sides of the horst are likely to interfere, resulting in lesser apparent footwall uplift along the flanks of the horst. Basement horsts that are wider than about 30 km and bounded by oppositely dipping normal faults should show uplift along their flanks, such that the horst margins will appear to have been uplifted in the vicinity of the bounding faults and the center of the horst will appear to have "sagged" (Figs. 10, 11). Incremental uplift at the footwall crest should be on the order of several centimeters to several meters, but the cumulative uplift caused by many earthquake cycles could be greater than several hundred meters.

Cumulative displacement on normal faults that bound basement horsts create stratal geometries within the platform that suggest that platform-margin facies were repeatedly elevated above inner-platform facies tracts. The elevated nature of platform-margin facies tracts may be recognizable only on regional seismic profiles, because angular relationships within strata from platform-margin to platform-interior settings would be very gently dipping and difficult to recognize, even in large mountainside exposures. Internal growth stratal patterns within syn-rift horst platforms is further complicated if secondary synthetic or antithetic faults break interior parts of the horst. There may be few syn-rift horst platforms in the rock record that have the appropriate size and thickness to document growth patterns related to footwall uplift, but a Neogene example from offshore Vietnam shows that this is a viable model for the growth stratigraphy of syn-rift horst platforms (Fig. 11).

Thus, for long-lived syn-rift platforms that are at least 30 km wide, syn-rift growth strata should dip toward the platform interior from platform-margin locations along normal-fault boundaries. Dips should increase down-section, due to the cumulative effects of many fault-slip events. Incremental displacement events are probably more frequent and have larger effects during early syn-rift phases. In younger parts of the stratigraphy of a horst platform, dips should progressively flatten and become more horizontal as the cumulative effects of many displacement events diminish, or as stretching rate decreases and the basin gradually transitions into its post-rift phase of evolution. If rifting is episodic, with several cycles of tectonic activity and quiescence that last several million years each, very complicated internal stratal patterns will develop and reflect the "pulsed" character of the rifting events.

Small syn-rift pinnacle buildups that form in some half-graben depocenters show growth patterns where the crest of the buildups may appear to "curve" away from the border fault to the hanging wall as the hanging-wall dip slope rotated (Fig. 12). The curvature of the aggrading pinnacles indicates that they could keep pace with accommodation increase as the hanging-wall fault block rotated during stretching.

Other tectonic factors that might influence growth stratal patterns.—Other tectonic factors must be considered when interpreting the growth stratigraphy of a syn-rift isolated platform. For example, steeply dipping, crustal-scale faults should exhibit the greatest amounts of coseismic slip and interseismic rebound because even minor amounts of extension cause significant unloading along these faults. In contrast, low-angle normal faults that flatten at shallow levels in the upper crust should show lesser differential uplift or subsidence during an earthquake cycle for an equivalent amount of stretching because the amount of footwall unloading and hanging-wall rotation is much less.

The magnitude of incremental coseismic slip and corresponding interseismic rebound also can vary over time as tectonic activity waxes and wanes, as normal faults grow and link with neighboring faults, or as extensional strain becomes preferentially accommodated along fewer and larger faults across a rift zone. In active rift systems, as late-stage extension is accommodated along fewer and larger faults, the amount of incremental coseismic slip increases as earthquake recurrence interval decreases. In other words, extension is accommodated on fewer but larger faults during fewer but larger earthquakes.

Transfer or accommodation zones represent special settings for syn-rift carbonate sedimentation. Transfer zones commonly are entry points for siliciclastic sediment into a rift network, so shallow-marine carbonate sedimentation may be rare there. If they are not siliciclastic entry points, however, and they are at appropriate water depths, carbonate platforms may build across a transfer zone. Ramp-like carbonate systems are likely to mimic the structural topography of relay ramps and may link steeper-sided platform profiles that were constructed on adjacent footwall highs. If the transfer zone is broken by numerous faults, however, then narrow, steep-sided or step-like platform margins may link platforms that were built on neighboring footwall highs. The platform margins that are constructed across accommodation and transfer zones are strongly controlled by displacement on these linking faults.

General Tectono-Stratigraphic Models for Syn-Rift Carbonate Platforms

The generalized tectono-stratigraphic models presented in this section are attempts to incorporate the effects of most factors that can influence stratigraphic development within syn-rift platforms. Infinite variations on these models are possible, but the platform models presented here emphasize the influence of syn-rift deformation on platform growth (Figs. 6–11). Accurate interpretation of actual growth stratal patterns within syn-rift platforms will also require independent, high-resolution data on eustatic fluctuations during platform growth.

In general, stratigraphies of syn-rift carbonate platforms begin with an initial transgressive succession, which consists of siliciclastic to mixed carbonate–siliciclastic facies that overlie a basal erosional unconformity. The basal unconformity surface develops across faulted pre-rift basement rocks that comprise the substrates upon which the platform is initiated. Basal siliciclastic facies are locally derived from the hanging wall and grade upward into progressively more carbonate-dominated facies as the remnant basement topography is flooded and onlapped by younger syn-rift strata. Basal sandstone facies may also form important aquifers for migration of petroleum into overlying carbonate-platform reservoirs.

Syn-rift isolated platforms that form on half-graben highs develop the characteristic stratal patterns that record hanging-wall rotation, namely recurrent onlap toward the footwall crest of the half-graben and divergent strata down the hanging-wall dip slope (cf. Cross et al., 1998). The point of stratal onlap may climb up the hanging-wall dip slope if this surface is gradually submerged over time (Figs. 6, 8, 9) or where tilting results in repeated footwall uplift. More complex patterns of onlap result where eustatic fluctuations interact with tectonic tilting. In contrast to siliciclastic systems, however, which generally show thickening across the entire hanging-wall depocenter, carbonate strata within syn-rift isolated platforms may thicken down the hanging-wall dip slope and then abruptly thin farther down the hanging-wall dip slope, well before reaching the adjacent bounding fault. This is because water depths in the hanging-wall depocenter may be too great for shallow-water carbonate deposition and the shallow-water factory is limited to those parts of the hanging-wall dip slope that are within photic depths (cf. Fig. 9). In addition, the flux of either platform-derived carbonate sediment or pelagic sediment is apparently too low to fill the adjacent depocenter and allow extensive progradation of the platform margin.

Syn-rift carbonate strata are also strongly influenced by paleowind direction. Windward–leeward effects influence facies locations and growth stratal patterns on half-graben highs, depending on where sea level intersects the half-graben surface and whether wave approach is toward the footwall crest or the hanging-wall dip slope (Fig. 6). Windward platform-margin facies tracts may have higher accumulation rates than inner-platform facies, so aggradational to backstepping platform-margin geometries may dominate hanging-wall dip slopes that face prevailing wind directions.

Syn-rift platforms that form on horsts will have growth stratal patterns and facies distributions different from those of half-graben isolated platforms (Figs. 10, 11). Opposite dips on horst-bounding faults do not lead to major rotation of the horst block unless the faults are active at different times for extended periods of time, perhaps on the order of a million years or more. That is, in order for growth strata to record alternate tilting of the entire horst, one bounding fault must be active for a million years or more while the other fault remains locked. This is probably a rare kinematic history for normal faults that bound individual horst blocks in most extensional settings.

Differential uplift of horst footwall highs might be possible, however, if the dips of opposing bounding faults have different steepness. More steeply dipping faults may show greater footwall uplift, whereas gently dipping bounding faults do not result in significant footwall uplift. The combined displacements on the two bounding faults may cause differential uplift and subsidence across the top of the horst (Figs. 10, 11).

Horst width is also important for recognizing internal growth stratal patterns. Isolated syn-rift platforms that grow on narrow horsts may have little internal growth stratal patterns. If the horst is over 30 km wide, however, the cumulative footwall uplift that is predicted for deformation around fault zones should be expressed as uplifted platform-margin facies that dip toward the platform interior. Syn-rift inner-platform facies should develop broad "sagging" geometries if opposing bounding faults are active at the same time. Strata along opposite platform margins should dip toward the platform interior and the effects of syn-rift footwall uplift should diminish in progressively younger strata as displacement on the bounding faults diminishes over time (Figs. 10, 11).

These basic styles of growth stratal patterns are further complicated if secondary antithetic or synthetic faults develop within the half-graben or horst (Figs. 10, 11). Secondary faults can develop at about the same time as the main bounding faults form, or they can develop much later, after the isolated syn-rift platforms on the basement highs are fully developed. The secondary faults may not be rooted at deep crustal levels, and they may be unrelated to syn-rift deformation. For example, secondary faults may develop along the margins of steep-sided isolated platforms due to gravitational failure along the margins and slopes of the platform. These secondary faults may have little or no relationship to faults rooted in basement rocks.

Incremental and cumulative displacement on bounding faults can also influence depositional facies and stratigraphic patterns in other ways. Seismic events can trigger gravitational failure and mass-wasting events along platform margins and slopes. Toe-of-slope deposits at the base of footwall fault scarps are progressively down-faulted and rotated toward the bounding fault as hanging-wall depocenters subside during the syn-rift phase. Long histories of syn-rift deformation can lead to stacked talus aprons at the foot of syn-rift platform margins with complex internal stratification patterns (Figs. 6–11). During tectonically quiescent intervals, when there is little active fault displacement, slope and platform-margin facies may prograde variable amounts across the hanging-wall depocenter. Progradation is more likely to occur where relief along the fault scarp and platform margin is relatively low and where the adjacent depocenter is nearly filled with sediment. Hanging-wall depocenters that are nearly filled decreases the amount of platform-derived carbonate sediment that is required to allow progradation of platform slopes (cf. Harris, 1991).

With regard to sequence stratigraphic relationships within syn-rift platforms, unconformity-bounded transgressive to early highstand systems tracts should dominate footwall highs, whereas late highstand to lowstand strata should be more common down hanging-wall dip slopes of half-graben elements (Fig. 8). Similar stratal patterns should characterize horst platforms, although lowstand systems tracts may not be well developed within platform interiors and are preferentially developed on leeward slopes (Fig. 10). Fault scarps are typically too steep for major accumulation of any carbonate sediment, except for thin veneers of sediment bound to the fault scarps by encrusting organisms.

Syn-Rift Unconformities

In practical terms, the onset of the syn-rift phase of basin evolution is usually interpreted to begin when the first brittle faults cut the upper crust. Local differential subsidence rates and rotation of basement blocks about subhorizontal poles of rotation are most rapid during this stage and exert strong controls on the areal extent of syn-rift unconformities. The areal extent of an unconformity across the top of a rotating fault block records the combined effects of eustatic sea-level fluctuations, block rotation, and flexural adjustments related to fault displacement, erosion of footwall highs, and sediment accumulation in hanging-wall depocenters.

For fully aggraded isolated carbonate platforms on a half-graben footwall high, the topology of the bounding faults and the incremental and cumulative displacements on those faults strongly influence the lateral extent of syn-rift unconformities (Figs. 8, 9). If eustatic sea level does not change during deformation, with each increment of displacement on a bounding fault, progressively more of the platform should become subaerially exposed, with the greatest uplift at the footwall crest. Uplift then progressively diminishes away from the footwall crest (Figs. 8, 9). Ultimately, the greatest lacuna should develop on the footwall crest.

Of course, fluctuations of eustatic sea level also influence the development of syn-rift unconformities in marine rift basins. Eustatic effects on unconformity development should be greater during global "icehouse" climatic conditions, when the amplitude of high-frequency, glacio-eustatic (i.e., Milankovitch time scales) sea-level changes may be > 100 m and rates of change are on the order of 10 m/kyr (cf. Read, 1995). Thus, during icehouse conditions, eustatic sea-level falls are likely to outpace any tectonically generated accommodation increase across the top of an aggraded platform and subaerial exposure surfaces develop. In contrast, high-frequency sea-level changes during "greenhouse" times are thought to be < 20 m, with rates of change on the order of 1 m/kyr (Read, 1995). Thus, rates of eustatic change during greenhouse times are comparable to average, short-term displacement rates on normal faults in modern rift systems (Leeder and Gawthorpe, 1987; Leeder, 1995) and fault-controlled increases in accommodation may match or outpace eustatic sea-level falls during greenhouse times. Subaerial exposure surfaces that develop across syn-rift carbonate platforms during greenhouse times might then be subdued or limited to footwall crests.

Interactions with Syn-Rift Siliciclastic Depositional Systems

Syn-rift carbonate platforms interact in complex ways with coeval siliciclastic systems depending on the sources and amounts of siliciclastic sediment supplied to a rift depocenter and the types of dispersal systems that distribute the sediment. Source areas for siliciclastic sediment include local fault scarps and footwall highs, regional rift-flank uplift, and hinterland fluvial drainage networks that are captured and diverted into the rift network. Fault scarps are not major sediment sources to rift depocenters, although progressive knick point erosion can create base-of-slope sediment aprons and wedges that might locally provide substrates for carbonate sedimentation. Footwall highs can be important source areas for siliciclastic sediment until they are covered by carbonate strata and are then no longer eroded by surface processes. Rift-flank uplifts are not prominent source areas for siliciclastic sediment during syn-rift stages unless drainage is somehow diverted around these highs and into the evolving rift network, which can occur in major accommodation zones. Similarly, hinterland drainage networks must also somehow be captured and diverted into the rift axis.

In general, then, siliciclastic systems have the most profound effects on syn-rift carbonate platform systems in rift depocenters that are located closer to the rift flank (Fig. 13). Progressive filling of fault-bounded rift depocenters by siliciclastic sediment first helps carbonate platforms to prograde into individual depocenters. Complex mixing of siliciclastic and carbonate facies follows in time, and ultimately, progradational siliciclastic systems can completely overwhelm any carbonate systems.

These patterns of progressive basin filling may be repeated as individual depocenters are filled farther offshore. Thus, carbonate platforms located closer to the rift flank generally are thinner and shorter-lived than more offshore platforms simply because they are overwhelmed sooner by prograding siliciclastic strata. In contrast, carbonate platforms located farther offshore are thicker and much longer lived because it may take many millions of years to fill up-dip depocenters and regional accommodation space.

SUMMARY

Hopefully, this review shows the complexity and significance of syn-rift carbonate platforms and the stratigraphic successions that are characteristic of these systems. Accurate stratigraphic analysis of syn-rift carbonate facies requires simultaneous, detailed analysis of local fault systems and regional tectonic history. Carbonate-platform facies also provide critical stratigraphic markers for reconstructing the history of base-level change across a rifted region.

Some important observations and conclusions about syn-rift carbonate platforms include:

(1) Syn-rift carbonate platforms typically form during later syn-rift stages, after significant fault linkage has already occurred, the amount of extension has progressed enough so that the evolving rift system can link to an adjacent ocean basin and becomes flooded with seawater, and fault-bounded depocenters are generally underfilled. Local underfilled depocenters "trap" siliciclastic sediment, allowing shallow-water carbonate platforms to grow on adjacent fault-bounded highs.

(2) Syn-rift carbonate platforms generally first develop within a rift system where greater amounts of stretching and related syn-rift tectonic subsidence has occurred. These areas of higher stretching are generally the first areas to be flooded by seawater as the evolving rift network connects to an adjacent ocean basin. Thus, it should be anticipated that the oldest, thickest, and longest-lived syn-rift carbonate platforms will be found where the amount of stretching is greatest. Siliciclastic strata also must first fill more updip depocenters before they can prograde to more highly stretched and more subsided regions, which allows carbonate platforms in the more highly stretched regions to grow without the deleterious effects of siliciclastic sediment.

(3) Syn-rift carbonate depositional systems typically form isolated platforms that are constructed on fault-bounded topography. Platforms may coalesce into larger composite platforms during later syn-rift stages, especially where initial horst or half-graben isolated platforms are closely spaced or where there is a large flux of siliciclastic sediment that fills fault-bounded depocenters, which allows carbonate-platform slopes to prograde and coalesce with neighboring platforms. Local pinnacle reefs may form on volcanic highs or in hanging-wall depocenters. Land-attached syn-rift platforms are rare in rift systems, perhaps because rift-flank uplift creates steep depositional gradients that prevent land-attached platforms from forming until the post-rift stage.

(4) Typical growth stratal patterns within syn-rift platforms on half-graben highs include increasing stratal dips downsection,

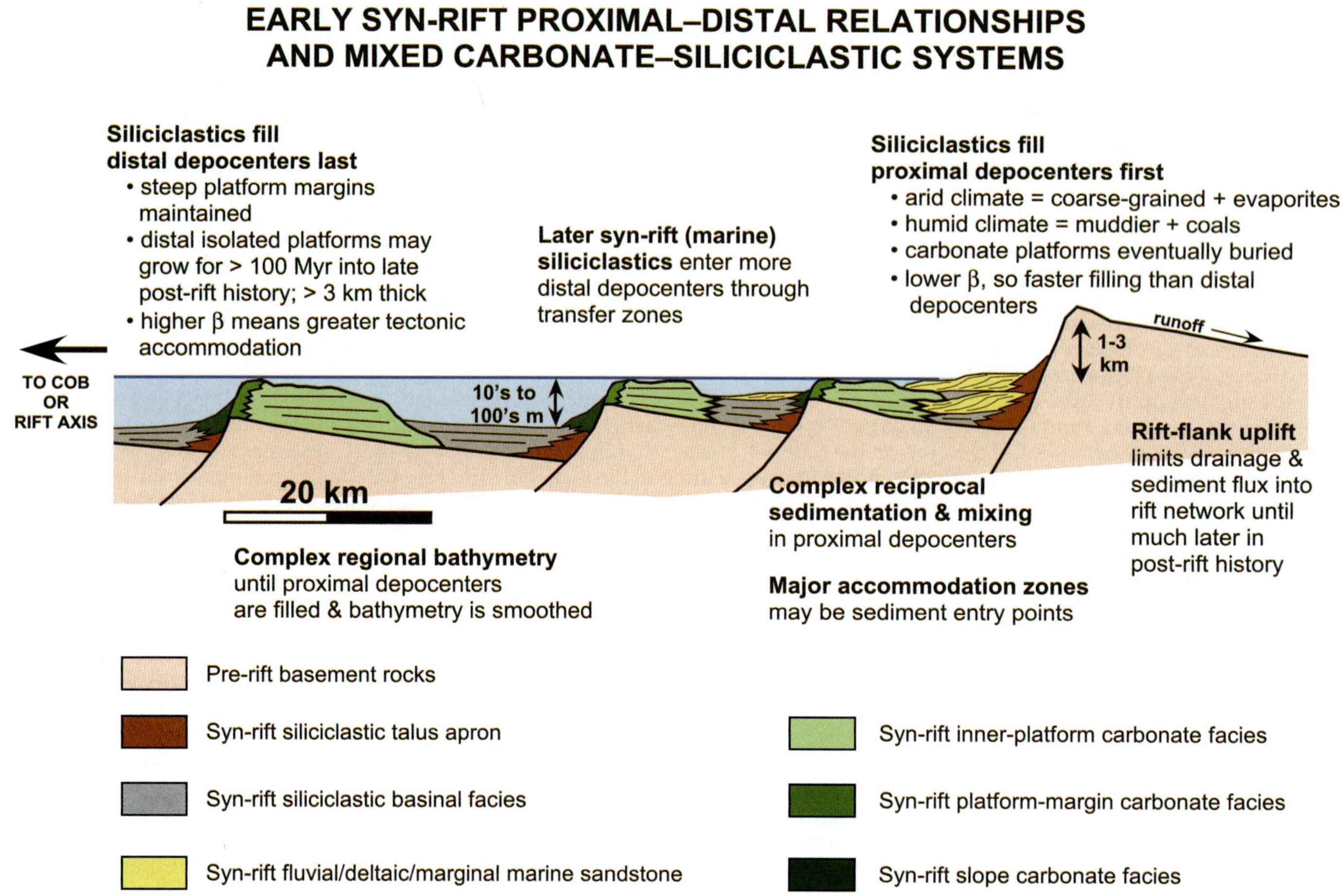

FIG. 13.—Schematic cross section showing generalized proximal-to-distal facies relationships for syn-rift carbonate–siliciclastic systems. Rift-flank uplifts may limit siliciclastic flux to rift depocenters until they have been eroded and no longer serve as topographic barriers. Local siliciclastic input along major accommodation or transfer zones may influence carbonate systems near the rift flank. Up-dip depocenters fill first, whereas more distal depocenters do not receive large volumes of siliciclastic sediment until later stages of basin evolution. Thus, up-dip platforms show complex patterns of reciprocal sedimentation, whereas more distal platforms form first, grow over longer periods of time, and may not interact with coarse-grained siliciclastic facies until late in their history.

greater unconformity development and related subaerial diagenesis near the footwall crest, complex onlap patterns toward the footwall crest, and platform-margin and slope facies that are cut by rift-related faults during deposition. It is often difficult to identify the relative influence of eustatic fluctuations on syn-rift carbonate stratigraphy, although the growth stratal patterns described here cannot be created without fault-block rotation or footwall uplift. It is important to recognize, however, that incremental coseismic surface deformations can accrue at time scales that are similar to glacio-eustatic fluctuations. This further emphasizes that 3D data and high-resolution age constraints are required for proper analysis of stratigraphic patterns within syn-rift carbonate platforms.

(5) Growth stratal patterns within syn-rift platforms on horsts depend on the spacing and topology of the horst's main bounding faults. If horst highs are more than about 30 km in width and bounding faults are steeply dipping, accrued footwall uplift creates syn-rift growth strata that are characterized by apparently uplifted platform margins and "sagged" inner-platform strata. These growth stratal patterns should become more subdued upsection. Where horsts are narrow (< 20 km wide) or where bounding faults have very shallow dips and flatten at shallow crustal levels, footwall uplift is minor and growth stratal patterns within inner platform strata are not obvious.

(6) Growth stratal patterns within segments of syn-rift platforms that develop above transfer zones are poorly understood, with few reliable observations from which to construct conceptual models. Analog rift models and detailed interpretation of 3D seismic data across transfer zones, however, suggest that carbonate-platform facies likely have platform-to-basin transitions that mimic the ramp-like to terraced topography of transfer zones.

With regards to sequence stratigraphic relationships, unconformity-bounded transgressive to early highstand systems tracts should dominate footwall highs, whereas late highstand to lowstand strata should be more common down hanging-wall dip slopes. Fault scarps are typically too steep for major accumulation of any carbonate sediment, except possibly for thin veneers of sediment bound to the fault scarps by encrusting organisms.

Paleowind direction is also an important control on stratigraphic development within syn-rift platforms. In general, higher-energy facies belts develop wherever rift-generated topography faces paleowind directions. Greater flux of platform-derived sediment is delivered to leeward slopes.

Interactions between syn-rift siliciclastic and carbonate depositional systems are more common along the margins of rift systems because lower amounts of stretching in these regions mean that there is less syn-rift subsidence and depocenters are filled more rapidly. Once updip depocenters are filled, siliciclastic facies can prograde toward the main rift axis and then interact with progressively more outboard carbonate-platform systems.

ACKNOWLEDGMENTS

Most of the research for this paper was completed while I was on leave from Texas A&M University during the 2000–2001 academic year. ExxonMobil Upstream Research Company supported the research during my leave and provided an excellent work environment. I am indebted to colleagues at EMURC, both past and present, but most notably Jim Markello and Randy Ewasko, who supported this research in every way possible, from start to finish. Seismic sections of syn-rift isolated platforms from offshore Vietnam were provided by TGS-NOPEC. Maersk Oil and Gas is gratefully acknowledged for providing funds to cover publication costs. Kate Giles and Finn Surlyk are thanked for providing useful, critical reviews that improved the manuscript. I also thank the editors of this volume, Jeff Lukasik and Toni Simo, for their patience and polite persistence during preparation of the manuscript.

REFERENCES

ACKERMANN, R.V., SCHLISCHE, R.W., AND WITHJACK, M.O., 2001, The geometric and statistical evolution of normal fault systems: an experimental study of the effects of mechanical layer thickness on scaling laws: Journal of Structural Geology, v. 23, p. 1803–1819.

ALLEN, P.A., AND ALLEN, J.R., 1990, Basin Analysis: Principles and Applications: Oxford, U.K., Blackwell Science, 451 p.

ALVAREZ, F., VIRIEUX, J., AND LEPICHON, X., 1984, Thermal consequences of lithosphere extension over continental margins: the initial stretching phase: Royal Astronomical Society, Geophysical Journal, v. 78, p. 389–411.

ANDERS, M.H., AND SCHLISCHE, R.W., 1994, Overlapping faults, intrabasin highs, and the growth of normal faults: Journal of Geology, v. 102, p. 165–180.

BARNETT, J.A.M., MORTIMER, J., RIPPON, J.H., WALSH, J.J., AND WATTERSON, J., 1987, Displacement geometry in the volume containing a single normal fault: American Association of Petroleum Geologists, Bulletin, v. 71, p. 925–937.

BEN-AVRAHAM, Z., HARTNADY, C.J.H., AND KITCHIN, K.A., 1997, Structure and tectonics of the Agulhas–Falkland fracture zone: Tectonophysics, v. 282, p. 83–98.

BOSENCE, D., 1998, Stratigraphic and sedimentological models of rift basins, *in* Purser, B.H., and Bosence, D.W.J., eds., Sedimentation and Tectonics of Rift Basins: Red Sea-Gulf of Aden: London, Chapman & Hall, p. 9–25.

BOSWORTH, W., 1985, Geometry of propagating continental rifts: Nature, v. 316, p. 625–627.

BOSWORTH, W., AND MCCLAY, K., 2001, Structural and stratigraphic evolution of the Gulf of Suez Rift, Egypt: A synthesis: Mémoires du Muséum National d'histoire Naturelle, Peri-Tethys Memoir 6: Peri-Tethyan Rift/Wrench Basins and Passive Margins, v. 186, p. 567–606.

BRAUN, J., AND BEAUMONT, C., 1989, A physical explanation of the relationship between flank uplifts and the breakup unconformity at rifted continental margins: Geology, v. 17, p. 760–764.

BURKE, K., 1977, Aulacogens and continental breakup: Annual Review of Earth and Planetary Sciences, v. 5, p. 371–396.

CHEN, W.P., AND MOLNAR, P., 1983, Focal depths of intracontinental and intraplate earthquakes and their implications for the thermal and mechanical properties of the lithosphere: Journal of Geophysical Research, v. 88, p. 183–4214.

COCHRAN, J.R., 1983, Effects of finite rifting times on the development of sedimentary basins: Earth and Planetary Science Letters, v. 66, p. 289–302.

CONTRERAS, J., AND SCHOLZ, C., 2001, Evolution of stratigraphic sequences in multisegmented continental rift basins: Comparison of computer models with the basins of the East African rift system: American Association of Petroleum Geologists, Bulletin, v. 85, p. 1565–1581.

CONTRERAS, J., SCHOLZ, C.H., AND KING, G.C.P., 1997, A model of rift basin evolution constrained by first-order stratigraphic observations: Journal of Geophysical Research, v. 102, p. 7673–7690.

COWIE, P.A., ATTAL, M., TUCKER, G.E., WHITTAKER, A.C., NAYLOR, M., GANAS, A., AND ROBERTS, G.P., 2006, Investigating the surface process response to fault interaction and linkage using a numerical modelling approach: Basin Research, v. 18, p. 231–266.

COWIE, P.A., GUPTA, S., AND DAWERS, N.H., 2000, Implications of fault array evolution for synrift depocentre development: insights from a numerical fault growth model: Basin Research, v. 12, p. 241–261.

COWIE, P.A., AND ROBERTS, G.P., 2001, Constraining slip rates and spacings for active normal faults: Journal of Structural Geology, v. 23, p. 1901–1915.

COWIE, P.A., AND SCHOLZ, C.H., 1992a, Displacement-length scaling relationship for faults: Data synthesis and discussion: Journal of Structural Geology, v. 14, p. 1149–1156.

CROSS, N.E., PURSER, B.H., AND BOSENCE, D.W.J., 1998, The tectono-sedimentary evolution of a rift margin carbonate platform: Abu Shaar, Gulf of Suez, Egypt, *in* Purser, B.H., and Bosence, D.W.J., eds., Sedimentation and Tectonics of Rift Basins: Red Sea-Gulf of Aden: London, Chapman & Hall, p. 271–295.

DAVISON, I., 1994, Linked fault systems; extensional, strike-slip and contractional, *in* Hancock, P.L., ed., Continental Deformation: Oxford, U.K., Pergamon Press, p. 121–142.

DAWERS, N.H., ANDERS, M.H., AND SCHOLZ, C.H., 1993, Fault length and displacement: scaling laws: Geology, v. 21, p. 1107–1110.

DOROBEK, S.L., 1995, Tectonic controls on carbonate platform evolution: Selected examples from the South China Sea region, *in* Ross, G.M., ed., University of British Columbia, 1995 Alberta Basement Transects Workshop, Lithoprobe Report #47, Lithoprobe Secretariat, p. 165–180.

ELLIS, P.G., AND MCCLAY, K.R., 1988, Listric extensional fault systems—results of analogue model experiments: Basin Research, v. 1, p. 55–70.

FALVEY, D.A., 1974, The development of continental margins in plate tectonic theory: Australian Petroleum Exploration Association, Journal, v. 14, p. 95–106.

FROSTICK, L.E., AND REID, I., 1989, Is structure the main control of river drainage and sedimentation in rifts?: Journal of African Earth Sciences, v. 8, p. 165–182.

GAWTHORPE, R.L., AND LEEDER, M.R., 2000, Tectono-sedimentary evolution of active extensional basins: Basin Research, v. 12, p. 195–218.

GAWTHORPE, R.L., SHARP, I., UNDERHILL, J.R., AND GUPTA, S., 1997, Linked sequence stratigraphic and structural evolution of propagating normal faults: Geology, v. 25, p. 795–798.

GILLESPIE, P.A., WALSH, J.J., AND WATTERSON, J., 1992, Limitations of dimension and displacement data from single faults and the consequences for data analysis and interpretation: Journal of Structural Geology, v. 14, p. 1157–1172.

Gouyet, S., Unternehr, P., and Mascle, A., 1994, The French Guyana margin and the Demerara Plateau: Geological history and petroleum plays, *in* Mascle, A., ed., Hydrocarbon and Petroleum Geology of France: European Association of Petroleum Geoscientists, Special Publication 4, p. 411–422.

Gupta, S., and Cowie, P., 2000, Processes and controls in the stratigraphic development of extensional basins: Basin Research, v. 12, p. 185–194.

Harris, M.T., 1991, Modeling carbonate progradation geometry and sediment accumulation rates: a comparison of "MARGIN" results with outcrop data, *in* Franseen, E.K., Watney, W.L., Kendall, C.G.St.C., and Ross, W.C., eds., Sedimentary Modeling; Computer Simulations and Methods for Improved Parameter Definition: Kansas Geological Survey, Bulletin 233, p. 509–517.

Jackson, J.A., 1987, Active normal faulting and crustal extension, *in* Coward, M.P., Dewey, J.F., and Hancock, P.L., eds., Continental Extensional Tectonics: Geological Society of London, Special Publication 28, p. 3–18.

Jackson, J.A., and McKenzie, D.P., 1983, Rates of active deformation in the Aegean Sea and surrounding areas: Basin Research, v. 1, p. 121–128.

Jackson, J.A., and White, N.J., 1989, Normal faulting in the upper continental crust: observations from regions of active extension: Journal of Structural Geology, v. 11, p. 15–36.

Jankhof, K., 1945, Changes in ground level produced by the earthquakes of April 14 to 18, 1928 in southern Bulgaria, *in* Treblements de Terre en Bulgarie, Nos. 29-31: Institut Méterologique Central de Bulgaria, Sofia, p. 131–136.

Jarvis, G.T., 1984, An extensional model of graben subsidence—the first stage of basin evolution: Sedimentary Geology, v. 40, p. 13–31.

Jarvis, G.T., and McKenzie, D.P., 1980, Sedimentary basin formation with finite extension rates: Earth and Planetary Science Letters, v. 48, p. 42–52.

King, G.C.P., Stein, R.S., and Rundle, J.B., 1988, The growth of geological structures by repeated earthquakes, 1. Conceptual framework: Journal of Geophysical Research, v. 93, p. 13,307–13,318.

Kinsman, D.J.J., 1975, Rift valley basins and sedimentary history of trailing continental margins, *in* Fischer, A.G., and Judson, S., eds., Petroleum and Global Tectonics: Princeton, New Jersey, Princeton University Press, p. 83–126.

Klitgord, K.D., Hutchinson, D.R., and Schouten, H., 1988, U.S. Atlantic continental margin : structural and tectonic framework, *in* Sheridan, R.E., and Grow, J.A., eds., The Geology of North America, v. I-2, The Atlantic Continental Margin, U.S.: Geological Society of America, p. 19–56.

Kusznir, N.J., Marsden, G., and Egan, S.S., 1991, A flexural-cantilever simple-shear / pure-shear model of continental lithosphere extension: applications to the Jeanne d'Arc Basin, Grand Banks and Viking Graben, North Sea, *in* Roberts, A.M., Yielding, G., and Freeman, B., eds., The Geometry of Normal Faults: Geological Society of America, Special Publication 56, p. 41–60.

Lambiase, J.J., and Bosworth, W., 1995, Structural controls on sedimentation in continental rifts, *in* Lambiase, J.J., ed., Hydrocarbon Habitat in Rift Basins: Geological Society of America, Special Publication 80, p. 117–144.

Larsen, P.-H., 1988, Relay structures in a Lower Permian basement-involved extension system, East Greenland: Journal of Structural Geology, v. 10, p. 3–8.

Leeder, M.R., 1995, Continental rifts and proto-oceanic rift troughs, *in* Busby, C.J., and Ingersoll, R.V., eds., Tectonics of Sedimentary Basins: Oxford, U.K., Blackwell Science, p. 119–148.

Leeder, M.R., and Gawthorpe, R.L., 1987, Sedimentary models for extensional tilt-block / half-graben basins: Geological Society of London, Special Publication 28, p. 139–152.

Lister, G.S., and Davis, G.A., 1989, The origin of metamorphic core complexes: Journal of Structural Geology, v. 11, p. 65–94.

Lister, G.S., Etheridge, M.A., and Symonds, P.A., 1986, Detachment faulting and evolution of passive continental margins: Geology, v. 14, p. 246–250.

Marrett, R., and Allmendinger, R.W., 1991, Estimates of strain due to brittle faulting: sampling of fault populations: Journal of Structural Geology, v. 13, p. 735–738.

Matthews, S.J., Fraser, A.J., Lowe, S., Todd, S.P., and Peel, F.J., 1997, Structure, stratigraphy and petroleum geology of the SE Nam Con Son Basin, offshore Vietnam, *in* Fraser, A.J., Matthews, S.J., and Murphy, R.W., eds., Petroleum Geology of Southeast Asia: Geological Society of America, Special Publication 126, p. 89–106.

McClay, K.R., 1990a, Deformation mechanics in analogue models of extensional fault systems, *in* Knipe, R.J., and Rutter, E.H., eds., Deformation Mechanisms, Rheology and Tectonics: Geological Society of America, Special Publication 54, p. 445–453.

McClay, K.R., 1990b, Physical models of structural styles during extension, *in* Tankard, A.J., and Balkwill, H., eds., Tectonics and Stratigraphy of the North Atlantic Margins: American Association of Petroleum Geologists, Memoir 46, p. 341–344.

McClay, K.R., 1995a, 2D and 3D analogue modelling of extensional fault structures: templates for seismic interpretation: Petroleum Geoscience, v. 1, p. 163–178.

McClay, K.R., and Ellis, 1987a, Analogue models of extensional fault geometries, *in* Coward, M.P., Dewey, J.F., and Hancock, P.L., eds., Continental Extensional Tectonics: Geological Society of America, Special Publication 28, p. 109–125.

McClay, K.R., and Ellis, 1987b, Geometries of extensional fault systems developed in model experiments: Geology, v. 15, p. 341–344.

McClay, K.R., and White, M.J., 1995, Analogue modelling of orthogonal and oblique rifting: Marine and Petroleum Geology, v. 12, p. 137–151.

McKenzie, D.P., 1978, Some remarks on the development of sedimentary basins: Earth and Planetary Science Letters, v. 40, p. 25–32.

McLeod, A.E., Dawers, N.H., and Underhill, J.R., 2000, The propagation and linkage of normal faults: Insights from the Strathspey–Brent–Statfjord fault array, northern North Sea: Basin Research, v. 12, p. 263–284.

Molnar, P., and Lyon-Caen, H., 1988, Some simple physical aspects of the support, structure and evolution of mountain belts: Geological Society of America, Special Paper 218, p. 179–207.

Montadert, L., Roberts, D.G., de Chapral, O., and Guennoc, P., 1979, Rifting and subsidence of the northern continental margin of the Bay of Biscay: Initial Reports of the Deep Sea Drilling Project, v. 48, p. 1025–1060.

Morewood, N.C., and Roberts, G.P., 2002, Surface observations of active normal fault propagation: implications for growth: Geological Society of America, Journal, v. 159, p. 263–272.

Morley, C.K., 1995, Developments in the structural geology of rifts over the last decade and their impact on hydrocarbon exploration, *in* Lambiase, J.J., ed., Hydrocarbon Habitat in Rift Basins: Geological Society of America, Special Publication 80, p. 1–32.

Morley, C.K., 1999a, Influence of preexisting fabrics on rift structure, *in* Morley, C.K., ed., Geoscience of Rift Systems—Evolution of East Africa: American Association of Petroleum Geologists, Studies in Geology, no. 44, p. 151–160.

Morley, C.K., 1999b, ed., Comparison of hydrocarbon prospectivity in rift systems, *in* Geoscience of Rift Systems—Evolution of East Africa: American Association of Petroleum Geologists, Studies in Geology, No. 44, p. 233–242.

Myers, W.B., and Hamilton, W., 1964, Deformation accompanying the Hegben Lake earthquake of August 17, 1959: U.S. Geological Survey, Professional Paper 435, p. 55–98.

Nicol, A., Watterson, J.J., Walsh, J., and Childs, C., 1996, The shapes, major axis orientations and displacement patterns of fault surfaces: Journal of Structural Geology, v. 18, p. 235–248.

OLSEN, K.H., AND MORGAN, P., 1995, Introduction: Progress in understanding continental rifts, *in* Olsen, K.H., ed., Continental Rifts: Evolution, Structure, Tectonics: Amsterdam, Elsevier, Developments in Geotectonics 25, p. 3–26.

OMAR, G.I., AND STECKLER, M.S., 1995, Fission track evidence on the initial rifting of the Red Sea: two pulses, no propagation: Science, v. 270, p. 1341–1344.

PEACOCK, D.C.P., AND SANDERSON, D.J., 1991, Displacements, segment linkage and relay ramps in normal fault zones: Journal of Structural Geology, v. 13, p. 721–733.

RABINOWITZ, P.D., AND LABRECQUE, J., 1979, The Mesozoic South Atlantic Ocean and evolution of its continental margin: Journal of Geophysical Research, v. 84, p. 5973–6002.

RAVNÅS, R., AND BONDEVIK, K., 1997, Architecture and controls on Bathonian–Kimmeridgian shallow-marine synrift wedges of the Osegberg–Brage area, northern North Sea: Basin Research, v. 9, p. 197–228.

RICHTER, C.F., 1958, Elementary Seismology: San Francisco, W.H. Freeman, 768 p.

RIHM, R., AND HENKE, C.H., 1998, Geophysical studies on early tectonic controls on Red Sea rifting, opening and segmentation, *in* Purser, B.H., and Bosence, D.W.J., eds., Sedimentation and Tectonics of Rift Basins: Red Sea–Gulf of Aden: London, Chapman & Hall, p. 29–49.

RIPPON, J., 1985, Contoured patterns of the throw and hade of normal faults in the Coal Measures (Westphalian) of north-east Derbyshire: Yorkshire Geological Society, Proceedings, v. 45, p. 147–161.

ROBERTS, A., AND YIELDING, G., 1994, Continental extensional tectonics, *in* Hancock, P.L., ed., Continental Deformation: Oxford, U.K., Pergamon Press, p. 223–250.

ROBERTS, A.M., YIELDING, G., KUSZNIR, N.J., WALKER, I., AND DORN-LOPEZ, D., 1993, Mesozoic extension in the North Sea: constraints from flexural backstripping, forward modelling and fault populations, *in* Parker, J.R., ed., Petroleum Geology of Northwest Europe: 4th Conference: The Geological Society of London, Proceedings, p. 1123–1136.

ROSENDAHL, B.R., AND GROSCHEL, B.H., 2000, Architecture of the continental margin in the Gulf of Guinea as revealed by reprocessed deep-imaging seismic, *in* Mohriak, W.U., and Talwani, M., eds., Atlantic Rifts and Continental Margins: American Geophysical Union, Geophysical Monograph 115, p. 85–103.

ROYDEN, L., AND KEEN, C.E., 1980, Rifting processes and thermal evolution of the continental margin of eastern Canada determined from subsidence curves: Earth and Planetary Science Letters, v. 51, p. 343–361.

SALVESON, J.O., 1978, Variations in the geology of rift basins: A tectonic model (abstract): Rio Grande Symposium, Santa Fe, NM: Los Alamos National Laboratory, 7487C, Proceedings, p. 82–86.

SCHLISCHE, R.W., 1991, Half-graben basin filling models: New constraints on continental extensional basin development: Basin Research, v. 3, p. 123–141.

SCHLISCHE, R.W., AND ANDERS, M.H., 1996, Stratigraphic effects and tectonic implication of the growth of normal faults and extensional basins, *in* Beratan, K.K., ed., Reconstructing the History of Basin and Range Extension Using Sedimentology and Stratigraphy: Geological Society of America, Special Paper 303, p. 183–203.

SCHLISCHE, R.W., WITHJACK, M.O., AND EISENSTADT, G., 2002, An experimental study of the secondary deformation produced by oblique-slip normal faulting: American Association of Petroleum Geologists, Bulletin, v. 86, p. 885–906.

SCHLISCHE, R.W., YOUNG, S.S., ACKERMANN, R.V., AND GUPTA, A., 1996, Geometry and scaling relations of a population of very small rift-related normal faults: Geology, v. 24, p. 683–686.

SCHOLZ, C.H., AND CONTRERAS, J.C., 1998, Mechanics of continental rift architecture: Geology, v. 26, p. 967–970.

SCHOLZ, C.H., AND COWIE, P.A., 1990, Determination of total geologic strain from faulting: Nature, v. 346, p. 837–839.

SENGÖR, A.M.C., 1995, Sedimentation and tectonics of fossil rifts, *in* Busby, C.J., and Ingersoll, R.V., eds., Tectonics of Sedimentary Basins: Oxford, U.K., Blackwell Science, p. 53–117.

SONDER, L.J., ENGLAND, P.C., WERNICKE, B.P., AND CHRISTIANSEN, R.L., 1987, A physical model for Cenozoic extension of western North America, *in* Coward, M.P., Dewey, J.F., and Hancock, P.L., eds., Continental Extensional Tectonics: Geological Society of London, Special Publication 28, p. 187–202.

STEIN, R.S., AND BARRIENTOS, S.E., 1985, Planar high-angled faulting in the Basin and Range: geodetic analysis of the 1983 Borah Peak, Idaho earthquake: Journal of Geophysical Research, v. 90, p. 11,355–11,366.

STEIN, R.S., KING, G.C.P., AND RUNDLE, J.B., 1988, The growth of geological structures by repeated earthquakes, 2. Field examples of continental dip-slip faults: Journal of Geophysical Research, v. 93, p.

SURLYK, F., 1990, Timing, style and sedimentary evolution of Late Paleozoic–Mesozoic extensional basins of East Greenland, *in* Hardman, R.F.P., and Brooks, J., eds., Tectonic Events Responsible for Britain's Oil and Gas Reserves: Geological Society of London, Special Publication 55, p. 107–125.

VENDEVILLE, B., AND COBBOLD, P.R., 1988, How normal faulting and sedimentation interact to produce listric fault profiles and stratigraphic wedges: Journal of Structural Geology, v. 10, p. 649–659.

VENDEVILLE, B., COBBOLD, P.R., DAVY, P., BRUN, J.P. AND CHOUKROUNE, P., 1987, Physical models of extensional tectonics at various scales, *in* Coward, M.P., Dewey, J.F., and Hancock, P.L., eds., Continental Extensional Tectonics: Geological Society of America, Special Publication 28, p. 95–107.

WALSH, J.J., AND WATTERSON, J., 1987, Distribution of cumulative displacement and of seismic slip on a single normal fault surface: Journal of Structural Geology, v. 9, p. 1039–1046.

WALSH, J.J., AND WATTERSON, J., 1988, Analysis of the relationship between displacements and dimensions of faults: Journal of Structural Geology, v. 10, p. 239–247.

WALSH, J.J., AND WATTERSON, J., 1989, Displacement gradients on fault surfaces: Journal of Structural Geology, v. 11, p. 307–316.

WATTERSON, J., 1986, Fault dimensions, displacements and growth: Pure and Applied Geophysics, v. 124, p. 365–373.

WERNICKE, B., 1985, Uniform sense, normal simple shear of the continental lithosphere: Canadian Journal of Earth Sciences, v. 22, p. 108–125.

WHITTEN, C.A., 1957, Geodetic measurements in the Dixie Valley area: Seismological Society of America, Bulletin, v. 47, p. 321–325.

WINN, R.D., JR., CREVELLO, P.D., AND BOSWORTH, W., 2001, Lower Miocene Nukhul Formation, Gebel el Zeit, Egypt: Model for structural control on early synrift strata and reservoirs, Gulf of Suez: American Association of Petroleum Geologists, Bulletin, v. 85, p. 1871–1890.

WITHJACK, M.O., AND JAMISON, W.R., 1986, Deformation produced by oblique rifting: Tectonophysics, v. 126, p. 99–124.

WITHJACK, M.O., SCHLISCHE, R.W., AND OLSEN, P.E., 1998, Diachronous rifting, drifting, and inversion on the passive margin of central eastern North America: An analog for other passive margins: American Association of Petroleum Geologists, Bulletin, v. 82, p. 817–835.

TECTONO-SEDIMENTARY MODELS FOR RIFT-BASIN CARBONATE SYSTEMS

NIGEL E. CROSS
BG Group 100 Thames Valley Park Drive, Reading, Berkshire RG6 1PT
Present address: Rashid Petroleum Company, New Maadi, Cairo, Egypt
email: nigel.cross@hotmail.co.uk
AND
DAN W. J. BOSENCE
Geology Department, Royal Holloway University of London, Egham, Surrey, TW20 0EX, U.K.

Abstract: Detailed outcrop study of Miocene platform carbonates along the southwestern margin of the Gulf of Suez rift, together with an extensive review of comparable systems elsewhere, has generated new tectono-sedimentary models for marine rift-basin carbonate systems. Both syntectonic and post-tectonic platforms are described for various extensional settings in rift basins. Syntectonic platforms are defined as systems deposited during periods associated with active faulting in rift basins. Post-tectonic platforms are defined as carbonate systems deposited after periods of active extensional faulting that develop over preexisting, fault-generated rift topography. Carbonate systems that occupy transfer zones are also defined, which display the effects of active tectonism and intervening phases of tectonic quiescence. These platform types produce recurring tectono-sedimentological signatures recognizable throughout the stratigraphic record.

The footwall margins to syntectonic platforms generally develop through the tectonic modification of existing platforms and display features associated with relative sea-level fall such as emergence and, depending upon the prevailing climate, meteoric diagenesis. In addition, fault activity truncates platforms that have prograded across the fault line during preceding phases of tectonic quiescence. Hanging-wall dip-slope margins can be complex, inasmuch as they display the effects of hanging-wall subsidence and up-dip footwall uplift. Down-dip of the fault-block fulcrum, depositional sequences are more likely to be bounded by flooding surfaces and display retrogradational to backstepping depositional sequences, while up-dip of the fault-block fulcrum, sequences may form as a result of forced regression driven by relative sea-level fall. In rift-margin areas, clastic influx is likely to play a more important part in tectonically active hanging-wall dip-slope stratigraphies because of the rejuvenation of adjacent footwall highs.

Post-tectonic footwall platforms develop on the uplifted highs of fault blocks during periods of relative tectonic quiescence. Such areas reflect favorable shallow-water sites for carbonate production, because commonly they are isolated from significant clastic drainage fairways. The geometry of a footwall margin is highly variable, because it is inherited from the morphology of the faulted margin. Low-relief footwall margins produce progradational platforms that can bridge the block-bounding fault zone. High-relief blocks produce smaller, aggradational platforms in which the platform margin corresponds to the position of the previous fault scarp. Hanging-wall dip-slope margins commonly have large progradational geometries and typically evolve from ramp-type to rimmed-shelf morphologies through time. Platforms are more likely to form on rotated fault blocks that dip toward the rift margin or upon intrabasinal blocks that are set well away from rift-margin clastic input.

In areas of low clastic flux, transfer-zone platforms occur with either faulted (hard linkage) or low-relief and progradational (soft linkage) margins. Such platforms commonly pass laterally into rift-parallel footwall margins to produce complex platform-margin geometries.

Hydrocarbon reservoirs in platforms from these settings have a range of tectonic, sedimentary, and diagenetic controls, but the 3D models presented form templates that can be used as analogues for exploration and/or production from fault-block carbonate platforms.

The active (syntectonic) and passive (post-tectonic) platform models presented can be used at scales varying from individual depositional sequences to large-scale rift-basin stratigraphies characterized by episodes of tectonic activity and intervening periods of tectonic quiescence.

INTRODUCTION

Carbonate platforms are commonly developed on passive continental margins, where they represent some of the largest and most widespread sedimentary units known (e.g., Jansa, 1981; Read, 1982; Gamboa et al., 1985). They have yielded significant amounts of the world's hydrocarbon production (e.g., Tethyan carbonates of the Middle East; Alsharan and Nairn, 1997) and continue to hold major reserves. The study of carbonate platforms on rifted continental margins is therefore of importance because they represent the foundations on which later passive-margin platforms develop. Fault-block carbonate platforms also represent important hydrocarbon producing reservoirs themselves (e.g., Erlich et al., 1990; Erlich et al., 1993; Grötsch and Mercadier, 1999; Fournier et al., 2005). The very nature of passive-margin systems also means that rift-related carbonate platforms are less well exposed or more poorly imaged in seismic data because of subsequent burial under passive-margin stratigraphies.

Although the development of carbonate platforms in extensional settings has long been recognized, little in-depth information has been available on the distribution and evolution of platform facies and depositional sequences. In many cases the absence of detailed descriptions means that it is not possible to differentiate between carbonate systems that are truly syntectonic from those that are founded on existing, rift-generated, antecedent topography and therefore are largely post-tectonic. We use the term fault-block carbonate platform (cf. Bosence, 2005) here to describe a carbonate platform or buildup whose geometry and facies distribution is controlled by an underlying extensional fault block.

To date, the description of fault-block carbonate platforms has been linked to a few selected areas, most notably the Dinantian of northern England (e.g., Gawthorpe, 1986, 1987; Gutteridge, 1987;

Controls on Carbonate Platform and Reef Development
SEPM Special Publication No. 89, Copyright © 2008
SEPM (Society for Sedimentary Geology), ISBN 978-1-56576-130-8, p. 83–105.

Grayson and Oldham, 1987; Gawthorpe et al., 1989; Fraser and Gawthorpe, 1990; Ebdon et al., 1990; Collier, 1991) and Ireland (e.g., Somerville et al., 1992; Strogen et al., 1995; Pickard et al., 1994), the Jurassic of the Tethyan province (e.g., Bernoulli and Jenkyns, 1974; Bosellini et al., 1981; Winterer and Bosellini, 1981; Santantonio, 1994; Ruiz-Ortiz et al., 2004), the Cretaceous of northern Spain (e.g., Reitner, 1986; Vera et al., 1988; Garcia-Mondejar, 1990; Rosales et al., 1994), the Tertiary of southeast Asia (van de Weerd and Armin, 1992; Wilson and Bosence, 1996; Wilson, 2000; Wilson et al., 2000) and the Miocene of the Gulf of Suez and the Red Sea rift (Burchette, 1988; James et al., 1988; Cross et al., 1998; Purser et al., 1998).

The tectonic controls on fault-block carbonate platforms have been demonstrated in a number of tectonically well-constrained settings ranging from the Cambrian (Bechstedt and Boni, 1989) to the present day (Gulf of Suez, Burchette, 1988, Roberts and Murray, 1988; Southeast Asia, Wilson et al., 2000). At a larger scale, it is often implied that many of the major present-day carbonate provinces in the world have their foundations upon rifted basement or individual fault blocks (e.g., Bahamas, Belize, and Great Barrier Reef, Australia; Enos et al., 1979; Sheriden et al., 1988; Mullins and Lynts, 1977; Mullins, 1983; Davies et al., 1989).

Fault-block carbonate platforms exhibit a range of scales from isolated footwall buildups, associated with synsedimentary faults (e.g., The Permian of the Barents Shelf, Gérard and Buhrig, 1990), to extensive, and sometimes amalgamated, carbonate platforms covering hundreds of square kilometers (South China Sea, Fulthorpe and Schlanger, 1989). Their importance as successful petroleum plays is highlighted by a number of published examples (van de Weerd and Armin, 1992; Montgomery, 1996; Mayall et al., 1997; Grötsch and Mercadier, 1999).

Leeder and Gawthorpe (1987) originally presented three-dimensional tectono-sedimentary facies models for extensional fault blocks / half graben. Their coastal / shelf basins with carbonate facies model ("model D") outlines the distribution of carbonates facies expected in an active rift system which is both in a low-latitude region and where clastic input is low. This model has served as a basic template for the study of carbonate facies in extensional basins, and some aspects were further developed to illustrate prerift, synrift, and postrift settings by Bosence (1998). Complementary tectono-sedimentary models for rifted carbonate platforms in deep-water pelagic settings have been discussed by Santantonio (1994). In a recent review of carbonate platforms from various tectonic settings (Bosence, 2005) fault-block platforms are reviewed as one specific class of carbonate platform. This paper expands on those earlier studies by integrating three-dimensional outcrop data from the Gulf of Suez rift (Cross, 1996; Cross et al., 1998) with other published examples of rift-related carbonate platforms in order to generate models that explain the diversity of stratigraphies that are developed in these settings. The focus in this paper is the three-dimensional geometry of rift-basin platforms that develop in response to fault-block rotation. These 3D qualitative models aim to facilitate the prediction of facies in this type of platform both in poorly exposed basins and in the subsurface. Despite the range of platform geometries and facies displayed by rift-basin fault-block platforms, we demonstrate reoccurring tectono-sedimentological signatures that can be related to the timing of fault movement in relation to the development of the carbonate stratigraphy; this allows differentiation of syntectonic and post-tectonic platforms. Syntectonic platforms are defined as systems deposited during periods associated with active faulting in rift basins. Post-tectonic platforms are defined as cabonate systems deposited after periods of active extensional faulting that develop over preexisting fault-generated rift topography. A model to account for carbonate systems developed in transfer zones is also presented.

It should be emphasized that the models presented focus on shallow-marine rift basins and in particular those that occur at low latitudes in a position favorable for tropical carbonate production. We are not aware of any temperate climate platforms formed in rift settings, and we suspect that this relates to a combination of the higher rates of clastic supply in these settings coupled with lower rates of carbonate production. The principal controls on the development of carbonate systems in rift-basin settings is discussed later in this paper.

The Gulf of Suez Rift

The uplifted and well exposed Upper Oligocene to Middle Miocene synrift stratigraphy of the failed Gulf of Suez rift is an excellent setting for the study of tectono-sedimentological relationships (Fig. 1; Garfunkel and Bartov, 1977; Patton et al., 1994; Steckler et al., 1988). In many cases, depositional geometries within the rift stratigraphy are well preserved (e.g., Purser et al., 1998; Cross et al., 1998; Young et al., 2000).

Structurally, the Gulf of Suez rift is characterized by northwest–southeast-trending extensional faults that bound rotated fault blocks (Fig. 1; Colletta et al., 1988; Bosworth, 1995). In the southwestern part of the Gulf of Suez rift these fault blocks are overlain by an early to late rift (Aquitanian to Langhian) stratigraphy, comprising a mixed carbonate–clastic marine to marginal marine succession that includes fault-block carbonate platforms (James et al., 1988; Burchette, 1988; Evans, 1988; Richardson and Arthur, 1988; Bosworth et al., 1998; Purser et al., 1998; Cross et al., 1998). Along the margins of the Gulf of Suez rift, where there is such a close interplay between clastic and carbonate regimes, clastic deposition dominates during the main rift phase when fault activity and subsidence was at a maximum (e.g., the Rudeis Formation). Once efficient drainage systems are established, topographic relief at the rift margin produces clastic influx and systems which display a strong degree of tectonic control (e.g., Gupta et al., 1999; Sharp et al., 2000a; Sharp et al., 2000b; Young et al., 2000). Carbonates are present during this early phase of rifting but are short-lived and spatially restricted (e.g., Gawthorpe et al., 1990; Gupta et al., 1999; Winn et al., 2001), In the late rift phase of the Gulf of Suez (Serravalian), fault-driven subsidence slowed, particularly along the rift margins, and became more focused towards the rift axis. At this time, fault blocks like the Esh el Mellaha Fault Block became less active and provided ideal shallow marine substrates for the development of carbonate complexes like the Abu Shaar Platform. Although platforms such as Abu Shaar display unequivocal evidence for synsedimentary tectonism and therefore an active tectonic control, such events appear to be relatively short-lived within a largely quiescent phase of late rift-basin development. Reefs and associated carbonate facies continue to play an important role in the present-day depositional system. The partially submerged rotated fault block of Gebel Zeit, 40 km north of the Abu Shaar Platform, accumulates mixed carbonate and clastic sediments, which are passively influenced by the position of this fault-block (Burchette, 1988; Roberts and Murray, 1988). Offshore, in the tectonically active central part of the Gulf of Suez rift, small carbonate platforms also develop on the present-day uplifted footwall of the Zeit–Shadwan Fault Block (Roberts and Murray, 1988; Jackson et al. 1988).

THE ESH EL MELLAHA FAULT BLOCK AND THE ABU SHAAR PLATFORM

Gebel Esh el Mellaha is the uplifted, footwall edge of a large synthetically (southwesterly) rotated fault block developed along the southwest margin of the Gulf of Suez rift (Figs. 1, 2). The

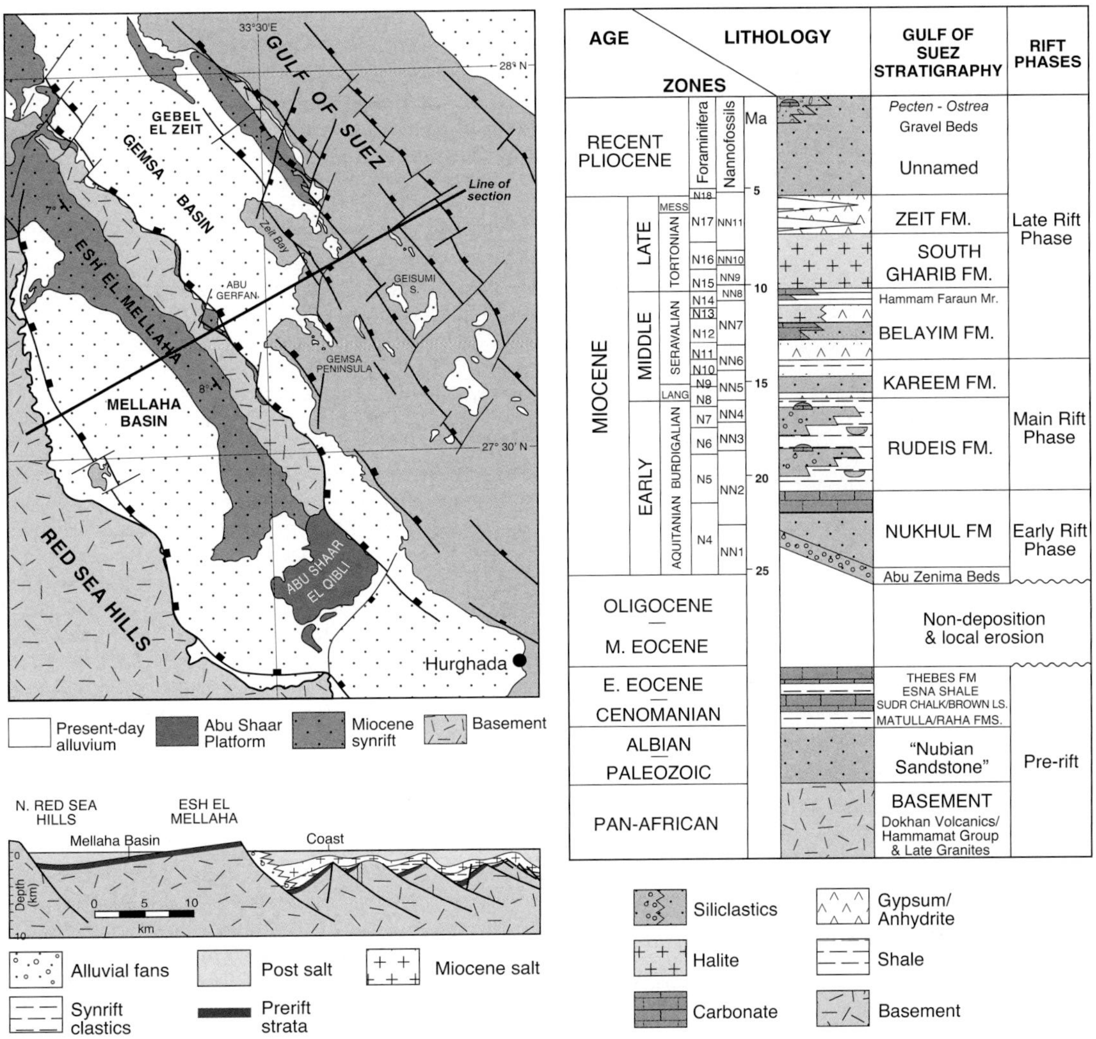

FIG. 1.—Structural location map and cross section of the Esh el Mellaha Fault Block and a schematic stratigraphy of the Gulf of Suez rift (modified after Bosworth, 1995). The Abu Shaar Platform is probably part of the Serravalian Belayim Formation (Bosworth et al., 1998).

northeast escarpment of the fault block represents a retreated fault-line scarp that borders and dips (60°) into the Gemsa half graben to the northeast. Fault displacement along this block-bounding fault is approximately 4–6 km (Bosworth, 1995). To the southwest the Esh el Mellaha Fault Block dips into the subsurface with dips of less than 5° and passes into the Esh el Mellaha half graben. The southwestern limit is marked by the Red Sea Hills, which represent the position of the rift border fault and the rift shoulder. Displacement on the rift border fault is approximately 3 km (Bosworth, 1995). Towards the southeast the Esh el Mellaha block loses relief with reduced fault displacement and is onlapped by Middle Miocene carbonate and clastic deposits of the Abu Shaar Platform (Figs. 2, 3). Along the northeast margin of the Esh el Mellaha Fault Block, several northeast–southwest to north–south oriented jogs in the fault-line scarp suggest that its broad rift-parallel trend is complicated by local cross faults (Fig. 1). This southeast end of the Esh el Mellaha Fault Block is delineated by a more significant cross-fault where fault displacement is transferred across to the Red Sea Hills border fault (Fig. 1).

The Abu Shaar Platform is an exceptionally well-exposed carbonate and mixed carbonate–clastic platform situated on the footwall and hanging-wall dip slope of the Esh el Mellaha Fault Block. The carbonate facies are now all preserved as a fine texture-preserving dolomite (James et al., 1988). In the Abu Shaar area the underlying fault block consists entirely of Precambrian igneous and metamorphic basement. Pre-rift strata have been eroded prior to Middle Miocene platform deposition, although it outcrops several kilometers along strike on the hanging-wall dip slope of the Esh el Mellaha Fault Block (Fig 1). 3D exposure of the Abu Shaar Platform is facilitated by a number of deeply incised wadis that dissect the platform perpendicular to the rift axis (Fig. 2). These wadis provide closely spaced sections through the footwall of the platform. At the northeast end of the Abu Shaar Platform a major wadi system (Wadi Bali'h) provides a single section through part of the hanging-wall dip slope of the fault block where platform strata onlap the underlying fault block in a northeast direction. The facies and sequence interpretation of the Abu Shaar Platform is aided by the fact that it is a passively exhumed platform so that its present-day tectono-sedimentary geometry preserves its original Miocene setting and morphology. Detailed sedimentological and sequence stratigraphical interpretations of the Abu Shaar Platform have been published by

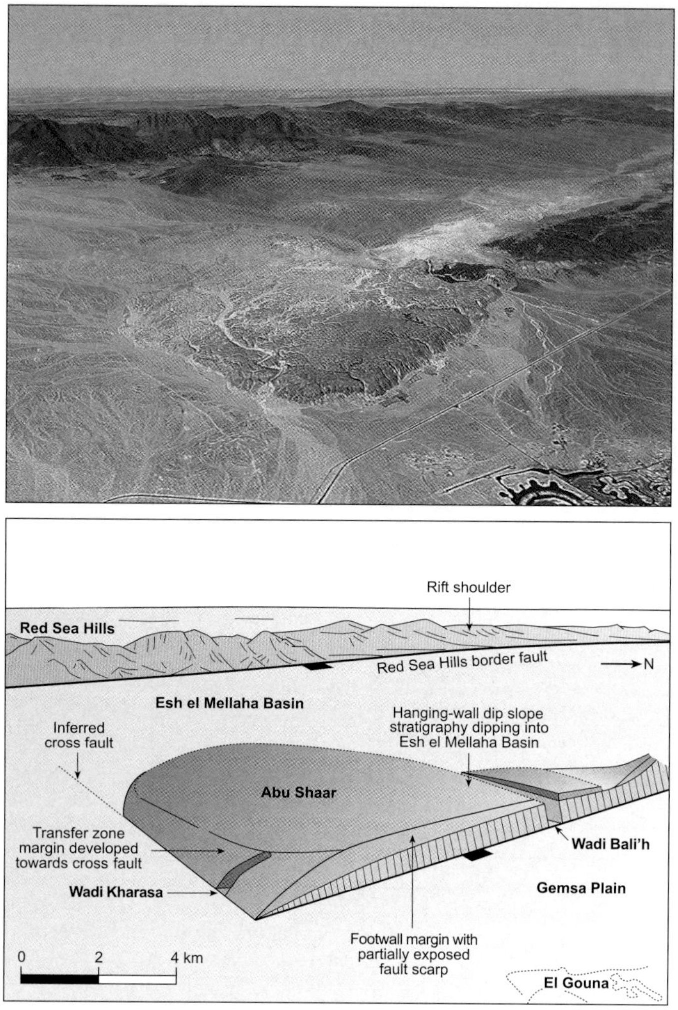

FIG. 2.—Oblique satellite image (from Google Earth™) looking west-northwest, and schematic block diagram of the southeastern end of the Esh el Mellaha Fault Block, illustrating the tectono-sedimentary context of the Abu Shaar Platform.

Burchette (1988), James et al. (1988), and Cross et al. (1998). Here we highlight its 3D morphology and facies and the contrasting tectono-sedimentological relationships that can be derived from different margins of the fault block. Abu Shaar probably represents the best-exposed fault-block carbonate platform in the world (Fig. 2).

Sedimentology and Stratigraphy

The Abu Shaar Platform stratigraphy was divided into three informal members by James et al. (1988) (Kharasa, Esh el Mellaha, and Bali'h members), which are broadly consistent with the three sequences of Burchette (1988). This stratigraphy was further divided by Cross et al. (1998), who defined nine depositional sequences (DS1 to DS9; Fig 4). DS 1–3 correspond to Sequence 1 of Burchette (1988) and the Kharasa member of James et al. (1988), DS 4–6 correspond to Sequence 2 of Burchette (1988) and the Esh el Mellaha member of James et al. (1988), and DS 7–9 correspond to Sequence 3 of Burchette (1988) and the Bali'h member of James et al. (1988). For a comparison of the detailed correlations between these three stratigraphic schemes, see Cross et al. (1998). At the broadest scale the stratigraphy of the Abu Shaar Platform initially onlaps the rifted basement and then shows an overall regressional succession of reefal to peritidal carbonates that also become progressively more clastic on the hanging-wall dip slope of the fault block towards the southwest (Fig. 4). The stratigraphy and sedimentology of the Abu Shaar Platform can be best considered in terms of the following tectono-sedimentary settings:

Footwall Margin.—

The northeast footwall margin to the Abu Shaar Platform is characterized by a relatively thin (0–100 m) carbonate succession (Fig. 4). It comprises an open-marine facies assemblage of high-energy coralgal reefs with associated fore-reef talus on the northeastern slopes and passes into low-energy micritic platform-interior facies to the southwest (Figs. 4, 5). The subhorizontal platform-top strata onlap the footwall of the block from northeast to southwest and from southeast to northwest with a thin alluvial to beach conglomeratic fringe.

The base of DS 1 represents the major prerift–synrift unconformity between the platform and the underlying basement fault block (Figs. 3, 4, 6). The base of DS 2 is an inclined erosion surface that can be traced through the wadi sections along the footwall platform margin. The base of DS 3 is inclined away from the platform margin at a very shallow angle (1–2° towards the southwest) and implies a small amount of block rotation. DS 1–3 (Kharasa member of James et al., 1988) are largely aggradational platform-interior-dominated deposits, which are truncated by either the inclined erosion surfaces at the base of reefal wedges of DS 4 and 5 (Esh el Mellaha member of James et al., 1988; Sequence 2 of Burchette, 1998) or by the footwall-margin fault zone itself (Fig. 4). DS 4 and 5 are interpreted as proximal slope and reef-repair phases following episodes of platform-margin collapse and are local to the high-relief footwall platform margin (Fig 6A). These sequences are overlain by a platform-wide biostrome unit (DS 6) up to 3 m in thickness (Fig. 4).

The upper-platform stratigraphy of the footwall area of the Abu Shaar Platform (DS 7–9) comprises a variety of intertidal facies including stromatolites, thin low-diversity coral biostromes, micrites, and thin peloid to low-diversity skeletal grainstone sandsheets (Fig. 7). Along the footwall margin of the Abu Shaar Platform these facies comprise DS 7 to 9 and have a combined thickness of 10 m (Fig. 4).

Transfer-Zone Margin.—

To the southeast of the Esh el Mellaha Fault Block, fault displacement decreases significantly and extension is transferred across to the Red Sea Hill border fault via a northeast–southwest-trending cross fault, which downthrows to the southeast (Fig 2; Bosworth, 1995). It is difficult to constrain the amount of offset on this platform-bounding cross fault, because it is not exposed at the surface, but judging by historical 2D seismic data it is probably several hundred meters (Bill Bosworth, personal communication). Footwall platform margin deposits can be correlated through closely spaced wadi sections to Wadi Kharasa where a southeast-facing platform margin is well exposed (Fig. 6B). In this position on the fault-block platform, carbonates are considerably thicker (over 120 m thick) than at the footwall margin, display a progradational geometry with lower angles of slope than the footwall, and contain more coralgal boundstone facies (Figs. 5, 6).

In Wadi Kharasa, DS 1 is poorly exposed and truncated by the erosion surface at the base of DS 2, while DS 2 comprises a prograding reef unit, up to 50 m thick, which, unlike the footwall counterpart, shows no sign of truncation at the platform margin. Instead,

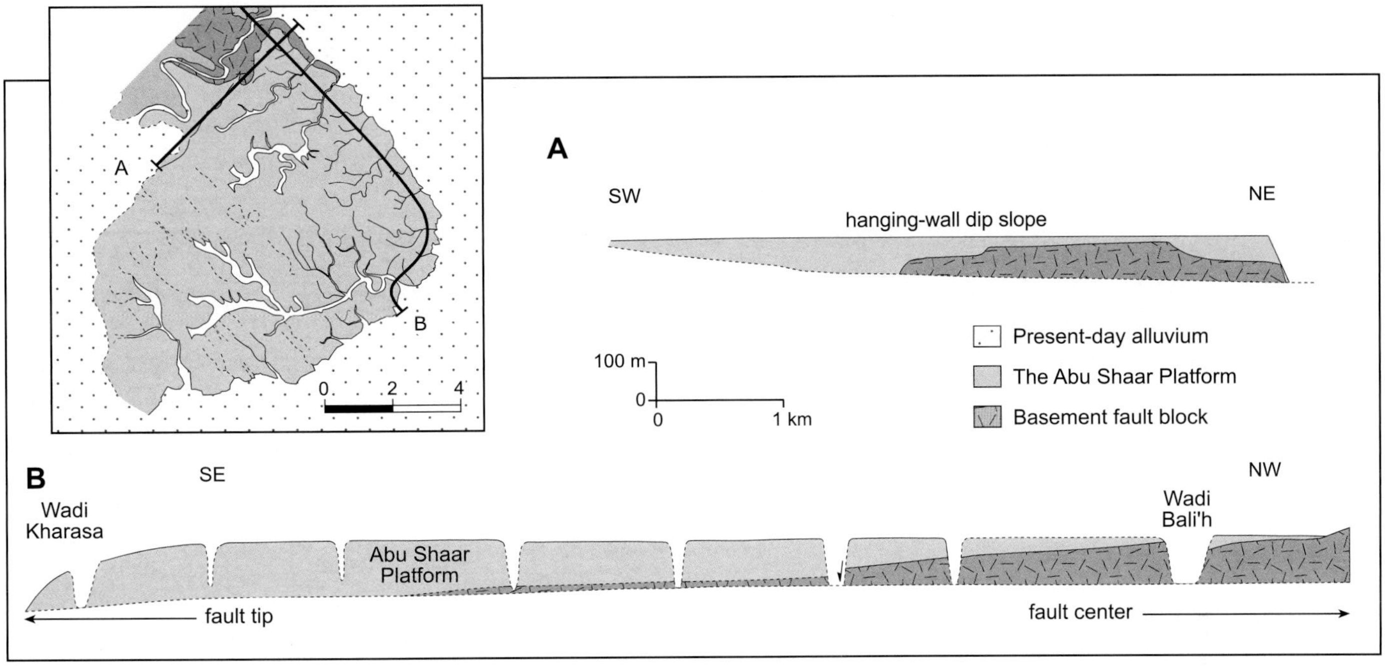

FIG. 3.—Map and schematic **A)** dip and **B)** strike cross sections of the Abu Shaar Platform, illustrating the overall platform geometry and its relationship with the underlying fault block.

it passes eastward and southeastward into well-developed progradational fore-reef talus slopes (Fig. 6B). There is no unequivocal evidence for a correlative equivalent of DS 3, or the reefal and slope wedges of DS 4 and 5. The reef biostrome unit of DS 6 is similar to DS 6 in the footwall margin, whereas DS 7, 8, and 9 are considerably thicker (up to 40 m) than their footwall equivalents.

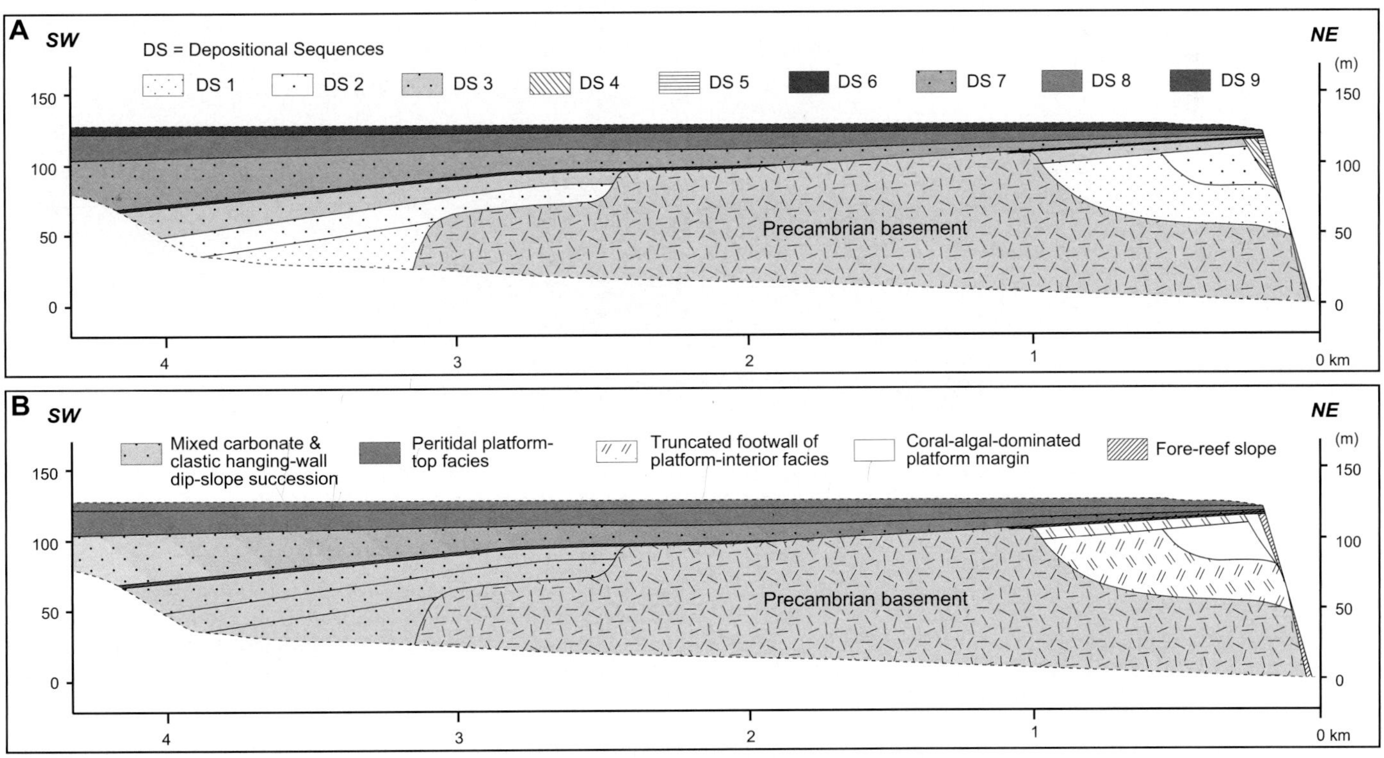

FIG. 4.—Schematic southwest–northeast cross sections of the Abu Shaar Platform, showing the distribution of **A)** depositional sequences and **B)** facies from the footwall of the Esh el Mellaha Fault-Block in the northeast to the hanging-wall dip slope in the southwest. (After Cross et al., 1998.)

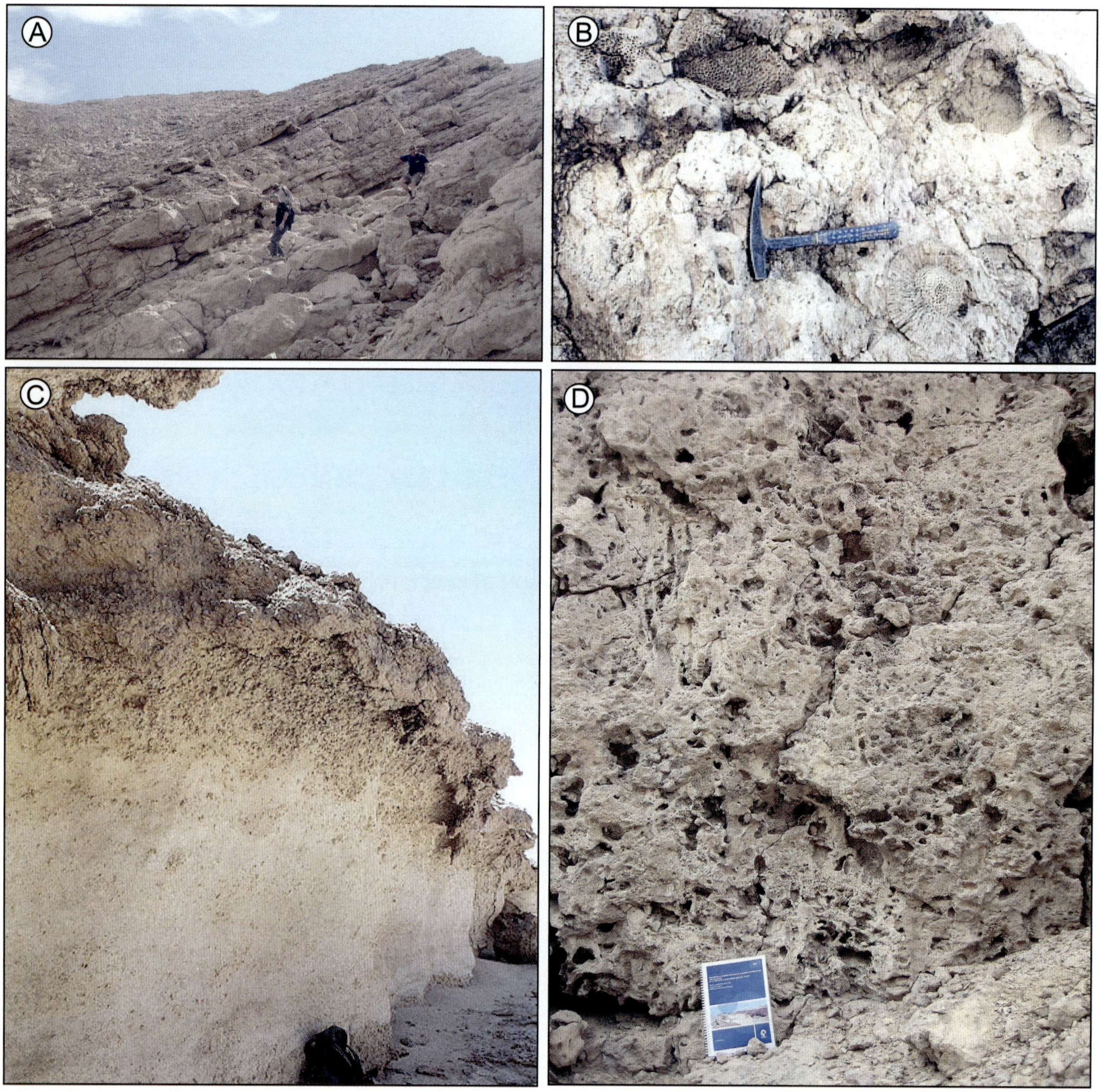

FIG. 5.—Selected facies from the footwall and transfer-zone platform margin of Abu Shaar. **A)** Steeply dipping fore-reef slope deposits comprising coarse-grained, redeposited coralgal material (DS 5). **B)** Represents a coarsening-upward parasequence of micritic algal wackestones to rudstones (DS 2; rucksack for scale) and **C)** *in situ* coral framestone with well developed *Porites* and faviids (DS 2; A4 field guide for scale).

Hanging-Wall Dip-Slope Margin.—

The hanging-wall dip slope of the Abu Shaar Platform is well exposed only in Wadi Bali'h, where it comprises over 120 m of mixed carbonate–clastic facies with a low-order regressive trend similar to that present along the footwall (Fig. 4). Platform strata display a largely aggradational geometry and onlap the inclined surface of the underlying fault block towards the northeast (Fig. 6C). Between the footwall and the hanging-wall dip-slope sites of the platform, direct correlation is hampered by high-relief basement outcrops, which represented fault-block islands that were onlapped during platform accumulation.

In detail the lower part of the platform stratigraphy (DS 1–3) comprises stacked units of both clastics-rich and carbonate-rich facies. Tabular sandbodies of planar cross-bedded arkoses are common and are cyclically interbedded with more carbonate-rich facies. They are made up of alternations of fine and coarse sand that are organized into foresets with a consistent dip orientation towards the southeast (Fig. 7). These arkoses are interpreted to be the products of tidal sandsheets that were reworked

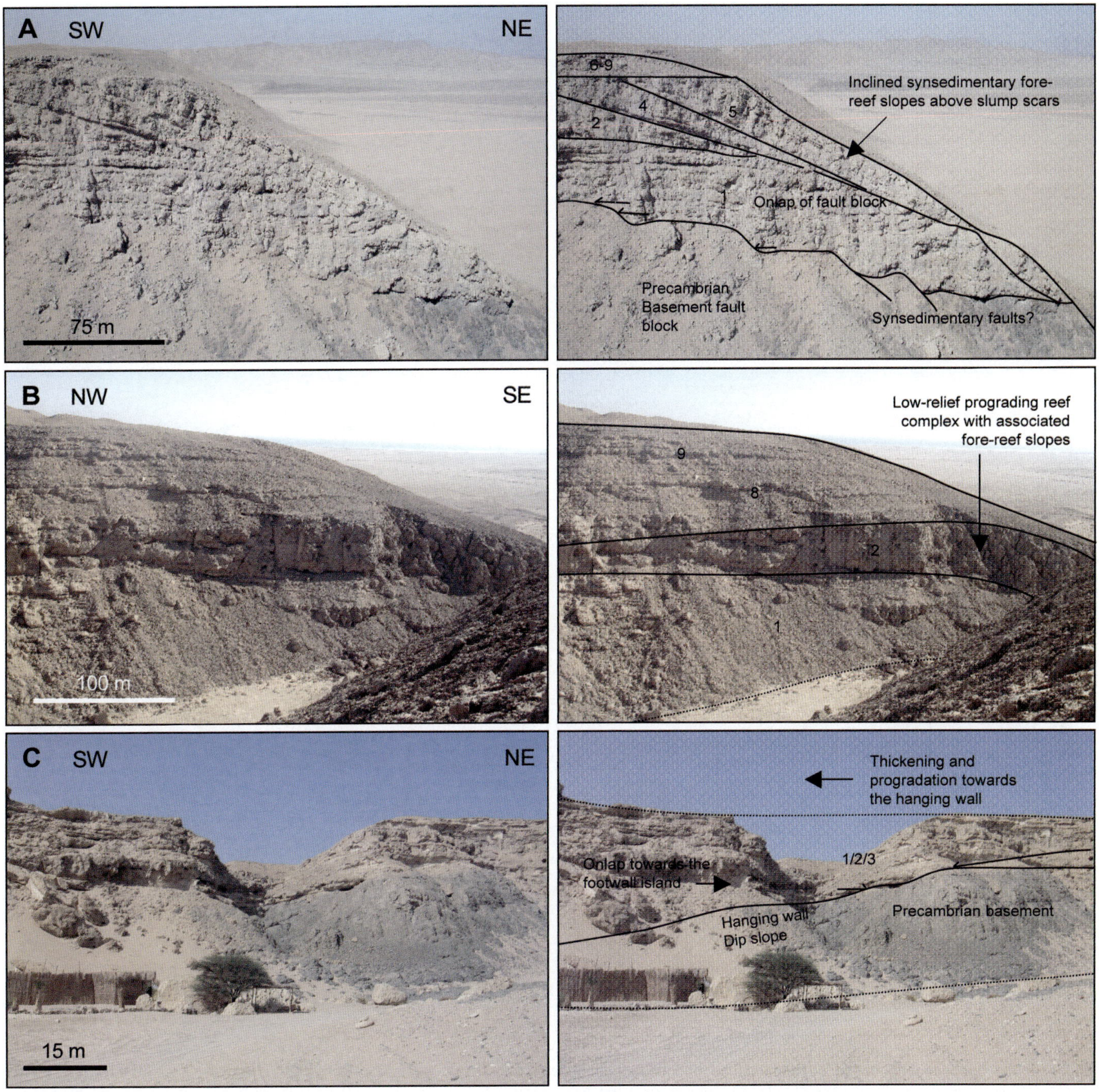

FIG. 6.—Three wadi sections through Abu Shaar Platform, highlighting the contrasting tectono-sedimentary settings present. **A)** A footwall section southeast of Wadi Bali'h through the platform margin, illustrating the wedge-shaped platform that onlaps the underlying fault block, together with the slump scars and high-relief margin. **B)** The transfer-zone margin exposed in Wadi Kharasa, illustrating the thick platform stratigraphy, low-relief and largely progradational platform margin developed in this setting. **C)** The platform interior of Wadi Bali'h, showing an aggradational-to-progradational mixed carbonate and clastic succession developed in the hanging-wall dip-slope setting. The numbers on the interpreted sections refer to the depositional sequences of Cross et al. (1998).

by currents oriented along strike and parallel to the rift axis. They are interbedded with more micritic intervals, also with a strong arkosic component, and are either sharp-based or organized into upward-shoaling cycles up to 4 m thick. Towards the top of DS 3, carbonate-rich facies become more dominant and several laterally extensive coralgal biostromes (up to 3 m thick) are developed (Fig. 7).

The lower platform section is capped by DS 6, a plateau-forming biostrome, which can be physically traced around the entire platform. Overlying this marker sequence, a subtidal sand-

FIG. 7.—Selected facies from the platform interior and hanging-wall dip slope of Abu Shaar. **A)** Stromatolites and **B)** trough cross-stratified, ooid–peloid grainstones from the upper, peritidal part of the platform stratigraphy (DS 8), whilst mixed carbonate and clastic facies are present in the hanging-wall dip slope; **C)** algal (rhodolith)-rich arkosic sandstone (DS 2), **D)** *Pecten*-rich arkosic sandstone (DS 3). **E)** Erosional contact between a coralgal biostrome dominated by faviids and coralline algae and an overlying alluvial basement-derived conglomerate. **F)** Cross-stratified arkose displaying tidally induced reactivation of cross beds (arrowed; pen for scale).

rich micrite dominates DS 7 and displays block-wide thickening from the footwall in the northeast to the hanging-wall dip-slope section in the southwest (Fig. 4). This sequence locally displays large-scale cross-beds oriented towards the northeast. Above, the intertidal deposits of DS 8 show a similar thickening towards the southwest indicating the infill of tectonically generated accommodation space following the rotation of the originally subhorizontal DS 6 biostrome.

Controls on the Development of the Abu Shaar Platform

The controls on the development of marine-rift-basin stratigraphies have been discussed at length in the literature (e.g., Ravnås and Steel, 1998; Gawthorpe and Leeder, 2000). These discussions focus primarily on the development of clastic systems within evolving rift basins, and whilst many of these factors are relevant to their carbonate counterparts, the environmentally sensitive nature of carbonate systems imparts significant variation. Rift-basin stratigraphies are controlled mainly by sediment supply (clastic input and carbonate production) in relation to accommodation changes driven by tectonism and sea-level changes. Carbonate stratigraphies are impacted strongly by climate and are also likely to exhibit the effects of local factors such as nutrient supply and water circulation. The following section highlights the principal controls that influenced the development of the Abu Shaar Platform and factors which influence the development of carbonate systems in rift basins generally.

Accommodation.—

Accommodation space for carbonate platform development in rift basins results from an interplay between tectonism, global sea-level changes, and sediment supply:

Tectonism.—Tectonically induced accommodation is a product of basin-wide subsidence and more local movement on individual faults. Vertical movement on normal faults produces coseismic hanging-wall subsidence and footwall uplift (Jackson and McKenzie, 1983; Stein and Barrientos, 1985), causing intervening fault blocks to rotate. Data from modern actively extending basins indicate that fault-related subsidence occurs as instantaneous seismic events, resulting in throws of up to about 5 m (Leeder, 1995). Slip rates averaged over thousands of years from the Gulf of Patras in Greece are 2–5 m/ky (Chronis et al., 1991), from Utah, U.S.A., are 2–5 m/ky (Machette et al., 1991), and from the Pisia fault, Greece, are 0.5 m/ka (Roberts et al., 1993; Machette et al., 1991). Uplift in the footwall immediately adjacent to extensional faults is around 10–15% of the total displacement (Jackson and McKenzie, 1983).

Tectonics played an important role in the evolution of the Abu Shaar Platform. On this platform, an active tectonic control on accommodation can be separated from post-tectonic effects whereby platform sequences are passively developed upon a preexisting tectonically generated topography:

- Synsedimentary fault-block rotation produces tilting and relative deepening of the hanging-wall dip slope down-dip of the rotational axis (fulcrum). This results in the generation of wedge-shaped accommodation space and stratigraphic fanning of dip-slope strata from the footwall towards the hanging-wall basin (Figs. 4, 8). Tilting of strata can be recognized across the dip slope of the Abu Shaar Platform and produces tectonically driven sequences and sequence boundaries (DS 3, 7, and 8) that do not appear to be present in correlative transfer-zone platform-margin strata. On the Abu Shaar Platform stratal dip towards the southwest is subtle and generally less than 2°, but it can be identified by the measuring of closely spaced vertical sections and through the use of panoramic sections, and on a kilometer scale they result in significant thickening.

- Fault-block rotation also produces spatial variation in the nature of bounding surfaces. Footwall uplift results in erosion of footwall platform sequences, which may become emergent and affected by meteoric diagenesis. These surfaces can be correlated down depositional dip to coeval flooding surfaces within dip-slope stratigraphies and do not have equivalents that can be readily identified in the transfer zone (e.g., the base of DS 7 on the Abu Shaar Platform; Fig. 8). This spatial variation in time-equivalent surfaces contrasts markedly with surfaces generated by more regional relative sea-level changes, which have a similar polarity across the platform and its margins (e.g., Pomar, 1991; see below).

- Fault displacement that takes place during platform evolution may result in the truncation of footwall sequences. On the Abu Shaar Platform, fault displacement along the edge of the footwall platform margin is demonstrated by DS 1 and 2 with platform-interior facies exposed along the platform margin. This contrasts with younger carbonate sediments that drape across these sequences and locally seal the basement fault plane. In isolation, features such as local slump scars along the footwall platform margin do not indicate a tectonically active margin, inasmuch as they also occur with sedimentary oversteepening on unfaulted margins (James and Ginsburg, 1979).

Sea-Level Changes.—Ravnås and Steel (1998) provide a comparison of rates of sea-level change versus fault displacement, and conclude that, with the exception of glacial periods, structurally driven accommodation changes are considerably higher than those generated by eustatic sea-level change. On the Abu Shaar Platform the correlation of erosion surfaces (e.g., the base of DS 2) between footwall and hanging-wall dip-slope settings of the same fault block suggest a more regional control on accommodation. Because the growth of the Abu Shaar Platform probably occurred during the latter stages of rifting (Serravalian), when fault-driven subsidence had waned significantly, it is probably not surprising to see the effects of more regional sea-level fluctuations recorded in the platform stratigraphy.

Carbonate Production and Clastic Supply.—The effect of rift-basin morphology, and in particular the complex controls offered by synsedimentary tectonism on clastic depositional systems, are well documented in the literature (e.g., Leeder and Gawthorpe, 1987; Dart et al., 1994; Gawthorpe and Hurst, 1993; Gawthorpe et al., 1994; Ravnås and Steel, 1998; Young et al., 2000, 2003; Sharp et al., 2000a; Sharp et al., 2000b; Gawthorpe and Leeder, 2000). Ravnås and Steel (1998) summarize a range of controls that affect sediment supply, including climatic factors, hinterland size and proximity (e.g., intra-basinal highs versus rift-margin source areas), and the nature of the pre-rift substrate.

On the hanging-wall dip slope of the Abu Shaar Platform, clastic supply was particularly important in controlling the nature of the platform facies and their geometries. In addition, the Esh el Mellaha Fault Block is attached to the rift margin and as a result was particularly susceptible to clastic influx during the Miocene. Clastic input to the Abu Shaar Platform was derived principally from the rift margin and axially along the Esh el Mellaha half graben. Minor amounts of sediment were also delivered from the footwall of the fault block itself.

The rate of carbonate production, particularly with respect to rates of fault movement, is a key controlling factor in rift-basin carbonate systems. At the broadest scale, maximum rates of present-day shallow-water carbonate production on carbonate platform margins are similar to those quoted for average normal-fault slips described above (1–4 m/ky; Bosence and Waltham, 1990; Bosence et al., 1998). However, production rates vary dramatically across carbonate platforms, with maximum rates attributed to reef-dominated margins. Carbonate production rates have also varied through geological time due to the evolution of carbonate-producing organisms (for summary, see Wright and Burchette, 1996).

Because the footwall carbonates of the Abu Shaar platform accumulated close to sea level, platform growth must have occurred during phases of tectonic quiescence, when accommodation reflected a balance between carbonate production and relative sea-level rise. Relative sea-level rise was probably driven by basin subsidence and/or eustasy. During episodes of faulting,

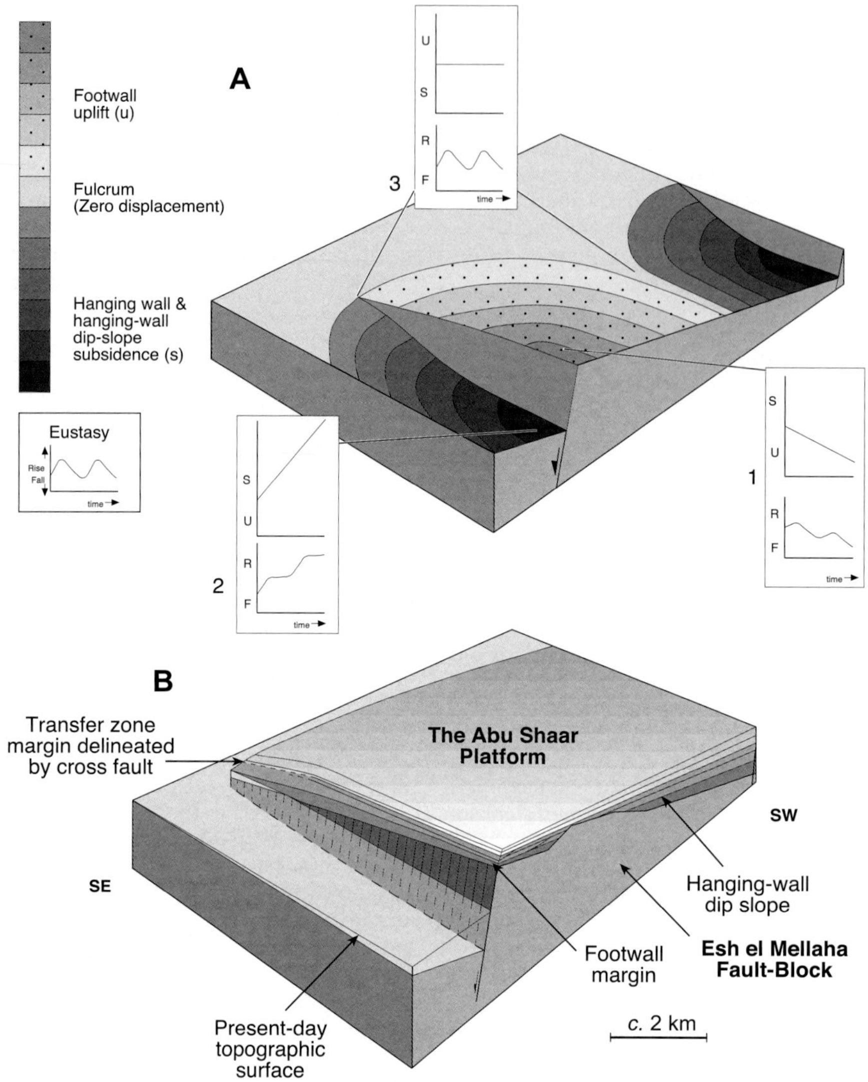

FIG. 8.— **A)** Schematic block diagram of the spatial variation in tectonically driven accommodation developed around a schematic extensional fault segment. The upper graph for locations 1 to 3 illustrates schematic variation in uplift (u) or subsidence (s) (modified from Gawthorpe et al., 1994). The lower graphs illustrate fault displacement superimposed on a third-order eustatic sea-level cycle, thus illustrating effective accommodation development. **B)** How the spatial variations in deformation associated with the Esh el Mellaha Fault Block produce 3D variations in footwall accommodation, which are manifested by changes in the geometry of platform sequences.

footwall uplift resulted in relative sea-level fall and would have caused at least the platform margin to become emergent and subject to erosion. More significant faulting induced platform-margin collapse. In the hanging-wall dip slope of the Abu Shaar Platform, fault-block rotation which generated accommodation down-dip of the fulcrum was filled by both carbonate and clastic sediments that were able to keep up with relative sea-level rise and produce a largely aggradational succession which infilled fault-block topography.

The coexistence of carbonate and clastic facies in the Abu Shaar Platform stratigraphy is primarily a function of the prevailing arid Miocene climate. Highly analogous facies relationships exist around the margins of the present-day Gulf of Suez (e.g., Roberts and Murray, 1988). In this setting the close proximity of carbonate and clastic facies is probably due to the ephemeral nature of clastic flux in such an arid environment. In addition, the rift-related topography of the basin produces high-relief hinterland areas adjacent to the main depocenters. The relatively short transport distances result in the deposition of mainly coarse-grained clastic sediment, which has a less detrimental effect on water clarity, nutrient input, and therefore efficiency of carbonate-producing organisms. The development of fault-block carbonate platforms in humid climates is more likely to be impacted by the perennial input of fine-grained clastic sediment, especially on fault blocks situated close to the rift margin. Many examples of fault-block carbonate platforms are described from the Tertiary of Southeast Asia, in which a more humid climate prevailed (e.g., Fulthorpe and Schlanger, 1989). Many of these examples occur on relatively isolated fault blocks where the effects of clastic input are more subdued or absent.

The nature of the pre-rift substrate on the Esh el Mellaha Fault Block also had an important role to play in the development of the

Abu Shaar Platform. Platform carbonates onlap crystalline Precambrian basement through an alluvial–beach fringe over a relatively short distance (10–20 m). Although basement-derived conglomerates and the onlapping of footwall islands indicate the basement to have been emergent during the growth of the Abu Shaar Platform, coarse-grained fan deposits are not a volumetrically significant component of the platform stratigraphy. The hard crystalline basement was reasonably durable to erosional processes, and as a result, relatively clastics-free conditions prevailed with carbonate production focused around the footwall high. The development of carbonate platforms on fault blocks where the prerift substrate is less crystalline or sedimentary in nature is likely to produce more clastic-prone platform stratigraphies.

Water Circulation.—Water circulation through both wind-driven and tidal currents produce some degree of lateral variability in most shallow marine carbonate systems (e.g., see examples in Harris and Kowalik, 1994). On the Abu Shaar Platform it is difficult to access the along-strike variability in facies because the platform has been truncated along much of its footwall margin and outcrop of the platform interior is limited. In the hanging-wall dip-slope section of the platform, cross-stratified arkoses testify to strong tidal currents oriented parallel to the rift structure. In this setting, axial currents were probably amplified by the narrow marine gulfs developed by the morphology of the underlying half graben (cf. Leeder and Gawthorpe, 1987; Gawthorpe and Leeder, 2000).

In the present-day Gulf of Suez, persistently strong winds (8–15 m/s) and accompanying waves are directed towards the southeast by the combined topographic steering of regional (shamal) winds, augmented by diurnal heating and cooling of the basin-facing mountain slopes (Roberts and Murray, 1988). Roberts and Murray (1988) describe the effects of these prevailing currents on the northwest–southeast-oriented Ashrafi Platform, a present-day syntectonic fault-block carbonate platform situated in the southern Gulf of Suez. Unlike the Abu Shaar Platform, this syntectonic footwall platform is situated in a more central part of the Gulf of Suez rift and therefore away from the effects of rift-margin clastic input. The Ashrafi Platform comprises a northwest windward margin that is reef-dominated and a leeward margin that includes elongate carbonate sand shoals. This platform is also dissected by northwest–southeast-oriented, rift-parallel tidal channels.

TECTONO-SEDIMENTARY MODELS

A comparison of the 3D outcrops of the Middle Miocene Abu Shaar Platform, and analogous rift-related platforms published in the literature, reveals recurring tectono-sedimentary platform geometries and facies distributions. These characteristics form the basis for the tectono-sedimentary models presented in this paper. The models are divided into those systems that are syntectonic from those that are post-tectonic. Syntectonic platforms are defined as systems deposited during periods associated with active faulting within rift basins. Post-tectonic platforms are defined as systems deposited after periods of active extensional faulting that develop over preexisting fault-generated rift topography. A model to account for carbonate systems that occupy transfer zones is also presented.

Syntectonic Platforms

Footwall Margins.—

During extensional faulting, fault blocks undergo coseismic footwall uplift and coeval hanging-wall subsidence (Jackson and McKenzie, 1983; Stein and Barrientos, 1985), with rotation occurring about a horizontal axis (fulcrum), leading to fault-block tilting. These relatively instantaneous events lead to distinct stratigraphic signatures in footwall margins, uplift and emergence, fault-related truncation, and platform-margin collapse (Fig. 9).

Fault blocks that are highly rotated, such as those that form in late rift settings, tend to form relatively narrow areas of substrate on which carbonate platforms can develop, and this is emphasized further if the fault block becomes emergent during footwall uplift (Fig. 9). In this latter case, narrow fringing platforms can develop immediately on the footwall scarp (e.g., present-day Gebel Zeit, Gulf of Suez rift). The amount of footwall relief is controlled by a number of factors: fault displacement, the water depth of the adjacent hanging-wall basin, fault-block lithology, and erosion rates. It can also be controlled by the nature of the fault system that develops: segmented/stepped fault geometries produce lower relief substrate over which platforms can develop, whilst single block-margin faults can produce steep fault-block margins. The sedimentary expression of footwall uplift includes all of the features associated with relative sea-level fall at a platform margin, and in particular subaerial erosion and meteoric diagenesis (Leeder and Gawthorpe, 1987). Such features are especially evident in carbonate shelves, which commonly build up to sea level, although the degree of meteoric diagenesis depends on the prevailing climate (Sun and Esteban, 1994). In humid climates, soils, karstic surfaces, and cave systems with associated fluvial deposits might be expected to occur. In contrast, more arid environments may result in calcretes and rhizocretion formation, or evaporites with associated eolian facies (see summary in Tucker and Wright, 1990).

The degree of footwall uplift also varies spatially on a footwall margin depending on the local geometry of the fault block and the bounding fault segments. Gawthorpe et al. (1994) predict that development of composite type I sequence boundaries (where emergence can be demonstrated) is favored at the centers of fault segments, where uplift is greatest, but because most platform margins aggrade to sea level even a minor relative sea-level fall may result in a type I sequence boundary (Fig. 8).

On the Abu Shaar Platform, erosional truncation of depositional sequences along the platform margin is generally the only record of a relative sea-level fall. James et al. (1988) attribute calcite mineralization to meteoric diagenesis related to footwall emergence. However, these were later demonstrated by Clegg et al. (1998) to be hydrothermal in origin and related to the block-bounding fault zone. Clegg et al. (1998) describe various vadose speleothems, together with evaporite residues along the footwall boundary of the Kharasa and Esh el Mellaha members, which may relate to episodes of footwall uplift during platform growth.

In the Albian of Northern Spain, Rosales et al. (1994) describe the formation of unconformities related to platform exposure and karstic diagenesis in the Castro Urdiales footwall platform. These are interpreted to be the results of footwall lowstands controlled by fault-block rotation. Associated lowstand deposits include the sedimentary fill of paleo-karst structures, incised valleys, and canyons. In the late Chadian (Dinantian) of the Dublin Basin, Pickard et al. (1994) also describe incision and the deposition of fluvial conglomerates (Smuggler's Cave Formation) on the footwall of the Balbriggan platform as a response to footwall uplift. A paleo-karst surface in the Frasnian–Famennian (Late Devonian) reef complexes of the Canning Basin, Australia, has also been linked to a phase of fault-block rotation and footwall uplift (Chow et al., 2004). In the Tonasa platform of South Sulawesi, Wilson and Bosence (1996) invoke episodes of footwall uplift using the occurrence of reworked platform-top speleothems in redeposited facies of the adjacent hanging-wall basin.

The location of footwall carbonate margins, adjacent to fault zones, means that they are particularly susceptible to platform-margin collapse. In tectonically active areas this can occur either through seismic triggering of collapse along already steep platform margins or by propagation of block-margin faults through the platform itself, generating a submarine fault scarp along the margin of the platform (Fig. 9). Such events can involve underlying fault-block basement and therefore be separated from platform-margin collapse of oversteepened platform margins. The distribution and stacking of redeposited carbonates containing reworked basement clasts along a footwall scarp has been linked to episodes of enhanced tectonic activity (e.g., Wilson and Bosence, 1996). However, the mass wasting of contemporaneous escarpment-type margins into adjacent basins is commonly found in tectonically stable platform settings, in both ancient and present-day carbonate systems (Mullins and Neumann, 1979; James and Ginsburg, 1979).

The truncation of footwall platform-margin sequences can be a direct indicator of active faulting in fault-block carbonate platforms. The most prominent relict of fault activity comes from footwall platforms that become truncated so that platform-interior deposits occur at the newly generated platform margin (erosional shelf margin of Read, 1982; Fig. 9). The Kharasa member (DS 1 to DS 3) of the Abu Shaar Platform is probably the best example of a truncated footwall margin (James et al., 1988; Burchette 1988; Cross et al., 1998), although similar margins are interpreted in the Dinantian from well and seismic data in northern England (Ebdon et al., 1990; Fraser and Gawthorpe, 1990) and from well and outcrop data in Ireland (Pickard et al., 1994).

The response of footwall carbonate-platform margins to fault movement depends on the intensity and magnitude of faulting coupled with the geometry of the platform-margin fault zone. Where fault displacement is limited, basinward thickening of footwall sequences can be enhanced by small-scale synsedimentary faults. This is locally observed in the Abu Shaar Platform (Fig. 6A), where small-scale faults cut both the basal DS 1 and the underlying fault block. Small fault displacements can also lead to downstepping platform margins and is common when fault throw is accommodated on several block-bounding faults (Fig. 9). Bosellini et al. (1981) imply that in the Lower Jurassic of northern Italy, basinward thickening of the Trento and Friuli footwall margins is accommodated on a series of downstepping faults. Hurst and Surlyk (1984) also recognize "stepped margins" as a distinct platform-margin type in the Silurian of North Greenland. Synsedimentary faults have been interpreted in the margin of the Late Permian Capitan Platform, U.S.A., (e.g., Kosa et al., 2003; Kosa and Hunt, 2006). They describe numerous steep extensional faults and associated fracture systems in the outer 5–6 km of the platform margin. These extend 33 km along strike parallel to the margin and have throws of up to 30 m. Most of the faults tip out upward within the platform succession and are associated with significant karst development. Synsedimentary fractures and joints have also been described in present-day and Pleistocene reefs of the Gulf of Suez rift (e.g., Strasser et al., 1992; Strasser and Strohmenger, 1997). Fractures develop parallel to the principal fault trend, are diagenetically modified, and also terminate upwards towards younger reefs which are not faulted (e.g., Ras Mohammad, Sinai; Ian Sharp, personal communication).

Hanging-Wall Dip-Slope Margins.—

Platforms developed in hanging-wall dip-slope areas are controlled primarily by their position relative to the rift margin. Platforms, like Abu Shaar, that are attached to the rift margin are

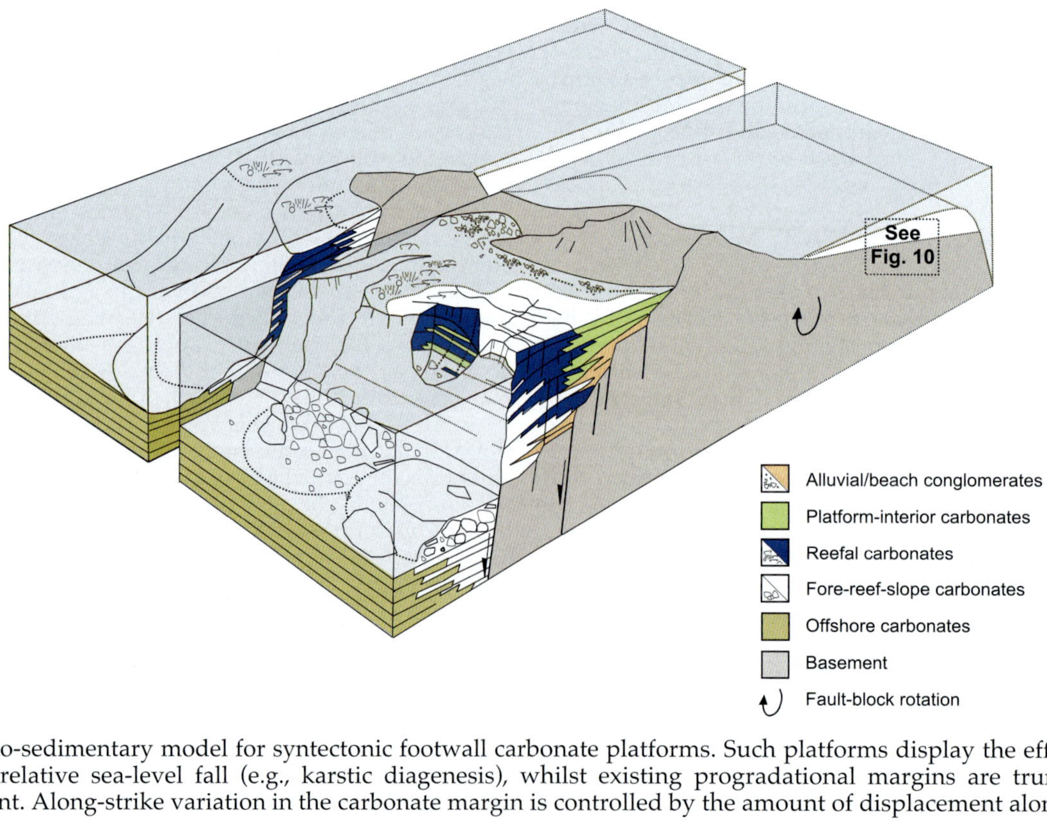

FIG. 9.—Tectono-sedimentary model for syntectonic footwall carbonate platforms. Such platforms display the effects of footwall uplift and relative sea-level fall (e.g., karstic diagenesis), whilst existing progradational margins are truncated by fault displacement. Along-strike variation in the carbonate margin is controlled by the amount of displacement along the fault.

likely to have mixed carbonate–clastic successions (Fig. 10B), whereas platforms that form on isolated blocks which are detached from the margin are more likely to have carbonate-dominated stratigraphies (e.g., Rosales et al., 1994; Fig. 10A). Complexities in this broad model are driven by the nature of the underlying pre-rift or basement substrate. Although fault blocks that are detached from the rift margin, and therefore the primary hinterland source areas, are likely to be relatively starved of clastic input, erosion of clastic bedrock in local footwall highs also contributes to the stratigraphy of fault-block carbonate platforms that occupy a more central position in the basin (cf. Ravnås and Steel, 1998).

During fault activity, rotating fault blocks experience relative sea-level rise on the hanging-wall dip slope and coeval relative sea-level fall in adjacent up-dip footwall areas. The transition between down-dip hanging-wall subsidence and up-dip footwall uplift occurs at a pivot point (fulcrum) about which the fault-block rotates (e.g., Leeder and Gawthorpe, 1987; Ravnås and Steel, 1998).

The size and geometry of a rift-margin-detached, hanging-wall dip-slope carbonate platform, which is developed on an actively rotating fault-block, is likely to be controlled primarily by the interplay between fault-block widths, the magnitude of

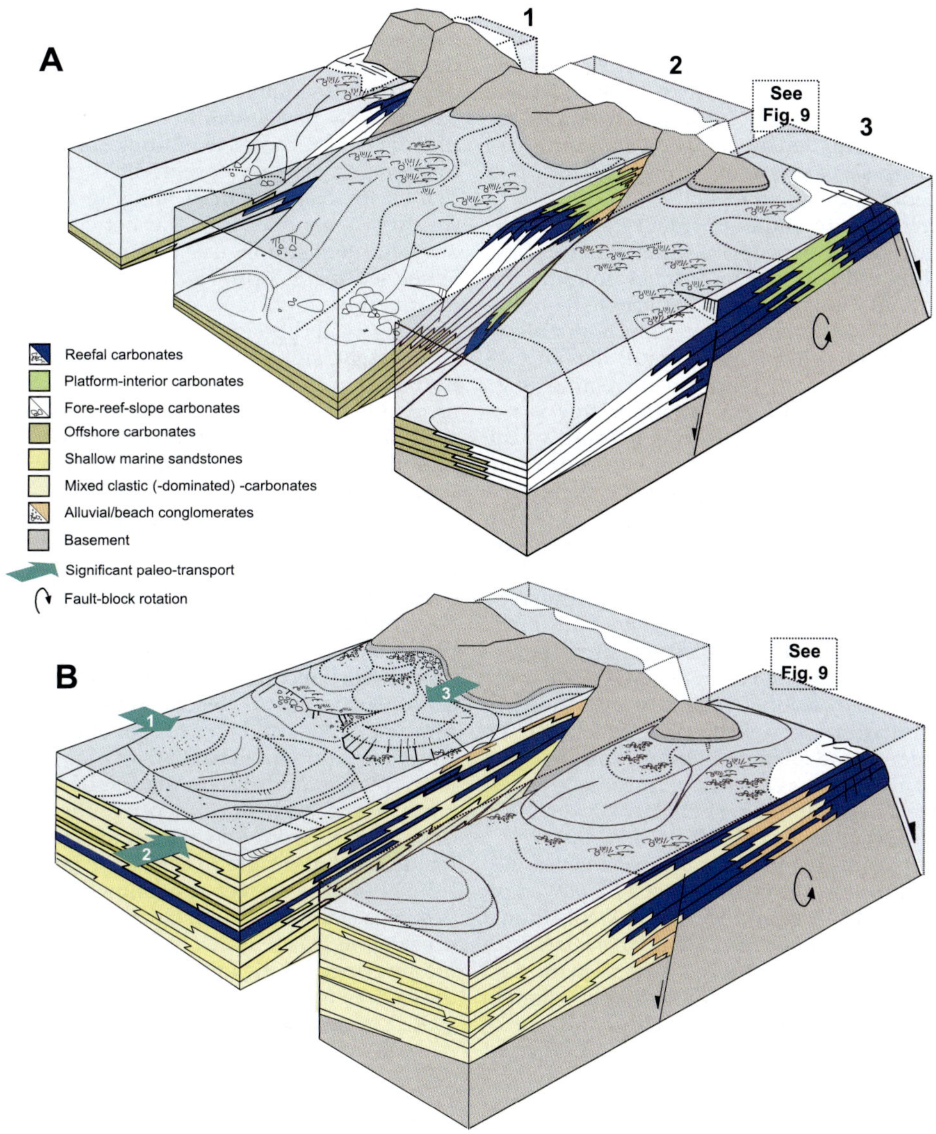

FIG. 10.—Tectono-sedimentary models for active hanging-wall dip-slope platform margins. **A)** Dip-slope platform that is detached from the rift margin. The geometry and distribution of facies in these platforms is dominated by tilting and fault-block rotation. Such platforms are likely to be complicated, because they are likely to exhibit the effects of hanging-wall dip-slope deepening and coeval sea-level fall caused by footwall uplift. Such platforms can show low-relief geometries (2) with broad retrogradational stacking patterns and depositional sequences bounded by flooding surfaces where they are outpaced by relative sea-level rise. If block rotation is significant, the platform margin may continue to backstep to form a narrow fringe around the uplifted high (1). Antithetic faults can also locally uplift dip-slope strata (3). **B)** The effects of block rotation on a margin-attached mixed carbonate–clastic platform. High clastic influx is generated by rejuvenated footwall highs, with sediment supply derived from along the axis of the hanging wall (1), from the rift margin (2), and locally from the adjacent footwall high (3).

fault displacement, the position of the depositional system in relation to the fault-block fulcrum, and carbonate production and/or clastic supply. Individual fault displacements along block-bounding faults generally result in shallow rotational tilting of hanging-wall dip-slope areas. The relatively high production rates of tropical shallow-water carbonate systems means that dip-slope platforms are more than likely to compensate for subsidence and tilting and commonly generate aggradational stacking patterns where carbonate production is able to keep up with relative sea-level rise, or progradation if this is exceeded (Fig. 10A). In this setting, the topographical effects of individual fault episodes are likely to be minimal. Nevertheless, successive episodes of fault displacement result in repeated thickening of platform strata into neighboring hanging-wall basins, with the development of sequences bounded by down-dip flooding surfaces that exhibit onlap, whilst up-dip coeval relative sea-level fall and erosion occurs. The evolutionary trend from carbonate ramps to rimmed shelves, which can occur in passive dip-slope platforms, is likely to be accentuated by the tilting and continued deepening in an active dip-slope platform setting (Fig. 10A; Bott and Johnson, 1967; Leeder and Gawthorpe, 1987).

During episodes of high differential subsidence, when hanging-wall dip-slope settings may experience significant rotation, platforms that form down-dip of the fault-block fulcrum are likely to generate retrogradational stacking patterns in response to relative sea-level rise if carbonate production is unable to keep pace (Fig. 10A). Although high rates of fault-driven subsidence have been associated with platform drowning (e.g., Schlager, 1981), with simple fault-block rotations this is not likely to cause the demise of entire platforms because of the counteraction of adjacent footwall uplift. In the case of rapid subsidence of a hanging-wall dip-slope platform, continued backstepping occurs toward the footwall shoreline and results in the development of a narrow fringing platform in this setting (Fig. 10A).

Retrogradational stacking patterns in hanging-wall dip-slope platforms, which have developed in response to fault-block rotation, have been documented prior to platform drowning in some Tethyan platforms (e.g., Ruiz-Ortiz et al., 2004).

In contrast, hanging-wall dip-slope systems that develop up-dip of the fault-block fulcrum experience uplift, which may be accompanied by emergence and erosion. In this setting, marked basinward facies shifts may occur as a result of forced regression during relative sea-level fall (Plint, 1988; Hunt and Tucker, 1992; Gawthorpe et al., 1997; Gawthorpe et al., 2000).

Fault-block rotations produce wedge-shaped geometries with growth fanning of platform-top strata; this is well displayed on the Abu Shaar Platform in DS 7, 8, and 9 (Fig. 3). Depositional sequences are bounded by flooding surfaces which correlate with coeval erosional surfaces developed at footwall sites. On the Abu Shaar platform the large proportion of clastic facies deposited in the hanging-wall dip-slope stratigraphy is probably a function of relative sea-level fall and erosion of rift margin and adjacent footwall highs. In particular, the sharp-based and locally erosional nature of some arkosic sandbodies may be the result of forced regression during relative sea-level fall.

Hanging-wall dip-slope areas that are dissected by small-scale, antithetic, or synthetic faults can result in abrupt changes in subsidence and water depth. Faulting of the hanging-wall dip slope produces small-scale local highs, which can become the sites of carbonate buildups (Fig. 10A; Burchette and Wright, 1992; Watchorn et al., 1998).

Burchette and Wright (1992) show how the progradation of ramps on hanging-wall dip slopes can be limited by such synsedimentary faulting. If the differential subsidence associated with these faults is significant, then the aggradation of rimmed shelf margins may be promoted in a manner similar to that seen along footwall margins (Fig. 10A). In the Dinantian of the Dublin Basin, Pickard et al. (1994) show how antithetic faults on the northern part of the Balbriggan fault block aided the preferential subsidence of the Drogheda area, accounting for the anomalously thick sequence of late Chadian hanging-wall dip-slope platform carbonates.

An important feature of hanging-wall dip-slope margins, which are proximal to the rift border, is the effect of clastic input (Fig. 10B). Uplift of the adjacent fault-block footwall high and rift margin leads to rejuvenation of hinterland source areas. Clastics derived from uplifted areas are accommodated through deepening of the dip-slope areas, which provides space for syntectonic sediment accumulation. On the Abu Shaar Platform this is well demonstrated by DS 7, which is undoubtedly a syntectonic depositional sequence, in which lagoonal micrites in the footwall stratigraphy pass laterally into a thickened dip-slope stratigraphy of coarse arkoses exhibiting southeast-directed cross-stratification and are therefore derived from the rift margin (Fig. 4). The clastic component of hanging-wall dip-slope margins is a result of the complex interplay between relative sea-level changes and sediment supply (see discussion in Ravnås and Steel, 1998). When sediment supply is outpaced by relative sea-level rise that is driven by high rates of hanging-wall dip-slope subsidence, hanging-wall dip-slope platforms may become clastics starved. Conversely, slower rates of subsidence, which are equalled or outpaced by clastic supply, can result in more clastics-rich platforms. On the Abu Shaar Platform clastic deposits in the hanging-wall dip-slope stratigraphy are sourced from southwestern rift margin, locally from the footwall high of the Esh el Mellaha Fault Block, and axially from the northwest.

Hanging-wall dip-slope platforms may become drowned by clastic influx during fault-block rotation when adjacent footwall hinterland areas are rejuvenated by uplift. This is most likely to occur when sediment supply is able to outpace accommodation development. Such platforms maybe characterized by mixed post-tectonic carbonate phases and syntectonic clastic stratigraphies, which record the alternation of tectonic activity and quiescence. Such a case is described by Bosence et al. (1996) from the Miocene of the northern margin of the Gulf of Aden. In this case study from one small basin with a steep rift-margin fault there is little rotation and differentiation of footwall and hanging-wall areas. Here, fault movement results in erosion and increased clastic supply from the rift margin in the form of alluvial fans and fan deltas that prograde across the fault block. Subsequent subsidence and sea-level rise allows marine carbonate facies to become established, which develop over the morphology of the preexisting clastic units. Similar transgressive carbonates commonly fringe clastic fan-delta deposits in the Lower Miocene of the eastern margin of the Gulf of Suez rift (e.g., Gawthorpe et al., 1990; Young et al., 2000).

Post-Tectonic Platforms

Footwall Margins.—

Post-tectonic footwall platform-margin geometries are governed by the preexisting (antecedent) morphology of the underlying fault block and are therefore likely to be variable. Where margins develop above strongly bevelled footwall scarps, wedge-shaped platforms develop that seal previous faults and thicken towards the platform margin (Fig. 11). Alternatively, long-lived eroded scarps, typical of mature late-rift fault blocks, can result in relatively tabular flat-topped platforms (Fig. 11).

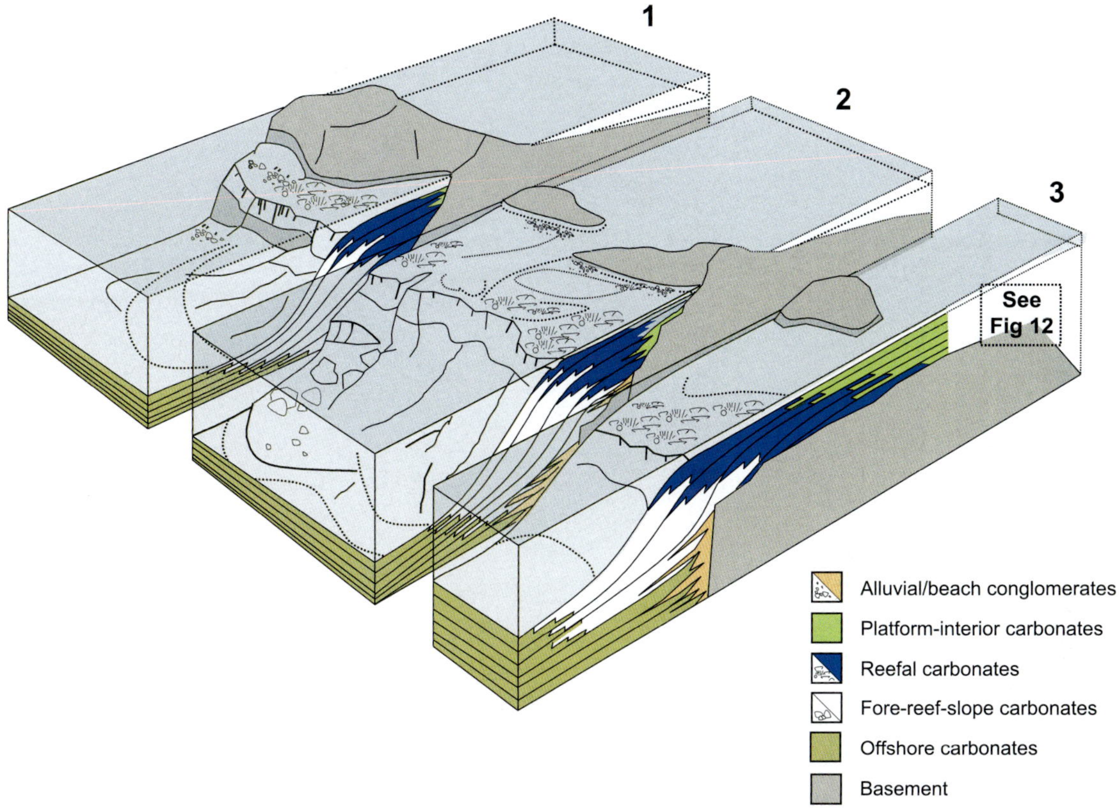

FIG. 11.—Tectono-sedimentary model for post-tectonic footwall carbonate platforms. The model shows a progradational and aggradational margin developed across a low-relief block-bounding fault zone. Such models apply to platforms developed along fault zones where displacement is minor or where the fault zone is bridged by aggrading hanging-wall sediments (e.g., alluvial and marine fans). High-relief footwall settings maintain aggradational platform margins (1). Beveled footwall highs result in narrow wedge-shaped platforms (2), whereas long-lived eroded footwalls result in more tabular footwall platforms (3).

The size of a footwall platform depends upon several related factors, including the size of the fault block, the amount of previous rotation, and the relief at the footwall-to-hanging-wall transition. Large fault blocks that were not rotated to a high degree, and fault blocks where erosion has modified earlier rift-generated topography may develop into broad shallow-marine areas. It has been shown that as footwall margins evolve and fault-generated topography is progressively buried, an initial perched footwall platform evolves into a more expansive, block-wide platform (e.g., Pickard, et al., 1994) or a mixed carbonate–clastic shelf (e.g., the Abu Shaar Platform).

In post-tectonic footwall margins, two end members can be recognized: firstly, along steep footwall scarps, where the adjacent half graben represents a relatively deep-water setting, the platform margin is confined to the footwall crest. In this case steep, strongly aggradational margins (bypass shelf margin, escarpment-type of Read, 1982, 1985) are produced where the platform margin corresponds to the position of the block-margin fault scarp (e.g., Rosales et al., 1994). The alternative is where footwall margins develop in low-relief areas and the platform is not confined to the footwall crest but is able to prograde across the footwall-to-hanging-wall transition (Fig. 11). This type of platform is facilitated in areas where the hanging-wall basin is filled with sediment and water depths are shallow enough to permit the bridging of the block-to-basin transition (Fig. 11). The progradation of the footwall platform into the hanging-wall setting can occur only while the block-margin fault zone remains inactive and basinal accommodation is infilled to shallow water depths. This model has been advocated in the Abu Shaar Platform, where footwall-derived conglomerate fans that reduce the relief of the block margin form a substrate over which later platform carbonates prograde (James et al., 1988; Burchette, 1988; Cross et al., 1998). This platform-margin type is also interpreted by Fraser and Gawthorpe (1990) in the Carboniferous of the U.K., where the same relationship is interpreted from seismic data.

Hanging-Wall Dip-Slope Margins.—

The models outlined here are based on carbonate and mixed carbonate–clastic platforms that develop during tectonic quiescence in a hanging-wall dip-slope setting (Fig. 12). The nature of such platforms is determined by the polarity of the underlying fault block and their position within the rift basin. These factors give rise to a range of possible platform types and geometries, of which two contrasting models are considered here: one for fault blocks that are attached to the rift border fault, such as the Esh el Mellaha Fault Block, and one for blocks that are detached or isolated from the border fault (e.g., South China Sea, Fulthorpe and Schlanger, 1989).

Margin-Attached.—Margin attached hanging-wall dip-slope platforms can be greatly affected by clastic influx derived either from the rift shoulder or axially along the hanging-wall half graben (Fig. 12A). On fault blocks that slope to the rift margin, like the Esh

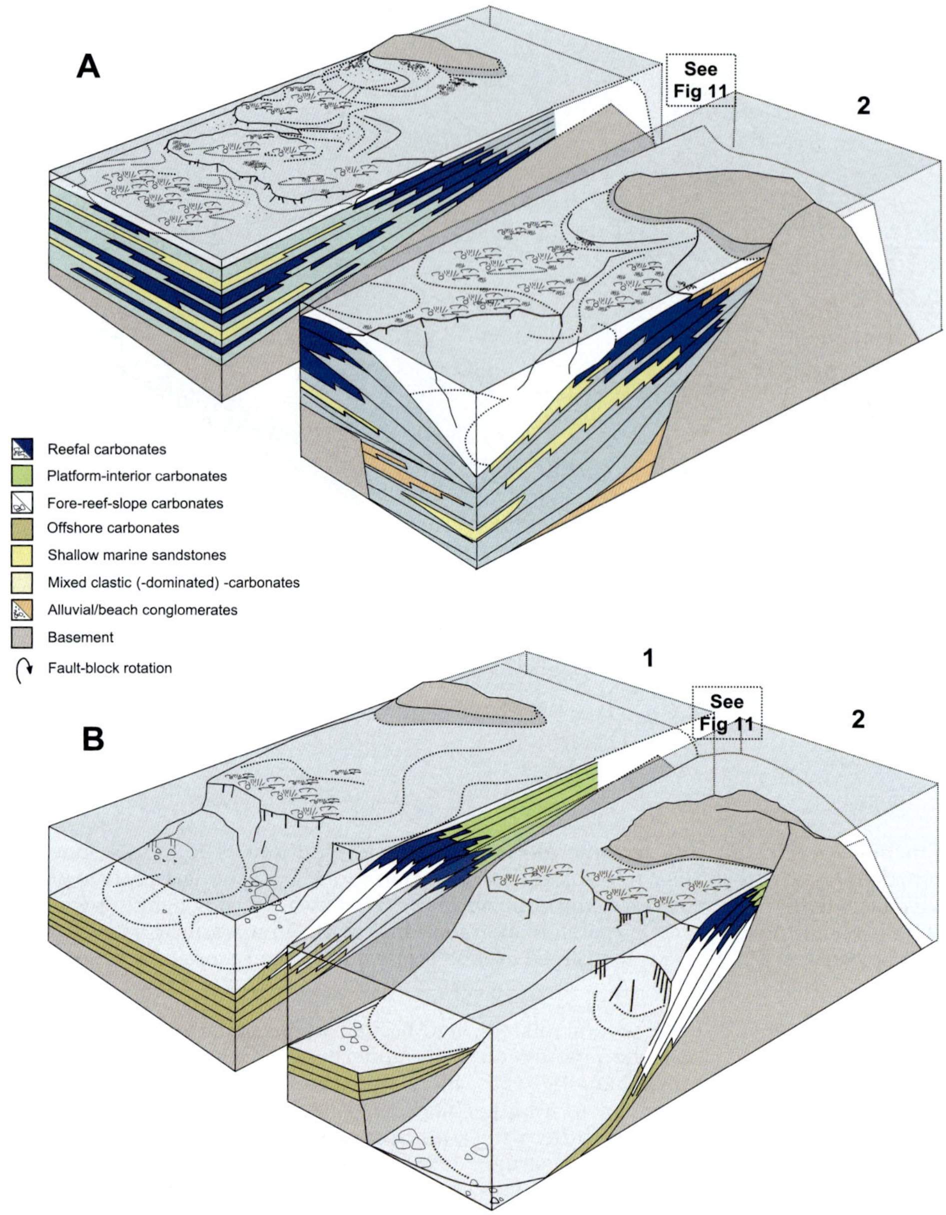

FIG. 12.—Tectono-sedimentary models for post-tectonic dip-slope carbonate platforms. **A)** A mixed clastic–carbonate platform developed on a fault block that is attached to the rift margin and is therefore clastics-influenced. If carbonate and clastic sedimentation is high, then aggradational profiles develop (1), while in deeper-water settings associated with steeply dipping fault blocks, a shelf margin may develop (2). **B)** A platform developed on a fault block that is detached from the rift margin and is therefore carbonate-dominated. Such platforms share similarities with transfer-zone platforms: they have low-relief geometries, which have the potential to evolve steeper profiles as they aggrade (1). Alternatively, if the underlying fault block is steeply rotated, then narrow fringing platforms can develop (2).

el Mellaha Fault Block, mixed carbonate–clastic platforms are more likely to develop. This is in contrast to flexural margins, where clastic runoff is likely to be reduced and dip slopes are basin-facing, therefore favoring the development of carbonate systems.

During deposition of DS 7 of the Abu Shaar Platform the influx of clastic sediments to the Esh el Mellaha half graben filled much of the accommodation space and therefore kept water depths relatively shallow through its entire growth. In this case a major platform margin and slope did not develop and the stratal geometries have more ramp-like geometries and are largely aggradational with proximal dip-slope carbonates interbedded with more distal clastics. Carbonate-dominated deposition is characterized by laterally extensive biostromes, as in DS6, and an abundance of decameter-scale, peritidal, upward-shallowing cycles.

As with footwall platforms, the landward extent of both platform types described above is controlled by the inherited relief and therefore the previous rotation of the underlying fault block and the amount of erosion. High-relief footwall islands are onlapped by platform-interior strata and have a local extent, whereas low-relief blocks are blanketed by extensive shallow-water carbonates, and hanging-wall dip-slope carbonates may

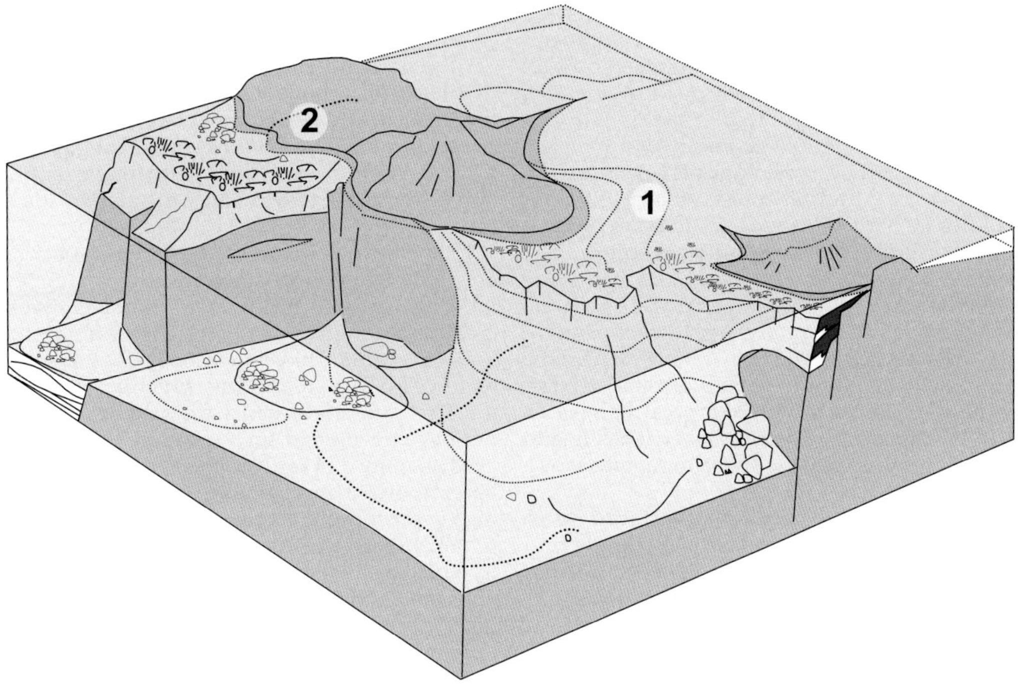

FIG. 13.—Tectono-sedimentary model for transfer-zone carbonate platforms. Such platforms can have complex geometries, depending on the underlying structural configuration. Area 1 shows the soft linkage between two fault segments and is reflected by the progradational nature of the transfer-zone platform margin. This margin has evolved from a low-relief ramp-type margin to a steeper escarpment-type margin through higher carbonate production rates in shallow-water sites or continued fault propagation and therefore steepening of the intervening relay ramp. Area 2 shows an aggradational margin associated with a steep, high-relief, hard-linkage cross fault (see Figures 9–12 for key to facies).

become amalgamated with the opposing footwall carbonates to form one large post-tectonic platform. The initial geometries of post-tectonic dip-slope carbonates are onlap and infill of existing topography.

Margin-Detached.—Fault blocks that are detached from the rift margin are generally characterized by carbonate-dominated stratigraphies (Fig. 12B), because they are not directly associated with uplifted areas capable of delivering significant clastic material to the platform. The clastic component of such platforms is generally limited to erosion of source areas provided by local footwall highs (cf. Ravnås and Steel, 1998). During tectonic quiescence the low-angle structural geometry of this setting is reflected in the geometry of the platforms developed there. The polarity of the fault block in this setting is not as critical as for margin-attached platforms, although platform growth is governed by other factors such as water circulation patterns. Initially, hanging-wall dip-slope margin stratigraphies onlap and infill relict topography and then are characterized by low-angle progradational geometries which reflect the low-gradient slope on which they form (Leeder and Gawthorpe, 1987; Burchette and Wright, 1992). As a result, a trend of platform-margin types may steepen up from depositional to bypass margins (Leeder and Gawthorpe, 1987). Gawthorpe (1986, 1987) describes the evolution from a hanging-wall dip-slope ramp margin to a rimmed shelf in the Dinantian of the Bowland Basin. He also emphasized that, at an advanced stage in basin evolution, neighboring footwall and hanging-wall dip-slope platforms may show a symmetrical profile, irrespective of the original underlying basement morphology. However, although the overall geometries of these two types may share many similarities, internal stratal geometries, sequence boundaries, and facies distribution exhibit major differences if carbonate development has any syntectonic component. Hanging-wall dip-slope platforms developed on relatively isolated fault blocks are widely developed in the Tertiary of southeast Asia (e.g., Fulthorpe and Schlanger 1989; see below, Reservoir Implications and Examples).

Transfer-Zone Margins

The distribution of rift-basin facies and sequence geometries is clearly responsive to the underlying three-dimensional structural architecture (Gawthorpe and Leeder, 2000). However, published examples of fault-block carbonate systems often consider only their two-dimensional profile in the context of the classic fault block-half graben (e.g., Rosales et al., 1994). This is mainly because of the paucity of outcrop examples and examples based on seismic-reflection profiles. The southeastern end of the Abu Shaar Platform appears to be the best-exposed example of a transfer-zone margin to a carbonate platform.

Transfer zones are important transverse or oblique structural elements in rift basins, which permit displacement transfer between fault blocks and segments (Rosendahl et al., 1986; Rosendahl, 1987; Morley et al., 1990). They occur on a variety of scales, from those which accompany the complex basin-defining polarity switches that occur along the length of most rift systems (e.g., Moustafa, 1976; Colletta et al., 1988; Patton et al., 1994), to those which occur locally and switch displacement between individual fault segments. Gawthorpe and Hurst (1993) highlighted in particular how topographic changes associated with intrabasinal transfer zones can offer a major control on basin drainage and sediment transport processes in rift systems. Commonly, transfer zones are local clastic entry points to half grabens and as such are not usually favorable sites for carbonate-platform development. However, as these authors describe, in areas of low clastic influx carbonate-platform development is facilitated by the low-angle, ramp geometry of a transfer zone (Fig. 13).

As with footwall margins, the initial geometry of a transfer-zone carbonate margin reflects the morphology of the underlying fault block. Simplistically, local transfer zones can form either as soft-linkage relay ramps or hard-linkage zones that are defined by rift-perpendicular or rift-oblique cross faults. In soft-linkage examples, the decrease in displacement on a block-bounding fault as it passes into a transfer zone is accompanied by a progressive reduction in relief along the footwall scarp towards a point of zero displacement or fault tip point. Hard linkage, where the displacement on primary block-bounding faults is transferred by cross faults, can result in more abrupt topographic changes. In reality, both kinds of linkage may be present in a rift basin, and there may be an evolution between these two end members as a fault system develops.

Initially the low-relief topography associated with intrabasinal transfer zones favors the relatively low-angle progradational platform margins; these can develop perpendicular to the main rift trend and the footwall-margin platform (Fig. 13). Provided that accommodation space can be maintained by relative sea-level rise, transfer-zone platforms are expected to evolve progressively steeper profiles, eventually developing into steep-sided shelves similar to the footwall margin (Read, 1982, 1985; cf. Wright and Faulkner, 1990; Barnaby and Read, 1990). This process is achieved by either repeated aggradation of platform strata during relative sea-level rise or through higher rates of carbonate production in shallower-water sites, and could probably be amplified by the downthrowing of minor transfer faults, although to our knowledge this has not been documented. In active rift basins, continued extension along block-bounding faults might produce fault propagation through transfer areas that were originally characterized by low-relief flexure. In such settings, transfer-zone margins may develop faulted margins through time (e.g., Hurst and Surlyk, 1983; Surlyk and Hurst, 1984).

Bosworth et al. (1998) describe a mixed carbonate–clastic system associated with a series of hard-linkage faults at Gebel Naqara (15 km south of Safaga) on the southwest margin of the Gulf of Suez rift. In this area, rift-trending fault blocks are dissected by several cross faults oriented perpendicular to the main rift axis. A number of Middle Miocene reef complexes with associated fore-reef slopes crop out on the uplifted footwall crests of rotated fault blocks and show geometries and facies very similar to those in the Abu Shaar Platform. In addition, platform carbonates are also developed upon the up-thrown sides of cross faults (Fig. 13). During Miocene deposition, the combination of rift-parallel and rift-perpendicular cross faults produced a tectonically active embayment which was fringed by a series of carbonate reefs. The cross-faults also acted as sediment conduits, channeling clastic input derived from the rift margin to the adjacent hanging-wall half graben in front of the reef complexes. In this case, sediment inputs occurred at the intersection between the rift-parallel faults and the cross faults, thus bypassing the carbonate-platform margins developed on the fault-block highs.

Although transfer zones are most commonly seen as important entry points for clastic input in rift basins, Wilson et al. (2000) and Cassabianca et al. (2002) also show that they may also control the redeposition of shallow-water carbonates into adjacent hanging-wall basins, especially when they are associated with relatively detached fault-block carbonate platforms.

RESERVOIR IMPLICATIONS AND EXAMPLES

The variability in geometry and facies distribution exhibited by rift-basin carbonate platforms has important implications for their subsurface prediction. Platforms developed on the uplifted crests of rotated fault blocks are instinctively attractive exploration targets. However, their performance as producing hydrocarbon reservoirs is also strongly dictated by their tectono-sedimentary setting.

The reservoir quality of syntectonic footwall platform carbonates may be enhanced significantly by meteoric diagenesis related to footwall uplift in a humid climatic setting. Porosity creation through hydrothermal leaching along active fault zones is also possible. This varies markedly when compared with post-tectonic footwall platforms, which are developed on tectonically inactive fault blocks which may not have been emergent and may not have developed such secondary porosity.

The nature of fault-related truncation of footwall platforms has important implications for the preservation of potentially high-reservoir-quality platform-margin reservoir facies. Platforms that have been truncated have poorer-quality margins than those in which a shoal or reef-dominated footwall margin has simply aggraded and not prograded across the margin fault zone. In the latter case, redeposited carbonates situated in the hanging wall adjacent to the footwall platform margin may form higher-quality reservoirs than remnant faulted footwall margins.

Small-scale faults, fractures, and joints associated with syntectonic footwall platforms are potentially significant reservoir heterogeneities. Not only do they have the potential to increase reservoir quality themselves, but they can also channel pore-enhancing diagenetic fluids within footwall reservoirs.

In hanging-wall dip-slope settings, platforms can be characterized by a range of carbonate and clastic facies. Facies prediction in such settings is problematic but is governed primarily by the proximity of the rift margin and therefore significant clastic flux, together with accommodation space or water depth across the hanging-wall dip slopes. Reservoirs vary from those that are carbonate-dominated and pass down-dip into non-reservoir basinal shales, where fault blocks are isolated and sediment starved, to laterally extensive mixed carbonate and clastic reservoirs where clastic input is high close to the rift margin and water depths are shallow.

Reservoir architecture is driven by sequence stacking patterns produced by the interplay of relative sea-level changes and clastic input/carbonate production. Hanging-wall dip-slope platforms in post-tectonic settings produce laterally extensive and highly layered reservoirs in shallow, mixed carbonate–clastic systems. These reservoirs are in contrast to areas where sedimentation is unable to keep pace with episodes of significant fault-block rotation, and these syntectonic platforms may backstep and produce spatially restricted reservoirs that rapidly shale-out across the hanging-wall dip slope.

Published examples of hydrocarbon-producing fault-block platforms come mainly from Southeast Asia, and their origin relates to the early rifting of the South China Sea; the most detailed publications come from the Liuhua Platform, the Pearl River Mouth Basin (Erlich et al., 1990; Ehrlich et al., 1993), and the Malampaya Platform, offshore Palawan (Grötsch and Mercadier, 1999; Fournier et al., 2005). These two examples are both detached from the rift margin and contrast with each other in that they appear to represent end members of the syntectonic and post-tectonic platform types outlined in this paper. However, a detailed comparison is unwise, because they are based on very different datasets, the Liuhua example being based on 1986–87 2D seismic (Erlich et al., 1993) and the Malampaya example on higher-resolution 3D seismic data and a greater number of wells (Fournier et al., 2005). The Malampaya Platform stratigraphy develops on an east-tilted fault block bounded to the west by a northeast–southwest-oriented extensional fault. It has a trapezoidal shape in plan and is wedge shaped in cross section (5 km long x 1–2 km wide). The platform is characterized by small northeast–southwest graben, synsedimentary tilting, unconformities, and onlap surfaces. In general the western (footwall) margin comprises shallower, higher-energy facies and the platform interior and hanging-wall dip slope to the east a range of different carbonate facies indicating deepening and shallowing episodes. For this example, the models outlined in Figures 9 and 10A would appear to be good analogues for reservoir prediction for this type of field. By contrast, the Liuhua field is a much larger, oblong-shaped structure (100 km x 20–30 km across) and with a less pronounced wedge-shaped cross section than Malampaya. The platform is sited on a block-faulted basement high with a steep (windward) southwest margin with reefs and high-energy shallow-water carbonates with some minor extensional faults and an extensive, gently sloping platform interior with some northeast-prograding clinoforms (Erlich et al., 1993). No syntectonic geometries or structures have been recorded, but the platform appears to have inherited the morphology and accommodation space typical of a post-tectonic platform (cf. Figs. 11, 12B).

CONCLUSIONS

- A number of factors control the development of carbonate systems in rift basins. Stratigraphy is controlled mainly by sediment supply (carbonate production and clastic input) in relation to accommodation changes driven by tectonism and sea-level changes. Carbonate stratigraphies are impacted strongly by climatic factors and are also likely to exhibit the effects of local factors such as nutrient supply and water circulation.

- Fault-block carbonate platforms in marine rift basins occur as two main classes: syntectonic platforms, where a tectonic driver imparts an imprint on the stratigraphy, and post-tectonic platforms that passively exploit preexisting rift topography and accommodation. Platforms developed in transfer zones display complex geometries, which can also be related to episodes of fault activity and intervening periods of quiescence.

- Data from 3D outcrops in the Gulf of Suez rift, and a review of published examples, indicate 3D variability in the morphology and stratigraphy of fault-block carbonate platforms; different tectono-sedimentary settings such as footwalls, hanging-wall dip slopes, and transfer zones reflect variations in platform geometry and facies distribution.

- Syntectonic platforms have footwall margins characterized by fault-truncated footwall sequences, shallow marine platform-margin facies cut by sequence boundaries with evidence of subaerial exposure, narrow margins with one dominant basin-bounding fault, or wider margins with downstepping faults.

- Hanging-wall dip slopes on syntectonic platforms accumulate wedge-shaped stratigraphies with stratigraphic fanning away from the footwall. These platforms are likely to be complex, inasmuch as they may show the effects of both down-dip flooding events and backstepping related to hanging-wall subsidence, and the up-dip effects of coeval relative sea-level fall caused by footwall uplift. On platforms that are attached to the rift margin, dip-slope stratigraphies interfinger with margin-sourced clastics, whose supply can also be controlled by fault movement.

- Post-tectonic platforms have footwall-margin successions whose accommodation space is controlled by the degree of previous fault-block rotation and subsequent erosion. These platforms show basal onlap and may develop into extensive flat-topped platforms. Where previous faulting has been minor or alluvial fans and/or fan deltas bridge the previously active fault zone, such platforms can prograde across the block-bounding fault into the adjacent hanging-wall basin.

- Hanging-wall dip slopes on post-tectonic platforms are characterized by strata that blanket preexisting fault-generated topography. They are more likely to develop parallel-sided sequences, which do not display the effects of fault-block rotation. Facies are dependent on the proximity of source areas such as the rift margin, with carbonate stratigraphies likely to dominate on isolated fault blocks.

- Transfer-zone carbonate margins vary depending on whether they occur in soft-linkage or hard-linkage zones. In soft-linkage relay zones, platforms typically have ramp profiles that steepen up to rimmed shelf margins. In hard-linkage zones, platforms share similarities with footwall margins but are more likely to form perpendicular to the main rift-basin axis. Carbonate platforms developed in transfer zones are less common, because such areas tend to act as major clastic entry points to adjacent basins.

- 3D facies models are presented for rift-related carbonate systems that may be useful analogues for the interpretation of less well exposed or subsurface seismically imaged examples.

- Hydrocarbon reservoir and/or exploration targets are known in rift-related carbonate platforms. However, predicting their flow behavior is a complex combination of tectonic, stratigraphic, and diagenetic controls. The 3D models presented in this paper enable these different controls to be assessed and applied to exploration and production for such reservoirs.

ACKNOWLEDGMENTS

This paper is derived largely from NEC's PhD research carried out at Royal Holloway University of London (1992–1996), which was funded from the EC SCIENCE grant "REDSED" and supported by BP Exploration and a London University Scholarship. These grants are gratefully acknowledged. Bill Bosworth (Apache), Trevor Burchette (BP Exploration), and Bruce Purser (Université de Paris) are thanked for much encouraging and constructive discussion on this work. We are indebted to Lynne Blything (Schlumberger) for her support in generating the figures. The authors are also grateful to the two reviewers, Ian Sharpe and Mario Coniglio, as well as the editors Jeff Lukasik and Toni Simo, for the suggestions and comments that improved the final version of this publication.

REFERENCES

Alsharan, R.J., and Nairn, J.F., 1997, Sedimentary Basins and Petroleum Geology of the Middle East: Amsterdam, Elsevier, 843 p.

Barnaby, R.J., and Read, J.F., 1990, Carbonate ramp to rimmed shelf evolution: Lower to Middle Cambrian continental margin, Virginia

Appalachians: Geological Society of America, Bulletin, v. 102, p. 391–404.

BECHSTÄDT, T., AND BONI, M., 1989, Tectonic control of the formation of a carbonate platform: the Cambrian of southwest Sardinia, *in* Crevello, P.D., Wilson, J.L., Sarg, J.F., and Read, J.F., eds., Controls on Carbonate Platform and Basin Development: SEPM, Special Publication 44, p. 107–122.

BERNOULLI, D., AND JENKYNS, H.C., 1974, Alpine, Mediterranean, and Central Atlantic Mesozoic facies in relation to the early evolution of the Tethys, *in* Dott, R.H., and Shaver, R.H., eds., Modern and Ancient Geosynclinal Sedimentation: SEPM, Special Publication 19, p. 129–160.

BOSELLINI, A., MASETTI, D., AND SARTI, M., 1981, A Jurassic "Tongue of the Ocean" infilled with oolitic sands: the Belluno trough, Venetian Alps, Italy: Marine Geology, v. 44, p. 59–95.

BOSENCE, D.W.J., 1998, Stratigraphic and sedimentological models of rift basins, *in* Purser, B.H., and Bosence, D.W.J., eds., Sedimentation and Tectonics of Rift Basins: Red Sea–Gulf of Aden: London, Chapman & Hall, p. 9–25.

BOSENCE, D.W.J., 2005, A genetic classification of carbonate platforms based on their basinal and tectonic setting in the Cenozoic: Sedimentary Geology, v. 175, p. 49–72.

BOSENCE, D.W.J., CROSS, N.E., AND HARDY, S., 1998, Architecture and depositional sequences of Tertiary fault-block carbonate platforms: an analysis from outcrop and computer modelling: Marine and Petroleum Geology, v. 15, p. 203–221.

BOSENCE, D.W.J., NICHOLS, G., AL-SUBBARY, M., AL-THOUR, A-K., AND REEDER, M., 1996, Syn-rift continental–marine depositional sequences, Tertiary, Gulf of Aden, Yemen: Journal of Sedimentary Research. v. 66, p. 766–777.

BOSENCE, D.W.J., AND WALTHAM, D.A., 1990, Computer modelling and interval architecture of carbonate platforms: Geology, v. 18, p. 26–30.

BOSWORTH, W., 1995, A high strain rift model for the southern Gulf of Suez, Egypt, *in* Lambiase, J.J., ed., Hydrocarbon Habitat in Rift Basins: Geological Society of London, Special Publication 80, p. 75–102.

BOSWORTH, W., CREVELLO, P., WINN, R.D., JR., AND STEINMETZ, J., 1998, Structure, sedimentation, and basin dynamics during rifting of the Gulf of Suez and north-western Red Sea, *in* Purser, B.H., and Bosence, D.W.J., eds., Sedimentation and Tectonics of Rift Basins: Red Sea–Gulf of Aden: London, Chapman & Hall, p. 77–96.

BOTT, M.H.P., AND JOHNSON, G.A.L., 1967, The controlling mechanism of Carboniferous cyclic sedimentation: Geological Society of London, Quarterly Journal, v. 122, p. 421–441.

BURCHETTE, T.P., 1988, Tectonic control on carbonate facies distribution and sequence development, Miocene, Gulf of Suez: Sedimentary Geology, v. 59, p. 179–204.

BURCHETTE, T.P., AND WRIGHT, V.P., 1992, Carbonate ramp depositional systems: Sedimentary Geology, v. 79, p. 3–57.

CASSABIANCA, D., BOSENCE, D., AND BECKETT, D., 2002, Reservoir potential of Cretaceous platform-margin breccias: Central Italian Apennines: Journal of Petroleum Geology, v. 25, p. 179–202.

CHOW, N., GEORGE, A.D., AND TRINAJSTIC, K.M., 2004, Tectonic control on development of a Frasnian–Famennian (Late Devonian) palaeokarst surface, Canning Basin reef complexes, northwestern Australia: Australian Journal of Earth Sciences, v. 51, p. 911–917.

CHRONIS, G., PIPER, D.J.W., AND ANAGNOSTOU, C., 1991, Late Quaternary evolution of the Gulf of Patras, Greece: Tectonism, deltaic sedimentation and sea-level change: Marine Geology, v. 97, 191–209.

CLEGG, N., HARWOOD, G., AND KENDALL, C., 1998, The dolomitisation and post- dolomite diagenesis of Miocene platform carbonates: Abu Shaar, Gulf of Suez, Egypt, *in* Purser, B.H., and Bosence, D.W.J., eds., Sedimentation and Tectonics of Rift Basins: Red Sea–Gulf of Aden: London, Chapman & Hall, p. 390–405.

COLLETTA, B., LE QUELLEC, P., LETOUZY, J., AND I., MORETTI, 1988, Longitudinal evolution of the Suez rift structure (Egypt): Tectonophysics, v. 153, p. 221–233.

COLLIER, R.E.Ll., 1991, The Lower Carboniferous Stainmore Basin, N. England: extensional basin tectonics and sedimentation: Geological Society of London, Journal, v. 148, p. 379–390.

CROSS, N.E., 1996, Sedimentology and sequence stratigraphy of a Miocene fault-block carbonate platform, Gulf of Suez, Egypt: unpublished Ph.D. thesis, University of London, 268 p.

CROSS, N.E., BOSENCE, D.W.J., AND PURSER, B.H., 1998, The tectono-sedimentary evolution of a rift margin carbonate platform: Abu Shaar, Gulf of Suez, Egypt, *in* Purser, B.H., and Bosence, D.W.J., eds., Sedimentation and Tectonics of Rift Basins: Red Sea–Gulf of Aden: London, Chapman & Hall, p. 271–295.

DART, C., COLLIER, R.E.LL., GAWTHORPE, R.L., KELLER, J.V.A., AND NICHOLS, G., 1994, Sequence stratigraphy of (?)Pliocene–Quaternary syn-rift Gilbert-type fan deltas, northern Peleponesos, Greece: Marine and Petroleum Geology, v. 11, p. 545–560.

DAVIES P.J., SYMONDS, P.A., FEARY, D.A., AND PIGRAM, C.J., 1989, The evolution of the carbonate platforms of northeast Australia, *in* Crevello, P.D., Wilson, J.L., Sarg, J.F., and Read, J.F., eds., Controls on Carbonate Platform and Basin Development: SEPM, Special Publication 44, p. 234–258.

EBDON, C.C., FRASER, A.J., HIGGINS, A.C., MITCHENER, B.C., AND STRANK, A.R.E., 1990, The Dinantian stratigraphy of the East Midlands: a seismostratigraphic approach: Geological Society of London, Journal, v. 147, p. 519–536.

ENOS, P., KOCH, W.J., AND JAMES, N.P., 1979, The geophysical anatomy of the southern Belize continental margin and adjacent basins, *in* James, N.P., and Ginsburg, R.N., eds., The Seaward Margin of Belize Barrier and Atoll Reefs: International Association of Sedimentologists, Special Publication 3, p. 15–23.

ERLICH, R.N., BARRETT, S.F., AND GUO, B.J., 1990, Seismic and geological characteristics of drowning events on carbonate platforms: American Association of Petroleum Geologists, Bulletin, v. 74, p. 1523–1537.

ERLICH, R.N., LONGO, A.P., AND HYARE, S., 1993, Response of platform margins to drowning: evidence of environmental collapse, *in* Loucks, R.G., and Sarg, J.F., eds., Carbonate Sequence Stratigraphy: Recent Developments and Applications: American Association of Petroleum Geologists, Memoir 57, p. 241–266.

EVANS, A.L., 1988, Neogene tectonic and stratigraphic events in the Gulf of Suez: Insights into synrift unroofing and uplift history: American Association Petroleum of Geologists, Bulletin, v. 74, p. 1386–1400.

FOURNIER, F., BORGOMANO, J., AND MONTAGGIONI, L.F., 2005, Development patterns and controlling factors of Tertiary carbonate build-ups: new insights from high-resolution 3-D seismic and well data from the Malampaya gas field (Offshore Palawan, Philippines): Sedimentary Geology, v. 174, p. 189–215.

FRASER, A.J., AND GAWTHORPE, R.L., 1990, Tectono-stratigraphic development and hydrocarbon habitat of the Carboniferous in northern England, *in* Hardman, R.F.P., and Brooks, J., eds., Tectonic Events Responsible for Britain's Oil and Gas Reserves: Geological Society of London, Special Publication 55, p. 49–86.

FULTHORPE, C.S., AND SCHLANGER, S.O., 1989, Paleo-oceanographic and tectonic settings of early Miocene reefs and associated carbonates of offshore south-east Asia: American Association of Petroleum Geologists, Bulletin, v. 73, p. 729–756.

GAMBOA, L.A., TRUCHAN, M., AND STOFFA, P.L., 1985, Middle Jurassic and Upper Jurassic depositional environments at outer shelf and slope of Baltimore Canyon Trough: American Association of Petroleum Geologists, Bulletin, v. 69, p. 610–621.

GARCIA-MONDEJAR, J., 1990, The Aptian–Albian carbonate episode of the Basque–Cantabrian Basin (northern Spain): general characteristics, controls and evolution, *in* Tucker, M.E., Wilson, J.L., Crevello, P.D., Sarg, J.F., and Read, J.F., eds., Carbonate Platforms: Facies Sequences and Evolution: International Association of Sedimentologists, Special Publication 9, p. 257–290.

Garfunkel, Z., and Bartov, Y., 1977, The tectonics of the Suez rift: Geological Survey of Israel, Bulletin, v. 71, p. 1–44.

Gawthorpe, R.L., 1986, Sedimentation during carbonate ramp-to-slope evolution in a tectonically active area: Bowland Basin (Dinantian), northern England: Sedimentology, v. 33, p. 185–206.

Gawthorpe, R.L., 1987, Tectono-sedimentary evolution of the Bowland Basin, N. England, during the Dinantian: Geological Society of London, Journal, v. 144, p. 59–71.

Gawthorpe, R.L., Fraser, A.J., and Collier, R.E.Ll., 1994, Sequence stratigraphy in active extensional basins: implications for the interpretation of ancient basin fills: Marine and Petroleum Geology, v. 11, p. 642–658.

Gawthorpe, R.L., Gutteridge, P., and Leeder, M.R., 1989, Late Devonian and Dinantian basin evolution in northern England and North Wales, *in* Arthurton, R.S., Gutteridge, P., and Nolan, S.C., eds., The Role of Tectonics in Devonian and Carboniferous Sedimentation in the British Isles: Yorkshire Geological Society, Special Occasional Publication 6, p. 1–23.

Gawthorpe, R.L., Hall, M., Sharp, I.R., and Dreyer, T., 2000, Tectonically enhanced forced regressions: examples from growth folds in extensional and compressional settings, the Miocene of the Suez rift and the Eocene of the Pyrenees, *in* Hunt, D., and Gawthorpe, R.L., eds., Sedimentary Responses to Forced Regressions: Geological Society of London, Special Publication 172, p. 177–191.

Gawthorpe, R.L., and Hurst, J.M., 1993, Transfer zones in extensional basins: their structural style and influence on drainage development and stratigraphy: Geological Society of London, Journal, v. 150, p. 1137–1152.

Gawthorpe, R.L., Hurst, J.M., and Sladen, C.P., 1990, Evolution of Miocene footwall-derived, coarse-grained deltas, Gulf of Suez, Egypt: Implications for exploration: American Association of Petroleum Geologists, Bulletin, v. 74, p. 1077–1086.

Gawthorpe, R.L., and Leeder, M.R., 2000, Tectono-sedimentary evolution of active extensional basins: Basin Research, v. 12, p. 195–218.

Gawthorpe, R.L., Sharp, I.R., Underhill, J.R., and Gupta, S., 1997, Linked sequence stratigraphic and structural evolution of propagating normal faults: Geology, v. 25, p. 795–798.

Gérard, J., and Buhrig, C., 1990, Seismic facies of the Permian section of the Barents shelf: analysis and interpretation: Marine and Petroleum Geology, v. 7, p. 234–252.

Grayson, R.F., and Oldham, L., 1987, A new structural framework for the Northern British Dinantian as a basis for oil, gas and mineral exploration, *in* Miller, J., Adams, A.E., and Wright, V.P., eds., European Dinantian Environments: Chichester, U.K., John Wiley, p. 33–59.

Grötsch, J., and Mercadier, C., 1999, Integrated 3-D reservoir modelling based on 3-D seismic: the Tertiary Malampaya and Camargo buildups, offshore Palawan, Philippines: American Association of Petroleum Geologists, Bulletin, v. 83, p. 1703–1728.

Gupta, S., Underhill, J.R., Sharp, I.R., Gawthorpe, R.L., 1999, Role of fault interactions in controlling synrift sediment dispersal patterns: Miocene Abu Alaqa Group, Suez Rift, Egypt: Basin Research, v. 11, p. 167–189.

Gutteridge, P., 1987, Dinantian sedimentation and the basement structure of the Derbyshire Dome: Geological Journal, v. 22, p. 25–41.

Harris, P.M., and Kowalik, W.S., 1994, Satellite Images of Carbonate Depositional Settings: American Association of Petroleum Geologists, Methods in Exploration Series, 10, 147 p.

Hunt, D., and Tucker, M.E., 1992, Stranded parasequences and the forced regressive wedge systems tract: deposition during base-level fall: Sedimentary Geology, v. 81, p. 1–9.

Hurst, J.M., and Surlyk, F., 1983, Initiation, evolution and destruction of an early Paleozoic carbonate shelf, eastern North Greenland: Journal of Geology, v. 44, p. 97–117.

Hurst, J.M., and Surlyk, F., 1984, Tectonic controls of Silurian carbonate-shelf margin morphology and facies, North Greenland: American Association of Petroleum Geologists, Bulletin, v. 68, p. 1–17.

Jackson, J.A., and McKenzie, D., 1983, The geometrical evolution of normal fault systems: Journal of Structural Geology, v. 5, p. 471–482.

Jackson, J.A., White, N.J., Garfunkel, Z., and Anderson, H, 1988, Relations between normal-fault geometry, tilting and vertical motions in extensional terrains: an example from the southern Gulf of Suez: Journal of Structural Geology, v. 10, p. 155–170.

James, N.P., Coniglio, M., Aissaoui, D.M., and Purser, B.H., 1988, Facies and geological history of an exposed Miocene rift margin carbonate platform: Gulf of Suez, Egypt: American Association of Petroleum Geologists, Bulletin, v. 72, p. 555–572.

James, N.P., and Ginsburg, R.N., 1979, The Seaward Margin of the Belize Barrier and Atoll Reefs: International Association of Sedimentologists, Special Publication 3, 191 p.

Jansa, L., 1981, Mesozoic carbonate platforms and banks of the eastern North America margin: Marine Geology, v. 44, p. 97–117.

Kosa, E., and Hunt, D., 2006, Heterogeneity in fill and properties of karst-modified syndepositional faults and fractures: Upper Permian Capitan Platform, New Mexico, U.S.A.: Journal of Sedimentary Research, v. 76, p. 131–151.

Kosa, E., Hunt, D., Fitchen, W.M., Bockel-Rebelle, M.-O., and Roberts, G., 2003, The heterogeneity of paleocavern systems developed along syndepositional faults, Upper Permian Capitan platform, New Mexico, USA, *in* Ahr, W.M., Harris, P.M., Morgan, W.A., and Somerville, I.D., eds., Permo-Carboniferous Carbonate Platforms and Reefs: SEPM, Special Publication 78, and American Association of Petroleum Geologists, Memoir 83, p. 291–322.

Leeder, M.R., 1995, Continental and proto-oceanic troughs, *in* Busby, C.J., and Ingersoll, R.V., eds., Tectonics of Sedimentary Basins: U.S.A., Blackwell Science Inc., p. 119–148.

Leeder, M., and Gawthorpe, R.I., 1987, Sedimentary models for extensional tilt-block/half-graben basins, *in* Coward, M.P., Dewey, J.F., and Hancock, P.L., eds., Continental Extensional Tectonics: Geological Society of London, Special Publication 28, p. 139–152.

Machette, M.N., Persounis, S.F., and Nelson, A.R., 1991, The Wasatch Fault Zone Utah, segmentation and history of Holocene earthquakes: Journal of Structural Geology, v. 13, p. 137–149.

Mayall, M.J., Bent, A., and Roberts, D.M., 1997, Miocene carbonate build-ups offshore Socialist Republic of Vietnam, *in* Fraser A.J., Mathews, S.J., and Murphy, R.W., eds., Petroleum Geology of Southeast Asia: Geological Society of London, Special Publication 126, p. 117–120.

Montgomery, S.L., 1996, Cotton Valley lime pinnacle reef play; Branton Field: American Association of Petroleum Geologists, Bulletin, v. 80, p. 617–629.

Morley, C.K., Nelson, R.A., Patton, T.L., and Munn, S.G., 1990, Transfer zones in the East African rift system and their relevance to hydrocarbon exploration in rifts: American Association of Petroleum Geologists, Bulletin, v. 74, p. 1234–1253.

Moustafa, A.R., 1976, Block faulting in the Gulf of Suez: 5th Egyptian General Petroleum Organisation Exploration Seminar, 19 p.

Mullins, H.T., 1983, Structural controls on contemporary carbonate continental margins: Bahamas, Belize, Australia, *in* Cook, H.E., Hine, A.C., and Mullins, H.T. eds., Platform Margin and Deep-Water Carbonates: SEPM, Short Course 12, p. 2.1–2.52.

Mullins, H.T., and Lynts, G.W., 1977, Stratigraphy and structure of the Northeast providence Channel, Bahamas: American Association of Petroleum Geologists, Bulletin, v. 60, p. 1073–1053.

Mullins, H.T., and Neumann, A.C., 1979, Deep carbonate bank margin structure and sedimentation in the northern Bahamas, *in* Doyle, L.J., and Pilkey, O.H., eds., Geology of Continental Slopes: SEPM, Special Publication 27, p. 165–192.

Patton, T.L., Moustafa, A.R., Nelson, R.A., and Ardine, S.A., 1994, Tectonic evolution and structural setting of the Suez Rift, *in* Landon, S.M., ed., Interior Rift Basins: American Association of Petroleum Geologists, Memoir 59, p. 7–55.

PICKARD, N.A.H., REES, J.G., STROGEN, P., SOMMERVILLE, I.D., AND JONES, G.Ll., 1994, Controls on the evolution and demise of Lower Carboniferous carbonate platforms, northern margin of the Dublin Basin, Ireland: Geological Journal, v. 29, p. 93–117.

PLINT, A.G., 1988, Sharp-based shoreface sequences and 'offshore bars' in the Cardium Formation of Alberta: their relationships to relative sea-level, *in* Wilgus, C.K., Hastings, B.S., Kendall, C.G.St.C., Posamentier, H.W., Ross, C.A., and Van Wagoner J.C., eds., Sea-Level Changes: An Integrated Approach: SEPM, Special Publication 42, p. 357–370.

POMAR, L., 1991, Reef geometries, erosion surfaces and high-frequency sea-level changes: Upper Miocene Reef Complex, Mallorca, Spain: Sedimentology, v. 38, p. 243-269.

PURSER, B.H., BARRIER, P., MONTENAT, C., ORSZAG-SPERBER, F., OTT D'ESTEVOU, P., PLAZIAT, J.-C., AND PHILOBBOS, E., 1998, Carbonate and siliciclastic sedimentation in an active tectonic setting: Miocene of the northwestern Red Sea rift Egypt, *in* Purser, B.H., and Bosence, D.W.J. eds., Sedimentation and Tectonics of Rift Basins: Red Sea–Gulf of Aden: London, Chapman & Hall, p. 239–270.

RAVNÅS, R., AND STEEL., R.J., 1998, Architecture of marine rift basin successions: American Association of Petroleum Geologists, Bulletin, v. 82, p. 110–146.

READ, J.F., 1982, Carbonate platforms of passive (extensional) continental margins: types, characteristics and evolution: Tectonophysics, v. 81, p. 195–212.

READ, J.F., 1985, Carbonate platform facies models: American Association of Petroleum Geologists, Bulletin, v. 69, p. 1-21.

REITNER, J., 1986, A comparative study of the diagenesis in diapir influenced reef atolls and a fault-block reef platform in the late Albian of the Vasco-Cantabrian basin, *in* Schroeder, J.H., and Purser, B.H., eds., Reef Diagenesis: Berlin, Springer-Verlag, p. 189–209.

RICHARDSON, M., AND ARTHUR, M.A., 1988, The Gulf of Suez–northern Red Sea Neogene rift: a quantitative basin analysis: Marine and Petroleum Geology, v. 5, p. 247–270.

ROBERTS, G.P., GAWTHORPE, R.L., AND STEWART, I., 1993, Surface faulting in active normal fault zones: examples from the Gulf of Corinth fault system: Zeitschift für Geomorphologie, N.F., Supplement v. 94, p. 303–328.

ROBERTS, H.H., AND MURRAY, S.P., 1988, Developing carbonate platforms: Southern Gulf of Suez, Northern Red Sea: Marine Geology, v. 59, p. 165–185.

ROSALES, I., FERNÁNDEZ-MENDIOLA, P.A., AND GARCÍA-MONDÉJAR, J., 1994, Carbonate depositional sequence development on active fault-blocks: the Albian in the Castro Urdiales area, northern Spain: Sedimentology, v. 41, p. 861–882.

ROSENDAHL, B.R., 1987, Architecture of continental rifts with special reference to East Africa: Annual Review of Earth and Planetary Sciences, v. 15, p. 445–503.

ROSENDAHL, B.R., REYNOLDS, D.J., LORBER, P.M., BURGESS, C.F., MCGILL, J., SCOTT, D., LAMBIASE, J.J., AND DERKSON, S.J., 1986, Structural expression of rifting: lessons from Lake Tanganyika, Africa, *in* Frostick, L.E., Renaut, R.W., and Reid, I., eds., Sedimentation in African Rifts: Geological Society of London, Special Publication 25, p. 29–44.

RUIZ-ORTIZ, P.A., BOSENCE, D.W.J., REY, J., NIETO, L.M., CASTRO, J.M., AND MOLINA, J.M., 2004, Tectonic control of facies architecture, sequence stratigraphy and drowning of a Liassic carbonate platform (Betic Cordillera, southern Spain): Basin Research, v. 16, p. 331–372.

SANTANTONIO, M., 1994, Pelagic carbonate platforms in the geologic record: their classification, and sedimentary and paleotectonic evolution: American Association of Petroleum Geologists, Bulletin, v. 78, p. 122–141.

SCHLAGER, W., 1981, The paradox of drowned reefs and carbonate platforms: Geological Society of America, Bulletin, v. 92, p. 197–211.

SHARP, I.R., GAWTHORPE R.L., ARMSTRONG B., AND UNDERHILL, J.R., 2000a, Propagation history and passive rotation of mesoscale normal faults: implications for syn-rift stratigraphic development: Basin Research, v. 12, p. 285–296.

SHARP, I.R., GAWTHORPE R.L., UNDERHILL J.R., AND GUPTA, S., 2000b, Fault-propagation folding in extensional settings: Examples of structural style and synrift sedimentary response from the Suez rift, Sinai, Egypt: Geological Society of America, Bulletin, v. 112, p. 1877–1899.

SHERIDEN, R.E., CROSBY, J.T., BRYON, G.M., AND STOFFA, P.L., 1988, Stratigraphy and structure of the southern Blake plateau, northern Florida Straits and northern Bahamashy Platform from recent multichannel seismic reflection data: American Association of Petroleum Geologists, Bulletin, v. 65, p. 2571–2591.

SOMERVILLE, I.D., PICKARD, N.A.H., STROGEN, P., AND JONES, G.Ll., 1992, Early to mid-Visean shallow water platform build-ups, north Co. Dublin, Ireland: Geological Journal, v. 27, p. 151–172.

STECKLER, M.S., BERTHELOT, F., LYBERIS, N., AND LE PICHON, X., 1988, Subsidence in the Gulf of Suez: implications for rifting and plate kinematics: Tectonophysics, v. 153, p. 249–270.

STEIN, R.S., AND BARRIENTOS, S.E., 1985, Planar high-angle faulting in the Basin and Range: Geodetic analysis of the 1983 Borah Peak, Idaho, Earthquake: Journal of Geophysical Research, v. 90, p. 11,355–11,366.

STRASSER, A., AND STROHMENGER, C., 1997, Early diagenesis in Pleistocene coral reefs, southern Sinai, Egypt: response to tectonics, sealevel and climate: Sedimentology, v. 44, p. 537–558.

STRASSER, A., STROHMENGER, C., DAVAUD, E., AND BACH, A., 1992, Sequential evolution and diagenesis of Pleistocene coral reefs (South Sinai, Egypt): Sedimentary Geology, v. 78, p. 59–79.

STROGEN, P., SOMMERVILLE, I.D., PICKARD, N.A.H., AND JONES, G.Ll., 1995, Lower Carboniferous (Dinantian) boreholes from west Co. Meath, Ireland: Geological Journal, v. 25, p. 103–137.

SUN, S.Q., AND ESTEBAN, M., 1994, Paleoclimatic controls on sedimentation, diagenesis and reservoir quality: lessons from Miocene carbonates: American Association of Petroleum Geologists, Bulletin, v. 78. p. 519–543.

SURLYK, F., AND HURST, J., 1984, The evolution of the early Paleozoic deep-water basin of North Greenland: Geological Society of America, Bulletin, v. 95, p. 131–154

TUCKER, M.E., AND WRIGHT, V.P., 1990, Carbonate Sedimentology: Oxford, U.K., Blackwell Scientific Publications, 482 p.

VAN DE WEERD, A.A., AND ARMIN, R.A. 1992, Origin and evolution of the Tertiary hydrocarbon bearing basins in Kalimantan (Borneo), Indonesia: American Association of Petroleum Geologists, Bulletin, v. 76, p. 1778–1803.

VERA, J-A., RUIZ-ORITZ, P.A., GARCIA-HERNANDEZ, M., AND MOLINA, J.M., 1988, Palaeokarst and related pelagic sediments in the Jurassic of the Sub-betic zone, Southern Spain, *in* James, N.P., and Choquette, P.W., eds., Paleokarst: Berlin, Springer-Verlag, p. 364–383.

WATCHORN, F., NICHOLS, G.J., AND BOSENCE, D.W.J., 1998, Rift-related sedimentation and stratigraphy, southern Yemen (Gulf of Aden), *in* Purser, B.H., and Bosence, D.W.J., eds., Sedimentation and Tectonics of Rift Basins: Red Sea–Gulf of Aden: London, Chapman & Hall, p. 165–189.

WILSON, M.E., 2000, Tectonic and volcanic influences on the development and diachronous termination of a Tertiary tropical carbonate platform: Journal of Sedimentary Research, v. 70, p. 310–324.

WILSON, M.E., AND BOSENCE, D.W.J., 1996, The Tertiary evolution of South Sulawesi: a record in re-deposited carbonates of the Tonasa Limestone Formation, *in* Hall, R., and Blundell, D., eds., Tectonic Evolution of Southeast Asia: Geological Society of London, Special Publication 106, p. 365–389.

WILSON, M.E., BOSENCE, D.W.J., AND LIMBONG, A., 2000, Tertiary syntectonic carbonate platform development in Indonesia: Sedimentology, v. 47, p. 395–419.

WINN, R.D., JR., CREVELLO, P.D., AND BOSWORTH, W., 2001, Lower Miocene Nukhul Formation of Gebel el Zeit, Egypt: sedimentation and structural movement during early Gulf of Suez rifting: American Association of Petroleum Geologists, Bulletin, v. 85, p. 1871–1890.

WINTERER, E.L., AND BOSELLINI, A., 1981, Subsidence and sedimentation on Jurassic passive continental margin, Southern Alps, Italy: American Association of Petroleum Geologists, Bulletin, v. 65, p. 394–421.

WRIGHT V.P., AND BURCHETTE, T.P., 1996, Shallow-water carbonate environments, *in* Reading, H.G., ed., Sedimentary Environments; Processes, Facies and Stratigraphy: Oxford, U.K., Blackwell, p. 325–394.

WRIGHT, V.P., AND FAULKNER, T.J., 1990, Sediment dynamics of early Carboniferous ramps: a proposal: Geological Journal, v. 25, p. 139–144.

YOUNG, M.J., GAWTHORPE, R.L., AND SHARP, I.R., 2000, Sedimentology and sequence stratigraphy of a transfer zone coarse-grained delta, Miocene Suez Rift, Egypt: Sedimentology, v. 47, p. 1081-1104.

YOUNG, M.J., GAWTHORPE, R.L., AND SHARP, I.R., 2003, Normal fault growth and early syn-rift sedimentology and sequence stratigraphy: Thal Fault, Suez Rift, Egypt: Basin Research, v. 15, p. 479–502.

CONTROLS ON UPPER CRETACEOUS (MAASTRICHTIAN) HETEROZOAN CARBONATE PLATFORMS DEVELOPED ON SALT DIAPIRS, LA POPA BASIN, NE MEXICO

KATHERINE A. GILES
Institute of Tectonic Studies, New Mexico State University, Las Cruces, New Mexico 88001-8003, U.S.A.
e-mail: kgiles@nmsu.edu
DOMINIC C. DRUKE
Shell Exploration and Production Company, 700 Milam, PNT, 16023, Houston, Texas 77002, U.S.A.
e-mail: dominic.druke@shell.com
AND
DAVID W. MERCER AND LELA HUNNICUTT-MACK
Institute of Tectonic Studies, New Mexico State University, Las Cruces, New Mexico 88001-8003, U.S.A.

ABSTRACT: In the distal part of the Late Cretaceous Hidalgoan foreland basin in NE Mexico three isolated carbonate platforms nucleated on seafloor topography created by rising passive diapirs. Carbonate facies type and architecture of each platform was distinctly influenced by a combination of both short-term local conditions surrounding individual diapirs and by long-term regional conditions that affected the entire shelf. Local conditions included windward–leeward platform paleogeography, possible elevated nutrient levels at the salt–sediment interface, and halokinesis. Regional conditions included eustatic sea-level changes, foreland-basin tectonism, and siliciclastic sediment supply to the shelf.

Maastrichtian carbonate-platform facies are distributed asymmetrically across individual diapirs, reflecting windward-margin versus leeward-margin paleogeographic setting and differential minibasin subsidence related to salt withdrawal. Southern (windward) margins are dominated by steep-sided sponge, coral, and red algal reefs displaying minor fore-reef progradation (< 1.5 km) into the adjacent minibasin, thick carbonate debris-flow beds containing diapir-derived detritus, and pervasive near-diapir halokinetic fracturing. In contrast, northern (leeward) margins are dominated by foraminifera, red algal grainstone banks displaying major progradation (3–4 km) into the adjacent minibasin and lack debris-flow beds or halokinetic fracturing. Carbonate facies at all the diapirs are primarily sand-prone, heterozoan faunal assemblages that are unusual for this period of time and paleogeographic location. The presence of heterozoan faunal assemblages may be in response to high nutrient levels from local methane seeps forming at the salt–sediment interface. Carbonate facies form the bases of angular-unconformity-bounded carbonate–siliciclastic cycles called halokinetic sequences. The cycles reflect local variations in net diapiric-rise rates versus local sediment accumulation rates. Halokinetic sequences vary in number and character between the different diapirs and between the windward and leeward margins of each diapir. On leeward margins, halokinetic sequences are more numerous and carbonate facies are dominated by grainstone banks, whereas on windward margins halokinetic sequences are amalgamated and carbonate facies are dominated by fore-reef debris and debris-flow facies.

The isolated carbonate platforms are best developed within the transgressive systems tracts (TST) of third-order deltaic siliciclastic depositional sequences within the regionally marine foreland-basin depositional system. Late Cretaceous to Paleogene Hidalgoan shortening of La Popa foreland basin formed large-wavelength (> 10 km) NW–SE trending salt-cored detachment folds. Diapirs that lie in the hinges of folds were shortened or "squeezed" significantly more than diapirs that lie on the limbs of folds. Squeezed diapirs generated much higher and broader topographic relief in response to higher diapiric rise rates and are correspondingly dominated by extensive, thick, shallow-water (< 15 m deep) sponge, red algal reef and grainstone-bank facies with carbonate strata extending more than 4 km away from the diapir. Age-equivalent carbonate strata on limb diapirs contain thin, deeper-water (> 30 m deep) silty, red algal packstone facies that extend < 2 km from the diapir, reflecting lower carbonate production rates in a deeper-water setting.

INTRODUCTION

Understanding the intrinsic controls on type, distribution, and development of carbonate-platform facies forms the basis of petroleum exploration and production strategies in carbonate reservoir systems and has been the subject of much debate over the past few decades (see Crevello et al., 1989; James and Clarke, 1997; Harris et al., 1999). At the center of the discussion is the relative importance of a particular control amongst the myriad of possible variables that affect the deposition of carbonates and how to isolate the effects of a particular variable when looking at rocks. Most studies have concentrated on documenting facies type and stratal architecture of platforms that have fewer possible variables at work or that have tight constraints on one or more variables.

In this study, we document and compare and contrast the facies types and stratal architecture of three coeval, isolated carbonate platforms that grew on salt-diapir topographic highs. The platforms display a suite of characteristics, some of which are unique to the individual platform and some that share a common theme between all three platforms. We interpret that these characteristics reflect the complex interplay of both short-term local conditions surrounding individual diapirs and long-term regional conditions that affected the entire shelf. Local conditions include windward–leeward platform paleogeography, possible elevated nutrient levels related to methane seeps at the salt–sediment interface, and halokinesis. Regional conditions include eustatic sea-level change, foreland-basin tectonism, and siliciclastic sediment supply to the shelf. We conclude that no single variable dominated the system at either the local or the

Controls on Carbonate Platform and Reef Development
SEPM Special Publication No. 89, Copyright © 2008
SEPM (Society for Sedimentary Geology), ISBN 978-1-56576-130-8, p. 107–124.

regional scale, but that all of them are intimately interrelated in cause and effect. However, the combined result of the unified variables creates a predictable pattern that can be applied to carbonate platforms of any age that are growing on passive diapir topography.

GEOLOGIC SETTING OF LA POPA BASIN CARBONATE PLATFORMS

La Popa basin of NE Mexico contains exceptional exposures of isolated carbonate platforms that grew on salt diapirs and are encased in Lower Cretaceous to Paleogene dominantly siliciclastic strata. The basin is part of the Upper Cretaceous to Paleogene Hidalgoan foreland basin system, which includes the Parras and Sabinas basins, and the adjacent Sierra Madre Oriental fold belt to the south (Fig. 1; DeCelles and Giles, 1996, Soegaard et al., 1996; Ye, 1997). Stratal thicknesses in the foreland generally increase to the southwest (Ye, 1997) toward the Sierra Madre Oriental fold belt. Hidalgoan shortening in La Popa basin started in Late Cretaceous (Maastrichtian) time (Hon, 2001; Weislogel, 2001; Druke, 2005). It resulted in partial inversion of the former Jurassic rift basin (Lawton et al., 2001) and in formation of a series of NW–SE trending long-wavelength (> 10 km) salt detachment folds that deform all strata exposed in the basin (Fig. 2).

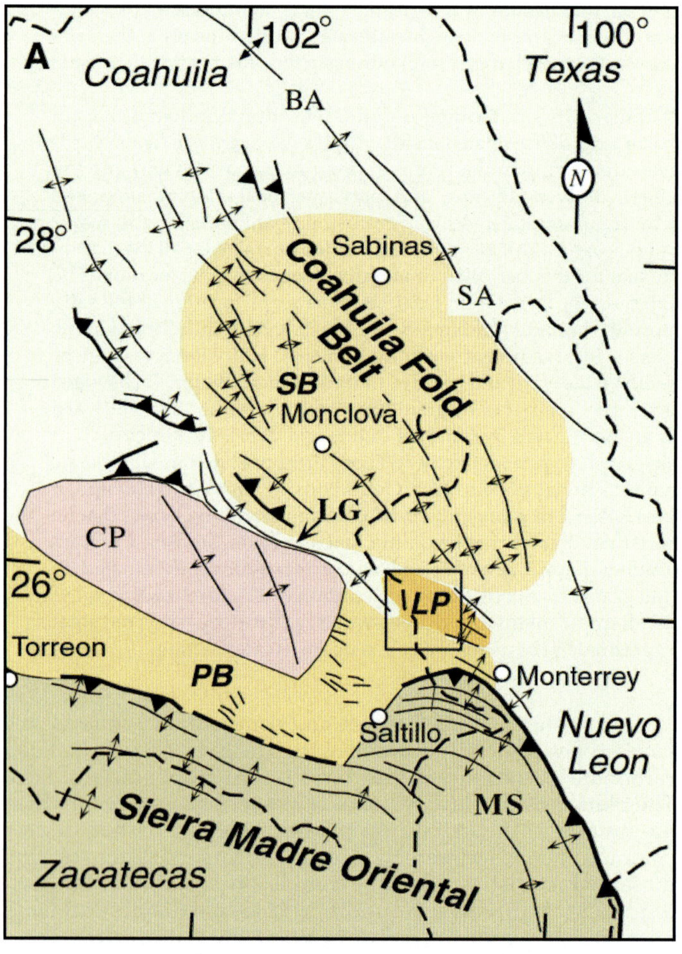

FIG. 1.—Map showing location of La Popa basin and the geologic elements of the Mexican states of Coahuila and Nuevo León, northeastern Mexico. Major Hidalgoan tectonic features are: CP, Coahuila platform; MS, Monterrey salient; Sierra Madre Oriental fold belt, which includes east-trending structures between Torreon and Monterrey; and the Coahuila fold belt, which occupies the Upper Cretaceous–Tertiary Sabinas Basin. Sedimentary basins of the Hidalgoan foreland region are: LP, La Popa; PB, Parras; and SB, Sabinas.

Salt diapirs in La Popa Basin are preserved at the surface as either subcircular gypsum caprock overlying salt stocks (El Papalote and El Gordo diapirs), which are 1.5–3 km in diameter, or as a fault-like structure called a salt weld (La Popa weld). La Popa weld represents a former 25-km-long salt wall (Fig. 2; Giles and Lawton, 1999) where salt was almost entirely evacuated, resulting in the fault-like weld structure. There is still a diapir (La Popa diapir) present at the western termination of the La Popa weld where salt evacuation was not complete (Fig. 2). The diapiric salt was sourced from the Upper Jurassic Minas Viejas Formation (Fig. 3) and contains clasts, as large as 200 m across, of metamorphosed intrusive and extrusive igneous rocks (Garrison and McMillan, 1999). The clasts referred to as meta-igneous blocks by Garrison and McMillan (1999) are interpreted to represent a rift-related magmatic assemblage that formed coevally with deposition of the rift-basin evaporite assemblage. The volcanic clasts represent rift magmas that were extruded into the salt basin and became interbedded with the salt beds. The plutonic clasts represent rift magmas that were intruded into the rift-basin assemblage. Both the plutonic and the volcanic rocks were subsequently carried upward within the salt as it became diapiric and were metamorphosed to greenschist facies by thermal wicking within the salt (Lawton et al., 2001). Adjacent to the diapirs are synclinal minibasins that have formed partially in response to salt withdrawal and partially related to Hidalgoan shortening (Rowan et al., 2003).

All exposed stratigraphic units in La Popa Basin show syndepositional deformation related to halokinesis and deformation due to Hidalgoan shortening. This indicates that passive diapirism occurred throughout deposition of the exposed stratal units (i.e. Lower Cretaceous–middle Eocene) and that shortening in the area ended after the middle Eocene (Hon, 2001; Weislogel, 2001; Druke, 2005).

The Parras Shale and the overlying Difunta Group are the primary exposed stratigraphic units that form the sedimentary fill of the Parras and La Popa basins (Fig. 3). Strata are shallow-marine, fine- to coarse-grained siliciclastics that shallow upward to marginal marine and fluvial deposits. The Difunta Group is as much as 4500 m thick within mini-basins and is divided into five formations (Murray et al., 1962), which, in ascending order, are the Muerto, Potrerillos, Adjuntas, Viento, and Carroza formations. The Muerto and Potrerillos formations are dominantly marine deltaic systems that transported sediment to the southeast in La Popa basin, judging by facies distribution and paleocurrent data (Weidie and Murray, 1967; McBride et al., 1974; McBride et al., 1975). The Potrerillos Formation contains finer-grained and more offshore facies than the underlying Muerto Formation. Punctuating the Potrerillos Formation stratigraphy are lenticular carbonate buildups, as much as 350 m thick that are termed lentils (McBride et al., 1974). The lentils are spatially restricted to areas directly adjacent to diapirs (McBride et al., 1974; Laudon, 1975, 1984) and are stratigraphically confined to prodeltaic and offshore shale units that represent periods of time when regional siliciclastic influx into the basin was low (Fig. 3; Giles and Lawton, 2002). In the Potrerillos Formation carbonate lentils are developed only in the Lower Mudstone, Middle Siltstone, and Upper Mudstone members (Laudon, 1975; Mercer, 2002; Druke, 2005).

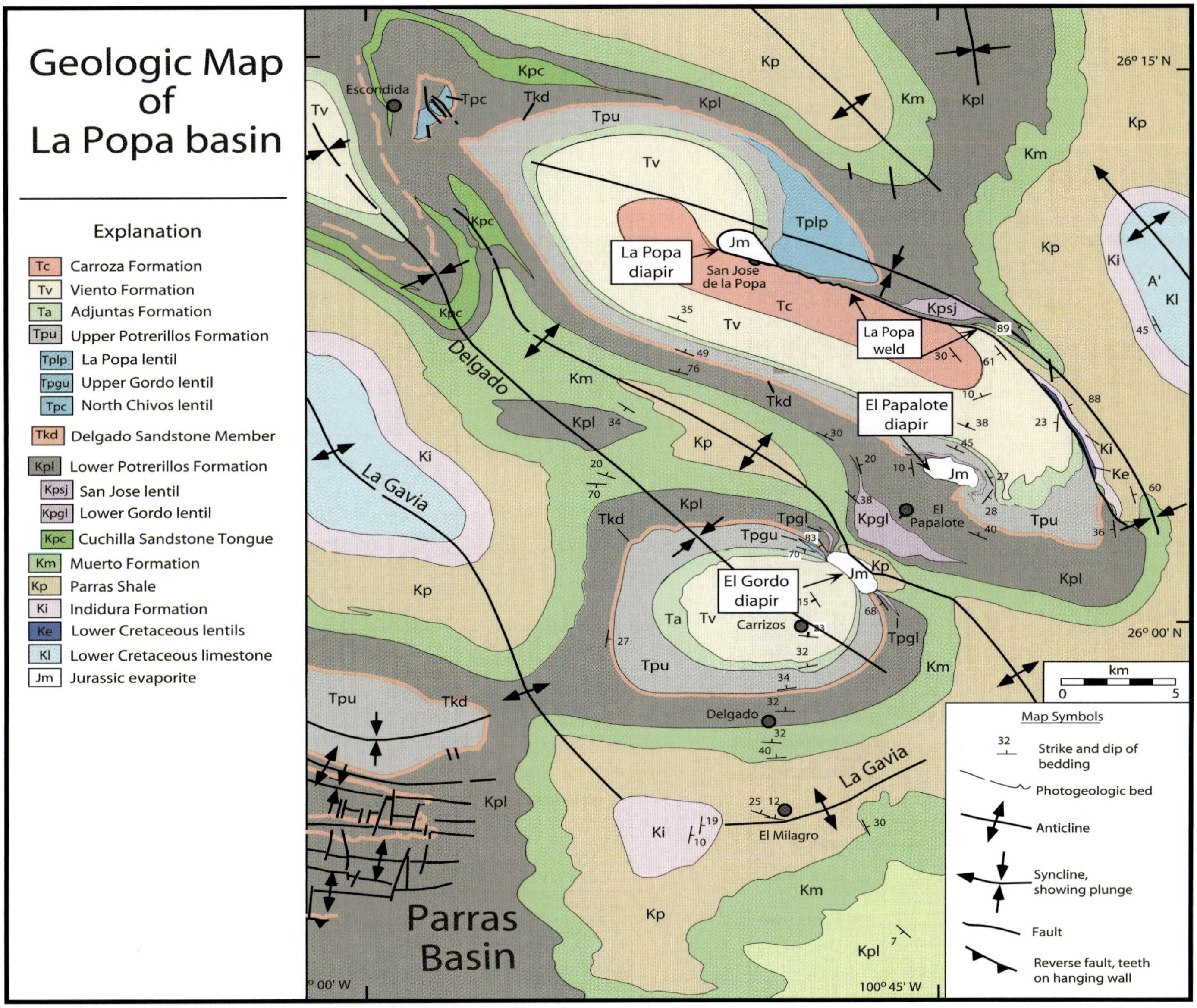

FIG. 2.—Geologic map of La Popa Basin.

This study documents the facies architecture and interprets the controls on Maastrichtian-age lentil systems present in the Lower Mudstone Member of the Potrerillos Formation at three different diapirs (El Gordo diapir, El Papalote diapir, and La Popa weld; Figs. 2, 3). The Lower Mudstone lentil system found at El Gordo diapir is referred to as the Lower Gordo Lentil (McBride et al., 1974; Laudon, 1975; Mercer, 2002); that found at El Papalote diapir is referred to as El Papalote lentil 1 (Hunnicutt, 1998; Giles and Lawton, 2002); and those present at La Popa weld are named the San Jose Lentils 1 and 2 (McBride et al., 1974; Druke, 2005). The coeval nature of the lentil systems was established by both lithostratigraphic and biostratigraphic correlations (McBride et al., 1974; Vega-Vera and Perrilliat, 1990). At El Gordo diapir and La Popa weld, additional Maastrichtian lentils are present (referred to as Middle Gordo Lentil and San Jose Lentils 3–8; Mercer, 2002; Druke, 2005) that are developed in the overlying Middle Siltstone Member of the Potrerillos Formation; however, they are not considered in this paper.

CARBONATE-PLATFORM FACIES AND DEPOSITIONAL ENVIRONMENTS

Detailed mapping of facies and measuring of stratigraphic sections of the Upper Cretaceous (Maastrichtian) carbonate lentils within the Lower Mudstone Member of the Potrerillos Formation indicates that they are actually composed of a series of carbonate horizons separated by wedges of siliciclastics (Mercer, 2002; Druke, 2005). The carbonates thin and lap out onto offshore marine black shale facies with distance from the diapir. Conversely, the intercalated siliciclastics thin toward the diapir, where they eventually pinch out by both onlap and erosion. At the siliciclastic pinchout points, the carbonate horizons amalgamate into thick continuous carbonate cliffs on outcrop, where they are mapped as a single carbonate lentil.

In general, the carbonate-lentil faunal assemblage is heterozoan (e.g., James, 1997) in character, with minor photozoan elements, and consists dominantly of red algal grains, benthic foraminifera,

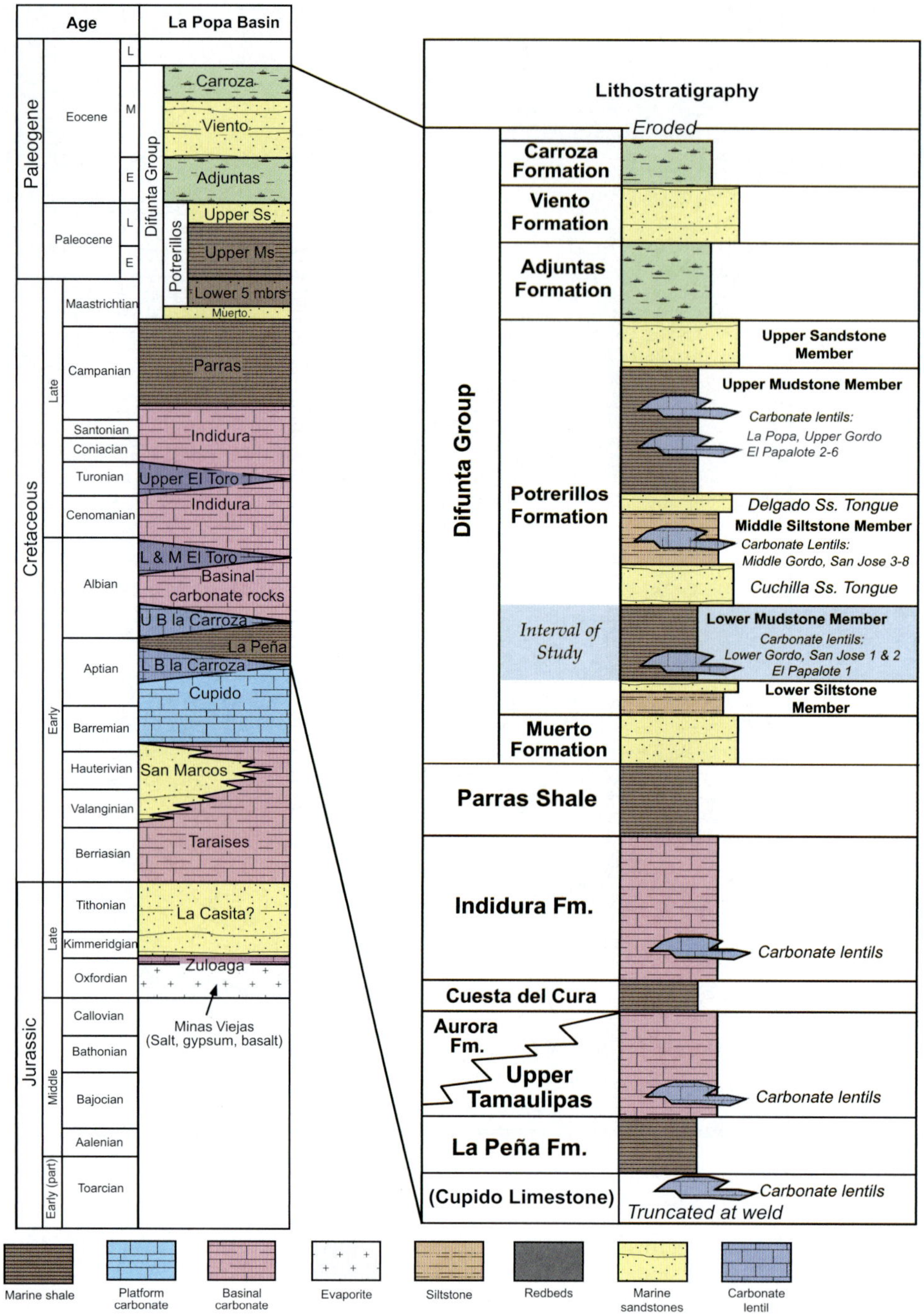

FIG. 3.—Stratigraphic column of La Popa basin (Modified from Giles et al., 2004).

rhodoliths, sponges, bivalves, and echinoids. Rudistid bivalves, which are characteristic of the shallow-water and reefal parts of other Late Cretaceous carbonate buildups, are relatively rare components of the lentils in La Popa basin. For each lentil system we describe the primary carbonate facies present and display their spatial and temporal distribution on a series of schematic cross sections constructed for each diapir.

Lower Gordo Lentil

The type and distribution of Maastrichtian Lower Mudstone Member carbonate facies at El Gordo diapir is markedly asymmetric from the southern to northern margin of the diapir (Figs. 4, 5). For this reason we have split the description of facies into southern-margin and northern-margin facies. The southern margin is dominated by a single thick (100 m+) continuous carbonate lentil (Lower Gordo Lentil) that extends only 1.5 km from the diapir. The facies comprise mainly reef boundstone, forereef packstone to rudstone, and amalgamated debris-flow units (Mercer, 2002). In contrast, the lentil on the northern margin forms a continuous cliff approximately 100 m thick near the diapir that splits into two resistant ledges separated by a recessive interval with distance from the diapir. The facies are dominated by red algal, foraminifera grainstone horizons, each about 20 m thick,

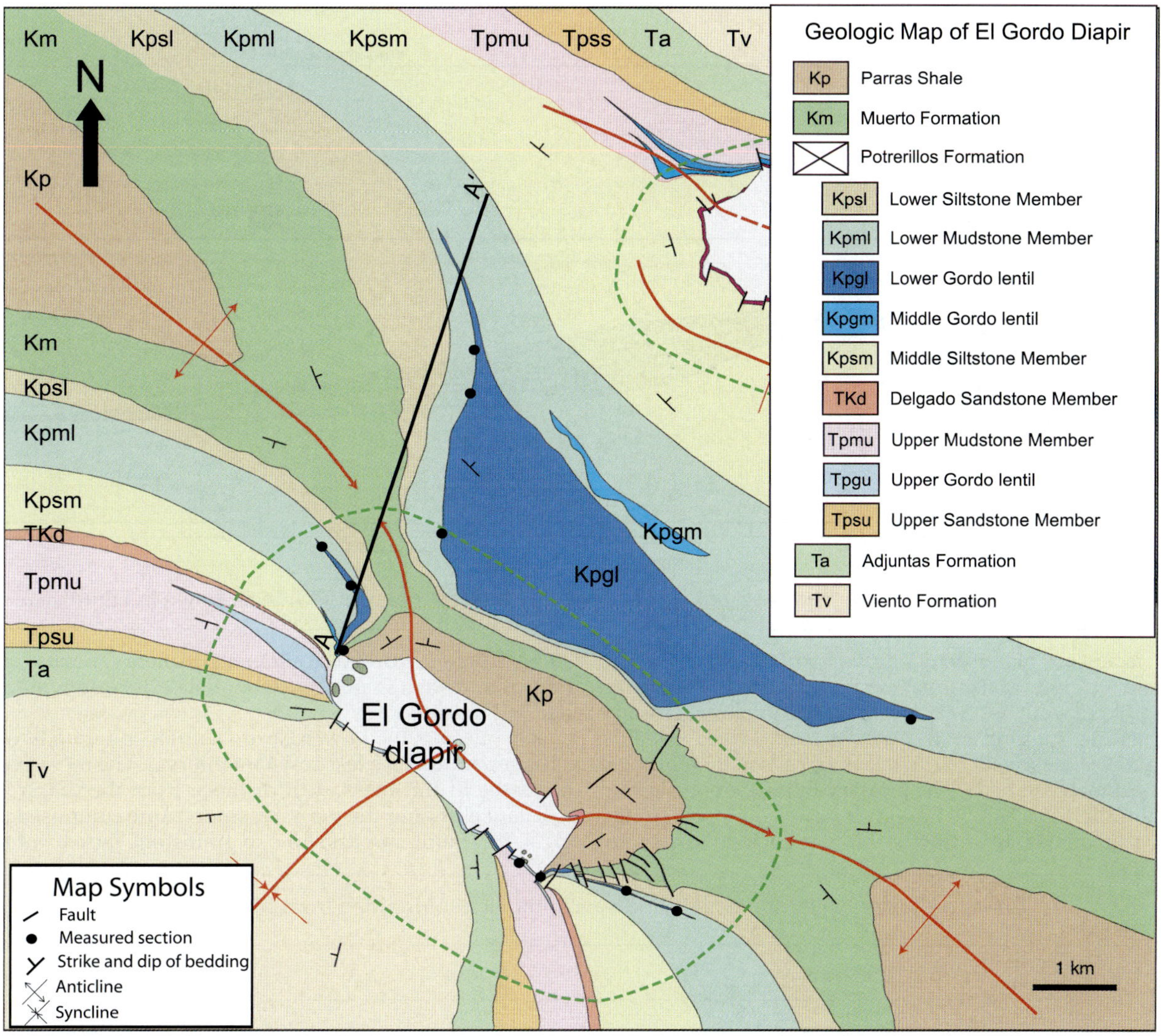

FIG. 4.—Geologic map of El Gordo diapir, showing distribution of Lower Gordo Lentil facies.

separated by a recessive wedge of silty, bivalve, red algal packstone. The northern-margin lentil extends more than 4 km from the diapir out into the surrounding minibasin.

Southern-Margin Facies.—

Carbonate lentil facies on the southern margin (Fig. 5) can be divided into five discrete lithofacies, whose characteristics and depositional setting are summarized in Table 1. A detailed description of each lithofacies and interpretation of depositional environment can be found in Mercer (2002). The facies follow a standard wave-dominated-reef model (James, 1983) with the heterozoan faunal assemblage following the shallow, moderate-energy, cool-water temperate to subtropical, carbonate shelf depth zonation model of James (1997). The lithofacies are: (1) restricted-fauna wackestone (backreef grass banks; Fig. 6A), (2) Sponge, coral, red algal boundstone (reef core; Fig. 6B), (3) Peloidal, coral, red algal packstone to rudstone (forereef; Fig. 6C), (4) Foraminifera, echinoid, red algal wackestone to grainstone (distal-slope calciturbidites; Fig. 6D), and (5) Meta-igneous-clast breccia to conglomerate (reef and diapir-derived detritus within debris flows; Fig. 6E). The units of meta-igneous-clast conglomerates are several meters thick, massively bedded, and extend more than 1 km from the diapir. The conglomerates contain abundant clasts of diapir-derived detritus that were shed by mass wasting of the diapir cap. The presence of the clasts in the surrounding sediments indicates periods of time when the diapir was exposed at the surface.

The geometry of the carbonate lentil shows wedged-shaped thickening of the lentil toward the diapir until within approximately 100 m of the diapir the carbonate lentil thins again, imparting a doubly tapering cross-sectional shape to the package. Internal facies stacking patterns show two progradational packages (parasequences) separated by a marine flooding surface. The upper parasequence displays best development (thickest and most widespread) of reefal facies. The upper surface of the lentil is bounded by a major marine flooding surface overlain by black shale beds.

Northern-Margin Facies.—

Carbonate facies on the northern margin of El Gordo diapir can be divided into three discrete lithofacies, whose characteristics and depositional setting are summarized in Table 2. The most

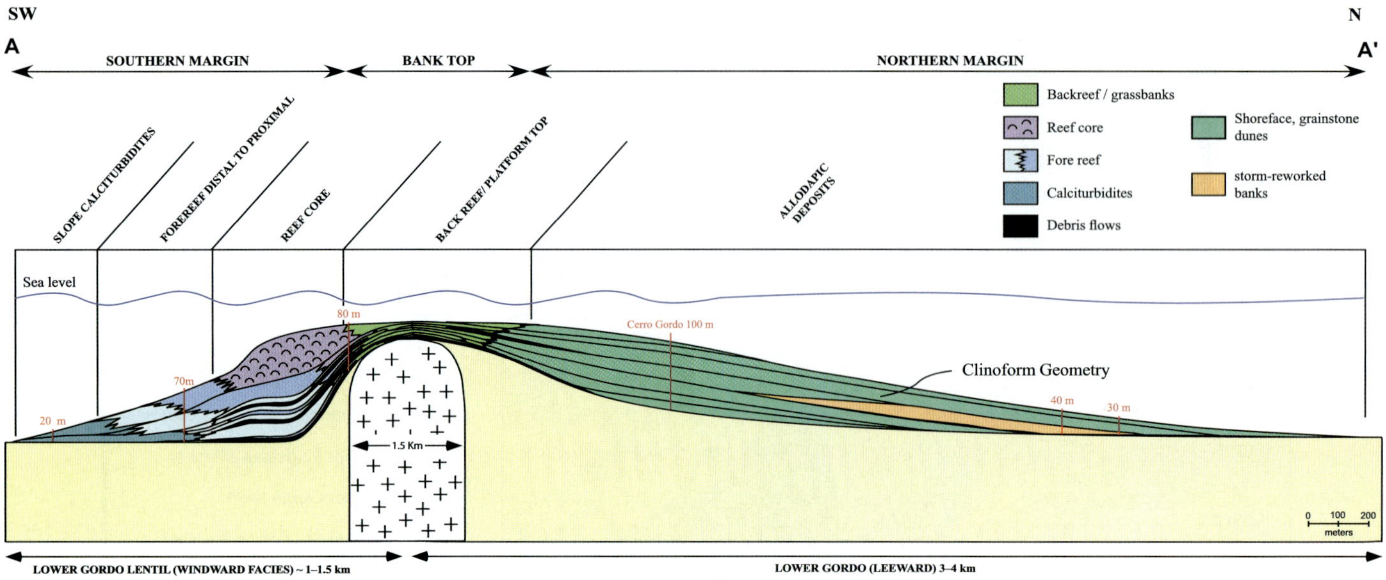

FIG. 5.—Schematic depositional cross section A–A' of Lower Gordo Lentil. Trace of cross section is shown in Figure 3.

abundant facies are planar-bedded, well sorted, red algal and foraminifera granule grainstones and packstones. The grain-dominated facies were deposited in shoreface bank systems in a back-reef setting, judging by the lack of reef-derived detritus, coarse grain size, degree of sorting, lack of carbonate mud, and clinoform geometry to beds. The primary lithofacies are: (1) red algal, foraminifera packstone and grainstone (shoreface dunes; Figs. 7A, B), (2) quartz silt, intraclast, bivalve packstone to grainstone (storm-reworked banks), and (3) meta-igneous clast, fossiliferous, red algal grainstone (shoreface / storm-reworked diapir-derived detritus).

On outcrop, the Lower Gordo lentil is split into two resistant cliffs separated by a less resistant interval. The recessive interval increases in thickness with distance from the diapir. Both the resistant cliffs and the recessive interval are composed of the red algal, foraminifera packstone to grainstone facies, but the reces-

TABLE 1.—Lower El Gordo Lentil–Windward Lithofacies (Modified from Mercer, 2002)

Depositional Facies Designation	Lithofacies Name	Description	Depositional Environment
Back Reef	Restricted-fauna wackestone	Thin- to massively bedded with pervasive bioturbation; contains benthic foraminifera, gastropods, bivalves, serpulid worms, and quartz silt within a carbonate mud matrix. Lacks red algae or other normal marine indicators.	Low energy, restricted to normal marine, shallow water above fair-weather wave base
Reef Core	Sponge, coral, and red algal boundstone	Massively bedded; constituents consist of coralline red algal (encrusting, fingers, and rhodolites), in situ large (1 m +) coral and sponge heads, green algae, and forams. Contains some carbonate mud. Contains halokinetic fractures.	Moderate to high-energy normal marine, shallow water, between storm and fair-weather wave base
Fore Reef	Peloidal, coral, red algal packstone to rudstone	Massively bedded; constituents consist of red algal fingers and rhodolites, peloids, and corals. Minor amounts of bryozoans and echinoids are present. Contains halokinetic fracturing.	Moderate to high energy, normal marine subtidal (deeper than the reef-core facies)
Slope Calciturbidites	Foraminifera, echinoid, red-algal wackestone to grainstone	Thin to massively bedded with localized bioturbation; constiuents include: red algae, echinoids, and foraminifera. Minor amounts of green algae, oysters, bryozoans, corals, and rudists are present.	Low to moderate energy with evidence of areas of high-energy deposition, normal marine, moderately deep water with near storm wave base
Subaqueous Debris Flows	Metaigneous-clast breccia to conglomerate	Massively bedded; contains lithoclasts and bioclasts of lentil debris with a high proportion of diapir-derived meta-igneous clasts. Contains halokinetic fractures.	Debris flow, low to moderate energy, deep water near storm wave base

FIG. 6.—Photomicrographs of Lower Gordo Lentil windward lithofacies. **A)** Thin-section photomicrograph of restricted fauna wackestone / back-reef grass banks. Scale bar = 0.5 mm. **B)** Outcrop photograph of sponge, red-algal reef-core facies. Sharpie pen cap = 5 cm. **C)** Outcrop photograph of peloidal, sponge, red-algal wackestone–rudstone fore-reef debris. Rudstone clasts are approximately 5 cm across. **D)** Outcrop photograph of foraminifera, echinoid, red-algal wackestone–grainstone / slope calciturbidites. **E)** Outcrop photograph of meta-igneous-clast conglomerate in subaqueous debris-flow facies. Rock hammer is approximately 31 cm long.

sive interval has thinner beds, more carbonate mud admixed with quartz silt, and more bivalve bioclasts, and grains display an overcompaction fabric with sutured grain boundaries. The recessive horizons were deposited in slightly deeper water with less winnowing (only during major storms) and no early marine cementation. Internal facies stacking patterns show two progradational parasequences forming a progradational parasequence set. Progradation was into the adjacent minibasin to the north. The upper parasequence displays the best development (thickest and most widespread) of grainstone-bank facies. The upper surface of the parasequence set is bounded by a major marine flooding surface overlain by black shale.

San Jose Lentils

The San Jose lentils only crop out on the northern side of the La Popa weld (Figs. 2, 8). During deposition of the San Jose lentils the area of the La Popa weld was occupied by salt and formed a roughly 25-km-long salt wall of unknown width (Giles and Lawton, 1999). During evacuation of salt from the salt wall, rocks on the south side of the salt wall were differentially downdropped approximately 2.5 km more than the north side, as estimated by stratigraphic offset across the weld (Loera and Giles, 2006). This process resulted in juxtaposition of Upper Cretaceous rocks against Paleocene rocks across the weld (Fig. 8).

TABLE 2.—Lower Gordo and San Jose Lentils–Leeward lithofacies (modified from Druke, 2005)

Depositional Facies Designation	Lithofacies Name	Description	Depositional Environment
Shoreface grainstone dunes	Red algal, foram packstone and grainstone	Found in thin to massive beds in clinoform geometries. Constitutes a majority of the Lower Gordo and San Jose north margin facies and is located at proximal to distal positions from the diapir. Major constituents are encrusting and finger red algae, foraminifera (*Vaughnina*), fragments of gastropods, echinoids, bivalves, green algae, corals, and bryozoans.	Lower to upper shoreface; normal marine conditions; leeward allodapic platform debris derived from grass banks
Storm-reworked banks	Quartz silt, intraclast, bivalve packstone to grainstone	Found in thin to medium nodular beds separated by argillaceous seams. The primary constituent is intraclasts that contain foraminfera (*Vaughnina*), bivalve fragments, red algal fragments, and other foraminifera. Lacks diapir-derived debris.	Offshore; normal marine conditions with increased turbidity; reworked platform facies.
Shoreface-reworked diapir-derived debris	Meta-igneous-clast fossiliferous grainstone to conglomerate	Identical to *Shoreface Subaqueous dunes* except it contains a significant increase in meta-igneous clast size and occurence. It is also found in broken up and nodular beds in the stratigraphy (usually at the basal part of the carbonate succession).	Lower to upper shoreface; normal marine conditions; storm-reworked eroded diapir-derived debris.
Rhodolite pavement	Rhodolith rudstone	Found at localized intervals, in medium tabular bedding near the salt weld. Primary consitutents consist of rhodolites in fossil hash matrix. The matrix is composed of red aglae, green algae, bivalve fragments, and gastropod fragments.	Foreshore; rhodolite pavement
Patch reef	Sponge–red algal boundstone	Thick to massive beds, laterally discontinuous, encased in red algal, foram grainstone facies. Primary constituents consist of encrusting red algae with interlayers of encrusting foraminifera (bindstone), whole sponge and fragments (bafflestones), foraminifera (*Vaughnina*), bivalves, green algal fragments, coral, and bryozoans.	Back reef, bank top; normal marine conditions; patch-reef facies

Therefore, the equivalent San Jose lentil rocks on the south side of La Popa weld are predicted to be present approximately 2.5 km in the subsurface. Facies on the northern side are very similar in type (Fig. 7A) and geometry (Fig. 7B) to those of the Lower Gordo Lentil present on the northern margin of El Gordo diapir and most likely formed under similar depositional conditions. For this reason, we have grouped the Lower Gordo and San Jose northern-margin facies together in Table 2. Two additional facies were documented in the San Jose Lentils that were not observed in the Lower Gordo Lentil northern-margin facies: (1) rhodolith framestone and rudstone facies (backreef rhodolite pavement; Fig. 7C) and (2) sponge, red algal boundstone (back-reef patch reef; Fig. 7D).

Near the diapir the San Jose lentils form a relatively continuous cliff face. With distance from the diapir it splits into two discrete cliffs (referred to as San Jose lentil 1 and lentil 2) separated by a wedged-shaped recessive horizon (Fig. 9). The recessive interval is similar to the overcompacted red algal, foraminifera packstone present at El Gordo but is thicker and more argillaceous.

El Papalote Lentil 1

El Papalote diapir is a cylindrical salt stock that is similar in size and shape to El Gordo diapir (Fig. 10). There is only one lentil (Lentil 1) developed within the Lower Mudstone Member. Lentil 1 is present on outcrop in vertical to overturned beds surrounding nearly two-thirds of the diapir (Fig. 10). The lentil is thin (< 12 m), with thick (1 m) to massive, rhodolith-bearing red-algal packstone beds and is consistent in thickness and facies types around the diapir (Hunnicutt, 1998). Two discrete facies were observed, whose characteristics and depositional setting are summarized in Table 3. Because the lentil has been deformed into vertical beds around the entire diapir (Figs. 10, 11A), all facies exposed are diapir-proximal; hence it was not possible to document a depositional-dip facies profile. In general, the facies are dominated by grains of coralline red algal sand similar to the northern-margin facies of Lower Gordo Lentil and San Jose Lentil. The two primary lithofacies present are: (1) red-algal packstone and minor grainstone (lower-shoreface subaqueous dunes; Fig. 11B) and (2) meta-igneous-clast breccia to conglomerate (reefal sponge and red-algal boundstone clasts and diapir-derived detritus within debris flows; Fig. 6E). Lenses within the red-algal packstone facies become dominated by peloids and benthic foraminifera, and contain sparse bioclasts of rudistid clams, sponges, echinoderms, bryozoans, and serpulid worm tubes. The shallowest-water depositional facies at El Papalote diapir are in general lower shoreface, containing minor hummocky cross-stratification indicating storm-wave reworking and winnowing. The facies do not show any asymmetry or circum-diapir variation as documented in the Lower Gordo and San Jose lentils.

DISCUSSION OF CONTROLS ON FACIES TYPE AND ARCHITECTURE

We have divided the controls on type and architecture of carbonate-platform facies into local controls present at individual

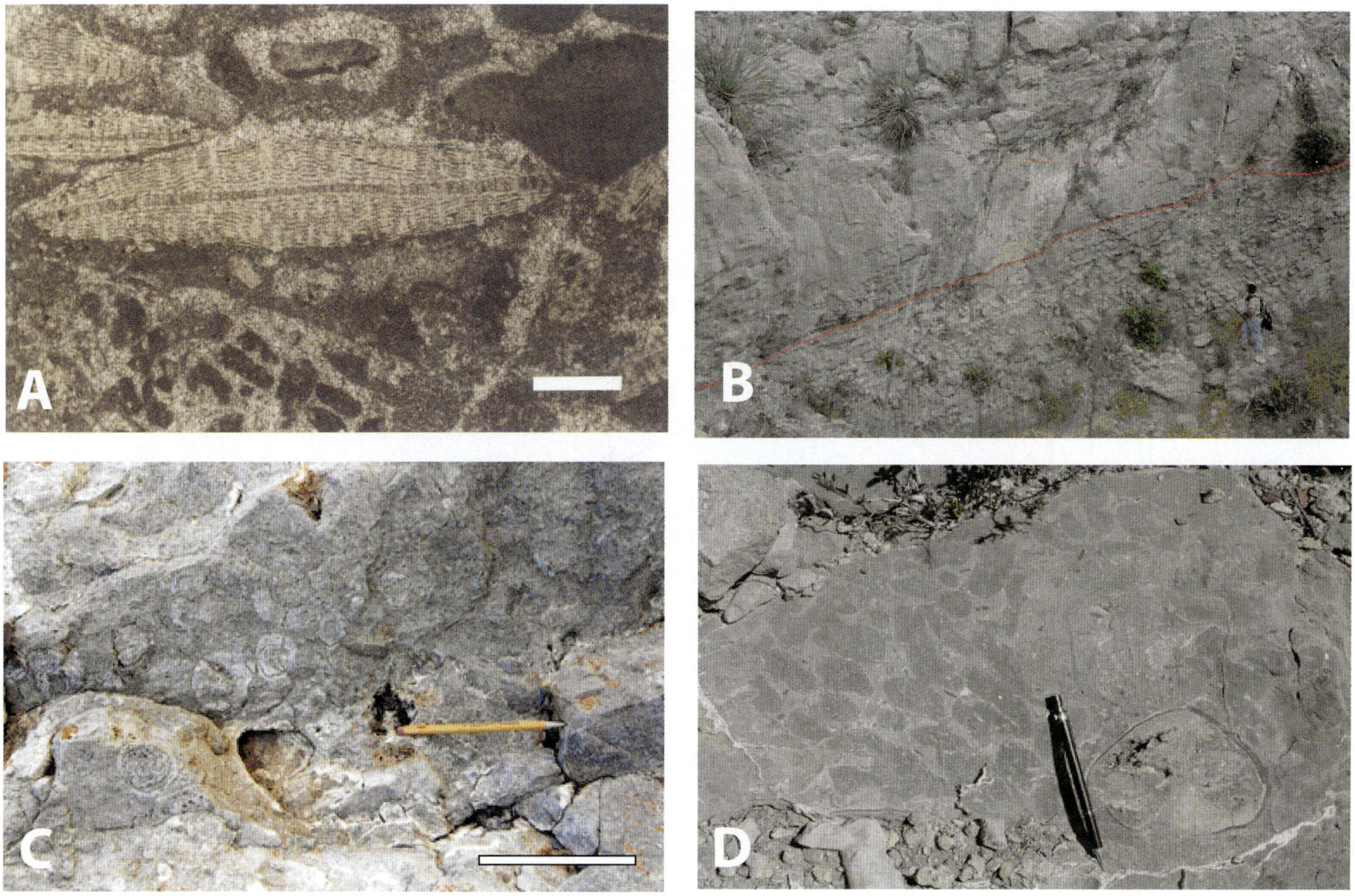

FIG. 7.—Photomicrographs of Lower Gordo Lentil and San Jose Lentil leeward lithofacies. **A)** Thin-section photomicrograph of foraminifera, red-algal packstone to grainstone. Scale bar = 0.5 mm. **B)** Outcrop photograph of foraminifera, red-algal subaqueous dune facies. **C)** Outcrop photograph of rhodolith rudstone/rhodolite pavement facies. Scale bar = 15 cm. **D)** Outcrop photograph of sponge, red-algal patch-reef facies. The mechanical pencil is approximately 15 cm long.

diapirs and more regional controls occurring across the entire shelf. The local controls reflect the local paleogeographic setting such as windward versus leeward margins in an outer-shelf setting and variations in halokinesis such as minibasin subsidence rates versus sediment accumulation rates. The sedimentologic response to local controls was highly variable between diapirs and from one side of a diapir to another. Regional controls reflect sedimentation patterns associated with eustatic sea-level change and subsidence/uplift variations generated by Hidalgoan tectonism and caused a consistent sedimentologic response in all lentil systems.

Local Paleogeographic Controls

La Popa Basin carbonate buildups are isolated platforms associated exclusively with penecomtemporaneous passive salt diapirs. The buildups have been shown to develop on topographic relief generated by the diapirs, which provided an elevated depositional setting locally sheltered from influx of terrigenous clastic sediment (Hunnicutt, 1998; Mercer, 2002; Druke, 2005).

During the Late Cretaceous (Maastrichtian) the three diapirs (El Gordo, El Papalote, and La Popa salt wall) were located at roughly the same position on a shelf-to-basin depositional profile. The dip trend of the Maastrichtian shelf was northwest to southeast, with shore-parallel trends running roughly northeast to southwest (Fig. 12). Depositional position on the shelf was determined by trends in depositional environments and current indicators in the overlying and underlying sandstone units (Muerto Formation and Delgado Sandstone Member of the Potrerillos Formation; Fig. 3) and interpretation of the Lower Mudstone Member of the Potrerillos Formation (Weislogel, 2001; Hon, 2001; Mercer, 2002; Aschoff, 2003; Aschoff and Giles, 2005; Shipley, 2004). The diapirs sat in a position slightly up dip from a change in slope from lowermost-shoreface deposition to outer-shelf deposition of black shale and siltstone facies in inter-diapir areas (Mercer, 2002; Shipley, 2004). The diapirs formed local steep-sided relief where the tops of the El Gordo and La Popa diapirs sat in shallow water above fair-weather wave base (Fig. 13) and the top of El Papalote diapir sat lower but still above storm wave base.

The Lower Gordo and San Jose lentil systems show a distinct asymmetry in type and distribution of carbonate facies. The southern margins consist of sponge–red algal boundstones deepening abruptly down depositional slope to forereef debris and foraminifera, echinoderm, red-algal packstone calciturbidites that toe out into black shale. The southern margin was very steep sided, with carbonates forming a sigmoidal geometry with great-

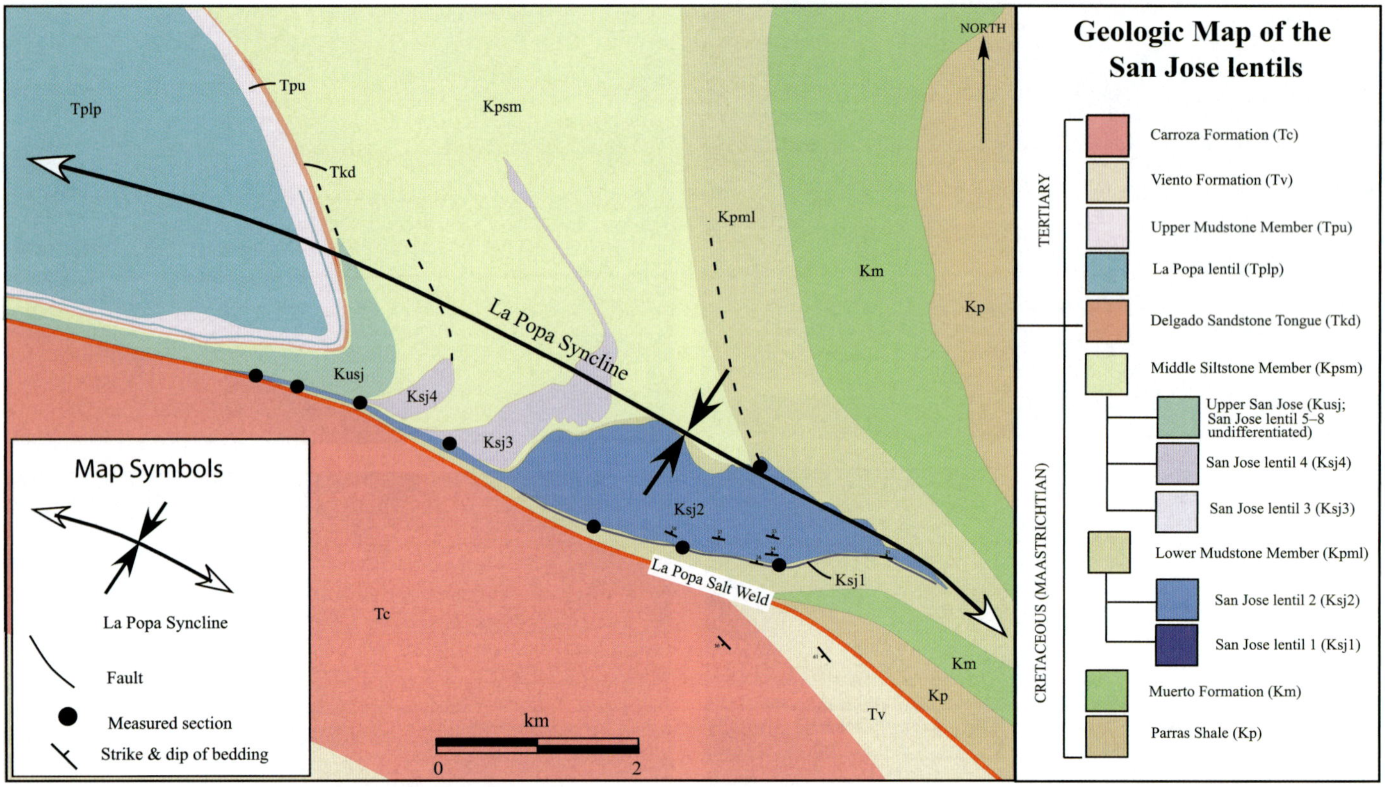

FIG. 8.—Geologic Map of La Popa salt weld and Upper Cretaceous San Jose lentils. Modified from Druke (2005).

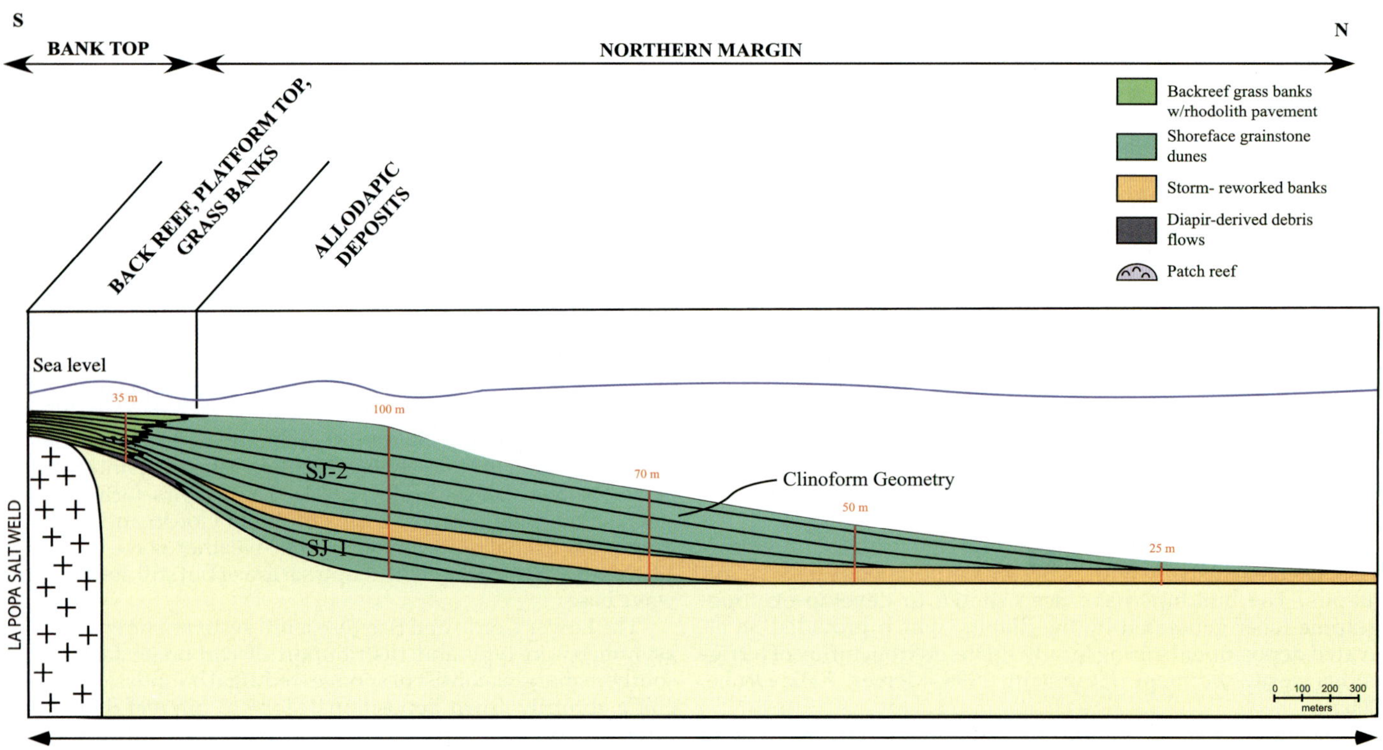

FIG. 9.—Schematic depositional cross section of San Jose Lentils 1 and 2 (leeward facies).

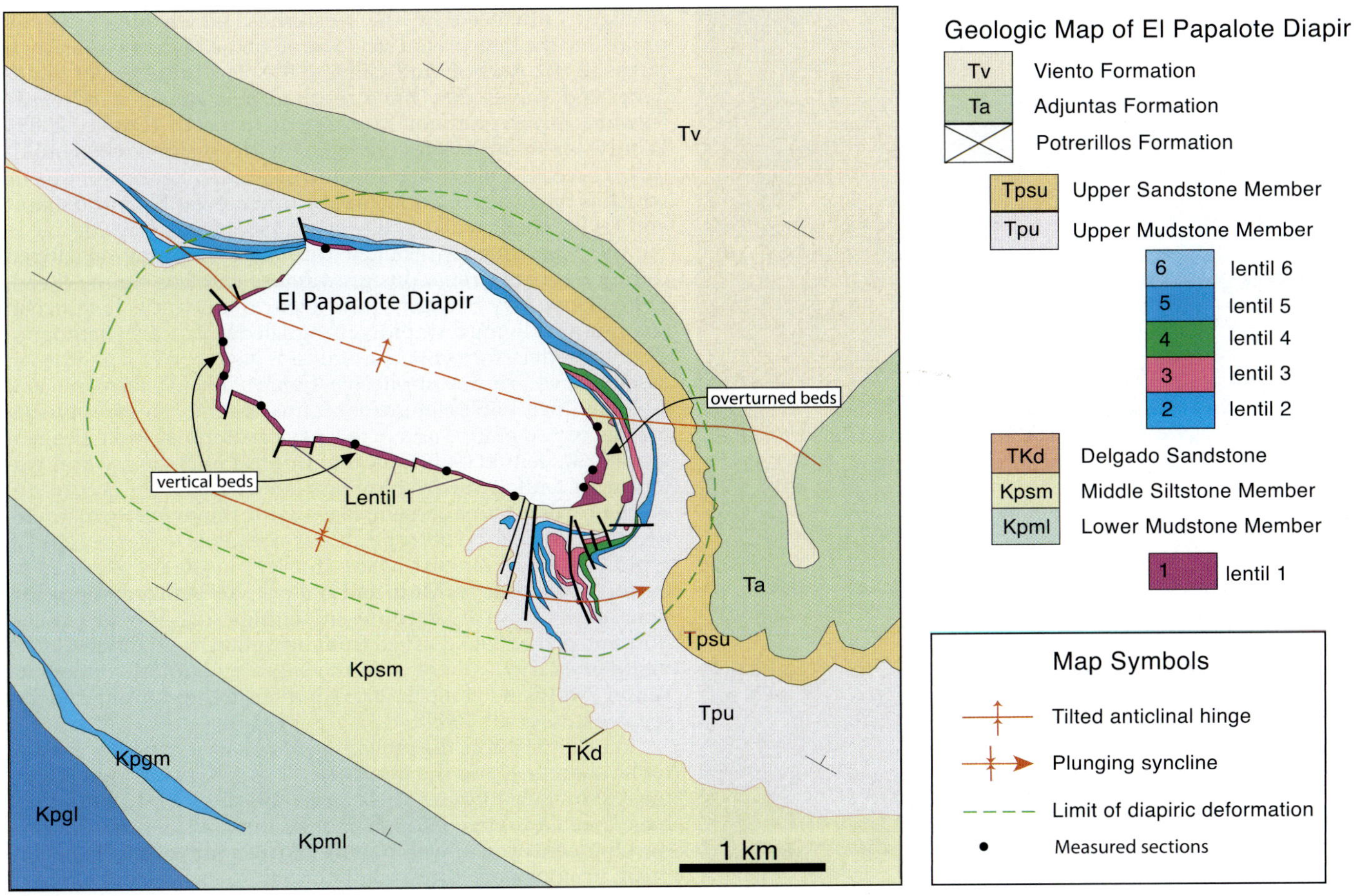

FIG. 10.—Geologic map of El Papalote diapir.

est lentil thickness a few hundreds of meters from the diapir and tapering away rapidly in both directions. Southern-margin carbonate facies extend < 1.5 km from the diapir. Northern-margin facies consist of echinoderm, foraminifera, red algal grainstones arranged into small-scale (10 m) clinoforms that prograded down the northern margin of the diapirs and the northern limbs of major Hidalgoan folds. The northern margin had relatively gentle slopes, with carbonates also forming a sigmoidal geometry that was more laterally extensive than on the southern margin. The thickest part of the lentils is significantly outboard from the diapir (1 km) and tapers gently in both directions, terminating at a distance of 3–4 km from the diapir.

The distinct facies asymmetry found at El Gordo diapir and La Popa salt wall and not at El Papalote diapir may be the result of several different local factors, including: differential subsidence related to salt withdrawal, basement-related antecedent topography, and/or windward–leeward variations in wave action. The southern margins of El Gordo diapir and La Popa salt wall face into major minibasins (Figs. 2, 12, 13, Delgado syncline and Carroza syncline) that contain substantially thicker Difunta Group

TABLE 3.—El Papalote Lentil 1 Lithofacies (modified from Hunnicutt, 1998).

Depositional Facies Designation	Lithofacies Name	Description	Depositional Environment
Lower-shoreface bank	Red-algal packstone–grainstone	Thick to massively bedded dominated by disarticulated and abraded coralline red algal grains. Locally rich in echinoids, peloids, and forams. Rare sponge, bryozoan, and serpulid worm tubes. Size sorting of grains indicates current or wave action.	Below normal wave base, but within storm wave base where minor grain reworking occurs.
Subaqueous debris flow	Meta-igneous-clast breccia to conglomerate	Massively bedded; contains lithoclasts and bioclasts of lentil debris with a high proportion of diapir-derived meta-igneous clasts. Contains halokinetic features.	Debris flow, low to moderate energy, restricted to normal marine.

FIG. 11.—Photographs of Lentil 1 facies at El Papalote diapir. **A)** Outcrop photograph of El Papalote diapir showing vertical lentils surrounding the diapir. La Popa and San Jose lentils in distant background. **B)** Photomicrograph of red-algal packstone to grainstone facies. Scale bar = 0.5 mm.

sections than minibasins on the northern side (La Popa syncline). This suggests that subsidence rates were greater in the southern-margin minibasins than in the northern-margin minibasins. Minibasin subsidence rates were controlled by both salt withdrawal (downbuilding) and Hidalgoan shortening and may have strongly influenced sedimentary facies (Rowan et al., 2003). Lawton et al. (2001) suggest that there is a basement step (down to the south) present at the La Popa weld / salt wall, on the basis of contours on magnetic basement (Eguiluz de Antunano, 1994) and stratigraphic reconstructions. This could have resulted in antecedent topographic highs on the northern margins and a steep escarpment on the southern margin, which could have affected the distribution of carbonate facies.

During the Maastrichtian, La Popa basin was located at approximately 30° N latitude (Roberts and Kirshbaum, 1995). This would place it in the modern transition zone between the northeast trade winds and the westerlies and near the northern limit of standard tropical carbonate deposition. However, during Maastrichtian greenhouse conditions the position of the subtropical high, which separates the northeast trade winds from the westerlies, may have been positioned farther to the north than it is today in the modern icehouse climate. The mapped distribution of the southern-margin and northern-margin carbonate facies suggests that the area may have been strongly influenced by the westerlies, which blew roughly parallel to the interpreted strike trend of the Maastrichtian shelf (Figs. 12, 13; Aschoff and Giles, 2005). The influence of wind-generated waves may have resulted in a strong windward–leeward facies gradient from south to north (Druke, 2005). Windward-margin facies are typically steep-sided reefal buildups, whereas leeward facies are lower-gradient grainstone belts, which is the facies relationship documented at El Gordo diapir and La Popa salt wall (Halley et al., 1983; Druke, 2005).

The Maastrichtian isolated carbonate buildups developed during the Late Cretaceous greenhouse climatic regime (Read, 1995). However, because these were dominantly heterozoan communities living on rapidly rising salt diapirs, carbonate depositional models of greenhouse buildups developed for photozoan assemblages are not applicable (James, 1997; Lehrmann and Goldhammer, 1999). Heterozoan faunal assemblages are unusual for this period of time and in this paleolatitudinal and paleogeographical position (subtropical to tropical intrashelf, warm waters in the Gulf of Mexico province; James, 1997). The occurrence of heterozoan faunal assemblages with minor photozoan elements in tropical to subtropical, warm shallow-water regimes is typical of environments where nutrient levels are elevated to eutrophic conditions (Mutti and Hallock, 2003). Eutrophic conditions may occur in several different settings, such as where cooler, nutrient-rich waters derived from upwelling onto the shelf prevail (James, 1997), there is an increased amount of continental runoff (Mutti and Hallock, 2003) or there are significant cold gas seeps (Suess et al., 1985).

In La Popa basin, the prevailing winds were blowing parallel to the shoreline, and the interpreted break in slope from lowermost shoreface to outer-shelf facies (Shipley, 2004) would not have created a large enough change in water depth to permit development of an upwelling zone. To the south of the diapirs the break in slope deepened rapidly into the Sierra Madre Oriental foreland basin, where changes in water depth may have been sufficient to generate upwelling (Ye, 1997). However, the onshore prevailing wind direction interpreted from the windward–leeward facies distribution is inconsistent with a southern source of upwelling to explain the presence of the heterozoan facies assemblage. Also, there are no complementary facies that typify upwelling zones, such as phosphorites, glauconite, or radiolarian cherts (O'Brien and Veeh, 1983; Brasier, 1995).

Nutrient flux related to terrestrial runoff sources can increase to eutrophic levels where river deltas build into enclosed basins (Mutti and Hallock, 2003) and especially during wetter climatic regimes (Lukasik et al., 2000). Throughout deposition of the Difunta Group in the enclosed La Popa salt basin, the primary dispersal mechanism of siliciclastic sediments was via river deltas (McBride et al., 1974). However, the carbonate lentils formed only in an outer-shelf environment when deltas were not present in La Popa basin but were forming updip in northern Mexico. Carbonate lentils did not form on the diapirs when large delta complexes such as those found in the Muerto Formation or the Delgado Sandstone Tongue were being deposited (Fig. 3). This suggests that terrestrial sources of nutrients were not controlling factors in the deposition of the carbonate lentils.

An alternative explanation for the localized high nutrient levels in warm, shallow waters on the shelf is to have local cold seeps (Suess et al., 1985). Cold seeps are common features associated with shelfal passive diapirs in the Gulf of Mexico today, where they provide nutrients to a diverse heterozoan faunal assemblage dominated by bivalves and serpulid worms (Roberts and Aharon, 1994; Sassen et al., 2004). Presumably this same process could have lead to the heterozoan carbonate buildups on salt diapirs in La Popa basin during the Maastrichtian. Source

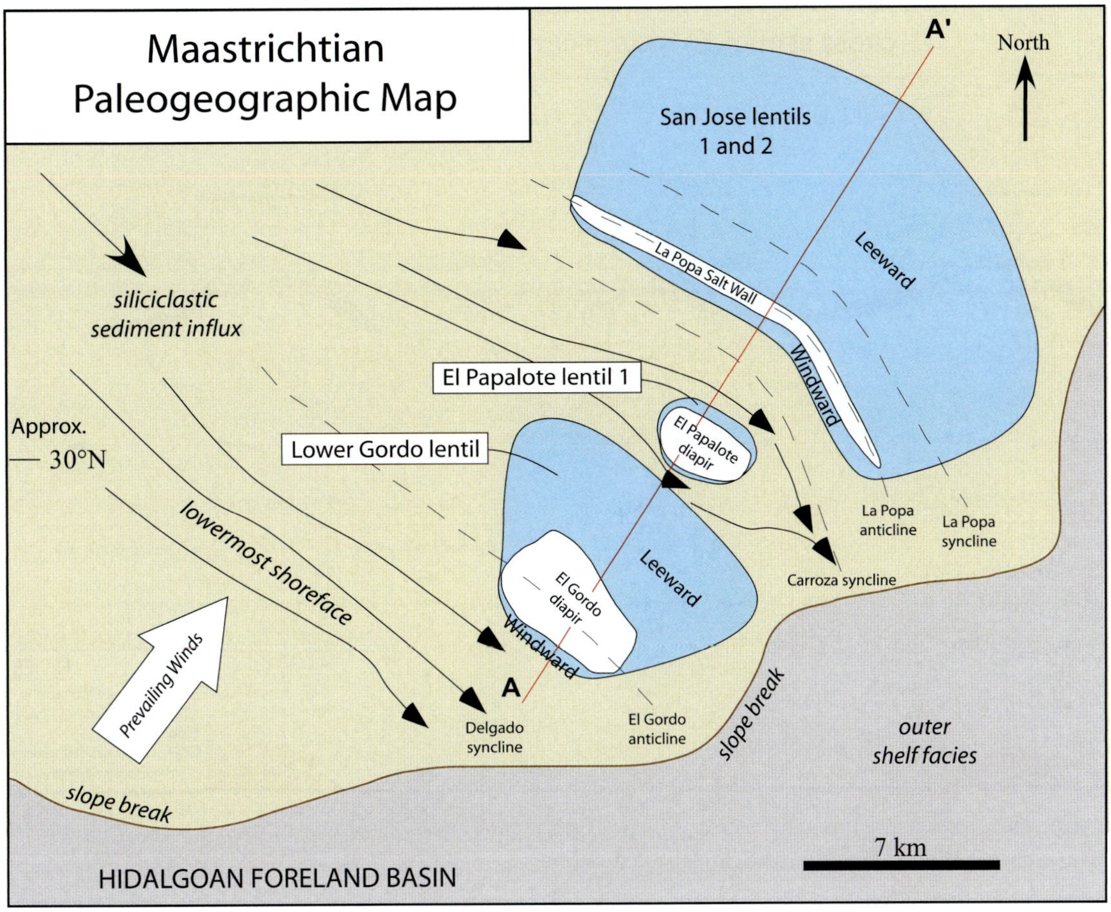

FIG. 12.—Maastrichtian paleogeographic map of the El Gordo, El Papalote, and La Popa diapirs.

rocks and evidence of migrating hydrocarbons associated with diapirs have been identified in La Popa basin (Hudson, 2004), suggesting that this mechanism is feasible. However, preliminary carbon isotope analysis of cements and matrix from the lentils did not yield characteristic signatures of methane-related precipitation (Ganging Jiang, personal communication, 2006).

Halokinetic Controls

Near-diapir strata in La Popa basin are arranged into halokinetic sequences (Giles and Lawton, 2002). Halokinetic sequences are conformable successions of strata genetically influenced by near-surface diapir or extrusive salt rise, and bounded near the diapir by angular unconformities that become conformable with distance from the salt body (Giles and Lawton, 2002). Halokinetic sequence boundaries are readily identified by pronounced near-diapir angular unconformities. Halokinetic sequences display internal stratal onlap and thinning in the direction of the diapir, indicating concurrence of halokinesis and deposition.

Halokinetic sequences are generated in response to local variations in the rise rate of passive diapirs relative to sediment accumulation rates adjacent to the diapir (Giles and Lawton, 2002). Angular unconformities form during or following periods when rates of diapir rise greatly exceed local rates of sediment accumulation. The imbalance of relative rates leads to topographic "inflation" of the diapir, creating steep, unstable slopes that eventually fail. Slope failure and mass wasting create the angular unconformities and debris-flow beds, which are further steepened by drape-folding deformation during continued rise of the diapir and subsidence of the adjacent minibasin (Rowan et al., 2003). In La Popa basin, carbonate lentil facies developed on diapir topographic relief during periods of decreased siliciclastic influx (slow net sediment accumulation rates) when diapir inflation was at its greatest (Fig. 14, Phase 1 and 2). Increased sediment accumulation rates that substantially exceeded diapir rise rates led to infill of the adjacent minibasin followed by onlap and eventual overlap of the diapir. Minibasin infill resulted in burial of the diapir and suppression of diapir topographic relief (Fig. 14, Phase 3). The angularity of the unconformity was substantially increased by subsidence-related drape folding following deposition during continued halokinesis. As a result, the unconformity was reactivated as a fault surface near the diapir (Fig. 14, Phases 3 and 4).

All of the halokinetic sequences in the Lower Mudstone Member display intense near-diapir deformation with major 90° angular unconformities at the base and top of the sequence. These are attributes of Type A halokinetic sequences (Giles et al., 2004), which form during periods of regionally slow sediment accumulation rates on the shelf relative to diapir rise rates. The halokinetic sequence stratigraphy varies in character between diapirs and between the southern/windward and northern/leeward margins of each diapir. The differences reflect local variations in net diapir rise rates and net sediment accumulation rates.

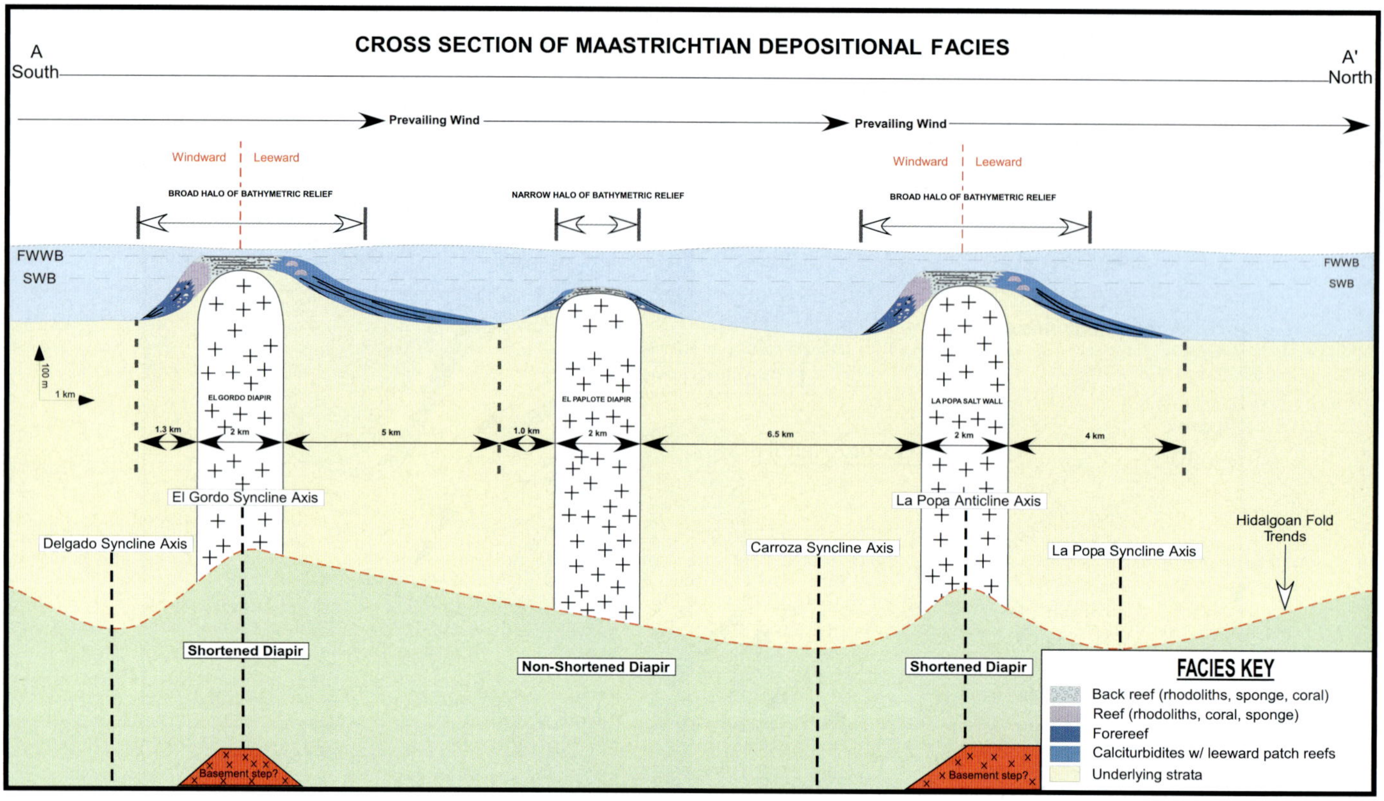

FIG. 13.—Schematic depositional profiles of each of the three lentil systems showing the position of fair-weather wave base (FWWB) and storm wave base (SWB), Hidalgoan anticlines and synclines.

Halokinetic sequences at El Gordo diapir and the former La Popa salt wall are more numerous and contain thicker (by an order of magnitude) basal carbonate buildup facies than halokinetic sequences at El Papalote diapir. This suggests that siliciclastic sediment accumulation rates around El Gordo diapir and La Popa salt wall were lower than rates at El Papalote diapir and / or that diapiric rise rates were higher at El Gordo diapir and La Popa salt wall than at El Papalote diapir. El Gordo diapir and La Popa salt wall occupy the anticlinal axes of Hidalgoan folds (Fig. 13) that were forming during deposition of the Lower Mudstone Member (Druke, 2005). El Papalote diapir occupies a position on the limb of a Hidalgoan fold. Diapirs positioned on the axis of a fold would have experienced greater shortening than diapirs situated on the limbs (Rowan et al., 2003). Tectonic squeezing of the El Gordo and La Popa diapirs may have led to higher salt rise rates in those diapirs than limb-positioned diapirs like El Papalote. Siliciclastic sediment influx across the shelf would have preferentially deposited sediment in topographic lows (i.e., Delgado and Carroza synclines), resulting in less sediment accumulation over the anticlines. These factors could account for variations in character of halokinetic sequences developed at El Gordo and La Popa diapirs compared to those developed on fold-limb diapirs (El Papalote).

Overlying the basal angular unconformity or halokinetic sequence boundary on the southern / windward margin of El Gordo diapir are thick carbonate-debris-flow facies containing abundant clasts of diapir-derived detritus. The thick debris flows grade upward into massive sponge, red-algal reef facies that are abruptly overlain by black shale. In the Lower Mudstone Member there is a single halokinetic sequence developed on the southern side of the El Gordo diapir, which contains thick carbonate lentils dominated by reefal and debris-flow facies. In contrast, on the northern margins halokinetic sequences have siliciclastic wedges in between two thin carbonate lentil horizons. The northern carbonate lentils have minor, small (sand- and pebble-size) diapir-derived detritus at their bases, indicating that two halokinetic sequences are developed on the northern / leeward margins. El Papalote lentil 1 is dominated by debris-flow facies derived from slumping of material from the top of the diapir. The lentil is relatively thin (< 13 m) and extends < 1 km from the diapir and represent a single halokinetic sequence.

Eustatic-Sea-Level Controls

The eustatic sea-level curve for the Maastrichtian generated by Haq et al. (1988) correlates well with previously interpreted third-order depositional sequence stratigraphic trends in La Popa basin (Fig. 15), suggesting that eustasy strongly influenced the timing of regional siliciclastic sediment influx into La Popa basin. The third-order depositional sequence stratigraphic framework presented here represents a compilation of a series of previously published detailed studies that documented regional relationships of siliciclastic units in the Difunta Group. Eustatic sea-level changes influenced fluctuations in sediment influx into the basin and local net sediment accumulation rates near diapirs, which in turn generated halokinetic sequences near the passively rising diapirs. Carbonate lentil facies are found only near diapirs and form an important component of Type A halokinetic sequences (Giles et al., 2004) when net siliciclastic sediment accumulation on the shelf was very low (deposition of black shale and mudstone). Low net siliciclastic sediment accumulation on the shelf corresponds with the transgressive systems tract (TST) of the previ-

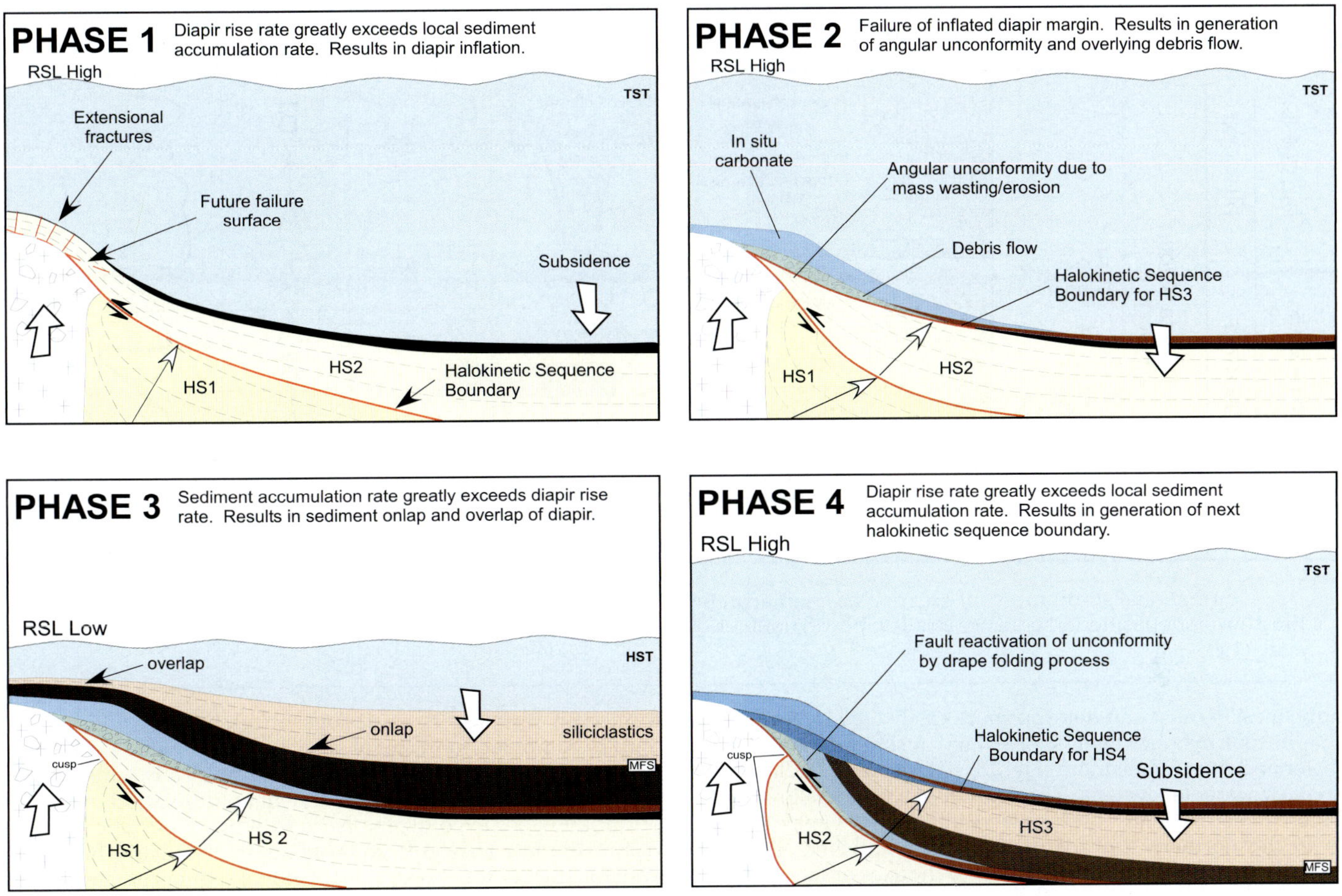

FIG. 14.—Generalized progressive model for the formation of halokinetic sequences. HS = halokinetic sequence. Modified from Druke (2005).

ously interpreted third-order siliciclastic depositional sequences and major rises in eustatic sea level (Fig. 15).

The Maastrichtian began with major eustatic sea-level fall, which resulted in progradation of the Muerto Formation deltas across La Popa basin during the late highstand systems tract (HST) of a third-order depositional sequence (Fig. 15, Ye, 1997; Hon, 2001; Weislogel, 2001; Soegaard et al., 2003). The Muerto siliciclastics onlapped and eventually completely buried the three diapirs, effectively eliminating all diapir-related topographic relief (Hon, 2001; Weislogel, 2001). Type B halokinetic sequences (Giles et al., 2004), which lack carbonate lentil and debris-flow facies, formed around the diapirs at this time (Hon, 2001). The Lower Siltstone Member along with the overlying Lower Mudstone Member of the Potrerillos Formation forms the trangressive systems tract (TST) of the next third-order depositional sequence (Ye, 1994; Mercer, 2002; Soegaard et al., 2003). The Lower Mudstone Member contains the significant carbonate lentil buildups described in this study. The lentils formed during times of low siliciclastic sediment accumulation where diapir inflation generated topographic highs that extended into clear, shallow water, permitting local carbonate deposition. The Cuchilla Sandstone Tongue, which locally forms the base of the Middle Siltstone Member, forms the HST of that sequence (Ye, 1994; Soegaard et al., 2003). The Cuchilla Sandstone Tongue is not present in the area near the three diapirs (Fig. 2), suggesting that the eustatic fall in sea level shown by Haq et al. (1988) was not significant enough to force the Cuchilla sands to prograde across the entire basin or that basin-wide subsidence rates were high. The next sequence formed entirely within the Middle Siltstone Member. Carbonate lentils formed within the TST (Druke, 2005), and deltaic facies of the Delgado Sandstone Tongue form the HST of that depositional sequence (Aschoff and Giles, 2005; Shipley, 2004). Increased siliciclastic input in prograding Delgado Sandstone deltas onlapped and buried the three diapirs, removing the relief required for carbonate sedimentation (Aschoff and Giles, 2005).

Regional Tectonic Controls

La Popa basin lies within the foreland-basin system of the Sierra Madre Oriental fold-and-thrust belt (Fig. 1), which is part of a retroarc foreland basin or collapsing remnant ocean basin related to subduction of the Farallon Plate off the west coast of Mexico (Dickinson and Lawton, 2001). Hidalgoan shortening in La Popa Basin started in the early Maastrichtian (Druke, 2005) and corresponds to a period of rapid subsidence (Hudson, 2004) related either to foreland flexural subsidence or to basin-wide salt withdrawal due to tectonic and sediment loading. Rapid subsidence in conjunction with major eustatic sea-level rise severely decreased siliciclastic sediment influx into the basin, which in turn led to major diapir inflation and concomitant deposition of carbonate buildups on diapir topographic highs.

Hidalgoan shortening may have differentially enhanced diapir topographic relief at the three diapirs (Rowan et al., 2004). El Gordo diapir and La Popa salt wall lie on the axes of Hidalgoan

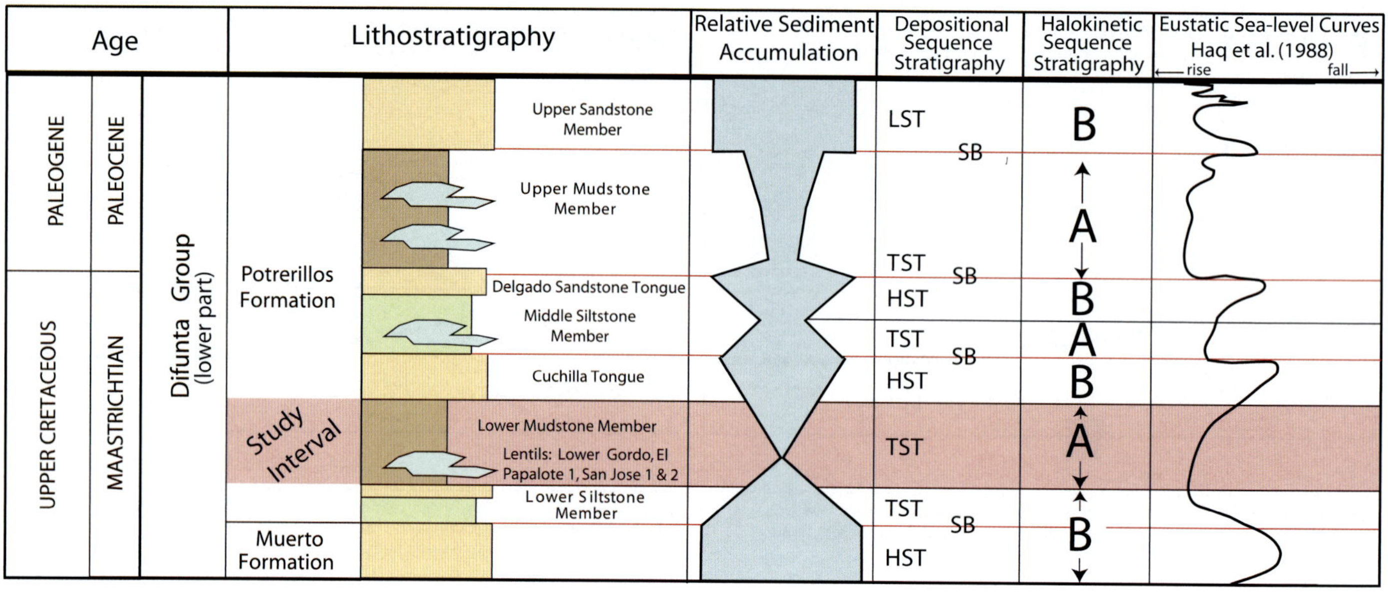

FIG. 15.—Formational stratigraphy of La Popa basin interpreted with regard to relative sediment accumulation rates on the shelf, third-order depositional sequence stratigraphy, halokinetic sequence stratigraphic type, and eustatic sea-level trends from Haq. et al. (1998).

anticlines. Because salt deforms more easily than strata, the salt diapirs took up much of the shortening, causing the diapirs to rise faster and thus inflate more (Fig. 13). These diapirs inflated into shallow water (a few tens of meters deep) where shallow-water, heterozoan carbonate buildups were initiated. El Papalote diapir is in the northern limb of the El Gordo anticline. This area experienced less shortening relative to the axes of the anticlines: since El Papalote diapir was in the downflexed limb (i.e., deeper water, 40–70 m) and was not "squeezed" as much, the diapir did not develop the topographic relief that the other two diapirs did.

Rowan et al. (2004) demonstrated that shortened diapirs rise higher and have a broader zone of bathymetric relief (Fig. 13). The shortened diapirs (El Gordo and La Popa) have carbonate lentils that formed in shallower water than the non-shortened diapir (El Papalote) and the carbonates extended over a much greater distance (< 4 km) with gentler depositional slopes than observed at the El Papalote lentil (< 1 km). Thus, El Papalote diapir would have been in deeper water and experienced less shortening because of its position on the fold. Siliciclastic-sediment-dispersal studies of slightly older and younger packages show that sediment followed a NE–SW trend, focused in the synclinal lows and away from the diapir highs (Weislogel, 2001; Hon, 2001; Aschoff, 2005).

CONCLUSIONS

Isolated Maastrichtian heterozoan carbonate buildups (lentils) developed on diapir-related topographic relief in La Popa basin. Carbonate buildups at El Gordo and La Popa salt wall consist of shallow-water heterozoan reef communities that display well-developed windward–leeward facies relationships and extensive leeward / backreef progradation of grainstone facies. In contrast, carbonate buildups at El Papalote are deeper-water, mud-prone heterozoan packstone facies that display uniform lithofacies and laterally restricted distribution around the diapir.

Carbonate buildups or lentils are present only at the bases of Type A halokinetic sequences, above angular unconformities developed adjacent to the diapirs. The halokinetic sequences formed in response to variations in relative rise rates of the diapirs compared to influx of siliciclastic sediment. Carbonates formed when the diapirs were topographically inflated and diapir rise rates far exceeded sediment accumulation rates. Diapir inflation occurred during periods of major eustatic sea-level rise in the late transgressive systems tract to early highstand systems tract of third-order depositional sequences, and during periods of major flexural subsidence and salt withdrawal.

Differential relative rise rates of the diapirs is related to position with respect to shortening structures, where diapirs on the axes of anticlines have higher rise rates and a broader halo of bathymetric relief. Carbonates that developed on these diapirs formed in shallower water, have greater lateral extent, and have gentler depositional slopes than carbonates that developed on non-shortened diapirs.

ACKNOWLEDGMENTS

This summary of our research on Late Cretaceous carbonate buildups in La Popa Basin has benefited greatly from our interactions and discussions with Tim Lawton and Mark Rowan, and numerous graduate students and colleagues at NMSU, UNAM, and UNLV over the past decade. Insightful reviews by Eric Blanc and Brian Coffey significantly improved the manuscript. Acknowledgment is made to the donors of the Petroleum Research Fund, administered by the American Chemical Society for grant AC8-33339 to Giles and Lawton, American Association of Petroleum Geologists Grants-in-Aid Awards to Dominic Druke, David Mercer, and Lela Hunnicutt, and to the sponsors of the La Popa Basin Joint Industry Consortium, administered by the Institute of Tectonic Studies at New Mexico State University. Sponsors include: Amerada Hess Corporation, Anadarko Petroleum Corporation, BHP Billiton Petroleum Inc., BP America Production Company, ChevronTexaco Exploration and Production Company, ConocoPhillips, Inc., Devon Energy Production Company, ENI-S. P. A., Agip Division, GEBA, ExxonMobil Exploration, Kerr-McGee Oil & Gas Corporation, Marathon Oil Company, Maxus (U.S.) Exploration Co., Nexen Petroleum U.S.A. Inc.,

Norsk Hydro ASA, Samson, Shell Offshore Inc., Statoil, Total, and Union Oil of California.

REFERENCES

ASCHOFF, J.L, 2003, Sedimentation patterns in a salt-diapir influenced foreland basin: Upper Cretaceous to Lower Tertiary Delgado Sandstone tongue, Potrerillos Formation, La Popa basin, Nuevo Leon, Mexico: Unpublished M.S. Thesis, New Mexico State University, Las Cruces, 259 p.

ASCHOFF, J.L., AND GILES, K.A., 2005, Salt diapir–influenced, shallow marine sediment dispersal patterns: Insights from outcrop analogs: American Association of Petroleum Geologists, Bulletin, v. 89, p. 447–469.

BRASIER, M.D., 1995, Fossil indicators of nutrient level. 1. Eutrophication and climate change, *in* Bosence, D.W.G., and Allison, P.A., eds., Marine Palaeoenvironmental Analysis from Fossils: Geological Society of London, Special Publication 83, p. 113–132.

CREVELLO, P.D., WILSON, J.L., SARG, J.F., AND READ, J.F., eds., 1989, Controls on Carbonate Platform and Basin Development: SEPM, Special Publication 44, 405 p.

DECELLES, P.G., AND GILES, K.A., 1996, Foreland basin systems: Basin Research, v. 8, p. 105-123.

DICKINSON, W.R., AND LAWTON, T.F., 2001, Carboniferous to Cretaceous assembly and fragmentation of Mexico: Geological Society of America, Bulletin, v. 113, p. 1142–1160.

DRUKE, D.C., 2005, Sedimentology and stratigraphy of the San Jose Lentil, La Popa basin, Mexico and implications for carbonate development in a tectonically influenced salt basin: Unpublished M.S. Thesis, New Mexico State University, Las Cruces, 132 p.

EGUILUZ DE ANTUNANO, S., 1994, La formación Carbonera y sus implicaciones tectónicas, estados de Coahuila y Nuevo León: Sociedad Geológica Mexicana, Boletín, v. 50, p. 3–4.

GARRISON, J.M., AND MCMILLAN, N.J., 1999, Evidence for Jurassic continental rift magmatism in northeast Mexico: allogenic meta-igneous blocks in El Papalote diapir, La popa basin, Nuevo Leon, Mexico, *in* Bartolini, C., Wilson, J.l., and Lawton, T.F., eds., Mesozoic Sedimentary and Tectonic History of North-Central Mexico: Geological Society of America, Special Paper 340, p. 319–332.

GILES, K.A., AND LAWTON, T.F., 1999, Attributes and evolution of an exhumed salt weld, La Popa Basin, northeastern Mexico: Geology, v. 27, p. 323–326.

GILES, K.A., AND LAWTON, 2002, Halokinetic sequence stratigraphy adjacent to the El Papalote diapir, northeastern Mexico: American Association of Petroleum Geologists, Bulletin, v. 86, p. 823–840.

GILES, K.A., LAWTON, T.F., AND ROWAN, M.G., 2004, Summary of halokinetic sequence characteristics from outcrop studies of La Popa salt basin, northeastern Mexico, *in* Post, P.J., Olson, D.L., Lyons, K.T., Palmes, S.L., Harrison, P.F., and Rosen, N.C., eds., Salt–Sediment Interactions and Hydrocarbon Prospectivity: Concepts, Applications, and Case Studies for the 21st Century: SEPM, Gulf Coast Section, 24th Annual Bob F. Perkins Research Conference, Program and Abstracts, 16 p.

HALLEY, R.B., HARRIS, P.M., AND HINE, A.C., 1983 Bank margin environment, *in* Scholle, P.A., Bebout, D.G., and Moore, C.H., eds., Carbonate Depositional Environments: American Association of Petroleum Geologists, Memoir 33, p. 464–506.

HAQ, B.U., HARDENBOL, J., AND VAIL, P.R., 1988, Mesozoic and Cenozoic chronostratigraphy and cycles of sea-level change, *in* Wilgus, C.K., Hastings, B.S., Ross, C.A., Posamentier, H.W., Van Wagoner, J.C., and Kendall, C.G.St.C., eds., Sea-Level Changes: An Integrated Approach: SEPM, Special Publication 42, p. 72–108.

HARRIS, P.M., SALLER, A.H., AND SIMO, J.A.T., eds., 1999, Advances in Carbonate Sequence Stratigraphy: Application to Reservoirs, Outcrops, and Models: SEPM, Special Publication 63, 421 p.

HON, K.D, 2001, Salt-influenced growth-stratal geometries and structure of the Muerto Formation adjacent to an ancient secondary salt weld, La Popa Basin, Nuevo Leon, Mexico: Unpublished Master's Thesis, New Mexico State University, Las Cruces, 97 p.

HUDSON, S.M., 2004, Hydrocarbon occurrences within the La Popa basin: Potential source rocks, thermal maturation, and hydrocarbon migration along the La Popa salt weld, La Popa basin, NE Mexico, Unpublished M.S. Thesis, University of Nevada, Las Vegas, 140 p.

HUNNICUTT, L.A., 1998, Tectonostratigraphic interpretation of Upper Cretaceous to Lower Tertiary limestone lentils within the Potrerillos Formation surrounding El Papalote diapir, La Popa basin, Nuevo Leon, Mexico: Unpublished M.S. Thesis, New Mexico State University, Las Cruces, 181 p.

JAMES, N.P., 1983, Reef environment, *in* Scholle, P.A., Bebout, D.G., and Moore, C.H., eds., Carbonate Depositional Environments: American Association of Petroleum Geologists, Memoir 33, p. 346–462.

JAMES, N.P., 1997, The cool-water carbonate depositional realm, *in* James, N.P., and Clarke, A.D., eds., Cool-Water Carbonates: SEPM, Special Publication 56, p. 1–20.

JAMES, N.P., AND CLARKE, J.A.D., eds., 1997, Cool-Water Carbonates: SEPM, Special Publication 56, 440 p.

LAUDON, R.C., 1975, Stratigraphy and petrology of the Difunta Group, La Popa and eastern Parras basins, northeastern Mexico: Unpublished Ph.D. Dissertation, University of Texas, Austin, 294 p.

LAUDON, R.C., 1984, Evaporite diapirs in the La Popa basin, Nuevo León, México: Geological Society of America, Bulletin, v. 95, p. 1219–1225.

LAWTON, T.F., VEGA, F.J., GILES, K.A., AND ROSALES-DOMINGUEZ, C., 2001, Stratigraphy and origin of the La Popa Basin, Nuevo León and Coahuila, Mexico, *in* Bartolini, C., Buffler, R.T., and Cantu-Chapa, A., eds., The Western Gulf of Mexico Basin: Tectonics, Sedimentary Basins, and Petroleum Systems: America Association of Petroleum Geologists, Memoir 75, p. 219–240.

LEHRMANN, D.J., AND GOLDHAMMER, R.K., 1999, Secular variation in parasequence and facies stacking patterns of platform carbonates: A guide to application of stacking-patterns analysis in strata of diverse ages and settings, *in* Harris, P.M., Saller, A.H., and Simo, J.A.T., eds., Advances in Carbonate Sequence Stratigraphy: Application to Reservoirs, Outcrops, and Models: SEPM, Special Publication 63, p. 187–225.

LOERA, D.M., AND GILES, K.A., 2006, Attributes and history of the Cretaceous–Eocene La Popa salt weld derived from halokinetic sequence stratigraphy (abstract): American Association of Petroleum Geologists, 2006 Annual Convention, Abstracts Volume, p. 65.

LUKASIK, J.J., JAMES, N.P., MCGOWRAN, B., AND BONE, Y., 2000, An epeiric ramp: low energy, cool-water carbonate facies in a Tertiary inland sea, Murray basin, South Australia: Sedimentology, v. 47, p. 851–881.

MCBRIDE, E.F., WEIDIE, A.L., AND WOLLEBEN, J.A., 1975, Deltaic and associated deposits of the Difunta Group (Late Cretaceous to Paleocene), Parras and La Popa Basins, northeastern Mexico, *in* Broussard, M.L.S., ed., Deltas: Houston Geological Society, p. 485–522.

MCBRIDE, E.F., WEIDIE, A.L., WOLLEBEN, J.A., AND LAUDON, R.C., 1974, Stratigraphy and structure of the Parras and La Popa basins, northeastern Mexico: Geological Society of America, Bulletin, v. 84, p. 1603–1622.

MERCER, D.A., 2002, Analysis of the growth strata of the Upper Cretaceous to Lower Paleogene Potrerillos Formation adjacent to El Gordo salt diapir, La Popa Basin, Nuevo Leon, Mexico: Unpublished M.S. Thesis, New Mexico State University, Las Cruces, 229 p.

MURRAY, G.E., WEIDIE, A.E., JR., BOYD, D.R., FORDE, R.H., AND LEWIS, P.D., JR., 1962, Formational subdivisions of the Difunta Group, Parras basin, Coahuila and Nuevo Leon, Mexico: American Association of Petroleum Geologists, Bulletin, v. 46, p. 374–383.

MUTTI, M., AND HALLOCK, P., 2003, Carbonate systems along nutrient and temperature gradients: some sedimentological and geochemical constraints: International Journal of Earth Science (Geologische Rundschau), v. 92, p. 465–475.

O'Brien, G.W., and Veeh, H.H., 1983, Are phosphorites reliable indicators of upwelling?, *in* Suess, E., and Thiede, J., eds., Coastal Upwelling; Its Sedimentary Record: New York, Plenum Press, p. 399–419.

Read, J.F., 1995, Overview of carbonate platform stratigraphy and reservoirs in greenhouse and ice-house worlds, *in* Read, J.F., Kerans, C., and Weber, L.J., eds., Milankovitch Sea Level Changes, Cycles and Reservoirs in Greenhouse and Icehouse Worlds: SEPM, Short Course Notes, no. 35, part 1, p. 1–102.

Roberts, H.H., and Aharon, P., 1994, Hydrocarbon-derived carbonate buildups of the northern Gulf of Mexico continental slope—A review of submersible investigations: Geo-Marine Letters, v. 14, p. 135–148.

Roberts, L.N.R., and Kirschbaum, M.A., 1995, Paleogeography of the Late Cretaceous of the Western Interior of middle North America—Coal distribution and sediment accumulation: U.S. Geological Survey, Professional Paper 1561, 115 p., 1 pl.

Rowan, M.G., Lawton, T.F., Giles, K.A., and Ratliff, R.A., 2003, Near-salt deformation in La Popa basin, Mexico, and the northern Gulf of Mexico: A general model for passive diapirism: American Association of Petroleum Geologists, Bulletin v. 87, p. 733–756.

Sassen, R., Sweet, S.T., Defreitas, D.A., Eaker, N.L., Roberts, H.H., and Zhang, C., 2004, Brine vents on the Gulf of Mexico Slope: Hydrocarbons, carbonate–barite–Uranium mineralization, red beds, and life in an extreme environment, *in* Post, P.J., Olson, D.L., Lyons, K.T., Palmes, S.L., Harrison, P.F., and Rosen, N.C., eds. Salt–Sediment Interactions and Hydrocarbon Prospectivity: Concepts, Applications, and Case Studies for the 21st Century (abstract): SEPM, Gulf Coast Section, 24th Annual Bob F. Perkins Research Conference, Program and Abstracts, p. 21.

Shipley, K.W., 2004, Ejecta-bearing deposits at the Cretaceous–Tertiary boundary and their implications for timing of Hidalgoan (Laramide) folding, La Popa basin, Nuevo Leon, Mexico: Unpublished M.S. Thesis, New Mexico State University, Las Cruces, 146 p.

Soegaard, K., Ye, H., Daniels, A., and Halik, N., 1996, Late Cretaceous–early Tertiary evolution of foreland to Sevier–Laramide fold–thrust belt, northeast Mexico (abstract): Geological Society of America, Abstracts with Programs, v. 28, no. 7, p. A115.

Soegaard, K., Ye, H., Halik, N., Daniels, A.T., Arney, J.W., and Garrick, S., 2003, Stratigraphic evolution of the latest Cretaceous to early Tertiary Difunta Group foreland basin in northeast Mexico: Influence of salt withdrawal on tectonically induced subsidence by the Sierra Madre Oriental fold and thrust belt, *in* Blickwede, J., ed., Sequences, Stratigraphy, Sedimentology: Surface and Subsurface: Canadian Society of Petroleum Geologists, Memoir 15, p. 401–416.

Suess, E., Carson, B., Ritger, S.D., Moore, J.C., Jones, M.L., Kulm, L.D., and Cochrane, G.R., 1985, Biological communities at vent sites along the subduction zone off Oregon, *in* Jones, M.L., ed., The Hydrothermal Vents of the Eastern Pacific—An Overview: Biological Society of Washington, Bulletin no. 6, p. 475–484.

Vega-Vera, F.J., and Perrilliat, M.C., 1990, Molúscos del Maastrichtiano de la Sierra el Antrisco, estado de Nuevo León: Universidad Nacional Autónoma de México, Paleontología Mexicana, no. 55, 65 p.

Weidie, A.E., and Murray, G.E., 1967, Geology of Parras Basin and adjacent areas of northeastern Mexico: American Association of Petroleum Geologists, Bulletin, v. 51, p. 678–695.

Weislogel, A.L., 2001, The influence of diapirism and foreland evolution on the depositional system, stratigraphy, and petrology of the Maastrichtian Muerto Formation, La Popa Basin, Mexico: Unpublished M.S. Thesis, New Mexico State University, Las Cruces, 310 p.

Ye, H., 1997, Sequence stratigraphy of the Difunta Group in the Parras–La Popa foreland basin, and tectonic evolution of the Sierra Madre Oriental, NE Mexico: Unpublished Ph.D. Dissertation, University of Texas, Dallas, 197 p.

EVALUATING STRUCTURAL CONTROLS ON THE FORMATION AND PROPERTIES OF CARBONIFEROUS CARBONATE RESERVOIRS IN THE NORTHERN MIDCONTINENT, U.S.A.

W. LYNN WATNEY, EVAN K. FRANSEEN, AND ALAN P. BYRNES
Kansas Geological Survey, The University of Kansas, 1930 Constant Avenue—Campus West, Lawrence, Kansas 66047, U.S.A.
e-mail: lwatney@kgs.ku.edu, evanf@kgs.ku.edu, abyrnes@kgs.ku.edu
AND
SUSAN E. NISSEN
Kansas Geological Survey, The University of Kansas, 1930 Constant Avenue—Campus West, Lawrence, Kansas 66047, U.S.A.
Present Address: 5761 Wellman Road, Mc Louth, Kansas 66054, U.S.A.
e-mail: nissen_susan@yahoo.com

ABSTRACT: During the Carboniferous a carbonate shelf covered areas of the central U.S, including Kansas, with the outer shelf and shelf margin intermittently extending through southern Kansas and northern Oklahoma. The regional setting resulted in deposition of relatively laterally continuous carbonate and siliciclastic facies belts. Areally sparse subsurface well data and surface exposures have led to the interpretation that most structures on the craton are simply shallow draped anticlines and associated synclines primarily reflecting general aspects of regional structure and depositional and erosional heterogeneity. Models that assume a broad continuous shelf relegate local structure to being minor or non-existent. However, our recent examination of subsurface data, 3-D seismic, and rock properties for oil fields from the Middle Mississippian shelf margin, Middle Pennsylvanian mid-shelf, and Late Pennsylvanian lower shelf indicate that regional- and kilometer-scale structures (e.g., faults, fractures, lineaments) segmented the shelf and shelf-margin areas in Kansas, primarily along Precambrian structures that were reactivated throughout the Phanerozoic. Movement on faults resulted in segmentation expressed as rhombic-shaped structural blocks (1–10s km) with subtle variations in relief (generally meter to ~ 70 m) and slope (near zero to upwards of 2–3 m/km). Regional, down-to-basin block faulting produced linear shelf edges and segmentation of the ramp and shelf profile repeatedly during the Carboniferous. The association of stratal packages and rock properties with structural elements argues that structure exerted continued, but episodic, influence and affected sediment accommodation, depositional patterns, paleotopography, weathering intensity, diagenesis, and later fluid movement, including hydrocarbon emplacement. Results from our study of the "stable" shelf carbonates of the Midcontinent indicate that tectonic events may have had far-reaching effects and caused structural deformation in the interiors of cratons. Sedimentologic and stratigraphic analyses in such settings can benefit by evaluating the possible influence of subtle faulting and fault reactivation on depositional and diagenetic patterns that can significantly influence rock properties and reservoir development.

INTRODUCTION

During the Carboniferous a carbonate shelf covered broad areas of the central U.S., including Kansas, with the outer shelf and shelf margin extended through southern Kansas (Lane and De Keyser, 1980; Witzke, 1990; Witzke and Bunker, 1996). This regional setting resulted in deposition of extensive, laterally continuous carbonate and siliciclastic facies belts deposited in stacked stratal packages. Kansas is characterized by broad basins and domes covering thousands of square kilometers with prominent uplift and subsidence occurring episodically throughout the Phanerozoic (Newell et al., 1987).

Based on available well and outcrop data, a common working hypothesis is that most structures on the craton are simply shallow draped anticlines and associated synclines reflected primarily by gentle regional fold structures and depositional and erosional heterogeneity. Given this paradigm, sea level, climate, and other depositional processes are thought to be predominant controls on facies and stratigraphic architecture. Though the influence of depositional environment and facies on rock properties have been well studied for Mississippian (e.g., Watney et al., 2001; Franseen, 2006) and Pennsylvanian (e.g., Watney et al., 1995) strata in the Midcontinent, there are patterns to facies distribution, field locations, and rock properties that are not adequately explained by a simple scenario of laterally extensive and uninterrupted facies belts and stratal packages occurring on a broad, continuous shelf. Examination of the juxtaposition of facies and rock properties with structures, 3-D seismic, and hydrocarbon accumulations leads to an alternate working hypothesis. Our ongoing studies indicate that local structures (e.g., faults, fractures, lineaments) segmented the broad shelf and shelf-margin areas in Kansas, and were reactivated throughout the Phanerozoic. Structural reactivation appears to have influenced depositional patterns, paleotopography, weathering intensity, diagenesis, and later fluid movements. Correlation between structural features and rock properties argues that reactivated, in some cases syndepositional, faulting and structural movement may have been a primary control on development of rock properties. The location of changes in rock properties and faults argues that reactivated faults may be linked to basement structural elements. Therefore, basement configuration and structural elements may serve as a template for understanding and predicting areas of segmentation, facies and stratigraphic patterns, and hydrocarbon accumulation. This paper reviews the structural setting of the "stable" carbonate-dominated shelf and examines the role of structure in several Carboniferous shelf–shelf-margin examples (Figs. 1, 2).

Controls on Carbonate Platform and Reef Development
SEPM Special Publication No. 89, Copyright © 2008
SEPM (Society for Sedimentary Geology), ISBN 978-1-56576-130-8, p. 125–145.

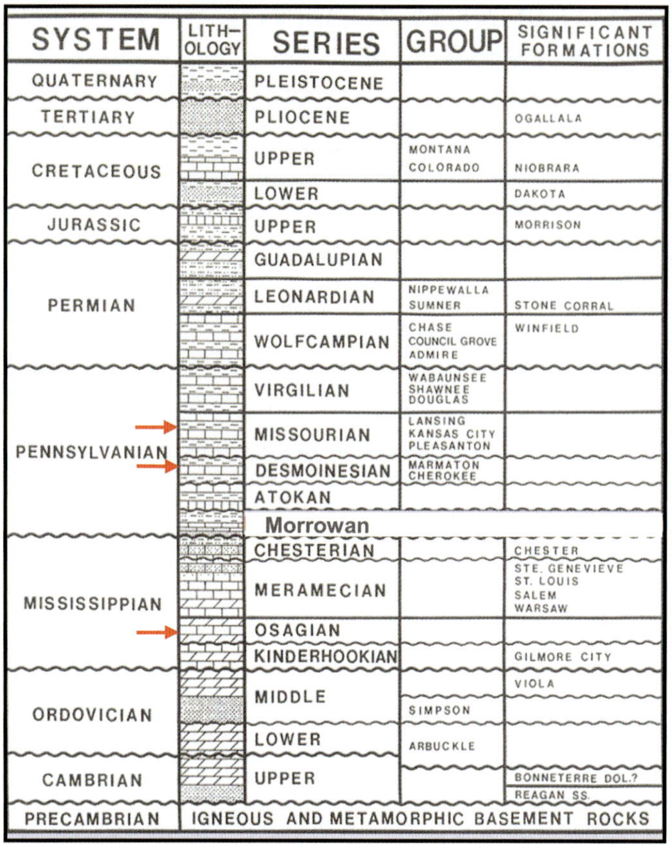

FIG. 1.—Stratigraphic column in Kansas following Zeller (1968) with arrows identifying intervals examined in this study.

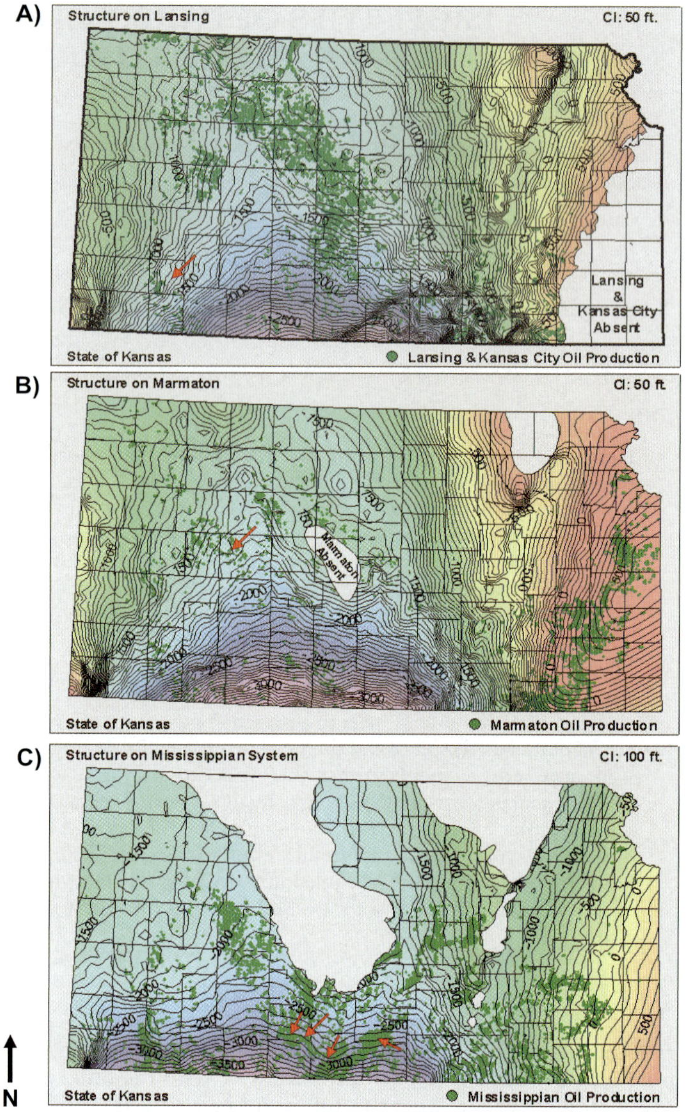

FIG. 2.—Structural maps for three stratigraphic horizons annotated with arrows showing fields examined in the paper: **A)** Upper Pennsylvanian Lansing and Kansas City Groups (Victory Field), **B)** configuration of the Middle Pennsylvanian Marmaton Group (Dickman Field), and **C)** configuration on top of Mississippian System (Nichols, Glick, Donald, and Spivey–Grabs fields delineated from east to west on map). Horizontal distance across Kansas is 400 mi (644 km). (modified from Gerlach, 1988).

REGIONAL FAULTS, BASEMENT STRUCTURE, AND LINEAMENTS

Geology and Structure of Precambrian Basement

Precambrian basement rocks of the upper Midcontinent have been studied using well cuttings, cores, limited outcrops, and potential fields and seismic geophysics. Though major structural elements and provinces of the Precambrian basement are well defined (Van Schmus et al., 1987; Gerhard, 2004; Sims et al., 2005), the basement configuration in the Midcontinent and Kansas has been variously depicted with only a few major faults (Cole, 1976) (Fig. 3) to many faults (Berendsen and Blair, 1996; Baars and Watney, 1991). This range in interpretation results from: (1) few well bores that penetrate basement, (2) varied utilization of basement compositional data, (3) nearly vertical orientation of many faults and lack of offset in individual wells, (4) assumption of continuity of subsurface data, and (5) lack of continuous surface exposures. Recent 3-D seismic imaging is leading to a clearer picture of basement and Phanerozoic horizons, lineaments, and faults. In addition, integration of basement structure and on reduced-to-pole aeromagnetic anomaly maps reveals correlations between anticlines, synclines, and abrupt linear changes in dip vectors with magnetically inferred compositional and structural features of the basement (Yarger, 1983; Van Schmus et al., 1987; Kruger, 1996) (Fig. 3).

Ancestral Rocky Mountain and Ouachita Tectonism and Reactivation of Precambrian Structures

Many prior structural studies of the U.S. craton have investigated deformation associated with Ancestral Rocky Mountain (ARM) and Ouachita tectonism, recognizing a west-northwest and north-northeast fault orientation defining a rhombic-shaped grid (Fig. 4; Marshak et al., 2000) with the same orientation in the present Rocky Mountains and central U.S. (Dorobek, 1995; Yang and Dorobek, 1995; Brown, 1978; Gerhard et al., 1982; Gerhard et al., 1987; Brown and Brown, 1987; Berendsen et al., 1983; Baars

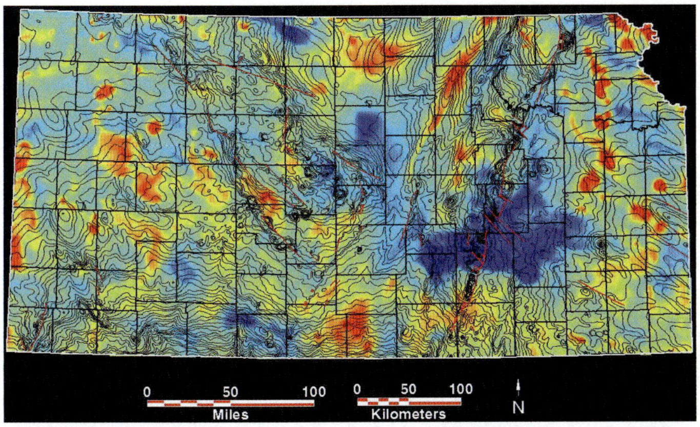

FIG. 3.—Aeromagnetic map of Kansas acquired on 1/4 mi (0.4 km) intervals along east–west flight lines spaced 2 mi (3.2 km) apart. North–south tie lines flown approx. 20 mi apart. Reduced to pole using 65 degrees and declination of 7 degrees. Shaded relief with vertical illumination with lowest magnetic values in blue. Steeper gradients are darker shading (Kruger, 1996). Contours are configuration of the Precambrian surface with mapped basement faults shown by red lines (Cole, 1976).

and Watney, 1991; Berendsen and Blair, 1996; Watney et al., 1997; Gay, 1999; Guo, 1999; Gerhard, 2004; Merriam, 2005; Stevenson and Baars, 1986; Thomas and Baars, 1988; Stone, 1999; Thomas, W.A., 1983; Jacobi and Fountain, 2002; Duchek et al., 2004; Kluth and Coney, 1981; Baars et al., 1995; Yang and Dorobek, 1995; Macfarlane and Wilson, 2006).

Geometries of associated uplifts have led to interpretations of faults and folds forming by inversion of two sets of Proterozoic extensional faults, originating at 1.1–1.3 Ga and 0.7–0.9 Ga, that were reactivated by stress transmitted during Phanerozoic compressional orogenies associated with ARM, Ouachita, and Laramide tectonism (Marshak et al., 2000). Reactivated faults are reverse, oblique, and transpressional / strike-slip (Marshak et al., 2000). Some west-trending Proterozoic faults are interpreted as NW-trending syn-rift accommodation zones crossing the NE-trending Midcontinent Rift System (MRS), including at least one that crosses the Nemaha Uplift at the Kansas–Nebraska border and others along the upper reaches of the MRS in the north-central U.S. (Van Schmus and Hinze, 1985; Hinze et al., 1997). The accommodation zones are recognized by apparent tens of kilometers of strike-slip offset of the MRS. Subsequent to rifting, Late Precambrian and Paleozoic reactivation, including that associated with ARM and Ouachita deformation, led to structural inversion of the MRS. Movement along NE-trending faults bounding the east side of the MRS led to formation of the Nemaha Uplift, and NW-trending cross faults are linked, in part, to reactivation of Precambrian accommodation zones (Berendsen and Blair, 1996; Berendsen et al., 1988; Baars and Watney, 1991). Vertical, high-angle, and reverse faulting of ~ 3000 ft (914 m) (in reverse sense from original Precambrian MRS faulting) occurred along the Humboldt Fault System during ARM-related deformation of the Nemaha Uplift in Kansas (Berendsen and Blair, 1996). Regional erosion of the Precambrian is estimated to have removed approximately 10 km of section prior to the Phanerozoic (Timmons et al., 2001). This contributes to uncertainties in assessing magnitudes of vertical movement associated with basement structures, particularly since rift-basin sedimentary fills have been removed.

Diachronous subsidence of ARM basins was also coincident with sequential closure of the Ouachita suture bordering the

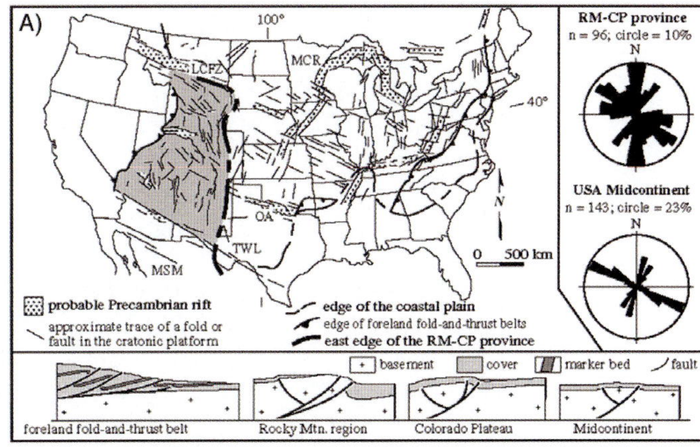

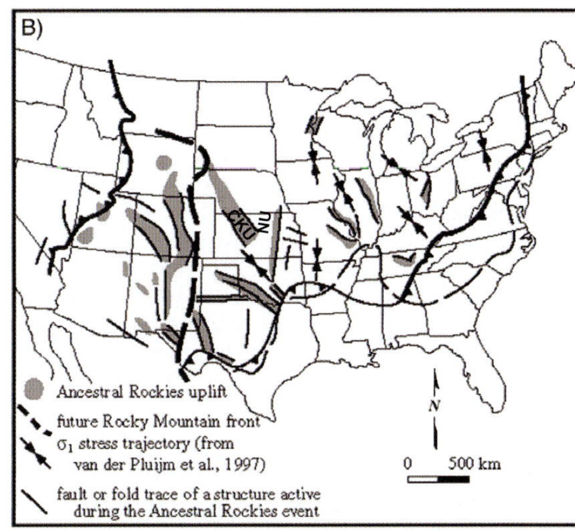

FIG. 4.—**A)** Map of faults and folds on US craton that highlights the Midcontinent Rift (MCR), Oklahoma aulacogen (OA) bordering the southern Midcontinent, and Rocky Mountain–Colorado Plateau province (RM-CP) forming western border of Midcontinent. The rose diagram of fault trends shows dominant NW and NE trends. (from Marshak et al., 2000). **B)** Ancestral Rockies structures including northwest-trending Central Kansas Uplift (CKU) in western Kansas and N-NE trending Nemaha Uplift (NU) in eastern Kansas. Stress trajectories, σ_1, range from N–S to NW–SE trends, which extend inboard into Midcontinent from Ouachita–Marathon orogenic suture zone (from Marshak et al., 2000, modified from Kluth and Coney, 1981, and McBride and Nelson, 1999).

southern portion of the Midcontinent (Fig. 5; Marshak et al., 2000; Dickinson and Lawton, 2003). The combined ARM and Ouachita orogenies began in the Chesterian and extended through the Leonardian (335 Ma to 275 Ma). Downflexure of the foreland basins along the flanks of the Ouachita orogen migrated west and southwest at the rate of ~ 35 km/My (Dickinson and Lawton, 2003). As a consequence of this migration, stress trajectories across the craton would have evolved from NW to NE trending, and a likely scenario would have been for sequential and systematic strain release by reactivation of optimally aligned preexisting faults (Thatcher, 1995, 2003).

In addition to the migration of stress-field directions, van der Pluijm et al. (1997) indicate that the compressional tectonic stress

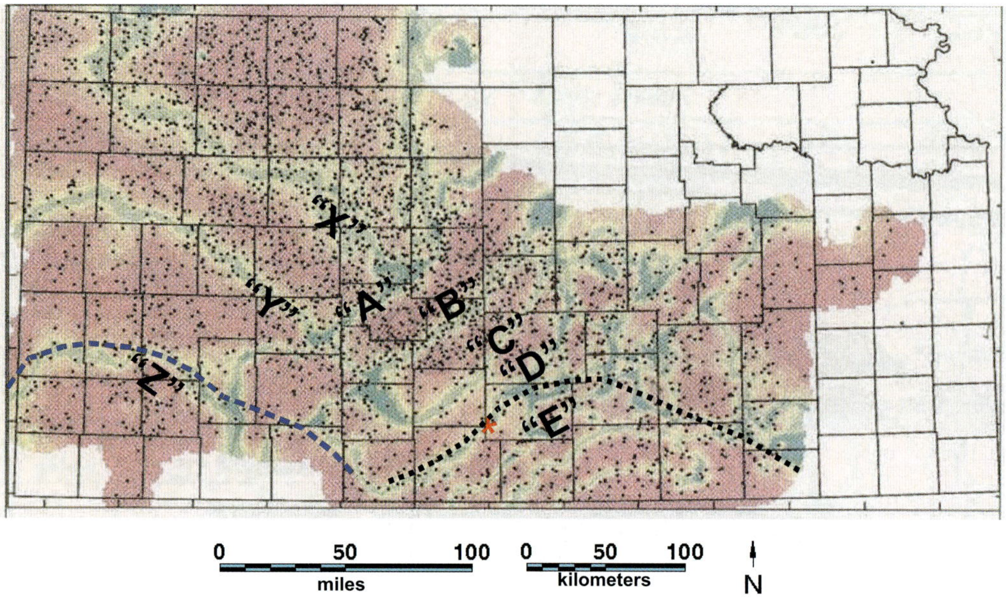

FIG. 5.—Results of regionalization analysis on six upper Pennsylvanian genetic units shown here as maximum probability of correct classification in an assigned group within 15 regions of similar thickness variation (Watney et al., 1997). Dots are wells, totaling more than 3700, that provide control. Lineaments are labeled A, B, ...G and X, Y, Z. Areas of low probability of wells being part of a region form narrow green-colored strips, whereas areas of high probability for wells being part of the region form red-colored subregional-scale, rhombic-shaped areas generally trending NW and NE. Blue-colored dashed line along SW Kansas shelf is structural flexure for Late Pennsylvanian carbonate ramp, and black dotted line in south-central Kansas is Late Pennsylvanian shelf margin (as mapped by Watney et al., 1995). Both flexure and margin were episodically active. Red star denotes general location along lineament "D", which corresponds to the common location of Mississippian and Late Pennsylvanian shelf margins.

was transmitted over 1,000 km onto the craton during the Phanerozoic and conclude that intra-plate fault reactivation was dependent mainly on the orientation of weaker fault zones in relation to plate-margin configuration and associated stress orientation. McBride and Nelson (1999) document monoclines and asymmetric anticlines overlying deep-seated high-angle reverse faults in the Illinois Basin that they interpret as distal mid-Carboniferous deformation related to ARM tectonism. Watney et al. (1997) documented the evolution of a set of active structural blocks in Kansas that formed in the late Pennsylvanian in response to variations in stress trajectory that reactivated basement faults. Ohlmacher and Berendsen (2005) document Mesozoic–Cenozoic contractional stepover along the *en echelon* NE-trending strike-slip basement faults in east-central Kansas that they attribute to stress from the Laramide orogeny, thereby illustrating reactivation of preexisting faults in the Mesozoic–Cenozoic as a result of distant tectonism.

Faults and Lineaments

Data collected in conjunction with studies related to oil production have been critical in delineating the influence of faulting and fault reactivation on depositional and stratigraphic patterns, diagenesis, and hydrocarbon accumulation in Paleozoic shelf and shelf-margin carbonate strata in Kansas. Six oil fields are used in this paper as examples: Early Carboniferous (Osagean) shelf and ramp margin (Spivey–Grabs, Glick, Donald, and Nichols fields), early Late Carboniferous (Desmoinesian) mid-shelf (Dickman Field), and Late Carboniferous (Missourian) lower shelf (Victory field) (Fig. 2). At the field scale, fracture sets and localized faults delineate kilometer-scale rhombic-shaped blocks. Structural movement of these blocks concurrent with deposition and early diagenesis significantly influenced rock properties and reservoir development, and fracture sets (joints) locally affect past and present reservoir connectivity and permeability.

Watney et al. (1997) identified major northeast- and northwest-trending faults that they interpreted to have been active during deposition of six Upper Pennsylvanian genetic units over an area covering more than 100,000 km^2 (Fig. 5). Overall, the thickness of each genetic unit increases southward toward the Anadarko Basin. However, using statistical classification of thickness similarity, Watney et al. (1997) noted vertical changes in thickness between genetic units in specific areas that they interpreted to reflect episodic spatial and temporal modifications to sediment accommodation caused predominantly by tectonically driven differential basin subsidence. Importantly, their analyses identified rectangular-shaped areas of similar thickness change, tens of kilometers in size, delimited by narrow, linear to curvilinear corridors with low similarity of thickness changes among the genetic units (Fig. 5).

The narrow corridors that separate areas of similar thickness change are identified as lineaments, often delimiting pronounced changes in shelf configuration (Fig. 5). A portion of lineament "D" (Fig. 5) represents the location of the carbonate shelf margin developed initially in the Early Carboniferous and again in the Late Carboniferous. This shelf-margin location corresponds with what appears to be a major lineament on magnetic data and an inferred change in basement composition (Figs. 5, 6). The linear nature of the shelf margin (and associated nearby parallel lineaments), the consistent shelf-margin location developed in the Early Carboniferous and again in the Late Carboniferous, and correspondence with basement features suggest a reactivated-fault origin for the shelf margin along a portion of lineament "D".

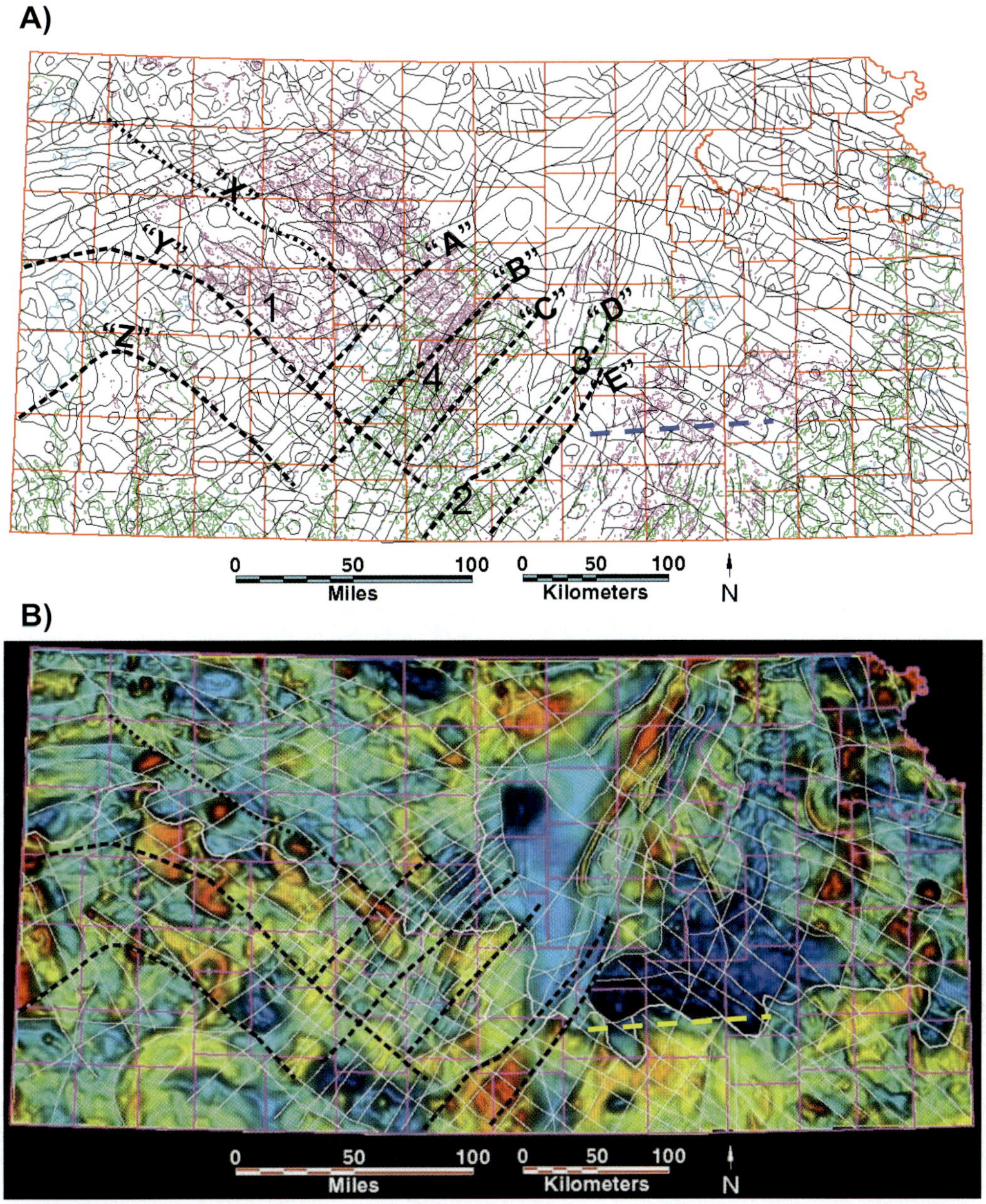

FIG. 6.—**A)** Aeromagnetic lineations derived from aeromagnetic map (Fig. 3) overlaid with oil (purple), gas (light blue), and oil and gas fields (green) showing nested set of NE and NW trends. Control of lineaments on oil and gas accumulation is evident throughout the map: area 1, between lineaments "X" and "Y"; area 2, concentration of fields between lineaments "D" and "E"; area 3, oil and gas field extending along lineament "D"; area 4, concentration of oil fields between lineaments "B" and "C". **B)** Aeromagnetic map with "relief" created by vertical illumination (Kruger, 1996) overlaid by regional lineaments as heavy black dashed lines as delimited in Fig. 5. Dashed line (blue in Part A and yellow in Part B) denotes southern boundary of a large magnetic low located in southeastern Kansas. This southern boundary also coincides with the eastern extension of shelf margin during episodes in Early and Late Carboniferous (e.g., Fig. 17).

Other lineaments, e.g., lineament "A" (Fig. 5), separate regions of lesser change in thickness and related lithofacies on the shelf. Although more subtle, the monoclinal flexure along these other lineaments is parallel to lineament "D". This parallel alignment, the extent of lineaments ("A" is at least 200 km long), offset of approximately 120 km to the northwest, and correspondence with northeast-trending magnetic lineaments in basement rocks (Fig. 6) strongly suggest that lineaments A, B, C, and E are also structural hinges that were active during the Late Carboniferous. Farther to the west, additional lineaments delineate the western margin of the CKU ("X"), the general location of a significant Precambrian terrane boundary (Yarger, 1983) ("Y"), and the central area of the Hugoton Embayment (extension of the Anadarko Basin) ("Z"). The combination of lineaments defines separate blocks including north of lineament "A"—a platform, between "A" and "D" - a ramp, and south of "D", near the

Anadarko Basin—an episodically drowned shelf margin. In addition, it is apparent that the structure at this regional scale played a role in the locations of reservoirs, inasmuch as the lineaments define distinctly separate areas of hydrocarbon accumulation (Fig. 6A).

FIELD-SCALE EXAMPLES OF STRUCTURAL CONTROL ON DEPOSITION AND DIAGENESIS

Examination of subsurface data, 3-D seismic, and rock properties for oil fields from the Early Carboniferous shelf margin, the Early Late Carboniferous middle shelf, and the Late Carboniferous lower shelf indicate that regional- and kilometer-scale structures (e.g., faults, fractures, lineaments) segmented the shelf and shelf-margin areas in Kansas, primarily along Precambrian structures that were reactivated throughout the Phanerozoic. The association of stratal packages and rock properties with structural elements illustrated in the examples argues that structure exerted continued, but episodic, influence and affected sediment accommodation, depositional patterns, paleotopography, weathering intensity, diagenesis, and later fluid movement, including hydrocarbon emplacement.

Lower Carboniferous

Lower Carboniferous strata examined in this study (Fig. 1) are situated on the southern and southwestern gently dipping flanks of the Central Kansas Uplift (CKU; Fig. 2). This uplift represents the southeastward extension of the Transcontinental Arch (see Goebel, 1968; Lane and DeKeyser, 1980) and is one of many large northwest- and northeast-trending structures that are commonly flat-topped, bordered by outward verging monoclinal folds, and are associated with ARM tectonism (Marshall et al., 2000). Lower Carboniferous strata are dominated by shallow-shelf and shelf-margin carbonates that were deposited over broad shelf to gently dipping (south-southwest) ramp profiles. Lower Carboniferous rocks, subcropping beneath the Pennsylvanian, are successively younger in a southwestward direction away from the CKU on which Mississippian rocks are absent. The Lower Carboniferous strata in the study area were differentially eroded at the post–Early Carboniferous unconformity, resulting in significant variable paleotopography along the unconformity (e.g., Schaben Field in Ness County, Kansas; Montgomery et al., 2000).

Examples in this study include Lower Carboniferous Warsaw limestone and Osagean strata, which are typically cherty limestone and dolomite. Lithofacies are dominated by heterozoan assemblages (e.g., echinoderms, bryozoa, gastropods, foraminifera, brachiopods, ostracods, solitary corals) and siliceous sponges. Generally, shelf facies comprise mudstone to grainstone limestones, dolomites, and cherts whereas basin facies are predominantly argillaceous lime mudstones and shales with minor amounts of nodular chert (Selk, 1968; Lane and DeKeyser, 1980; Gutschick and Sandberg, 1983). During the Osagean, Kansas was located at approximately 20° S latitude, within the tropical to subtropical latitude belt. The predominance of Lower–Middle Mississippian heterozoan carbonate and biosiliceous (spiculitic) deposits, and lack of photozoan deposits, in the mid-latitude shallow-shelf setting in Kansas and surrounding areas was likely due to abundant nutrients and dissolved silica derived predominantly from basinal sources (Franseen, 2006). The Osagean Cowley Formation, noted for an abundance of chert, accumulated in lower-shelf to shelf (ramp)–margin locations in southern Kansas bordering the Anadarko Basin (Watney et al., 2001). Along its subcrop beneath the Upper Carboniferous in southern Kansas, the Cowley Formation becomes highly enriched in chert (commonly referred to as "chat" when highly weathered with microporosity), in part, coinciding with localized sponge-dominated buildups. Cycles are apparent in some Osagean strata and may be tied to relative sea-level fluctuations, a component of which may include glacio-eustatic fluctuations (Watney et al., 2001). More sedimentologic information can be found in Franseen (2006), Watney et al. (2001), Montgomery et al. (1998), and Rogers et al. (1995).

Lower Carboniferous Shelf Margin: Nichols, Glick, Donald, and Spivey–Grabs Fields

Early Carboniferous "chat" hydrocarbon reservoirs in south-central Kansas developed along the updip, northern depositional edge and subcrop of the Cowley Formation, a shelf-margin lithofacies deposited in southern Kansas (Watney et al., 2001). The Cowley Formation is primarily sponge spicule–rich dolomitic limestone containing variable amounts of bedded chert believed to have been derived from dissolution of monaxion sponge spicules (Thomas, 1982). The Early Carboniferous fields in this area are closely aligned with northeast- and northwest-trending lineaments "D" and "Y" (Fig. 7; Watney et al., 2001). The "chat" reservoirs consist of very porous, intensely weathered, autoclastically brecciated, porous chert (Rogers et al., 1995; Watney et al., 2001). Four moderate to large fields (Fig. 8; Nichols, Glick, Donald, and Spivey–Grabs fields) illustrate the role of structure in field formation. Three fields are bounded on the north by the "Y" lineament and one by the "D" lineament. All four fields are interpreted as being composed predominantly of localized buildups of sponge spicules, up to 150 ft (46 m) thick, that were deposited along the shelf margin and the updip edge of the Cowley Formation (e.g., Montgomery et al., 1998). In Nichols field the pattern of net thickness of residual (low-resistivity) chert reveals a rhombic-shaped thick area surrounded by abrupt thinning along northwest- and northeast-trending linear features (Fig. 8A) that are interpreted as faults. The structure map closely follows the isopach map with the same two dominant trends, where maximum chert thickness is bounded by parallel edges of a set of northwest- and northeast-trending structural highs.

The isopach map of the Lower Carboniferous interval in Glick field indicates both NW- and NE-trending patterns of thickening within the field, which is dominated by weathered chert (sponge buildup) along the updip fringe of the Cowley Formation (Montgomery et al., 1998) (Fig. 8B). The chat is layered, and Glick Field exhibits differential preservation of the chat layers across lineaments.

The lithofacies and structure maps associated with the chat reservoir in Donald Field show chat located in a southwest-dipping graben bounded by a fault on the east flank of the field. Limestone and shale occur on either side of the chat reservoir, suggesting fault-bounded preservation of the chat (Fig. 8C).

Spivey–Grabs Field shows a general NE-dominated trend of thicker chat delineated by lineaments that are interpreted from structural and production data. In particular, production is distinctly segmented into localized sites of thicker chat (Fig. 8D; Watney et al., 2001). Structural blocks in the Spivey–Grabs field preserve the chat along a prominent northeast-trending subcrop trend located on the southeastern edge of Lineament "D" (Fig. 7). Well locations, representing the sites of porous chert, are generally clustered in the northern corners of the fault-bounded rhombic blocks. The southern corners of these blocks have little or no porous chert. Instead, strata in these locations are composed of mud-dominated sponge-spicule-poor wackestones with less porosity.

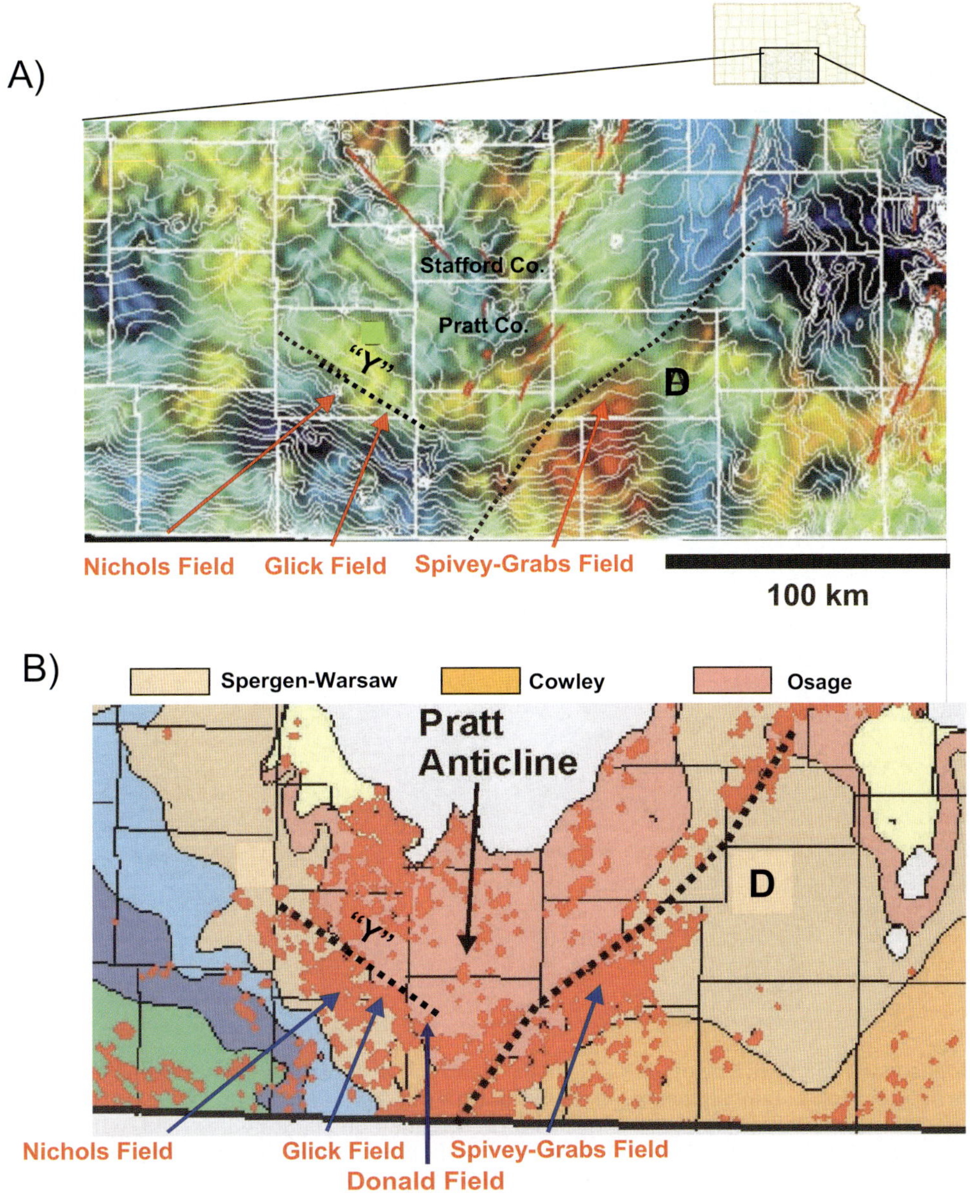

FIG. 7.—**A)** Aeromagnetic map of south-central Kansas with lineaments "Y" and "D" and locations of Early Carboniferous fields shown with arrows (subset of Fig. 3). White lines are contours depicting configuration of the Precambrian surface (after Cole, 1976). Red lines are basement faults recognized by Cole. **B)** Subcrop map of Lower Carboniferous strata beneath basal Pennsylvanian unconformity with lineaments "Y" and "D", gas fields shown in red, and color key to subcrops of Lower Carboniferous strata (modified from Gerlach, 1998). Note petroleum accumulations relative to lineaments. Fields discussed in text are shown by arrows.

The coincidence of magnetic and structural lineaments in conjunction with abrupt changes in depositional and facies patterns in association with lineaments is interpreted to indicate reactivation of deep-seated faults during Osagean deposition that resulted in the formation of rhombic-shaped structural blocks (several kilometers on a side) (Watney et al., 2001).

Upper Carboniferous

The Upper Carboniferous in Kansas was dominated by warm-water tropical photozoan carbonates, including abundant phylloid algal accumulations and extensive, cyclic accumulations of bioclastic and oolitic grainstones on the Midcontinent shelf areas. Onset of icehouse conditions started in the late Early Carboniferous (late Visean–Chesterian; Smith and Read, 2000) and continued through the Late Carboniferous, resulting in high-frequency, high-magnitude sea-level oscillations that created fourth-order (0.1–1 My) depositional sequences (cyclothems) consisting of alternating shale and carbonate in Kansas (Feldman et al., 2005; Watney et al., 1995).

The interaction of subsidence and sea-level position greatly influenced facies distribution and stratal geometries across the

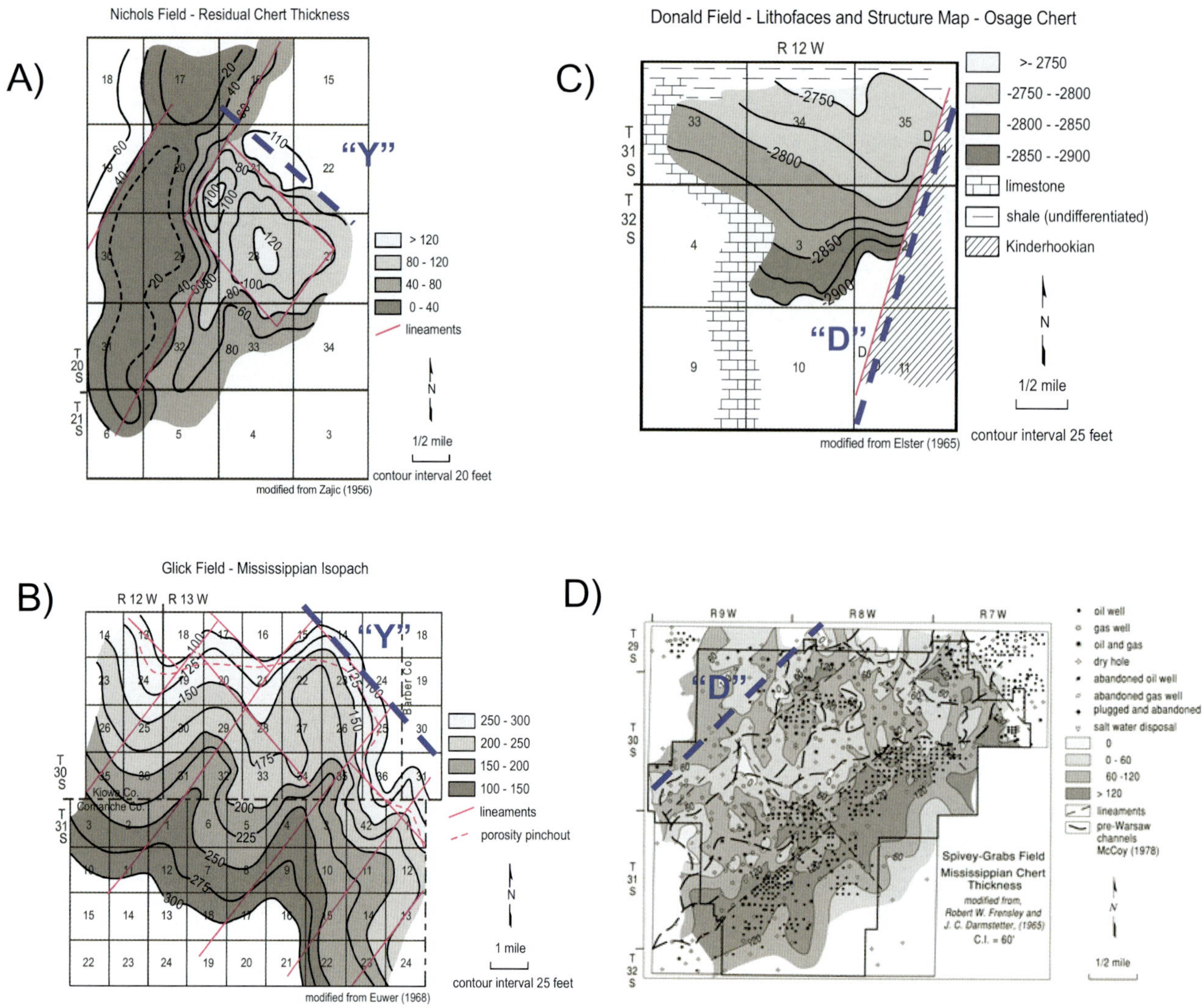

FIG. 8.—Maps showing isopach and structure of gas fields in Osagean "chat" illustrating association with superimposed basement lineaments shown in pink. Major regional lineaments shown as blue dashed lines. **A)** Isopach map of low-resistivity chert ("chat") in Nichols Field, showing proximity of lineament "Y". **B)** Structure map of chat for Glick field, with blue dashed line showing proximity of lineament "Y". **C)** Isopach map for Donald Field, with blue dashed line showing proximity of lineament "D". **D)** Isopach map for Spivey–Grabs Field with blue dashed line showing proximity of lineament "D". Field locations are shown in Figure 7.

upper Midcontinent shelf. The control of climate and general tectonics has also been documented (Feldman et al., 2005; Olszewski and Patzkowsky, 2003; Rankey, 1997; Soreghan, 1994).

Late Carboniferous Mid-Shelf: Dickman Field.—

Dickman Field (Ness County) lies in a mid-shelf setting (Figs. 2, 9) and illustrates the role of reactivated basement faults in creating paleotopography that controlled development of a Late Carboniferous oolite reservoir. Hydrocarbon-producing intervals include upper Lower Carboniferous Spergen and Warsaw cherty dolomites, an Upper Carboniferous (Cherokee Group) incised-valley-fill sandstone that locally unconformably overlies the Lower Carboniferous, and a thin (< 10 ft, 3 m) oolitic grainstone of the Upper Carboniferous Fort Scott Limestone, lying approximately 100 ft (30 m) above the top of the Lower Carboniferous (Fig. 10).

Dickman Field is located along the southern edge of a well-defined 12-mile-long (19 km) northeast-trending aeromagnetic lineament, lies approximately 30 km north of lineament "Y" (Figs. 9, 5), and is situated along the Lower Carboniferous subcrop bordering the CKU. The subcrop coincides with the northwest-trending portion of a state-wide (400-mile/644-km-long) Precambrian basement terrane boundary that separates the Central Plains Province from the Western Granite–Rhyolite Province (Yarger, 1983; Gerhard, 2004). Identification of structural elements at Dickman Field is aided by a 2 mi^2 (5 km^2) 3-D seismic survey (Figs. 11–14).

Seismic horizons were tied to key stratigraphic tops (Fig. 10) using synthetic seismograms. Subsea-depth structure maps of the horizons were constructed using time–structure maps and velocity grids computed from data at well locations. The deepest horizon illustrated in a series of 3-D seismic images is the inferred Precambrian basement surface, followed by younger

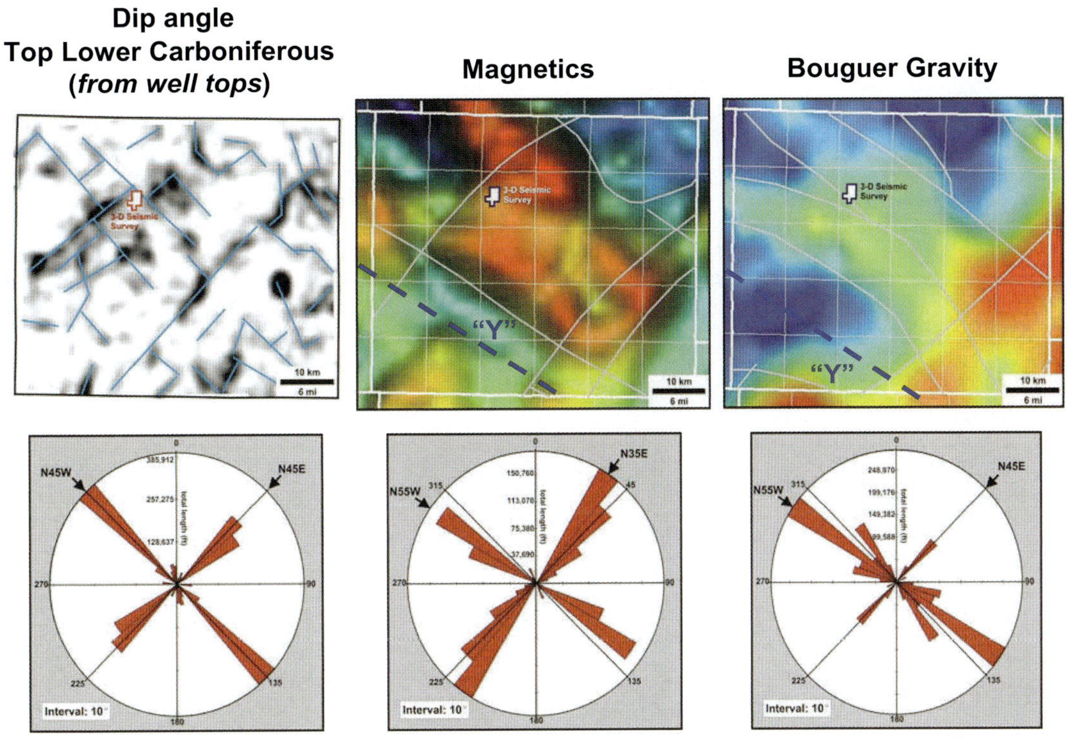

FIG. 9.—From left to right, dip-angle map (darker = higher angle) for top of Lower Carboniferous, magnetics map and Bouguer gravity map, and associated rose diagrams for total length and orientation of interpreted lineaments illustrate regional structure trends for Ness County, Kansas. Dickman Field seismic survey is also located in northeast Ness County (also identified in Figure 2). Approximate location of regional lineament "Y", active during the Carboniferous, is shown as blue dashed line.

surfaces including the top of the Mississippian, the top of Fort Scott, and the top of the Lower Permian Stone Corral Formation, 2500 ft (0.76 km) above the Fort Scott Formation (Figs. 11, 12). Major persistent features include a down-to-the-north, northeast trending fault at the northwest corner of the seismic survey, which bounds a high, upthrown rhombic-shaped block (~ 1.5 mi x 0.7 mi, 2.4 km x 1 km, in size). The upthrown structural block to the south of the fault dips generally south-southeast at 100 ft/mile (19 m/km) at the top of the Precambrian (Fig. 11A), decreasing to 25 ft/mile (4.8 m/km) at the top of the Lower Permian Stone Corral Formation (Fig. 12B). At the south end of the structural block, an incised valley in the top of the Lower Carboniferous is partially filled with Cherokee Group sandstone, and corresponds to a low in the Precambrian surface (Fig. 11 A, B). The sinuous valley pattern persists into overlying horizons, including the immediately overlying Fort Scott Limestone horizon (Fig. 12A).

Additional general observations include: (1) the Top Lower Carboniferous-to-basement interval thins abruptly to the north of the northern bounding fault, whereas the post–Lower Carboniferous isopachs consistently show abrupt thickening to the north of the fault. This suggests possible post–Early Carboniferous reversal of motion along the fault; (2) there is consistent post–Lower Carboniferous thinning over the upthrown, angular southeast-dipping block at least through the Lower Permian Stone Corral Formation. This thinning, along with radiating surface drainage, suggests further surface deformation on Upper Cretaceous bedrock.

The Fort Scott Limestone is porous and permeable in four wells in the greater Dickman Field area, with sample cuttings and core description indicating an oomoldic, vuggy limestone lithology. Local structure may have played an important role in creating inferred local shoal-water conditions and porosity development. The Fort Scott to Lower Carboniferous isopach indicates in excess of 50 ft (15 m) of thinning over the angular upthrown block, suggesting a southward (basinward) tilting of a topographic high during Fort Scott deposition (Fig. 13A). The NW-trending fault on the north (downthrown side to north) and the structural low flanking the topographic high on the south are still sharply defined, indicating continued differential subsidence/uplift that is consistent with movement prior to and following deposition of the Ft. Scott Limestone.

Seismic attributes were used to visualize subtle structural features in the 3-D seismic data and to infer rock properties. Curvature attributes, which can be used to identify folds, subtle faults, and fractures, were calculated for every sample in the 3-D seismic data volume using the technique of Chopra and Marfurt (2006). Data extracted from the resultant most negative curvature volume along the Fort Scott Limestone horizon shows NW- and NE-trending, discontinuous, high-curvature "wrinkles" on that surface that have orientations paralleling regional structural trends (Fig. 9) and likely reflect joints and other fractures (Fig. 13B). A similar joint system in the Lower Carboniferous carbonate in Dickman Field is solution enhanced and, in part, shale-filled, compartmentalizing that reservoir (Nissen et al., 2004). In the Fort Scott Limestone, two cored wells (labeled "F" in Fig. 13B) contain abundant fractures that correspond closely to locations of these "wrinkles" derived from the seismic data, thus suggesting that the wrinkles reflect oriented joint systems. The persistent arcuate-shaped structural low in the southern portion of the imaged area (corresponding to the location of the incised valley) contains a series of separate small linear joint sets (Fig. 13B).

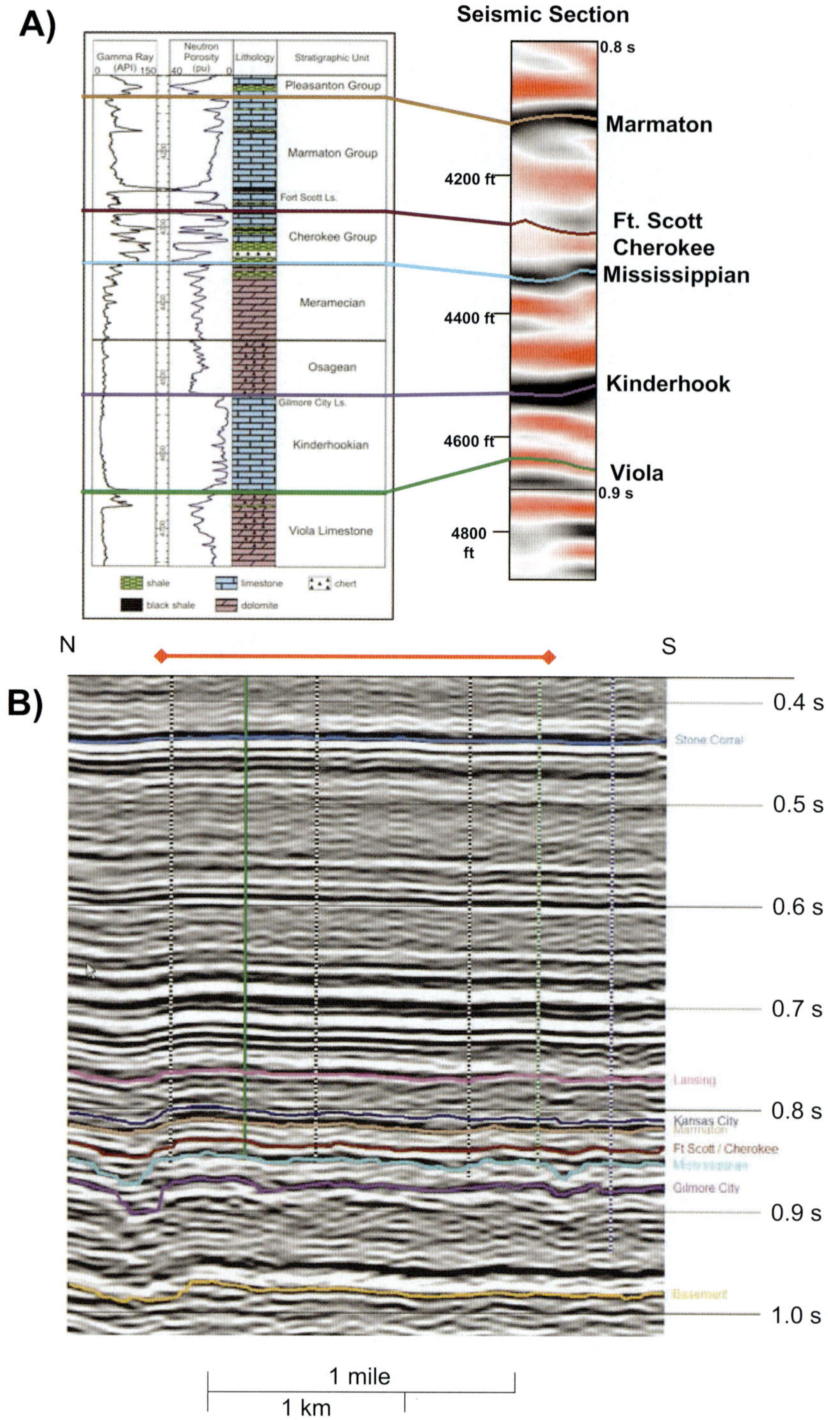

FIG. 10.—**A)** A vertical seismic section, approximately 1000 ft (305 m) long, through a section of the 3D seismic data at Dickman Field. Section is shown with corresponding formation tops and gamma-ray and neutron-porosity logs from the Stiawalt #3. The top of the Fort Scott Limestone is a small positive seismic reflection immediately above the Cherokee Group (Nissen et al., 2004). **B)** Seismic section running north south through Dickman Field showing key stratigraphic intervals described in text. Stiawalt #3 well is located on south side, identified as blue dotted line. Extent of the structural block described in text is noted by red line at top of the seismic section, bounded by structural lows.

Model-based inversion was used to generate an impedance volume from the 3-D seismic data. Average impedance was extracted for a 4 millisecond window approximating the Fort Scott Limestone interval. This attribute resolves a patch (1 mi x 0.5 mi, 1.6 km x 0.8 km) of low velocity (yellow) on the southeastern edge of the structural block corresponding to areas of higher porosity and/or lithology change (Fig. 14). Calibration of these seismic results was aided by lithologic descriptions of cuttings and core. Sample data confirm: (1) higher velocities (blues/greens) are generally low-porosity rock, and (2) lower velocities (red/yellow) have good to excellent porosity in vuggy oomoldic and fossil-mold lithofacies.

Subsurface mapping of the Fort Scott Limestone using well logs coupled with cuttings descriptions further refines the nature of the oolitic carbonate reservoir interval. The top of the Fort Scott Limestone mapped from subsurface well control closely resembles the configuration of the equivalent seismic surface (Fig. 15A). A map of gamma-ray departure between the uppermost Fort Scott Limestone reservoir interval and a zone of uniform gamma-ray in the lower part of the overlying Pawnee Limestone reveals an arcuate-shaped pod of greatest gamma-ray departure (cleanest carbonate) (Fig. 15B), corresponding to a portion of the area of low seismic impedance (Fig. 14). Sample descriptions of three wells included on a west-to-east oriented cross section across this gamma-ray low indicate that the upper Fort Scott Limestone is bioclastic and peloidal with minor oomoldic limestone on either side of the gamma-ray low (Fig. 16). The center well of the cross section, within the gamma-ray low, is a carbonate unit, 4.5 ft (1.4 m) thick, exhibiting higher, oomoldic, intercrystalline, and vuggy porosity.

The localized arcuate-shaped oolitic unit (low gamma ray) accumulated along the southeastern edge of the structural block with a narrow (0.25 mi [0.4 km] wide) lobe extending northward (landward) along a persistent but more subtle structural sag that defines the eastern margin of the structural block (Fig. 15). Since the isopachous map of the Ft. Scott to top of Lower Carboniferous interval from seismic data (Fig. 13A) also shows slight thickening and similar geometry at the same location, it is inferred that the structural sag was also a bathymetric low during upper Ft. Scott Limestone deposition in which clean, porous oolitic grainstone accumulated,

Late Carboniferous Lower Shelf: Victory Field.—

A final example of structural reactivation comes from Victory Field (Haskell County). Subsurface mapping demonstrates that structural movement persisted throughout much of the Paleozoic, influencing the depositional patterns of a series of Upper Pennsylvanian carbonate units (Fig. 17). Victory Field lies on a prominent northwest-trending basement ridge approximately 3 mi x 50 mi (5 km x 80 km) in dimension and resides on the western edge of an 18 mi x 12 mi (29 km x 19 km)

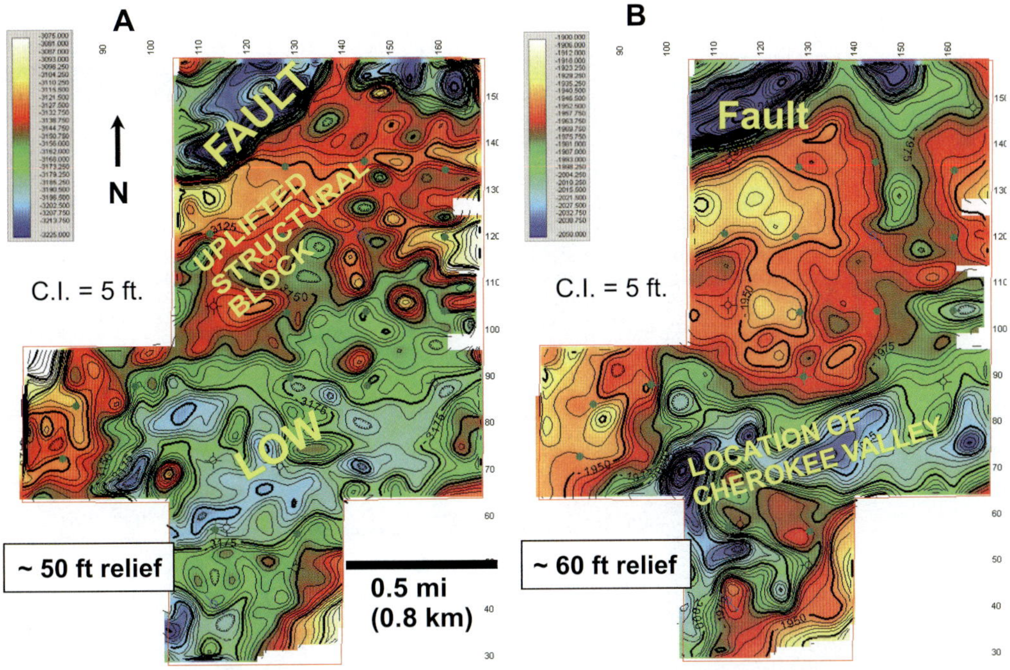

FIG. 11.—**A)** Configuration of the Precambrian basement converted to depth from a 3D seismic time horizon using a constant 19,000 ft/s (5791 m/s) velocity below the top of Lower Carboniferous. **B)** Top of the Lower Carboniferous surface from 3D seismic and well data. Both maps have the same contour interval. Both maps show an uplifted, NE-trending structural block, bounded on the north by a fault and on the south by a structural sag. Structural sag is occupied by an incised valley and is Cherokee in age.

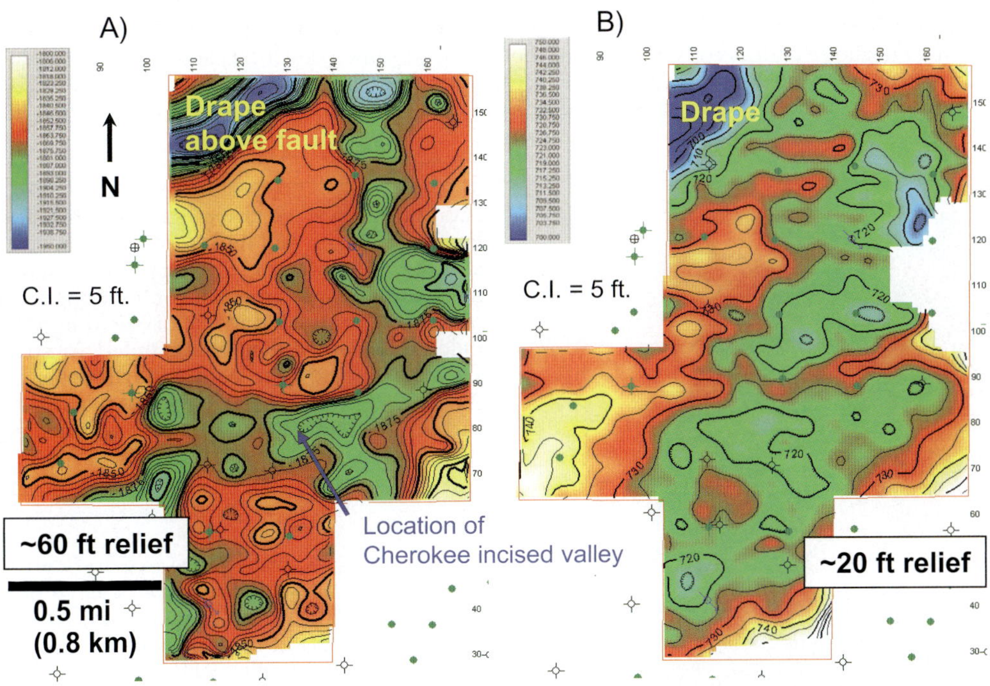

FIG. 12.—**A)** Structural elevation on top of the Fort Scott Limestone and **B)** structural map on top of the Lower Permian Stone Corral Formation from 3D seismic and well data. Both maps have the same contour interval. The geometry and relief on the structural block at the Fort Scott Limestone horizon closely resembles that on stratigraphically lower surfaces. Configuration of the Stone Corral Formation, 2500 ft (791 m) above the Fort Scott Limestone, indicates drape of fault, and a subtle structural nose running NE at the location of the underlying faulted structural block. Also, the structural low on the south side of the structural high that is shown in Figure 11 persists in shape but has lower relief because of decreased cumulative deformation at shallower depths.

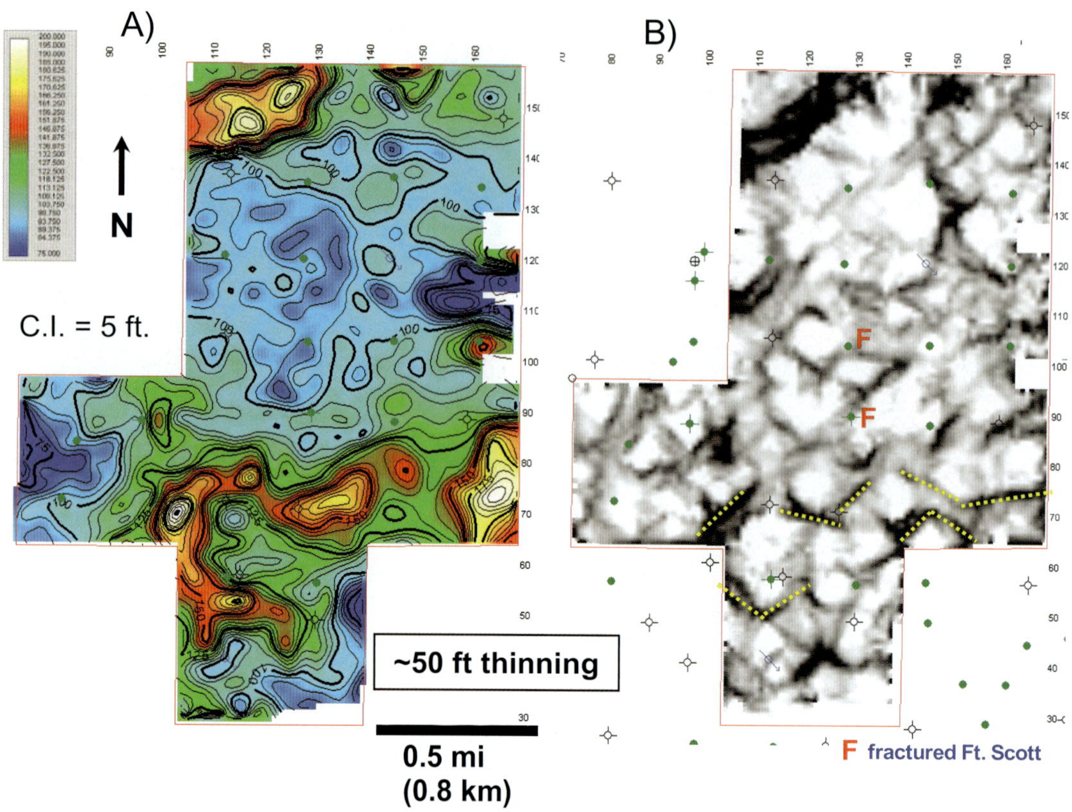

FIG. 13.—**A)** Isopach map of the Fort Scott to Lower Carboniferous interval obtained from the 3D seismic volume. Contour interval is 5 ft (1.5 m). Thicker in Cherokee incised valley and north of fault. Thinning over structural high, implying that location exhibited positive relief. **B)** Most negative volumetric curvature of the Fort Scott Limestone interval of the 3D seismic volume. Two wells containing fractures as observed in cores are labeled "F", corresponding with close proximity to curvature "wrinkles" that are both NW- and NE-trending. Yellow dotted lines delineate local joint sets that correspond to the location of the incised valley.

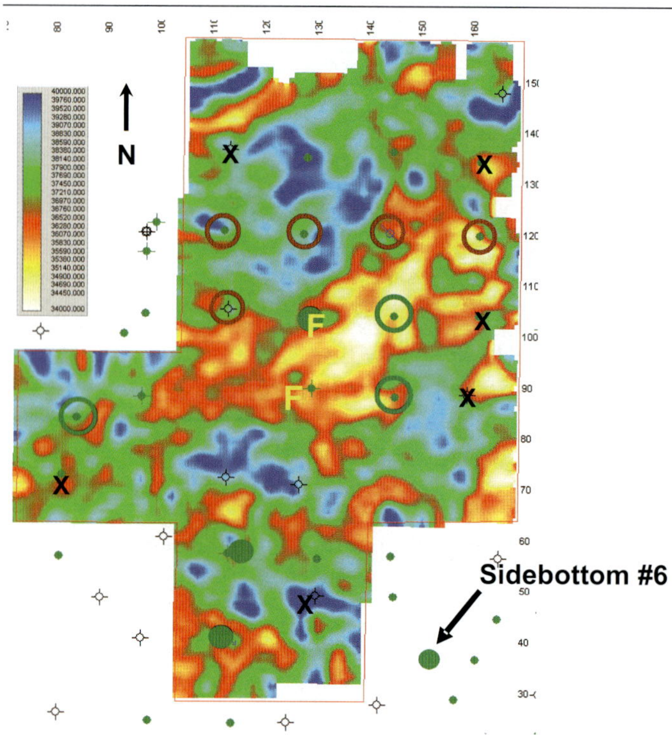

rhombic-shaped, northwest- and northeast-trending area of low Bouguer gravity and high aeromagnetic intensity (Fig. 17). Two stratigraphic intervals, the Upper Pennsylvanian Dewey Limestone and Swope Limestone of the Kansas City Group, are compared and contrasted in Victory field in relation to evidence for paleotopography that is attributed to structural reactivation. These carbonate intervals are fourth-order genetic units (Fig. 18) (Watney et al., 1995). Each of these two intervals produce hydrocarbon in Victory field, but they contrast significantly in areal distribution.

←

FIG. 14.—Seismic impedance in Fort Scott Limestone interval. This seismic attribute helps resolve porosity if the matrix velocity of the rock is constant. Impedance is shown as colors that correspond to variations in velocity. In this example, porous low-velocity rocks = yellow, and tight high-velocity rocks = blue. Map is annotated for areas of tight wackestone (X), low porosity–permeability (red circle), good porosity–permeability and hydrocarbon (green circle), very good porosity–permeability oomoldic, vuggy carbonate (solid green circle), fractures (as described in core, yellow F). Main area of apparently porous and permeable oomoldic reservoir resides on southern edge of the reactivated structural block.

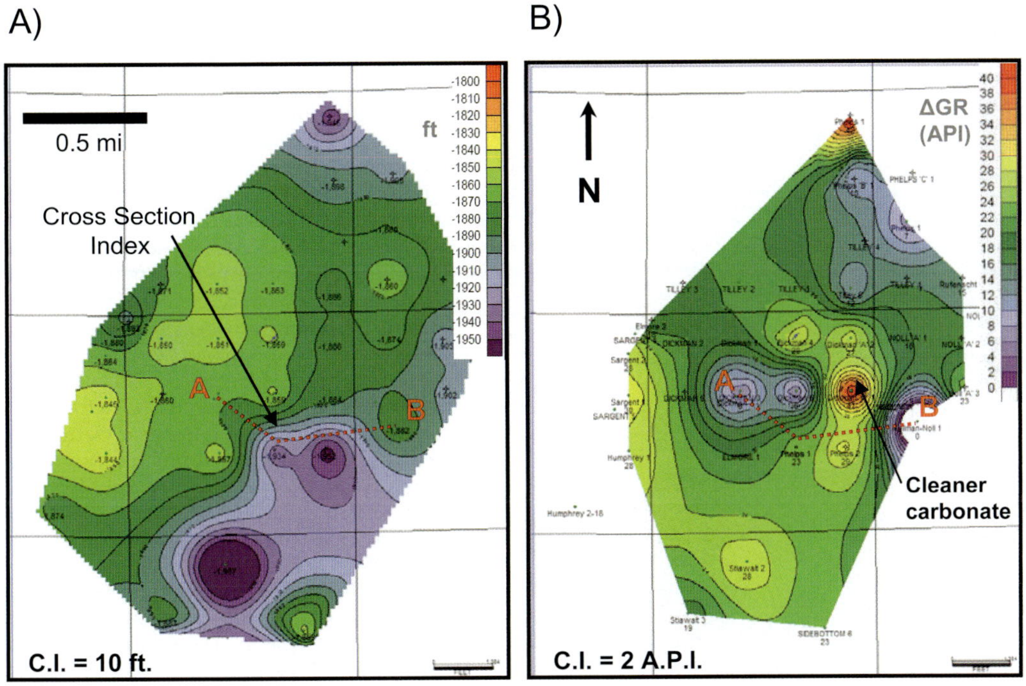

FIG. 15.—**A)** Structure contour map on top of Fort Scott Limestone based on well-log data. Elevations range from -1950 to -1800 ft (-594 to -549 m) (subsea). **B)** Isochore of gamma-ray departure between upper Fort Scott Limestone reservoir and lower zone in Pawnee Limestone, immediately overlying Fort Scott Limestone. Higher values suggest cleaner carbonate. Index line A–B is shown in map in Part A for cross section illustrated in Figure 16.

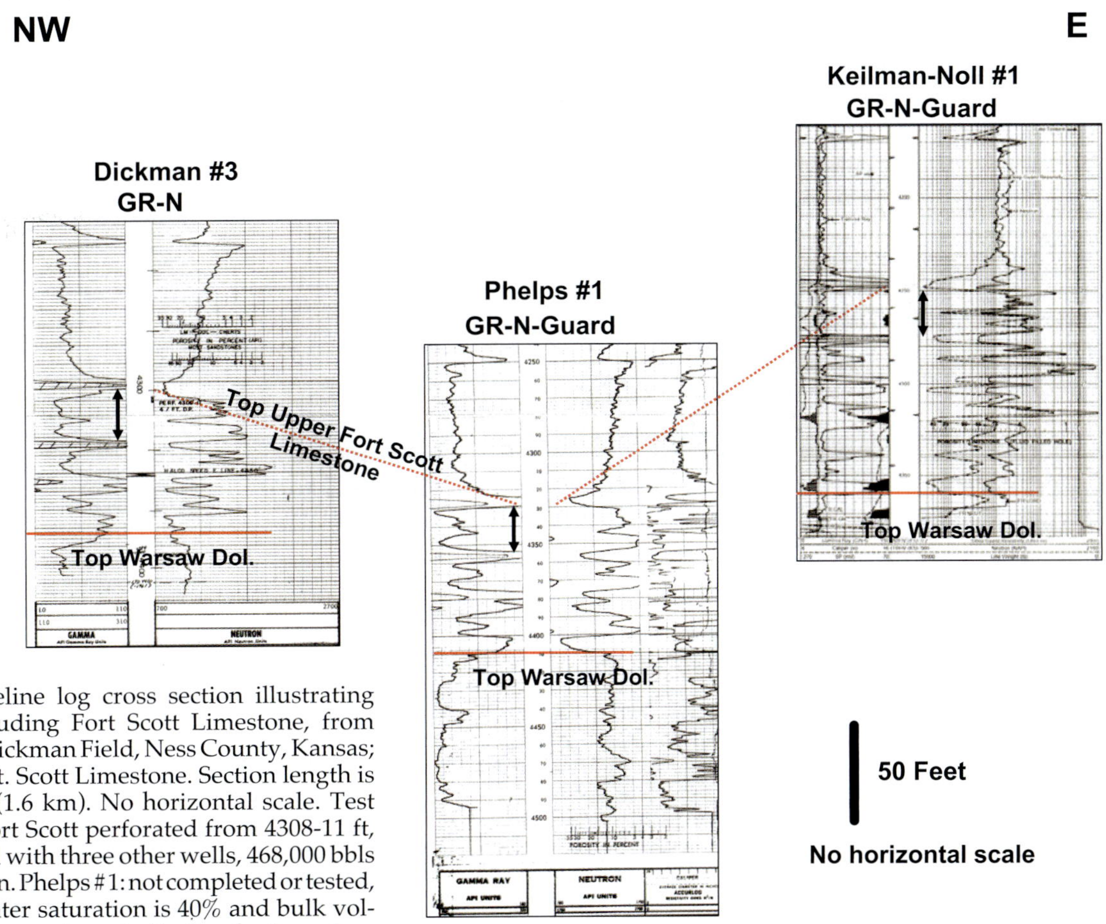

FIG. 16.—Structural wireline log cross section illustrating nature of units, including Fort Scott Limestone, from northwest to east in Dickman Field, Ness County, Kansas; datum, Base Upper Ft. Scott Limestone. Section length is approximately 1 mi (1.6 km). No horizontal scale. Test data: Dickman #3: Fort Scott perforated from 4308-11 ft, production combined with three other wells, 468,000 bbls cumulative production. Phelps #1: not completed or tested, but log calculated water saturation is 40% and bulk volume water is 0.02, suggestive of effective hydrocarbon pay. Keilman-Noll #1: DST over Fort Scott: 30′ GIP+30′ SOCM, FP 58-58#, suggesting interval is tight.

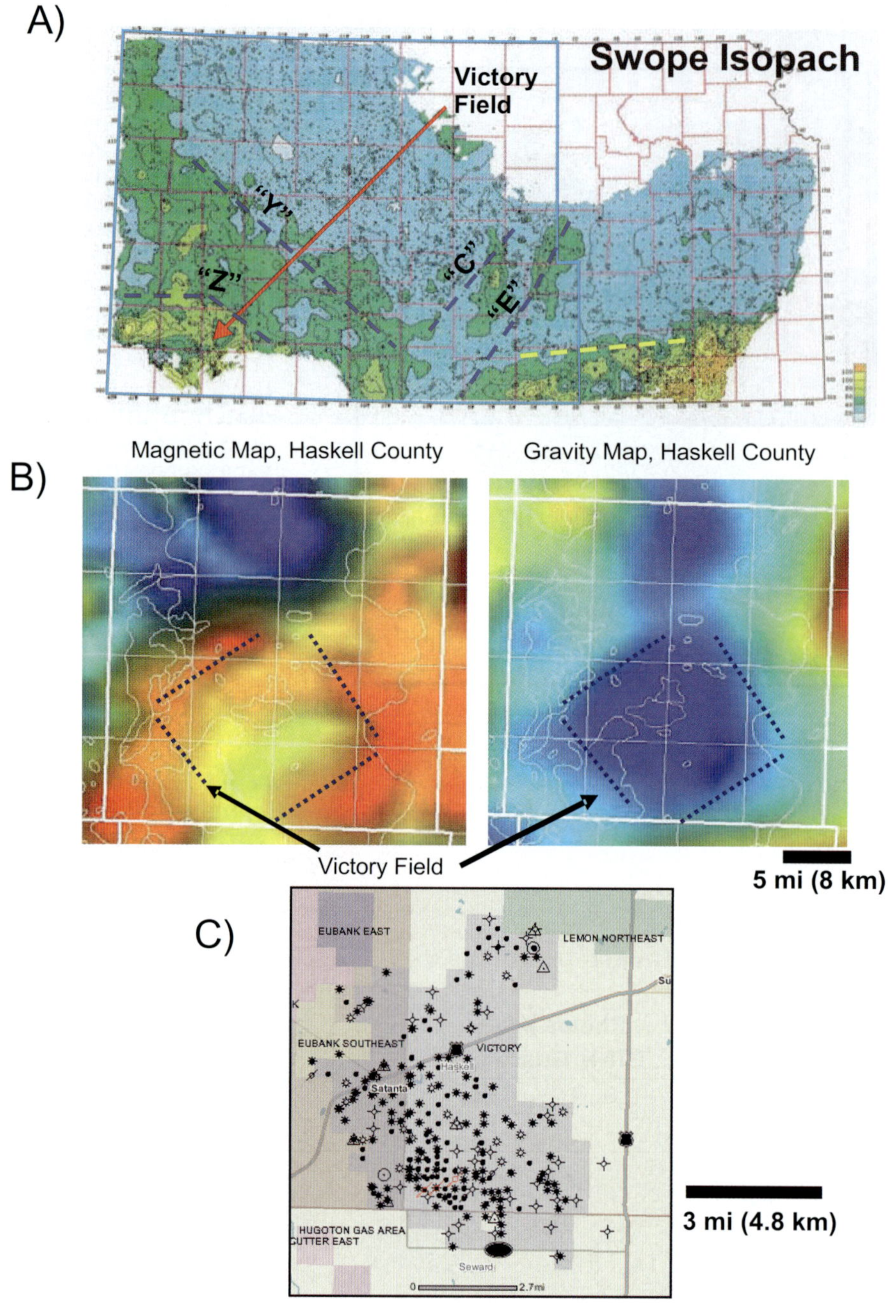

FIG. 17.—**A)** Isopach map of an Upper Pennsylvanian fourth-order genetic unit (Swope Limestone) that is one of six primary reservoirs in Victory Field (A). Horizontal distance across Kansas is 400 mi (643 km). Thicknesses vary between 6 and 37 m, thin across the upper shelf containing thin bioclastic grainstones to thick of oolitic grainstone (green to yellow, 12–30 m) across the lower southern shelf that borders the Anadarko Basin immediately to the south. Major lineaments are shown by blue dashed lines. Yellow dashed line in southeastern Kansas delimits approximate southern boundary of large prominent aeromagnetic low (see Fig. 6). **B)** Aeromagnetic (left) and Bouguer gravity (right) maps for Haskell County showing outlines of oil and gas fields and identification of Victory Field located on the west edge of NW- and NE-trending potential fields. Lineaments are denoted by dark blue dotted lines, which define a rhombic-shaped basement structure. **C)** Base map showing well locations in Victory Field.

The top of the Lower Permian Chase Group occurs 1500 ft (457 m) above the top of the Upper Pennsylvanian Lansing Group, yet structure maps of each group are remarkably similar in configuration and relief of the structure, suggesting either late movement or recurring fault reactivation (Fig. 19). An isopach of the Chase to Lansing interval (Fig. 19C) indicates thinning (up to 100 ft [30 m]) over current structural highs, matching the structure in form, which indicates either recurrent deformation with a strong northwestern trend, or differential compaction, with less compaction over the structural high. However, differential compaction seems improbable due to the maintenance of structural relief and pattern extending over both a large temporal and spatial scale.

Previous studies have shown that differences in facies and thickness within and between the Swope and Dewey limestones are tied to: (1) conditions favorable for shoal-water sediment accumulation, and (2) third-, fourth-, and fifth-order relative sea-level fluctuations (Watney et al., 1995; Watney et al., 1996). The Swope Limestone (near the base of the Nuyaka Creek third-order genetic set; Fig. 18) shows a backstepping-shelf-margin stacking pattern suggesting rising sea level and increased accommodation. In contrast, the Dewey Limestone, found at the top of the same third-order genetic set, was deposited during a time when fourth-order genetic units were prograding and downlapping, thereby indicating a loss of accommodation on the shelf (Watney et al., 1995). Our continuing studies of these units, as exemplified in Victory Field, indicate that reactivated structural elements also are discernible as controls on depositional patterns. The Dewey Limestone, isopached as a genetic unit from flooding surface to flooding surface, generally thins over the current structurally high areas of Victory Field (Fig. 20A). Similarly, the porous, mainly grainstone carbonate interval (Watney et al., 1997) thickens on the flanks of the highs (Fig. 20B). Both maps of the Dewey Limestone suggest limited sediment accommodation over the crest of the structure as it is currently expressed today.

In contrast, the crestal areas of the current structural high are sites of thick Swope Limestone and its porous oomoldic grainstone (Watney et al., 1997) interval (Fig. 21A, B). The porous grainstone accounts for much of the thickening of the genetic unit. Sediment accommodation was apparently sufficient to allow buildup of in excess of 30 ft (9 m) of oolitic grainstone over the structural crest, contrasting with the Dewey Limestone. The dominant depositional trend is northwest, conforming closely with the current structure.

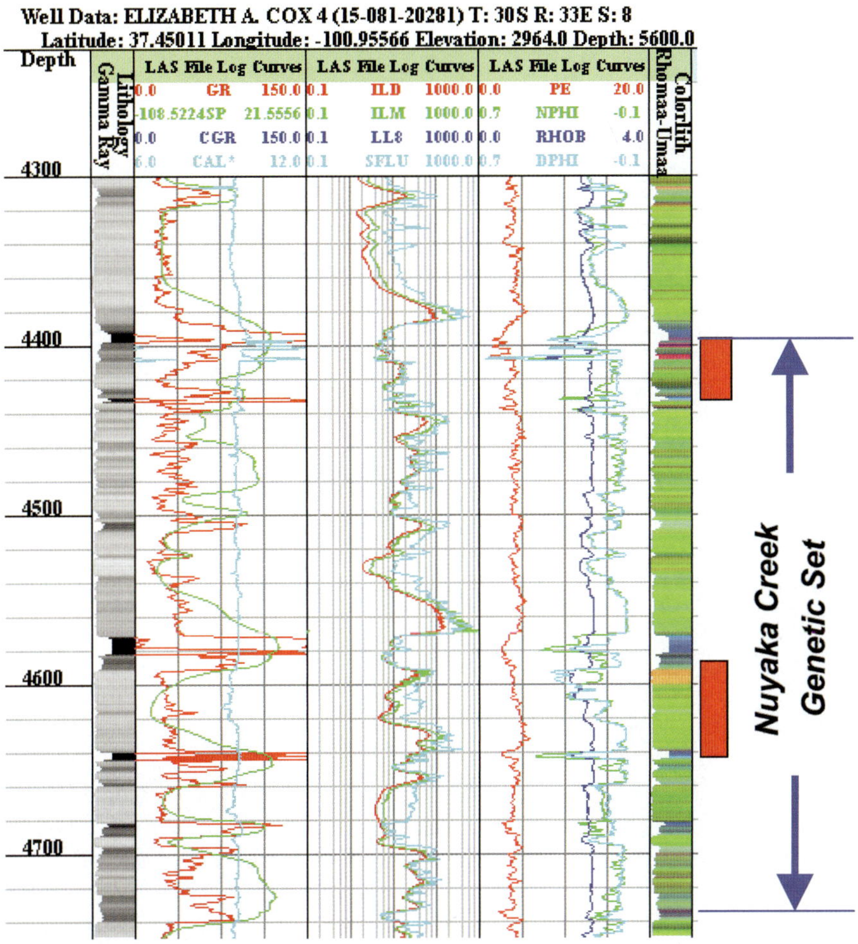

FIG. 18.—Profile of cored well in Victory Field, Amoco Cox A-4 showing log suites, lithology solution from well logs, and stratigraphic nomenclature. Well log suite including gamma ray (GR) in left log track, resistivity curves (SFLU, ILM, and ILD) in middle log track, porosity curves (NPHI, DPHI, RHOB) and photoelectric curve (PE) in right log track. Gray-scale gamma ray on left side of figure. ColorLith rhomma-umma lithology solutions in central area of figure (olive to green = carbonates, blue = shale, orange = siliceous carbonate). Fourth-order genetic units discussed in the text are identified by red rectangles defined by flooding unit boundaries) for the Dewey and Swope Limestones. These 4th-order genetic units represent a lower middle and upper-most portion of the 3rd-order Nuyaka Creek genetic set (blue arrows, again defined at the base of a flooding unit boundary).

Further evidence of paleotopographic influence on sedimentation and diagenesis is conveyed by maps of ϕ•ft (porosity x ft, ϕ x ft) and percent effective porosity for the Bethany Falls and Dewey limestones (Fig. 22A, B) and crosssection (Fig. 23). Lobate-shaped areas of percent effective oomoldic porosity and ϕ•ft are closely coincident in the Bethany Falls Limestone (Fig. 22A), which in turn closely correlate with present-day structurally high positions (Fig. 23). In contrast, the higher ϕ•ft values in the Dewey Limestone corresponding to buildups of the Dewey Limestone grainstone are on the flanks of the structural high (due to decreased accommodation as described earlier) whereas the percent effective porosity (heavy contours) remain on top of the structure. Although it is thinner on the structural crest, the Dewey grainstone still has a greater percentage of effective porosity (Fig. 22B), consisting of mainly connected oomoldic porosity, probably related to increased (e.g., intensity, duration) diagenesis over this structural high (Watney et al, 1995). Outlined areas of current production shown in Figure 22 correspond to limited areas where cuttings and well logs indicate reservoir rock behind the production casing in wells producing from other zones, suggesting opportunities for recompletion of wells into these two intervals.

DISCUSSION

Neotectonic studies provide useful insight for understanding the nature of deformation documented in our study. For example, the fault-bounded blocks defined by analysis of successive fourth-order sequences in our study are believed analogous in scale and mode of formation to rigid crustal blocks ranging from tens to hundreds of kilometers separated by narrow fault zones in actively deforming areas that are observed in neotectonic studies that utilize GPS, InSAR, seismic, and geologic data (e.g., McCaffery, 2005; Hammond and Thatcher, 2005; Thatcher, 2003). Neotectonic studies show that under a given stress trajectory (operating over a limited time frame) strain is released along a small fraction of existing faults that have proper orientation, strength, and proximity to the stress (Thatcher, 2003). Additionally, the distance of strain propagation onto the craton documented from neotectonic studies is on the order of hundreds of

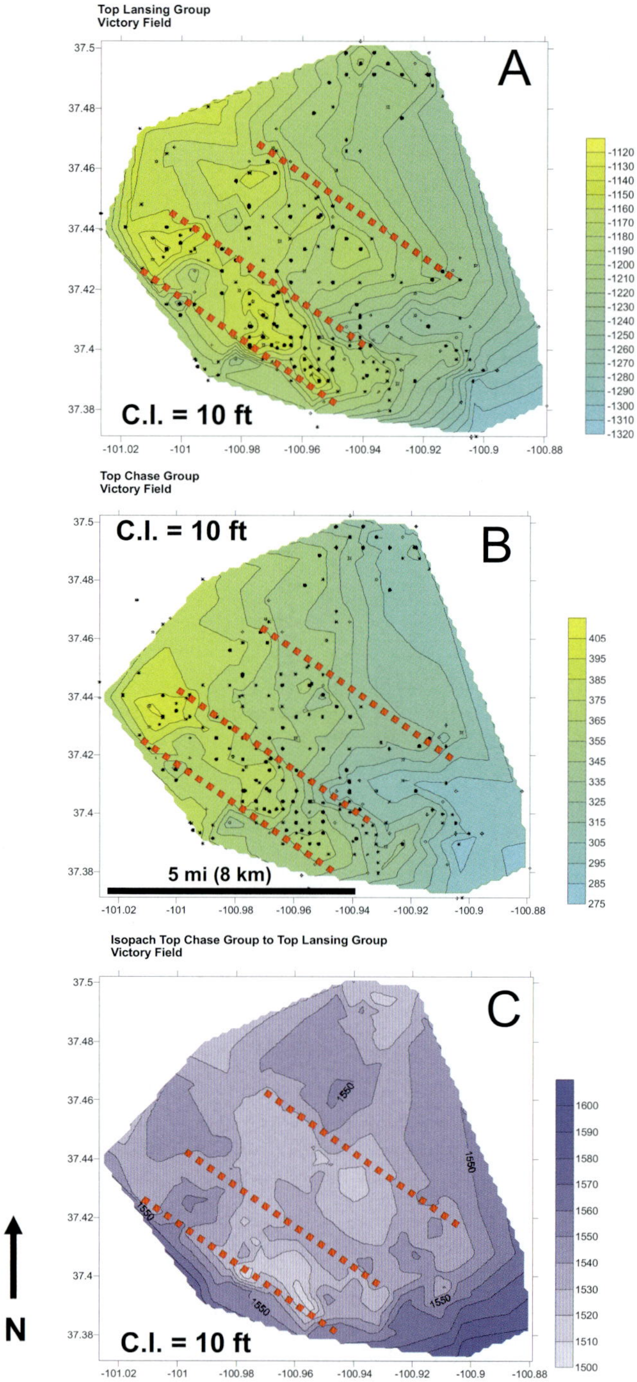

FIG. 19.—**A)** Structure map on the top of the Lansing Group in Victory Field. **B)** Structure map of the top of the Chase Group in Victory Field. **C)** Isopach map of the interval from top of the Chase Group to top of the Lansing Group. Three local structural lineaments are annotated on the three maps to aid in making comparisons between the maps. Structural configuration is very similar for intervals separated by some 450 meters. Isopach between the two intervals shows thinning conforming to the structural highs, suggesting that structural uplift was consistent and probably dominated by deep-seated structure, i.e., suggesting basement reactivation. Coordinates are latitude and longitude.

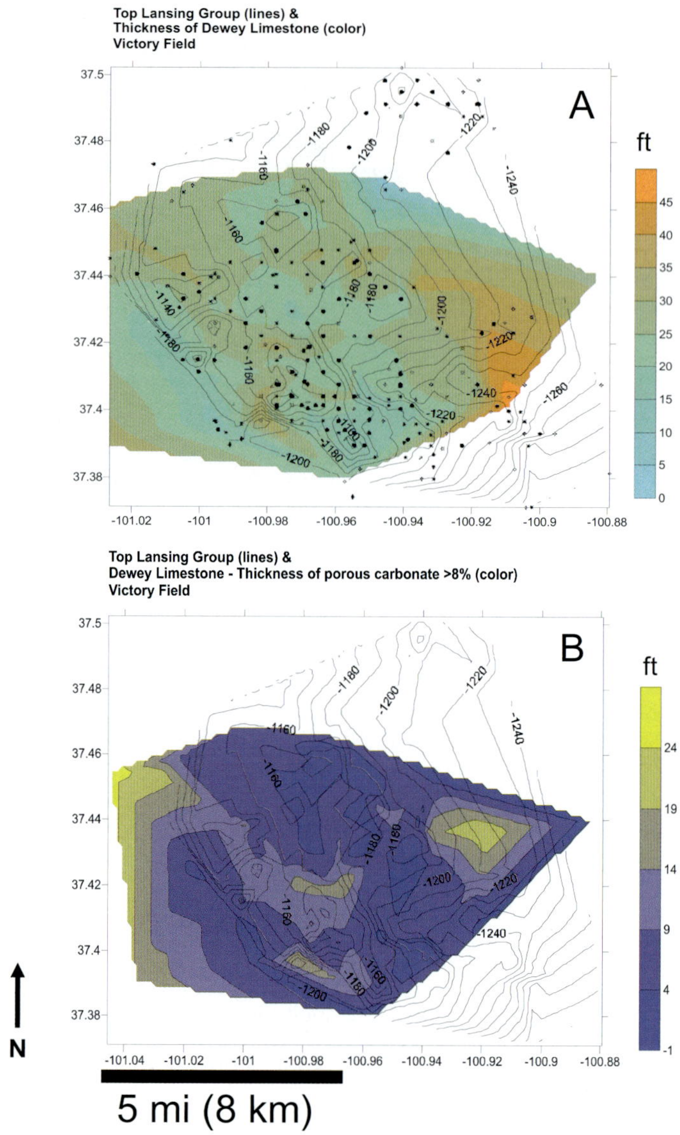

FIG. 20.—**A)** Structure contour map on top of the Lansing Group (contour lines), with color shading depicting thickness of the Dewey Limestone fourth-order genetic unit for the Victory Field area. **B)** Same structure contour map as Part A, but color shading depicts thickness of porous (exceeding 8% porosity) carbonate grainstone in Dewey Limestone. Coordinates are latitude and longitude. Thick porous carbonate grainstone lies on flanks of structure in the Dewey Limestone (Part A), suggesting limited sediment accommodation over an inferred bathymetric high.

kilometers up to 1000 kilometers (Thatcher, 2003), which is comparable to the 400–500+ kilometer distance from the major orogenic zone that existed for the central Kansas portion of the Paleozoic shelf.

Based on the similarities with observations in neotectonic studies, it is envisioned that the strain created from major areas of orogenic activity located to the south and west was propagated onto the craton and released along preexisting faults, which resulted in the segmented blocks documented in our study. Changes in the stress field eventually led to activation of different sets of lineaments and quiescence of others. In the process,

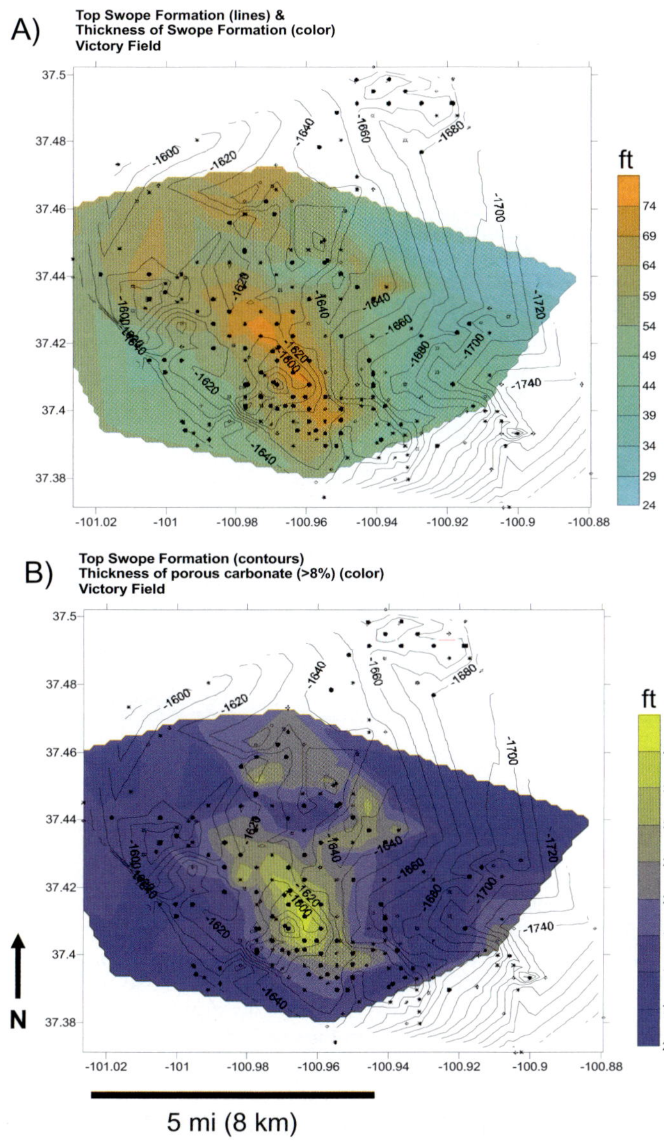

FIG. 21.—**A)** Structure contour map on the top of the Swope Formation (contour lines) overlaid with color shading depicts thickness of the Swope Limestone fourth-order genetic unit for Victory Field area. **B)** Same structure contour map as Part A, but color shading depicts thickness of porous (exceeding 8% porosity) carbonate grainstone in the Swope Limestone. Thick, porous carbonate grainstone lies on flanks of structure in the Swope Limestone (Part A), suggesting sufficient sediment accommodation to permit buildup over an inferred bathymetric high.

reactivation of basement faults resulted in the crosscutting of major Paleozoic structures, like the CKU, indicating variations in tectonic stress field that led to structural modification, albeit subtle, of even the larger structures developed under a different stress regime.

As is apparent in our study, the ability to define high-resolution temporal frameworks (e.g., at the scale of fourth- and fifth-order sequences) is especially critical for identifying syndepositional structural movement at scales that approach the scales of observation in present-day neotectonic studies. Although the time resolution will always be coarser in the ancient, it appears that the integration of observations from neotectonic studies can be useful in aiding studies in the ancient for identifying and understanding patterns and timing of subtle structures that could be important for facies distribution, stratal geometries, and fluid migration.

Implications for Petroleum Systems

As demonstrated in the examples provided in this study, reactivated structures played a role in reservoir formation during the Carboniferous by influencing facies distribution and stratal geometries. Importantly, reactivated structures appear to have had an important, long-lived role in reservoir formation beyond the Carboniferous time period. This is exemplified in an area that shows a high density of intersecting NE- and NW-trending lineaments on the aeromagnetic map in south-central Kansas above where lineaments "Y" and "D" meet in southern Stafford and northern Pratt Counties (Fig. 7A). The documented structural reactivation along some of these lineaments coincides with a region that has the most stacked hydrocarbon-productive intervals in Kansas (eight zones) that span an interval from the Cambro-Ordovician Arbuckle Group through the Upper Pennsylvanian Lansing Group (Newell et al., 1987).

In addition to playing a role in reservoir formation, the timing and distribution of structural reactivation presumably would have had a significant effect on the migration of hydrocarbons, particularly through providing pathways for vertical movement between carrier beds and charging reservoirs. Fluid inclusions, source-rock to reservoir-oil correlations, and the increase in gas–oil ratio in a southward direction strongly suggest that fluids migrated northward out of the Anadarko Basin onto the Kansas Shelf during the late Paleozoic (Newell, 1998). Deep-seated faults that were variously active through time, including during Laramide orogeny, may have encouraged the long-distance lateral fluid migration and provided local conduits for vertical migration. Comparison of the regional distribution of hydrocarbon accumulations with structural features (Figs. 2, 6, 7) shows the close correspondence between structural features, lineaments, and oil and gas fields.

CONCLUSIONS

The regional and local-scale results of this study indicate that Carboniferous (Mississippian and Pennsylvanian) shelf–shelf-margin carbonate settings were influenced through time by recurring, in some cases syndepositional, structural movement and faulting associated with reactivation of basement structural elements. An evolving stress field derived from compressional deformation along margins of the craton is interpreted to have led to systematic fragmentation of the craton, focused along selected, preferentially oriented, preexisting basement faults and lineaments that influenced depositional patterns, paleotopography, weathering intensity, diagenesis, and later fluid movements, including hydrocarbon emplacement. Several important conclusions concerning the influence of recurring fault movement can be inferred from the trends presented: (1) Mississippian and Pennsylvanian carbonate settings were affected by block-fault movement at a scale of 1 km to tens of kilometers, (2) Deformation can be closely linked to reactivation of deep-seated basement faults evident on potential-fields geophysical data, (3) Regional, down-to-basin block faulting controlled development of linear shelf edges and caused segmentation of the ramp and shelf profile, often repeated during the Mississippian and Pennsylvanian along select, probably weaker fault systems, (4) Distinct localized differences in lithofacies, thicknesses, and stratal archi-

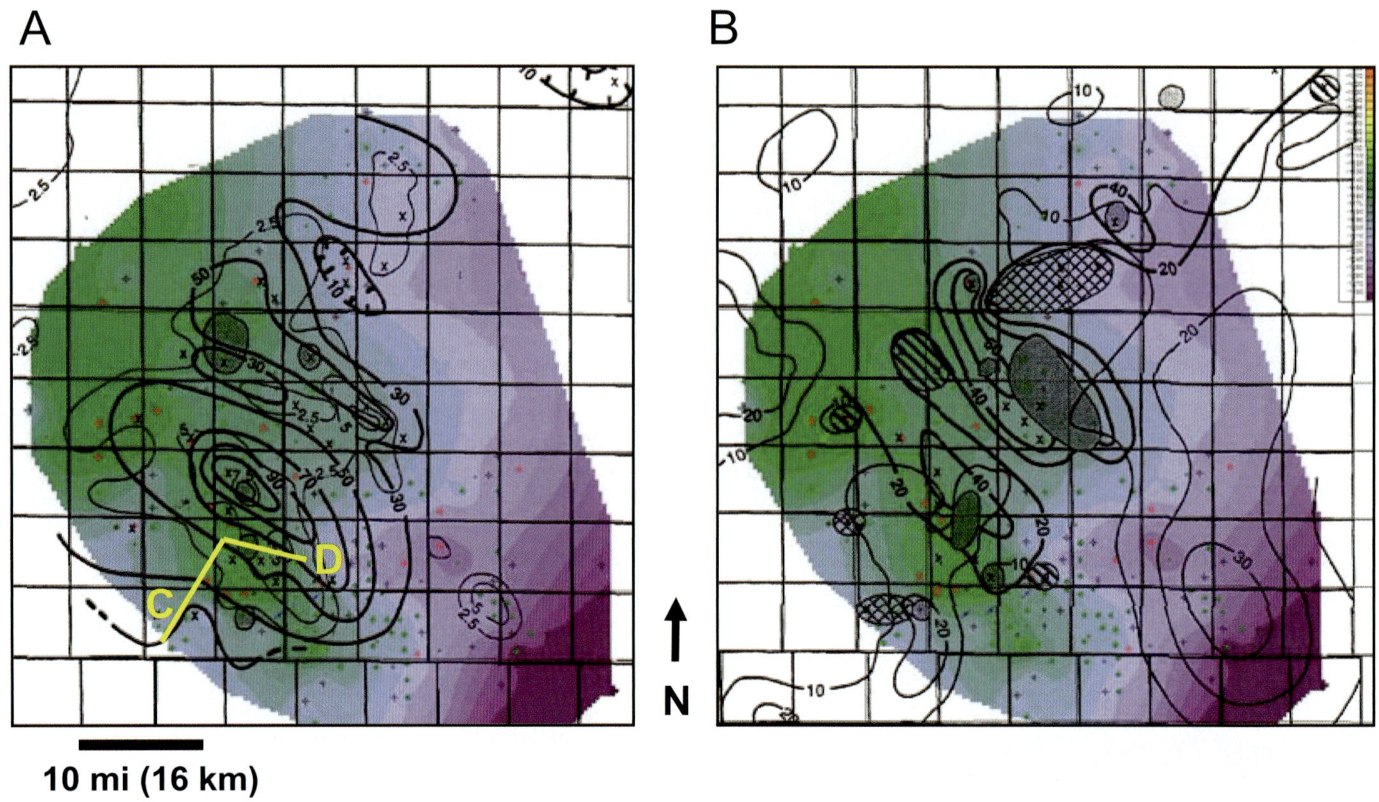

FIG. 22.—Overlay maps for **A)** Bethany Falls and **B)** Dewey limestones showing: (1) color structural elevation on the top of the Lansing Group (green high [~ -1120 ft], blue low [-1280 ft]); (2) producing areas from zone (fine stippled pattern); (3) finer-lined isochores of Φ*ft (Φ > 16% for Bethany Falls [Part A] and Φ > 8% for Dewey [Part B]; and (4) heavy-lined isochores of percent effective porosity (permeable connected pores) estimated from point counts of thin sections of well cuttings for Bethany Falls (map A) and visual estimates from cuttings examined under low-power binocular microscope for Dewey (map B) (Watney et al., 1994). Effective porosity cutoff, Φ cutoff, in the fully oomoldic Bethany Falls Limestone is assigned 16% (> 1 md permeability), whereas the Φ cutoff in the Dewey Limestone containing more preserved ooids and interooid porosity is assigned 8%. Also the Dewey Limestone (map B) shows areas with well-cemented ooid grainstone (cross-hachured lines) and no ooid grainstone (crossing-hachured lines) as determined from cuttings. Index for cross section C–D (Fig. 23) is shown in Figure 22A.

tecture across segments could be anticipated due to changes in dip, strike, and elevation of structural blocks, (5) Reversal of block-fault movement appears to have created areas for thick accumulations of sponge spiculite and oolite reservoir deposits, (6) Regional and local block-fault uplift created topographic highs that tended to be subjected to more intense diagenesis as well as more fracturing and migration of subsurface fluids during renewed uplift, (7) Sea level, climate, depositional setting, and regional tectonics interacted with local contemporaneous structural movement to establish site-specific conditions favorable for reservoir development, (8) The recurring nature of structural activity often led to paleotopographic expression acting to focus deposition (e.g., carbonate grainstones on highs and flanks; incised-valley formation in lows). Some areas of recurrent activity provided environments conducive to stacked accumulation of reservoir facies.

Based on our initial results as presented in this study, full understanding of rock properties at the regional and local scale should include analysis of the possible influence of structural movement and its synergistic influence on the many other paleoenvironmental variables. This analysis may prove more important than we previously thought in understanding the "stable" shelf carbonates of the Midcontinent.

The working model developed from our study suggests that: (1) True stratigraphic hydrocarbon accumulations are probably fewer than believed, (2) Lithofacies belts and stratal packages may not be as laterally continuous as is commonly thought. Local variations in accommodation due to recurrent and syndepositional structural deformation are important elements in predictive sedimentary modeling, (3) Combining sequence stratigraphy, sedimentary process, paleoenvironment, and diagenetic analyses with detailed structural analyses is necessary to improve sedimentary modeling, (4) Further study is needed to refine basement structure, identify key fault systems and determine timing of their activity, and develop the kinematics of major structural blocks on the craton in relation to petroleum reservoir development, hydrocarbon migration, and distribution of aquifer systems.

ACKNOWLEDGMENTS

Grand Mesa Operating in Wichita, Kansas, is thanked for providing access to the 3-D seismic data from Dickman Field. Appreciation is extended to the U.S. Department of Energy in support of GEMINI (Cooperative Agreement No.: DE-FC26-00BC15310). Seismic Micro-Technology, Inc., provided access to

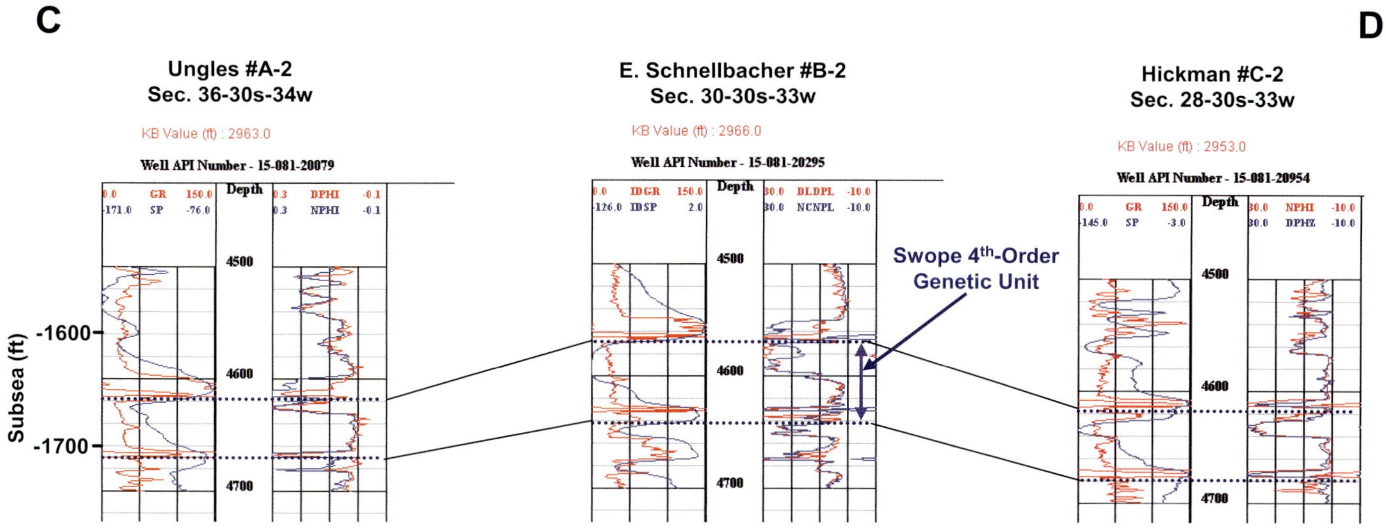

FIG. 23.—West-to-east structural well-log cross section depicting doubling of thickness of the highly porous, low-gamma-ray, oomoldic lithofacies of Swope Limestone fourth-order genetic unit across a current structural high in the southern portion of Victory Field, Haskell County, Kansas. Left track of well logs contains gamma ray (red) and spontaneous potential (blue) curves. Right track of well logs has neutron Φ (red) and neutron Φ (blue). Porosity increases to left (up to 30% or 0.3). Index for cross section is provided in Fig. 22. No horizontal scale. Total length of cross section is 5.6 km. Cross section further illustrates the close tracking of the development of the porous oomoldic pay zone with the location of the structurally higher positions of the field as noted in Fig. 21. The coincidence of current and inferred structural highs developed during time of deposition is believed to be due to structural reactivation as discussed in text.

seismic interpretation and mapping software, and Hampson-Russell Limited Partnership provided access to seismic inversion software. GEMINI, a web-based analysis, www.kgs.ku.edu/GEMINI, was used to compile and analyze subsurface well data and create Figure 18 in this paper. Appreciation is extended to Rick Sarg and Art Saller for their careful and constructive reviews and to Pat Acker for assisting with graphics.

REFERENCES

BAARS, D.L., THOMAS, W.A., DRAHOVZAL, J.A., AND GERHARD, L.C., 1995, Preliminary investigations of basement tectonic fabric of the conterminous USA., *in* Ojakangas, R.W., Dickas, A.B., and Green, J.C., eds., Basement Tectonics 10, Proceedings Tenth International Conference on Basement Tectonics held in Duluth, Minnesota, USA, August 1992: Dordrecht, The Netherlands, Kluwer Academic Publishers, p. 149–158.

BAARS, D.L., AND WATNEY, W.L., 1991, Paleotectonic control of reservoir facies, *in* Franseen, E.K., Watney, W.L., Kendall, C.G.St.C., and Ross, W., eds, Sedimentary Modeling: Computer Simulations and Methods for Improved Parameter Definition: Kansas Geological Survey, Bulletin 223, p. 253–262.

BERENDSEN, P., AND BLAIR, K.P., 1996, Structural configuration of the Precambrian basement, Kansas: Kansas Geological Survey, Map Series, no. 45-1, 4 pages, 1 sheet, scale 1:500,000.

BERENDSEN, P., WILSON, F.W., YARGER, H.L., AND STEEPLES, D.W., 1983, New data on major basement fractures in the tectonic development of eastern Kansas: 3rd Basement Tectonics International Conference, Proceedings, p. 227–240.

BERENDSEN, P., BORCHERDING, R.M., DOVETON, J., GERHARD, L., NEWELL, K.D., STEEPLES, D. AND WATNEY, W.L., 1988, Texaco Poersch #1, Washington County, Kansas–preliminary geologic report of the pre-Phanerozoic rocks: Kansas Geological Survey, Open-File Report 88-22, 116 p.

BROWN, D.L., 1978, Wrench-style deformation patterns associated with a meridional stress axis recognized in Paleozoic rocks in parts of Montana, South Dakota, and Wyoming, *in* Willison Basin Symposium: Montana Geological Society, Billings, Montana, Twenty-Fourth Annual Conference, p. 17–31.

BROWN, D.L., AND BROWN, D.L., 1987, Wrench-style deformation and paleostructure influence on sedimentation in and around a cratonic basin, *in* Peterson, J.A., Kent, D.M., Anderson, S.B., Pilatzke, R.H., and Longman, M.W., eds., Williston Basin: Anatomy of a Cratonic Oil Province: Rocky Mountain Association of Geologists, 1987 Symposium, p. 57–70.

CHOPRA, S., AND MARFURT, K., 2006, Seismic attributes—Promising aid for geological prediction: CSEG Recorder 2006 Special Edition, p. 111–121.

COLE, V.B., 1976, Configuration of the top of Precambrian rocks in Kansas: Kansas Geological Survey, Map Series, no. M-7, 1 sheet, scale 1:500,000

DICKINSON, W.R., AND LAWTON, T.F., 2003, Sequential intercontinental suturing as the ultimate control for Pennsylvanian Ancestral Rocky Mountains deformation: Geology, v. 31, p. 609–612.

DOROBEK, S.L., 1995, Synorogenic carbonate platforms and reefs in foreland basins: Controls on stratigraphic evolution and platform/reef morphology, *in* Dorobek, S.L., and Ross, G.M., Stratigraphic Evolution of Foreland Basins: SEPM, Special Publication 52, p. 127–147.

DUCHEK, A.B., MCBRIDE, J.H., NELSON, W.J., AND LEETARU, H.E., 2004, The Cottage Grove fault system (Illinois Basin): Late Paleozoic transpression along a Precambrian crustal boundary: Geological Society of America, Bulletin, v. 116, p. 1465–1484.

ELSTER, T., 1965, Donald Field: Kansas Geological Society, Kansas Oil and Gas Fields, v. 4, p. 68–77.

EUWER, R.M., 1968, Glick field, Kiowa and Comanche counties, Kansas, *in* Beebe, B.W., and Curtis, B.F., eds., Natural Gases of North America, part 3, vol. 2: American Association of Petroleum Geologists, Memoir 9, p. 1576–1582.

FELDMAN, H.R., FRANSEEN, E.K., JOECKEL, R.M., AND HECKEL, P.H., 2005, Impact of longer-term modest climate shifts on architecture of high-frequency sequences (cyclothems), Pennsylvanian of Midcontinent U.S.A.: Journal of Sedimentary Research. v. 75, p. 350–368.

FRANSEEN, E.K., 2006, Mississippian (Osagean) Shallow-water, mid-latitude siliceous sponge spicule and heterozoan carbonate facies: An example from Kansas with implications for regional controls and distribution of potential reservoir facies: Kansas Geological Survey, Bulletin 252, part 1, Current Research in Earth Sciences: 23 p.; available online at http://www.kgs.ku.edu/Current/2006/franseen/index.html .

FRENSLEY, R.W. AND J.C. DARMSTETTER, 1965, Spivey–Grabs field: Kansas Geological Society, Kansas Oil and Gas Fields, Kansas Geological Society, v. 4, p. 221–228.

GAY, S.P., JR., 1999, Strike-slip, compressional thrust-fold nature of the Nemaha system in eastern Kansas and Oklahoma, *in* Merriam, D.F., ed., Transactions of the 1999 American Association of Petroleum Geologists Mid-Continent Section Meeting: Kansas Geological Survey, Open-File Report 99-28, p. 39–50.

GERHARD, L.C., 2004, A New Look at an Old Petroleum Province: Kansas Geological Survey, Current Research in Earth Sciences, Bulletin 250, part 1, 27 p. http://www.kgs.ku.edu/Current/2004/Gerhard/index.html

GERHARD, L.C., ANDERSON, S. B., AND LEFEVER, J., 1987, Structural history of the Nesson anticline, North Dakota, *in* Peterson, J.A., Longman, M.W., Anderson, S.B., Pilatzke, R.H., and Kent, D.M., eds., Williston Basin—Anatomy of a Cratonic Oil Province: Rocky Mountain Association of Geologists, Denver, p. 337–353.

GERHARD, L.C., ANDERSON, S.B., LEFEVER, J.A., AND CARLSON, C.G., 1982, Geological development, origin, and energy mineral resources of Williston Basin, North Dakota: American Association of Petroleum Geologists, Bulletin, v. 66, p. 989–1020.

GERLACH, P.M., 1998, Statewide-production maps: Kansas Digital Petroleum Atlas, http://www.kgs.ku.edu/DPA/Plays/ProdMaps/miss_gas.html.

GOEBEL, E.D., 1968, Mississippian rocks of western Kansas: American Association of Petroleum Geologists, Bulletin, v. 52, p. 1732–1778.

GUO, G., 1999, An analysis of surface and subsurface lineaments in fractures for oil and gas exploration in the Mid-Continent region: Department of Energy, DOE/PC/91008-0223, 36 p.

HAMMOND, W.C., AND THATCHER, W., 2005, Contemporary tectonic deformation of Basin and Range province, western United States: 10 years of observation with Global Positioning System: Journal of Geophysical Research, v. 109, B08403, 21 p.

HINZE, W.J., ALLEN, D.J., BRAILE, L.W., AND MARIANO, J., 1997, The Midcontinent Rift System: A major Proterozoic continental rift, *in* Ojakangas, R.W., Dickas, A.B., and Green, J.C., eds., Middle Proterozoic to Cambrian Rifting, Central North America: Geological Society of America, Special Paper 312, p. 7–35.

JACOBI, R.D., AND FOUNTAIN, J.C., 2002, The character and reactivation history of the of the southern extension of the seismically active Clarendon–Linden Fault System, western New York State, *in* Fakundiny, R.H., Jacobi, R.D., and Lewis, C.F.M., eds., Neotectonics and Seismicity in the Eastern Great Lakes Basin: Tectonophysics, v. 353, p. 215–262.

KLUTH, C.F., AND CONEY, P.J., 1981, Plate tectonics of the Ancestral Rocky Mountains: Geology, v. 9, p. 10–15.

KRUGER, J.M., 1996, On-line gravity and magnetic maps of Kansas: Kansas Geological Survey, Open-File Report, no. 96-51, http://www.kgs.ku.edu/PRS/PotenFld/potential.html

LANE, H.R., AND DE KEYSER, T.L., 1980, Paleogeography of the late Early Mississippian (Tournaisian 3) in the central and south-western United States, *in* Fouch, T.D., and Magathan, E.R., eds., Paleozoic Paleogeography of West-Central United States: SEPM, Rocky Mountain Section, p. 149–159.

MACFARLANE, P.A., AND WILSON, B.B., 2006, Enhancement of the bedrock-surface-elevation map beneath the Ogallala portion of the High Plains Aquifer, Western Kansas: Kansas Geological Survey, Technical Series, no. 20, 28 p., 2 plates.

MARSHAK, S., KARLSTROM, K., AND TIMMONS, J.M., 2000, Inversion of Proterozoic extensional faults: An explanation for the pattern of Laramide and Ancestral Rockies intracratonic deformation, United States: Geology, v. 28, p. 735–738.

MCBRIDE, J.H., AND NELSON, W.J., 1999, Style and origin of mid-Carboniferous deformation in the Illinois Basin, USA: Ancestral Rockies deformation?: Tectonophysics, v. 305, p. 275–286.

MCCAFFREY, R., 2005, Block kinematics of the Pacific–North American plate boundary in the southwestern United States from inversion of GPS, seismological, and geologic data: Journal of Geophysical Research, v. 110, B07401, 27 p.

MCCOY, J.R., 1978, Character of the Mississippian formation in south-central Kansas: Kansas Geological Survey, Open-File Report 78-9, 10 p.

MERRIAM, D.F., 2005, Origin and development of plains-type folds in the Mid-Continent (United States) during the late Paleozoic: American Association of Petroleum Geologists, Bulletin, v. 89, p. 101–118.

MONTGOMERY, S.L., FRANSEEN, E.K., BHATTACHARYA, S., GERLACH, P., BYRNES, A.P., GUY, W., AND CARR, T.R., 2000, Schaben Field, Kansas: Improving performance in a Mississippian shallow-shelf carbonate: American Association of Petroleum Geologists, Bulletin, v. 84, p. 1069-1086.

MONTGOMERY, S.L., MULLARKEY, J.C., LONGMAN, M.W., COLLEARY, W.M., AND ROGERS, J.P., 1998, Mississippian "chat" reservoirs, South Kansas; low-resistivity pay in a complex chert reservoir: American Association of Petroleum Geologists, Bulletin, v. 82, p. 187–205.

NEWELL, K.D., 1998, Comparison of maturation data and fluid-inclusion homogenization temperatures to simple thermal models: Implications for thermal history and fluid flow in the Midcontinent: Kansas Geological Survey, Bulletin 240, part 2, Current Research in Earth Sciences, p. 13–27.

NEWELL, K.D., WATNEY, W.L., CHANG, S.W.L., AND BROWNRIGG, R.L., 1987, Stratigraphic and spatial distribution of oil and gas production in Kansas: Kansas Geological Survey, Subsurface Geology Series 9, 86 p.

NISSEN, S.E., MARFURT, K.J., AND CARR, T.R., 2004, Identifying Subtle Fracture Trends in the Mississippian Saline Aquifer Unit Using New 3-D Seismic Attributes: Kansas Geological Survey, Open File 2004-56, http://www.kgs.ku.edu/PRS/publication/2004/2004-56/index.html

OHLMACHER, G.C., AND BERENDSEN, P., 2005, Kinematics, mechanics, and potential earthquake hazards for faults in Pottawatomie County, Kansas, USA: Tectonophysics, v. 396, p. 227–244.

OLSZEWSKI, T.D., AND PATZKOWSKY, M.E., 2003, From cyclothems to sequences: The record of eustasy and climate on an icehouse epeiric platform (Pennsylvanian–Permian, North American Midcontinent): Journal of Sedimentary Research, v. 73, p. 15–30.

RANKEY, E.C., 1997, Relations between relative changes in sea level and climate shifts: Pennsylvanian–Permian mixed carbonate–siliciclastic strata, western United States: Geological Society of America, Bulletin, v. 109, p. 1089–1100.

ROGERS, J.P., LONGMAN, M.W., AND LLOYD, R.M. 1995, Spiculitic chert reservoir in Glick Field, south-central Kansas: The Mountain Geologist, v. 32, p. 1–22.

SIMS, P.K., SALTUS, R.W., AND ANDERSON, E.D., 2005, Preliminary Precambrian basement structure map of the continental United States—An interpretation of geologic and aeromagnetic data: U.S. Geological Survey, Open-File Report 2005-1029, 31 p. (text) + plate. http://pubs.usgs.gov/of/2005/1029/pdf/OFR-1029.pdf.

SMITH, L.B., JR., AND READ, J.F., 2000, Rapid onset of late Paleozoic glaciation on Gondwana: Evidence from Upper Mississippian strata of the Midcontinent, United States: Geology, v. 28, p. 279–282.

SOREGHAN, G.S., 1994, The impact of glacioclimatic change on Pennsylvanian cyclostratigraphy, *in* Embry, A.F., Beauchamp, B., and Glass,

D.J., eds., Pangea; Global Environments and Resources: Canadian Society of Petroleum Geologists, Memoir 17, p. 523–543.
Stevenson, G.M., and Baars, D.L., 1986, The Paradox: A pull-apart basin of Pennsylvanian age, *in* Peterson, J.A., ed., Paleotectonics and Sedimentation in the Rocky Mountain Region: American Association of Petroleum Geologists, Memoir 41, p. 513–539.
Stone, D.S., 1999, Foreland basement-involved structures: Discussion: American Association of Petroleum Geologists, Bulletin, v. 83, p. 2006–2016.
Thatcher, W., 1995, Continuum versus microplate models of active continental deformation: Journal of Geophysical Research, v. 100, p. 3885–3894.
Thatcher, W., 2003, GPS constraints on the kinematics of continental deformation: International Geology Review, v. 45, p. 191–212.
Thomas, M.A., 1982, Petrology and diagenesis of the Lower Mississippian, Osagian Series, western Sedgwick basin, Kansas: Master's thesis, University of Missouri–Columbia, 87 p.
Thomas, W.A., 1983, Continental margins, orogenic belts, and intracratonic structures: Geology, v. 11, p. 270–272.
Thomas, W.A., and Baars, D.L., 1988, Synsedimentary tectonics: Geology, v. 16, p. 190-191.
Timmons, J.M., Karlstrom, K.E., Dehler, C.M., Geissman, J.W., Heizler, M.T., 2001, Proterozoic multistage (ca. 1.1 and 0.8 Ga) extension recorded in the Grand Canyon Supergroup and establishment of northwest- and north-trending tectonic grains in the southwestern United States: Geological Society of America, Bulletin, v. 113, no. 2, p. 163–181.
van der Pluijm, B.A., Craddock, J.P., Graham, B.R., and Harris, J.H., 1997, Paleostress in cratonic North America: implications for deformation of continental interiors: Science, v. 277, p. 794–796
Van Schmus, W.R., Bickford, M.E., and Zietz, I., 1987, Early and Middle Proterozoic provinces in the central United States, *in* Kroner, A., ed., American Geophysical Union, Proterozoic Lithospheric Series, v. 17, p. 43–68.
Van Schmus, W.R., and Hinze, W.J., 1985, The midcontinent rift system. Annual Review of Earth and Planetary Sciences, v. 13, p. 345–383.
Watney, W.L., Baars, D., Carlson, R., Feldman, H., French, J., Franseen, E., Guy, B., Meham, T., Stevenson, G., Verma, A., Wong, J.C., and Youle, J., 1994, Depositional sequence analysis and sedimentologic modeling for improved prediction of Pennsylvanian reservoirs, final report: U.S. Department of Energy, Bartlesville, OK, Report, no. DOE/BC/14434-13, 589 pages, (NTIS no. DE95000116/HDM).
Watney, W.L., Davis, J.C., Olea, R. A., Harff, J., and Bohling, G.C., 1997, Modeling of sediment accommodation realms by regionalized classification: Geowissenschaften, v. 15, p. 28–33.
Watney, W.L., French, J.A., Doveton, J.H., Youle, J.C., and Guy, W.J., 1995, Cycle hierarchy and genetic stratigraphy of Middle and Upper Pennsylvanian strata in the upper Mid-Continent, *in* Hyne, N., ed., Sequence Stratigraphy in the Mid-Continent: Tulsa Geological Society, Special Publication 3, p. 141–192.
Watney, W.L., French, J.A., and Guy, W.J., 1996, Modeling of petroleum reservoirs in Pennsylvanian strata of the Midcontinent, USA, *in* Forester, A., and Merriam, D.F., eds., Spatial Modeling of Geologic Systems: New York, Plenum Press, p. 43–77.
Watney, W.L., Guy, W.J., and Byrnes, A.P., 2001, Characterization of the Mississippian Osage chat in south-central Kansas: American Association of Petroleum Geologists, Bulletin, v. 85, p. 85–114.
Witzke, B.J., 1990, Paleoclimatic constraints for Paleozoic paleolatitudes of Laurentia and Euramerica, *in* McKerrow, W.S., and Scotese, C.R., eds., Palaeozoic Palaeogeography and Biogeography: Geological Society of London, Memoir 12, p. 57–73.
Witzke, B.J., and Bunker, B.J., 1996, Relative sea-level changes during Middle Ordovician through Mississippian deposition in the Iowa area, North American craton, *in* Witzke, B.J., Ludvigson, G.A., and Day, J., eds., Paleozoic Sequence Stratigraphy: Views from the North American Craton: Geological Society of America, Special Paper 306, p. 307–330.
Yang, K.-M., and Dorobek, S.L., 1995, The Permian basin of West Texas and New Mexico: History of a "composite" foreland basin and its effects on stratigraphic development, *in* Dorobek, S.L., and Ross, G.M., eds., Stratigraphic Evolution of Foreland Basins: SEPM, Special Publication 52, p. 149–174.
Yarger, H.R., 1983, Regional interpretation of Kansas aeromagnetic data: Kansas Geological Survey, Geophysics Series, no. 1, 35 p.
Zajic, W.E., 1956, Nichols pool: Kansas Geological Society, Kansas Oil and Gas Pools, vol. 1 (South Central Kansas), p. 69–73.
Zeller, D.E., ed., 1968, The Stratigraphic Succession in Kansas: Kansas Geological Survey, Bulletin 189, 81 p.

UPPER MISSISSIPPIAN ANTLER FORELAND BASIN CARBONATE AND SILICICLASTIC ROCKS, EAST-CENTRAL IDAHO AND SOUTHWESTERN MONTANA, U.S.A.: DISTINGUISHING TECTONIC AND EUSTATIC CONTROLS ON DEPOSITION

LISELLE S. BATT

Department of Geological Sciences, University of Idaho, Moscow, Idaho 83844, U.S.A.
Present Address: ExxonMobil Producing Company, 233 Benmar, Houston, Texas 77060, U.S.A.
e-mail: liselle.s.batt@exxonmobil.com

MICHAEL C. POPE

School of Earth and Environmental Sciences, Washington State University, Pullman, Washington 99164, U.S.A.

PETER E. ISAACSON

Department of Geological Sciences, University of Idaho, Moscow, Idaho 83844, U.S.A.

ISABEL MONTAÑEZ

Department of Geology, University of California at Davis, Davis, California 95616, U.S.A.

AND

JASON ABPLANALP

Department of Geological Sciences, University of Idaho, Moscow, Idaho 83844, U.S.A.

Abstract: A regional study, based on detailed descriptions of 17 outcrops across east-central Idaho and southwestern Montana, provides a dip-oriented cross section in which to better understand the distribution of Upper Mississippian (Chesterian) stratigraphy on the distal margin of the Antler foreland basin. Chesterian strata constitute an eastward-thinning wedge of mixed carbonate and siliciclastic rocks that formed on a west-facing ramp. Foreland-basin tectonism subdivided the ramp into three distinct depositional settings: the western, central, and eastern ramp. The western ramp records nearly continuous Chesterian deposition, whereas the central and eastern ramps have significant unconformities. Mud-rich subtidal carbonate predominates on the western ramp, but this interfingered during the late Chesterian with tidally influenced siliciclastics. The central ramp contains an intraramp basin with deep subtidal siliciclastics and carbonate that formed adjacent to shallower-water facies to the west and east. The eastern ramp has mostly peritidal carbonate and shallow marine to fluvial siliciclastics, but a transgression from the north during the late Chesterian inundated this portion of the ramp with open marine carbonate.

New conodont biostratigraphic constraints indicate that these Chesterian strata are a second-order megasequence (10–12 My duration) composed of more than seven third-order depositional sequences (S0–S7), each having a duration of 1–5 My. The sequences are grouped into three composite sequences (I, II, and III) that define long-term changes in accommodation controlled by syndepositional tectonism. Composite sequence I was deposited during a period of tectonic loading that partitioned the ramp via subsidence loading and extension. Composite sequence II records a period of tectonic stabilization and deposition during the most extensive eustatic flooding, whereas composite sequence III is dominated by a localized subsidence event in the Big Snowy Trough. Higher-frequency (fourth- and fifth- order) parasequences are common throughout the study interval, but they are only locally correlative. A change from thick-bedded carbonate-dominated parasequences in the early and middle Chesterian to thinner-bedded mixed carbonate and siliciclastic parasequences in the late Chesterian likely reflects the onset of moderate- to high-amplitude, high-frequency eustatic fluctuations caused by the initiation of Gondwanan glaciation.

INTRODUCTION

Carbonate and siliciclastic successions in foreland-basin settings form complex lateral relationships that record the interaction of differential, sometimes episodic tectonic subsidence, eustatic sea-level changes, and sediment supply derived from both sides of the basin, commonly including cratonic and allochthonous thrust-sheet sources, as well as short-lived bathymetric highs within the basin (Dorobek et al., 1991a; Pope and Read, 1997; Smith and Read, 2001). Combining basin-wide stratigraphic chronologies with meter-scale facies stacking patterns enables distinction between regional tectonic and global eustatic controls in such foreland deposits (DeCelles and Giles, 1996). In east-central Idaho and southwestern Montana, a succession of Upper Mississippian carbonates and mixed carbonates and siliciclastics were deposited on a gently sloping ramp during the formation and subsequent fill of the Antler foreland basin (Fig. 1). Deposition of this assemblage occurred during the mid-Carboniferous transition from a global greenhouse to icehouse climate (González, 1990; Isbell et al., 2003a; Isbell et al., 2003b). Chesterian meter-scale cycles (parasequences) in east-central Idaho indicate increases in glacio-eustatic frequency and amplitude of relative sea-level changes on the ramp (Butts, 2005), providing a proxy for the timing of onset of glaciation that may help constrain the wide range (from mid-Visean to late Serpukhovian) of estimates based on direct glacial indicators. The purposes of this study are foremost to correlate the Chesterian stratigraphy from the western, central, and eastern ramps using new conodont biostratigraphic zones and secondly to understand the relative influence of regional tectonics and global eustatic processes on deposition in the distal part of the Antler foreland basin.

Previously the Chesterian stratigraphy in east-central Idaho (Fig. 2; Table 1) was described as an eastward-shallowing homoclinal carbonate ramp that formed in the distal part of the

Controls on Carbonate Platform and Reef Development
SEPM Special Publication No. 89, Copyright © 2008
SEPM (Society for Sedimentary Geology), ISBN 978-1-56576-130-8, p. 147–170.

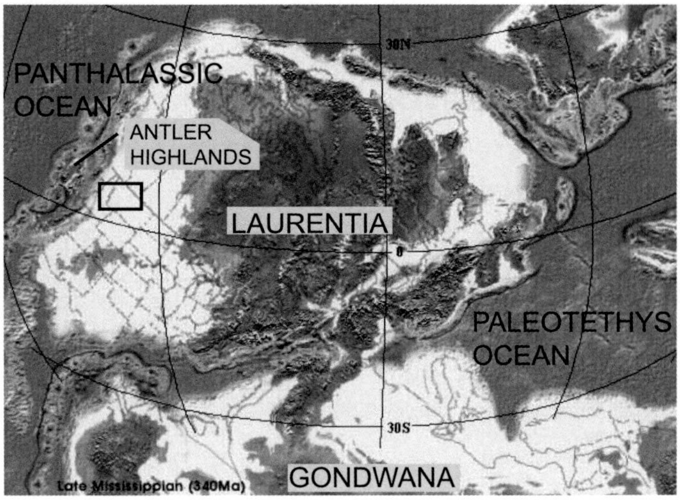

FIG. 1.—Paleogeographic reconstruction of the northern hemisphere during the late Mississippian illustrating the equatorial position of the study area (black box) between the Antler Highlands and the cratonic margin. Light shading indicates shallow marine depositional environments. Adapted from Blakey, unpublished map from http://jan.ucc.nau.edu/~rcb7/RCB.html.

Antler foreland basin (Rose, 1976; Skipp et al., 1979a; Skipp et al., 1979b; Skipp and Hall, 1980; Poole and Sandberg, 1991). Macrofaunal (corals; Sando et al., 1969) and microfaunal (foraminifera; Mamet et al., 1971) biostratigraphy enabled construction of a preliminary chronology for regional events, including the onset of subsidence and the timing of basin fill. Updip the Upper Mississippian stratigraphy of southwestern Montana records coeval shallow-water deposition on the cratonic side of the ramp (Table 1; Rose, 1976; Smith and Gilmour, 1979; Peterson, 1981; Sando et al., 1985; Wardlaw and Pecora, 1985; Key, 1986). The occurrence of laterally heterogeneous facies in Montana resulted in a complex stratigraphic nomenclature for the region that is challenging to correlate with the better exposed Upper Mississippian strata in Idaho, particularly given the widespread use of lithostratigraphic correlation techniques and limited biostratigraphic data. Southwestern Montana paleogeographic reconstructions indicate the presence of geographically coincident bathymetric highs and lows that became inverted throughout the Late Mississippian (Peterson, 1981; Poole and Sandberg, 1977, 1991), but the timing of these regional inversions, their mechanistic link to Antler foreland subsidence, and their temporal relationship to the mostly carbonate succession in east-central Idaho remain unconstrained. Our high-resolution framework links the subsidence and carbonate-ramp deposition in Idaho to the formation of an intraramp basin and adjacent peritidal environment in Montana. Resolving timings of local processes controlling lateral facies variations in Idaho and Montana enables a more realistic evaluation of the control that tectonics and eustasy imparted upon this foreland-basin succession.

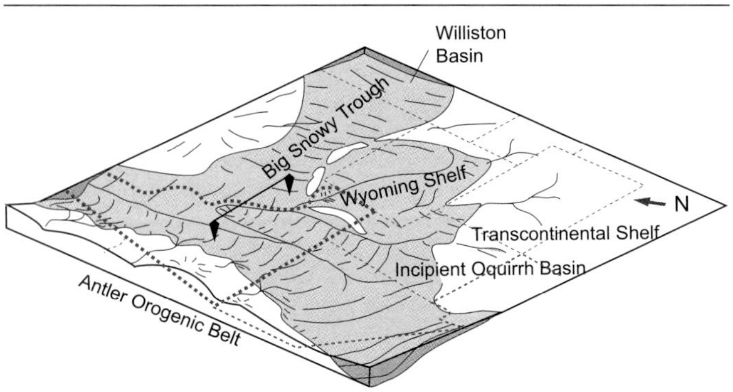

FIG. 2.—Regional paleogeography of the western margin of Laurentia (adapted from P. Rose, unpublished data). The black line indicates the dip-oriented cross section (Figs. 4 and 8) constructed for this study. Gray shading represents submarine environments. Dotted lines indicate present-day state borders.

REGIONAL GEOLOGIC SETTING

In the Late Devonian, western Laurentia transitioned from a passive continental margin to a convergent margin with the onset of Antler orogenesis, which continued episodically throughout the late Paleozoic (Speed andSleep, 1982; Poole and Sandberg, 1991; Miller et al., 1992; Dickinson, 2004; Isaacson et al., 2007). Recent models of Antler convergence involve development of an east-dipping subduction zone in the back-arc basin of the Klamath–Sierran arc to the west (Burchfiel and Royden, 1991; Dickinson, 2004). Direct and indirect evidence for tectonism occurs in an arcuate belt extending from the Canadian Cordillera south to Nevada (Poole and Sandberg, 1977, Nilsen and Stewart, 1980; Dover, 1980; Smith et al., 1993; Trexler et al., 1995). Direct evidence of subduction is missing from Idaho, but detrital-zircon geochronology in Pioneer Range indicates a 340–380 Ma arc situated to its west (Link and Fanning, 2004). An Antler allochthon in western Idaho is further inferred from a thick Mississippian basinal succession of fine-grained turbidites and fan conglomerates (Skipp et al., 1979b; Dover, 1980; Link et al., 1996; Link and Janecke, 1999) that likely formed on the proximal side of this asymmetric north–south trending foreland basin (Fig. 2), which shallowed eastward onto the craton (Poole and Sandberg, 1977, 1991; Skipp et al., 1979b; Dover, 1980).

Antler basin infill occurred within 5° of the paleoequator (Fig. 1; Scotese, 1997) and resulted in shallowing and the development of a west-facing subtidal carbonate ramp (Fig. 2) in the distal foreland region of east-central Idaho during the Meramecian (Skipp et al., 1979b; Smith and Gilmour, 1979; Dover, 1980; Buoniconti et al., 2004). Subtidal carbonates of the Meramecian–Chesterian Scott Peak Formation (Crawford, 1976; Gutschick et al, 1980; Stamm, 1985; Evans, 1989) formed adjacent to emergent topography across most of southwestern Montana (Table 1; Rose, 1976; Dorobek et al., 1991a; Dorobek et al., 1991b). Subsequent deposition of the Chesterian South Creek Formation in east-central Idaho coincides with an episode of increased subsidence and carbonate drowning, followed by progradation of Surrett Canyon Formation carbonate (Galvin, 1981), which is overlain by mixed carbonate and siliciclastic rocks of the uppermost Mississippian Arco Hills Formation (Griffing, 1987; Butts, 2005) and the Bluebird Mountain Formation (Skipp and Hall, 1980). Sandstones are rare in the lower and middle Chesterian ramp deposits from east-central Idaho but become increasingly abundant in the late Chesterian.

In Montana, emergent topography during the Late Mississippian resulted in the development of a widespread unconformity above the Madison Group, which was overlain by deposition of thin-bedded tidally influenced mixed carbonates and siliciclastics. Chesterian units in southwestern Montana are summarized in Table 1 and include the Kibbey Sandstone, the Lombard Limestone (Pecora, 1984; Sando et al., 1985; Wardlaw and Pecora,

TABLE 1.—Chesterian lithostratigraphy of east-central Idaho and southwestern Montana.

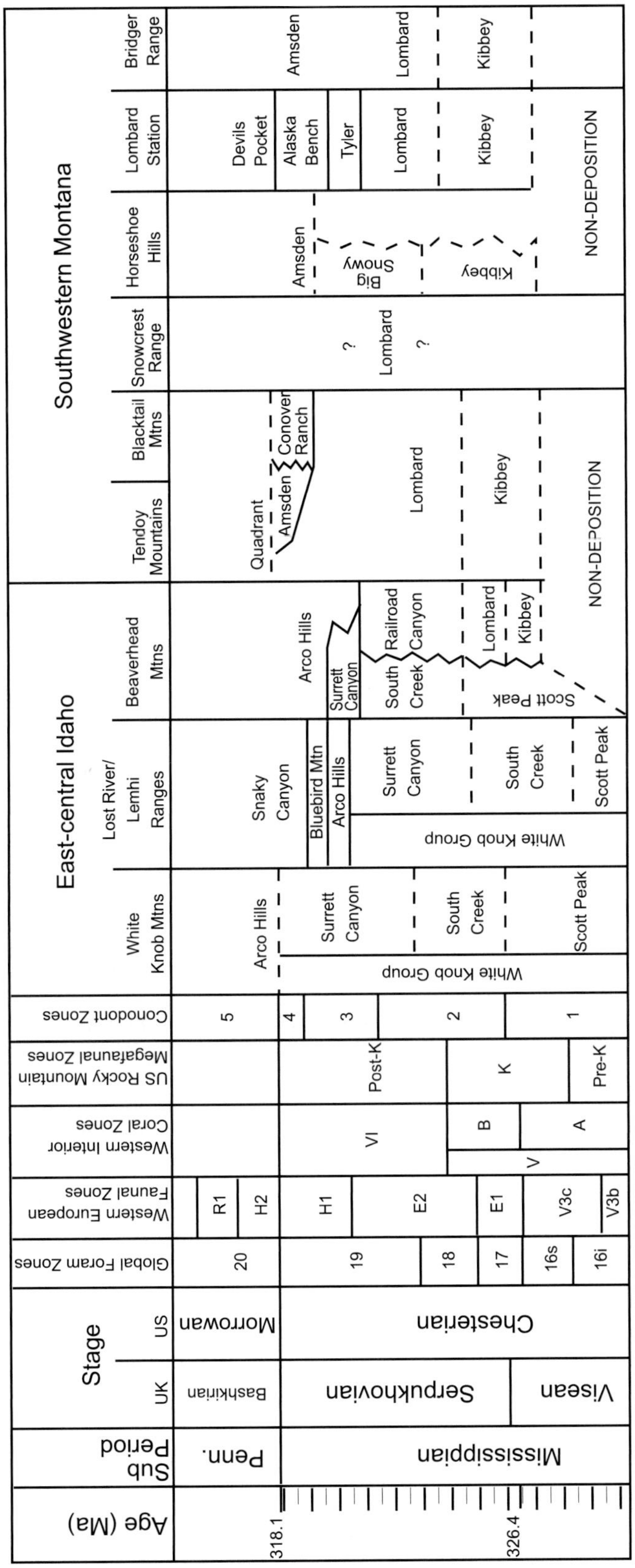

1985) and the Big Snowy Formation (Brewster, 1984; Byrne, 1985; Hildreth, 1981; Guthrie, 1984), as well as time-equivalent formations farther east in central Montana (Shepard, 1993; Easton, 1962). The Big Snowy Trough was an east–west trending depositional basin located in southwestern Montana that formed perpendicular to the Antler foreland basin during the Late Mississippian (Fig. 2). This elongate basin is the northern extension of the Central Montana Trough, which formed during late Proterozoic rifting and persisted as an episodically subsiding trough throughout the Paleozoic (Peterson, 1981). Subsidence of the Big Snowy Trough likely occurred along reactivated Proterozoic normal faults in response to horizontal stresses associated with tectonic loading of the Antler allochthon (Winston, 1986; Dorobek et al., 1991a; Dorobek et al., 1991b).

Laramide and Sevier crustal shortening and Eocene extension complicate estimates of the original width of the Antler foreland basin (Oldow et al., 1989). It was proposed that Cenozoic extension may approximate Mesozoic crustal shortening (Levy and Christie-Blick, 1989); thus, generalized palinspastic reconstructions based on field-determined structural relationships estimating a Mississippian basin–shelf width of less than 500 km (Peterson, 1981) may be roughly accurate.

METHODOLOGY

Seventeen Chesterian outcrops ranging in thickness from 80 to 520 m were measured in detail along an east–west, dip-oriented transect (Fig. 3). The transect spans ~ 320 km from south of Mackay, Idaho, to north of Bozeman, Montana, and includes outcrops from nine mountain ranges. The Chesterian stratigraphy measured for this study is summarized in Table 1, and latitude and longitude coordinates for each outcrop are in Batt (2006). Bed-by-bed outcrop descriptions included bed thickness, weathering profile, weathered-surface and fresh-surface color, grain type and size, fossils, sedimentary structures, and the character of bedding contacts. Sample collection accompanied the stratigraphic log and included samples for: (1) conodont biostratigraphy from every section, sampled every 15 meters for a total of 148 samples, (2) carbon and oxygen isotopes from three sections, sampled every 0.5 meters for a total of 640 samples and discussed in Batt (2006) and Batt et al. (2007), (3) strontium isotopes from one section, sampled wherever whole unsilicified brachiopods occurred for a total of 30 samples, and (4) lithofacies samples from every section. Select samples were thin sectioned and studied petrographically under transmitted light for more detailed analysis.

BIOSTRATIGRAPHY

Conodont biostratigraphy provides the basis for regional correlations between east-central Idaho and southwestern Montana (Fig. 4). Five applicable conodont zones for the Upper Mississippian were obtained from a continuous stratigraphic assemblage in east-central Idaho and extended eastward to southwestern Montana (Fig. 4; Abplanalp, 2006). Two zones are new and three are modifications of previous work from the U.S. Mid-continent and western interior (Dunn, 1966; Lane and Straka, 1974; Sando et al., 1981; Basemann and Lane, 1985). Zone 1 is the first appearance of *Hindeodus cristula* and / or *Vogelgnathus campbelli* and spans the uppermost Meramecian to the lower Chesterian. This newly proposed zone was first observed by Stamm (1985), in the Scott Peak Formation, and we extend it into the South Creek Formation. Zone 2 is based on the first occurrence of *Cavusgnathus naviculus* and overlaps with the Mamet's foraminiferal zones 17, 18, and the base of 19. Zone 3 is based on

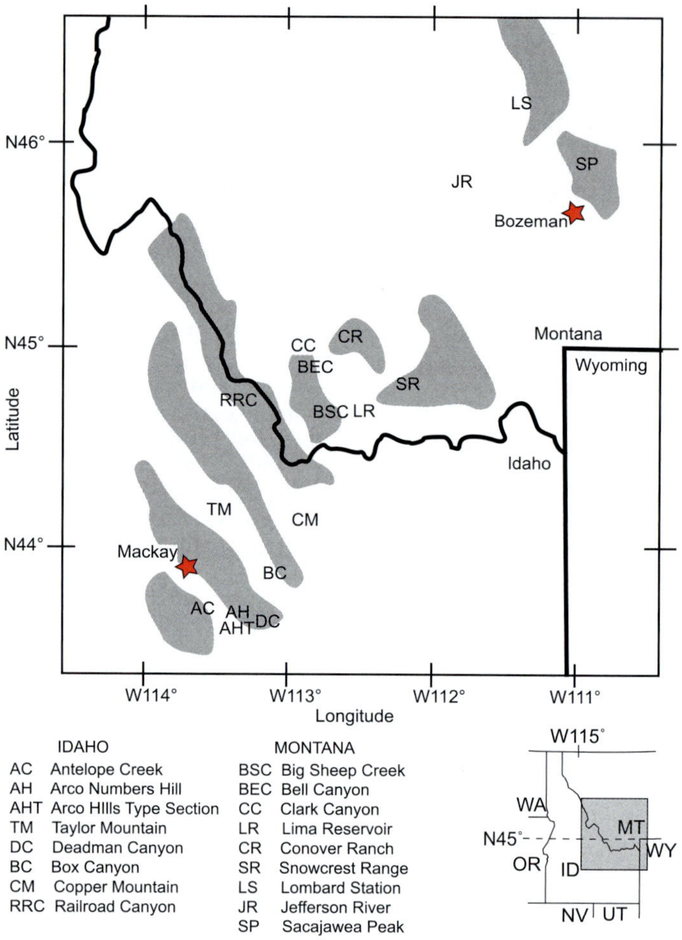

FIG. 3.—Study area showing the location of 17 measured sections in nine mountain ranges (gray shading) from Mackay, Idaho, to north of Bozeman, Montana. For bed-by-bed descriptions and detailed section location coordinates see Batt (2006).

the appearance of *Adetognathus unicornis* and correlates with the middle of foraminiferal zone 19. The range of Zone 3 is extended for the northern Rocky Mountains due to the rare occurrence of *Rhachistognathus muricatus* in the Surrett Canyon and Arco Hills formations. Zone 4 has the first appearance of *Adetognathus lautus* and is a newly proposed biostratigraphic division based on the upper subdivision of the *R. muricatus* zone of Basemann and Lane (1985). This stratigraphically thin zone correlates with the uppermost Mamet zone 19 and occurs within the Surrett Canyon, Arco Hills, and Bluebird Mountain formations. Zone 5 (Pennsylvanian) is based on the first occurrence of Pennsylvanian forms and is limited by the full range of *R. primus* (Abplanalp et al., 2006).

LITHOFACIES AND DEPOSITIONAL SETTING

The Chesterian succession (330–318 Ma; Gradstein et al., 2004) in the study area defines an eastward-thinning wedge of carbonate and mixed carbonate–siliciclastic sediments. Chesterian stratigraphy is bounded at its base by the Madison unconformity in Montana and correlative carbonate in Idaho and it terminates at the Mississippian–Pennsylvanian boundary, which has a stratigraphic expression across the region that varies from conformable mixed carbonate–siliciclastic deposition in Idaho to nondeposition, erosion, or deposition of siliciclastics across much of southwestern Montana.

Chesterian strata in east-central Idaho and southwestern Montana were deposited on a ramp that is subdivided into three distinct depositional settings, informally termed the western, central, and eastern ramp (Fig. 5). The western ramp occurs in east-central Idaho and is composed predominantly of muddy skeletal carbonates formed in deep subtidal to intertidal water depths. The central ramp occurs along the Idaho–Montana boundary where predominantly shale and calcisiltstone formed in deep subtidal settings. The eastern ramp occurs throughout most of southwestern Montana and is composed predominantly of peritidal carbonate interbedded with fluvial–estuarine siliciclastics. Detailed descriptions of the 14 lithofacies observed in the study area are in Table 2. These descriptions form the basis for interpretations of water depth and depositional setting (Fig. 5).

Deep Subtidal Carbonate

Bedded mudstone (Fig. 6A) and laminated calcisiltstone (Fig. 6B) lithofacies are thick (10–50 m), dark gray, generally unfossiliferous, mud- or silt-size carbonates interbedded with chert that form recessive ledges or partially covered slopes in outcrop. The thickest accumulations are on the western and central ramp. The thin-bedded and laminated nature of these facies and the absence of skeletal fragments indicates deposition below storm wave base (Schlager, 2005). Additional evidence for a deep subtidal depositional setting is the dark gray color of the carbonates, which indicates lower oxygen conditions and preservation of organic material (Pederson and Calvert, 1990), and the abundance of interbedded chert. The pinch-and-swell structures in the laminated calcisiltstone (Fig. 6B) resulted from soft-sediment deformation associated with slope instability and syndepositional extensional tectonism (J. Watkinson, personal communication).

Deep Subtidal Shales

Shale intervals occur in the central and eastern ramp (Table 2). The shale facies is black, dark gray, or green, fissile, with rare orbiculoid brachiopods (Fig. 6C) and ranges in thickness from 1 to 75 m. A more fossiliferous shale facies occurs only at the Lombard Station, Montana section as thick (15–25 m) laminated deposits of dark gray, organic-rich, calcareous shale with abundant whole articulated brachiopods interbedded with skeletal packstone. The dark color and the laminated nature of the nonfossiliferous shale indicates deposition below storm wave base and the organic-rich character, lack of bioturbation, and orbiculoid brachiopods (Butts, 2005) suggest that water circulation was restricted. That the fossiliferous shale facies contains thin, irregular skeletal packstone beds with corals, bryozoa, and pelmatozoan fragments indicates deposition in moderate to shallow depths where water conditions ranged from restricted marine during shale deposition to open marine during carbonate deposition.

Intermediate-Depth Subtidal Carbonate

The skeletal wackestone and bioturbated wackestone lithofacies occur in the western and central ramp (Table 2). In the skeletal wackestone facies, open marine fauna including brachiopods, solitary corals, and crinoid ossicles occur as whole, unfragmented fossils or as fossil fragments in thin discontinuous skeletal lags. The occurrence of both nonfragmented and

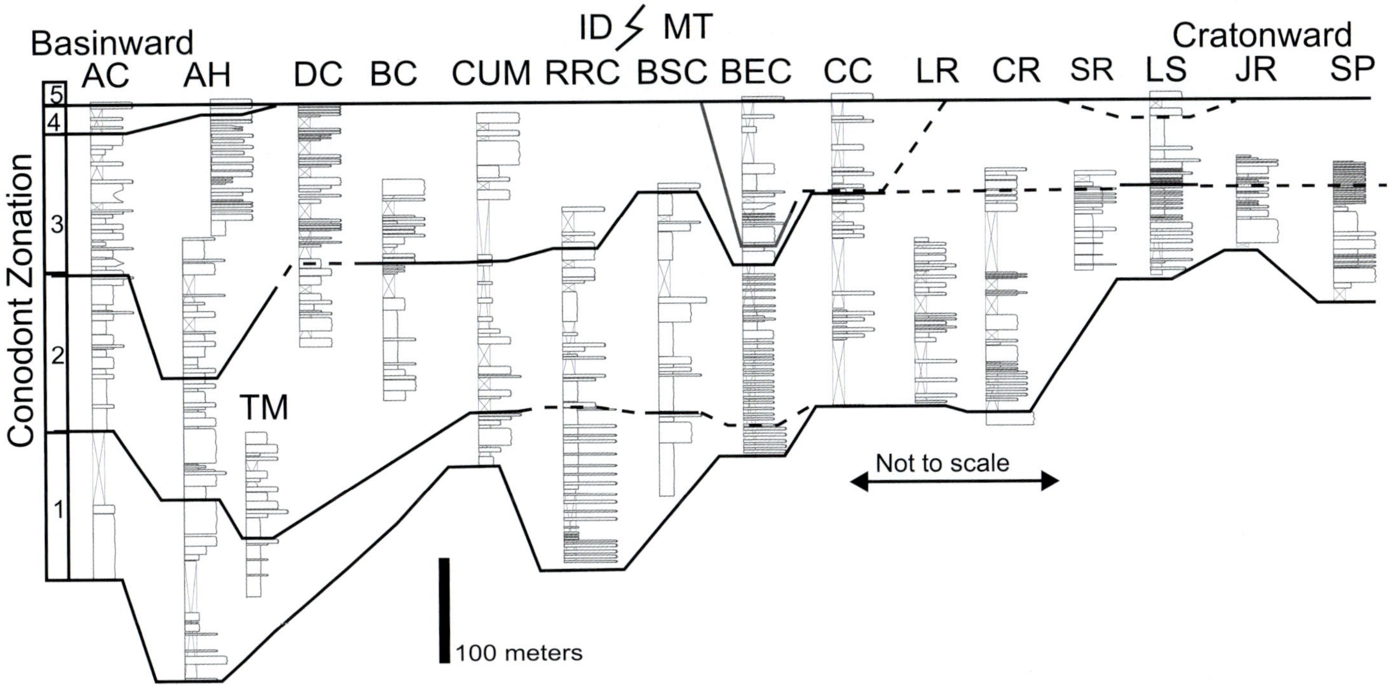

FIG. 4.—Biostratigraphic zonations across the study area provided the basis for sequence correlation (adapted from Abplanalp, 2006). Solid lines represent good biostratigraphic control. Dashed lines indicate where stratigraphic relationships were used in conjunction with lower-resolution biostratigraphy to determine age relationships. Stratigraphic columns based on grain size are generalized from Figure 8.

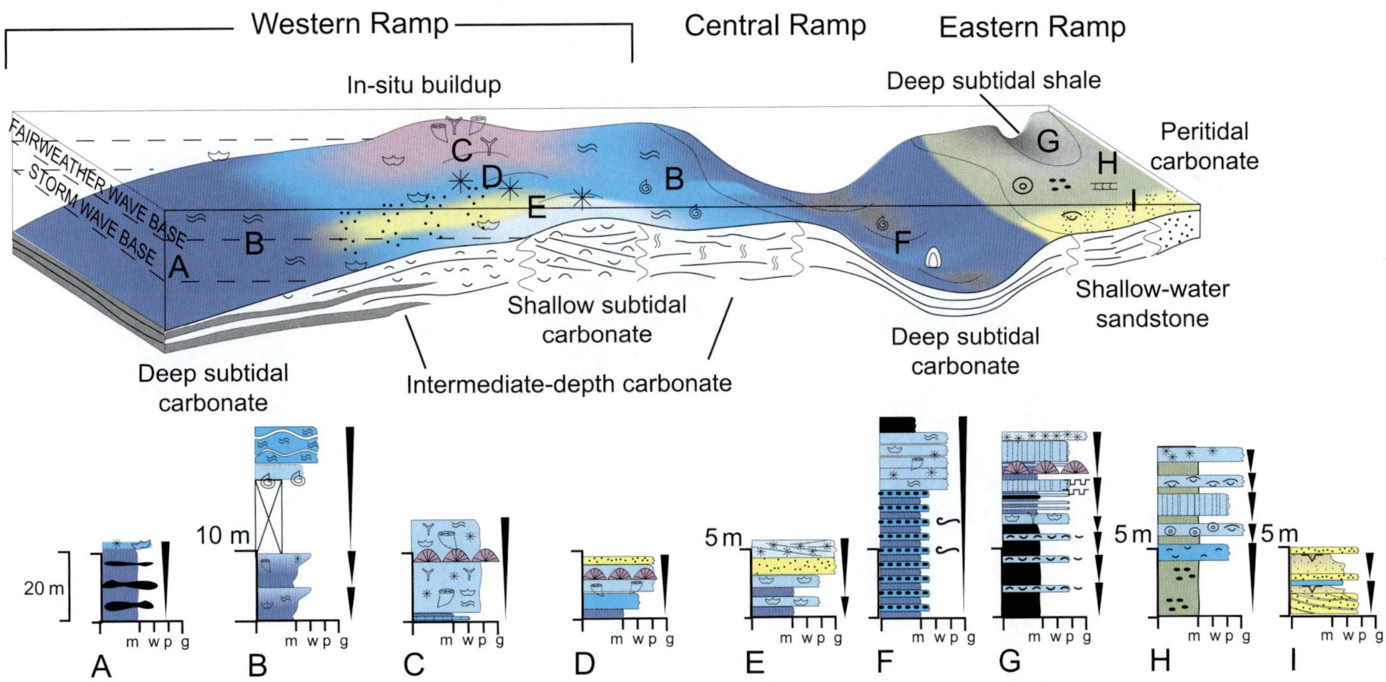

FIG. 5.—Schematic block diagram illustrating the range of depositional settings on the east-central Idaho to southwestern Montana Chesterian ramp. No horizontal scale implied. Representative parasequences A–I (individual parasequences indicated by inverted black triangles) illustrate facies stacking geometries according to position on the ramp. **A)** Bedded mudstone–based parasequences; **B)** Bioturbated wackestone parasequences; **C)** Skeletal carbonate parasequences; **D)** Skeletal carbonate parasequences; **E)** Grainstone–sandstone parasequences; **F)** Laminated-calcisiltstone parasequences; **G)** Shale-based parasequences; **H)** Peloidal packstone–based parasequences; **I)** Fine sandstone parasequence. Vertical twenty-meter bar scale unless otherwise noted in figure. Lithofacies color and symbol key is in Figure 8.

FIG. 6.—Photographs of Chesterian marine carbonate and mixed carbonate and siliciclastic lithofacies. **A)** Bedded mudstone and bedded chert, South Creek Formation, Antelope Creek, Idaho. **B)** Laminated to thin-bedded calcisiltstone; note the pinching and swelling of individual beds, Railroad Canyon Formation, Railroad Canyon, Idaho. **C)** Fossiliferous shale, Lombard Limestone, Lombard Station, Montana. **D)** Rugose and tabulate coral assemblage, Surrett Canyon Formation, Arco Hills, Idaho. **E)** Colonial rugose corals, Surrett Canyon Formation, Antelope Creek, Idaho. **F)** Abrupt transition from fine-grained quartz arenite to pelmatozoan grainstone, Bluebird Mountain Formation, Arco Hills type section, Idaho. **G)** Same sandstone–grainstone contact, larger scale.

TABLE 2.—Lithofacies description of upper Mississippian–lowermost Pennsylvanian strata in east-central Idaho and southwestern Montana.

Lithofacies	Stratigraphic Expression	Sedimentary Structures	Fossils/ Allochems	Depositional Environment; Estimated Water Depth	Cycle Expression	Occurrence
Deep Subtidal Carbonate					ps = parasequences	
Bedded mudstone	Bedded mudstones,gray, form massive parasequences 10–50 meters thick in recessive slopes and occasional cliff-forming outcrop. Chert occurs as laterally continuous beds or as discontinuous lenses and nodules.	Thin beds with bedded, lensoidal, and nodular chert up to 20–30% by unit thickness	Rare brachiopods, globular foraminifera	Outer ramp; below storm wave base; > 50 m	Base of bedded mudstone ps	Western ramp
Laminated calcisiltstone	Laminated calcisiltstones, gray, form recessive slopes and platy ledges.	Planar laminae may be deformed by pinch-and-swell structures	Brachiopods, bivalves	Outer ramp; near storm wave base; 25–50 m	Base of laminated calcisiltstone–based ps	Central ramp
Deep Subtidal Sandstone						
Shale	Shale, black, green or red, is fissile and up to 75 m thick, and may be interbedded with fine sandstone. It forms recessive slopes.	Planar lamination and fissile weathering	Orbiculoid brachiopods	Outer ramp or restricted shallow-water lagoon; 10–50 m	Depends on context. Top of laminated calcisiltstone–based ps or base of shale-based ps	Central and Eastern Ramp
Fossiliferous shale	Fossiliferous shale, dark gray, forms thin platy beds (1 cm or less) with organic carbon and abundant marine fossils. It forms recessive slopes where whole fossils weather out of the matrix. Occasional thin (< 5 cm) beds of skeletal wackestone/packstone occur within the fossiliferous shale.	Planar and wavy laminae containing thin, dark-gray to black bands (< 1 cm) of organic carbon	Brachiopods (productids with spines intact), bryozoa, solitary rugose corals, bivalves	Semi-restricted embayment; intermediate depth; 15–25 m	Base of shale-based ps	Eastern Ramp
Intermediate-depth Subtidal Carbonate						
Skeletal wackestone	Skeletal wackestones, gray, are medium (0.5–2 m thick) to massively bedded (up to 10 m thick).		Brachiopods, bryozoa, corals, pelmatozoa, gastropods	Open marine ramp; below fair-weather wave base, 10–50 m	Top of massive calcisiltstone–based ps or base of skeletal carbonate ps and bioturbated carbonate ps	Western Ramp
Bioturbated wackestone	Bioturbated wackestones, gray, are medium (0.5–2 m thick) bedded and they form ledges or cliffs that resist weathering.	Bioturbation, burrowing	Skeletal fragments filling burrows, brachiopods, bryozoa, corals, rare pelmatozoa	Semi-restricted ramp; lagoonal; below fair-weather wave base, 10–50 m	Top of bioturbated carbonate ps	Western and Central Ramp

TABLE 2 (continued).—Lithofacies description of upper Mississippian–lowermost Pennsylvanian strata in east-central Idaho and southwestern Montana.

Lithofacies	Stratigraphic Expression	Sedimentary Structures	Fossils/ Allochems	Depositional Environment; Estimated Water Depth	Cycle Expression	Occurrence
Shallow subtidal carbonate						
Skeletal packstone	Skeletal packstones are thin to medium (0.1 to 2 m thick) bedded, and they form ledges or cliffs resistant to weathering.		Fragmented and whole brachiopods, pelmatozoa, corals (rugose and tabulate), bryozoa	Open marine ramp; above fair-weather wave base, < 10 m	Top of skeletal carbonate ps	Western Ramp, Eastern Ramp (rare)
Skeletal grainstone	Skeletal grainstones are thin to medium (0.1 to 2 m thick) bedded, and they form ledges or cliffs resistant to weathering.	Planar cross bedding (rare)	Pelmatozoa, brachiopods (fragmented), subrounded fine quartz sand	Inner ramp shoals; above fair-weather wave base, < 10 m	Base or Top of grainstone-sandstone ps	Western Ramp, Eastern Ramp (rare)
Peritidal carbonate						
Carbonate conglomerate	Carbonate conglomerates, red or gray, have undulating thicknesses that average < 5 cm thick and may be grain or matrix supported, with subrounded granule- to pebble-size chert and carbonate mudstone lithoclasts, disarticulated brachiopod fragments, and carbonate matrix.	Possibly scoured base or top resulting in non-uniform lateral thickness variation	Chert lithoclast, brachiopods	Tidal channel, but may also indicate a deep marine flooding surface; < 5 m	Top of laminated calcisiltstone-based ps	Central Ramp
Peloidal packstone	Peloidal packstones, tan, are thin bedded (< 1 m thick) and form laterally continuous ribs of outcrop in otherwise recessive slopes.		Peloids, rare foraminifera	Subtidal, semi-restricted marine; < 15 m	Base of peloidal packstone ps	Western Ramp, Eastern Ramp
Ostracod packstone	Ostracod packstones, tan, are thin bedded (< 1 m thick) and form laterally continuous ribs of outcrop in otherwise recessive slopes.		Ostracods, rare foraminifera	Intertidal, semi-restricted marine; < 15 m		
Oncoid Packstone	Oncoid packstones, gray, are thin bedded (< 1 m thick) and form low ledges in otherwise recessive slopes.		Oncoids (2–4 cm diameter)	Intertidal marine; 5–15 m	Local occurrence at top of peloidal packstone ps	Eastern Ramp, one occurrence at Clark Canyon, Montana
Algal Mudstone	Algal mudstones, tan, are thin bedded (~ 10 cm thick) and form a laterally continuous rib of outcrop.	Domed stromatolite forms	Stromatolitic	Intertidal marine; 5–15 m	Local occurrence at base of one skeletal carbonate ps	Eastern Ramp, one occurrence at Sacajawea Peak, Montana

TABLE 2 (continued).—Lithofacies description of upper Mississippian–lowermost Pennsylvanian strata in east-central Idaho and southwestern Montana.

Lithofacies	Stratigraphic Expression	Sedimentary Structures	Fossils/ Allochems	Depositional Environment; Estimated Water Depth	Cycle Expression	Occurrence
Shallow water sandstone						
Fine sandstone	Thin-bedded (< 1 m thick) fine-grained quartz sandstone, yellow or red, may be calcareous or interbedded with carbonates. It forms recessive slopes.	Pedogenic features, interbedded with "B" and "C" soils, trough cross bedding, ripple marks	Plant impressions	Fluvial-estuarine; Shallow water < 15 meters	Comprise the fine sandstone ps, become interbedded with carbonates up-section	Central Ramp, Eastern Ramp
Significant Surfaces						
Flooding Surface	Flooding surfaces are regionally continuous thin (< 10 cm thick) intervals characterized by organic carbon, conglomeritic lags with phosphate nodules, pyrite, and chert.	Nodular or bedded chert, pyrite	Phosphate nodules	Depth depends on ramp location; typically > 50 m	Bases of sequence sets and most sequences	Western, Central, Eastern Ramp
Exposure Surface	Exposure surfaces are disconformable, identifiable as "B" and "C" soil horizons, or as discontinuous surfaces with collapse breccia filling dissolution cavities.	Karst, "B" and "C" soils, pedogenic features, erosional base		Subaerial 0 m	Bases of some sequences	Eastern Ramp

fragmented fossils, the abundance of carbonate mud matrix, and the medium gray color all suggest deposition in intermediate depths between fair-weather and storm wave base (Schlager, 2005). The bioturbated wackestone lithofacies is darker gray and contains mostly sand-size fossil allochems and restricted marine fauna (gastropods and rare ostracods) in a highly bioturbated matrix. These characteristics indicate deposition in a more restricted, quiet-water, perhaps hypersaline marine setting of intermediate depth (Schlager, 2005).

Shallow Subtidal Carbonate

Skeletal packstone and skeletal grainstone lithofacies (Table 2) formed predominantly on the western ramp, and they contain fragmented brachiopods, crinoids, corals, and bryozoa (Fig. 6D). The diverse skeletal assemblage and the fragmented nature of the allochems indicate deposition of skeletal packstones above fair-weather wave base (Schlager, 2005). The occurrence of low-relief *in situ* coral and bryozoan buildups (average of 10–50 cm; Fig. 6E) in the skeletal packstone lithofacies also supports deposition above fair-weather wave base, where winnowing of sea-floor muds by active currents would have enabled coral and bryozoan growth (Lasemi et al., 2003).

Skeletal grainstone contains syntaxial-cemented crinoid ossicles, with variable amounts (up to 20%) of subrounded quartz grains. Grainstone beds are limited to the upper Chesterian on the western ramp, where they are interbedded with fine-grained calcareous sandstone to form grainstone–sandstone parasequences (Figs. 6F, 6G). Skeletal grainstone formed just below the tidal zone, in a wave-agitated setting where active water circulation winnowed fines and promoted grain cementation (Harris et al., 1985). The wide spacing between outcrops prevents a quantitative understanding of the geometry of the grainstone facies; they may have formed as laterally extensive sheets with low or no relief, or they may have formed high-energy skeletal banks that baffled current energy in a landward direction.

Peritidal Carbonate

Peloidal packstone, ostracod packstone, oncoid packstone, algal mudstone, and carbonate conglomerate (Table 2) all formed in the tidal zone. They are best developed on the central and eastern ramp. All of these lithofacies contain thin, laterally discontinuous beds that are poorly exposed in the study area (Fig. 7A). The algal mudstone (Fig. 7B) and oncoid packstone (Fig. 7C) occur in only one section each. The formation of algal mats (stromatolites) and coated grains (oncoids) occurs in the intertidal zone where algae bind carbonate sand grains transported by tidal currents (Tucker and Wright, 1990). Deposition of this facies occurred in the supratidal to intertidal environment. Peloidal mudstone and ostracod packstone record semi-restricted circulation in a shallow marine setting. Peloids are excrement from burrowing organisms; they, along with the presence of ostracods, suggest a partly restricted environment (Tucker and Wright, 1990). Carbonate conglomerates occur as beds 0–5 cm thick containing sand- to pebble-size clasts of carbonate and chert (Table 2). The presence of pebble-size grains indicates bedload transport in a high-energy environment. Those carbonate conglomerates occur interbedded with claystone and soil surfaces (Fig. 7E), suggesting highly variable depositional energy conditions in a marginal marine environment. Estuary or tidal channels may be appropriate analogs for deposition of the carbonate conglomerate lithofacies (Tucker and Wright, 1990).

FIG. 7.—Photographs of Chesterian peritidal carbonate and terrestrial siliciclastic lithofacies. **A)** Thin-bedded, poorly exposed peloidal- and ostracod-rich packstones typical of peritidal deposits on the eastern ramp, Lombard Limestone, Clark Canyon Reservoir, Montana. **B)** Stromatolite mudstone, Big Snowy Formation, Sacajawea Peak, Montana. **C)** Oncoid packstone, Lombard Limestone, Clark Canyon Reservoir, Montana. **D)** Carbonate conglomerate with pebble- to cobble-size subrounded skeletal wackestone and chert clasts bound by a skeletal mudstone–packstone matrix, Lombard Limestone, Lombard Station, Montana. **E)** Soil formation above medium-grained siltstone, Conover Ranch Formation, Conover Ranch, Montana. **F)** Cross-bedded calcareous siltstone, Kibbey Sandstone, Sacajawea Peak, Montana. **G)** Plant fossils, Kibbey Sandstone, Bell Canyon, Montana. **H)** Very fine-grained carbonate-cemented sandstone, Surrett Canyon Formation, Antelope Creek, Idaho.

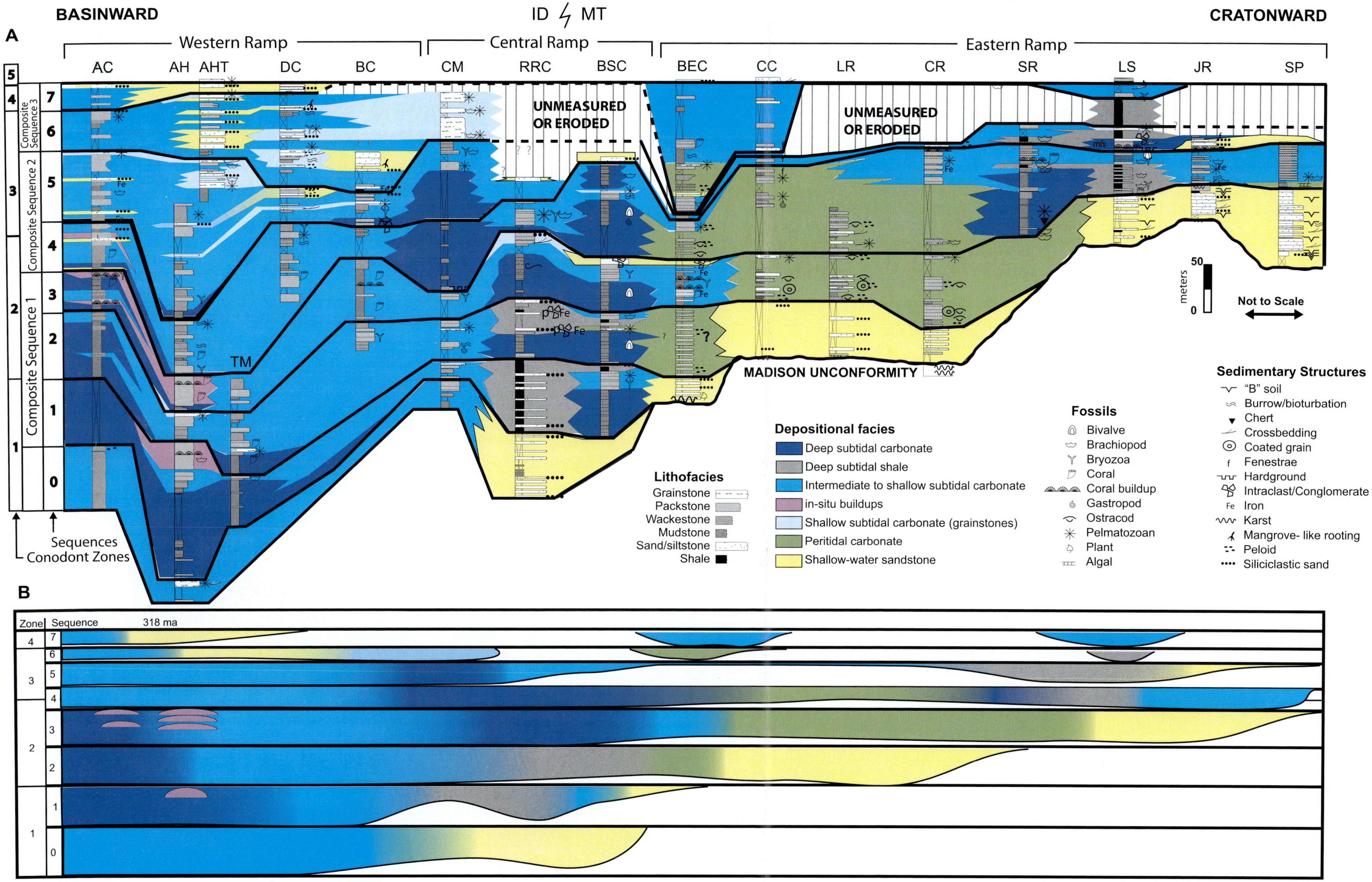

FIG. 8.—**A)** Regional cross section of Chesterian strata from east-central Idaho across southwest Montana. Abbreviated outcrop locations as in Figure 3. No hor **B)** Chronostratigraphic chart showing the generalized distribution of sediment on the ramp during each sequence.

Shallow-Water Sandstones

The fine-grained sandstone lithofacies (Fig. 7F) occurs on all parts of the ramp. In southwestern Montana it overlies the Madison unconformity during the lower and middle Chesterian. In east-central Idaho, fine-grained sandstone is common only in the upper Chesterian, where fine-grained calcite-cemented quartz sandstone is interbedded with skeletal grainstone (Fig. 6G). The association of fine-grained sandstones with skeletal grainstone indicates that they formed between fair-weather base and the base of the tidal zone. Siliciclastics on the east-central Idaho ramp occur only rarely in the early and middle Chesterian as thin, laterally extensive drapes within the bedded mudstone lithofacies (Fig. 7H). Here siliciclastic sand is fine grained and usually occurs adjacent to *in situ* coral and bryozoan buildups. The environment of deposition of these siliciclastic sands is most likely intermediate depth. The absence of up-dip equivalents makes their origin difficult to determine. In Montana sections, the fine-grained sandstone lithofacies contains soil horizons, tangential and planar cross-bedding (Fig. 7F), and ripple marks. Lycopod plant impressions also occur in this lithofacies at Bell Canyon, MT (Fig. 7G). These sandstones are texturally distinct from those on the western ramp, and they are 1–5 m thick.

Significant Surfaces

Subaerial exposure surfaces are common on the eastern ramp but they are rare to absent on the western and central ramps. The most prominent subaerial exposure surface is a regionally correlative unconformity that forms a karst on the upper surface of the Madison Formation. A well developed karst surface also occurs at Jefferson River, Montana (JR, Fig. 8) during the middle Chesterian. Here a disconformable surface penetrates 1.5 m into underlying strata, and the karst cavities are filled with clay and solution-collapse breccia. Elsewhere on the eastern ramp paleosols are abundant during the middle Chesterian, but they are locally discontinuous and only rarely correlate between sections. Soil surfaces commonly are 1–3 cm of dark red, waxy clay that often contains carbonate nodules. On the western and central ramp, unconformities are rare. The only definitive evidence of subaerial exposure on the western ramp is an upper Chesterian horizon of centimeter-size, carbonate-filled root cavities in a sandy matrix.

Carbon isotopes collected at 0.5 m intervals at Antelope Creek, Idaho, do not show a trend of decreasing $\delta^{13}C$ values from the bases to the tops of shallowing-upward parasequences (Batt et al., 2007). This would be the expected trend if meteoric water associated with subaerial exposure had affected parasequence caps (Allan and Mathews, 1982; Rush and Chafetz, 1990).

Flooding surfaces are locally developed on each section of the ramp. Characterized by an abrupt transition from shallow-water facies to deep-water facies, or by condensed intervals of organic matter, flooding surfaces are locally to regionally correlative across the study area. Regionally correlative surfaces are labeled maximum flooding zones (MFZ) in Figure 8. They form the bases of composite sequences, and they correspond with large-scale changes in depositional settings on the eastern ramp. On the eastern ramp, flooding surfaces form gradational transitions to deeper-water facies, or they may occur as distinct surfaces marked by organic matter, coral buildups, nodular chert, or iron oxide nodules. Phosphatic pebble lags are distinctive features within or just below flooding surfaces that formed only on the central ramp.

CHESTERIAN CYCLICITY

The Chesterian succession in east-central Idaho and southwestern Montana is 10–12 million years in duration and constitutes the latter stage of the Kaskaskia second-order sea-level cycle (Ross and Ross, 1985, 1987). Lithofacies in this Chesterian stratigraphic succession are grouped into a hierarchical set of parasequences, sequences, and composite sequences that record fourth-order, fifth-order, third-order, and second-order relative changes in sea level, respectively. The cycle hierarchy used in this study is modified from Mitchum and Van Wagoner (1991), Weber et al. (1995), Kerans and Tinker (1997), and Sarg et al. (1999).

Parasequences

Parasequences represent the shortest duration of genetically related vertical lithofacies assemblages, and they are bounded by local flooding surfaces (Fig. 5). The stratigraphic expression of parasequences varies according to depositional setting, but most commonly they are coarsening-upward cycles less than 20 m thick. Parasequences may result from autocyclic or allocyclic processes, depending on their position on the ramp, and because of the large distances between outcrops they are usually only locally correlative. The duration of parasequences remains equivocal because biostratigraphic zonations provide only third-order chronological constraints for the region. In addition, significant differences in sedimentation rate and accommodation likely existed from the western to eastern ramp. We instead consider the duration of parasequences in a relative sense. In some settings parasequences may be equivalent to the duration of an entire sequence defined elsewhere. During conodont zones 1 and 2 (lower to middle Chesterian), individual sequences on the western and central ramps contain between one and six parasequences. However, the upper Chesterian sequences contain more than 10 parasequences. Thus, the number of parasequences within individual sequences increases from the western and central ramps to the eastern ramp, and also through time.

Chesterian parasequences range in thickness from less than 5 m in peritidal facies up to greater than 20 m in deep-water facies. The thickness, geometry, lateral extent, and lithology of parasequences vary across depositional settings on the ramp. Representative meter-scale parasequences, arranged in order of decreasing water depth, include: bedded mudstone–based, laminated calcisiltstone–based, shale-based, skeletal wackestone–based, bioturbated wackestone–based, interbedded skeletal grainstone–sandstone, peloidal packstone–based, and fine sandstone (Fig. 5).

Bedded mudstone–based parasequences are between 20 and 75 meters thick, and they occur in lower to middle Chesterian strata on the western ramp (Fig. 5A). These parasequences commonly are asymmetric with thick, interbedded mudstone and chert bases that are capped by one or more thin beds of skeletal wackestone–packstone. These parasequences formed dominantly below storm wave base in an open marine, deep-water setting. The skeletal wackestone–packstone parasequence caps indicate shallowing to storm wave base.

Thin laminated calcisiltstone–based parasequences are between 20 and 50 meters thick, and they occur in lower to middle Chesterian strata from the central ramp. Laminated calcisiltstone–based parasequences are similar to bedded mudstone–based parasequences because they are asymmetric, composed mostly of calcisiltstone with thin skeletal wackestone–packstone caps that locally are phosphatic (Fig. 5F). The laminated calcisiltstone–based parasequences are different from the bedded mudstone–based parasequences because they are laminated, they contain little or no chert, and they locally contain abundant organic carbon (Fig. 6B). They record shallowing from deposition of laminated calcisiltstone below storm wave base to deposition of skeletal packstones above fair-weather wave base.

Shale-based parasequences are 10–20 m thick and are uniform deposits of organic-rich, highly fissile black or green shale (Fig. 5G). Shale units contain abundant *in situ* productid and orbiculoid brachiopods, and the abundance of thin carbonate beds in each parasequence increases up-section. Medium-bedded, irregular nodular skeletal wackestone–packstone (< 1 m) caps the shale-based parasequences. Shale-based parasequences occur in lower Chesterian strata from the central ramp and in upper Chesterian strata from the eastern ramp. The transition from shale to carbonate in these parasequences may indicate shallowing water depths, or alternatively it may record the transition from a dysoxic to open marine environment with little or no change in water depth.

Skeletal carbonate parasequences range from 5 to 20 m thick, with thin-bedded calcisiltstone or skeletal wackestone bases that coarsen and thicken upward into massive skeletal packstone caps (Figs. 5C, 5D, 9A). *In situ* colonial rugose and bryozoan buildups (< 1 m thick) may form parasequence capping marker beds that are laterally traceable between sections (Fig. 6E). Skeletal carbonate parasequences occur in middle to upper Chesterian strata from the western ramp. The skeletal carbonate parasequences indicate shallowing from storm wave base to fair-weather wave base across much of the ramp.

Bioturbated carbonate parasequences are 5–10 m thick (Fig. 5B). They are medium-bedded throughout, with skeletal wackestones or packstones that become bioturbated and increasingly peloidal up-section. These parasequences occur in lower Chesterian strata on the central ramp. These trends of increasing infaunal activity and mud-dominated nature indicate a transition from open marine shallow subtidal conditions into quieter-water, more restricted marine conditions (Tucker and Wright, 1990).

Interbedded skeletal grainstone–sandstone parasequences are 2–10 m thick and are commonly bimodal, with thin- to medium-bedded (0.5–2 m thick), massive or planar cross-bedded skeletal grainstone at their bases and calcite-cemented fine-grained quartz arenite at their tops (Fig. 5E). These facies may also stack in the reverse order, with sandstone grading upward into skeletal grainstone (Figs. 6G, 9B). Pelmatozoan fragments, with lesser fractions of brachiopod fragments, constitute the grainstone. The contact between carbonate and siliciclastic facies in these cycles may be abrupt or gradational, and internal planar cross-bedding is common. Interbedded skeletal grainstone–sandstone parasequences occur in upper Chesterian strata from the western and central ramp. These mixed carbonate–siliciclastic parasequences form above fair-weather wave base in a high-energy, shallow-water, open marine setting where fine material was preferentially swept away. They record lateral migration of skeletal shoals and sandy bars in response to changes either in relative sea level or sediment supply of fine siliciclastic material to the ramp.

Peloidal packstone–based parasequences are 1–5 m thick and form as a series of recessive ledges within otherwise covered outcrop (Figs. 5H, 7A). Parasequence bases are interbedded peloidal or massive tan carbonate mudstone that commonly are poorly exposed. Carbonate mud interbeds decrease towards the top of the parasequence, concomitant with an increase in either skeletal grains or oncoids, which form packstones that cap these parasequences. The peloidal packstone–based parasequence occur in lower Chesterian strata from the western ramp and in middle to upper Chesterian strata on the eastern ramp. Peloidal packstone–based parasequences record the transition from shallow subtidal to intertidal environments in a shallow-water, semirestricted-ramp setting.

Fine-grained sandstone parasequences are recessive and commonly form covered slopes in outcrop, where they consist of thin interbedded fine-grained calcite-cemented quartz sandstone and sandy mudstone (Fig. 5I). Fine-grained sandstone parasequences form in early to middle Chesterian strata on the eastern ramp, directly above the regional Madison Group unconformity, where they contain numerous dark red paleosols with rooting structures (Fig. 7E). Basal sandstones may be planar bedded or have small trough cross-bedding (Fig. 7F). Fine-grained sandstone parasequences become interbedded with carbonate up-section. The depositional environment of these sandstones is difficult to infer, partly because of limited exposure, but also because of the wide range of water depths possible for sandstone deposition and overall paucity of sedimentary structures or fossil depth indicators. Where they are interbedded with whole-fossil-bearing calcareous facies, the sandstone beds may be lowstand sands in a marine setting of moderate depth. Alternatively, the presence of soils, plant fossils (Fig. 7G), and cross bedding in some facies is suggestive of a coastal-plain setting involving sand deposition in channels or marine inlets and nondeposition and soil formation on adjacent floodplains or tidal marshes.

Third-Order Sequences

Sequences are composed of stacked parasequences and are bounded by regionally correlative flooding zones (Fig. 8A). The Chesterian strata in the study area are subdivided into the upper half of a sequence (S0) that is preserved only downdip in Idaho, and seven full sequences (S1–S7; Fig. 8A). The sequences are numbered sequentially in order, with S0 being the oldest and S7 being the youngest. The average duration of the seven sequences is ~ 1.5 Myr, on the basis of the duration of the Chesterian ~ 12 Myr (330–318 Ma), divided by the seven sequences. However, their likely range is 1–5 Myr; thus these sequences are considered third-order (Weber et al., 1995). These Chesterian sequences are equivalent in duration to the third-order unconformity-bounded sequences defined by Van Wagoner (1991). However, the downdip location of sections in east-central Idaho and lateral discontinuity of exposure horizons in updip locations preclude the use of unconformities to define all sequence boundaries in this study area, so some sequence boundaries are downdip conformable surfaces.

Sequence 0 is the upper part of an incomplete sequence that occurs in the lower half of conodont biozone 1 (Figs. 8A, 8B). Deposition in this sequence is restricted to locations west of Bell Canyon, Idaho, on the western ramp. S0 is composed of thin- to medium-bedded peloidal packstone parasequences that thin eastward into bioturbated-wackestone parasequences. Eastward the shallow-water carbonate passes into interbedded siltstone and mudstone at Railroad Canyon, Idaho. Farther updip, S0 onlaps against the widespread unconformity formed above the Madison Group.

Sequence 1 was deposited during the upper half of conodont biozone 1 and records deeper-water carbonate deposition to the west and siliciclastic deposition only near Bell Canyon, Montana (Figs. 8A, 8B). The base of S1 is defined in east-central Idaho by the occurrence of the deepest-water facies in the Chesterian. Here massive calcisiltstone facies overlie the thin-bedded peloidal packstone parasequences of S0. The thickest deposits in S1 occur in the western ramp, where calcisiltstone–based parasequences with locally abundant chert (South Creek Formation) shallow upward into skeletal carbonate parasequences capped

FIG. 9.—Photographs of parasequences and sequences in Chesterian strata. **A)** Middle Chesterian thick-bedded parasequences, Surrett Canyon Formation, Antelope Creek, Idaho. **B)** Late Chesterian thin-bedded heterolithic parasequences, Arco Hills Formation, Arco Hills type section, Idaho.

by *in situ* rugose corals. Calcisiltstone–based parasequences thin and become grain-rich to the east, passing into a thin skeletal carbonate parasequence on the central ramp at Copper Mountain, Idaho. Northward at Railroad Canyon, Idaho, multiple shale-based parasequences occur, with the abundance of shale increasing up-section. Eastward, a thick asymmetric calcisiltstone-based parasequence, capped by skeletal packstone, passes into fine sandstone parasequences in the eastern ramp, which pinch out against the Madison Group unconformity to the east and north.

Sequence 2 occurs in the lower half of conodont biozone 2 and records nearly equal distribution of carbonate to the west and siliciclastics to the east (Figs. 8A, 8B). Deposition of calcisiltstone–based parasequences continues in the westernmost part of the ramp, but at Arco Hills this interval of calcisiltstone is capped by pelmatozoan grainstone. Additionally, muddy coral banks began to develop for the first time at Taylor Mountain, Idaho. In the central ramp, multiple, thinner skeletal carbonate parasequences transition to bioturbated carbonate parasequences at Copper Mountain, ID. North and east, but still in the central ramp, multiple calcisiltstone-based parasequences are capped by thin phosphatic conglomerates or skeletal packstone. These deeper-water parasequences pass abruptly eastward into poorly exposed peloidal packstone parasequences at Bell Canyon, Montana. Farther north, these carbonates become fine sandstone parasequences, which truncate against the Madison unconformity east of Conover Ranch, Montana.

Sequence 3 records carbonate deposition throughout most of the study area during the middle of conodont biozone 2 (Figs. 8A, 8B). This sequence generally has uniform thickness from west to east. On the western ramp, skeletal carbonate continues to dominate, and muddy carbonate banks expand throughout this area (Fig. 9A). Laminated calcisiltstone–based parasequences continue to accumulate at Railroad Canyon and Big Sheep Creek, Montana, transitioning to multiple thin skeletal carbonate parasequences to the north and east with corals at Bell Canyon, Montana. On the eastern ramp, peloidal packstone parasequences are deposited above sandstone parasequences from sequence 2. The facies transition to siliciclastics occurs east of Conover Ranch, where multiple soils are interbedded with fine-grained calcareous sandstones.

Sequence 4 records carbonate deposition across the entire study area during upper conodont biozone 2 and lowermost conodont biozone 3 (Figs. 8A, B). On the western ramp, skeletal carbonate forms thick, coarsening-upward parasequences locally with siliciclastic bases that thin eastward, disappearing toward Box Canyon, Idaho. Deposition of deeper-water calcisiltstone-based parasequences occurred in two isolated areas, Copper Mountain and Big Sheep Creek, on either side of skeletal carbonate parasequences that formed coevally at Railroad Canyon. Peloidal packstone parasequences formed from Bell Canyon to Conover Ranch, and they pass eastward and northward into more open marine calcisiltstone-based and shale-based parasequences. Carbonate content increases upward in the shale-based parasequences at Lombard Station and becomes fully carbonate at the top of the sequence. To the east, shale passes into carbonate facies that include skeletal packstone and stromatolitic mudstone.

Sequence 5 occurs in conodont biozone 3 and records significant depositional thicknesses on the western ramp that correlate with thin or isolated sediment accumulations east of Copper Mountain (Fig. 8A, 8B). The S5 basal flooding surface is regionally correlative. Skeletal carbonate parasequences on the western ramp have muddy bases and coarsen upwards. Skeletal grainstone and calcareous sandstones increasingly cap parasequences up-section, and interbedded grainstone–sandstone parasequences dominate the upper part of the sequence at Arco Hills and Deadman Canyon. Calcisiltstone deposition continued at Copper Mountain during the basal portion of S5, shallowing upward to skeletal packstones at the top of the sequence. On parts of the eastern ramp, S5 deposition was limited to a thin wedge of skeletal packstones. East of Conover Ranch, however, shale-based parasequences occur above the basal flooding surface and shallow upward to interbedded siliciclastics and skeletal grainstone at the top of the sequence.

Sequence 6 occurs during uppermost conodont biozone 3, and records deposition only on the western ramp and at Lombard Station, Montana (Figs. 8A, 8B). Thick skeletal carbonate parasequences (15–20 m) on the western most ramp pass eastward into multiple thin-bedded (5–10 m) grainstone–sandstone cycles (Fig. 9B). At Copper Mountain, stacked skeletal packstone and grainstone parasequences of S6 lack siliciclastics. East of Copper Mountain S6 is a thin carbonate interval recorded at Bell Canyon and Clark Canyon. In contrast, a 30-meter-thick shale interval was deposited at Lombard Station during S6.

Sequence 7 was deposited during conodont biozone 4 and includes deposition of both carbonate and siliciclastics to the west, whereas the few outcrops to the east are completely carbonate (Figs. 8A, 8B). Interbedded grainstone–sandstone parasequences on the western ramp pass upward into poorly exposed skeletal packstone at Antelope Creek, Idaho. Few time-equivalent strata east of Deadman Canyon, Idaho, were measured because of their poor exposure or structural complexity. This interval was described at Bell Canyon, Clark Canyon, and Lombard Station in Montana. At Bell Canyon and Clark Canyon, a succession of thin-bedded peloidal packstone parasequences transition up-section to poorly exposed skeletal carbonate parasequences. This carbonate succession is significantly thicker than coeval accumulations east or west on the ramp. At Lombard Station, skeletal carbonates of S7 occur above the shale succession from S6, passing upward into hummocky cross-bedded skeletal packstone across the Mississippian–Pennsylvanian boundary (Wardlaw and Pecora, 1985).

Composite Sequences

Composite sequences group genetically related sequences and differentiate major changes in the style of regional deposition and sequence stacking patterns (Mitchum and Van Wagoner, 1991). Systems tracts are not identified in this study, and so the terminology is a modification of Mitchum and Van Wagoner (1991). There are three composite sequences (I, II, and III) defined for the Chesterian strata in east-central Idaho and southwestern Montana (Fig. 8A). The composite sequences are numbered sequentially, with I being the oldest and III being the youngest. The duration of composite sequences varies significantly, but each is less than 10 Myr.

Composite Sequence I (S0–S3) formed in the early and middle Chesterian (Fig. 8A). In east-central Idaho its basal boundary is the deepest-water flooding event observed in the study, which is correlative updip with the Madison unconformity in Montana. The upper boundary of composite sequence I is defined in east-central Idaho by a transition from sequences composed of thick-bedded parasequences to sequences composed of thin-bedded parasequences; in Montana this upper boundary is defined by the maximum landward position of peritidal lithofacies relative to underlying sequences. Parasequence stacking patterns in composite sequence I vary from progradational on the western ramp, to aggradational on the central ramp, and retrogradational on the eastern ramp. Carbonate deposition dominates composite sequence 1 on the western ramp, and corresponds to a transition from siliciclastic to carbonate deposition on the central and eastern ramp.

Composite Sequence II (S4 and S5) formed in the middle and late Chesterian (Fig. 8A). This composite sequence records several significant changes in sedimentation across the ramp. On the western ramp, sequences are composed of thinner parasequences with grain-rich caps. On the central ramp, the intraramp basin appears to have broadened and become segmented (Figs. 8, 10, 11). On the eastern ramp, carbonate-dominated lithofacies occur above the siliciclastics-dominated lithofacies of composite sequence I. In composite sequence II, carbonate dominates across the entire ramp, and the relative dearth of peritidal facies and

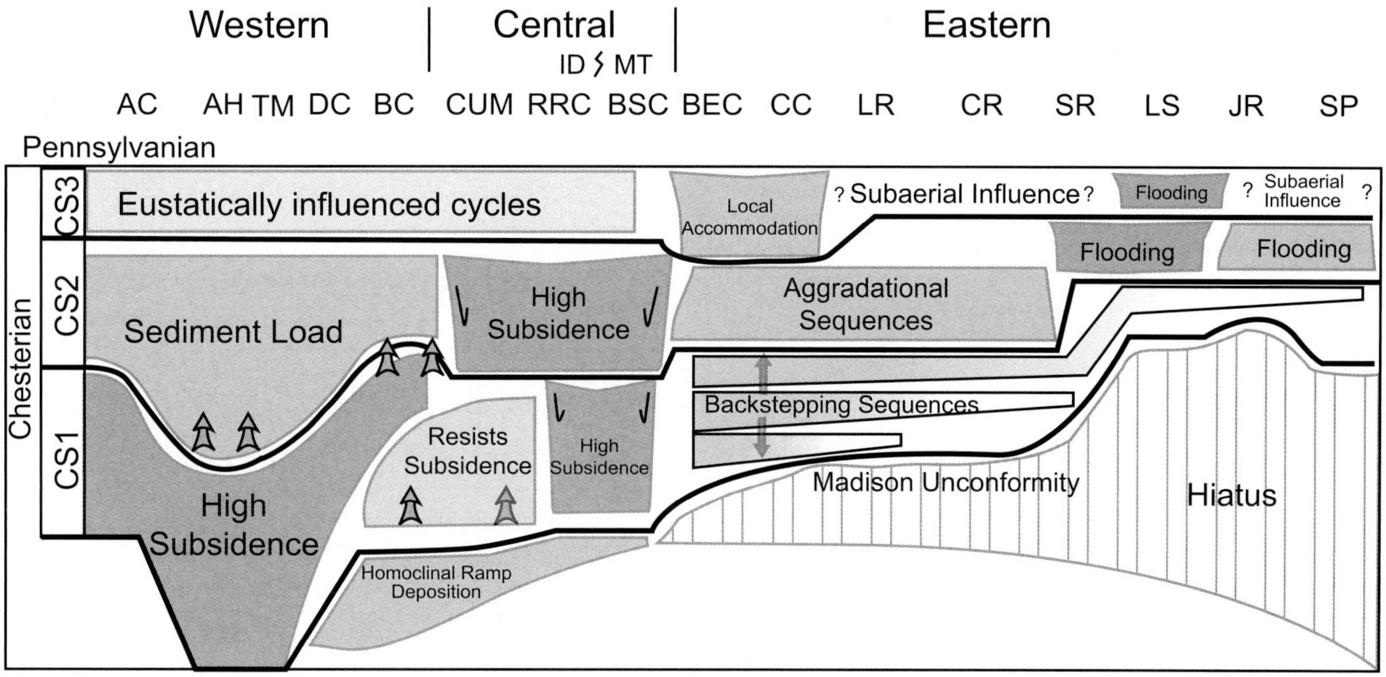

FIG. 10.—Schematic cross section generalizing processes affecting accommodation on the ramp during the Chesterian. Tectonic loading during composite sequence I segmented the ramp into three unique settings: western, central, and eastern ramp. Composite sequence I records tectonic loading in the west, establishment of an extensional intraramp basin on the central ramp and transgressive sedimentation on the eastern ramp. Composite sequence II records continued muddy-carbonate-bank deposition on a broad bathymetric high that was resisting tectonic subsidence on the western ramp, expansion then infilling of the intraramp basin on the central ramp, and initiation of flooding onto the eastern ramp from the Big Snowy Trough. Composite sequence III records eustatically driven parasequences on a homoclinal ramp, on the western ramp, peritidal carbonate deposition on the central ramp and continued deeper subtidal carbonate deposition on the eastern ramp.

nearshore to fluvial siliciclastics on the eastern ramp indicates that relative sea level was at its local maximum during this period. The depositional thicknesses in composite sequence II are relatively uniform, being thickest on the western ramp.

Composite Sequence III (S6 and S7) formed in the late Chesterian (Fig. 8A). Parasequence stacking patterns in composite sequence III are progradational overall and record the increased influx of siliciclastics to the western ramp. Continuous deposition on the western ramp contrasts with minimal stratal accumulation, nondeposition, or erosion on the central ramp. Outcrops on the eastern ramp are biostratigraphically unconstrained except for Lombard Station, where conodont assemblages in shallow subtidal lithofacies indicate continuous deposition across the Mississippian–Pennsylvanian boundary.

DISCUSSION

The three composite sequences (I, II, and III) have unique facies distribution and parasequence stacking patterns that record changes in accommodation resulting from regional tectonics, differential subsidence, and eustasy changes in the Antler basin during the Chesterian (Fig. 10). Composite sequence I (S0–S3) involves a stage of active tectonic subsidence during the lower and middle Chesterian (330 to 322 Ma). Tectonic subsidence during this interval overlaps eustatic sea-level changes that may otherwise have controlled stratal distributions. Composite sequence II (S4 and S5) is aggradational and records tectonic stabilization of the ramp during the local maximum Chesterian flooding event, both of which occurred in the middle and late Chesterian (322 to 320 Ma). Composite sequence III (S6 and S7) is progradational and records the combination of reduced accommodation on the ramp and increased siliciclastic input. Composite sequence III also records the initiation of high-amplitude glacio-eustatic sea-level changes in this area, produced by Gondwanan glaciation during the transition into a late Paleozoic icehouse climate.

Siliciclastic versus Carbonate Sedimentation

Carbonate and siliciclastic sand deposition across the study area occurs contemporaneously but in spatially distinct areas. Little or no mixing occurred between siliciclastic sand and peritidal carbonate on the eastern ramp (Fig. 8). Any interplay between siliciclastic and carbonate would likely inhibit productivity and result in thinner carbonate accumulations (Budd and Harris, 1990; Mount, 1984). Third-order-sequence thickness differences are insignificant between peritidal carbonate adjacent to siliciclastic sand on the eastern ramp and open marine carbonate far removed from siliciclastic sources on the western ramp. Uniform thicknesses and the near-absence of siliciclastic sand in carbonate-bearing strata argue for reduced siliciclastic sediment influx to the ramp during the lower and middle Chesterian. High rates of sediment input typically result in progradational sandstones (Posamentier and Allen, 1999). During the lower to middle Chesterian, siliciclastics on the eastern ramp backstep cratonward,

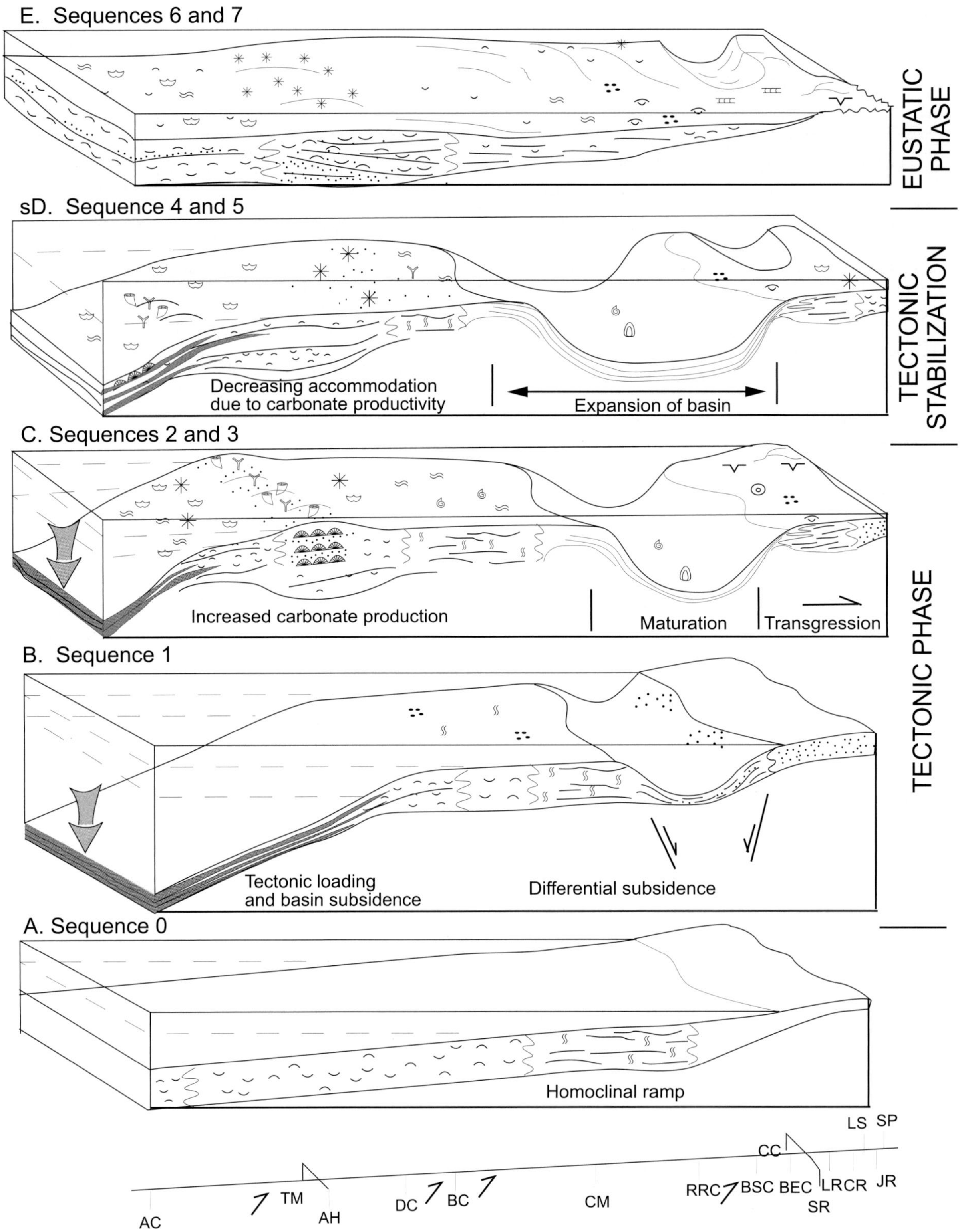

FIG. 11.—Stacked paleogeographic reconstructions for the Chesterian strata of east-central Idaho and southwestern Montana representing significant time intervals. Symbols as in Figure 8. **A)** Sequence 0, homoclinal ramp conditions. **B)** Sequence 1, differential subsidence and the formation of an incipient intraramp basin during tectonic loading. **C)** Sequences 2 and 3, increased carbonate productivity on the western ramp, formation of a mature intraramp basin on the central ramp, and transgression on the eastern ramp. **D)** Sequences 4 and 5, decreasing accommodation on the western ramp, expansion of the intraramp basin and deepening on the eastern ramp that may reflect peripheral deepening of the Big Snowy Trough to the north. **E)** Sequences 6 and 7, a return to homoclinal-ramp conditions with continued deepening from the Big Snowy Trough to the north and east.

indicating that relative sea level controlled deposition through its effect on base level. The limited lateral and vertical resolution in this study is insufficient to differentiate systems tracts within third-order sequences, particularly on the eastern ramp. It is not possible, therefore, to evaluate whether siliciclastics formed across the eastern ramp in response to lowered base level during higher-frequency (third-order or fifth-order) lowstand systems tracts, or highstand systems tracts.

Within the intraramp basin, shale and carbonate are interbedded. Shale deposits are thick and correspond to episodes of tectonic loading which may have simultaneously increased accommodation in the intraramp basin and mobilized (by isostatic uplift) siliciclastics from the craton (cf. Smith and Read, 2001). Carbonate forms thin beds at the tops of third-order cycles and is associated with shallowing to wave base and reduced influx of siliciclastics to the basin.

On the western ramp, both siliciclastic sand and carbonate were deposited during the late Chesterian. This sedimentation pattern was possible because of the textural and compositional maturity of the quartz-rich fine-grained sands. Bed-load transport rather than suspended-load transport dominated, enabling continued growth of filter-feeding organisms (Schlager, 2005). Periods of moderate input of clastics may have facilitated carbonate development by providing sandy substrate and bathymetric highs on sand bodies for marine colonization (Mount, 1984). Seasonal climate may have played a leading role in the progradational nature of siliciclastics during the late Chesterian. A change from arid to perhumid conditions documented for the Appalachian Basin (Al-Tawil et al., 2003) and the Illinois Basin (Smith and Read, 1999) during the late Visean was attributed to rotation of North America from sub-equatorial to equatorial latitudes (Scotese, 1997). The change from arid to per humid conditions in the Late Mississippian has also been attributed to increased seasonal rainfall associated with the onset of late Paleozoic glaciation (Cecil et al., 2004). Increased humidity increases erosion and delivery of sediment to basins (Posamentier and Allen, 1999). The extent to which climate rather than relative sea level may have controlled the distribution of sandstones on the western ramp warrants further study.

The provenance of late Chesterian fine sandstones on the western ramp remains unconstrained, because coeval coarse sandstone beds are absent from contemporaneous sections to the east. Either sediment bypass occurred in submarine channels not observed in outcrop, or clastic input to the western ramp occurred from another direction (e.g., Mount, 1984). Possible sources of siliciclastics to the western ramp include (1) reworked sands from the Alberta Shelf to the north, transported along south-flowing coastal currents, (2) the Antler Highlands to the west, (3) the Lemhi–Beaverhead Arch in southern Idaho, which was episodically emergent throughout the Paleozoic (Peterson, 1981), and (4) the Transcontinental Arch to the southeast (Fig. 2). However, the Antler highlands to the west are an unlikely source of siliciclastics to the ramp during the middle Chesterian, given the occurrence of an intervening deep basin to the west (see Link et al., 1996). Thus, the sandstones were likely derived from a northern or southern source region outside the study area.

Composite Sequence I: Tectonic Subsidence

Composite sequence I formed during an episode of tectonic loading in east-central Idaho and southwestern Montana during the early Chesterian (Figs. 10, 11). Several stratigraphic relationships support a tectonic control on the development of S1–S3. Increases in accommodation during composite sequence I were not uniform across the entire ramp, as would occur during a eustatically driven sea-level rise. The juxtaposition of deep-water calcisiltstone above peloidal packstones requires greater than 50 meters of depth change associated with the flooding event marking the base of S1 (Fig. 8A). The lithofacies of S0 indicates that homoclinal-ramp conditions persisted prior to flooding, and therefore a eustatically driven flooding event of this magnitude should have inundated the entire ramp. However, restriction of this flooding to the western and central ramp suggests flexural downwarping of the lithosphere in response to tectonic loading. Differential subsidence and flexural accommodation was also documented for an earlier phase of Antler tectonism, resulting in similar lateral variations in facies for strata in Nevada and Utah (Giles and Dickinson, 1995). Additional support for a tectonic control on composite sequence 1 is early Chesterian segmentation of the ramp (Figs. 11A, B), including the formation of an incipient intraramp basin along the present-day border between Idaho and Montana during S1 (Fig. 11B). Separated from the deep ramp to the west by shallow marine bioturbated and skeletal wackestone and packstone, this incipient basin provided a proximal locus for deposition of over 50 m of shale and siltstone (Fig. 8A). Limited stratigraphic exposure and the presence of the Medicine Lodge thrust complicate estimates of the lateral extent and paleogeographic trend of this basin. The abrupt lateral facies heterogeneity observed in this study suggests that this intraramp basin was narrow in extent compared to contemporaneous adjacent shallow-water environments and that its depocenter may have migrated during the early and middle Chesterian. This bathymetric feature was described as a northern precursor of the Pennsylvanian Oquirrh Basin (P. Rose, personal communication, 2003). The shallow-water carbonate that formed west of the intraramp basin represent a bathymetric high that resisted subsidence throughout composite sequence 1 (Figs. 8A, 10). Limited outcrop coverage and subsequent tectonic deformation in the study area prohibits a quantified understanding of the tectonic mechanisms controlling the formation of the intraramp basin and adjacent bathymetric high. Both lithospheric flexure and extensional block faulting are possible explanations for the abrupt lateral facies distribution on the central ramp, and both these processes can form by regional tectonic loading of the lithosphere (Quinlan and Beaumont, 1984; Flemings and Jordan, 1990; Dorobek et al., 1991a; Dorobek et al., 1991b). Pinch-and-swell structures (Fig. 6B) in the laminated calcisiltstone lithofacies provide sedimentological support for syndepositional extension during the formation of the intraramp basin.

A similar depositional model of a foreland-basin ramp with an offshore peloidal-skeletal bank was postulated for Lower Mississippian strata in southern Idaho and northern Utah (Chen and Webster, 1994) and would be, in part, correlative with the shallow-water facies of the Scott Peak Formation in east-central Idaho. If so, the shallow-water bank was a long-lived feature likely produced by tectonic uplift within the otherwise rapidly subsiding foreland basin.

Eastward backstepping of lithofacies in S1–S3, above the regional Madison unconformity, provides additional support for increased accommodation during composite sequence I (Figs. 8B, 10). Sea-level curves from western Europe, South China, Donets, and the US Mid-Continent all document a transgression beginning at the Visean–Serpukhovian boundary that culminated in the mid-Serpukhovian (Izart et al., 2003). The geographically isolated, episodic increases in accommodation space across the study area began in the late Visean and continued at least through Sequence 3, suggesting that factors besides eustasy played a role in generating accommodation. Facies backstepping on the eastern ramp corresponds to aggradational parasequence stacking in the central ramp. Laminated calcisiltstone–based parasequences are thick, asymmetrical, shallowing-upward parasequences that correlate to thinner shallow-ramp parasequences on the eastern ramp, suggesting that a long-term relative sea-level rise maintained accommodation within the intraramp basin and increased accommodation on the eastern ramp. The transition from an incipient to a mature intraramp basin occurs in sequence 2 with deposition of laminated calcisiltstone above siliciclastic shales (Fig. 8A). The stratigraphy indicates a deeper, wider basin and a reduction in siliciclastic input. Coeval development of a peritidal marine setting above the fluvial deposits on the eastern ramp likely facilitated the transition from shale to calcisiltstone sedimentation within the intraramp basin by trapping fluvial siliciclastics updip, far from the intraramp basin. The persistent location of deep-water deposition within the intraramp basin during S1–S3 suggests that subsidence, possibly from crustal extension related to increased loading to the west, was focused in this area during the formation of these sequences. Maximum relative sea-level rise of composite sequence I occurred in sequence 3, with deposition of extensive intraramp calcisiltstones and the maximum backstepping of facies on the eastern ramp.

We qualitatively assess the extent to which subsidence on the central ramp is related to tectonic loading west of the foreland basin by comparing parasequence stacking patterns on the western ramp to those on the eastern and central ramp. On the western ramp, the transition from thick asymmetrical calcisiltstone-based parasequences to thick-bedded skeletal carbonate parasequences indicates an overall decrease in accommodation space. However, calcisiltstone-based parasequences also persist in the basal part of the westernmost ramp during S1–S3. We propose that sustained tectonic subsidence of the foreland basin resulted in retrogradational sequences in Montana, but that establishment of a productive carbonate factory on the shallow ramp west of the intraramp basin coupled with a decrease in the subsidence rate in this area enabled the rate of carbonate accumulation in east-central Idaho to keep pace with the rate of tectonic subsidence during S2 and S3 (Figs. 8A, 11C). Environmental conditions that favored carbonate accumulation in this setting were circulation with the open ocean and trapping of craton-derived siliciclastics in the intraramp basin (cf. Pigram et al., 1989). Therefore, initial drowning of the western ramp in S1–S3 occurred by flooding associated with tectonic subsidence but was followed almost immediately by prolific carbonate growth in response to decreased subsidence in this area (cf. Kendall and Schlager, 1981). Coeval extension produced the intraramp basin in the central ramp, whereas regional tectonic subsidence coupled with eustatic sea-level rise produced backstepping parasequences on the eastern ramp.

Composite Sequence II: Tectonic Stabilization and Basin Fill

Several significant changes in the geometry of parasequences and sequences differentiate composite sequence I from composite sequence II, indicating a change in the processes controlling sedimentation across the ramp. We suggest that composite sequence II formed during regional tectonic abeyance and an interval of relative sea-level rise.

On the western ramp, reduced tectonism is inferred from a change in the character of parasequences between composite sequences I to II. Composite sequence I parasequences at Arco Hills, Idaho, (AH, Fig. 8) are thick, asymmetrical accumulations that represent abrupt flooding events involving at least 50 m of sea-level change. By contrast, parasequences in composite sequence II are thinner and more heterolithic accumulations, with nearly equivalent portions of calcisiltstone, wackestone, and packstone in each parasequence (compare Fig. 5B with Fig. 5D). The uniform composition and facies thickness distribution in these parasequences likely reflect uniform relative sea-level changes of low to moderate amplitude.

On the central ramp, a shift in the position of the intraramp basin may provide additional evidence for a reduction in tectonic subsidence during composite sequence II. This shift occurred at the base of sequence 4 and resulted in deposition of calcisiltstone-based parasequences above the thick, uniform succession of skeletal carbonate parasequences at Copper Mountain, Idaho (Fig. 8A). It is unclear whether the depositional center of the intraramp basin shifted laterally or expanded outward (Figs. 11C, 11D). The shallowest-water facies on the central ramp appears to have shifted westward from Copper Mountain to Box Canyon from composite sequence I to composite sequence II. These bathymetric adjustments may reflect crustal relaxation associated with the cessation of tectonic loading (Beaumont and Quinlan, 1984; Flemings and Jordan, 1990). A similar sequence was reported for the Antler foreland basin in Nevada and Utah during the Early Mississippian (Giles and Dickinson, 1995). The associated cratonward migration of the forebulge during the Early Mississippian was linked to episodes of active subsidence. Quantifying the relationship between the polarity of facies migration and subsidence events is not possible for this study, because no chronology of regional subsidence events has been developed for the Antler foreland basin in Idaho and because this dip-oriented cross section does not address strike-oriented changes in intrashelf or forebulge dimensions.

During S5, all remaining accommodation in the intraramp basin was filled and shallow-subtidal carbonate facies prograded over the top of the preexisting deep-water basin (Figs. 8A, 11E). This bathymetric inversion reestablished homoclinal-ramp conditions across most of southwestern Montana and east-central Idaho by the end of composite sequence II.

On the eastern ramp, a flooding event occurred that may reflect either a change in the style of regional tectonism or a eustatically driven sea-level rise during composite sequence II. In composite sequence II, extensive peritidal carbonates on the western side of the eastern ramp pass eastward into shale-based and skeletal carbonate parasequences deposited above fluvio-estuarine siliciclastics of composite sequence I. These facies relationships indicate a regional deepening event that appears to have originated from a cratonward direction. It is possible that this flooding event on the eastern ramp reflects a change in the style of tectonism from high to low differential subsidence. If antecedent faults controlled the differential subsidence observed in composite sequence I, then a change in the orientation or magnitude of the stress field may have reduced the response of antecedent faults to lithospheric flexure, resulting in uniform subsidence rather than differential subsidence (cf. Dorobek et al., 1991a).

Alternatively, a eustatic rise in sea level during composite sequence II could explain the flooding event on the eastern ramp. The east–west trending Big Snowy Trough included the northeast end of the study area and extended into central Montana during the Chesterian. It was likely in communication with the open ocean during this time (Peterson, 1981; Peterson and Smith, 1986; Poole and Sandberg, 1991), and a rise in relative sea level during the middle Chesterian may have resulted in peripheral deepening of the Big Snowy Trough to its north and south, resulting in flooding on the eastern ramp during the early part of composite sequence II. Deposition of shallow-water siliciclastics on the eastern ramp in the upper part of composite sequence II coincides with basinward progradation of shallow subtidal carbonate on the western ramp. Together these trends indicate that maximum relative sea-level rise occurred during sequence 4, followed by a regional decrease in accommodation on the central and western ramp (Figs. 8, 11). The long-term transgression beginning at the Visean–Serpukhovian boundary and culminating in the middle Serpukhovian may explain these regional observations (Ross and Ross, 1987; Izart et al., 2003). The lower Serpukhovian part of the transgression may have been enhanced by tectonic subsidence so that only the latter part of the transgression significantly impacted deposition in the Antler basin.

Composite Sequence III: Glacio-Eustasy

Regional correlation of sequences during composite sequence III is complicated in the study area by poor exposure, structural complications, and sub-Pennsylvanian erosion of Chesterian strata. Biostratigraphically well-constrained, nearly continuous deposition throughout the latest Chesterian was recorded on the western ramp and in localized areas of the eastern ramp. Interbedded carbonates and siliciclastics on the western ramp record an increase in the amplitude and frequency of sea-level change during the late Chesterian (middle to late Serpukhovian; Fig. 10) and a transition into an icehouse climate (Fig. 11; see also Butts, 2005). S6 and S7 contain parasequences whose biofacies indicate large-amplitude (> 50 m) sea-level fluctuations representative of an icehouse climate (Butts, 2005). Parasequences are thin (commonly < 5 m) and heterolithic, with calcisiltstone bases and skeletal grainstone or fine sandstone caps (Fig. 5E). No obvious regionally correlative unconformities can be traced, but parasequences shallow to depths above fair-weather wave base, some even showing evidence of tidal activity. The lack of evidence for unconformities may reflect the relative downdip position of the western ramp, or it may be that brief subaerial exposure events did not result in erosional unconformities. The frequency of parasequences is challenging to estimate on the western ramp. The number of upward-shallowing parasequences in each sequence varies by location on the ramp. For instance at Arco Hills, Idaho, the number of parasequences increases three-fold from the middle to the upper Chesterian (Compare Fig. 9A with Fig. 9B).

Marine proxy records based on sequence stratigraphic analyses indicate a late Visean age for the onset of middle Carboniferous that glaciation (Wright and Vanstone, 2001; Smith and Read, 1999, 2000). Direct glacial indicators suggest that glaciation began from the early Namurian (Veevers and Powell, 1987; Isbell et al., 2003a, 2003b) to the middle Serpukhovian (González, 1990). Global sea-level curves indicate that maximum drawdown during the Late Mississippian occurred in the late Chesterian and resulted from a protracted decrease in sea level beginning in the middle Serpukhovian (Ross and Ross, 1985, 1987; Izart et al., 2003). The middle to late Chesterian estimate of this study is later than most of the estimates based on direct glacial indicators, and it significantly postdates the estimate from the marine record in the Illinois Basin, USA (Smith and Read, 1999, 2000). As previously discussed, foreland-basin tectonics may have overprinted any eustatic signature in the Antler basin prior to the middle Serpukhovian. The late Chesterian estimate from this study thus provides a minimum age estimate for the onset of middle Carboniferous glacio-eustatic sea-level change.

Composite sequence III records a return to homoclinal-ramp conditions in the late Chesterian (Fig. 8A), and this homoclinal ramp prograded across the previously existing intraramp basin. Mixed siliciclastic–carbonate parasequences on the western ramp continue to record homoclinal-ramp conditions and the westward flux of craton-derived siliciclastics. The stratigraphically complete section at Lombard Station, Montana, likely indicates that drowning from north of the study area associated with subsidence of the Big Snowy Trough persisted through composite sequence III (Fig. 8).

CONCLUSIONS

Chesterian (330–318 Ma) carbonate and siliciclastic rocks in east-central Idaho and southwestern Montana form an eastward-thinning wedge of Upper Mississippian strata that were deposited on a tectonically segmented ramp in the Antler foreland basin. Syndepositional tectonism subdivided the ramp into three distinct areas (western, central, and eastern ramp), each of which records unique histories of subsidence and deposition. Detailed sequence stratigraphic analysis of these Chesterian rocks, constrained by new conodont biostratigraphy, indicates that the entire succession is a single second-order megasequence composed of three composite sequences (I, II, and III). Each composite sequence is composed of two or more third-order depositional sequences. Parasequences (meter-scale cycles) are ubiquitous in these rocks, but they are only locally correlative. The composite sequences record long-term accommodation changes across the ramp that were controlled by subsidence driven by foreland-basin tectonism. Nearly continuous Chesterian deposition is recorded in east-central Idaho, but the succession contains a number of hiatuses in southwestern Montana.

Composite sequence I (S1–S3) records transgressive deposition across the Chesterian ramp above the Madison unconformity and documents the sedimentary response to tectonically induced changes in accommodation. On the western ramp the base of composite sequence I is a thick succession of massive mudstone above older skeletal–peloidal carbonate, reflecting a significant increase in accommodation on the western ramp produced by tectonic loading. Immediately following the initial tectonic drowning, composite sequence I records the initiation of a muddy skeletal-rich carbonate bank on the western ramp that formed because of decreasing subsidence coupled with abundant carbonate productivity. An intraramp basin on the central ramp was produced by extensional subsidence, and it records predominantly deeper-water carbonate and siliciclastic deposition punctuated by pulses of shallow-water skeletal carbonate deposition. The eastern ramp preserves a record of transgressive peritidal carbonate to shallow marine and fluvial sedimentation during composite sequence I.

Composite sequence II (S4 and S5) records continued deposition of thick-bedded, carbonate-dominated parasequences on the western ramp, but there is an increase in fine sand to these parasequences during this period. On the central ramp the intraramp basin widened during S4 and was eventually filled with shallow-water sediments during deposition of S5. Throughout S4 and S5 deposition, open marine waters appear to have flooded the eastern ramp from the north, likely from the Big Snowy Trough, as indicated by the decrease in peritidal facies and the increase in open marine facies. Thus, composite sequence II records the regionally highest relative sea level in S4 followed by filling of much of the accommodation space that reflects tectonic quiescence and establishment of a homoclinal ramp.

Composite sequence III (S6 and S7) is poorly exposed on the central ramp and is marked by increasing amounts of siliciclastics interbedded with shallow-water and deep-water carbonate on the western ramp and continued deposition of subtidal carbonate from the Big Snowy Trough on the eastern ramp. The transition from thick-bedded, carbonate-dominated parasequences to thin-bedded mixed carbonate and siliciclastic parasequences in composite sequence III on the western ramp reflects the transition to high-amplitude, high-frequency eustatic fluctuations likely produced by the onset of Gondwanan glaciation during the late Chesterian.

ACKNOWLEDGMENTS

National Science Foundation EAR-0229369 provided the principal financial support for this research. Additional funding is from a 2005 Geological Society of America Graduate Student Award, from the University of Idaho Carl Savage and A.H. Featherstone Scholarships, and from the University of Idaho graduate teaching assistantship program. We thank everyone who assisted in the field and in the laboratory including: Bonny Archuleta, Renee Breedlove, Justin Murphy, Jennifer Clark, and Alex Boughamer. We are grateful to private landowners in Montana who granted us access to outcrops. This manuscript benefited from reviews of an earlier version by Art Saller, Jeff Lukasik, and Toni Simo.

REFERENCES

ABPLANALP, J.M., 2006, Late Mississippian (Chesterian) conodont biostratigraphy of east-central Idaho and southwest Montana: unpublished M.S. thesis, University of Idaho, Moscow, Idaho, 119 p.

AL-TAWIL, A., WYNN, T.C., AND READ, F., 2003, Sequence response of a distal-to-proximal foreland ramp to glacio-eustasy and tectonics: Mississippian, Applalachian basin, West Virginia–Virginia, U.S.A, *in* Ahr, W.M., Harris, P.M., Morgan, W.A., and Somerville, I.D., eds., Permo-Carboniferous Carbonate Platforms and Reefs: SEPM, Special Publication 78, p. 11–34.

BASEMANN, J.F., AND LANE, H.R., 1985, Taxonomy and evolution of the Genus *Rhachistognathus* Dunn (Conodonta; late Mississippian to early middle Pennsylvanian): Courier Forschungsinstitut Senckenberg, v. 74, p. 93–136.

BATT, L.S., 2006, Late Mississippian sequence and carbon isotope stratigraphy, east-central Idaho and southwest Montana: proxies for Gondwana glacio-eustasy: unpublished Ph.D. Dissertation, University of Idaho, Moscow, 451 p.

BATT, L.S., MONTAÑEZ, I.P., ISAACSON, P., POPE, M.C., BUTTS, S.H., AND ABPLANALP, J., 2007, Multi-carbonate component reconstruction of mid-Carboniferous (Chesterian) seawater ^{13}C: Palaeogeography Palaeoclimatology Palaeoecology, v. 256, p. 298–318..

BREWSTER, D.P., 1984, Stratigraphy, biostratigraphy, and depositional history of the Big Snowy and Amsden formations (Carboniferous) of southwestern Montana: unpublished M.S. Thesis, Indiana University, Bloomington, Indiana, 164 p.

BUDD, D.A., AND HARRIS, P.M., 1990, Carbonate–Siliciclastic Mixtures: SEPM, Reprint Series, no. 14, 272 p.

BUONICONTI, M.R., EBERLI, G.P., AND SMITH, L.B., JR., 2004, Quantifying the relationship between foreland basin processes and their prograding carbonate fill: tectonostratigraphic co-evolution of the Antler foreland basin and Lower–Middle Mississippian carbonates, SW Montana and east-central Idaho (abstract): Geological Society of America, Abstracts with Programs, v. 36, p. 545.

BURCHFIEL, B.C., AND ROYDEN, L.H., 1991, Antler Orogeny: A Mediterranean-type orogeny: Geology, v. 19, p. 66–69.

BUTTS, S.H., 2005, Latest Chesterian (Carboniferous) initiation of Gondwanan glaciation recorded in facies stacking patterns and brachiopod paleocommunities of the Antler foreland basin, Idaho: Palaeogeography, Palaeoclimatology, Palaeoecology, v. 223, p. 275–289.

BYRNE, D.J., 1985, Stratigraphy and depositional history of the Mississippian Big Snowy Formation in the Snowcrest Range, Beaverhead and Madison Counties, Montana: unpublished M.S. Thesis, Oregon State University, Corvallis, 417 p.

CECIL, C.B., BREZINSKI, D.K., AND DULONG, F.T., 2004, The Paleozoic record of changes in global climate and sea level: Central Appalachian Basin: U.S. Geological Survey, Report C 1264, p. 77–135.

CHEN, X., AND WEBSTER, G.D., 1994, Sedimentology, tectonic control and evolution of a Lower Mississippian carbonate ramp with offshore bank, central Wyoming to eastern Idaho and northeastern Utah, U.S.A., *in* Embry, A.F., Beauchamp, B., and Glass, D.J., eds., Pangea: Global Environments and Resources, Canadian Society of Petroleum Geologists, Memoir 17, p. 557–587.

CRAWFORD, S.D., 1976, Stratigraphy and faunal aspects of Upper Mississippian formations, eastern Idaho: unpublished M.S. Thesis, University of Iowa, Iowa City, Iowa, 41 p.

DECELLES, P.G., AND GILES, K.A., 1996, Foreland basin systems: Basin Research, v. 8, p. 105–123.

DICKINSON, W.R., 2004, Evolution of the North American Cordillera: Annual Review of Earth and Planetary Sciences, v. 32, p. 13–45.

DOROBEK, S.L., REID, S.K., AND ELRICK, M., 1991a, Antler foreland stratigraphy of Montana and Idaho: the stratigraphic record of eustatic fluctuations and episodic tectonic events, *in* Cooper, J.D., and Stevens, C.H., eds., Paleozoic Paleogeography of the Western United States II: Los Angeles: SEPM, Pacific Section, Fieldtrip 67, p. 487–507.

DOROBEK, S.L., REID, S.K., ELRICK, M., BOND, G.C., AND KOMINZ, M.A., 1991b, Subsidence across the Antler foreland of Montana and Idaho: tectonic versus eustatic effects, *in* Franseen, E.K., Watney, W.L., Kendall, C.G.St.C., and Ross, W., eds., Sedimentary Modeling: Computer Simulations and Methods for Improved Parameter Definition: Kansas Geological Survey, Bulletin 233, p. 231–250.

DOVER, J.H., 1980, Status of the Antler orogeny in central Idaho—clarifications and constraints from the Pioneer Mountains, *in* Fouch, T.D., and Magathan, E.R., eds., Paleozoic Paleogeography of West-Central United States: Denver, SEPM, West Central United States Paleogeography Symposium I, p. 371–385.

DUNN, D.L., 1966, New Pennsylvanian platform conodonts from southwestern United States: Journal of Paleontology, v. 40, p. 1294–1303.

EASTON, W.H., 1962, Carboniferous formations and faunas of central Montana: U.S. Geological Survey, Professional Paper 348, 126 p.

EVANS, R.D., 1989, Lithostratigraphy and depositional environments of the Upper Mississippian Scott Peak Formation, south-central Idaho: unpublished M.S. Thesis, University of Idaho, Moscow, 148 p.

FLEMINGS, P.B., AND JORDAN, T.E., 1990, Stratigraphic modeling of foreland basins: Interpreting thrust deformation and lithosphere rheology: Geology, v. 18, p. 430–434.

GALVIN, T.J., 1981, Stratigraphy, faunal assemblages and depositional environments of the Surrett Canyon Formation (Chesterian), Lost River Range, south-central Idaho: unpublished M.S. thesis, University of Idaho, Moscow, 116 p.

GILES, K.A., AND DICKINSON, W.R., 1995, The interplay of eustasy and lithospheric flexure in forming stratigraphic sequences in foreland settings: an example from the Antler foreland, Nevada and Utah, *in* Dorobek, S.L., and Ross, G.M., eds., Stratigraphic Evolution of Foreland Basins: SEPM, Special Publication 52, p. 187–211.

GONZÁLEZ, C.R., 1990, Development of the late Paleozoic glaciations of the South American Gondwana in western Argentina: Palaeogeography, Palaeoclimatology, Palaeoecology, v. 79, p. 275–287.

GRADSTEIN, F.M., OGG, J.G., SMITH, A.G., AGTERBERG, F.P., BLEEKER, W., COOPER, R.A., DAVYDOV, V., GIBBARD, P., HINNOV, L.A., HOUSE, M.R., LOURENS, L., LUTERBACHER, H.P., MCARTHUR, J., MELCHIN, M.J., ROBB, L.J., SHERGOLD, J., VILLENEUVE, M., WARDLAW, B.R., ALI, J., BRINKHUIS, H., HILGEN, F.J., HOOKER, J., HOWARTH, R.J., KNOLL, A.H., LASKAR, J., MONECHI, S., PLUMB, K.A., POWELL, J., RAFFI, I., RÖHL, U., SADLER, P., SANFILIPPO, A., SCHMITZ, B., SHACKLETON, N.J., SHIELDS, G.A., STRAUSS, H., VAN DAM, J., VAN KOLFSCHOTEN, T., VEIZER, J., AND WILSON, D., 2004, A Geologic Time Scale 2004: Cambridge, U.K., Cambridge University Press, 589 p.

GRIFFING, D.H., 1987, Petrography, stratigraphy, and depositional environments of the Arco Hills Formation (Chesterian) of east-central Idaho: unpublished M.S. Thesis, University of Idaho, Moscow, 116 p.

GUTHRIE, G.E., 1984, Stratigraphy and depositional environment of the Upper Mississippian Big Snowy Group in the Bridger Range, southwest Montana: unpublished M.S. Thesis, Montana State University, Bozeman, Montana, 105 p.

GUTSCHICK, R.C., SANDBERG, C.A., AND SANDO, W.J., 1980, Mississippian shelf margin and carbonate platform from Montana to Nevada, *in* Fouch, T.D., and Magathan, E.R., eds., Paleozoic Paleogeography of the West-Central United States: Denver, SEPM, Rocky Mountain Section, West-Central United States Paleogeography 1, p. 111–128.

HARRIS, P.M., KENDALL, C.G.ST.C., AND LERCHE, I., 1985, Carbonate cementation—a brief review, *in* Schneidermann, N., and Harris, P.M., Carbonate Cements: SEPM, Special Publication 36, p. 79–95.

HILDRETH, G.D., 1981, Stratigraphy of the Mississippian Big Snowy Formation of the Armstead anticline, Beaverhead County, Montana: Montana Geological Society, Field Conference, southwest Montana, p. 49–57.

ISAACSON, P.E., GRADER, G.W., POPE, M.E., MONTAÑEZ, I.P., BUTTS, S.H., BATT, L.S., AND ABPLANALP, J.A., 2007, Devonian–Mississippian Antler foreland basin carbonates in Idaho: Significant subsidence and eustasy events: The Mountain Geologist, v. 44, no. 3, p. 1–21.

ISBELL, J.L., LENAKER, P.A., ASKIN, R.A., MILLER, M.F., AND BABCOCK, L.E., 2003b, Reevaluation of the timing and extent of late Paleozoic glaciation in Gondwana: role of the Transantarctic mountains: Geology, v. 31, p. 977–980.

ISBELL, J.L., MILLER, M.F., WOLFE, K.L., AND LENAKER, P.A., 2003a, Timing of the late Paleozoic glaciation in Gondwana: was glaciation responsible for the development of Northern Hemisphere cyclothems?, *in* Chan, M.A., and Archer, A.W., eds., Extreme Depositional Environments; Mega End Members in Geologic Time: Geological Society of America, Special Paper 370, p. 5–24.

IZART, A., STEPHENSON, R., VAI, G.B., VACHARD, D., LE NINDRE, Y., VASLET, D., FAUVEL, P., SUESS, P., KOSSOVAYA, O., CHEN, Z., MASLO, A., AND STOVBA, S., 2003, Sequence stratigraphy and correlation of Late Carboniferous and Permian in the CIS, Europe, Tethyan area, North Africa, Arabia, China, Gondwanaland and the USA: Palaeogeography, Palaeoclimatology, Palaeoecology, v. 196, p. 59–84.

KENDALL, C.G.ST.C., AND SCHLAGER, W., 1981, Carbonates and relative changes in sea level: Marine Geology, v. 44, p. 181–212.

KERANS, C., AND TINKER, S.W., 1997, Sequence Stratigraphy and Characterization of Carbonate Reservoirs: SEPM, Short Course 30, 130 p.

KEY, C.F., 1986, Stratigraphy and depositional history of the Amsden and lower Quadrant formations, Snowcrest Range, Beaverhead and Madison Counties, Montana: unpublished M.S. Thesis, Oregon State University, Corvallis, 187 p.

LANE, H.R., AND STRAKA, J.J., II, 1974, Late Mississippian and Early Pennsylvanian Conodonts, Arkansas and Oklahoma: Geological Society of America, Special Paper 152, 144 p.

LASEMI, Z., NORBY, R.D., UTGAARD, J.E., FERRY, W.R., CUFFEY, R.J., AND DEVER, G.R., JR., 2003, Mississippian carbonate buildups and development of cool-water-like carbonate platforms in the Illinois Basin, Midcontinent, U.S.A., *in* Ahr, W.M., Harris, P.M., Morgan, W.A., and Somerville, I.D., eds., Permo-Carboniferous Carbonate Platforms and Reefs: SEPM, Special Publication 78, p. 69–95.

LEVY, M., AND CHRISTIE-BLICK, N., 1989, Pre-Mesozoic palinspastic reconstruction of the eastern Great Basin (western United States): Science, v. 245, p. 1454–1462.

LINK, P.K., AND FANNING, C.M., 2004, Detrital zircon studies of the Pioneer Mountains correctly summarize geologic relations mapped by Betty Skipp and colleagues over the last 40 years, (abstract): Geological Society of America, Abstracts with Program, v. 36, p. 547.

LINK, P.K., AND JANECKE, S.U., 1999, Geology of east-central Idaho: geologic road logs for the Big and Little Lost River, Lemhi, and Salmon River valleys, *in* Hughes, S.S., and Thackray, G.D., eds., Guidebook to the Geology of Eastern Idaho: Pocatello, Museum of Natural History, p. 295–334.

LINK, P.K., WARREN, I., PREACHER, J.M., AND SKIPP, B., 1996, Stratigraphic analysis and interpretation of the Mississippian Copper Basin Group, McGowan Creek Formation, and White Knob Limestone, south-central Idaho, *in* Longman, M.W., and Sonnenfeld, M.D., eds., Paleozoic Systems of the Rocky Mountain Region: Denver, Rocky Mountain Section, SEPM, p. 117–144.

MAMET, B.L., SKIPP, B., SANDO, W.J., AND MAPEL, W.J., 1971, Biostratigraphy of Upper Mississippian and associated Carboniferous rocks in south-central Idaho: American Association of Petroleum Geologists, Bulletin, v. 55, p. 20–33.

MILLER, E.L., MILLER, M.M., STEVENS, C.H., WRIGHT, J.E., AND MADRID, R., 1992, Late Paleozoic paleogeographic and tectonic evolution of the western US Cordillera, *in* Burchfiel, B.C., Lipman, P.W., and Zoback, M.L., eds., The Cordilleran Orogen: Boulder, Colorado, The Geological Society of America, The Geology of North America, v. G-3, Conterminous US, p. 57–97.

MITCHUM, R.M., JR., AND VAN WAGONER, J.C., 1991, High frequency sequences and their stacking patterns: sequence stratigraphic evidence of high-frequency eustatic cycles: Sedimentary Geology, v. 70, p. 131–160.

MOUNT, J.F., 1984, Mixing of siliciclastics and carbonate sediments in shallow shelf environments: Geology, v. 12, p. 432–435.

NILSEN, T.H., AND STEWART, J.H., 1980, Penrose conference report: The Antler Orogeny, mid-Paleozoic tectonism in western North America: Geology, v. 8, p. 298–302.

OLDOW, J.S., BALLY, A.W., AVE LALLEMANT, H.G., AND LEEMAN, W.P., 1989, Phanerozoic evolution of the North American Cordillera: United States and Canada, *in* Bally, A.W., and Palmer, A.R., eds., Geological Society of America, The Geology of North America, vol. A., The Geology of North America–An Overview, p. 139–232.

PECORA, W.C., 1984, Bedrock geology of the Blacktail Mountains, southwest Montana: unpublished M.S. thesis, Wesleyan University, Middletown, Connecticut, 203 p.

PEDERSON, T.F., AND CALVERT, S.E., 1990, Anoxia vs. productivity: what controls the formation of organic-carbon-rich sediments and sedimentary rocks?: American Association of Petroleum Geologists, Bulletin, v. 74, p. 454–466.

PETERSON, J.A., 1981, General stratigraphy and regional paleostructure of the western Montana over thrust belt, *in* Tucker, T.E., Aram, R.B., Brinker, W.F., and Grabb, R.F., Jr., eds., Montana Geological Society 1981 Field Conference, Southwest Montana, p. 5–35.

PETERSON, J.A., AND SMITH, D.L., 1986, Rocky Mountain paleogeography through geologic time, *in* Peterson, J.A., ed., Paleotectonics and Sedimentation in the Rocky Mountain region, United States: American Association of Petroleum Geologists, Memoir 41, p. 3–19.

PIGRAM, C.J., DAVIES, P.J., FEARY, D.A., AND SYMONDS, P.A., 1989, Tectonic controls on carbonate platform evolution in southern Papua New Guinea: passive margin to foreland basin: Geology, v. 17, p. 199–202.

POOLE, F.G., AND SANDBERG, C.A., 1977, Mississippian paleogeography and tectonics of the western United States, *in* Stewart, J.H., ed., Paleozoic Paleogeography of the Western United States: SEPM, Pacific Section, Pacific Coast Paleogeography Symposium I, p. 67–85.

POOLE, F.G., AND SANDBERG, C.A., 1991, Mississippian paleogeography and conodont biostratigraphy of the western United States, *in* Cooper, J.D., and Stevens, C.H., eds., Paleozoic Paleogeography of the Western United States II: Fieldtrip Guidebook for Pacific Section SEPM, Fieldtrip 67, p. 107–136.

POPE, M.C., AND READ, J.F., 1997, High-resolution surface and subsurface sequence stratigraphy of late Middle to Late Ordovician (Late Mohawkian–Cincinnatian) foreland basin rocks, Kentucky and Virginia: American Association of Petroleum Geologists, Bulletin, v. 81, p. 1866–1893.

POSAMENTIER, H.W., AND ALLEN, G.P., 1999, Fundamental concepts of sequence stratigraphy, *in* Siliciclastic Sequence Stratigraphy; Concepts and Applications: SEPM, Concepts in Sedimentology and Paleontology, no. 7, 191 p.

QUINLAN, G.M., AND BEAUMONT, C., 1984, Appalachian thrusting, lithospheric flexure, and the Paleozoic stratigraphy of the Eastern Interior of North America: Canadian Journal of Earth Sciences, v. 21, p. 973–996.

ROSE, P.R., 1976, Mississippian carbonate shelf margins, western United States: U.S. Geological Survey, Journal of Research, v. 4, p. 449–466.

ROSS, C.A., AND ROSS, J.R., 1985, Late Paleozoic depositional sequences are synchronous and worldwide: Geology, v. 13, p. 194–197.

Ross, C.A., and Ross, J.R.P., 1987, Late Paleozoic sea levels and depositional sequences, *in* Ross, C.A., and Haman, D., eds., Timing and Depositional History of Eustatic Sequence: Constraints on Seismic Stratigraphy: Cushman Foundation for Foraminiferal Research, Special Publication 24, p. 137–149.

SANDO, W.J., MAMET, B.L., AND DUTRO, J.T., JR., 1969, Carboniferous megafaunal and microfaunal zonation in the northern Cordillera of the United States: U.S. Geological Survey, Professional Paper 613-E, p. E1–E29.

SANDO, W.J., SANDBERG, C.A., AND GUTSCHICK, R.C., 1981, Stratigraphic and economic significance of Mississippian sequence at Georgetown Canyon, Idaho: American Association of Petroleum Geologists, Bulletin, v. 65, p. 1433–1443.

SANDO, W.J., SANDBERG, C.A., AND PERRY, W.J., JR., 1985, Revision of Mississippian stratigraphy, northern Tendoy Mountains, Southwest Montana, *in* Sando, W.J., ed., Mississippian and Pennsylvanian Stratigraphy in Southwest Montana and Adjacent Idaho: U.S. Geological Survey, Bulletin 1656, p. A1–A10.

SARG, J.F., MARKELLO, J.R., AND WEBER, L.J., 1999, The second-order cycle, carbonate-platform growth, and reservoir, source, and trap prediction, *in* Harris, P.M., Saller, A.H., and Simo, J.A., eds., Advances in Carbonate Sequence Stratigraphy: Application to Reservoirs, Outcrops and Models: SEPM, Special Publication 63, p. 11–34.

SCHLAGER, W., 2005, Carbonate Sedimentology and Sequence Stratigraphy: SEPM, Concepts in Sedimentology and Paleontology, no. 8, 200 p.

SCOTESE, C.R., 1997, Paleogeographic Atlas: Arlington, Texas, University of Texas at Arlington, Department of Geology, PALEOMAP Progress Report 90-0497, 37 p.

SHEPARD, W., 1993, Upper Mississippian biostratigraphy and lithostratigraphy of central Montana, *in* Hunter, L.D., ed., 1993 Field Conference Guidebook, Old Timers Rendezvous Edition: Energy and Mineral Resources of Central Montana: Montana Geological Society, p. 26–36.

SKIPP, B., AND HALL, W.E., 1980, Upper Paleozoic paleotectonics and paleogeography of Idaho, *in* Fouch, T.D., and Magathan, E.R., eds., Paleozoic Paleogeography of West-Central United States: Denver, SEPM, Rocky Mountain Section, Rocky Mountain Paleogeography Symposium I, p. 387–422.

SKIPP, B., HOGGAN, R.D., SCHLEICHER, D.L., AND DOUGLAS, R.C., 1979a, Upper Paleozoic carbonate bank in east-central Idaho—Snaky Canyon, Bluebird Mountain, and Arco Hills Formations, and the paleotectonic significance: U.S. Geological Survey, Bulletin 1486, 78 p.

SKIPP, B., SANDO, W.J., AND HALL, W.E., 1979b, The Mississippian and Pennsylvanian systems in the United States—Idaho: U.S. Geological Survey, Professional Paper 1110, p. AA1–AA42.

SMITH, D.L., AND GILMOUR, E.H., 1979, The Mississippian and Pennsylvanian (Carboniferous) systems in the United States—Montana: U.S. Geological Survey, Professional Paper 1110, 32 p.

SMITH, L.B., JR., AND READ, J.F., 1999, Application of high-resolution sequence stratigraphy to tidally influenced Upper Mississippian carbonates, Illinois Basin, *in* Harris, P.M., Saller, A.H., and Simo, J.A., eds., Advances in Carbonate Sequence Stratigraphy: Application to Reservoirs, Outcrops and Models: SEPM, Special Publication 63, p. 107–126.

SMITH, L., AND READ, J.F., 2000, Rapid onset of late Paleozoic glaciation on Gondwana: Evidence from Upper Mississippian strata of the Midcontinent, United States: Geology, v. 28, p. 279–282.

SMITH, L.B., JR., AND READ, J.F., 2001, Discrimination of local and global effects on Upper Mississippian stratigraphy, Illinois Basin, USA: Journal of Sedimentary Research, v. 71, p. 985–1002.

SMITH, M.T., DICKINSON, W.R., AND GEHRELS, G.E., 1993, Contractional nature of Devonian–Mississippian Antler tectonism along the North American continental margin: Geology, v. 21, p. 21–24.

SPEED, R.C., AND SLEEP, N.H., 1982, Antler orogeny and foreland basin: a model: Geological Society of America, Bulletin, v. 93, p. 815–828.

STAMM, R.G., 1985, Conodont biostratigraphy of the Scott Peak Formation (Upper Mississippian) of south-central Idaho: unpublished M.S. Thesis, University of Idaho, Moscow, Idaho, 105 p.

TREXLER, J.H., JR., SNYDER, W., SCHWARZ, D., KURKA, M.T., AND CROSBIE, R., 1995, An overview of the Mississippian Chainman shale, *in* Hansed, M.W., Walker, J.P., and Trexler, J.H., Jr., eds., Mississippian Source Rocks in the Antler Basin of Nevada and Associated Structural and Stratigraphic Traps: Reno, Nevada Petroleum Society, 1995 Fieldtrip Guidebook, 165 p.

TUCKER, M.E., AND WRIGHT, V.P., 1990, Carbonate Sedimentology: Oxford, U.K., Blackwell Science Ltd., 482 p.

VEEVERS, J.J., AND POWELL, C.M., 1987, Late Paleozoic glacial episodes in Gondwanaland reflected in transgressive–regressive depositional sequences in Euramerica: Geological Society of America, Bulletin, v. 98, p. 179–196.

WARDLAW, B.R., AND PECORA, W.C., 1985, New Mississippian–Pennsylvanian stratigraphic units in Southwest Montana and adjacent Idaho, *in* Sando, W.J., ed., Mississippian and Pennsylvanian Stratigraphy in Southwest Montana and Adjacent Idaho: U.S. Geological Survey, Bulletin 1656, p. B1–B9.

WEBER, L.J., SARG, J.F., AND WRIGHT, F.M., 1995, Sequence stratigraphy and reservoir delineation of the Middle Pennsylvanian (Desmoinesian), Paradox basin and Aneth field, southwestern USA, *in* Read, J.F., Kerans, C., and Weber, L.J., eds., Milankovitch Sea-Level Changes, Cycles, and Reservoirs on Carbonate Platforms in Greenhouse and Icehouse Worlds: SEPM, Short Course Notes 35, 81 p.

WINSTON, D., 1986, Stratigraphic correlation and nomenclature of the middle Proterozoic Belt Supergroup, Montana, Idaho and Washington, *in* Roberts, S.M., ed., Belt Supergroup; A Guide to Proterozoic Rocks of Western Montana and Adjacent Areas: Montana Bureau of Mines and Geology, Special Publication 94, p. 69–84.

WRIGHT, V.P., 1992, Speculations on the controls on cyclic peritidal carbonates: ice-house versus greenhouse eustatic controls: Sedimentary Geology, v. 76, p. 1–5.

WRIGHT, V.P., AND VANSTONE, S.D., 2001, Onset of late Paleozoic glacio-eustasy and the evolving climates of low latitude areas: a synthesis of current understanding: Geological Society of London, Journal, v. 158, p. 579–582.

TECTONIC CONTROLS ON OLIGOCENE CARBONATE PLATFORM–BASIN DEPOSITION, COSTA BLANCA, SOUTHEAST SPAIN

MICHELLE STOKLOSA

Department of Geosciences, Boise State University, 1910 University Dr., Boise, Idaho 83725, U.S.A.
e-mail: mstoklos@boisestate.edu

AND

J.A. (TONI) SIMO

Department of Geology and Geophysics, University of Wisconsin–Madison, Madison, Wisconsin 53706, U.S.A.
Present address: ExxonMobil Upstream Research Company, GW3 908B, P.O. Box 2189, Houston, Texas 77252, U.S.A.
e-mail: toni.t.simo@exxonmobil.com

ABSTRACT: A thick Oligocene succession of shallow-water carbonates exposed in southeast Spain provides an opportunity for studying the controls on platform and off-platform morphology and arrangement of facies successions, including compositional trends. The Oligocene platform of the Costa Blanca progressively onlaps a topographic high. It includes four high-frequency sequences that, together, show an overall upward-deepening trend and a vertical transition from cyclical inner-platform strata at the base to open-marine coralline algal-mounded morphologies at the top. The margin is interpreted to have been a steep-sided and possibly faulted margin that shed coarse-grained platformal and older material into the basin. The five defined facies associations (restricted inner shelf, open-marine inner shelf, open-marine middle shelf, open-marine mounded outer shelf, and basin) consist chiefly of heterozoan biotic assemblages (red algae, benthic foraminifera, bryozoans, mollusks) that are in contrast to most other coral-dominated Oligocene strata of the Tethys.

Deposition of the Oligocene platform and basin strata occurred contemporaneously with the opening of the Valencia Trough, and postdated and predated contractional events that folded the area. Faulting and increased subsidence associated with this extensional tectonic setting governed the overall platform morphology, deepening-upward trend, and slope–basin sedimentary wedges. The tectonic setting allowed the accumulation of a thick stratigraphic succession and subsequent drowning; it also overprinted the effects of global sea-level fall. The opening of the Valencia Trough may have brought deeper, cooler, and possibly nutrient-rich waters onto the shelf, controlling the heterozoan biotic assemblage. The tectonic setting, biotic assemblages, and deepening-upward characteristics of this Tethyan platform make it a good analog for Cenozoic carbonates of Southeast Asia.

INTRODUCTION

Tertiary Tethyan platforms were generated as the Eurasian plate interacted with the African and Arabian plates and various microplates, which created complex structural settings and constrained the size of the Tethys Sea. Because of this, these platforms have various geometries, ranging from isolated platforms with steep margins to low-gradient ramps attached to growing structural features (Buxton and Pedley, 1989; Esteban, 1996). They contain facies belts with compositions and arrangements that reflect not only the tectonic settings of the platforms but also their precursor substrates and variations in both sea level and oceanographic conditions to which the platforms were exposed during deposition (Fulthorpe and Schlanger, 1989; Esteban, 1996). For instance, some interpretations suggest that the architecture of Cenozoic carbonate platforms followed sea-level fluctuations, with transgressive systems dominated by red-algal and large-foraminifera facies and late transgressive, highstand, and early lowstand systems dominated by coral facies (Esteban, 1996; Franseen and Goldstein, 1996). However, rhodalgal carbonates have also been associated with estuarine-type circulation and areas with upwelling, strong winds, and intense vertical mixing (Esteban, 1996; Pedley, 1998).

Of all the Tethyan carbonate platforms, those from the Oligocene are relatively rare and limited to northeastern Italy (Frost, 1981; Nebelsick et al., 2005; Knoerich and Mutti, 2006), northern Slovenia, northwest Greece (Nebelsick et al., 2005), Malta (Buxton and Pedley, 1989), the German Alps (Nebelsick et al., 2005), and southeast Spain (Stoklosa, 2002).

This paper describes the characteristics of an Oligocene carbonate succession exposed along the Costa Blanca of southeast Spain and evaluates the factors controlling its characteristics. These factors include the role of tectonic activity versus eustasy in determining platform morphology, facies belts, and slope deposits, and the role of paleo-oceanography and climate in generating specific biota and facies types. The descriptions and interpretations presented in this study are useful as an analog for other platforms that have developed in complex tectonic settings, with overlapping extensional and contractional events in time and space that controlled platform morphologies and drowning. This may also serve as an analog for other platforms displaying similar facies successions, such those described from southeast Asia (Fulthorpe and Schlanger, 1989).

GEOLOGIC SETTING AND BIOSTRATIGRAPHY

The Costa Blanca of southeast Spain is in the eastern part of the External Zone (EZ) of the Betic Cordillera (Fig. 1). This zone is characterized by Mesozoic and Cenozoic sedimentary rocks that have been folded, thrusted, and deformed by salt diapirs (De Ruig, 1992). The EZ represents the former passive continental margin of southern Iberia during the Mesozoic and Paleogene (Azema et al., 1979; Baena Perez and Jerez Mir, 1982; Moseley, 1990; De Ruig, 1992). During most of the Mesozoic and Paleogene, lagoonal and shallow marine deposition was occurring to the north-northwest of the External Zone, while open marine, deeper-water deposition was occurring to the south-southeast. Terrigenous material was sourced from the northwest (Geel et al., 1992).

Controls on Carbonate Platform and Reef Development
SEPM Special Publication No. 89, Copyright © 2008
SEPM (Society for Sedimentary Geology), ISBN 978-1-56576-130-8, p. 171–184.

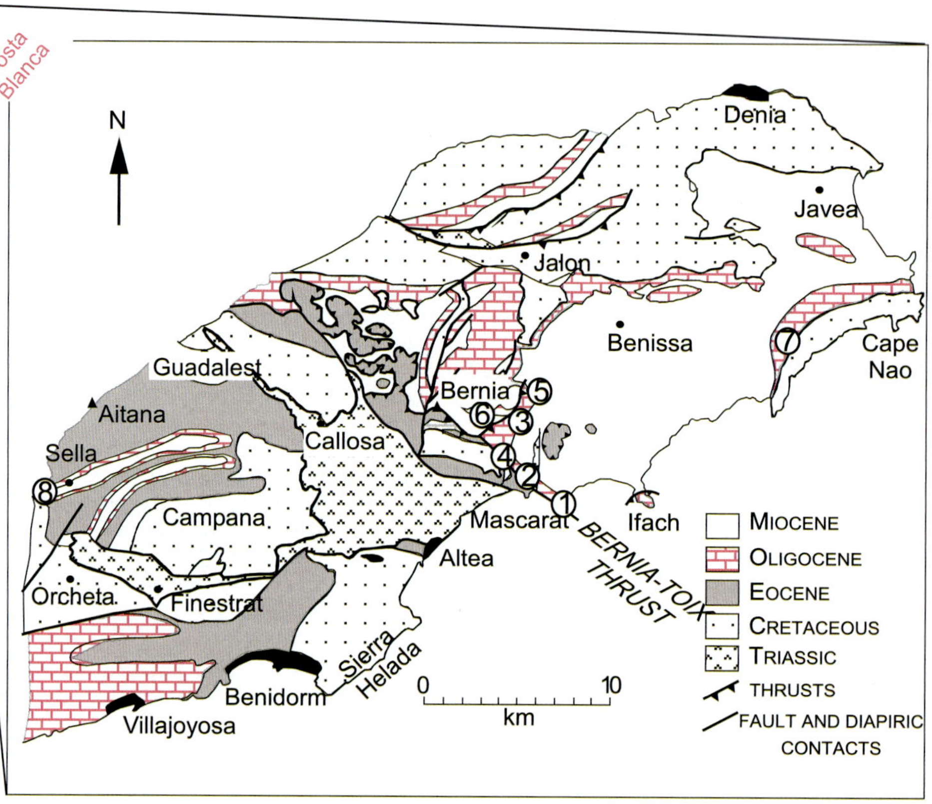

FIG. 1.—Location and structure map of field area. Inset shows the Costa Blanca located in the Prebetic Zone (former shelf) of the External Betics (which was the former continental margin of Iberia during the Mesozoic and Paleogene). The larger map on the bottom right shows the complex structure of the Costa Blanca and the locations of the measured sections: 1. Toix; 2. Pedrizas; 3. Estret; 4. Autopista; 5. Pinos; 6. Restaurant Bernia; 7. Cumbre del Sol; and 8. Relleu. Adapted from Gamble and Moseley (1975) and Moseley (1990).

During the Late Cretaceous and Paleogene, the platform experienced deformation and subsequent fragmentation of the passive margin, thought to have been caused by a peripheral bulge associated with the north–south-directed compression associated with the building of the Pyrenees (García-Hernández et al., 1980; De Ruig, 1992). Paleocene–Eocene sedimentary rocks consist of shallow-water carbonates with large foraminifera and algae on the shelf, and pelagic and hemipelagic marl with platform-derived turbidites in the basin. Eocene and Oligocene strata are separated by a major unconformity, and Oligocene strata transgressively overlie deformed Eocene and Cretaceous rocks (Geel, 1995). Deposition during the Oligocene–Early Miocene (Aquitanian) was contemporaneous with normal faulting related to rifting and opening of the Gulf of Valencia (De Ruig, 1992; Geel, 1995; Maillard and Mauffret, 1999). The Costa Blanca region changed from an extensional margin to a foreland basin during the Burdigalian–early Tortonian. Shallow-water coralline algal limestone strata are overlain by deep-water turbidites (McColgin, 2002). Foreland folding in the External Zone ceased in the late Tortonian to Messinian and continental sedimentation associated with emergence dominated thereafter (Megías et al., 1983; Pierson d'Autrey, 1987; Ott d'Estevou et al., 1988; De Ruig, 1992; Geel et al., 1992).

Correlation of Costa Blanca strata with global geochronology and with previously established regional tectonic events (Fig. 2) is based on the age constraints of biota in both shelf and basinal areas. Samples from the base of the shallow shelf section contain Late Eocene large benthic foraminifera genera (*Discocyclina* = Paleocene, Eocene; *Borelis* = Late Eocene) (Caus and Serra-Kiel, 1992; Serra-Kiel, personal communication). The remainder of the shallow shelf section is Early Oligocene (Rupelian) in age, an interpretation based primarily upon the appearance and abundance of *Lepidocyclina*, which first appears in the stratigraphic record in the Rupelian (biozone SB22A), and of *Asterigenina*, also common in the Rupelian (Serra-Kiel, personal communication). Chattian fauna are absent except in the uppermost meter of the shallow shelf section. The base of the basinal section is also Late Eocene, as indicated by the large benthic foraminifera assemblages (*Spiroclypeus, Biplanispira, Pellatispira*) and temporally significant planktonic foraminifera (*Globigerinatheca, Hantkenina, Turborotalia*, and *Catapsydrax*) (Everts, 1991). The planktonic foraminifera used to date the overlying basinal strata as Early Oligocene (or bichronozone P18–19) are *Chiloguembelina cubensis, Globigerina galarisi, Catapsydraz onicavus, Catapsydrax* sp., *Globigerina tripartita*, and *Globigerina eoceana*.

FACIES ASSOCIATIONS

The Costa Blanca carbonate succession was divided into facies based upon common sediment textures, biota, grain type,

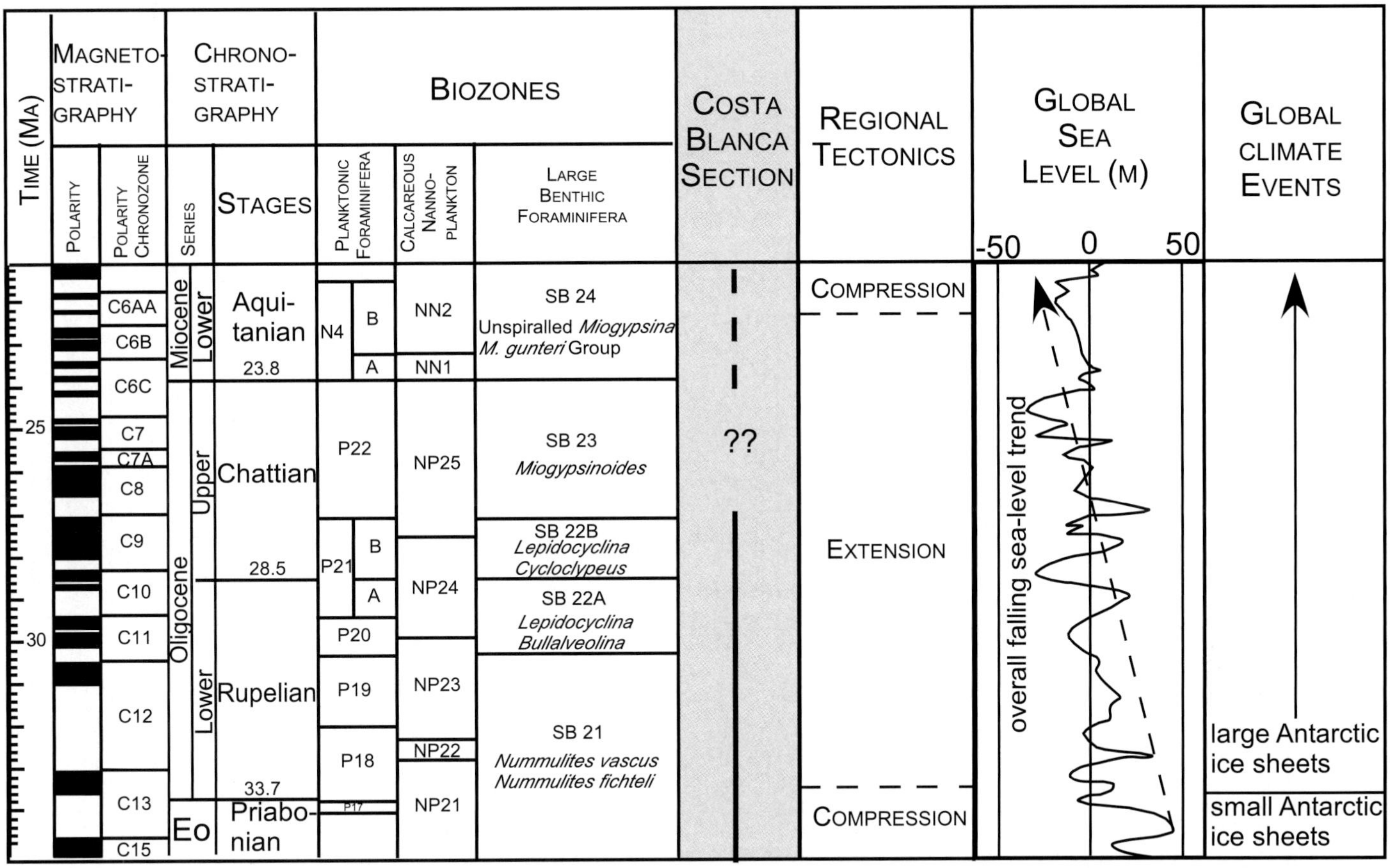

FIG. 2.—Position of the Costa Blanca stratigraphic section chronostratigraphically, biostratigraphically, and in the regional tectonic and global climate framework. Biostratigraphy and magnetostratigraphy data are from Cande and Kent (1992, 1995), Berggren et al. (1995), and Hardenbol et al. (1998). Tectonic events refer to the late Eocene–Early Oligocene compression associated with the building of the Pyrenees, extension associated with the opening of the Valencia Trough, and the compression associated with the transformation of southern Iberia into a foreland basin (De Ruig, 1992). Global sea level and climate events are from Miller et al. (2005).

grain size, sorting, bedding style, and sedimentary features of the rocks. Facies descriptions were refined based upon 635 representative samples collected from these sections and 105 thin sections. Flora, fauna, and other sedimentary features were used to determine depositional environments, and were compared to other Tertiary models (Frost, 1981; Buxton and Pedley, 1989; Carozzi, 1989; Geel, 1995; Stoklosa, 2002). Facies with similar interpretations of depositional environment and found in close proximity in vertical successions were grouped into an association. Each facies association was given an interpretive name based on a position, or facies belt, on an idealized carbonate platform (restricted inner shelf, open-marine inner shelf, open-marine middle shelf, open-marine mounded outer shelf, and basin). Table 1 lists the names and characteristics of these facies associations.

Restricted-Inner-Shelf Facies Association

The stratigraphic interval containing this association is thick (up to 150 m) and is found in the lowermost part of the two most complete sections (Toix and Pedrizas, Fig. 1). The association comprises three repetitive facies (Table 1): intraclast conglomerate, skeletal wackestone, and well-sorted fine-grained grainstone (Fig. 3). Intraclast conglomerate and well-sorted fine-grained grainstone are dominant (80%) in this association. The intraclast conglomerate is composed of rip-up clasts (mainly well-sorted fine-grained grainstone) and skeletal grains, and some of these grains types are black in color. Fenestral porosity and intergranular cracking are common features in the conglomerate and well-sorted fine-grained grainstone. Skeletal grains are the dominant allochem in both the wackestone and grainstone, and miliolids are by far the most abundant.

Blackened grains have been observed in modern and ancient carbonates, are commonly associated with features indicating subaerial exposure (e.g., caliche, root casts, fenestral structures), and have most often been interpreted to have formed in shallow subtidal, intertidal, and supratidal zones (Strasser, 1984; Tucker and Wright, 1990). Regardless of the specific geochemical environment and mode of impregnation by organic material and other (subordinate) minerals that cause the blackening, these clasts are more easily preserved when reworked in a shallow-water, even emergent, depositional environment (Strasser, 1984). The presence of blackened grains in association with rip-up clasts, intergranular cracking, and fenestral porosity in the intraclast conglomerate of the Costa Blanca strata have led to the interpretion that this conglomerate formed in a shallow marine (a few meters of water) depositional environment that experienced episodic exposure.

Repeating restricted, inner-shelf lithologies are interpreted to represent shallowing-upward cycles. Intraclast conglomerate forms the base of the typical cycle, skeletal wackestone, if present,

TABLE 1.—Description and interpretation of the facies of the Costa Blanca Oligocene strata.

Facies Association	Facies Name	Textures	Characteristic fauna	Sorting	Grain sizes	Environmental Interpretation
Restricted Inner Shelf	Intraclast conglomerate	Matrix-supported conglomerate	Miliolids, small benthic forams, bryozoans, echinoids	Poor to moderate	Fine, medium, coarse, very coarse, and pebble	Restricted, very shallow (few m), high energy, episodic exposure
	Skeletal wackestone	Wackestone to mud-rich packstone	Miliolids, small benthic forams, bryozoans, *Porites*, echinoids	Poor	Fine, medium and pebble-size	Semi-restricted, shallow, low energy
	Well-sorted fine-grained grainstone	Packstone to grainstone	Miliolids, small benthic forams, rare to moderate echinoids, mollusks	Well to moderate	Fine and medium	Restricted, shallow, high energy
Open-Marine Inner Shelf	Coarse skeletal grainstone	Packstones, grainstones	*Nummulites*, echinoids	Well to moderate	Coarse, medium, very coarse	Open marine, shallow, high energy
	Fine skeletal grainstone	Mud-lean packstone to grainstone	Small benthic forams, red algae	Well to moderate	Medium, fine, very fine	Open marine, shallow, high energy
Open-Marine Middle Shelf	Coralgal–bryozaon rudstone and bindstone	Muddy bindstone and rudstone	Small benthic and encrusting forams, bryozoans, *Porites*, red algae, *Nummulites*	Poor	Medium, very coarse, pebble	Open marine, shallow, moderate to low energy
	Bryozoan–coral floatstone	Floatstone to Mud-rich rudstone	Small benthic forams, encrusting forams, bryozoans, *Porites*, red algae, *Nummulites*	Poor	Medium, coarse, pebble	Open marine, shallow, moderate to low energy
	Lepidocyclinid mud-rich packstone	Wackestone to mud-rich packstone	*Lepidocyclina*, echinoids, red algae	Poor	Pebble, very coarse	Open marine, shallow, low to moderate energy
Open-Marine Mounded Outer Shelf	Biostromal red algal–large benthic foram boundstone	Boundstone	Red algae, rhodoliths, *Nummulites, Lepidocyclina*	Poor	Coarse, very coarse, pebble	Open marine, shallow, high energy
	Red algal–large benthic foram rudstone to grainstone	Pebble-size rudstone to grainstone	Red algae, rhodoliths, *Nummulites, Lepidocyclina*	Poor	Coarse, very coarse, pebble	Open marine, shallow, high energy
	Coarse red algal–skeletal grainstone	Coarse grainstone and packstone	Red algae, *Nummulites, Lepidocyclina*	Moderate	Medium, coarse, very coarse	Open marine, shallow, high energy
Basin	Marl	Argillaceous mudstone	Planktonic forams, occasional large benthic forams	N/A	very fine, fine, occasional very coarse	Open marine, deep, low energy
	Turbidite	Grainstone Clast-supported conglomerate	Large benthic forams, red algae, echinoids Large benthic forams, red algae, coral	Well to moderate Poor	coarse, very coarse, medium pebble, very coarse, medium	Open marine, deep, high energy transport
	Mass wasting	Clast- and matrix-supported conglomerate	Large benthic forams, red algae, echinoids, coral	Poor	Very fine, very coarse, boulder	Open marine, deep, closer to steep margin, sudden transport

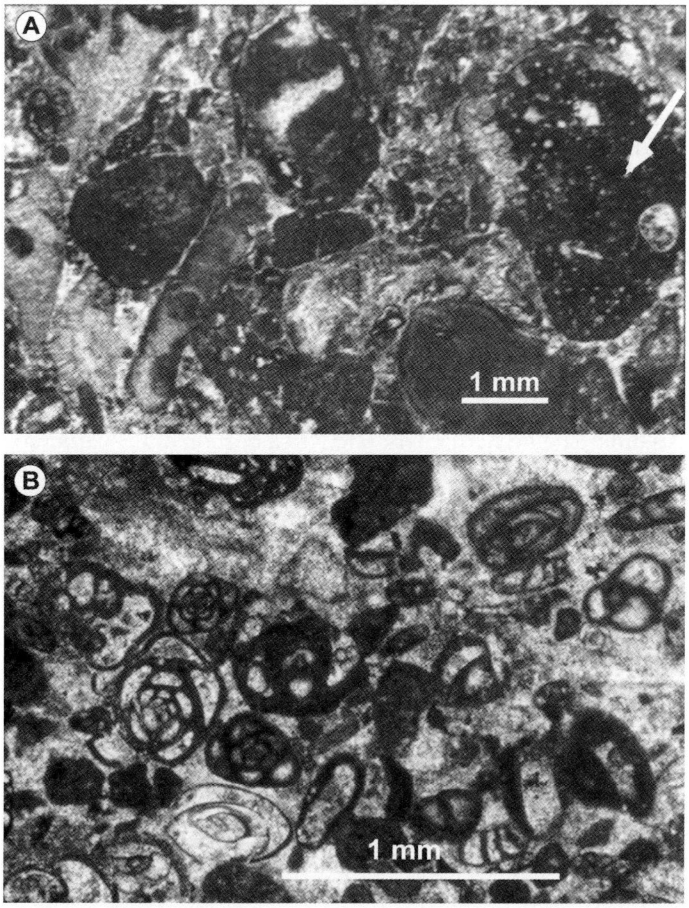

FIG. 3.—Photomicrographs of restricted inner-shelf facies. **A)** Intraclast conglomerate. Clasts are the dark rounded grains; some are micritic and some are fine skeletal grainstone (arrow). Bar scale is 1 mm. **B)** Well-sorted fine-grained grainstone. The grain sizes in this photo are primarily medium-sand size, the upper grain-size limit for this facies type. The dominant skeletal grains are miliolids. This sample is moderately sorted, and contains little micrite. Bar scale is 1 mm.

forms the middle, and well-sorted fine-grained grainstone forms the top. Cycle thicknesses range from 2 to 13 m, with an average thickness of 7.5 m. The cycle boundary is a truncation surface represented by intraclasts of the underlying well-sorted fine-grained grainstone in the overlying conglomerate.

Miliolids, a common biotic constituent in these rocks, are capable of living in waters with high salinities (40–56‰). Their dominance may indicate a restricted depositional environment (Chaproniere, 1975; Enos, 1983; Buxton and Pedley, 1989). Although echinoids are generally considered to be stenohaline, open-marine organisms (Sprinkle and Kier, 1987), the echinoid grains in these strata are generally broken, and most likely transported. Intact echinoid specimens are present only in the wackestone facies, which was most likely deposited during flooding of the platform, thus reflecting an open-marine influence.

Open-Marine Inner-Shelf Facies Association

The open-marine inner-shelf association is usually less than 50 m in any stratigraphic section. This association is present in upper Eocene and upper Rupelian strata of the succession. The two lithofacies that form this association, coarse and fine skeletal grainstone (Table 1, Fig. 4), alternate within the succession. Both lithologies are micrite-lean and contain red algae fragments. Large benthic foraminifera are dominant in the coarse skeletal grainstone, and small benthic foraminifera are dominant in the fine skeletal grainstone.

Alternations of these facies are interpreted to represent fining-upward and shallowing-upward cycles. The presence of rip-up clasts at the base of each coarse-grained grainstone marks the base of each cycle. Cycle thickness ranges from 3.5 to 17 m, with an average of 8.7 m.

This facies association is interpreted as having been deposited in an open-marine, shallow, inner-platform setting because of the dominance of organisms such as echinoids, nummulitid foraminifera, and red algae (Chaproniere, 1975; Wray, 1977; Hallock and Glenn, 1986; Sprinkle and Kier, 1987). Good sorting and lack of micrite suggest shallow-water, high-energy conditions. The rocks do not show evidence of exposure, but the presence of intraclasts indicates cementation prior to reworking.

Open-Marine Middle-Shelf Facies Association

The open-marine middle-shelf association corresponds to 40–75 m of the Pedrizas and Toix sections and is Rupelian in age. This association is composed of three facies types: coralgal–bryozoan rudstone and bindstone, bryozoan–coral floatstone, and *Lepidocyclina* mud-rich packstone (Table 1, Fig. 5). Rocks of this association include more micrite than those of the inner-shelf associations and form poorly defined cycles consisting of alternating floatstone and rudstone averaging 13 m in thickness. These rocks interfinger with inner-shelf and outer-shelf facies.

The biota (particularly the large benthic foraminifera and coralline algae) imply an open-marine depositional environment. Large benthic foraminifers are usually stenohaline, open-marine organisms (Hallock and Glenn, 1986). Abundant Melobesieae (crustose coralline algae) suggests constant circulation and normal–salinity waters (Johnson, 1961; Wray, 1977). Many studies have concluded that red algae with the morphology of those present in this association (not densely branched and with a leafy crust morphology) suggest low-energy environ-

FIG. 4.—Photomicrographs of coarse skeletal packstone to grainstone of the open-marine inner-shelf facies association. Nummulitids are the main skeletal component in this sample, although other rotaliid benthic forams are also present. Bar scale is 1 mm.

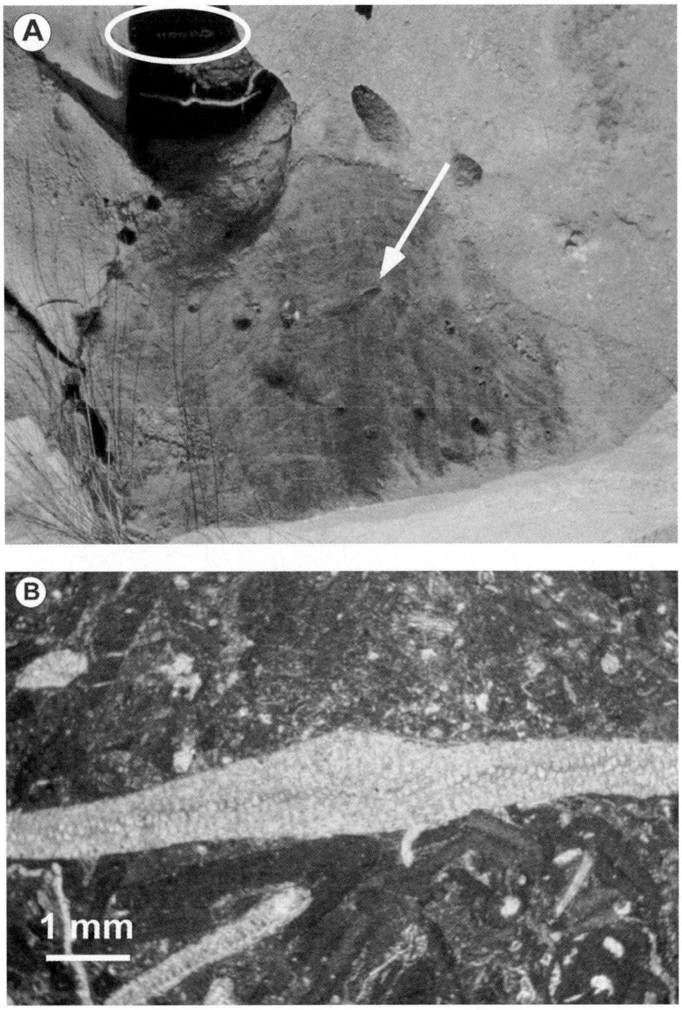

FIG. 5.—Photographs of open-marine middle-shelf facies. **A)** Outcrop of bryozoan–coral floatstone. This photomicrograph shows a hermatypic coral (arrow) "floating" in a fine skeletal packstone matrix. White circle shows the outline of a camera lens cap for scale (5 cm in diameter). **B)** Photomicrograph of lepidocyclinid mud-rich packstone. The white chambered foraminifera in the center is a *Lepidocyclina*. Bar scale is 1 mm.

ments (Bosellini and Ginsburg, 1971; Chaproniere, 1975; Bosence, 1983, 1991). Finger corals are also indicative of a low-energy, protected environment (James, 1979; Carozzi, 1989).

In addition to biotic content, the higher mud content, absence of subaerial exposure features, lack of sorting, thick bedding, and weakly cyclic nature also contribute to the interpretation of deposition in low-energy, deeper-water conditions. The lateral interfingering with other associations supports a mid-shelf interpretation.

Open-Marine Mounded Outer-Shelf Facies Association

The open-marine mounded outer-shelf association is found at the tops of most of the measured sections and is thus probably the most widespread of the shallow-shelf facies associations. It ranges in thickness from 25 to 80 m. This association consists of three facies: biostromal red algal–large benthic foraminiferal boundstone, red algal–large benthic foram rudstone to grainstone, and coarse red algal–skeletal packstone to grainstone (Table 1, Fig. 6). The thick bedding (meter scale to ten-meter scale) and mounded shape of the biostromal rocks are striking features distinguishing this association from others. Cyclicity is poorly defined; when present, it consists of thick (15–20 m) alternations of red algal–large benthic foraminiferal boundstone and rudstone capped by red algal–skeletal packstone to grainstone.

Biotic constituents (red algae, nummulitids, and lepidocyclinids) suggest deposition in open-marine waters (Johnson, 1961; Wray, 1977; Hallock and Glenn, 1986). The lack of both lime mud and exposure features indicates deposition where energy was concentrated, but exposure was uncommon. An environment that is seaward of the inner and middle shelf and closer to the shelf margin fits the features recognized in this association. An outer-shelf environment refers to a portion of the carbonate platform which receives much wave energy, and where thick organic buildups develop (James, 1979; Buxton and Pedley, 1989; Carozzi, 1989).

FIG. 6.—Photographs of open-marine mounded outer-shelf facies. **A)** Outcrop of massive, thick beds of the open-marine mounded outer-shelf facies association. The massive beds are biostromal rudstone and boundstone, and the recessive areas are the interbedded fine-grained packstone to grainstone. **B)** Photomicrograph of red algal–large benthic foram rudstone to grainstone. The dark, rounded and oblong areas are various morphologies of crustose coralline algae, and the lighter areas are large benthic forams (labeled "F") and intragranular cements. Bar scale is 1 mm.

Basin Facies Association

Marls, turbidites, and mass-wasting deposits are the three facies that compose this association (Table 1, Fig. 7), which is Late Eocene to Miocene in age. Marls are slightly argillaceous and contain planktonic foraminifera. Turbidites, composed principally of shallow-shelf material, show scoured bases, flute casts, and normal grading. Mass-wasting deposits consist of large (tens of meters in diameter) limestone blocks floating in a pebbly limestone conglomerate and slumped marls with limestone blocks. Clasts in the mass-wasting deposits and turbidites are Eocene to Oligocene in age. Mass-wasting deposits dominate closer to the platform whereas turbidites dominate distally.

FIG. 7.—Photographs of basin facies. **A)** Outcrop photo of the basin facies association. Younger strata are to the right. Marls are the recessive beds that make up most of the succession. Turbidites are the more resistant beds. **B)** Closer view of turbidite beds of the basin facies association. The resistant beds are coarse-grained with scoured bases and which fine upward into recessive marl. Hammer for scale (circled).

The very fine-grained texture and presence of planktonic foraminifera in the marl facies indicates a low-energy, deep-water depositional environment (Wilson, 1975; Buxton and Pedley, 1989; Carozzi, 1989; Geel, 1995). Turbidites represent gravity flows resedimenting shelf material downslope. The abundance of turbidites and mass-wasting deposits in this association suggests continuous reworking of the shelf.

STRATIGRAPHIC CORRELATION

Lateral relationships and correlation of the measured sections across the Costa Blanca are illustrated in Figure 8. Shallow-shelf sections were correlated using aerial photos, field tracing of unconformities and key beds, sudden appearance datums for distinctive allochems and fossils, and similarity in stratigraphic pattern. The basinal section was correlated to the shelf using broad age correlations based on biostratigraphy and on trends in allochem content. This approach has been applied successfully in other platform-to-basin correlations (Everts et al., 1999).

Both the shallow-shelf and basinal Oligocene strata are bounded by two major correlation surfaces. The Oligocene section rests unconformably on Cretaceous or Eocene strata. This unconformity is found at the Toix, Cumbre del Sol, Restaurante Bernia, and Relleu sections. At Cumbre del Sol and Restaurante Bernia, this surface is an angular unconformity between Cretaceous and Oligocene strata. At Toix and Relleu, this surface is an erosional unconformity marked by a conglomerate between Eocene and Oligocene strata. This unconformity has been linked to a compressional phase associated with the Pyrenees orogeny (De Ruig, 1992; Geel, 1995). Another major surface of correlation is an unconformity between Oligocene and Miocene strata found in the Toix, Autopista, Estret, Pinos, and Cumbre del Sol sections. This erosional unconformity is composed of a conglomerate commonly overlain by glauconitic sandstone. The Oligocene–Miocene boundary at Relleu is marked by a change from a dominance of fining-upward gravity-flow deposits to debris-flow deposits. Geel (1995) interprets this unconformity as having formed in the Chattian, when most of the area became emergent and eroded.

Some of the major compositional correlation lines are shown in Figure 8. For example, blackened clasts and grains are very common in a discrete interval in the shallow-water sections as well as in resedimented beds in the basinal section. Other correlations include the abrupt appearance of large benthic foraminifera such as nummulitids and giant lepidocyclinids in the succession.

The shallow-shelf succession is much thicker (369 m thick) towards the south, near the inferred shelf margin (Fig. 8), appearing to reflect the topography during the early Oligocene. The thickest sections overlie Eocene strata, and the thinnest overlie Cretaceous basement. This relation suggests a tectonic control on development of the sub-Oligocene unconformity (Geel, 1995) and suggests onlap of the Oligocene onto a fault-bounded high. The slope and basinal sections are today separated from the shallow-shelf sections by a salt diapir (Triassic salt, Fig. 1). Although not measured in detail, a thick chaotic succession of mass-wasting deposits, with large platform blocks of different ages, and gravity flows within hemipelagic sediments, exists between the shelf and basinal sections (Villajoyosa area, Figs. 1 and 8). This succession is interpreted as a toe-of-slope escarpment suggesting that the platform margin was steep and probably faulted, and exposed rocks of different ages. The mast-wasting succession thins basinward and grades into a distal slope–basin succession 175 m thick (Relleu, Fig. 1).

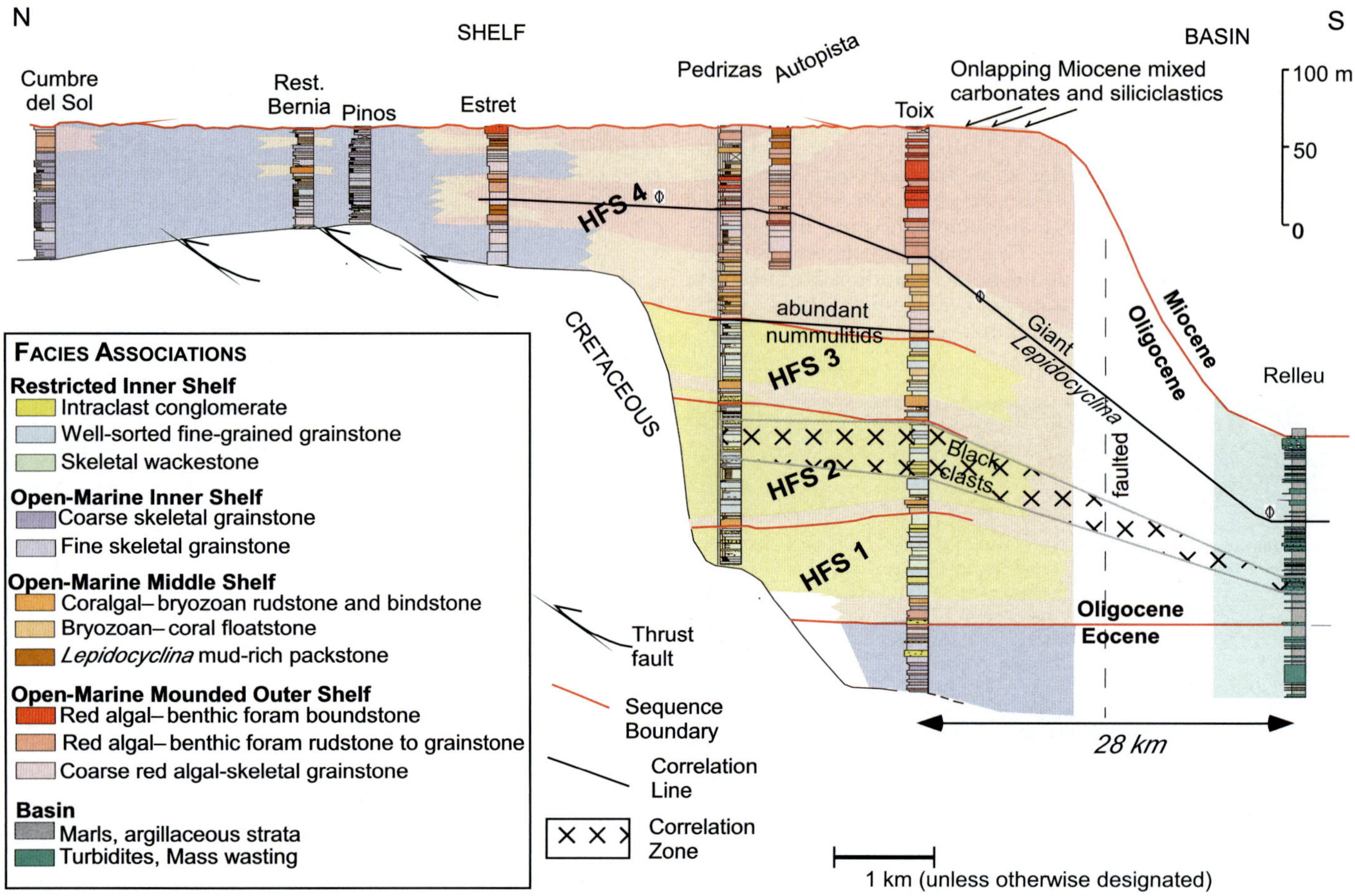

FIG. 8.— East-facing cross section of Oligocene strata of the Costa Blanca, showing the major correlation lines, facies associations, and sequence stratigraphic framework. The location of the stratigraphic sections can be found in Fig. 1. Cumbre del Sol and Relleu are farther from the section than the other locations, and their distances are not according to scale. The distance between Cumbre del Sol and Restaurante Bernia is 15 km, whereas 28 km separates Toix and Relleu. The facies types are indicated by color, as defined by the legend. Four high-frequency sequences are labeled, and sequence boundaries are in red.

SEQUENCE STRATIGRAPHY

The stratigraphic succession between the two regional unconformities described above was subdivided into four high-frequency sequences (HFS, Figs. 8, 9). The HFS are numbered from 1 to 4, from oldest (1) to youngest (4). Each successive HFS onlaps the depositional high, and the youngest platform appears to fossilize the positive relief.

HFS 1–3 include open-marine facies in their lower parts, overlain by restricted-marine facies (Fig. 9) that are interpreted to represent an initial deepening followed by an upward shallowing of local sea level. HFS 1 and 2 each show a facies succession of a thin (8–14 m) package of open-marine middle-shelf facies at their base overlain by a thicker (40–73 m) package of restricted inner-shelf facies (Figs. 8, 9). The open-marine middle-shelf facies at the base of HFS 1 and 2 show cycles of alternating coarse and fine skeletal rudstone and floatstone. These cycles range from 5 to 10 meters in thickness in both sequences. HFS 1 includes little evidence of exposure in these cycles, but some rip-up clasts are present in these cycles in HFS 2. In the restricted inner-shelf facies, or upper part, of HFS 1 and 2, the cyclicity is better defined. These cycles are successions of wackestone to packstone to fine-grained, well-sorted grainstone, interpreted as deepening- then shallowing-upward cycles with evidence of exposure. These cycles are composed of thin beds (0.1–0.5 m) and have an average cycle thickness of 7.5 m. Cycle bases are commonly conglomerates and breccias with blackened clasts and clasts with fenestral porosity, suggesting reworking of exposure surfaces.

The contact between HFS 1 and 2 is an abrupt transition from restricted inner-shelf facies to open-marine middle-shelf facies. Cycle thicknesses of HFS 1 show a broadly thinning-upward trend, from cycles 10 m thick at the base to cycles 2–3 m thick at the top. No clear upward-thickening trend in cycles is discernible in HFS 2.

HFS 1 and 2 are very similar in the Toix and Pedrizas sections (Figs. 8, 9), located 1.8 km apart. HFS 1 is complete only at Toix, the more basinward shallow-shelf section, so its lateral variability is difficult to assess. However, the upper part of HFS 1 is found in both sections, and the Pedrizas section includes more conglomerate beds and thinner cycles than the section at Toix. HFS 2 is about 30% thicker at Pedrizas compared to Toix, and Pedrizas includes proportionally less open-marine facies than Toix.

A traceable intraclast conglomerate truncating restricted facies and overlain by open-marine facies marks the boundary between HFS 2 and 3. HFS 3 shows a thick, open-marine middle-shelf package at its base (15–20 m) overlain by a restricted inner-

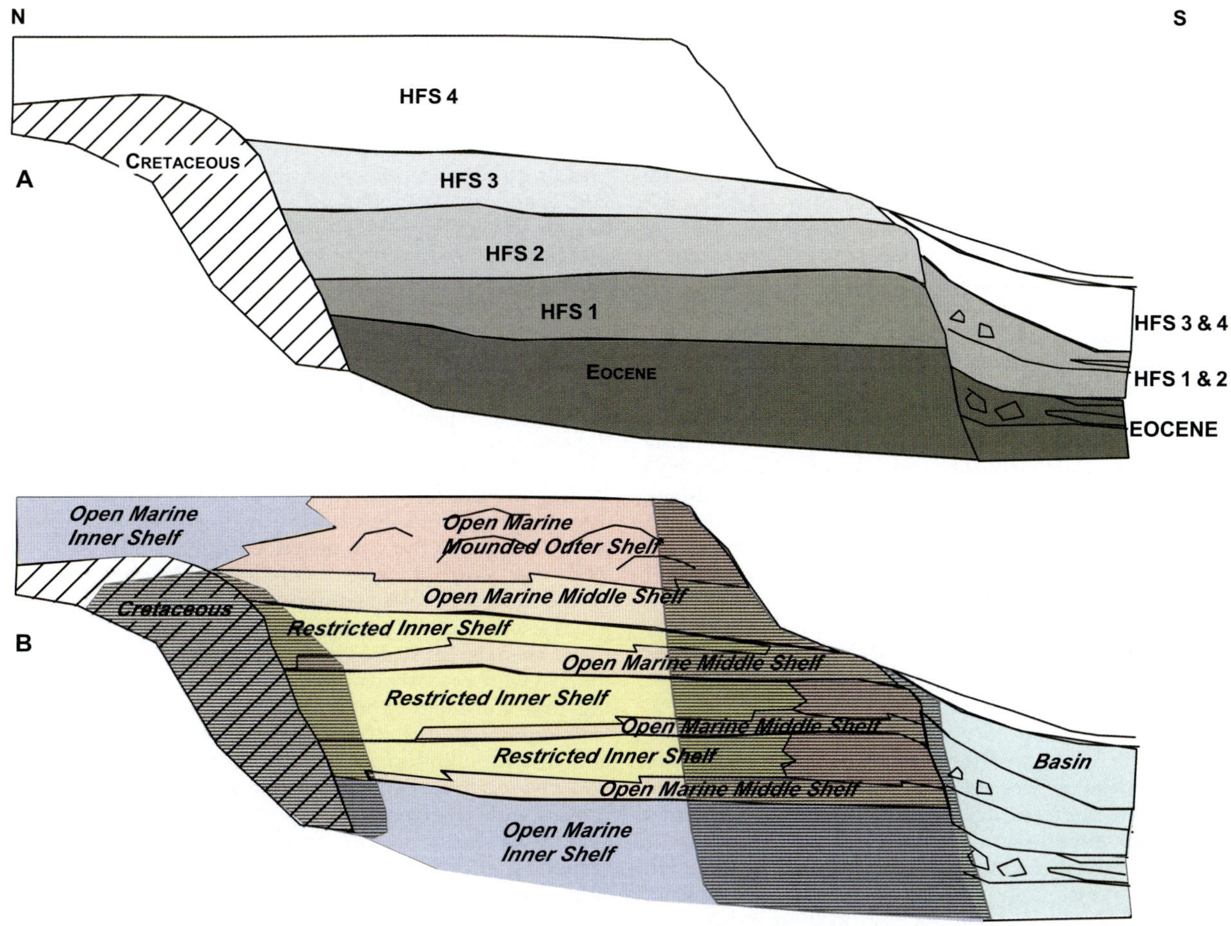

FIG. 9.— A more simplified interpretation of the cross section shown in Figure 8. Orientation is the same as the cross section. **A)** Position of the interpreted high-frequency sequences, and interpretation of the shelf margin as fault-controlled, with slumping and mass wasting on the slope and basin. **B)** The occurrence of the facies associations in the sequences, showing their aggradation and backstepping over time. Gray-shaded areas indicate that the platform margin is not actually preserved, but inferred, and that the onlap of the Oligocene onto Cretaceous strata is not seen in all sections, just the section farthest north.

shelf facies package (35–-40 m) comprising interbedded open-marine facies within the restricted regressive facies, especially in sections nearer the inferred margin (Figs. 8, 9). Specifically, the HFS 3 at Toix includes 36% open-marine facies, whereas 16% are open-marine facies at Pedrizas. HFS 3 at Toix also includes less intraclast-conglomerate facies (5%) compared to the section at Pedrizas (14%). The cyclicity of HFS 3 is similar to that of HFS 1 and 2, but cycles are thicker (5–16 m) and possess fewer exposure surfaces and fenestral porosity.

The base of HFS 4 corresponds to an abrupt change in skeletal and allochem components and overall facies distribution. This HFS is dominated by open-marine middle-shelf, mounded outer-shelf, and inner-shelf facies associations (Figs. 8, 9). Open-marine facies are found at all measured sections, indicating widespread platformal cover. Towards the inferred shelf margin (Toix and Pedrizas), this HFS is thick (125–150 m) and is composed of a thick package (60 m) of open-marine middle-shelf facies overlain by an even thicker package (90 m) of open-marine mounded outer-shelf facies. A comparison between these two sections shows that Pedrizas includes more middle-shelf facies (approximately 50% middle-shelf and 50% mounded outer-shelf facies) than Toix (30% middle-shelf and 80% mounded outer-shelf). HFS 4 is thinner (50–70 m) in the more landward and northward sections of Estret, Pinos, Bernia, and Cumbre del Sol. Middle-shelf facies decrease progressively in abundance towards the north, whereas open-marine inner-shelf facies increase in abundance. The cyclicity of this HFS is poorly defined in all sections, and evidence for subaerial exposure is rare. Where observed, the cycles are thick (15–20 m) and characterized as alternations between grainstone–rudstone and bindstone facies. The presence of open-marine facies over such a wide area of the platform during deposition of HFS 4 suggests that strong currents redistributed sediment and rarely allowed the available accommodation space to be filled. Thicker cycles suggest aggradation and possibly facies backstepping (Figs. 8, 9). The depositional profile and possible controls of HFS 4 are noticeably different than HFS 1–3 and are discussed below.

Overall, the HFS's show a successively thicker open-marine base and an increase in the intercalation of open-marine facies with the restricted facies. HFS 1 and 2 represent shallow-water and restricted conditions where sediment repeatedly filled the accommodation space. HFS 3, and especially 4, show evidence for deeper water and more open marine conditions, where accommodation space went unfilled. This trend suggests that accommodation increased through time, thus weakening the effects of higher-frequency sea-level variations and instead allowing thicker cycles. The increase in accommodation space may have induced platform backstepping. The platform was subsequently exposed, terraced, and eventually shut down by increasing subsidence and siliciclastic input during the Early Miocene.

The correlation between the platform and basin (Fig. 8) suggests that significant mass wasting and turbidite deposition in the basin was contemporaneous with the formation of the Eocene–Oligocene unconformity on the platform. HFS 1 and 2 coincide with a lower rate of basin sedimentation and the deposition of marls. An upward increase in the occurrence of channelized turbidites in the basinal section can be correlated to where HFS 1 and 2 have abundant exposure surfaces and black clasts showing that the platform was reworked. HFS 3 appears to correlate with a thin succession of marl and turbidites, whereas HFS 4 corresponds to a thicker basinal section, characterized by marl and turbidites, interbedded with rare mass-wasting deposits. The Oligocene–Miocene boundary, however, is characterized by conglomeratic turbidites and mass-wasting deposits. It is apparent that mass-wasting processes dominated during the development of HFS bounding unconformities. The variable-age clasts in the basinal deposits imply the presence of a faulted escarpment, with fault movement associated with platform evolution (Fig. 9).

DISCUSSION OF CONTROLS ON PLATFORM DEVELOPMENT

The Oligocene carbonate succession exposed along the Costa Blanca provides an example of the complex interplay of tectonics and oceanography on the morphology, facies, and biotic arrangement of a carbonate platform. It shows how tectonics overprinted the eustatic signal and allowed the platform to deviate from the Mediterranean Tertiary Model (Esteban, 1996; Franseen and Goldstein 1996) and the regional Oligocene photozoan assemblages (Frost, 1981) in terms of its biotic composition and arrangement.

Detailed study has shown that each facies association of the shallow-water carbonate platform consists of a distinct suite of organisms. The primary biotic constituents of the inner-shelf facies are small benthic forams, bryozoans, echinoids, and mollusks. The two inner-shelf facies associations show that variation between open marine and restricted environments existed in this succession, marked by the dominance (restricted) or lack (open marine) of miliolids. The middle-shelf community is characterized by both small and large benthic foraminifera, red algae, bryozoans, and finger corals. The mounded outer-shelf association is dominated by red algae and large benthic foraminifera, with minor corals. Abundant coral in the mass-wasting deposits of the slope to basinal sections suggest that during deposition of HFS 1 and 2, and possibly 3, the margin (not currently observed) was dominated by a combination of corals and red algae. The Costa Blanca succession thus shows a platform (with the exception of the margin) that lacked a typical photozoan assemblage with organisms such as green algae or abundant corals and other carbonate elements commonly found associated with a photozoan assemblage such as ooids and early marine cementation. The margin, however, appears to have had a more photozoan assemblage, reflecting the open marine Oligocene conditions interpreted from other areas of the Mediterranean (Frost, 1981).

The biotic assemblage for the entire platform is different for HFS 4, however, and can be classified as heterozoan (James, 1997; Halfar et al., 2004). One hypothesis to explain this difference is that during deposition of HFS 4 the platform was dominated by upwelling and strong currents, as shown by its upward increase in heterozoan biota and high-energy facies. Perhaps upwelling increased the trophic resource level and, combined with increasing water depth and cooler waters, caused a decrease in photozoan fauna. Following the model of Halfar et al. (2004), HFS 1 and 2, and their transitional heterozoan–photozoan associations, probably reflect deposition in shallow and warm waters with slightly mesotrophic conditions, whereas the heterozoan associations of HFS 3 and, in particular, HFS 4 probably reflect deposition in deeper and cooler waters with mesotrophic conditions. It may be that the western margin of the Mediterranean was affected by an increase in nutrients possibly associated with upwelling coupled with the opening of the Valencia Trough, which could have made nutrient-rich deeper waters available to the margin.

The temporal platform evolution from restricted to open-marine facies and the lack of intertonguing between all facies recognized suggest that the depositional model has evolved through time (Figs. 9, 10). HFS 1 and 2 show the interfingering of restricted inner-shelf and open marine middle-shelf facies. Although the platform margin is not preserved, the occurrence of open and restricted facies indicates the presence of a margin with relief (similar to the one described by Carozzi, 1989, and Geel, 1995) that reduced water circulation on the shelf (Fig. 10). Based on the time-equivalent components found in the basin, the margin was created by red algae, corals, and large benthic foraminifera. The basinal section indicates that both the inner shelf and margin were reworked and resedimented into the basin, suggesting that complete exposure of the platform during the high-frequency lowering of sea level occurred, followed by erosion and resedimentation during low and transgressive seas. HFS 1 and 2 have a more cyclic nature across all of the outcrops, and the presence of closely spaced exposure surfaces in these sequences gives support to the interpretation that this was a shallow, rimmed platform affected by high-frequency sea-level variations (Fig. 10).

The biota of HFS 3 and 4 indicates less restricted waters, and the facies suggest that high-energy, open-marine conditions culminated with a backstepped, broad, algal biostrome with no significant relief. The inner-shelf facies of HFS 3 and 4 show repetition indicative of fining and shallowing upward and an overall diminution of energy. For HFS 3 and 4 the depositional model suggests a submerged platform affected by strong currents, and without a margin to restrict circulation (Fig. 10).

Development of the Antarctic ice sheet in the Oligocene resulted in a eustatic sea-level fall of ~ 30 m (Miller et al., 2005). The deepening-upward and backstepping platform succession of the Costa Blanca is interpreted, therefore, to represent an increase in subsidence with continuing platform growth. Increasing subsidence could have placed the top of the platform at greater water depths, enhancing the effects of cooler and possibly more nutrient-rich water on the biota.

The regional tectonic setting appears to have been the overriding control on the morphology of the platform and even its components. The tectonic setting created subsidence that allowed a local increase in relative sea level, in contrast to global events. This subsidence could have allowed a heterozoan assemblage to dominate because of a decrease in temperature and/or an increase in nutrients, and for the margin facies to backstep, changing the margin morphology and character of the basinal and restricted sediments.

The overall deepening-upward succession is punctuated by exposure surfaces. Exposure surfaces do not show karsting and are difficult to identify in the field because they are often reworked during subsequent transgression. The most obvious exposure surface caps the deepening-upward and backstepping platform succession prior to burial by mixed carbonate and

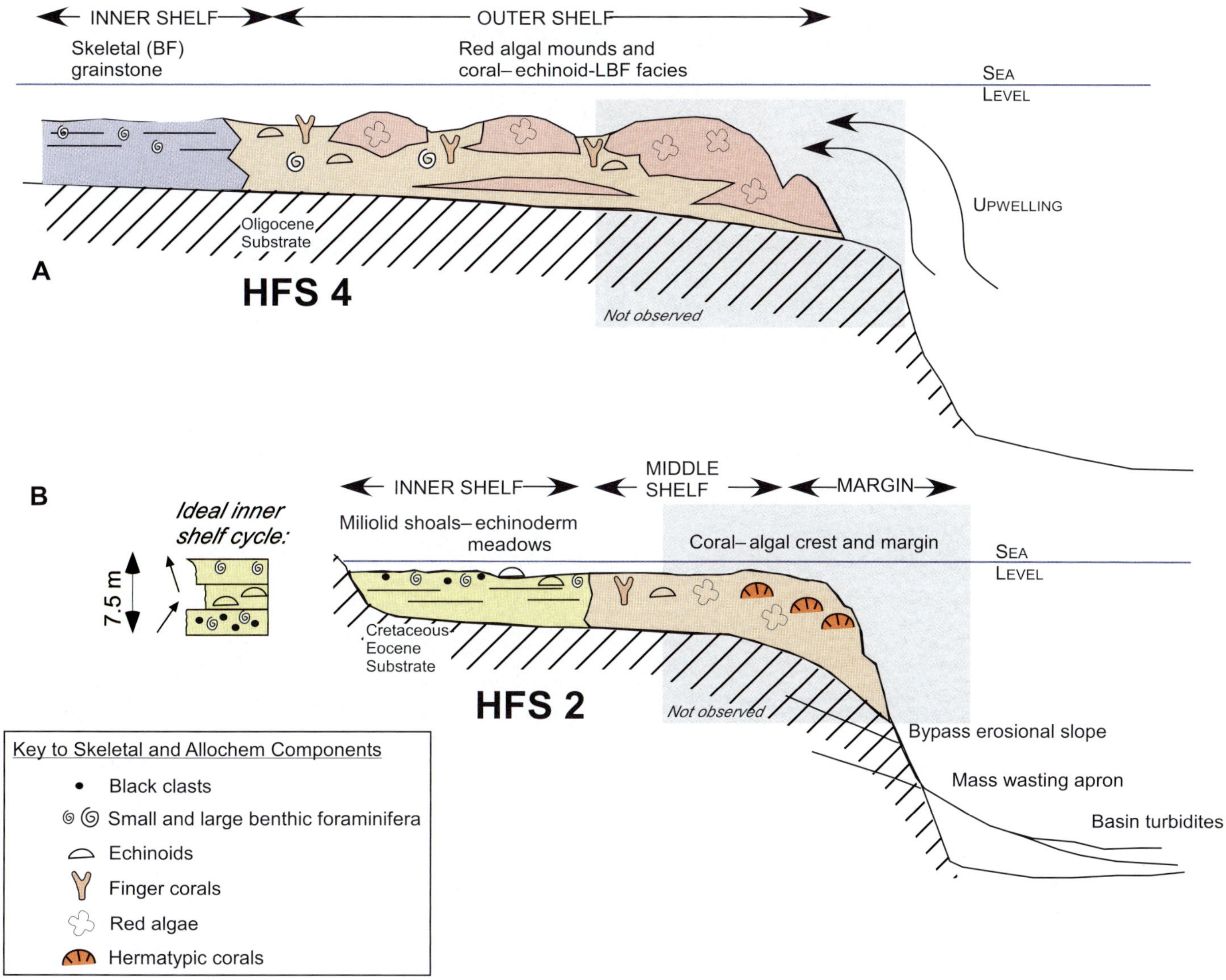

FIG. 10.—Two depositional models for the Oligocene Costa Blanca strata. **A)** Interpreted morphology and arrangement of facies for HFS 4. The model shows an outer shelf dominated by red-algal mounds, but without the relief that would restrict the inner shelf. Upwelling and high-energy waves sweep the platform. **B)** Interpreted platform morphology and facies arrangement for the high-frequency sequences (HFS) 1, 2, and possibly 3. The model includes a coral–algal crest and margin with relief that restricts the inner shelf. The inner shelf would be frequently exposed and show shallowing-upward cyclicity such as that shown to the left of the model: an intraclast conglomerate overlain by a skeletal wackestone, topped by a fine-grained skeletal grainstone. Upwelling and high-energy waves sweep the platform. The gray boxes in both Parts A and B indicate areas that are not directly observed in outcrop.

siliciclastic sediments. Potentially the gap between the Oligocene and Miocene strata across that boundary is on the order of 2–4 million years. The exposure may be a combination of lower global sea levels in the late Chattian (Miller et al., 2005) and a decrease in subsidence prior to another burst of subsidence in the Aquitanian due to compression from advancing thrust sheets.

Analogous Platforms

Subsurface studies of Cenozoic carbonate platforms offshore Southeast Asia show many similarities to the Costa Blanca platform of this study. Seismic and outcrop data from the Rembang Zone of the East Java Basin, for example, show an Oligocene–Miocene backstepping platform composed of red algae, large benthic forams, corals, and echinoids (Sharaf et al., 2005), similar to that described in this study (Fig. 11). The onshore Java example illustrates a platform that grew on a faulted block with a steep margin and shed sediment into the basin. The platform shows an early stage of pseudohorizontally bedded reflectors, which vertically show a backstepped and mounded platform. This seismic geometry has a strong similarity with the Costa Blanca exposures (Fig. 11). Backstepping Cenozoic carbonate platforms are very common in Southeast Asia. These patterns are generally related to increasing subsidence and/or eustatic rise, with platform demise due to drowning or to exposure and later drowning (Bachtel et al., 2004; Vahrenkamp et al., 2004; Sharaf et al., 2005). The biotic content is generally similar, as is the complexity of the

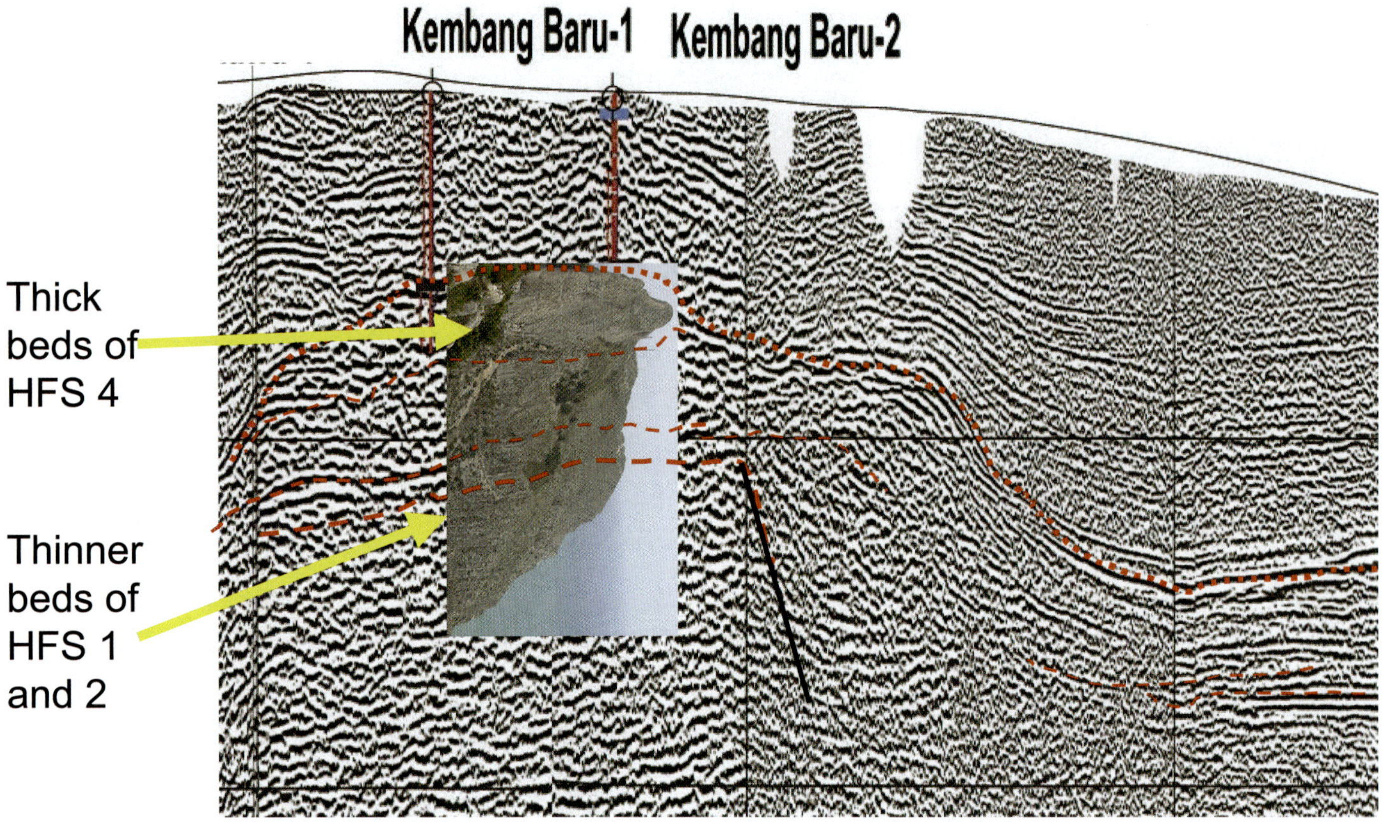

FIG. 11.—Comparison of the Oligocene Costa Blanca strata of this study (photo of outcrop at Toix) to the seismic data of Oligo-Miocene strata of the Rembang Zone of the East Java Basin. Red algae are a dominant skeletal grain in both successions. Note the thinner beds in the lower part of the succession for both examples, and the thicker, backstepped beds in the upper part of the successions.

tectonic settings, which also appears to exert a major control on the morphology of the platform. The exposed Oligocene platform should thus serve as a good analog for the subsurface Cenozoic platforms of Southeast Asia.

CONCLUSIONS

1. The Oligocene shallow-water carbonate succession of the Costa Blanca is unique with respect to other Tethyan Cenozoic platforms. Although the Tethys supported platforms of variable morphologies in a variety of tectonic settings, a platform similar to that of the Costa Blanca has yet to be found. The Costa Blanca Rupelian platform was an open shelf with a steep margin that formed in a post-compressional setting which exhibited increased subsidence, and possibly extensional faulting, with time.

2. The four high-frequency sequences identified in this succession show an overall deepening-upward, and possibly backstepping, trend. Increasing subsidence associated with the opening of the Valencia Trough in this region would account for this trend.

3. The platforms also evolved in terms of morphology and biotic assemblage. The platforms appeared to have changed from those with restrictive, high-relief margins to those with low-relief margins. Platform biota also shows change from a transitional to photozoan assemblage in the lower sequences to a heterozoan community in the upper sequences. Such changes are associated with deeper, possibly cooler, and more nutrient-rich waters. Tectonic activity thus appears to be the main factor controlling platform morphology, facies belts, cyclicity, and perhaps even biotic assemblages of the Oligocene Costa Blanca.

4. The study of this platform is significant for many reasons. It contributes to the continued evaluation of the controls on morphology and facies belts of carbonate platforms. Understanding the controls can be useful for the study and prediction of the characteristics (including morphology and porosity) of other subsurface carbonates, particularly those of the same tectonic setting. In particular, this study can be used as an analog for the subsurface Cenozoic carbonates of Southeast Asia.

ACKNOWLEDGMENTS

We would like to thank the University of Wisconsin–Madison Department of Geology and Geophysics for its financial support and the use of its facilities. Other financial support came from American Association of Petroleum Geologists Grants-in-Aid of Research, ARCO Field Research Grant, the Vilas Foreign Travel Award, and the Van Hise Fellowship. We would also like to thank Toni Barnolas for the maps and assistance in Spain and Carrie Gilliam for her field assistance. Thanks to Conxita Taberner, Jeff Lukasik, and an anonymous reviewer for their thorough and insightful comments and suggestions for this manuscript.

REFERENCES

Azema, J., Foucault, A., Fourcade, E., García-Hernández, M., Gonzalez-Donoso, J.M., Linares, A., López-Garrido, A.C., Rivas, P., and Vera, J.A., 1979, Las Microfacies de Jurásico y Cretácio de las Zonas Externas de las Cordilleras Beticas: University of Granada, 83 p.

Bachtel, S., Kissling, R.D., Martono, D., Rahardjanto, S.P., Dunn, P.A., and Macdonald, B.A., 2004, Seismic stratigraphic evolution of the Miocene–Pliocene Segitiga Platform, East Natuna Sea, Indonesia: The origin, growth and demise of an isolated carbonate platform, *in* Eberli, G.P., Masaferro, J.L., and Sarg, J.F., eds., Seismic Imaging of Carbonate Reservoirs and Systems: American Association of Petroleum Geologists, Memoir 81, p. 309–328.

Baena Perez, J., and Jerez Mir, L., 1982, Síntesis para un Ensayo Paleogeográfico entre la Meseta y la Zona Bética (sensu stricto): Madrid, Instituto Geológico y Minero de España, 256 p.

Berggren, W.A., Kent, D.V., Swisher, C.C., III, and Aubry, M.P., 1995, A revised Cenozoic geochronology and chronostratigraphy, *in* Berggren, W.A., Kent, D.V., Aubry, M.P., and Hardenbol, J., eds., Geochronology, Time Scales and Global Stratigraphic Correlation: SEPM, Special Publication 54, p. 129–212.

Bosellini, A., and Ginsburg, R.N., 1971, Form and internal structure of Recent algal nodules (rhodolites) from Bermuda: Journal of Geology, v. 79, p. 669–682.

Bosence, D.W.J., 1983, The coralline algae from the Miocene of Malta: Palaeontology, v. 26, p. 147–173.

Bosence, D.W.J., 1991, Chapter 5: Coralline Algae: Mineralization, taxonomy, and palaeoecology, *in* Riding, R., ed., Calcareous Algae and Stromatolites: Heidelberg, Springer, p. 98–113.

Buxton, M.W.N., and Pedley, H.M., 1989, Short Paper: A standardized model for Tethyan Tertiary carbonate ramps: Geological Society of London, Journal, v. 146, p. 746–748.

Cande, S.C., and Kent, D.V., 1992, A new geomagnetic polarity time scale for the Late Cretaceous and Cenozoic: Journal of Geophysical Research, v. 97, p. 13,917–13,951.

Cande, S.C., and Kent, D.V., 1995, Revised calibration of the geomagnetic polarity timescale for the Late Cretaceous and Cenozoic: Journal of Geophysical Research, v. 100, p. 6093–6095.

Carozzi, A.V., 1989, Carbonate Rock Depositional Models: A Microfacies Approach: Englewood Cliffs, New Jersey, Prentice Hall, 604 p.

Caus, E., and Serra-Kiel, J., 1992, Macroforaminífers: Estructura, Paleoecologia I Biostratigrafia: Servei Geològic de Catalunya, Publícacío, Monografies núm. 2, 211 p.

Chaproniere, G.H.C., 1975, Paleoecology of Oligo-Miocene larger Foraminiferida, Australia: Alcheringa, v. 1, p. 37–58.

De Ruig, M.J., 1992, Tectono-sedimentary Evolution of the Prebetic Fold Belt of Alicante (SE Spain): Amsterdam, Vrije Universiteit, 207 p.

Enos, P., 1983, Shelf environment, *in* Scholle, P.A., Bebout, D.G., and Moore, C.H., eds., Carbonate Depositional Environments: American Association of Petroleum Geologists, Memoir 33, p. 267–295.

Esteban, M., 1996, An overview of Miocene reefs from Mediterranean areas: General trends and facies models, *in* Franseen, E.K., Esteban, M., Ward, W.C., and Rouchy, J.M., eds., Models for Carbonate Stratigraphy from Miocene Reef Complexes of Mediterranean Regions: SEPM, Concepts in Sedimentology and Paleontology, no. 5, p. 3–53.

Everts, A.J.W., 1991, Interpreting compositional variations of calciturbidites in relation to platform-stratigraphy: an example from the Paleogene of SE Spain: Sedimentary Geology, v. 71, p. 231–242.

Everts, A.J.W., Schlager, W., and Reijmer, J.J.G., 1999, Carbonate platform-to-basin correlation by means of grain-composition logs: an example from the Vercors (Cretaceous, SE France): Sedimentology, v. 46,. 261–278.

Franseen, E.K., and Goldstein, R.H., 1996, Paleoslope, sea-level and climate controls on Upper Miocene platform evolution, Las Negras area, southeastern Spain, *in*, Franseen, E.K., Esteban, M., Ward, W.C., and Rouchy, J.M., eds., Models for Carbonate Stratigraphy from Miocene Reef Complexes of Mediterranean Regions: SEPM, Concepts in Sedimentology and Paleontology, no. 5, p. 159–176.

Frost, S.H., 1981, Oligocene reef coral biofacies of the Vicentin, northeast Italy, *in* Toomey, D.F., ed., European Fossil Reef Models: SEPM, Special Publication 30, p. 483–539.

Fulthorpe, C.S., and Schlanger, S.O., 1989, Paleo-oceanographic and tectonic settings of early Miocene reefs and associated carbonates of offshore Southeast Asia: American Association of Petroleum Geologists, Bulletin, v. 73, p. 729–756.

Gamble, H.J., and Moseley, F., 1975, Notes on the Pre-Betic geology of the Costa Blanca, SE Spain: London, Geologists' Association Handbook, 28 p. (unpublished).

García-Hernández, M., López-Garrido, A.C., Rivas, P., Sanz de Galdeano, C., and Vera, J.A., 1980, Mesozoic paleogeographic evolution of the External Zones of the Betic Cordillera: Geologie en Mijnbouw, v. 59, p. 155–168.

Geel, T., 1995, Oligocene to early Miocene tectono-sedimentary history of the Alicante region (SE Spain): implications for Western Mediterranean evolution: Basin Research, v. 7, p. 313–336.

Geel, T., Roep, T.B., Ten Kate, W., and Smit, J., 1992, Early–Middle Miocene stratigraphic turning points in the Alicante region (southeastern Spain): Sedimentary Geology, v. 75, p. 223–239.

Halfar, J., Godines-Orta, L., Mutti, M., Valdez-Holguín, J., and Borges, J.M., 2004, Nutrient and temperature controls on modern carbonate production: an example from the Gulf of California, Mexico: Geology, v. 32, p. 213–216.

Hallock, P., and Glenn, E.C., 1986, Larger foraminifera: a tool for paleoenvironmental analysis of Cenozoic carbonate depositional facies: Palaios, v. 1, p. 55–64.

Hardenbol, J., Thierry, J., Farley, M.B., De Graciansky, P., and Vail, P.R., 1998, Mesozoic and Cenozoic sequence chronostratigraphic framework of European basins, *in* De Graciansky, P., Hardenbol, J., Jacquin, T., and Vail, P.R., eds., Mesozoic and Cenozoic Sequence Stratigraphy of European Basins: SEPM, Special Publication 60, p. 3–13.

James, N.P., 1979, Reefs, *in* Walker, R.G., ed., Facies Models: Geological Association of Canada, Geoscience Canada Reprint Series 1, p. 121–133.

James, N.P., 1997, The cool-water carbonate depositional realm, *in* James, N.P., and Clarke, J.A.D., eds., Cool-Water Carbonates: SEPM, Special Publication 56, p. 1–20.

Johnson, J.H., 1961, Limestone-Building Algae and Algal Limestones: Boulder, Colorado, Johnson Publishing Company, 297 p.

Knoerich, A.C., and Mutti, M., 2006, Missing aragonitic biota and the diagenetic evolution of heterozoan carbonates: A case study from the Oligo-Miocene of the Central Mediterranean: Journal of Sedimentary Research, v. 76, p. 871–888.

Maillard, A., and Mauffret, A., 1999, Crustal structure and riftogenesis of the Valencia Trough (north-western Mediterranean Sea): Basin Research, v. 11, p. 357–379.

McColgin, K., 2002, The Development of a Mixed Carbonate and Siliciclastic Slope System in a Tectonically Active Setting, Southeastern Spain: Master's Thesis, University of Wisconsin–Madison, 194 p.

Megías, A.G., Leret, G., Martínez del Olmo, W., and Soler, R., 1983, La sedimentación Neógena en las Béticas: Análisis tectosedimentario: Mediterránea Serie de Estudios Geológicos, v. 1, p. 83–103.

Miller, K.G., Mountain, G.S., Katz, M.E., Sugarman, P.J., Cramer, B.S., Christie-Blick, N., Pekar, S.F., Kominz, M.A., Browning, J.V., and Wright, J.D., 2005, The Phanerozoic record of global sea-level change: Science, v. 310, p. 1293–1298.

Moseley, F., 1990, A Geological Field Guide to the Costa Blanca, Spain: London, The Geologists' Association, 79 p.

Nebelsick, J.H., Rasser, M.W., and Bassi, D., 2005, Facies dynamics in Eocene to Oligocene circumalpine carbonates: Facies, v. 51, p. 197–216.

OTT D'ESTEVOU, P., MONTENAT, C., LADURE, F., AND PIERSON D'AUTREY, L., 1988, Evolution tectonosédimentaire du prébétique oriental (Espagne) su Miocéne: Academie des Sciences (Paris), Comptes Rendes, v. 307 (II), p. 789–796.

PEDLEY, M., 1998, A review of sediment distributions and processes in Oligo-Miocene ramps of southern Italy and Malta (Mediterranean divide), *in* Wright, V.P., and Burchette, T.P., eds., Carbonate Ramps: Geological Society of London, Special Publication 149, p. 163–179.

PIERSON D'AUTREY, L., 1987, Sédimentation et structuration synsédimentaire dans le bassin Neogene d'Alcoy (Cordilléres bétiques externes orientales-Espagne): Ph.D. thesis, University of Paris XI–Orsay, Paris, 314 p.

SHARAF, E., SIMO, J.A., CARROLL, A.R., AND SHIELDS, M., 2005, Stratigraphic evolution of Oligocene–Miocene carbonates and siliciclastics, East Java basin, Indonesia: American Association of Petroleum Geologists, Bulletin, v. 89, p 799–819.

SPRINKLE, J., AND KIER, P.M., 1987, Chapter 18: Phylum Echinodermata, *in* Boardman, R.S., Cheetham, A.H., and Rowell, A.J., eds., Fossil Invertebrates: Cambridge, Massachusetts, Blackwell Scientific Publications, p. 550–611.

STOKLOSA, M.L., 2002, Evaluation of Controls on Carbonate Platfrom Morphology, Facies, and Diagenetic Variability: An Oligocene Example from Southeast Spain: Ph.D. Thesis, University of Wisconsin–Madison, 284 p.

STRASSER, A., 1984, Black-pebble occurrence and genesis in Holocene carbonate sediments (Florida Keys, Bahamas, and Tunisia): Journal of Sedimentary Petrology, v. 54, p. 1097–1109.

TUCKER, M.E., AND WRIGHT, V.P., 1990, Carbonate Sedimentology: Oxford, U.K., Blackwell Scientific Publications, 482 p.

VAHRENKAMP, V.C., DAVID, F., DUIJNDAM, P., NEWALL, M., AND CREVELLO, P., 2004, Growth architecture, faulting and karstification of a Middle Miocene carbonate platform, Luconia Province Offshore Sarawak, Malaysia, *in* Eberli, G.P., Masaferro, J.L., and Sarg, J.F., eds., Seismic Imaging of Carbonate Reservoirs and Systems: American Association of Petroleum Geologists, Memoir 81, p. 329–350.

WILSON, J.L., 1975, Carbonate Facies in Geologic History: Berlin, Springer-Verlag, 471 p.

WRAY, J.L., 1977, Calcareous Algae: Amsterdam, Elsevier, 185 p.

Environmental Controls
(Oceanography, Climate, Trophic Resources, Temperature)

ARCHITECTURE OF CARBONATE PLATFORMS: A RESPONSE TO HYDRODYNAMICS AND EVOLVING ECOLOGY

LUIS POMAR
Univeristat de les Illes Balears ctra., Valldemossa km 7.5, Palma de Mallorca 07122, Spain
e-mail: luis.pomar@uib.es

AND

CHRISTOPHER G. St. C. KENDALL
Department of Geological Sciences, University of South Carolina, Columbia, South Carolina 29208, U.S.A.
e-mail: kendall@geol.sc.edu

ABSTRACT: The spectrum of carbonate-platform types, their heterogeneities, and their architecture is complex. Each platform succession has a distinctive and unique character that is a response to the particular geotectonic context and the physical, chemical, and biological conditions to that specific Phanerozoic window. Each succession has a distinct depositional profile, facies-belt distribution, and platform architecture, and it is expressed by the order of the basic accretional units and their stacking patterns. Critical differences between platform types are often the result of differences in their ecological accommodation. End members include (A) low-relief carbonate ramps that match a shelf equilibrium profile and are composed of either loose, fine-grained sediments produced in shallow, well-illuminated areas but shed downdip, or sediment produced and accumulated (sometimes as a distal bulge) in the deeper part of the depositional profile (poor-light or no-light zones), (B) open-shelf platforms involving large-skeleton metazoans with a marked to moderate capacity to build a platform margin above the shelf equilibrium profile, (C) rigid rimmed platforms with biotic components capable of accumulating to sea level with a maximum ecological accommodation, and (D) platforms with steep, massive and thick marginal slopes characteristic of many Paleozoic and some Mesozoic settings.

Interpretation of carbonate platforms and prediction of their facies heterogeneities involves analyzing and integrating geometrically related data. Analysis involves iterative and successive backstripping of sediment accumulation from youngest to oldest. This is reassembled to determine the genetic character of the carbonate sequences, cycles, parasequences, and/or beds as products of changes in physical and ecological accommodation. This reassembly considers the evolution of the biota involved, and the resulting changes in ecological requirements, the hydrodynamic setting, the physical accommodation, and the ecological accommodation (capacity of building up above a certain hydrodynamic energy level). The limits to this analytical strategy are tied to the knowledge of the ecology of ancient biota, while its advantage is that it formulates new questions that lead to more realistic interpretations and enhanced predictions of lithofacies heterogeneities.

INTRODUCTION

This paper addresses Mesozoic and Cenozoic carbonate platforms in terms of the spectrum of their facies heterogeneity, and external and internal architecture. It demonstrates how the variable character of platforms is related to differences in accommodation (Jervey, 1988; Vail et al. 1991; Posamentier and James, 1993) tied to eustasy and total subsidence, and the changing character of carbonate production. This biological production has three broad environmental controls (Fig 1):

1) Ecological requirements that change as the biotas evolve.

2) Global temperature gradients, ocean and atmosphere chemistry.

3) Regional hydrodynamic setting, nutrient availability, salinity, and substrate.

These controls on carbonate systems can be linked to the principles of sequence stratigraphy and used to predict the lateral and vertical extent of carbonate facies geometry. Currently, sequence stratigraphy can be used to successfully interpret and predict facies architecture of siliciclastic depositional sequences generated by the combined products of changing accommodation and changing sediment flux. The framework of sequence stratigraphic surfaces is used to determine the positioning of the stratal patterns and facies architecture of the various clastic systems. These, in turn, are used in the interpretation of local depositional processes and their products through time.

The architectures of carbonate platforms can be subdivided as well, with one significant difference—stacked accretional units form a wider spectrum of geobody shapes than those of clastics (e.g., Read, 1982, 1985). For instance, carbonate geometries vary from homoclinal ramps to rimmed shelves, and their facies have a wide range of distribution and geometric heterogeneity. This variability is a direct response to the constantly evolving biological milieu of carbonate platforms. Thus, though hydrodynamic laws are unvarying through geological time and produce strong stratigraphic signals in both clastics and carbonates, carbonates often respond differently. This means that the kind of genetic analyses that works for clastics may fail to make accurate predictions for carbonates. This failure is often related to critical bio-physico-chemical factors that cause carbonate architectures to become less well defined the older the geologic section.

A major controlling factor of carbonates is the base level to which this sediment accumulates. Our central thesis is that for carbonate systems the base level for sediment accumulation, although influenced by hydrodynamics, is also strongly controlled by ecological factors (Pomar, 2001a, 2001b). Jervey's (1988) original definition of accommodation as "the space available for sediment accumulation" is extended in our paper by subdividing it into "physical" and "ecologic" accommodation in the sections below.

Controls on Carbonate Platform and Reef Development
SEPM Special Publication No. 89, Copyright © 2008
SEPM (Society for Sedimentary Geology), ISBN 978-1-56576-130-8, p. 187–216.

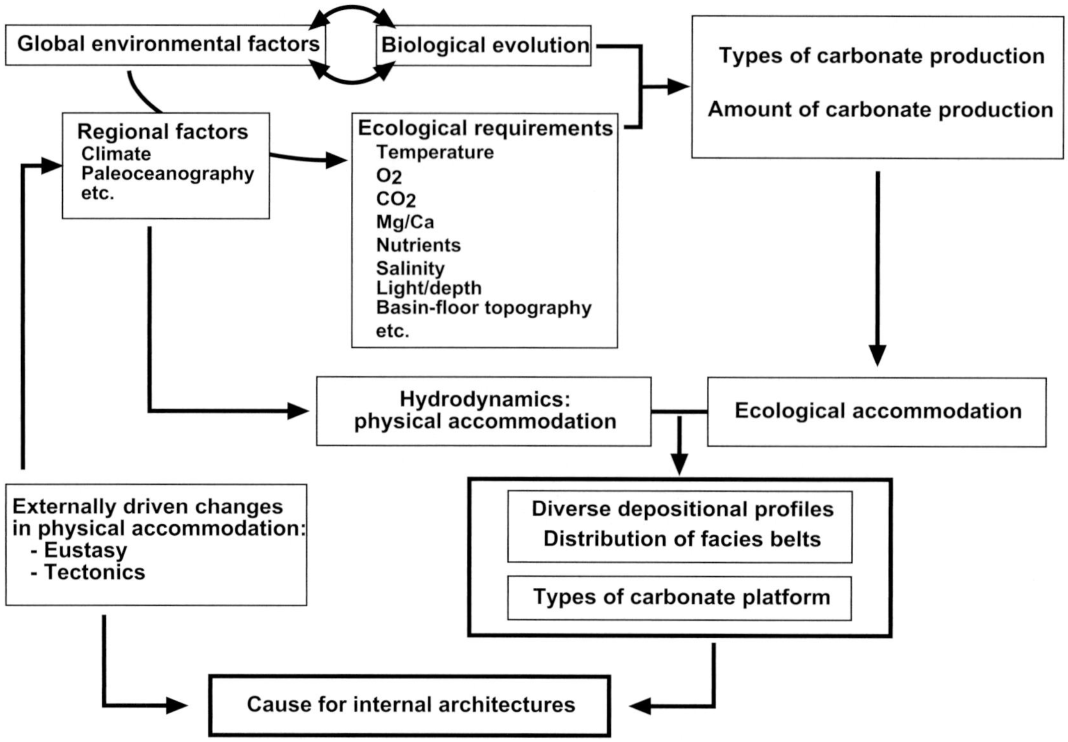

FIG. 1.—Flow diagram of the physical and biological relationships that control carbonate accumulation in carbonate platforms.

Physical Accommodation: the Base Level in Clastic Systems

We propose that the designation "physical accommodation" be used to describe the space available for sedimentary fill by a lithoclastic system in a setting dominated by the character of the local hydrodynamics. This definition is an outgrowth from the "shelf equilibrium profile" concept proposed by Swift and Thorne (1991). This has evolved from the recognition that in a marine system the shoreline is often the depositional setting that is exposed to the greatest wave energy, and thus forms a good datum as the shelf character changes seaward away from it. Shores are the conduits through which lithoclasts are commonly transported into the basin and redistributed in response to the hydrodynamic regime of the basin (Fig. 2). These sediments aggrade until they reach a depth exposed to waves and currents that can stir and move them down-shelf. These episodic processes progressively flatten the depositional profile. Grain size decreases basinward, reflecting the balance between input, redistribution, and accumulation of sediment (Johnson, 1919; Reineck and Singh, 1980; Allen, 1982; Swift et al., 1991), and defining the "shelf equilibrium profile" of Swift and Thorne (1991). This profile represents a conceptual surface of dynamic equilibrium that is close to the "marine profile of equilibrium" of Dietz (1963). For this he recognized that wave base was not a sharp surface but a generally fluctuating zone that bounds the marine profile of equilibrium. It matches the definition of Rich (1951) as a zone that coincides with the greatest depth to which the bottom is stirred by waves during storms.

Taking the above argument further, we propose that Jervey's (1988) accommodation be divided into physical and ecologic accommodation. The former, "physical accommodation", recognizes that this space is entirely dominated by the local hydrodynamics when associated with lithoclastic systems. In this case, accommodation comprises the space between sea floor and the "shelf equilibrium profile" of Swift and Thorne (1991). Often confused with sea level, the "base level" for sediment accumulation tends to match the "shelf equilibrium profile".

In some cases, the base-level datum can be seen as effectively equivalent to sea level when associated with overfed sedimentary systems like deltas. Conversely, the "dynamic equilibrium" between sediment input and hydrodynamic power means that deepening of the shelf equilibrium profile (delta retreat) results if there is a subsequent decrease in sediment supply.

At geological time scales, the dynamic equilibrium between the rate and character of sediment input and rate of sediment transport results in a depositional surface (Fig. 2). Sediments aggrade until they reach the equilibrium profile, and subsequently the rates of net deposition decrease when there is sufficient fluid power to sweep sediments down-shelf and offshore in response to intermittent storm and tidal currents. This results in a seaward shifting of the locus of deposition to form a slope dominated by gravity processes (Swift and Thorne, 1991). Thus the "shelf equilibrium profile" represents a base level for sediment to accumulate to, and defines the physical accommodation that is involved in the creation of the depositional shelf and slope in lithoclastic systems (Figs. 2, 3A). Eustasy and total sea-floor subsidence, as well as changes in hydrodynamic conditions, govern some of the changes in this accommodation space.

Ecological Accommodation in Carbonate Systems

In carbonate systems, controls on the space available for potential sediment accumulation are more complex than those for clastics. In addition to being affected by the hydrodynamics, accommodation space for carbonates is also strongly influenced by the capacity of organisms to produce and to accumulate sediments above the hydrodynamic thresholds associated with clastic systems. This is the ecological accommodation defined by Pomar (2001a, 2001b).

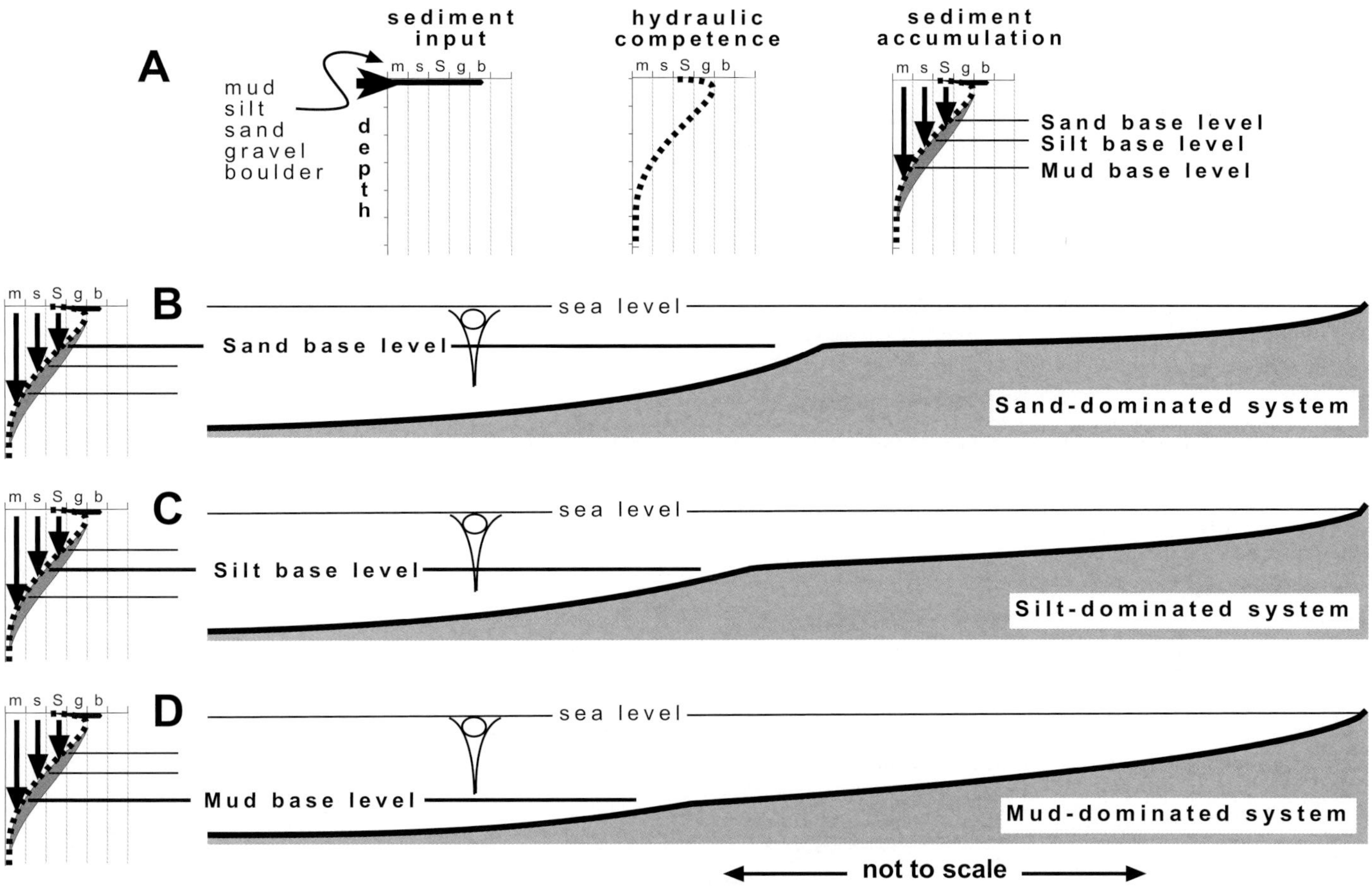

FIG. 2.—Using the concepts of Swift and Thorne (1991), the shelf equilibrium profiles displayed here are the products of a balance between sediment input and the effects of fluid motion. **A)** Sediment input: sediment is introduced at the seashore and ranges in size from mud to gravel. Hydraulic competence: the capacity of water (here, for the purposes of simplification, a wave-dominated system) to move sediments in terms of particle size and expressed as the diameter of the largest particle transported. "Gravel" is used to refer granule-, pebble-, and cobble-size particles, and it excludes boulders. Sediment accumulation: in a shelf dispersal system, sediments move downshelf and offshore in response to wave motion and intermittent storm and tidal currents. Grain size decreases in a seaward direction as the water deepens and bottom water stresses weaken. Textural gradient and facies differentiation originate from these large-scale diffusion mechanisms, which depend on the episodic nature of transport with a resulting seaward shift of the locus of deposition to form a slope dominated by gravity processes. **B–D)** shelf-to-basin profiles of settings dominated by sand-, silt-, and mud-size sediment.

In carbonate systems, the diversity of the component grains and their depositional textures define the lithofacies. They are the consequence of a wide variety of carbonate production processes and mechanisms that cause their redistribution within the basin. This production (sediment input) depends on the type, size, and efficiency of the carbonate depositional setting, which in turn depends on the area available to thriving carbonate-producing biota (basin-floor physiography), on intrabasinal conditions (nutrients, temperature, water energy, water transparency, salinity, oxygen, Ca^{2+} and CO_2 concentrations, Mg/Ca ratio, alkalinity, etc.), and biological evolution including ecological requirements (substrate, competitive displacement, etc.). Furthermore, sediment dispersal depends on the interaction between the physical characteristics of the different types of sediment being produced (grain size, bulk density as determined by porosity within the grain, e.g., intraskeletal porosity, etc.) and the hydraulic energy ambient to the production loci, and its modification by binding, trapping, baffling, and framework building (Ginsburg and Lowenstam, 1958) as well as by early cementation processes. Consequently, each different biotic system has a unique competence (ecological accommodation) for building above and below the hydrodynamic shelf equilibrium profile (Fig. 3).

Most carbonate (skeletal) production depends on primary organic production (directly for autotrophs and mixotrophs, indirectly for heterotrophs), and there are also major differences on depositional profiles (Fig. 3: B through G). The steep margins that characterize many carbonate platforms have commonly been related to the strong depth dependence of carbonate production, with higher production rates associated with shallow-water settings (Wilson, 1975; Kendall and Schlager, 1981; Schlager, 1981, 1992; Bosscher and Schlager, 1992; Pomar, 2001a, 2001b). In contrast, ramp geometries are seen as the products of decreased differentiation of depth-dependent production rates (Koerschner and Read, 1989; Wright and Faulkner, 1990; Burchette and Wright, 1992), depth-enhanced carbonate production (Brandano et al., 2001; Brandano and Corda, 2002; Pomar, 2001a, 2001b; Pomar et

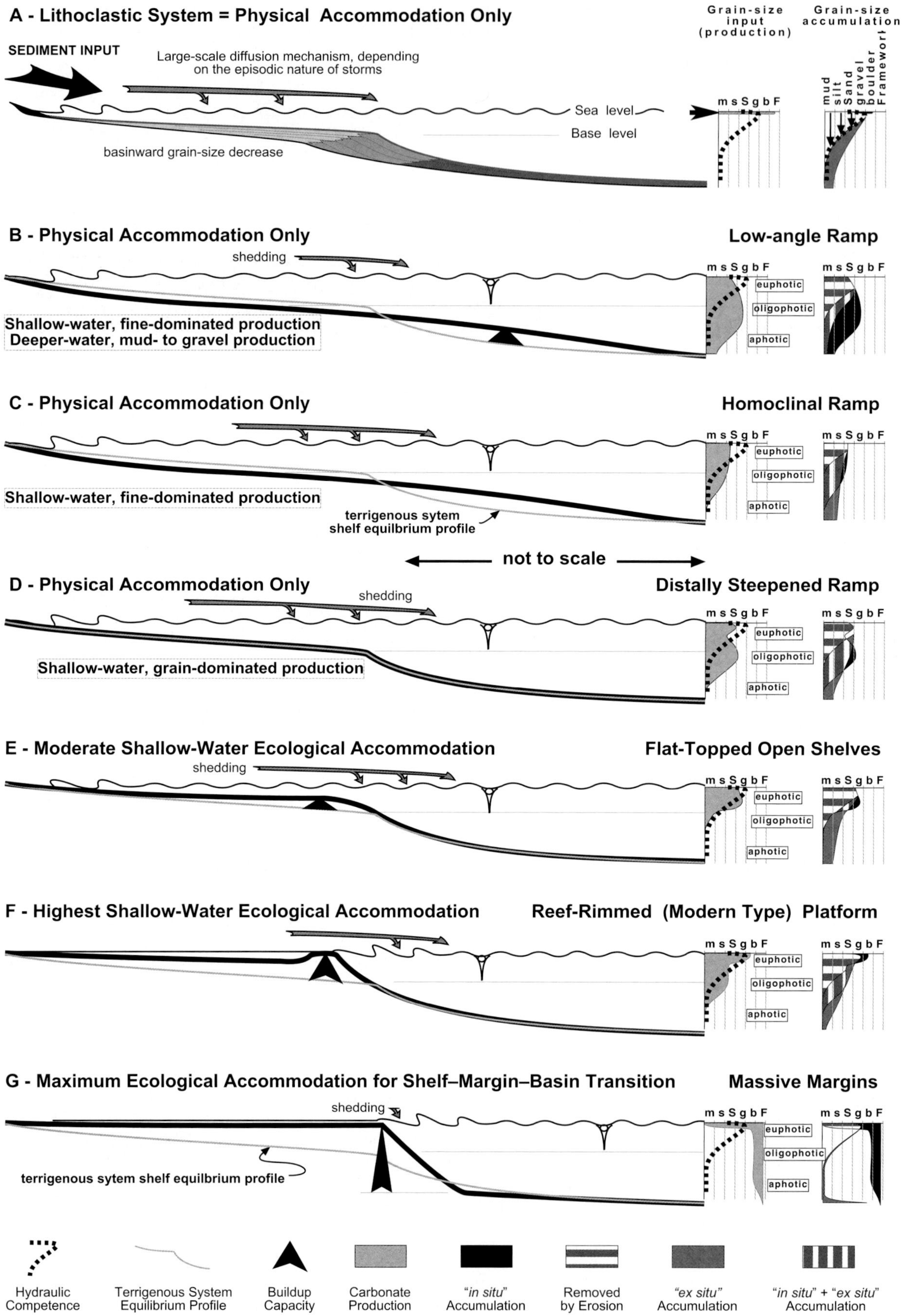

A - Lithoclastic System = Physical Accommodation Only
SEDIMENT INPUT
Large-scale diffusion mechanism, depending on the episodic nature of storms
Sea level
Base level
basinward grain-size decrease
Grain-size input (production)
Grain-size accumulation
mud silt Sand gravel boulder Framework
m s S g b F
B - Physical Accommodation Only
Low-angle Ramp
shedding
Shallow-water, fine-dominated production
Deeper-water, mud- to gravel production
euphotic
oligophotic
aphotic
C - Physical Accommodation Only
Homoclinal Ramp
Shallow-water, fine-dominated production
terrigenous sytem shelf equilbrium profile
not to scale
D - Physical Accommodation Only
Distally Steepened Ramp
shedding
Shallow-water, grain-dominated production
E - Moderate Shallow-Water Ecological Accommodation
Flat-Topped Open Shelves
shedding
F - Highest Shallow-Water Ecological Accommodation
Reef-Rimmed (Modern Type) Platform
G - Maximum Ecological Accommodation for Shelf–Margin–Basin Transition
Massive Margins
shedding
terrigenous sytem shelf equilbrium profile
Hydraulic Competence
Terrigenous System Equilibrium Profile
Buildup Capacity
Carbonate Production
"in situ" Accumulation
Removed by Erosion
"ex situ" Accumulation
"in situ" + "ex situ" Accumulation

al., 2004), and/or strong offshore transport (Aurell et al., 1995; Aurell et al., 1998; Bádenas et al., 2003; Wright and Burgess, 2005), as well as the lack of frame-building organisms capable of building steep platform margins (Beavington-Penney et al., 2005).

Interestingly, a more detailed analysis of the platform-margin–slope depositional profile, the distribution of skeletal components, and their associated textures enables the identification of the genetic processes involved. These, in turn, can be related to differences in the ecological accommodation of carbonate systems.

Carbonate systems where profound coarse-grained carbonate production predominates, either in the deeper part of the photic zone or in the aphotic zone, results in a low-angle ramp (Fig. 3B). Similar depositional profiles, although with different lithofacies organizations, result from systems dominated by production of mud-size sediment in the shallow-water zone that have a minimum capacity to infill the accommodation of a shallow-water shelf. Storms erode and suspend this fine material, so it is easily shed downshelf (Fig. 2). The angle of repose is very low, and the resulting depositional profile of the mud-dominated system is a homoclinal ramp (Fig. 3 C). Here, coarse-grained components often accumulate in shallow-water settings as beach or shoal deposits, reflecting the hydraulic regime and energy-dissipation processes in the wave-friction-dominated zone. In both cases, subtle topographic bulges result from the increased sediment accumulation at particular depths.

Physical accommodation that matches terrigenous systems prevails when loose-grain production dominates in the shallow-water, wave-agitated zone (Fig. 3 D). Skeletal components may accumulate only up to the shelf equilibrium profile, but then are swept seaward from this friction-dominated zone (Fig. 4A). The slope angle of this ramp depends on the sediment fabric (angle of repose) and the storm and current intensity and frequency (reworking processes). As with the red algae of many Neogene platform examples, gravel-size bioclasts may be produced in the deeper part of the photic zone. These coarse-grained bioclasts typically accumulate *in situ*, and are only episodically moved by currents or during exceptional storms. In these circumstances, the main locus of deposition is located down-shelf, at a particular water depth, and a slope forms as a result of an increased rate of sediment accumulation. The resulting depositional profile is a distally steepened ramp (Fig. 3D).

A marked to moderate capacity to build above the shelf equilibrium profile can be seen in open shelves dominated by soft-substrate platform dwellers, which, in some Mesozoic platforms, include rudists, corals, and stromatoporoids (Fig. 3E). If these large skeletons build up the elevation of the margin, they may also act as a hydrodynamic baffle along the margin of the outer shelf, often dissipating the hydraulic energy of this zone (Fig. 4B). Depending on the hydraulic regime, fine- to coarse-sand-size bioclasts may be shed from the shallow-water inner platform and transported onto a depositional slope. This then builds out, with the gravel-size skeletons mostly remaining on the platform top.

An enhanced capacity to build above the hydrodynamic shelf equilibrium profile is expressed by modern reef-rimmed platforms in which large corals and encrusting algae build a rigid, wave-resistant, framework up to sea level (Fig. 3F). The resulting rim may act as a "wave-energy barrier" that forms along the shelf margin and restricts circulation in the back-reef area (Fig. 4C). In the lagoon, which is protected from open-sea waves, the base level for sediment accumulation is much shallower than in open areas.

Paleozoic platforms, and some Mesozoic platforms, are interpreted to be related to depth-extended carbonate production (Fig. 3G). Many have characteristically steep, massive and thick marginal slopes that maintain a constant thickness and depositional dip. The origin of textures, including microbial boundstone, cementstone, and automicrite, are difficult to ascertain but are interpreted to be the result of photo-independent carbonate production. The scarcity of talus deposits in such systems suggests low rates of sediment export from the platform top. If this photo-independent boundstone factory nucleates on local relief on the sea floor (hard-bottom points tied to cementation or allochthonous blocks) of a ramp, mud mounds may form.

Thus, as has been demonstrated, in carbonate systems the base level to which sediment accumulates is influenced by hydrodynamics but may be controlled more strongly by ecological factors. So accommodation in carbonate systems (the space available for potential sediment accumulation) has two interrelated components: physical accommodation (hydrodynamic conditions) and ecological accommodation (competence to build up) (Pomar, 2001a, 2001b).

Factors Influencing Platform Architecture and Facies Distribution

Changes in physical accommodation, induced by high-frequency changes in relative sea level (para-cycles and cycles) determine the character of the basic building blocks (parasequences and basic sequences) and stacking of platform architecture. However, depending on ecological accommodation space, the bedding and internal facies architecture of sequences differ. Heterogeneities driven by high-frequency sea-level cycles are maximized in rimmed shelves dominated by shallow-water, euphotic, framework-dominated biota and minimized in low-angle ramps generated from aphotic and oligophotic carbonate production.

In addition to the impact of ecological accommodation and relative sea level, other processes contribute to the generation of

←

FIG. 3 (opposite page).—The depositional profiles that result from different types of carbonate production and dispersal. Sediment dispersal and accumulation depend on the interaction between "effective density" and grain size of clasts, production loci, and hydraulic competence. Finer-grained carbonate clasts produced in the shallow and agitated euphotic zone are "Removed by Erosion" and shed from the platform top to build out a talus ("*ex situ*" Accumulation). However, sediments coarser than erosional energy (hydraulic competence) build depositional relief above wave-base level ("*in situ*" Accumulation). Carbonate sediments produced deeper in the low-energy zones mostly remain in place ("*in situ*" Accumulation), and/or are mixed with sediments shed from shallow-water settings ("*in situ*" + "*ex situ*" Accumulation), building up a depositional relief until they reach the level at which, according to hydraulic competence, particles start to be moved and to accumulate while building outwards. Hydraulic competence: as in Figure 2, it is the capacity of water to move sediments according to their particle size as expressed as the diameter of the largest particle transported. It should be noted that with carbonates their intraskeletal porosity modifies the bulk grain density, so departures in textural differentiation (with respect to siliciclastics) have to be considered to be the result of their "effective" grain density. "Terrigenous Systems Equilibrium Profile", refers to the "theoretical" equilibrium profile of a lithoclastic system, as in Figure 2. "Build-Up Capacity": capacity of carbonate system to accumulate sediment above the terrigenous system equilibrium profile. "Carbonate Production": "effective" grain size of carbonate sediment production.

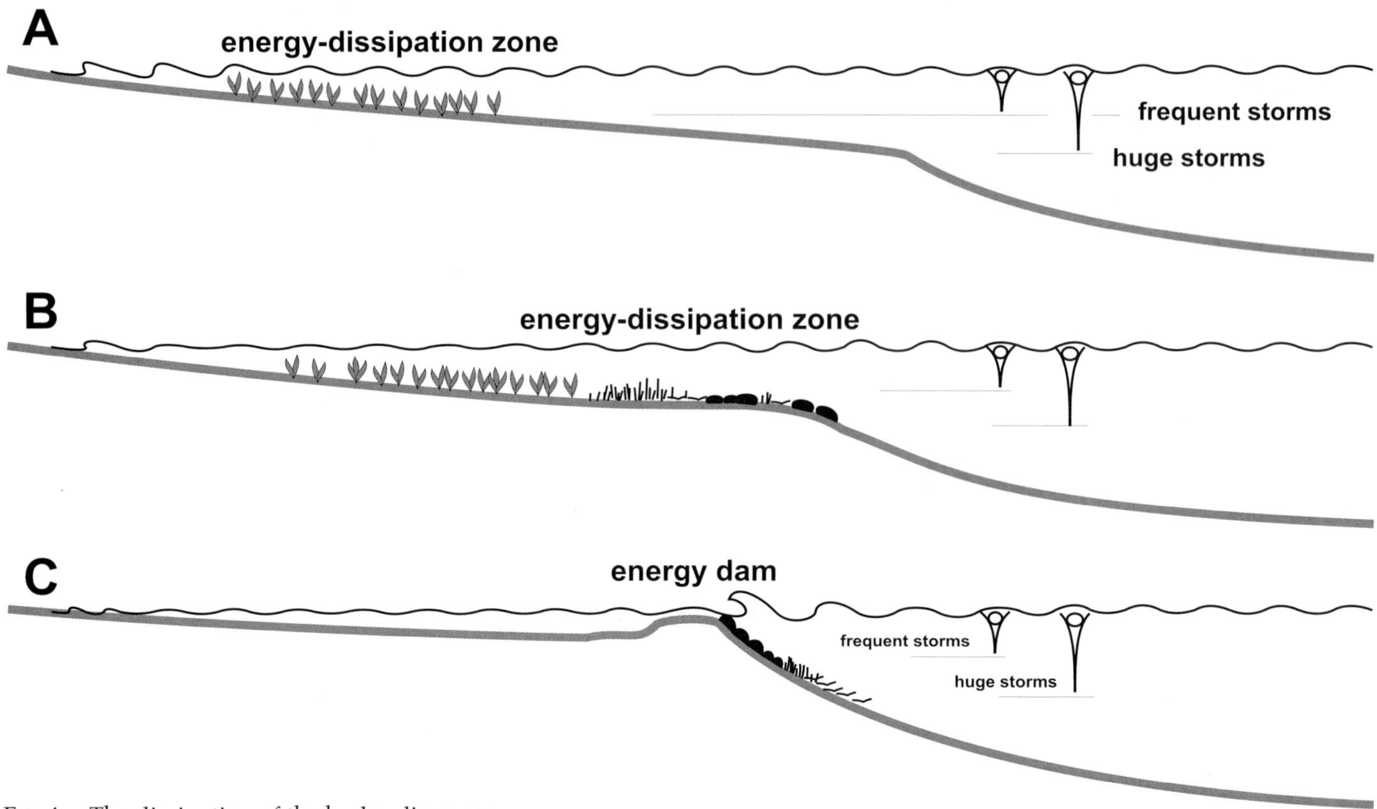

FIG. 4.—The dissipation of the hydraulic energy.

complexities in platform architecture and the distribution of their heterogeneities. Often several different biotic assemblages, or "carbonate factories", may coexist or alternate through time, reflecting changes in the conditions required for each type of carbonate biota to thrive (basin-floor physiography, substrate, nutrients, temperature, water energy, water transparency, salinity, oxygen, Ca^{2+} and CO_2 concentrations, Mg/Ca ratio, alkalinity, etc.). These changes may occur in phase with, out of phase with, or independent of sea-level change.

Moreover, carbonate accumulation may vary, as in lithoclastic systems, in tandem with a change in relative sea level (basinward/landward shifts) or they may turn on or off, thus resulting in flooding or drowning, etc. In modern reef-rimmed platforms, the lagoonal factory has a nonlinear response to sea-level change: turning on during sea-level rise and turning off during sea-level fall. Low-angle ramps tend to have a linear response (migration of facies belts) to sea-level changes (Calvet and Tucker, 1988; Calvet et al. 1990). Consequently, changes in oceanographic or climatic conditions (Lukasik et al., 2000; Lukasik and James, 2006) may cause larger effects on stratal patterns and facies architecture than sea-level changes by affecting the carbonate-producing biota.

TYPES OF CARBONATE PLATFORMS

The introduction to this paper explains how accommodation space for carbonates is affected by the capacity of organisms to produce and accumulate sediments above the hydrodynamic thresholds that control clastic settings. Thus, as the ecological requirements of organisms evolve, so the dimensions of the available accommodation may change reflecting the changing ecological accommodation (Pomar, 2001a, 2001b). This is traced in the section below. This systematically describes how the character of a variety of carbonate platforms responds to different physical and ecological accommodations (Table 1). Since the uncertainty related to interpreting the character of carbonate production increases with more ancient biotas, the examples selected for this analysis are restricted to Mesozoic–Cenozoic rocks. The inventory of examples begins with carbonate platforms for which the ecological controls on accommodation are slim to non-existent and physical accommodation is the dominant control (Figs. 3A–D). It then moves on to other examples for which ecological controls on accommodation become progressively more significant (Figs. 3E–F).

Lithoclastic Systems— Physical Accommodation Only (Fig. 3A)

The sedimentary fill for this example, the Calcarenite di Gravina, matches the physical accommodation, or the base level, of clastic systems. As summarized below, the "hydrodynamic base level" alone controls the architecture of the basic accretional units (basic sequences and parasequences). These are visible in the outcrops of the Upper Pliocene–Lower Pleistocene Calcarenite di Gravina in the vicinity of Matera (southern Italy). Here a very well exposed coastal transition from shoreline to offshore settings for a coarse-grained clastic system occurs.

The region lies within a small horst located at the south–southwest margin of the Murge, where, as part of the Apulian foreland of southern Italy, regional subsidence occurred during the Middle Pliocene to Middle Pleistocene (Iannone and Pieri, 1982). The Calcarenite di Gravina is formed of a mixed bioclastic–lithoclastic sediment body. This accumulated through the Late Pliocene–Early Pleistocene on the margins of an archipelago

Table 1.—Carbonate-platform types.

Type of accommodation (1)	Location	Age	Dominant sediment source	Carbonate sediment production: Euphotic zone	Carbonate sediment production: Mesophotic zone	Carbonate sediment production: Oligophotic zone	Carbonate sediment production: Aphotic zone	Platform type	Platform top shedding	Tectonic setting	Sea-level period (2)
Physical only (A)	Matera, South Italy	Plio-Pleistocene	Lithoclasts derived from wave abrasion	Seagrass epiphytes	None	Some gravel (red algae, bryozoans, etc.)	None	Infralittoral prograding wedges	From the shoreface zone	Tectonic subsidence	Not well determined. High frequency
Physical only (B)	Latium–Abruzzi, Apennines, Italy	Middle Miocene	Oligophotic and aphotic loose grains	Skeletal grainstone–packstone	Packstone to red-algae rudstone–floatstone	Red algae rudstone–floatstone	Abundant bryozoan floatstone	Low-angle ramp	Not abundant	Tectonic subsidence	2nd-order. Barely 3rd order **(H)**
Physical only (C)	NE Iberia	Kimmeridgian Late Jurassic	Shallow-water lime mud and grains	Lime mud and oolites	Microbial-coral pinnacles	None	Not significant	Low-angle ramp	Important from inner ramp	Tectonic subsidence	20 ky and 100 ky
Physical only (C)	High Atlas of Morocco	Toarcian–Aalenian, Mid Jurassic	Shallow-water lime mud and grains	Lime mud and oolites	Some microbial-coral pinnacles	?	Not significant	Low-angle ramp	Important from inner ramp	Tectonic subsidence	4th and 5th order cycles **(H)**
Physical only (D)	Menorca, Spain	Late Miocene	Shallow-water and deeper-water-depth loose grains	Seagrass epiphytic skeletal packstone	Low	Red algae rudstone/ floatstone	Not significant	Distally steepened ramp	Abundant from inner ramp	Tectonic stability	0.5–1 My 3rd order **(AD)**
Moderate ecological (E)	South Central Pyrenees, Spain	Late Cretaceous	Large metazoan skeletons	Foraminifera-rich grainstone–packstone	Hippuritid pillarstone	Coral and rudist mixstone	None?	Flat-topped platform	Not important (except fines)	Thermal subsidence	Milankovitch Few 10's ky?
Physical only (D)	South Central Pyrenees, Spain	Late Cretaceous	Protozoan-dominated loose grains	Foraminifera-dominated grainstone	?	?	None	Infra-littoral prism	Important from shallow platform top	Thermal subsidence	Milankovitch Few 10's ky?
Moderate ecological (E)	Mut basin, South Turkey	Middle Miocene	Shallow- and intermediate-water-depth loose grains	Seagrass epiphytic skeletal pack-grainstone	Red algae, LBF and coralgal rudstones and boundstones	Red algae, LBF and coralgal rudstones and boundstones	Not abundant	Flat-topped platform		Tectonic subsidence	Few 10's ky? 5th order **(AD)**
Maximum shallow-water ecological (F)	Mallorca, Spain	Late Miocene	Shallow water framestone – Loose grains on the lagoon	Coral framestone	Red algae	Red algae	None	Shallow reef-rimmed platform	Important from lagoon	Tectonic stability	Few ky 6th order **(AD)** 7th order **(H)**
Depth-extended ecological (D)	Western Pyrenees, Spain	Middle Cretaceous	Massive slope down to 120 m Depth—extended boundstone	Orbitolinid packstones–grainstones, coral–rudist wackestones	Rudist-*Bacinella* boundstones, wack, pack, and floatstones	Automicrite-coral–sponge boundstone and cement	? Automicrite–coral–sponge boundstone	Slope-controlled platforms "slope-forms"	Little	Tectonic subsidence	Unclear on the margin
Depth-extended ecological (D)	Dolomites, Southern Alps	Triassic	Massive slope down to 800 m Depth—extended boundstone	Lofer cyclothems (in subsiding platforms)	Massive microbial-automicrite boundstone and cementstone	Massive microbial-automicrite boundstone and cementstone	Massive microbial–automicrite boundstone and cementstone	Slope-controlled platforms "slope-forms"	Little	Tectonic subsidence	Milankovitch cyclicity on platform top. Not clear on the margins

(1) Character of accommodation listed in Figure 3.

(2) Shortest period (highest frequency) of sea-level changes involved in the creation of the basic accretional units. Third-order sea-level cyclicity for the Cenozoic, based on the oxygen isotope curve of Abreu and Haddad (1998) has different magnitude than third-order cyclicity of Haq et al. (1987). In this column, **(AD)** refers to Abreu and Haddad cycle order, while **(H)** refers to the Haq et al. cycle order.

during the progressive drowning of the Apulian foreland. Outcrops in both the dip and strike directions line the walls of canyons incised during the Middle–Late Pleistocene uplift of the Apulian foreland (Ciaranfi et al., 1983; Doglioni et al., 1994; Doglioni et al., 1996).

Details gathered from photomosaics of the canyon walls were used to map bedding patterns, facies architecture, and bounding surfaces. This established the basic accretional units, and enabled their interpretation with respect to high-frequency changes of relative sea level (Pomar and Tropeano, 2001). These high-frequency changes are interpreted to have controlled the type of accretional units formed (parasequences or simple sequences). Basic accretional units stack to form a transgressive systems tract that reaches a thickness of 50 m. In addition to tectonic subsidence, seafloor morphology controlled the stacking patterns of the basic accretional units and the overall thickness of the transgressive systems tract. The physical relationships that control shoreline trajectories down depositional-dip match those described by Helland-Hansen and Martinsen (1996).

The basic accretional units of the Calcarenite di Gravina are prism-shaped sigmoids that extend along strike parallel to the paleo-coastline. In proximal settings where these abut against the rocky coast they are conglomeratic, but they become progressively sandier basinward (Fig. 5). Topsets consist of crudely bedded to well-stratified gently seaward-dipping strata representing beach-face deposits that pass basinward into gently inclined shoreface beds of moderately sorted pebbles to granules. Foresets consist of basinward-dipping clinobedded granules and calcarenites, with a complex sigmoidal-oblique configuration (Fig. 5). Bottomsets are composed of subhorizontal finer-grained and bioturbated calcarenite, interpreted to represent offshore sediment formed in the deeper part of the photic zone (Pomar and Tropeano, 2001).

The parasequences and simple sequences were defined according to bedding patterns, facies architecture, bounding surfaces, and the inferred relationship to higher-frequency sea-level changes (Pomar and Tropeano, 2001). The parasequences consist of sigmoidal prisms, containing three facies belts that succeed one another vertically and laterally (Fig. 5 A and B). Uppermost and shoreward are topsets that overlie and pass basinward into prograding slope clinobeds, which, in turn, interfinger and pass downdip into fine-grained offshore deposits. They are interpreted to have prograded seaward during stillstands of relative sea level. The upper boundary is interpreted to be a flooding (transgressive) surface produced during a subsequent rise of relative sea level (Pomar and Tropeano, 2001).

Simple sequences are bounded by unconformities and contain two systems tracts stacked with a basinward-stepping configuration (Fig. 5A and C). Progradational highstand systems tract match the previously described parasequence, with the exception of the upper parts of the clinobeds and the shoreface topsets that have been removed by erosion. A forced-regressive systems tract (*sensu* Hunt and Tucker, 1992, 1995) occurs basinward of the highstand systems tract. This overlies an internal downlap surface (Fig. 5C), which merges seaward into the main downlap surface. Simple sequences formed during high-frequency cycles of relative sea-level change (Fig. 5C). The forced-regressive systems tract prograded during the regression. However, it expresses no evidence of subaerial exposure above the main lithosome. The increase in the dips of clinobeds and the sharp

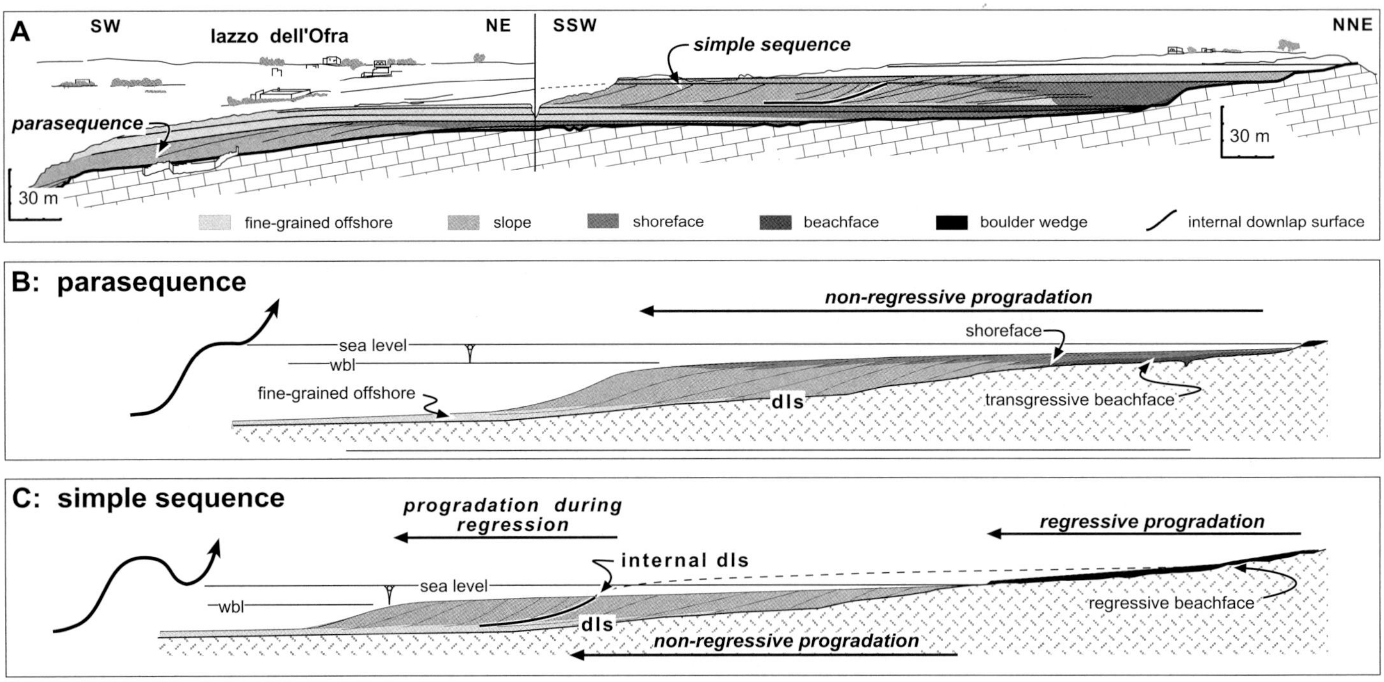

FIG. 5.—**A)** Cross section drawn from photomosaics of river-canyon cliffs and measured sections showing the stacking pattern of the accretional units with their internal facies architecture. Two types of accretional units are characteristic of the Calcarenite di Gravina. **B)** Parasequences consisting of prism-shaped units, bounded by flooding surfaces. They comprise three facies belts: subhorizontal shoreface topsets (inner belt) that overlie and pass into a middle belt of prograding slope lithofacies (foresets) that interfinger and downlap basinward onto fine-grained offshore deposits (outer belt). **C)** Simple sequences that also exhibit the same three facies belts as parasequences but are composed of two systems tracts. The transgressive lag deposits and the progradational highstand systems tract developed as in the parasequence, but the topsets are truncated. **dls** = downlap surface, **internal dls** = internal downlap surface, **wbl** = wave base level.

increase in grain size above the internal downlap surface reflect an increase in the seaward transportation of the sediments during a fall of relative sea level. This was induced by a lowering of the position of wave-base level and the concomitant erosion of the former shoreface and slope deposits of the highstand systems tract.

This example provides a means to better understand the mechanisms for transport of sand and gravel across the shelf. These sediments form elongate lithosomes parallel to and detached from the coeval coastline. They preserve both coarsening-upward sequences and seaward progradation (Fig. 3A).

The example illustrates the sedimentary expression of the "shelf equilibrium profile" in a system dominated by physical processes, where waves and currents stir the clasts, most of which are carried into the basin through the shoreline (Pomar and Tropeano, 2001). The shelf surface progressively flattens basinward, and is associated with a grain-size decrease that reflects the balance between input, redistribution, and accumulation of sediments (Johnson, 1919; Reineck and Singh, 1980; Allen, 1982; Swift et al., 1991) and defines the "shelf equilibrium profile" of Swift and Thorne (1991). Differences in the depositional profile and in the slope dip angle are related to grain size. Basic accretional units in coarse-grained dominated system, with low amount of fines, tend to have a flattened and shallower shelf surface and a steeper slope (see Fig. 2B). In contrast the accretional units are composed mostly of fine-grained sediments, and have the flattened shoreface with a deeper and a more gently dipping slope (Fig. 2C, D).

Additionally, this example illustrates how high-frequency changes in physical accommodation (sea-level fluctuations) control the internal architecture of the basic accretional units. The stacking patterns of the basic accretional units also capture how the overall subsidence-driven accommodation changed.

Low-Angle Ramps (Fig. 3B, C)

Examples of low-angle ramps include the depth-enhanced carbonate production associated with the "Calcari a Briozoi e Litotamni" of Italy, and the storm-driven resedimentation processes associated with the Tethyan ramp of northwestern Iberia and the Amellago Transect of the Moroccan High Atlas. These low-angle ramp geometries formed largely by physical accommodation and are likely the product of loose-grain sedimentation with accumulation rates that varied slightly down depositional dip. They illustrate that, despite similar depositional profiles, their facies differences reflect down-dip differences in the production/accumulation processes.

Enhanced Carbonate Production in the Aphotic Zone (Fig. 3B).—

Depth-enhanced carbonate production is interpreted to have occurred in the Upper Aquitanian–Lower Langhian "Calcari a Briozoi e Litotamni" (Brandano, 2001, 2003; Brandano and Corda, 2002; Brandano et al. 2001; Pomar et al. 2004). This 40–100-m-thick limestone body unconformably overlies Cretaceous and, locally, the Paleogene limestones of the Latium–Abruzzi Platform of the Central Apennines, Italy (Accordi et al., 1967; Bergomi and Damiani, 1976; Barbera et al., 1978; Damiani et al., 1992). This ramp formed during the diachronous marine transgression of the Late Oligocene to Langhian (Carannante et al., 1988), which was probably the result of tectonic subsidence. This tectonic event was produced by the east-northeastward migration of the Apennines fold-and-thrust belt, and is related to the eastward rollback of the Apennines subduction during the Neogene (Doglioni et al., 1999).

Inner-ramp lithofacies (Fig. 6 A and B) consist of balanid floatstone passing basinward into poorly sorted, mud-rich skeletal packstone to floatstone (seagrass-meadow deposits) interbedded with free-living, branching red algae rudstone to floatstone (Brandano, 2003). These pass, downslope, into crudely stratified, subhorizontal beds of red-algae bafflestone, rhodolithic rudstone, and red-algae, mollusk, and larger-foraminifera packstone (middle-ramp lithofacies). Small coral buildups are scattered within these lithofacies where corals are commonly encrusted by red algae and associated with oysters. There is an absence of wave-related structures and abundance of red algae and larger benthic foraminifera that include *Heterostegina, Operculina, Elphidium,* and *Amphistegina*. This places the depositional setting of this middle ramp in the oligophotic zone, below the base of wave action (Brandano, 2003; Pomar et al., 2004). Seaward, crudely stratified and structureless outer-ramp lithofacies are composed of: (1) coarse-grained echinoid–foraminiferal packstone and pectinid–bryozoan floatstone that passes downdip into (2) fine- to medium-grained echinoid–planktonic foraminiferal packstone. Seaward, in a basinal setting, limestone and marl alternate and contain siliceous sponge spicules, echinoid fragments, and planktonic foraminifera. The dominance of heterozoan skeletal components and the absence of light-related skeletal components suggest that outer-ramp carbonate production occurred in the aphotic zone (Brandano, 2003).

In this Miocene platform, enhanced carbonate production in the deeper-water aphotic zone contrasts with minor production in the shallow inner ramp. This difference produced a low-angle ramp profile. Heterozoan-dominated sediments (bryomol: Carannante et al., 1988; and molechfor: Nelson et al., 1988), produced in the aphotic outer ramp exceed the volume of up-dip oligophotic production. Red-algae production dominated the middle-ramp deposits, whereas sediment produced in the euphotic inner ramp was volumetrically minor (Brandano and Corda, 2002). This provides a good example of the departure from the "standard" decrease in grain size associated with increasing depth that characterizes terrigenous systems. Here, the increase in grain size with depth is tied to differences in production rates of carbonate components. In this example, production of gravel-size sediment in the deeper, low-hydraulic-energy zone resulted in floatstone-dominated textures in the middle-ramp and outer-ramp lithofacies. Moreover, packstone to floatstone textures occurring in the agitated, shallow-water lithofacies has resulted from seagrass baffling and sheltering, which prevented winnowing of fine-grained sediment.

Though sea-level changes were significant, the stratigraphic architecture of this carbonate ramp does not fully capture the effects of the sea-level cyclicity (Fig. 6C). Nevertheless, glacial eustasy was prominent during the Middle Miocene. By the Late Burdigalian, the East Antarctic ice sheet was already formed and the West Antarctic ice sheet was forming (Abreu and Anderson, 1998). Deposition on the Latium–Abruzzi ramp took place over a period of 5 My. The ramp is crudely stratified and structureless in the middle and outer portions. Here the most outstanding lithofacies barely reflects the third-order sea-level cyclicity (Fig. 6A). Only the bedding and lithofacies arrangement within the low-volume inner-ramp lithofacies (Fig. 6B) reflects cyclic changes in base level. Simultaneous backstepping and progradation of the outer-ramp facies belt after 17.5 Ma is interpreted to be a response to enhanced heterozoan carbonate production related to an increase in nutrient availability (Brandano and Corda, 2002).

Downslope Resedimentation Processes (Fig 3C).—

During the Late Jurassic, extensive low-angle carbonate ramps developed in northwestern Iberia (Fig. 7A). These exceeded 200 km in width in the dip direction and faced the Tethys Sea to the

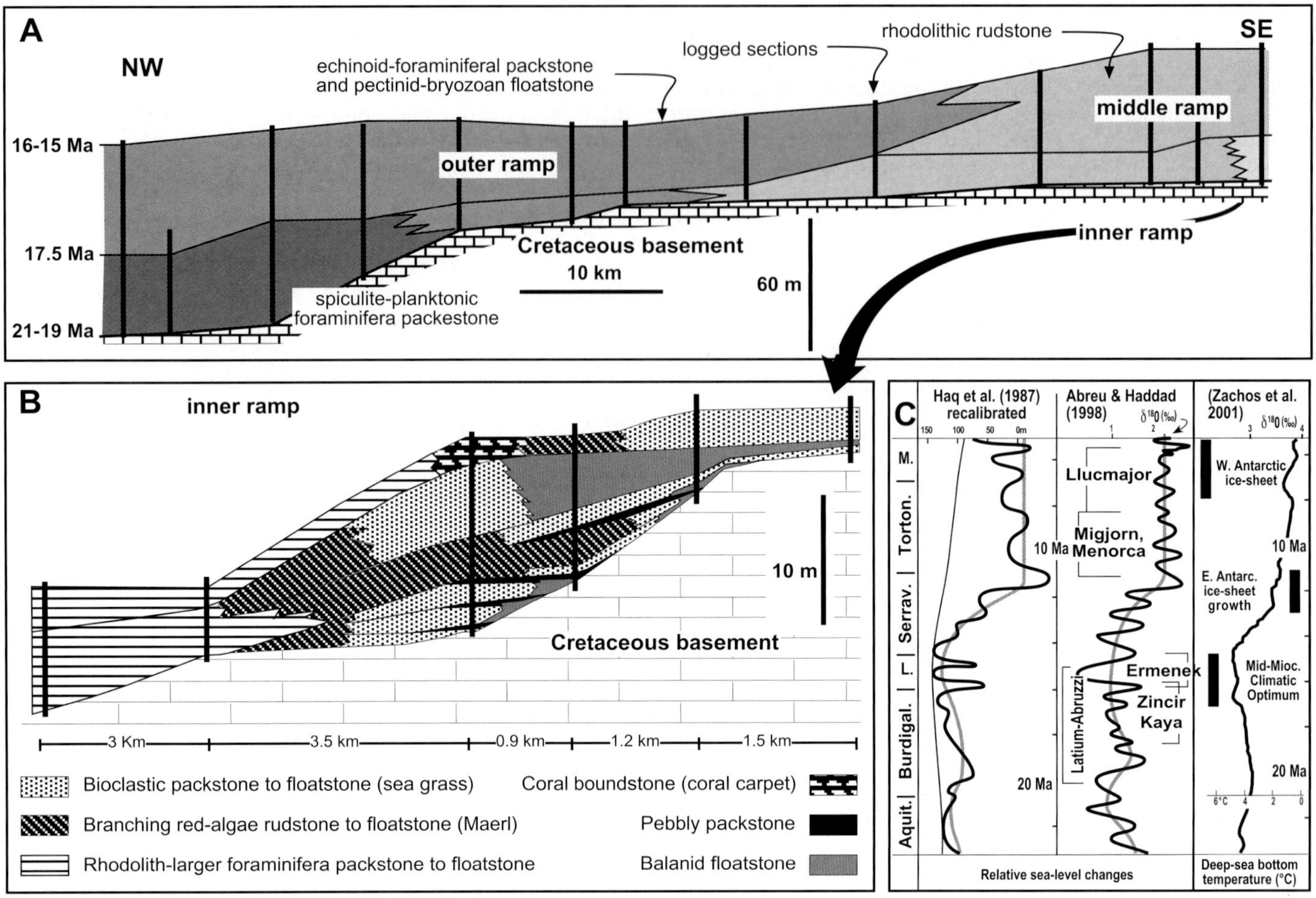

FIG. 6.—**A)** Middle Miocene Latium–Abruzzi platform, Italy (from Brandano, 2001). **B)** Inner–middle ramp facies architecture in the Latium–Abruzzi platform, Italy (modified from Brandano, 2003). **C)** Global sea-level changes (Haq et al., 1987). Composite smoothed short-term oxygen isotope record is from Abreu and Haddad (1998), as a proxy for sea-level changes. Occurrence of other Miocene carbonate platforms analyzed in this paper is shown. Global deep-sea oxygen isotope from Zachos et al. (2001), based on data compiled from bottom-dwelling deep-sea foraminifera from more than 40 DSDP and ODP sites, provides constraints on the evolution of deep-sea temperature as well as a time-averaged record of high-latitude sea-surface temperatures.

east (Bádenas and Aurell, 2001a and 2001b; Aurell et al., 2003; Bádenas et al., 2005). Local and temporary phases of fault-related subsidence affected these Tethyan ramps. They developed under tropical conditions facing a windward position. They undoubtedly were affected by winter storms from the west and summer hurricanes from the southeast (Price et al., 1995).

During the Kimmeridgian, inner-ramp lithofacies to the north consist of cross-bedded oolitic grainstone, with variable amounts of quartz sand, and burrowed skeletal packstone. These are interpreted to have accumulated in shallow-water settings, sporadically reworked by storms. Occasional low-angle planar cross beds (subaqueous dunes and megaripples) were formed by storm-generated currents. To the south, although cross-bedded oolitic grainstone are also present, inner-ramp lithofacies consist mostly of peloidal–oolitic grainstone and packstone, with variable amounts of skeletal fragments and lithoclasts. Middle-ramp lithofacies consist of sandy marl and skeletal sandy limestone, interbedded with graded beds of quartz sand interpreted as tempestite deposits (Bádenas and Aurell, 2001a). Patch reefs built by corals and chaetetid sponges, solenoporacean algae, and microbial crusts also occur in these middle-ramp settings. These are associated with skeletal and oolitic grainstone and well-bedded mudstone that pass basinward into mudstone with tempestite layers. Outer-ramp lithofacies are composed of laminated marl and mudstone–marl rhythmites, containing scattered ammonites and benthic fossils (brachiopods, bivalves, crinoids, serpulids, and benthic foraminifera).

All the facies associations of this Tethyan ramp suggest that resedimentation processes were dominant (Fig. 7B, C). Shallow-water skeletal, oolitic, and peloidal grains were shed down-ramp during winter storms and summer hurricanes (Bádenas and Aurell, 2001a). Forward modeling supports down-ramp transport (Aurell et al., 1995; Aurell et al., 1998). There is a good match between stratigraphic reconstructions of the real ramp and computer-generated models. They determine the thicknesses, large-scale stratigraphical geometries, amounts of aggradation and progradation, and location of sequence boundaries, flooding surfaces, and systems tracts. The computer simulations also suggest that though the origin of the percentages of lime mud cannot be inferred from rock analysis, a significant amount of this sediment (between 30% and 50%) may have been derived from erosion and redeposition from shallow-water areas. Neverthe-

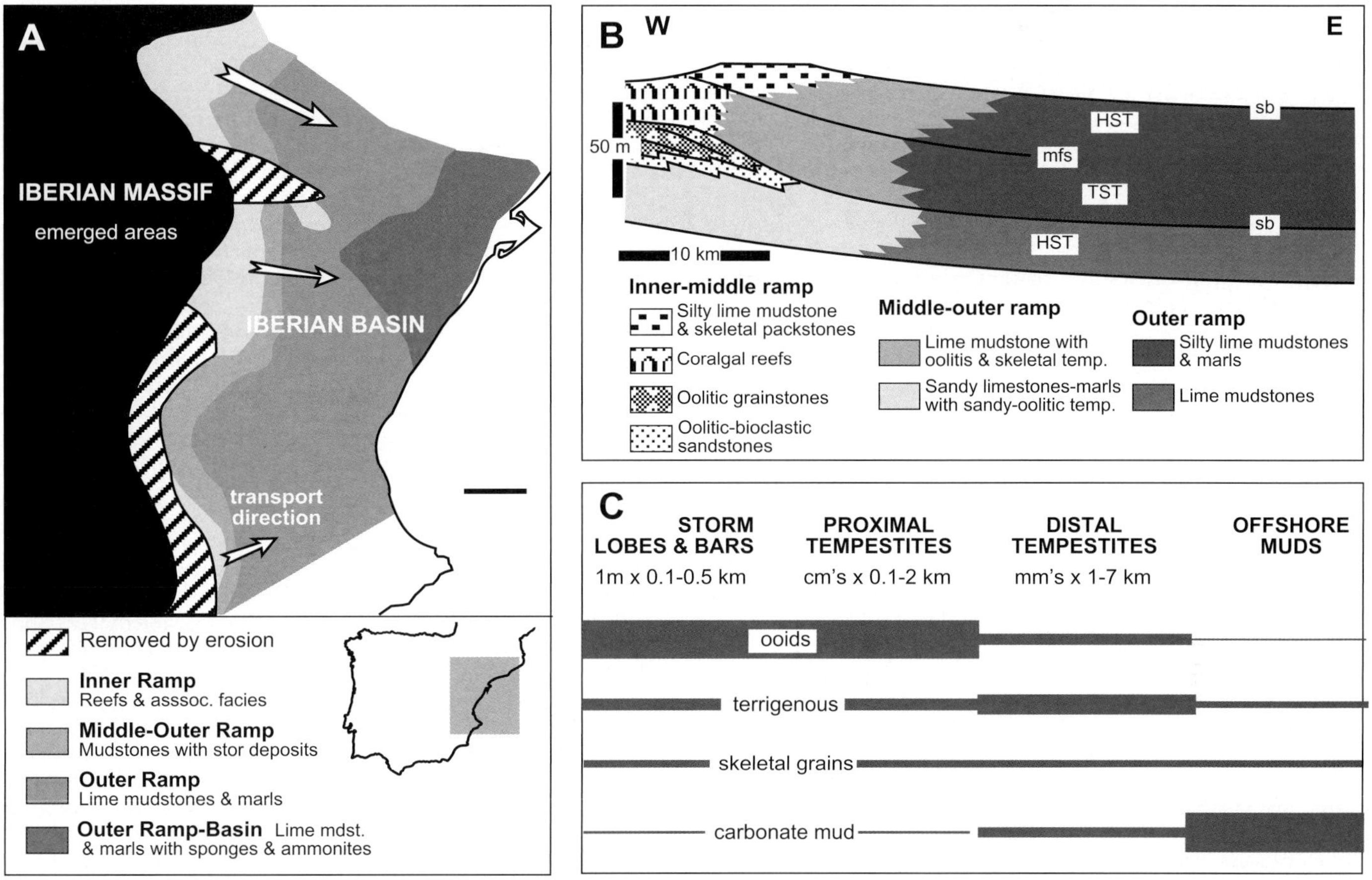

FIG. 7.—**A)** Facies distribution of the Late Kimmeridgian ramp in the Iberian basin (from Bádenas et al., 2005). **B)** Synthetic cross section showing facies architecture from shallow-water, inner-ramp to outer-ramp settings (from Bádenas et al., 2003). **C)** Relative proportions of grains in the storm deposits of the TST of Sequence 2 (from Bádenas and Aurell, 2001b).

less, the presence of some coccoliths in the muddy facies supports the contention that there was some pelagic carbonate production. Further evidence for downdip transport processes has recently been provided from facies analysis and bedding-pattern correlation between middle-ramp and outer-ramp lithofacies tied to variations in carbon isotopes (Bádenas et al., 2005).

Facies architecture analysis indicates that three third-order depositional sequences, with inherent regional variation, occur in the uppermost Oxfordian to lower Tithonian strata (Aurell and Bádenas, 2004). Shallow-water facies (inner and middle ramp) that collected on the western marginal areas of the Iberian Basin during the late Kimmeridgian have a hierarchy of higher-order depositional sequences and a periodicity of a few hundred thousand years. The amplitude of the relative sea-level fluctuations estimated from these high-frequency sequences range from 5 to 10 m (Aurell and Bádenas, 2004). In the middle-ramp and outer-ramp lithofacies, both deepening-upward and shallowing-upward high-frequency cycles are well constrained and are interpreted to reflect the amount of carbonate exported from the inner ramp (Bádenas et al., 2003). Changes in bed thickness and in bedding-plane character in the lime-mudstone lithofacies of the outer ramp enable the identification of bundles of beds and sets of bundles within the transgressive and highstand strata of the late Kimmeridgian third-order depositional sequence (Bádenas et al., 2003). Stratal bundles reflect changes in carbonate production and offshore shedding in response to sea-level cycles whose periodicity falls in the order of 20 ky. Sets of bundles, in turn, are estimated to reflect the 100 ky frequency band. Thicker sets of bundles occur in the late transgressive systems tract, while thinner sets occur in the highstand systems tract, thus indicating enhanced sediment export during the stages of a maximum gain in accommodation on the shallow inner ramp (Bádenas et al., 2003). Moreover, many bed bundles exhibit a thinning-upward trend, suggesting a progressive decrease in carbonate accumulation in the outer-ramp settings during periods of high-frequency sea-level rise.

Similar characteristics have also been documented in the Toarcian–Aalenian Ramp of the Amellago Transect, High Atlas of Morocco, a seismic-scale, 1000-m-thick, 37-km-long, quasi-continuous outcrop (Pierre et al., 2005; Pierre et al., 2006; Pierre, 2006). Here, precise physical correlation of beds and sedimentological logs using continuous photomosaics are tied to abundant biostratigraphic data. The fine-grained middle- to outer-ramp lithofacies are, volumetrically, the most conspicuous.

The above examples illustrate how, despite most of the sediment being produced in shallow-water settings, efficient downdip transport of abundant mud-size carbonate resulted in low-angle depositional profiles.

Distally Steepened Ramp (Fig. 3D)

Examples of distally steepened ramp geometries of the Late Tertiary and Holocene appear to have had high rates of sediment production and accumulation at the shelf-margin crest. These rates controlled the depositional profile and the location of a major break in the slope. An example of this kind of physical accommodation is represented by the Lower Tortonian sediments of Menorca in the Balearic Islands of Spain (Pomar, 2001a; Pomar et al., 2002; Pomar et al., 2004). The accumulation rates of these loose-grain sediments varied slightly down depositional dip, producing a characteristic distally steepened ramp.

Enhanced Carbonate Production in the Oligophotic Zone.—

The well-exposed example of Menorca forms a distally steepened carbonate ramp that crops out on the sea cliffs. It has a depositional profile and facies architecture that can be reconstructed along a complete dip section. This stretches from shoreline to basin. Bedding geometries, relative position of facies belts, sedimentary structures, and estimation of paleobathymetry from biotic evidence enable its interpretation.

The Lower Tortonian sediments prograded as a HST. They formed a prograding, high-energy, distally steepened ramp within a low-frequency third-order depositional sequence. Locally the tectonic setting was quiescent. In this carbonate ramp, siliciclastic basin-margin conglomerate and sandstone (fan-delta and pebbly beach deposits) pass basinward into crudely to well-stratified, subhorizontal to gently basinward-dipping inner-ramp carbonate lithofacies. These are composed of poorly sorted, structureless, foraminiferal, lithoclastic and mollusk dolopackstone. They are interpreted to have collected as seagrass-meadow deposits in the shallow euphotic zone (Pomar et al., 2002). Downdip, they interfinger with medium- to coarse-grained cross-bedded grainstone (middle ramp) containing fragments of red algae, echinoids, bryozoans, and molluscs, as well as a mix of large and small benthic foraminifera and scattered rhodoliths. These grainstone units are interpreted to have collected as medium-scale 2D subaqueous dunes. These formed under episodic, unidirectional, contour-parallel currents below wave base and were deeper than seagrass meadows (Pomar, 2001a; Pomar et al., 2002). Downdip, the middle ramp passes into 15–20° dipping clinobeds, composed of rhodolithic rudstone–floatstone with red-algae-rich grainstone and packstone matrix, interbedded with red-algal grainstone–packstone. The steepening of the slope indicates an increase in the sedimentation rate. This resulted from combined accumulation of "*in situ*" production of gravel-size skeletal components in the oligophotic zone, and of "*ex situ*" finer-grained sediment swept from the shallower euphotic and mesophotic zones. Lower in the slope, rhodolithic clinobeds interfinger with rudstone and floatstone to grainstone and packstone beds. These are either graded, crossbedded, or massive, and are interpreted to be mass-movement debris-flow and turbidity-flow deposits. The absence of "*in situ*" photic-dependent biota components (red algae) places this setting below the lower limit of the photic zone. Downdip, laminated fine-grained wackestone to packstone and grainstone rich in planktonic and small benthic foraminifera (outer ramp) occur. They are locally interbedded with 1–2 m thick, convex-up, coarse-grained, cross-laminated, red-algae-rich grainstone units. These grainstone units are interpreted to be 3D subaqueous dunes at the toe of slope (Fig. 8). These are composed of current-winnowed platform- and slope-derived sediment, and driven by contour-parallel currents (Pomar et al., 2002).

Despite the fact that glacio-eustatic changes were significant during the Miocene, the stratigraphic architecture of this ramp, deposited during a period of relative tectonic quiescence in the Balearic archipelago, is fairly simple. The inner- and middle-ramp lithofacies are stacked as an aggradational succession that expands progressively both landward and basinward, producing an expansive platform-top area. Shore-related lithofacies successively onlap onto the basement, whereas the ramp slope is progradational. Crudely stratified to well-stratified bed-bounding surfaces of the inner and middle ramp are marked by changes in grain size, but bed-set boundaries developed sharper surfaces and lithofacies changes across them, indicating a backstepping of the shoreline. Similarly, slight changes on skeletal components, with increased red-algae fragments and rhodoliths, provide evidence that sea-level cyclicity affected the middle-ramp lithofacies, though no sharp bed boundaries and bed-set boundaries developed.

Progradational ramp-slope lithofacies, however, record high-frequency sea-level cyclicity. This is reflected by alternating intervals of rhodolithic grainstone and rhodolithic rudstone–floatstone. The increase of inner-ramp-derived sediment components

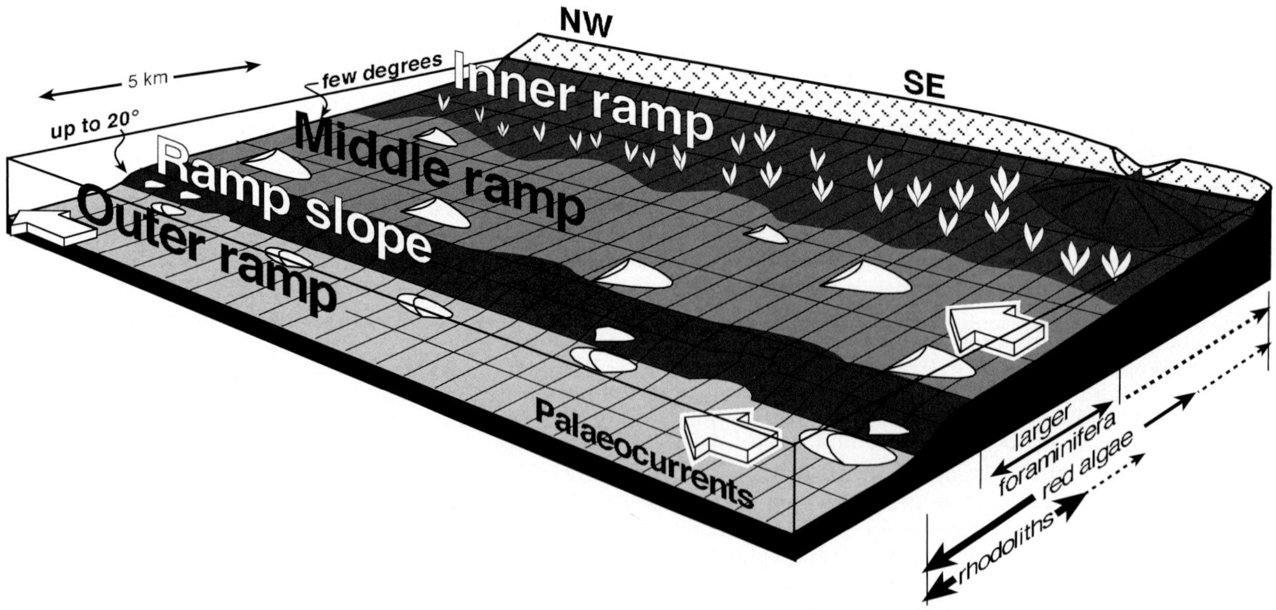

Fig. 8.—The Late Miocene distally steepened ramp of Menorca, Spain (modified from Pomar et al., 2002).

in the grainstone intervals relates them to an increase in sediment shedding when base level was lowered by falls in sea level.

This example demonstrates that sediment production/accumulation loci controlled the depositional profile and the location of the break in the slope. Loose foramol sediment production dominated the shallow-water, euphotic zone, and rhodalgal sediment production was preponderant deeper in the oligophotic zone. In this context, elevation of the base level during rises of sea level caused the inner- and middle-ramp lithofacies to aggrade. This was concomitant with a shoreline retreat and a landward expansion of the oligophotic factory (red algae) onto mid-ramp settings (Fig. 9). This increased the area available for red algae to produce and resulted in progradation of rhodolithic clinobeds (rudstone–floatstone). During sea-level falls, lowering of base level increased erosion and basinward-sweeping of grain-dominated sediments from the inner and middle ramp. This resulted in progradation of the grainstone–rhodolithic floatstone clinobeds.

Flat-Topped Open Shelves: Moderate Shallow-Water Ecological Accommodation (Fig. 3E)

Examples of these include the Late Cretaceous rudist buildups of the southern Central Pyrenees of Spain and the Early to Middle Miocene isolated banks and flat-topped carbonate shelves of the Mut basin of Turkey. The character of these bodies reflects the component large-skeleton soft-substrate-dwelling metazoans. These organisms are interpreted to have resisted moderate wave action and so contributed to sediment accumulation above the shelf equilibrium profile. Hydrodynamic baffling by these large skeletons contributed to the elevation of the outer edge of the platform and created a flatter platform top. Additionally, the dissipation of hydraulic energy at the margin affected the character of the sediment textures of the inner platform, which may be supplemented, as in many Tertiary platforms, by the ancillary baffling exerted by sea grasses (see Fig. 4). Finally the predominant grain size of the sediments shed from the platform determined the angle of repose for the fore-platform slope.

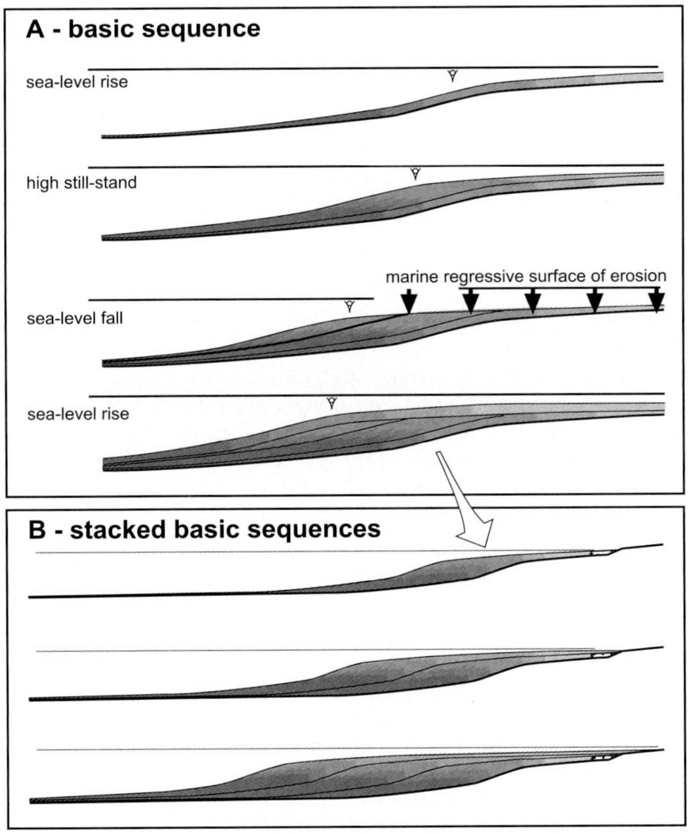

FIG. 9.—Architectural evolution of the Lower Tortonian distally steepened ramp in relation to changes of sea level. **A)** The basic sequences. **B)** Stillstand of sea level, punctuated by high-frequency cycles that result in basic sequences stacked in forestepping configuration.

Late Cretaceous Rudist Buildups.—

Late Cretaceous carbonate platforms crop out on the northern flank of Sant Corneli anticline, in the southern Central Pyrenees, Spain. There, the Santonian platform succession is 350 m thick, and its accumulation in the South Pyrenean Zone was strongly influenced by tectonic subsidence (Pomar et al., 2005). During the Mesozoic, the South Pyrenean Zone was part of the northern margin of the Iberian plate. Extensive carbonate-platform sequences accumulated here prior to the Late Santonian–Early Cenozoic basin inversion (Deramond et al., 1993, Berástegui et al., 1990; Berástegui et al., 1993; García-Senz, 2002). Excellent exposures enable the detailed mapping of the facies architecture along both depositional strike and dip. Contrasting facies changes across mappable surfaces allowed differentiation of the basic accretional units. The internal lithofacies arrangement and an inferred relationship to sea-level cyclicity (Pomar et al., 2005) enabled the differentiation of two kinds of carbonate platform in this setting. These are rudist buildups and calcarenite wedges (Fig. 10). They both contain "simple" sequences and parasequences (*sensu* Van Wagoner et al., 1990, Vail et al., 1991, Mitchum and Van Wagoner, 1991).

The Rudist Buildup Parasequences.—Rudist-buildup parasequences consist (Fig. 10A, B) of a basal coral–sponge–rudist "sheetstone" unit and an upper interval composed of coral–rudist "mixstone" rim at the platform margin, and an extensive back-rim lithosome of dense-hippuritid "pillarstone" landward. These match the growth-fabric nomenclature of Insalaco (1998) of "sheetstone", "mixstone", and "pillarstone". Marginal coral–rudist mixstone rims pinch out basinward into low-angle-dipping, mud-dominated packstones to wackestones. These are locally interbedded with small, scattered coral–rudist mounded colonies.

The recurrent facies arrangement, along with the textural and skeletal composition of the matrix, is the key to the genetic interpretation of the rudist-buildup parasequences. Corals and rudists started to grow on the basin floor, forming an extensive sheetstone biostrome in the lower photic zone, below the storm wave base. This is indicated by the presence of branching red algae, and the fine-grained wackestone matrix with calcispheres, planktonic foraminifera, and ostracods. Subsequently, a coral–rudist mixstone rim grew up, forming a margin seaward of an extensive pillarstone of slender hippuritids (Fig. 10B). This marginal rim provided some protection from frequent to episodic storm surges indicated by the common landward inclination of the hippuritids (Gili and LaBarbera, 1998). However, the common hippuritid-fragments floatstone beds occurring in some buildups suggest that occasional storms caused severe destruction of hippuritid meadows. This supports an interpretation that deposition was below the fair-weather wave base but above the depth of transient storm wave base. The upper boundary is a transgressive surface that, in aggradational suites, is overlain by a fine-grained basinal facies that indicates a subsequent rise in relative sea level (Fig. 11).

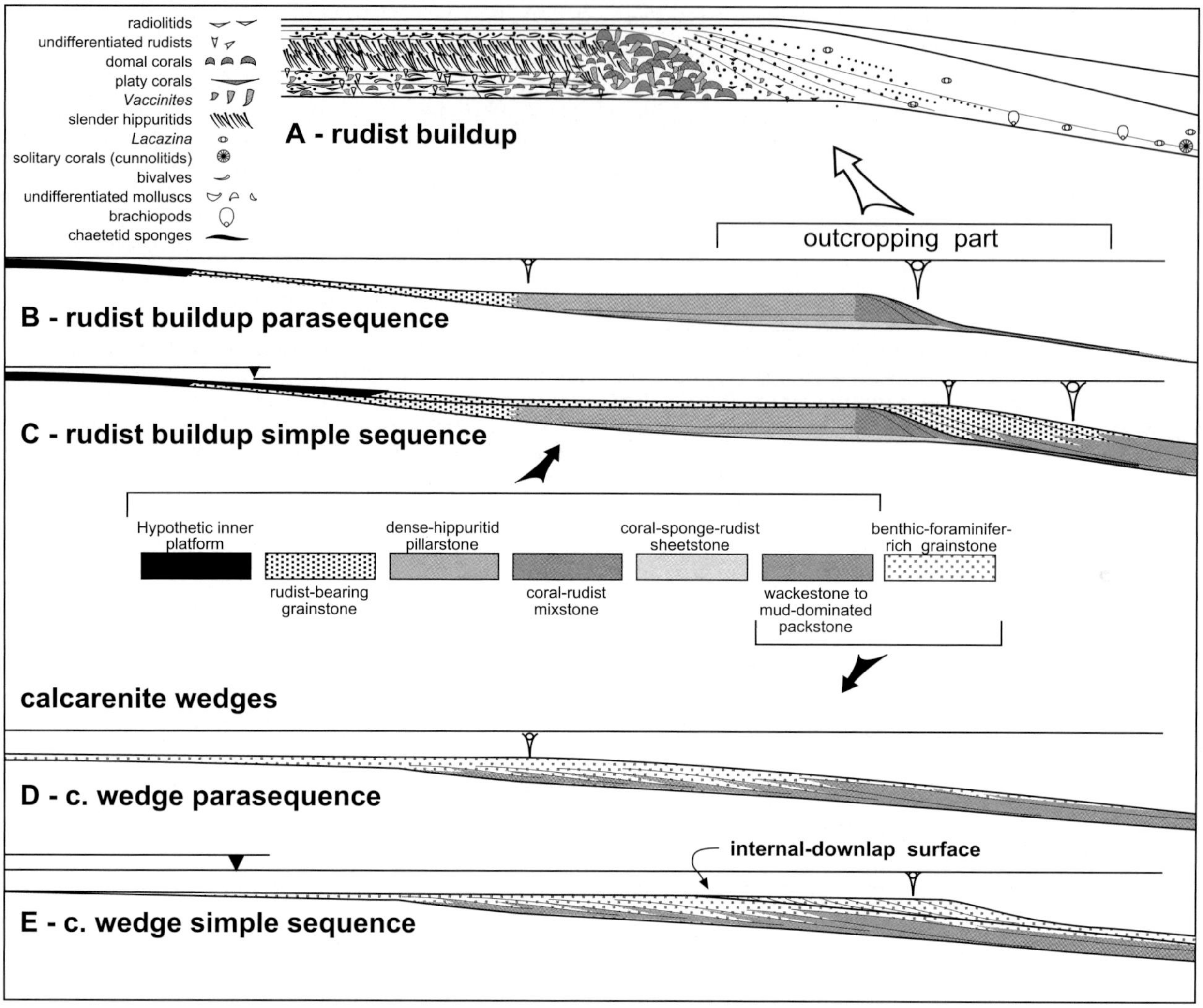

FIG. 10.—Two types of carbonate shelves are defined on the basis of on the persistent occurrence of a grouped assemblage of two facies: rudist buildups and calcarenite wedges (modified from Pomar et al., 2005). **A)** Fossil assemblages, along with rock textures, define the lithofacies in a rudist buildup. These are dominated by large sediment-dweller metazoans. **B, C)** Rudist buildup parasequences and simple sequences are defined according to lithofacies arrangement and inferred high-frequency sea-level fluctuation. Facies interpretation has helped reconstruct the development of the middle and inner platform. **D, E)** Calcarenite wedges are grain-dominated with foraminifera (protozoan) as the most conspicuous components. Here, parasequences and simple sequences are defined according to the lithofacies stacking patterns and the inferred sea-level changes.

Rudist-Buildup Simple Sequences.—Some rudist buildups are capped by rudist-bearing grainstone and packstone beds, with codiacean green algae and scattered rudists and corals, locally in living position. These grainstone beds overlie an erosion surface on top of the rudist buildups, which locally weakly truncates the slender *Hippurites*, and extend basinward to form apron-like clinobedded deposits on the platform margin (Fig. 10A, C). These grainstones are thought to have accumulated in the fair-weather wave-action zone: copious rounded rudist and coral bioclasts in the grainstone beds suggest high hydrodynamic energy, and presence of codiacean green algae denote shallow-photic conditions. A fall in sea level explains the increase in both hydrodynamic energy and light penetration needed for these sediments to form (Fig. 10C). These grainstones might represent mid-shelf facies belts, with profusion of larger and small benthic foraminifera, which migrated basinward during pulses of sea-level fall. Inner-platform lithofacies do not crop out in the Sant Corneli anticline. Consequently, as a parasequence, the rudist buildup represents the HST, and the overlying grainstone unit the regressive system tract.

Consequently, these rudist buildups tend to be flat topped as a result of the moderate ecological accommodation provided by large sediment-dweller metazoans (corals and large *Vaccinites* rudists) that were able to build up a storm-wave-resistant rim. This rim protected the wide meadows of slender *Hippurites*, where storm energy partially dissipated (see Fig. 4B) before impinging the foraminifera-dominated mid-shelf grainstone belt.

Alternating Rudist Buildups (Fig. 3E) and Calcarenite Wedges (Fig. 3A).—

Rudist buildups and calcarenite wedges alternate, although not rhythmically, in the upper Santonian carbonate-platform succession on the northern flank of Sant Corneli anticline (Figs. 11, 12).

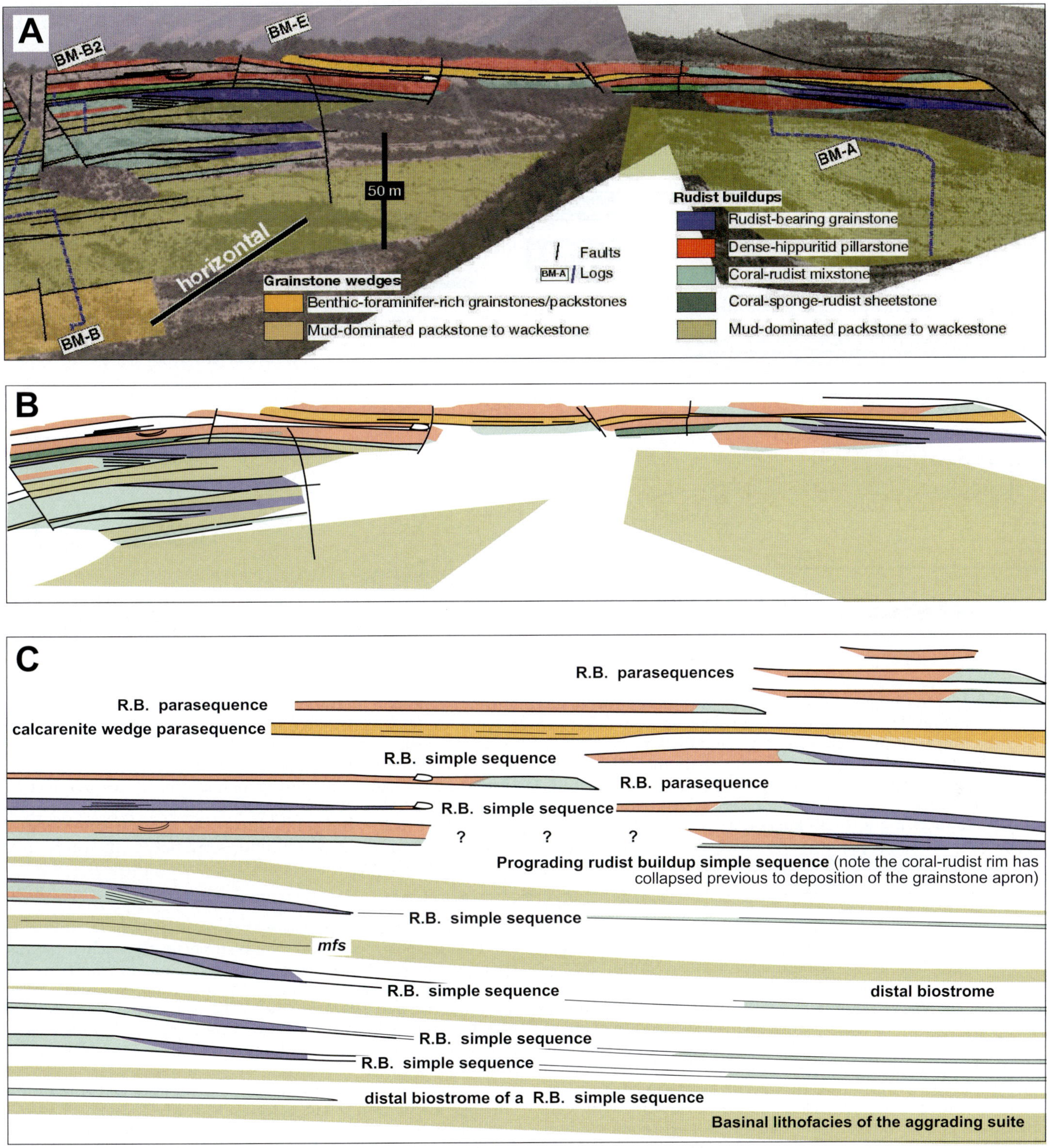

FIG. 11.—**A)** Photomosaic of the west rock face of the Montagut gully in the northern flank of the Sant Corneli anticline, showing the stacking of simple sequences and parasequences. The photomosaic has been tilted 40° anticlockwise (from Pomar et al., 2005). **B)** Accretional units as in Part A. **C)** Individualization of the accretional units in Part B, reconstructed by keeping thickness and position of facies belts and correcting deformation and faulting. The lowermost calcarenitic units have been omitted.

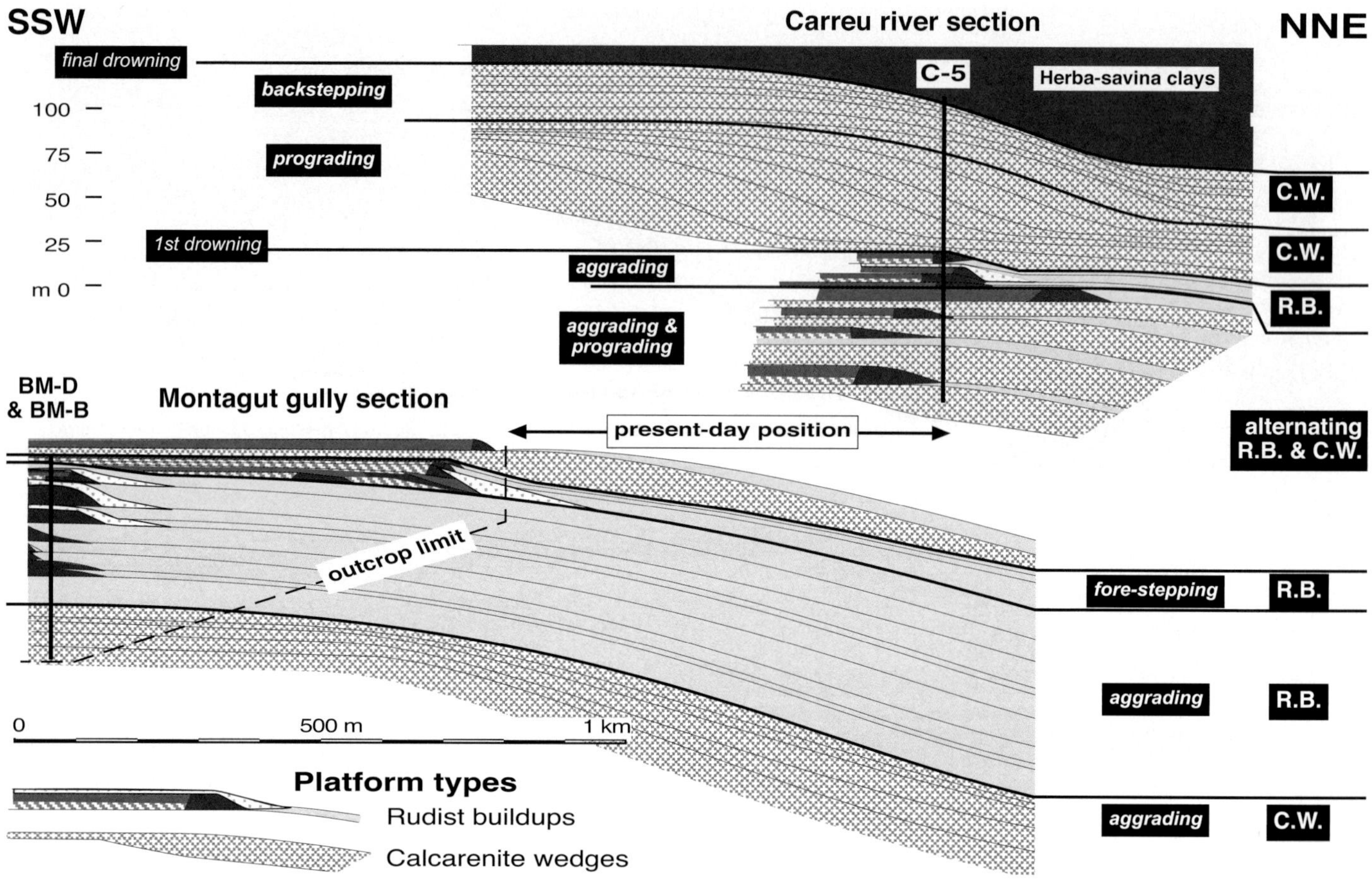

FIG. 12.—Architectural reconstruction of the studied carbonate-platform succession, in depositional dip direction. The Montagut gully section (Fig. 12) has been reconstructed from lithofacies and boundaries mapped on outcrop photomosaics, avoiding and correcting fault-induced repetitions and thickness changes. Two logged sections (BM-D and BM-B) have been used to estimate the thickness of the aggrading platform packages. The Carreu River platform architecture (cropping out along depositional strike) has been reconstructed by assembling basic accretional units (aggrading and prograding), fitting their internal facies architecture on the logged Carreu-5 section. The relative position of the Carreu River and Montagut Gully sections is shown at the present-day situation. Platform packages are defined according to the platform type (R.B. = rudist buildups; C.W. = calcarenite wedges) and their stacking patterns. The internal facies arrangement is shown only for the rudist buildups but not for the calcarenite wedges. (Modified from Pomar et al., 2005).

Calcarenite Wedges: Parasequences and Basic Sequences (Fig. 3A).— Calcarenite wedges form contour-parallel elongated wedges, roughly sigmoidal in dip section, and they are the product of physical accommodation only (Fig. 10 D, E). They consist of coarse grainstone, 2–5 mm thick, on the platform top that passes basinward into up to a 20-m-thick coarsening- and thickening-upward succession. This is composed of a mixed benthic-foraminifer-rich grainstone and grain-dominated packstone. This, in turn, grades basinward into and interfingers with basinal, fine-grained, mud-dominated packstone and wackestone with variable quartz sand-quartz silt and clay (Fig. 10D). In some wedges, an upper grainstone lithosome progrades with high-angle (up to 20°) clinobeds, downlapping onto an internal downlap surface and passing basinward into finer-grained grainstone and farther into mud-dominated packstone (Fig. 10E).

Characteristic constituents in grainstone and packstone intervals are abundant benthic foraminifera, echinoids, branching red algae, and ostracods. The miliolids are the most conspicuous of foraminifera present. Scattered small rudists and some small coral colonies may be present locally. Fragments of rudists, corals, bivalves, gastropods, and bryozoans are also common. In the basinal packstone and wackestone to clay-rich intervals, sponge spicules, calcispheres, ostracods, and brachiopods are present, as well as planktonic foraminifera, *Lacazina* larger benthic foraminifera, and solitary corals. The calcarenite wedges have been interpreted to correspond to infralittoral prograding prisms (Pomar et al., 2005). They are dominated by physical accommodation only (Fig. 3E).

Parasequences and simple sequences (Fig. 10D, E) are differentiated in the Santonian calcarenite wedges on the basis of bedding patterns, facies architecture, bounding surfaces, and an inferred relationship to the highest frequency of sea-level changes (Pomar et al., 2005). Key diagnostic features are the absence of topsets and the presence of the internal downlap surfaces in the basic sequences. The parasequences represent non-regressive progradation below the storm wave base during stillstands of relative sea level. The simple sequences are composed of two systems tracts bounded by an internal downlap surface (Fig. 10E). This internal downlap surface is a sharp boundary marked by an abrupt grain-size increase, but it has no evidence of erosion.

Erosional truncation of the topsets of the lower lithosome (HST) provided the increased sediment supply, a consequence of the lowering of wave-base level inherent to the decrease to accommodation space.

Alternation of Platform Types.—Alternation of platform types adds a factor of complexity to the analysis of the carbonate-platform architecture. In the upper Santonian platform succession of the Sant Corneli anticline, the calcarenite wedges alternate, although not rhythmically, with rudist buildups (Fig. 12). Following a package of aggrading calcarenitic wedges, an aggrading and then prograding suite of rudist buildups occurs. The later rudist buildups (see Figure 11) prograde in a forestepping configuration, to be succeeded by a thick group of alternating rudist buildups and calcarenite wedges stacked in aggrading and prograding patterns (Carreu river section in Figure 12). These in turn are overlain by a set of aggrading rudist buildups only. Overlying a third-order transgressive surface, a 65-m-thick package of calcarenite wedges progrades over the top of the rudist buildups. It is, in turn, overlain by a set of calcarenitic wedges, which are finally overlain by the deeper-water Herba-savina clays.

These changes in platform skeletal composition reflect changes in the dominant type of carbonate-producing biota independent of the changes in accommodation. Rudist buildups were dominated by large metazoans (large corals and rudists) that raised platform margins that never reached sea level, with wide meadows of slender hippuritids to their lee. These latter thrived close to the storm wave base, with rudist-bearing grainstones and packstones, with codiacean green algae, probably representing a mid-platform facies belt (Fig. 10B, C). In contrast the yellow-brown calcarenite wedges formed in a dominantly grain-producing carbonate factory, in which protozoans were abundant, and rudists and corals were scarce, subordinate biota. In these wedges, sediments accumulated up to a "shelf equilibrium profile" that was dominated by hydrodynamic accommodation alone.

This oscillation in the dominant type of carbonate-producing biota could be the result of shifts in trophic resources, temperature, and other factors controlling the carbonate-mineral stability. The calcarenite wedges represent calcitic benthic foraminifers, and loose-grain-dominated carbonate production, whereas the rudist-buildup carbonate production was dominated by large filter-feeder rudists (aragonitic and calcitic shells), aragonitic corals, and, to some extent, aragonitic green algae. Consequently, there is a clear shift in skeletal composition of the hypercalcifying organisms (*sensu* Stanley and Hardie, 1998, 1999). Rudist buildups with abundant corals and aragonitic green algae may represent "short-term" aragonite-facilitating episodes (*sensu* Sandberg, 1983), whereas calcarenite wedges may represent aragonite-inhibiting episodes.

Middle Miocene Flat-Topped Open Shelves (Fig. 3A).—

During the Early to Middle Miocene, isolated open banks and open flat-topped carbonate shelves prograded over a complex paleo-topography and a context of gentle basin-wide subsidence, in the Mut basin of south-central Turkey (Bassant, 1999; Bassant et al., 2004, 2005; Janson, 2002). Extraordinarily well-exposed outcrops mean that the 3D facies architecture and bounding surfaces can be documented in detail, and allow establishing their relationships to relative-sea-level cyclicity. Bassant et al. (2004) and Bassant et al. (2005) established that the character of the South Turkey Miocene carbonate platforms were affected by glacio-eustatic sea-level fluctuations of around 100–150 m for the 400-ky-period cycles (large-order sequences) and 18–30 m for the 100-ky-period cycles (medium-order sequences). However, they were uncertain as to the origin of the small-order sequences.

Isolated carbonate platforms, 1–4 km across and 100–120 m thick, developed on the northern margin of the Mut basin during transgression (Fig. 13A), but they were finally drowned when rates of sediment accumulation did not match rapid sea-level rise. Depositional sequences, interpreted to be the products of fourth–fifth order cycles (Fig. 13B), are composed of consecutive aggradational and a progradational successions (Bassant, 1999; Bassant et al., 2005). The sequences are flat-topped with no rimmed margin. Loose-grain sediment forms the bulk sediment that was dispersed and moved downslope by waves and currents. Foraminiferal packstone–wackestone to grainstone (inner platform) pass basinward to miliolid grainstones–packstones with larger benthic foraminifera (shallow open platform). Later these pass into clinobedded nummulitid–foraminifera grainstone–packstone with amphisteginids, and some planktonic foraminifera (platform and slope), and to planktonic foraminifera micro-packstones to marls with nummulitids (Bassant, 1999; Bassant et al., 2004; Bassant et al., 2005).

Within the platform top and the talus, coralgal (both clay free and muddy) boundstones and coralgal floatstones in a fine-grained (clean to muddy) matrix are conspicuous. Microbial coralgal boundstones are found locally on the platform top (Fig. 13A). Unlike in many modern reef examples, coral growth was never enough to sustain wave-resistant buildups that reached sea level. Instead, they built discrete "rigid" bioherms associated with red algae and microbial crusts. The discrete bioherms controlled the depth of the wave friction, which in turn had a strong influence on the position of base level across the flat-topped platforms. Terrigenous shale trapped between the corals suggests a turbid water column with relatively low hydrodynamic energy.

In the western part of the Mut basin, a 20-km-wide platform succession, the Ermenek platform, developed attached to a steep pre-Miocene substratum. Here Late Burdigalian to Serravallian carbonate platforms can be traced from the shoreline to the platform margin along several valleys (Janson, 2002). The beat of increasing and decreasing accommodation space tied to higher-frequency fourth–fifth order relative-sea-level cycles explain the alternating isopachous and sigmoidal clinothem geometries (Fig. 13C).

Skeletal components of the aggrading Burdigalian strata consist mostly of echinoderms fragments, oysters and other bivalves, larger benthic foraminifera (*Heterostegina*), bryozoans, and red algae. Few corals occur toward the top of this interval. During the Early Langhian, small isolated coralgal buildups started to aggrade and subsequently to prograde as steep-angle clinobeds forming flat-topped platform banks, with relief of 15 to 35 m above the surrounding depressions. These platforms consist mostly of coarse- to medium-grained red-algae grainstone and packstone, with large benthic foraminifera and molluscs and, locally, small coral buildups. Janson (2002) estimates that the bathymetry for platform-top settings in about 5 to 20 m. The intra-platform depressions were shallow and filled by wackestone to micro-packstones with planktonic and small benthonic foraminifera, echinoderms, and bivalves.

A change in the main skeletal components was contemporaneous with a subsequent change from progradation to aggradation (Fig. 13C). While the platform-margin clinobeds are still dominated by red-algae packstone, the lithofacies of the platform tops are composed mostly of a packstone of small foraminifera, and coral buildups increase. Coral patch reefs are small and numerous on the platform top, whereas they reach a few meters in height on the slope (Janson, 2002). They consist of branching

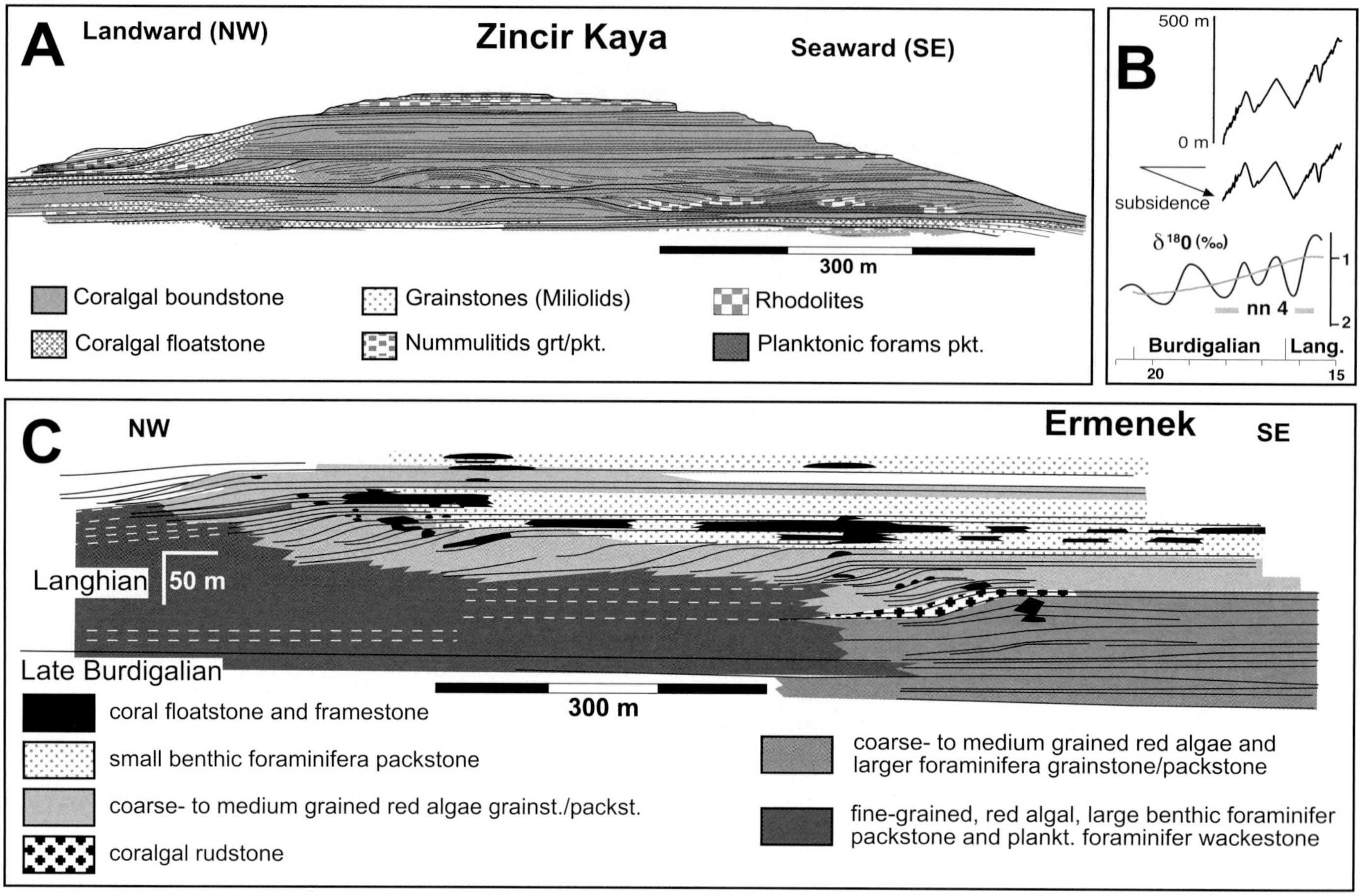

FIG. 13.—**A)** Bedding patterns and facies architecture of the Zincir Kaya isolated platform (Modified from Bassant, 1999). **B)** Accommodation changes inferred from 3D outcrop control of different platforms in South Turkey (modified from Bassant, 1999), subsidence correction of accommodation changes in Part A, and oxygen isotope curve from Abreu and Haddad (1998). **C)** Facies architecture of the prograding flat-topped platforms margin in the Bijou valley, based on facies mapping on photomosaics and logged sections (simplified from Janson, 2002).

and spherical corals encrusted by red algae. Locally, abundant encrusting red algae, larger benthic foraminifera, and small coral heads were able to develop at the toe of slope and out into the basin. By the end of the Langhian, the prograding flat-topped platform banks encroached to form a single large flat-topped platform (Janson, 2002).

Reef-Rimmed Modern Platforms: Highest Shallow-Water Ecological Accommodation (Fig. 3F)

Reef-rimmed platforms that match those of modern settings have the highest ecologically driven capacity to build above the hydrodynamic shelf equilibrium profile. Large corals and encrusting algae build a rigid, wave-resistant framework up to sea level. This acts as a "wave-energy dam" along the shelf margin, often restricting circulation in the back-reef area, so that a wide, low-energy lagoon develops (Figs. 3F, 4C). In the lagoon, the base level for sediment accumulation is much shallower than in the reef and forereef, and these latter are exposed to the erosive forces of waves coming from the open sea. An example of a platform of this kind is that of the Llucmajor reef of Mallorca, which exhibits high-frequency heterogeneous depositional sequences that have a high ecological accommodation.

The Llucmajor Platform, Mallorca.—

The sea cliffs of Mallorca in the Western Mediterranean exhibit continuous and complete exposures of the Late Tortonian–Early Messinian Llucmajor carbonate platform. These cliff exposures have enabled the detailed mapping of bedding patterns, facies architecture, and bounding surfaces on photomosaics. They provide a well-documented database for the basic accretional units of this carbonate buildup, displaying the hierarchy of their high-frequency stacking patterns that accumulated in a region of tectonic stability, and these sea cliffs have been used to determine the genetic relationship between carbonate production, sea-level changes, and internal facies architecture (Pomar, 1991, 1993; Pomar and Ward, 1994, 1995, 1999; Pomar et al., 1996).

The basic accretional unit, or building block, of the Llucmajor carbonate platform is the "*sigmoid*" (Pomar, 1991; Pomar and Ward, 1994). Each sigmoid is composed of a relatively conformable succession of genetically related stratified lithofacies bounded by unconformities or their correlative conformities. Updip and landward, this sigmoid unit is represented by a horizontal lagoonal bed that basinward passes into a sigmoidally bedded reef-core lithofacies belt and then farther seaward into clinoform-bedded forereef slope deposits and subhorizontal basinal lithofacies (Fig. 14A). Sigmoids stack into progressively larger-scale accretional units of sets, cosets, and megasets of sigmoids, which reflect hierarchical orders of sea-level cycles (Fig. 14B, C, D). The amplitudes of these cycles are less than 15 m, 20–30 m, 60–70 m, and ~ 100 m, respectively, and are calculated from the amplitude of the reef-crest line (Fig. 14), and from the successive positions of the reef crest (Pomar, 1991). These sea-level fluctuations are thought to be glacioeustatic in origin. Certainly there is a good match between the cyclicity of the megasets and the composite smoothed short-term oxygen isotope curve (Fig. 15) of Abreu and Haddad (1998), suggesting that both can be used as a proxy for sea-level change.

All these accretional units represent high-frequency depositional sequences. They have similar characteristic stratal geometries, bounding surfaces, and facies architecture. The relative volume of facies, along with the facies arrangement within these accretional units and across the erosional boundaries, is interpreted to reflect the interdependence between accommodation

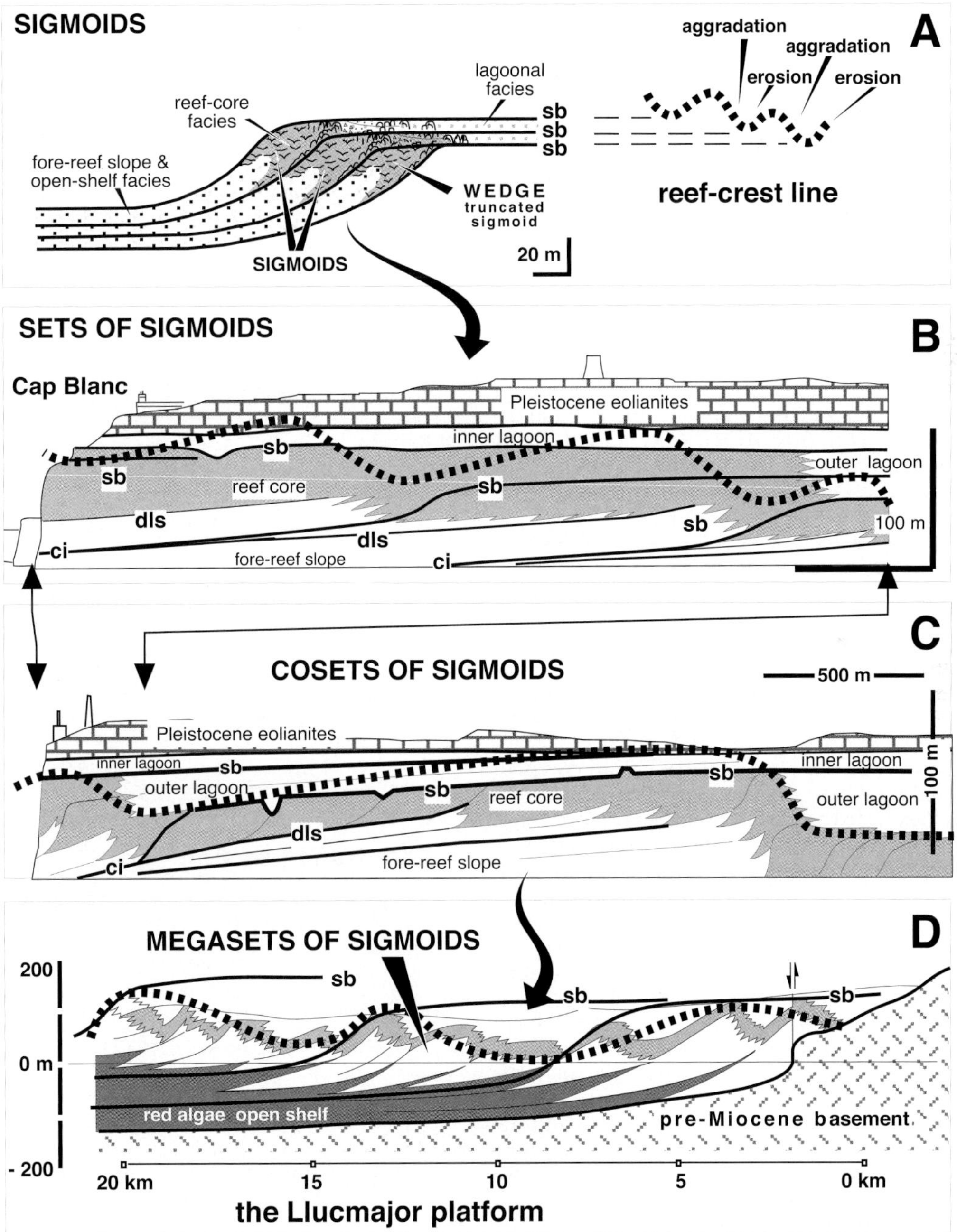

FIG. 14.—Hierarchical stacking of up to four orders of basic accretional units in the Llucmajor platform. The reef-crest line is an accurate proxy for sea-level changes. It traces the successive progradational position of the reef crest as a function of sea-level fluctuation and sediment accumulation through time (Modified from Pomar and Ward, 1994). **Sb**, sequence boundary; **dls**, downlap surface; **ci**, condensed interval.

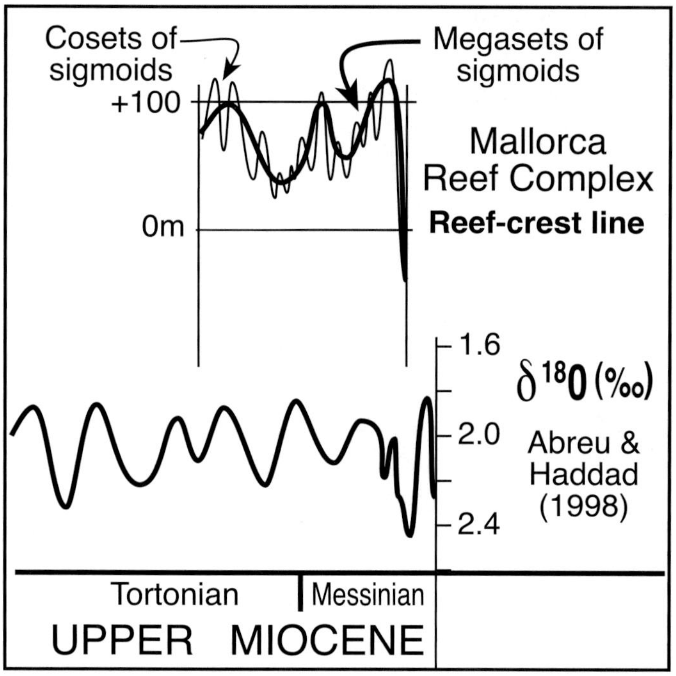

FIG. 15.—Comparison between the reef-crest line established for the Mallorca reef complex and the composite smoothed short-term oxygen isotope record from Abreu and Haddad (1998), as a proxy for sea-level changes. The reef-crest curve has been constructed from the cosets and megasets of sigmoids identified on the cross section in Figure 14C and D.

and carbonate production. The amount of carbonate production and sedimentation was directly related to accommodation changes driven by sea-level fluctuations (Figs. 14, 16).

During sea-level rise, the reefs aggraded vertically, the lagoonal sediment surface matched the highest position of the sea, and the greatest amount of sediment was shed, inducing the forereef talus to aggrade (Fig. 16). This aggrading systems tract is volumetrically the most important and differs from the transgressive systems tract of other models (Vail et al., 1977; Van Wagoner et al., 1988; Van Wagoner et al., 1990; Sarg, 1988) in that it lacks a condensed section basinward and backstepping is absent (Figs. 14, 16). The highstand systems tract is composed of the combination of (1) a thin progradational reef-core facies in a basinward-stepping trajectory, (2) a forereef-slope facies that wedges out basinward, and (3) a condensed open-shelf facies. Both indicate that there was a minor amount of sediment shed from the platform top. Lagoon beds are commonly absent (if they were deposited, they were eroded during the subsequent sea-level fall). The falling-stage, forced-regression, or offlapping systems tract has a trajectory that slopes seaward and is composed of thin prograding and downstepping reef-core facies, which downlap onto the open-shelf facies (of the previous systems tract), without significant forereef-slope facies. The lagoon facies is missing, and the open-shelf facies is condensed. The lowstand systems tract, like the HST, incorporates thin progradational reef-core facies and basinward-wedging forereef-slope deposits.

Basin-floor bathymetry controlled carbonate production. The areal extension of the lagoon was dependent upon the previous floor morphology and on the changes of relative sea level (Figs. 16, 17). Maximum lagoonal extension took place behind barrier reefs that developed over gently inclined surfaces, during rises of sea level, while areas with more steeply inclined surfaces had

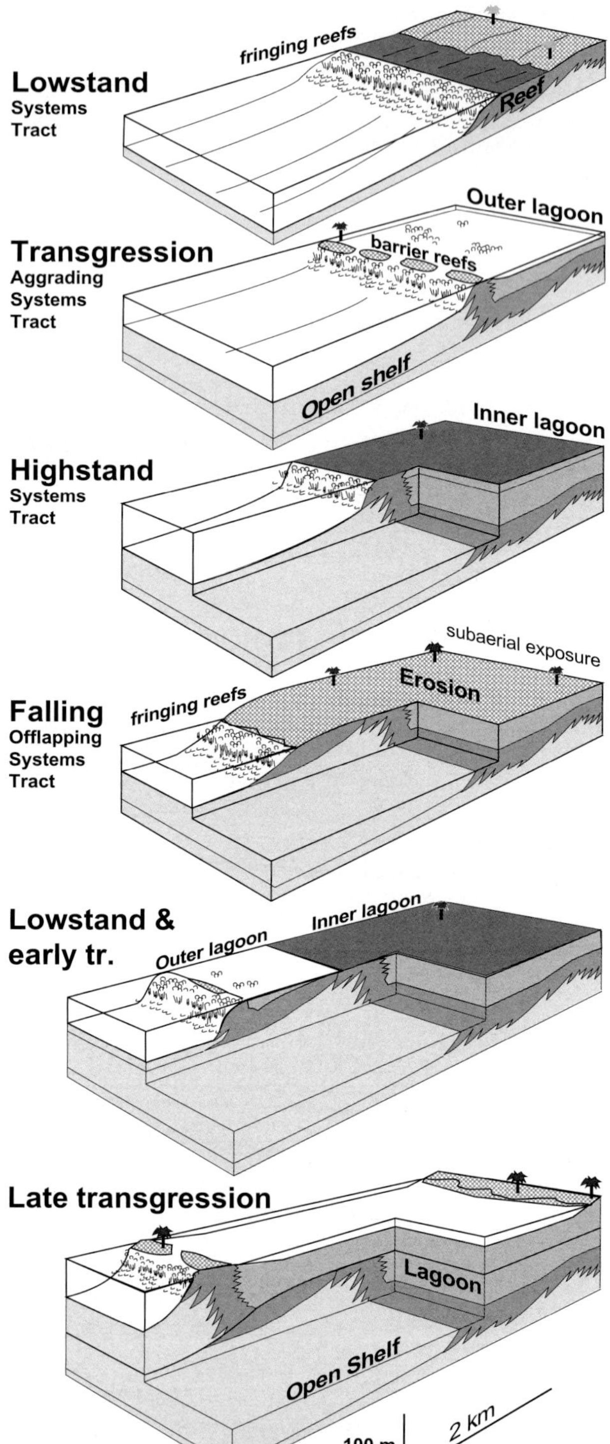

FIG. 16.—The stacking trajectories of systems tracts, composed of sets of sigmoids, which form cosets of sigmoids as seen in the sea-cliff exposures along the southern Llucmajor platform (modified from Pomar and Ward, 1994). Most sediment volume in the entire prograding platform corresponds to aggrading systems tracts deposited during rises of sea level. Highstand, falling, and lowstand systems tracts are composed of thin reefal facies passing basinward into a condensed interval.

narrower lagoons (Palma basin; Fig. 17). The volume of the forereef-slope sediments, however, was directly related to how extensive the existing lagoons were. Fringing reefs with little or no evidence of lagoons to their lee predominated in all areas during falls of sea level. The trajectory of the coral reefs shifted downward and basinward as result of decreasing accommodation.

During lowstands of sea level, carbonate production in shallow-basin settings was significant over a large area (Fig. 17). Only coarse red-algae-rich sediment remained on the shallow open shelf, and wave action winnowed finer material and transported it to the deeper open platform. The red-algal biostromes of this setting interfingered with the fore-reef slope of the LST on the shallower shelf. When sea level rose, most of the sediment production shifted to the shallower shelf, so extensive lagoons developed behind reefs, and red-algal deposits were not produced because the basin floor was too deep to allow light penetration. Maximum progradation rates occurred across gently inclined areas, not only because accommodation was less, but also because carbonate production was greater.

In this context, the rhodalgal unit existing at the base of the Llucmajor platform (Fig. 14D) represents amalgamated lowstands, produced in shallow-basin settings when the oligophotic factory was on, and the higher-frequency sea-level cyclicity produced textural fluctuations (rudstone–floatstone) within the lowstand units. The sigmoids, sets, cosets, and megasets of sigmoids reflect the changes induced in the euphotic factory by the high-frequency sea-level cyclicity (Figs. 14, 17).

This Late Miocene platform exemplifies how the heterogeneity of the lithofacies architecture in this type of prograding platform was controlled by the high-frequency changes both in accommodation and in carbonate production when there was an absence of significant compaction and subsidence. Carbonate production was not an independent factor with respect to changes in accommodation. In fact, production varied in response to changes in relative sea level and the depositional profile of the basin floor. The enhanced ecological accommodation makes the shallow-water carbonate factory very sensitive even to minor sea-level changes. This sensitivity allowed this platform to record up to sixth-order sea-level fluctuations with periods of a few ky (Figs. 14, 15).

The combined effect of sea-level cyclicity and the bathymetry of the forereef and the open shelf areas leads to two different carbonate factories that alternate in phase with sea-level changes: a euphotic chlorozoan factory that was efficient during sea-level rises and also during highstands; and an oligophotic rhodalgal factory that turned on during sea-level lowstands.

Massive Platform Margins: Depth-Extended Ecological Accommodation (Fig. 3G).—

Many Paleozoic and some Mesozoic platforms have steep, massive and thick marginal slopes that maintain a constant thickness and depositional dips (Fig. 3G). These accretional platform margins have depositional textures that are often difficult to interpret, but they share some common trends and differ from

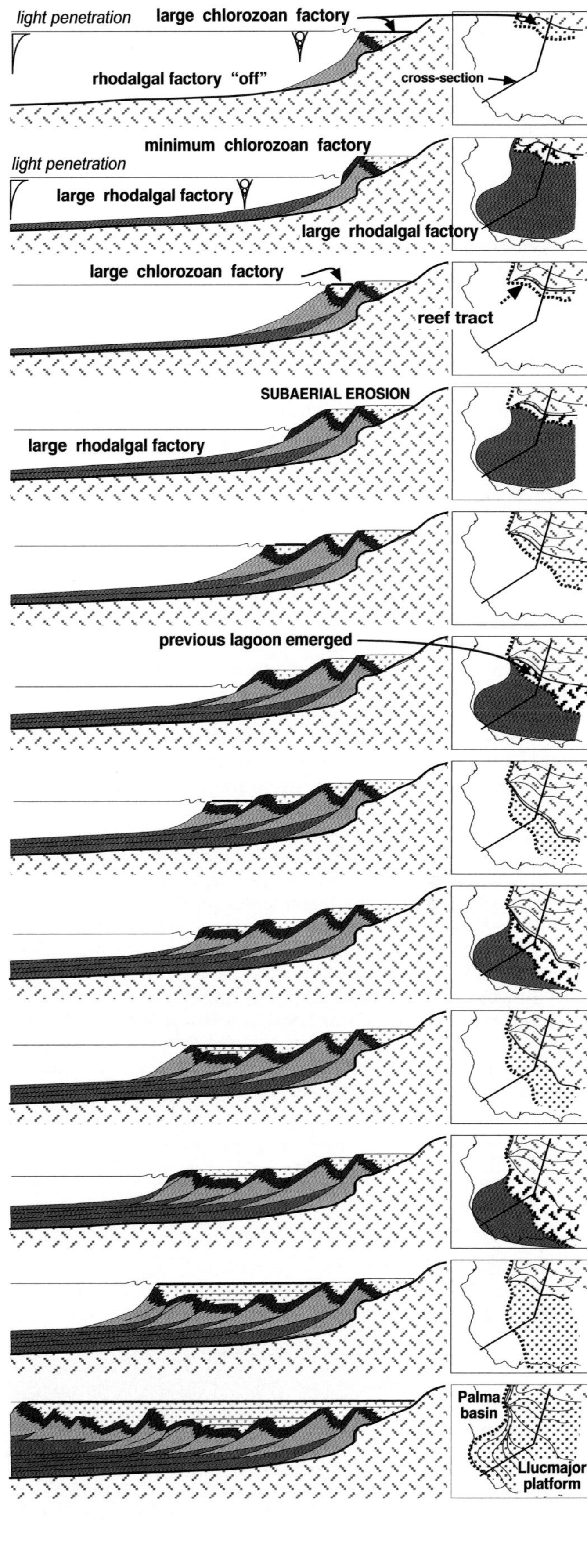

→

FIG. 17.—Reconstruction of the 20 km of reefal progradation to the south-southwest, during a period of about 2 My, based on borehole and outcrop data (modified from Pomar and Ward, 1995). The relatively fast progradation is interpreted to be the result of two carbonate factories that alternated in phase with sea-level fluctuations on a gently dipping area.

most other types of carbonate platforms. They contain abundant microbial boundstone and automicrite, along with heterotroph skeletons, that suggest an important light-independent, encrusting or binding carbonate factory. Calcification associated with on-ground and underground degradation of benthic microbes and metazoans communities, probably sponges, by heterotrophic organisms is the most likely origin of these biotically induced carbonates (Monty, 1995; Neuweiler et al., 1999). The scarcity of talus deposits at the toe of the slope indicates low rates of sediment export from the platform top. This depth-extended, organomineral-boundstone carbonate production on the slope often occurred in low-oxygen and sediment-starved basins, particularly in those with low metazoan diversity and density. The abundance of submarine cement is another common feature on many of these platform margins, and its origin is still poorly understood. Examples of these kinds of carbonate setting include the Lower Albian Gorbea carbonate platform succession of the Basque–Cantabrian Basin, eastern Pyrenees, during the Late Aptian, and Triassic carbonate buildups of the Dolomites in the Southern Alps.

The Early Albian Gorbea Platform.—

The Gorbea platform succession accumulated in the Basque–Cantabrian Basin, eastern Pyrenees, during the Late Aptian to Early Albian. It evolved consecutively from a mixed carbonate–siliciclastic ramp into a carbonate ramp and finally into a rimmed platform. The rimmed-platform succession changed from progradational into aggradational (Gómez-Pérez et al., 1999). The northwestern margin of the Gorbea platform developed a north–south trending steep margin facing the deeper Orozko basin. Present-day topography of this platform matches that of the Cretaceous, because no significant postdepositional faulting or major tectonic deformation has affected the paleo–platform edge. During the aggradational stage of the Gorbea platform succession, sedimentation kept pace with accelerated subsidence and little excess of sediment was transported off-bank. An elevated margin with no evidence of a wave-resistant organic framework separated the horizontal platform area from the steep and massive upper-slope beds, which maintained a nearly constant thickness and depositional dips up to 35° (Fig. 18). Small concave-up clinoforms, which split into thinner beds, are interbedded with marly limestone at the toe of slope (Gómez-Pérez et al., 1999).

Platform-interior lithofacies consist mostly of orbitolinid marly packstone and grainstone, coral-rudist and oyster-like *Chondrodonta* wackestone, *Bacinella* boundstone, and miliolid wackestone–packstone. Basinward, they pass to burrowed spiculitic marl, marly limestone, and skeletal packstone and grainstone, which grade laterally into coral–wackestone shelf mounds. Platform-margin lithofacies are massive to biostromal and consist of boundstone, wackestone, packstone, and floatstone, with local grainstone and rudstone with highly diverse fauna: rudists, *Bacinella*, *Chondrodonta*, rare corals, bivalves, brachiopods, and skeletal fragments. Fossils and textures suggest open marine conditions in the photic zone, with moderate to low energy levels. The upper slope consists of massive, up to 35° dipping clinoforms composed of rudists, algae, and *Chondrodonta* skeletal boundstone and floatstone, similar to that of the platform margin. These change downdip into a mid-slope massively bedded coral boundstone, with lenses of skeletal grainstone and packstone. Farther downdip, these massive slope clinobeds change into calcareous sponge (chaetetid) wackestone and boundstone, locally with meter- to decameter-scale coral–rudist bioherms. Depositional depth has been estimated to be in the range of 20 m down to 120 m (Gómez-Pérez et al., 1999). These bioherms were created by automicrite precipitation, a certain degree of frame binding, and early cementation on the slope (Gómez-Pérez et al., 1999). They suggest low rates of sediment export from the platform system.

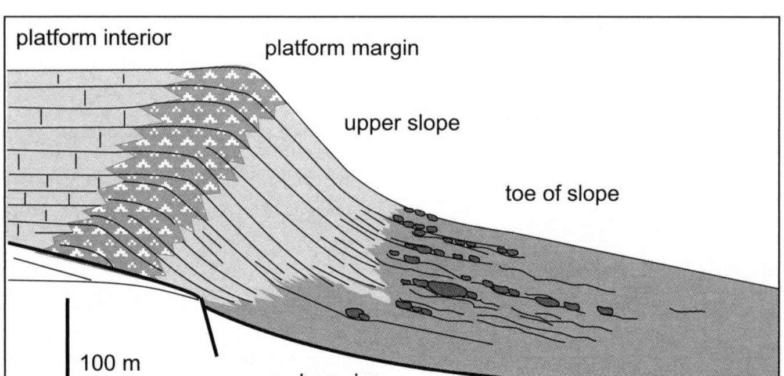

FIG. 18.—Cross section of the Early Albian aggradational stage of the Gorbea rimmed-platform succession based on field observations and logged sections (from Gómez-Pérez et al., 1999).

Downslope, the massive beds split into decimeter-scale beds that pinch out, and have a facies change into marl and marly limestone at the toe of slope, including some platform-margin-collapse megabreccias. Depositional dip decreases from 12° to 1–2°. Here, lithofacies consist of thin-bedded, fine-grained bioclastic marl and marly limestone interbedded with locally graded skeletal floatstone, and packstone, and matrix-supported breccia, and olistoliths. Deeper-water facies consist of basinal argillaceous limestone and marl. Cross-bedded grainstone and packstone suggest current action at the toe of slope. High-order depositional sequences in this platform succession express changes in relative sea-level cyclicity thought to be related to tectonic activity. These are superimposed on a major eustatic signal.

Triassic Examples from the Dolomites, Southern Alps.—

The Triassic was a period of biota readaptation after the end-Paleozoic extinction. During the Triassic, swarms of carbonate platforms developed in the Southern Alps in the northwestern part of the Tethyan embayment. These extensional basins were controlled by strike-slip tectonics (Doglioni, 1987).

Slope-Controlled Platforms.—In the South Alpine Tethyan embayment, although there were differences between different margins that changed through time, most platforms were dominated by automicritic boundstones (both peloidal and aphanitic micrite, mostly organized into stromatolitic laminites and thrombolites) with minor skeletal contribution (*Tubiphytes*, skeletal cyanobacteria, sponges) of the rock, and a considerable amount of early submarine cements. As established by Russo et al. (1997), these platform margins can be considered as microbial mud mounds (*sensu* Bosence and Bridges, 1995). They have an accretionary and organically controlled character of mostly micrites. Brandner et al. (1991) consider this carbonate-producing biota association as low-growing binding and baffling communities that thrived in the subtidal, upper to middle fore-slope. They did not grow on the platform-slope margin or at the platform top. The abundance of the megabreccia clasts and olistoliths in the lower slope, along with abundance of cements, indicates that the automicrite mud was produced in upper- to middle-slope settings. They were lithified early, and they frequently collapsed. All these platforms show a very low amount of shallow-water sediments redeposited at the toe of slope and in the basin (Brandner et al., 1991; Russo et al., 1998).

By the end of the Carnian, after diversification reef-building communities, carbonate platforms began to change style. Corals and sponges dominated some platforms, whereas others were controlled by serpulid–microbial cement. The platform-margin boundstones formed a narrow belt that might be indicative of a limitation of the binding organisms to thrive in the platform-slope outer edge. Frequent talus grainstone and rudstones, with abundant crinoids, foraminifers, *Tubiphytes*, bivalves, brachiopods, and serpulid fragments, and coarse- to finer-grained turbidites interbedded with dark, organic-rich micrites (Climaco et al., 1997; Gianolla et al., 2003), indicate an increment of platform-top sediment export, with respect to Middle Triassic platforms, into a poorly oxygenated basin.

Shallow-Water Platform Tops.—In the Lofer cyclothems (Fischer, 1964) that characterize the Upper Triassic platform-top lithofacies, individual cyclothems consists of a lower intertidal interval that records the flooding of the platform top, and an upper subtidal interval that captures the platform submergence. Truncation on top relates to subsequent platform-top emergence. This indicates base level to have been controlled by a shallow-water rim at the platform margin, and, consequently, minor falls in sea level resulted in exposure and erosion of the platform top, recording Milankovitch cyclicity, although Enos and Samankassou (1998) point out that not all of these Upper Triassic cyclothems have a clear upward-deepening trend.

In the Triassic platform slopes, with depth-extended, light-independent, carbonate production, the impact of high-frequency sea-level cyclicity is minimum, except when there is a general trend of increase in accommodation driven by subsidence. Nevertheless, the impact of this cyclicity is better recorded in the platforms top as the margin built closer to sea level.

DISCUSSION

Recognizing the diversity of carbonate platforms and their controls, this paper proposes a scheme to generate more accurate interpretations of carbonate platforms. This should then lead to an improved prediction of the distribution of the sedimentary geometries and associated facies heterogeneity.

Such a strategy recognizes that each group of carbonate rocks is unique and formed in its own distinctive way. Stratigraphic architecture of each is controlled by the physical, chemical, and biological rules for that specific Phanerozoic window. Successful interpreters develop their own and often unique depositional models to understand how their group of rocks formed. These models will explain the local capacity for production of carbonate sediment and the setting in which the local biota thrived. To this end the models would need to consider the ecological and physicochemical conditions that existed for that position!

The philosophy we outline is not new. It matches that used by Nils Steno (1669) to describe the origin of the layered character of rocks in Tuscany. Steno's principles of stratigraphy enabled him to focus directly and systematically on the geological processes responsible for the fabrics that caught his attention. He first established the order in which the rocks and their fabrics were formed and then integrated his understanding of the physical processes he thought were responsible for the character of the rocks he saw.

Today we use the same integrated analytical approach, though the details of the mechanics of such an approach are seldom expressed in what we write or read. We tend to skip the details when explaining models, assuming that the audience intuitively understands them. Our contention is that for the novitiate to understand the mechanics of how a particular suite of carbonate rocks formed, they should mentally backstrip these sediments in the order they were laid down. Next they should analytically rebuild them to match the rock fabrics they are examining.

The difference in our analysis, from that of Steno, is the much larger and increasing size of our collective data and rule base. This has grown and is constantly being improved. Preexisting models are used to synthesize ideas and build interpretations of sedimentary rocks. So long as these models do not contravene physical laws, we use them. We even modify them to interpret unique sedimentary stratigraphic situations, but we do so with critical thought and caution! Most novitiates have no choice but to work from models in the literature. Their limited experience tends to make them loath to iteratively change existing models. Nevertheless if these models do not meet the conditions found in an outcrop, then they should be either modified or discarded. We firmly believe that stratigraphic interpretation is impossible without critical thought tied to iterative process modeling.

Backstripping Carbonate Platform Architecture: Proposed Analytical Strategy

The analytical strategy we propose for carbonates formalizes common heuristic rules used to examine carbonate rocks and exchange information with peers and students. This strategy proposes involves establishing:

Geologic setting of sedimentary section	Within the basin: • Margin • Shelf • Outer shelf • Down slope • Basin Plate tectonic setting Age
Regional depositional setting using:	Identification and description of any widespread surfaces that subdivide the succession Characterization of the enveloped bedding patterns Characterization of the associated lithofacies.

Once the general the stratigraphic setting of a carbonate platform has been determined, a more specific genetic analysis establishes:

Lithofacies relationships within and across enveloping surfaces.
Sediment components and the interrelationships of: • Physico-chemical and biogenic components • Their textures • *In situ* versus *ex situ* forms
Processes involved in depositional setting: How these processes changed through time
Bathymetric position of the carbonate setting inferred from • Mix of geometry • Lithofacies components and properties from o Textures (hydrodynamic level) o Depth related components e.g. algae (light penetration) o Deep basin versus high on the margin (depositional geometries including basic sequences and parasequences or the building blocks)?

The geological section is now subdivided into sequences, cycles, parasequences, and/or beds.

Each of the subdivisions of stratigraphic sections is described in terms of:

Lithology
Grain type and size Fauna Sedimentary structures including • Unidirectional current structures • Bidirectional current structures • Wave structures • Biogenic structures • Ichnofacies) Contact character: gradational, abrupt, etc.

Using Walther's "Law", the stacked lithofacies and lithofacies assemblages should be mentally backstripped and related to conceptual models of the depositional setting. Further it should be determined if these stacked bodies are:

- Prograding (proximal strata overlying distal strata)
- Retrograding (distal strata overlying proximal strata)
- Aggrading

This paper considers the effects of paleoecology alone, though the recipe outlined above is flawed by important limitations that arise from the lack of information related to the ecological requirements of the ancient biota. If Neogene carbonates are analyzed, these limitations may not be dramatic. However, the uncertainty increases when older carbonates generated from often-extinct biotas are considered. These uncertainties can be expected to be quite significant when interpreting Paleozoic biotas. It is for this reason that we currently have avoided this debate by excluding a discussion of carbonates from this part of the geological section. We have also excluded analyzing how the beat of icehouse vs. greenhouse climatic conditions impacts the cyclic character of some carbonate sections, since this is beyond the scope of this paper.

The analytical strategy presented here follows a classic path to the most plausible answer. It offers a means for better evaluating the uncertainty involved in an interpretation. It also focuses attention on formulating questions similar to those in the sections below that are tied to the interpretation of the geological section and better-constrained geological data. The formulation of these questions often leads the way to more realistic interpretations and predictions of the extent of lithofacies geometries.

In the sections above we have shown how genetic analysis provides a better understanding of the factors that, in a complex interaction, control the architecture and distribution of heterogeneities of carbonates and helps one better understand and unravel apparent paradoxes. We now consider potential problems associated with the interpretation of carbonate platforms. These include "understanding of the controls on downslope carbonate sedimentation", "alternation of carbonate systems" as tied to their ecological accommodation, problems with "Holocene carbonate record as the key to the past", controls on "subaerial exposure of a platform top" determined from stacking patterns of basic accretional units, and finally, the causes for the differences in "unconformity-bounded accretional units".

Understanding Downslope Carbonate Sedimentation

The key to solving this question is the down-dip character of grain size and texture. Though grain size commonly decreases with depth as result of decreasing hydrodynamic energy, the studied Neogene platforms, for instance, document significant departures from this expectation. In some of these examples, the grain size increases down-dip and the sediment is poorly sorted. The dilemma of shallow water or deeper water promoted a closer examination of the texture and components of these rocks. As a result, it was recognized that the unsorted packstone beds of the upper portion of the section accumulated in shallower water. This, plus recognizing the role of the sheltering and baffling effect of seagrass meadows on the shelf, modified the expectations of a high-wave-energy setting to be below wave base. In contrast, the gravel-size rudstone and floatstone units were generated mostly in deeper water, though in some cases reworking by currents modified them. Moreover, in the Late Miocene ramp of Menorca the cross-bedded grainstone deposits are composed of shallow-water skeletal components. These collected well below wave-base level in settings different from those of the shore or shoal areas. Unidirectional currents that crossed the middle-ramp, ramp-slope, and toe-of-slope settings formed the cross-bedded structures associated with them. Without analysis of the intermeshing relationships of sediment composition, textures, bedding patterns, and the modern conditions offshore the Balearic Islands, a dip-inverted interpretation would result.

"Modern" reef-rimmed platforms tend to shed sediment as soon as the platform top is flooded. During the Pleistocene, this occurred during the highstand of sea level, when the top of the "basement-controlled" platform was flooded and the carbonate factory increased production. The Caribbean platforms provide good examples of this (Glaser and Droxler, 1991; Schlager et al., 1994). However, for the Late Miocene reefal platform of Mallorca, most shedding is documented to have occurred during rises in sea level when flooding created large lagoons that decreased during highstands and shut off during sea-level fall. Here, the generalized concept of carbonate-platform shedding predominating during highstand of sea level has to be modified when the carbonate factory may shed sediment and add sediment to a slope during other positions of sea level.

In fact, our contention is that carbonate systems dominated by physical accommodation may shed most sediment during sea-level falls or lowstands. For instance, in the case of the Menorca Early Tortonian ramp, sand-grade sediment was shed at increasing rates off platform when falls of sea level lowered the base level and sea-grass epiphytic components were transferred basinward and forced the slope to prograde. The Upper Miocene reef complex of Mallorca, in contrast, involves an example of an oligophotic factory where production of red algae increased significantly during lowstands when light reached the sea floor in fore-reef areas. In this latter case, the coarse-grained sediment remained *in situ*, whereas the fines were transported off farther downslope. Shallow-water, euphotic, chlorozoan production predominated when sea level rose again in the lagoonal areas.

When considering transport of skeletal carbonates, as in evaluating the angle of repose (Kenter, 1990) or estimating the sediment-transfer processes associated with waves and currents, it is absolutely essential to consider the bulk density (or effective density) of the skeletal grains, which is lowered by the intraskeletal pores. Aigner (1982) measured the porosity preserved in Eocene nummulitid tests to range between 1% and 54%. In recent *Amphisorus* foraminifera, this increases to 72%. Using these measurements, he estimated the bulk density for *Nummulites* to be as low as 1.28 g/cm^3 and for *Amphisorus* to be 0.305 g/cm^3. Similarly, Jorry (2004) and Jorry et al. (2006), measuring porosity in Eocene nummulitid tests, found higher values than Aigner (1982), 46 to 63% porosity and estimated bulk density from 1.48 to 2.61 g/cm^3. From these values, Jorry (2004) and Jorry et al. (2006) concluded that large *Nummulites* (2 cm diameter) with bulk (effective) densities of 1.7 cm^3 could have been transported when shear velocity had reached 3.3 cm/s. The low velocities needed to move nummulitid tests may explain the rarity of tractional sedimentary structures associated with them and the difficulty in recognizing talus deposits since these would have a low angle of repose. Ginsburg (2005) notes that disarticulated skeletons of crinoids had porosities that ranged between 29 to 69% for some Carboniferous species. Their resulting weights are equivalent to quartz grains with a tenth of their size (Savarese et al., 1996). Currents as low as 16 cm/s can entrain these highly porous skeletal grains.

Further, not all carbonate platforms shed sediment from the platform top. For instance, the amount of sediment shed from the platform top is very limited for those carbonate buildups that host organomineral boundstones on their margins. These include the Early Albian Gorbea platform and the Triassic examples from the Dolomites in the Southern Alps. This lack of platform-top shedding may be related to a low concentration of oxygen in seawater, which allowed microbes and sponges to thrive, but not other metazoans with higher oxygen concentration requirements. This is evident for the Triassic, a period with global suboxia (Berner, 1999; Berner et al., 2000; Falkowski et al., 2005; Huey and Ward, 2005). Coeval occurrence of mud-dominated ramps and the non-shedding platforms points to an environmental influence rather than to a biological evolutionary trend

Alternation of Carbonate Factories

Alternation of carbonate factories that build a carbonate-platform succession is another important issue. In the case of the Upper Cretaceous example referred in this paper, alternation of factories (rudist buildups and calcarenitic wedges) was independent of sea-level changes. Instead, alternation of factories was induced by changes in paleoceanographic conditions that affected the carbonate-producing biota. In the Upper Miocene reef complex of Mallorca, carbonate-factory alternation was in phase with sea-level changes. As the platform top was flooded during sea-level rises, the chlorozoan, euphotic factory enhanced production, and shut off when the lagoonal area emerged during sea-level fall and lowstand. The rhodalgal-dominated, forereef, oligophotic factory turned on as soon as light reached the basin floor during lowstands and shut off when sea level rose and basin floor became dark.

The lesson here is that changes in paleoceanographic conditions may induce more spectacular variations in platform architecture than sea-level changes. The Late Miocene platforms of the Balearic Islands dramatically illustrate this effect; the increase in accommodation space that allowed the Late Tortonian–Early Messinian reefal platform to prograde onto the Early Tortonian distally steepened ramp was not induced by a relative sea-level rise. It was instead induced by a change in the ecological accommodation with no significant change in physical accommodation. The base level for the ramp was controlled by the wave base and by the contour-parallel currents, whereas the reef–lagoon system was controlled by the sea-level position. The coral reef built a hydrodynamic-energy dam that influenced the base level for sediment to accumulate close to sea level.

Holocene Uniformitarianism

Genetic analysis of the intermeshing relationship of sediment composition, textures, and bedding patterns also highlights the need to deemphasize the application of Holocene uniformitarianism. This is particularly true of other widely accepted concepts based on Bahamian facies relationships that responded to changing physical conditions. Thus, it can be demonstrated that during the Jurassic, corals and microbes built mounds in mid-ramp settings. This contrasts with corals in the Late Cretaceous, which appeared to grow best below storm wave base and, although they built a margin, were not able to build a rigid framework up to sea level. Later, during the Early and Middle Miocene, they built mounds and biostromes from the toe of the slope to the platform top. It seems that it was not until the Late Miocene that corals started to thrive in the shallow-water, wave-agitated zone and constructed rigid frameworks up to sea level. This change from the oligophotic to the euphotic zone for thriving corals coincides with a period of diversification of *Symbiodinium* zooxanthellae, and is coeval with the global cooling that followed the Miocene climate optimum (Pochon et al., 2006). Moreover, this change also suggests a modification in the nutrient requirements for corals, shifting to oligotrophy. This coincided with the change in the dominance of red algae to coral reefs, and it may be associated with a global decrease in productivity (Halfar and Mutti, 2005). It also appears to match a shift by the Late Miocene from aragonite-inhibiting to aragonite-facilitating conditions, which allowed both corals to thrive and oolites to form in shallow-water settings. Furthermore, this raises the question of climatic influence controlling the skeletal composition. So-called temperate-type rhodalgal and foramol sediment associations predominated in the Mediterranean during the Middle Miocene. However, the climate at this time was warmer than during the Late Miocene, a time proposed for tropical-type chlorozoan sediment associations to form. The question is thus raised as to whether all of these changes in coral requirements were in fact induced by biological evolution. Or did this reflect the evolution of zooxanthellae? Was it triggered by oceanographic changes? Or, are all of these effects interrelated? All of these considerations illustrate how genetic analysis raises important questions that previous interpretations did not recognize and now drive one to formulate new questions and compel a search for more information.

Subaerial Exposure of a Platform Top

The potential for subaerial exposure of a platform top during high-frequency sea-level falls can be better constrained through genetic analysis. "Modern" reef-rimmed platforms can build a structural framework to sea level (maximum ecological accommodation). This allows a very shallow backreef area to develop, while minor sea-level falls result in exposure of the complete platform top (reef and previous lagoonal area). This situation can be used in the inverse order through backstripping the Upper Triassic Lofer cyclothems in Southern Alps. Here, shallow subtidal and intertidal lower intervals are overlain by an upper subtidal event. These are top truncated in a subaerial setting. These cyclic sediments are related to flooding followed by submergence and subsequent emergence of the platform top. This is interpreted to be driven by high-frequency, low-amplitude sea-level cycles. Similarly, the Middle Triassic cyclothems in the Latemar buildup are interpreted to require a shallow base level to accumulate and also to be the products of meter-scale orbitally forced sea-level oscillations (Preto et al., 2001). In both cases, the continuous subsidence provided the increase in accommodation needed for accumulation of a thick and continuous sequence of shallow-water cyclic carbonates. These architectures require the base level to have been controlled by a shallow-water rim at the platform margin. Consequently, minor falls in sea level resulted in exposure and erosion of the platform top. In contrast, grain-dominated ramps with minimum ecological accommodation, exposed to lowering of base level, are more likely to feature basinward sediment transfer within the middle ramp. This restricts the potential for subaerial exposure of the intertidal, or shallow subtidal, lithofacies. The potential for early diagenesis, when coupled to the amplitude and velocity of the sea-level fall, may be solved through better analysis, and perhaps to aragonite dominance vs. calcite dominance in the sediments.

Stacking Patterns of Basic Accretional Units

Tectonic subsidence (and lower-order sea-level cycles) changes accommodation at slower rates, and so a resulting increase or decrease of physical accommodation is reflected in the stacking patterns of basic accretional units. In the context of relative sea-level stability (tectonic stability, insignificant subsidence, and a stillstand of sea level), accretional units stack in a forward-stepping configuration. In contrast, in the context of relative sea-level rise (tectonic subsidence and/or sea-level rise), the basic accretional units aggrade or even backstep. This depends on the differences in rates of sediment production and accumulation and/or increase of physical accommodation. This is true for the Late Tortonian ramp of Menorca. Here it is difficult to see the prograding pulses induced by sea-level changes because the sedimentary units have a forward-stepping configuration. In the Cretaceous of the Pyrenees this is expressed more clearly in the aggrading packages than in the forward-stepping ones.

The examples analyzed in this paper emphasize the need to consider the effect of ecological accommodation on platform architecture. The hope is that the examples provided in this paper will enable others to recognize how the effect of ecological accommodation increases when the high-frequency, low-amplitude sea-level changes increase too (see Table 1).

Unconformity-Bounded Accretional Units

In the "modern" reefal platform (F platform type in Figure 3), minor falls in sea level produce unconformity-bounded accretional units expressed by individual beds. In terms of processes, these fit the definition of a basic depositional sequence. Although these are not composed of systems tracts bounded by physical surfaces, their internal sub-zonation allows the identification of sediment accretion in the different parts of a sea-level cycle. Correspondingly, parasequences are the basic accretional units bounded by flooding surfaces, independent of their rank in the hierarchy of bed accretion. Nevertheless, as ecological accommodation decreases, the amplitude of the sea-level fluctuation required to produce a recognizable physical surface increases. This can be seen in the Middle Miocene low-angle ramp that crops out in the Apennines (B platform type in Figure 3). Here it becomes difficult to recognize even the impact of third-order cycles.

In the case of flat-topped platforms with moderate ecological accommodation (E in Figure 3), the inner-platform facies belt commonly becomes subaerially exposed during falls of sea level, whereas these platforms are associated with submarine erosion surfaces in the outer-platform facies belt. The result is that sea-level fluctuations are evidenced by the platform-top lithofacies migration, basinward and/or landward.

In grain-dominated flat-topped platforms to distally steepened ramps (D in Figure 3), platform-top lithofacies are preserved only where there is a net gain in physical accommodation. In this case the sequence boundaries consist mostly of submarine surfaces of erosion over the platform-top lithofacies. These are amalgamated with subsequent flooding surfaces and pass basinward into the correlative conformities.

In low-angle ramps influenced by shallow-water sediment production and subsequent basinward shedding (C in Figure 3), subaerial erosion surfaces develop over the inner-ramp settings. The correlative submarine surfaces of erosion and the conformity surfaces basinward are recognized by a combination of subtle changes in grain size, skeletal components, and/or bed thicknesses. These are expressed by both basinward and landward migration of facies belts. Nevertheless, these subtle changes in lithofacies seem to be even fainter where ramps are dominated by deeper-water carbonate production.

GENERAL PHILOSOPHICAL CONCLUSIONS

The approach recommended by this paper, in particular the separation of physical and ecological accommodation and the use of backstripping analysis, lends an increasingly "apparent" complexity to the interpretation of carbonate platforms. However, if a better understanding of carbonate platform heterogeneity is to be developed, then the proposed analysis helps us push the envelope of our knowledge. Conversely, one could argue that, instead, this approach forces the observer to enter the realm of unfettered speculation. We clearly disagree with this position, arguing that as one examines all the models in the paper, the kind of analysis we propose forces us to use every bit of data we can to develop reasonable and the least speculative of models! These models require reassembly of the backstripped rock record, and consideration of the evolution of the biota involved coupled with the resulting changes in ecological requirements. This also requires an integration of many different properties from the outcrop, including an understanding of the hydrodynamic setting, the physical accommodation, and the "proposed" ecological accommodation. This lends to the strength of the proposed models, because they are, and will be, based on solid scientific observation and principles.

Thus, the diversity in platform architecture and the complexities in the stacking patterns can be better recognized as the products of the interaction of ecological accommodation, carbonate-producing biota, and changes in physical accommodation. The platform descriptions presented in this paper demonstrate how successful this approach proves to be in deciphering the key factors that operate in controlling platform architecture and distribution of heterogeneities. The paper argues that, at another level, the analytical philosophy traced in the text increases the potential for accurate prediction of heterogeneities. It involves a mix of cautiously applied uniformitarianism associated with process–product relationships. It also provides clues for improving an evaluation of uncertainties associated with predictions so that these can be applied in subsurface interpretations. Simultaneously, it initiates new questions that drive the attention of the interpreter to look for new answers.

Finally, the paper shows that biotas from different Phanerozoic windows may produce similar products, suggesting that these carbonate sediments accumulated in a similar manner. Nevertheless, different platforms do occur in the same Phanerozoic window. The paper proposes that these differences are the result of the carbonate being produced in different ecological accommodations. It also recognizes that changes in oceanographic conditions may often produce a larger impact on stratal patterns and facies architecture than sea-level changes, particularly where these conditions affect the carbonate-producing biota.

ACKNOWLEDGMENTS

Comments, criticisms, and suggestions from Michele Morsilli are much appreciated. In particular we thank Jeff Lukasik, Toni Simo, Wolfgang Schlager, and Heiko Hillgartner, who read our manuscript and offered us valuable advice on improving its style and content. Without their input it was no more than a flawed masterpiece. Funding from Spanish projects DGI n° BTE2001-0372-C02 and DGI n° CGL200500537/BTE are acknowledged.

REFERENCES

ABREU, V.S., AND ANDERSON, J.B., 1998, Glacial eustasy during the Cenozoic: sequence stratigraphic implications: American Association of Petroleum Geologists, Bulletin, v. 82, p. 1385–1400.

ABREU, V.S., AND HADDAD, G.A., 1998, Glacioeustatic fluctuations: the mechanism liking stable isotope events and sequence stratigraphy from the Early Oligocene to Middle Miocene, *in* De Graciansky, P.C., Hardenbol, J., Jacquin, T., and Vail, P.R., eds., Mesozoic and Cenozoic Sequence Stratigraphy of European Basins, SEPM, Special Publication 60, p. 245–259.

ACCORDI, B., DEVOTO, G., LA MONICA, G.B., PRATURLON, A., SIRNA, G., AND ZALAFFI, M., 1967, Il Neogene nell'Appennino laziale-abruzzese. Commitee on Mediterranean Neogene Stratigraphy, (1969), Proc. IV Session, Bologna: Giornale di Geologia, v. 35, p. 235–268.

AIGNER, T., 1982, Event-stratification in nummulite accumulations and in shell beds from the Eocene of Egypt, *in* Einsele, G., and Seilacher, A., eds., Cyclic and Event Stratification: New York, Springer-Verlag, p. 248–262.

ALLEN, J.R.L., 1982, A model of storm sedimentation in shallow waters, *in* Sedimentary Structures; Their Character and Physical Basis: Amsterdam, Elsevier, Development in Sedimentology, 30B, p. 487–506.

AURELL, M., AND BÁDENAS, B., 2004, Facies and depositional sequences evolution controlled by high-frequency sea-level changes in a shallow-water carbonate ramp (Late Kimmeridgian, NE Spain: Geological Magazine, v. 141, p. 717–733.

AURELL, M., BÁDENAS, B., BOSENCE, D.W.J., AND WALTHAM, D.A., 1998, Carbonate production and offshore transport on a Late Jurassic carbonate ramp (Kimmeridgian, Iberian basin, NE Spain): evidence from outcrops and computer modelling, *in* Wright, V.P., and Burchette, T.P., eds., Carbonate Ramps: Geological Society of London, Special Publication 149, p. 137–161.

AURELL, M., BOSENCE, D.W.J., AND WALTHAM, D.A., 1995, Carbonate ramp depositional systems from a late Jurassic epeiric platform (Iberian basin, Spain): a combined computer modelling and outcrop analysis: Sedimentology, v. 42, p. 75–94.

AURELL, M., ROBLES, S., BÁDENAS, B., QUESADA, S., ROSALES, I., MELÉNDEZ, G., AND GARCÍA-RAMOS, J.C., 2003, Transgressive/regressive cycles and Jurassic paleogeography of Northeast Iberia: Sedimentary Geology, v. 162, p. 239–271.

BÁDENAS, B., AND AURELL, M., 2001a, Kimmeridgian paleogeography and basin evolution of northeastern Iberia: Palaeogeography, Palaeoclimatology, Palaeoecology, v. 168, p. 291–310.

BÁDENAS, B., AND AURELL, M., 2001b, Proximal–distal facies relationship and sedimentary processes in a storm-dominated carbonate ramp (Kimmeridgian, northwest of the Iberian Ranges, Spain): Sedimentary Geology, v. 139, p. 319–340.

BÁDENAS, B., AURELL, M., AND GRÖCKE, D.R., 2005, Facies analysis and correlation of high-order sequences in middle–outer ramp successions: variations in exported carbonate on basin-wide $^{13}C_{carb}$ (Kimmeridgian, NE Spain): Sedimentology, v. 52, p. 1253–1275.

BÁDENAS, B., AURELL, M., RODRÍGUEZ-TOVAR, F.J., AND PARDO-IGÚZQUIZA, E., 2003, Sequence stratigraphy and bedding rhythms of an outer ramp limestone succession (Late Kimmeridgian, Northeast Spain): Sedimentary Geology, v. 161, p. 153–174.

BARBERA, C., SIMONE, L., AND CARANNANTE, G., 1978, Depositi circalittorali di piattaforma aperta nel Miocene Campano, Analisi sedimentologica e paleoecologica: Società Geologica Italiana, Bollettino, v. 97, p. 821–834.

BASSANT, P., 1999, The high-resolution stratigraphic architecture and evolution of the Burdigalian carbonate-siliciclastic sedimentary systems of the Mut Basin, Turkey: GeoFocus, v. 3, Université de Fribourg, Institut de Géologie et Paléontologie (Switzerland), 277 p.

BASSANT, P., VAN BUCHEM, F.S.P., STRASSER, A., AND GORUR, N., 2005, The stratigraphic architecture and evolution of the Burdigalian carbonate–siliciclastic sedimentary systems of the Mut Basin, Turkey: Sedimentary Geology, v. 173, p. 187–232.

BASSANT, P., VAN BUCHEM, F.S.P., STRASSER, A., AND LOMANDO, T., 2004, A comparison of two early Miocene carbonate margins: the Zhujiang carbonate platform (subsurface, South China) and the Pirinç platform (outcrop, Southern Turkey), *in* Grammer, M., and Harris, P.M., eds., Integration of Outcrop and Modern Analogs in Reservoir Modeling: American Association of Petroleum Geologists, Memoir 80, p. 153–170.

BEAVINGTON-PENNEY, S.J., WRIGHT, V.P., AND RACEY, A., 2005, Sediment production and dispersal on foraminifera-dominated early Tertiary ramps: the Eocene El Garia Formation, Tunisia: Sedimentology, v. 52, p. 537–569.

BERÁSTEGUI, X., LOSANTOS, M., AND MUÑOZ, J.A., 1990, Structure and sedimentary evolution of the Organyà basin (Central South Pyrenean Unit, Spain) during the lower Cretaceous: Societé Géologique de France, Bulletin, v. 8, p. 251–264.

BERÁSTEGUI, X., LOSANTOS, M., MUÑOZ, J.A., AND PUIGDEFABREGAS, C., 1993, Tall geològic del Pirineu Central 1/200.000, Institut Cartogràfic de Catalunya, Barcelona.

BERGOMI, G., AND DAMIANI, A.V., 1976, Diagenesi precoce nei depositi Serravalliano–Tortoniani del Lazio e considerazioni sulla evoluzione strutturale del Bacino di sedimentazione miocenico: Servizio Geologico d'Italia, Bollettino, v. 97, p. 35–66.

BERNER, R.A., 1999, Atmospheric oxygen over Phanerozoic time: National Academy of Sciences, U.S.A., Proceedings, v. 96, p. 10955–10957.

BERNER, R.A., PETSCH, S.T., LAKE, J.A., BEERLING, D.J., POPP, B.N., LANE, R.S., LAWS, E.A., WESTLEY, M.B., CASSAR, N., WOODWARD, F.I., AND QUICK, W.P., 2000, Isotope fractionation and atmospheric oxygen: implications for Phanerozoic O_2: Science, v. 287, p. 1630–1633.

BOSENCE, D.W.J., AND BRIDGES, P.H., 1995, A review of the origin and evolution of carbonate mud-mounds, *in* Monty, C.L.V., Bosence, D.W.J., Bridges, P.H., and Pratt, B.R., eds., Carbonate Mud-Mounds; Their Origin and Evolution: International Association of Sedimentologists, Special Publication 23, p. 3–9.

BOSSCHER, H., AND SCHLAGER, W., 1992, Computer simulations of reef growth: Sedimentology, v. 39, p. 503–512.

BRANDANO, M., 2001, Risposta fisica delle aree di piattaforma carbonatica agli eventi più significativi del Miocene nell'Appennino centrale: Ph.D. Thesis, Universitá di Roma "La Sapienza", 180 p.

BRANDANO, M., 2003, Tropical/subtropical inner ramp facies in Lower Miocene "Calcari a Briozoi e Litotamni" of the Monte Lungo area (Cassino Plain, Central Apennines, Italy): Società Geologica Italiana, Bollettino, v. 122, p. 85–98.

BRANDANO, M., CIVITELLI, G., AND CORDA, L., 2001, A foramol type carbonate ramp in tropical–subtropical setting: the link between tectonics and nutrients (abstract): American Association of Petroleum Geologists, Annual Meeting, p. A 24.

BRANDANO, M., AND CORDA, L., 2002, Nutrients, sea level and tectonics: constraints for the facies architecture of a Miocene carbonate ramp in central Italy: Terra Nova, v. 14, p. 257–262.

BRANDNER, R., FLÜGEL, E., AND SENOWBARI-DARYAN, B., 1991, Microfacies of carbonate slope boulders: indicator of the source area (Middle Triassic: Mahlknecht Cliff, Western Dolomites): Facies, v. 25, p. 279–296.

BURCHETTE, T.P., AND WRIGHT, V.P., 1992, Carbonate ramp depositional systems: Sedimentary Geology, v. 79, p. 3–57.

CALVET, F., AND TUCKER, M.E., 1988, Outer ramp carbonate cycles in the Upper Muschelkalk, Catalan Basin, NE Spain: Sedimentary Geology, v. 57, p. 185–198.

CALVET, F., TUCKER, M.E., AND HENTON, J.M., 1990, Middle Triassic carbonate ramp systems in the Catalan Basin, northeast Spain: facies, systems tracts, sequences and controls, *in* Tucker, M.E., Wilson, J.L., Crevello, P.D., Sarg, J.R., and Read, J.F., eds., Carbonate Platforms: Facies, Sequences and Evolution: International Association of Sedimentologists, Special Publication 9, p. 79–108.

CARANNANTE, G., ESTEBAN, M., MILLIMAN, J.D., AND SIMONE, L., 1988, Carbonate lithofacies as paleolatitude indicators: problems and limitations: Sedimentary Geology, v. 60, p. 333–346.

CIARANFI, N., GHISETTI, F., GUIDA, M., IACCARINO, G., LAMBIASE, S., PIERI, P., RAPISARDI, L., RICCHETTI, G., TORRE, M., TORTORICI, L., AND VEZZANI, L., 1983, Carta Neotettonica dell'Italia meridionale, Progetto Finalizzato Geodinamica, Carta Neotettonica dell'Italia meridionale, Pubblicazione 515 del Progetto Finalizzato Geodinamica del Consiglio Nazionale delle Ricerche, 62 p.

CLIMACO, A., BONI, M., IANNACE, A., AND ZAMPARELLI, V., 1997, Platform margins, microbial/serpulids bioconstructions and slope-to-basin sediments in the Upper Triassic of the 'Verbicaro Unit' (Lucania and Calabria, Southern Italy): Facies v. 36, p. 37-56.

DAMIANI, A.V., CHIOCCHINI, M., COLACICCHI, R., MARIOTTI, G., PAROTTO, M., PASSERI, L., AND PRATURLON, A., 1992, Elementi litostratigrafici per una sintesi delle facies carbonatiche meso-cenozoiche dell'Appennino centrale, *in* Studi preliminari all'acquisizione dati del profilo CROP (CROsta Profonda), 11 Civitavecchia-Vasto", Tozzi M., Cavinato G.P., Parotto M., eds., Studi Geologici Camerti, vol. spec. 1991/92, p. 187–213.

DERAMOND, J., SOUQUET, P., FONDECAVE-WALLEZ, M.-J., AND SPECHT, M., 1993, Relationship between thrust tectonics and sequence stratigraphy surfaces in foredeeps: model and examples from the Pyrenees (Cretaceous–Eocene, France, Spain), *in* Williams, G.D., and Dobb, A., eds., Tectonics and Seismic Sequence Stratigraphy, Geological Society of London, Special Publication 71, p. 193–219.

DIETZ, R.S., 1963, Wave-base, marine profile of equilibrium, and wave-built terraces: A critical appraisal: Geological Society of America, Bulletin, v. 74, p. 971–990.

DOGLIONI, C., 1987, Tectonics of the Dolomites (Southern Alps, Northern Italy): Journal of Structural Geology, v. 9, p. 181–193.

DOGLIONI, C., GUEGUEN, E., HARABAGLIA, P., AND MONGELLI, F., 1999, On the origin of west-directed subduction zones and applications to the western Mediterranean, *in* Durand, B., Jolivet, L., Horvath, F., and Seranne, M., eds., The Mediterranean Basins: Tertiary Extension within the Alpine Orogen: Geological Society of London, Special Publication 156, p. 541–561.

DOGLIONI, C., MONGELLI, F., AND PIERI, P., 1994, The Puglia uplift (SE-Italy): an anomaly in the foreland of the Apenninic subduction due to buckling of a thick continental lithosphere: Tectonics, v. 13, p. 1309–1321.

DOGLIONI, C., TROPEANO, M., MONGELLI, F., AND PIERI, P., 1996, Middle-Late Pleistocene uplift of Puglia: an "anomaly" in the Apenninic foreland: Società Geologica Italiana, Memorie, v. 51, p. 101–117.

Enos, P., and Samankassou, E., 1998, Lofer cyclothems revisited (Late Triassic, Northern Alps, Austria): Facies, v. 38, p. 207–228.

FALKOWSKI, P.G., KATZ, M.E., MILLIGAN, A.J., FENNEL, K., CRAMER, B.S., AUBRY, M.P., BERNER, R.A., NOVACEK, M.J., AND ZAPOL, W.M., 2005, The rise of oxygen over the past 205 million years and the evolution of large placental mammals: Science, v. 309, p. 2202– 2204.

FISCHER, A.G., 1964, The Lofer cyclothems of the Alpine Triassic, *in* Merriam, D.F., ed., Symposium on Cyclic Sedimentation: Kansas Geological Survey, Bulletin 169, p. 107–149.

GARCIA-SENZ, J., 2002., Cuencas extensivas del Cretácico inferior en los Pirineos centrales, formación y subsiguiente inversión: Ph.D. Thesis, University of Barcelona, 310 p.

GIANOLLA, P, DE ZANCHE, V, AND ROGHI, G, 2003, An Upper Tuvalian (Triassic) platform–basin system in the Julian Alps: the start-up of the Dolomia Principale (Southern Alps, Italy): Facies, v. 49, p. 19–23.

GILI, E., AND LA BARBERA, M., 1998, Hydrodynamic behaviour of hippuritid rudist shells: ecological consequences: Géobios, v. 22, p. 137–145.

GINSBURG, R., 2005, Disobedient sediments can feedback on their transportation, deposition and geomorphology: Sedimentary Geology, v. 175, p. 9–18.

GINSBURG, R.N., AND LOWENSTAM, H.A., 1958, The influence of marine bottom communities on the depositional environments of sediments: Journal of Geology, v. 66, p. 310–318.

GLASER, K.S., AND DROXLER, A.W., 1991, High production and highstand shedding from deeply submerged carbonate banks, northern Nicaragua Rise: Journal of Sedimentary Petrology, v. 61, p. 128–142.

GÓMEZ-PÉREZ, I., FERNÁNDEZ-MENDIOLA, P.A., AND GARCÍA-MONDÉJAR, J., 1999, Depositional architecture of a rimmed carbonate platform (Albian, Gorbea, western Pyrenees): Sedimentology, v. 46, p. 337–356.

HALFAR, J., AND MUTTI, M., 2005, Global dominance of coralline red-algal facies: A response to Miocene oceanographic events: Geology, v. 33, p. 481–484.

HAQ, B.U., HARDENBOL, J., AND VAIL, P.R., 1987, Chronology of fluctuating sea levels since the Triassic: Science, v. 235, p. 1156–1167.

HELLAND-HANSEN, W., AND MARTINSEN, O.J., 1996, Shoreline trajectories and sequences: descriptions of variable depositional-dip scenarios: Journal of Sedimentary Research, v. 66, p. 670–688.

HUEY, R.B., AND WARD, P.D., 2005, Hypoxia, global warming, and terrestrial Late Permian extinctions: Science, v. 308, p. 398–401

HUNT, D., AND TUCKER, M.E., 1992, Stranded parasequences and the forced regressive wedge systems tract: deposition during base-level fall: Sedimentary Geology, v. 81, p. 1-9.

HUNT, D., AND TUCKER, M.E., 1995, Stranded parasequences and the forced regressive wedge systems tract: deposition during base-level fall—reply: Sedimentary Geology, v. 95, p. 147–160.

IANNONE, A., AND PIERI, P., 1982, Caratteri neotettonici delle Murge: Geologia Applicata e Idrogeologia, v. 18, p. 147–159.

INSALACO, E., 1998, The descriptive nomenclature and classification of growth fabrics in fossil scleractinian reefs: Sedimentary Geology, v. 118, p. 159–186.

JANSON, X., 2002, Architecture and seismic expression of Miocene carbonate barrier–lagoon systems (Ermenek platform, Turkey and Zhujiang platform, South China Sea): Ph.D. Thesis, University of Miami, Comparative Sedimentology Laboratory, 240 p.

JERVEY, M.T., 1988, Quantitative geological modeling of siliciclastic rock sequences and their seismic expression, *in* Wilgus, C.K., Hasting, B.S., Kendall, C.G.St.C., Posamentier, H.W., Ross, C.A., and Van Wagoner, J.C., eds., Sea-Level Changes: An Integrated Approach: SEPM, Special Publication 42, p. 47–69.

JOHNSON, D.W., 1919, Shore Processes and Shoreline Development, First Edition: New York, Wiley, 584 p.

JORRY, S., 2004, The Eocene Nummulite Carbonates (Central Tunisia and NE Libya): Sedimentology, Depositional Environments, and Application to Oil Reservoirs: Ph.D. Thesis, Université de Genève: Terre and Environment, v. 48, p. 206.

JORRY, S.J., HASLER, C.A., AND DAVAUD, E., 2006, Hydrodynamic behaviour of Nummulites: implications for depositional models: Facies, v. 52, p. 221–235.

KENDALL, C.G.ST.C., AND SCHLAGER, W., 1981, Carbonates and relative changes in sea level: Marine Geology, v. 44, p. 181–212.

KENTER, J.A.M., 1990, Carbonate platform flanks: slope angle and sediment fabric: Sedimentology, v. 37, p. 777–794.

KOERSCHNER, W.F., III., AND READ, J.F., 1989, Field and modelling of Cambrian cycles, Virginia Appalachians: Journal of Sedimentary Petrology, v. 59, p. 654–687.

LUKASIK, J., AND JAMES, N.P., 2006, Carbonate sedimentation, climate change, and stratigraphic completeness on a Miocene cool-water epeiric ramp, Murray Basin, South Australia, *in* Pedley, H.M., and Carannante, G., eds., Cool-Water Carbonates: Depositional Systems and Palaeoenvironmental Controls: Geological Society of London, Special Publication 255, p. 217–244.

LUKASIK, J.J., JAMES, N.P., MCGOWRAN, B., AND BONE, Y., 2000, An epeiric ramp: low-energy, cool-water carbonate facies in a Tertiary inland sea, Murray Basin, South Australia: Sedimentology, v. 47, p. 851–881.

MITCHUM, R.M., AND VAN WAGONER, J.C., 1991, High-frequency sequences and their stacking patterns: sequence stratigraphic evidence of high-frequency eustatic cycles: Sedimentary Geology, v. 70, p. 131–160.

MONTY, C.L.V., 1995, The rise and nature of carbonate mud-mounds: an introductory actualistic approach, *in* Monty, C.L.V., Bosence, D.W.J., Bridges, P.H., and Pratt, B.R., eds., Carbonate Mud-Mounds; Their Origin and Evolution: International Association of Sedimentologists, Special Publication 23, p. 11–48.

NELSON, C.S., KEANE, S.L., AND HEAD, P.S., 1988, Non-tropical carbonate deposits on the modern New Zealand shelf: Sedimentary Geology, v. 60, p. 71–94.

NEUWEILER, F., GAUTRET, P., THIEL, V., LANGE, R., MICHAELIS, W., AND REITNER, J., 1999, Petrology of Lower Cretaceous carbonate mud mounds (Albian, N. Spain): insights into organomineralic deposits of the geological record: Sedimentology, v. 46, p. 837–859.

PIERRE, A., 2006, Un analogue de terrain pour les rampes oolitiques anciennes. Un affleurement continu à l'échele de la sismique (falaises jurassiques d'Amellago, Haut Atlas, Maroc): Ph.D. Thesis, Université de Bourgogne, 232 p.

PIERRE, A., DURLET, C., AND CHELAI, E.H., 2005, Spatial and temporal distribution of facies in a Jurassic oolitic ramp (Amellago transect, High Atlas of Morocco): 24th Internation Association of Sedimentologists, Meeting of Sedimentology, Muscat, p. 124.

PIERRE, A, DURLET, C, AND RAZIN, P, 2006, A seismic to reservoir oolitic ramp model (example from the Lias–Dogger transition, High Atlas, Morocco): Architecture of Carbonate Systems Through Time: Reference Models for the Mesozoic and Tertiary of Southern Europe, North Africa and the Middle East: American Association of Petroleum Geologists, European Region Conference, Mallorca, Spain.

POCHON, X., MONTOYA-BURGOS, J.I., STADELMANN, B., AND PAWLOWSKI, J., 2006, Molecular phylogeny, evolutionary rates, and divergence timing of the symbiotic dinoflagellate genus *Symbiodinium*: Molecular Phylogenetics and Evolution, v. 38, p. 20–30.

POMAR, L., 1991, Reef geometries, erosion surfaces and high-frequency sea-level changes, upper Miocene reef complex, Mallorca, Spain: Sedimentology, v. 38, p. 243–270.

POMAR, L., 1993, High-resolution sequence stratigraphy in prograding carbonates: application to seismic interpretation, *in* Loucks, B., and Sarg, J.F., eds., Carbonate Sequence Stratigraphy: Recent Developments and Applications: American Association of Petroleum Geologists, Memoir 57, p. 389–407.

POMAR, L., 2001a, Ecological control of sedimentary accommodation: evolution from a carbonate ramp to rimmed shelf, Upper Miocene, Balearic Islands: Palaeogeography, Palaeoclimatology, Palaeoecology, v. 175, p. 249–272.

POMAR, L., 2001b, Types of carbonate platforms, a genetic approach: Basin Research, v. 13, p. 313–334.

POMAR, L., BRANDANO, M., AND WESTPHAL, H., 2004, Environmental factors influencing skeletal-grain sediment associations: a critical review of Miocene examples from the Western-Mediterranean: Sedimentology, v. 51, p. 627–651.

POMAR, L., GILI, E., OBRADOR, A., AND WARD, W.C., 2005, Facies architecture and high-resolution sequence stratigraphy of an upper Cretaceous platform margin succession, Southern Central Pyrenees, Spain: Sedimentary Geology, v. 175, p. 339–365.

POMAR, L., OBRADOR, A., AND WESTPHAL, H., 2002, Sub-wavebase cross-bedded grainstones on a distally steepened carbonate ramp, upper Miocene, Menorca, Spain: Sedimentology, v. 49, p. 139–169.

POMAR, L., AND TROPEANO, M., 2001, The "Calcarenite di Gravina" Fm. in Matera (Southern Italy): new insights for large-scale cross-bedded sandbodies encased in offshore deposits: American Association of Petroleum Geologists, Bulletin, v. 84, p. 661–689.

POMAR, L., AND WARD, W.C., 1994, Response of a Miocene carbonate platform to high-frequency eustasy: Geology, v. 22, p. 131–134.

POMAR, L., AND WARD, W.C., 1995, Sea-level changes, carbonate production and platform architecture: the Llucmajor Platform, Mallorca, Spain, *in* Haq, B.U., ed., Sequence Stratigraphy and Depositional Response to Eustatic, Tectonic and Climatic Forcing: Dordrecht, The Netherlands, Kluwer Academic Press, p. 87–112.

POMAR, L., AND WARD, W.C., 1999, Reservoir-scale heterogeneity in depositional packages and diagenetic patterns on a reef-rimmed platform, Upper Miocene, Mallorca, Spain: American Association of Petroleum Geologists, Bulletin, v. 83, p. 1759–1773.

POMAR, L., WARD, W.C., AND GREEN, D.G., 1996, Upper Miocene reef complex of the Llucmajor area, Mallorca, Spain, *in* Franseen, E., Esteban, M., Ward, W.C., and Rouchy, J.M., eds., Models for Carbonate Stratigraphy from Miocene Reef Complexes of the Mediterranean regions: SEPM, Concepts in Sedimentology and Paleontology, no. 5, p. 191–225.

POSAMENTIER, H.W., AND JAMES, D.P., 1993, An overview of sequence-stratigraphic concepts: uses and abuses, *in* Posamentier, H.W., Summerhayes, C.P., Haq, B.U., and Allen, G.P., eds., Sequence Stratigraphy and Facies Associations: International Association of Sedimentologists, Special Publication 18, p. 3–18.

PRETO, N., HINNOV, L.A., HARDIE, L.A., AND DE ZANCHE, V., 2001, Middle Triassic orbital signature recorded in the shallow-marine Latemar carbonate buildup (Dolomites, Italy): Geology, v. 29, p. 1123–1126.

PRICE, G.C., SELLWOOD, B.W., AND VALDES, P.J., 1995, Sedimentological evaluation of general circulation model simulations for the "greenhouse" Earth: Cretaceous and Jurassic case studies: Sedimentary Geology, v. 100, p. 159–180.

READ, J.F., 1982, Carbonate platforms of passive (extensional) continental margins: types, characteristics and evolution: Tectonophysics, v. 81, p. 195–212.

READ, J.F, 1985, Carbonate platform facies models: American Association of Petroleum Geologists, Bulletin, v. 69, p. 1–21.

REINECK, H.E., AND SINGH, I.B., 1980, Depositional Sedimentary Environments: Berlin, Springer-Verlag, 549 p.

RICH, J.L., 1951, Three critical environments of deposition and criteria for recognition of rocks deposited in each of them: Geological Society of America, Bulletin, v. 62, p. 1–20.

RUSSO, F., MASTANDREA, A., AND NERI, C., 1998, Evoluzione degli organismi costruttori nelle piataforme Triassiche delle Dolomiti (Italia): Società Geologica Italiana, Memorie, p. 479–488.

RUSSO, F., NERI, C., MASTANDREA, A., AND BARACCA, A., 1997, The mud mound nature of the Cassian platform margins of the Dolomites. A case history: the Cipit boulders from Punta Grohmann (Sasso Piatto Massif, Northern Italy): Facies, v. 36, p. 25–36.

SANDBERG, P.A., 1983, An oscillating trend in Phanerozoic nonskeletal carbonate mineralogy: Nature, v. 305, p. 19–22.

SARG, J.F., 1988, Carbonate sequence stratigraphy, *in* Wilgus, C.K., Hastings, B.S., and Kendall, C.G.St.C., eds., Sea-level Changes: An Integrated Approach: SEPM, Special Publication 42, p. 155–182.

SAVARESE, M., DODD, J.R., AND LANE, N.G., 1996, Taphonomic and sedimentological implications of crinoid intraskeletal porosity: Lethaia, v. 29, p. 141–156.

SCHLAGER, W., 1981, The paradox of drowned reefs and carbonate platforms: Geological Society of America, Bulletin, v. 92, p. 197-211.

SCHLAGER, W., 1992, Sedimentology and Sequence Stratigraphy of Reefs and Carbonate Platforms: American Association of Petroleum Geologists, Continuing Education Course Note Series, no. 34, 71 p.

SCHLAGER, W., REIJMER, J.J.G., AND DROXLER, A., 1994, Highstand shedding of carbonate platforms: Journal of Sedimentary Research, v. A64, p. 270–281.

STANLEY, S.M., AND HARDIE, L.A., 1998, Secular oscillations in the carbonate mineralogy of reef-building and sediment-producing organisms driven by tectonically forced shifts in seawater chemistry: Palaeogeography, Palaeoclimatology, Palaeoecology, v. 144, p. 3–19.

STANLEY, S.M., AND HARDIE, L.A., 1999, Hypercalcification: paleontology links plate tectonics and geochemistry to sedimentology: GSA Today, v. 9, no. 2, p. 2–7.

STENO, N. (alias Nils Steno, Niels Stensen, and Nicolai Stenonis), 1669, De Solido Intra Solidium Naturaliter Contento Dissertationis Prodromus: Florence, Italy, Library of Grand Duke Ferdinand II, - V. iv, 131 p. English version: Stensen, Niels 1671, The prodromus to a dissertation concerning solids naturally contained within solids: London, J. Winter, 112 p.

SWIFT, D.J.P., PHILLIPS, S., AND THORNE, J.A., 1991, Sedimentation on continental margins, VI: lithofacies and depositional systems, *in* Swift, D.J.P., Oertel, G.F., Tillman, R.W., and Thorne, J.A., eds., Shelf Sand and Sandstone Bodies: International Association of Sedimentologists, Special Publication 14, p. 89–152.

SWIFT, D.J.P., AND THORNE, J.A., 1991, Sedimentation on continental margins, I: a general model for shelf sedimentation, *in* Swift, D.J.P., Oertel, G.F., Tillman, R.W., and Thorne, J.A., eds., Shelf Sand and Sandstone Bodies: International Association of Sedimentologists, Special Publication 14, p. 3–31.

VAIL, P.R., AUDEMARD, F., BOWMAN, S.A., EISNER, P.N., AND PEREZ-CRUZ, C., 1991, The stratigraphic signatures of tectonics, eustasy and sedimentology—an overview, *in* Einsele, G., Ricken, W., and Seilacher, A., eds., Cycles and Events in Stratigraphy: Berlin, Springer-Verlag, p. 617–659.

VAIL, P.R., MITCHUM, JR., R.M., TODD, R.G., WIDMIER, J.M., THOMPSON, S., III, SANGREE, J.B., BUBB, J.N., AND HATLELID, W.G., 1977, Seismic stratigraphy and global changes of sea level, *in* Payton, C.E., ed., Seismic Stratigraphy—Applications to Hydrocarbon Exploration: American Association of Petroleum Geologists, Memoir 26, p. 49–212.

VAN WAGONER, J.C., MITCHUM, R.M., JR., CAMPION, K.M., AND RAHMANIAN, V.D., 1990, Siliciclastic Sequence Stratigraphy in Well Logs, Cores and Outcrops: American Association of Petroleum Geologists, Methods in Exploration Series no. 7, p. 1–55.

VAN WAGONER, J.C., POSAMENTIER, H.W., MITCHUM, R.M., JR., R.M., VAIL, P.R., SARG, J.F., LOUTIT, T.T., AND HARDENBOL, J., 1988, An overview of the fundamentals of sequence stratigraphy and key definitions, *in* Wilgus, C.K., Hastings, B.S., Kendall, C.G.St.C., Posamentier, H.W., Ross, C.A., and Van Wagoner, J.C., eds., Sea-Level Changes: An Integrated Approach: SEPM, Special Publication 42, p. 39–45.

WILSON, J.L., 1975, Carbonate Facies in Geologic History: New York, Springer-Verlag, 470 p.

WRIGHT, V.P., AND BURGESS, P.M., 2005, The carbonate factory continuum, facies mosaics and microfacies: an appraisal of some of the key concepts underpinning carbonate sedimentology: Facies, v. 51, p. 17–23.

WRIGHT, V.P., AND FAULKNER, T.J., 1990, Sediment dynamics of Early Carboniferous ramps: a proposal: Geological Journal, v. 25, p. 139–144.

ZACHOS, J., PAGANI, M., SLOAN, L., THOMAS, E., AND BILLUPS, K., 2001, Trends, rhythms, and aberrations in global climate 65 Ma to present: Science, v. 292, p. 686–693.

LATERAL VARIABILITIES OF CYCLE STACKING PATTERNS IN THE LATEMÀR, TRIASSIC, ITALIAN DOLOMITES

ARNDT PETERHÄNSEL

Department of Earth Sciences, University of Cambridge, Downing Street, Cambridge, CB2 3EQ, United Kingdom
Present address: LUKOIL Overseas UK Limited, Charles House, 5–11 Regent Street, London, SW1Y 4LR, United Kingdom
e-mail: arndt@lukoiloverseasuk.com

AND

SVEN O. EGENHOFF

Colorado State University, Department of Geosciences, 322 Natural Resources Building,
Fort Collins, Colorado 80523-1482, U.S.A.
e-mail: sven@warnercnr.colostate.edu

ABSTRACT: The well-known cyclic carbonate succession of the Middle Triassic Latemàr Massif in the Italian Dolomites reveals significant lateral variability in cycle numbers in platform-interior strata. Within an interval of 60 m, a 25% increase in the number of marine flooding surfaces was detected when approaching the several-hundred-meters-wide tepee belt in the backreef area, which represents the maximum elevation of the isolated Latemàr buildup. The impact of high-frequency– low-amplitude sea-level fluctuations on this elevated zone resulted in the development of spatially restricted intermittent emergence and marine flooding surfaces bounding small-scale upward-shallowing cycles. It is postulated that these alternations of submergence and subaerial exposure have favored tepee formation. Sediment collecting in the saucer-shaped tepee megapolygons further expedited upward shallowing of small-scale cycles. Conversely, deeper parts of the lagoon remained largely unaffected by high-frequency, low-amplitude sea-level oscillations: marine flooding surfaces disappear and cycles amalgamate. It is concluded that tepee structures are generally confined to topographically elevated areas where low-amplitude sea-level fluctuations were recorded. Lateral variations in cycle stacking pattern should be commonplace in shallow carbonate buildups throughout the geological record, where paleorelief existed in the platform interior.

INTRODUCTION

In recent years, the spatial distribution of carbonate lithofacies organized in small-scale cycles and related bounding surfaces has received increasing attention given its implications for reservoir architecture, sequence stratigraphy, and cyclostratigraphic interpretations (e.g., Raspini, 2001; Della Porta et al., 2002; Goldhammer et al., 1993; Sonnenfeld and Cross, 1993; Strasser et al., 2005; Borkhataria et al., 2006). Lateral continuity of these cycles and their facies for up to several hundreds of kilometers has been claimed for many pelagic to hemipelagic calcareous sequences (e.g., Maurer and Schlager, 2003) but also for shallow-water carbonate platform environments (e.g., Grotzinger, 1986; Read, 1995; Lehrmann and Goldhammer, 1999; Strasser et al., 2000; Sandulli and Raspini, 2004). Meter-scale cycles with upward shallowing are characteristic features of greenhouse periods and correspond to low-amplitude sea-level fluctuations (Wright, 1992; Read, 1995; Satterley, 1996). However, only few studies have demonstrated this on regionally extensive outcrops where spatial dimensions of facies were determined by physical tracing of beds and surfaces (Sonnenfeld and Cross, 1993; Adams and Grotzinger, 1996; Lukasik et al., 2000). These studies do not rely on facies interpolation and composite sections but are based on walking out beds along outcrops. Egenhoff et al. (1999) applied this technique in the well-exposed Lower to Middle Triassic Latemàr platform of the Italian Dolomites. In the lagoonal succession of this small buildup of 10 km^2, they demonstrated that the complexity of facies architecture was controlled by subtle paleorelief. It was apparent that most cycles are laterally continuous across the buildup interior. However, reconnaissance observations revealed that cycle stacking patterns vary laterally, in particular when approaching the tepee-dominated platform margin. Lateral variations of cycle boundaries are controlled by the magnitude of accommodation changes (Grotzinger, 1986; Read et al., 1986; Adams and Grotzinger, 1996), which in turn may be linked to autocyclic sediment transport (e.g., Pratt and James, 1986; Goldhammer et al., 1990, p. 553–554; Satterley, 1996) and/or high-frequency sea-level changes (e.g., Goldhammer et al., 1990).

We attempt to identify the extent of lateral variations in cycle stacking pattern in the isolated Latemàr carbonate platform (Fig. 1) and the governing factors that account for them. In view of the controversy surrounding the Latemàr's role as a potential cyclostratigraphic measuring stick in the Milankovitch band, the observations presented here highlight the cyclostratigraphic complexity of that buildup.

GEOLOGICAL SETTING

In the Middle Triassic, the area of the Dolomites was located in the tropics at the northwestern margin of the Tethys Ocean, forming part of a continental shelf characterized by the extended carbonate platform of the Contrin Formation (Fig. 2; Masetti and Neri, 1980; Rüffer and Zühlke, 1995). Tectonic movements in the late Anisian (early Middle Triassic; Castellarin et al., 1988) led to the fragmentation of the ramp (Gaetani et al., 1981) with shallow-marine platforms evolving on structural highs. Persisting subsidence resulted in aggrading buildups such as the Latemàr (Schlern Formation) and intervening deep-marine, partly anoxic basins (Buchenstein or Livinallongo Formation; Bosellini and Rossi, 1974). The lagoonal interior of the Latemàr was characterized by subtle paleorelief, which is reflected by lateral facies changes within individual shallowing-upward cycles (Egenhoff et al., 1999). A tepee belt in the immediate backreef formed the topo-

Controls on Carbonate Platform and Reef Development
SEPM Special Publication No. 89, Copyright © 2008
SEPM (Society for Sedimentary Geology), ISBN 978-1-56576-130-8, p. 217–229.

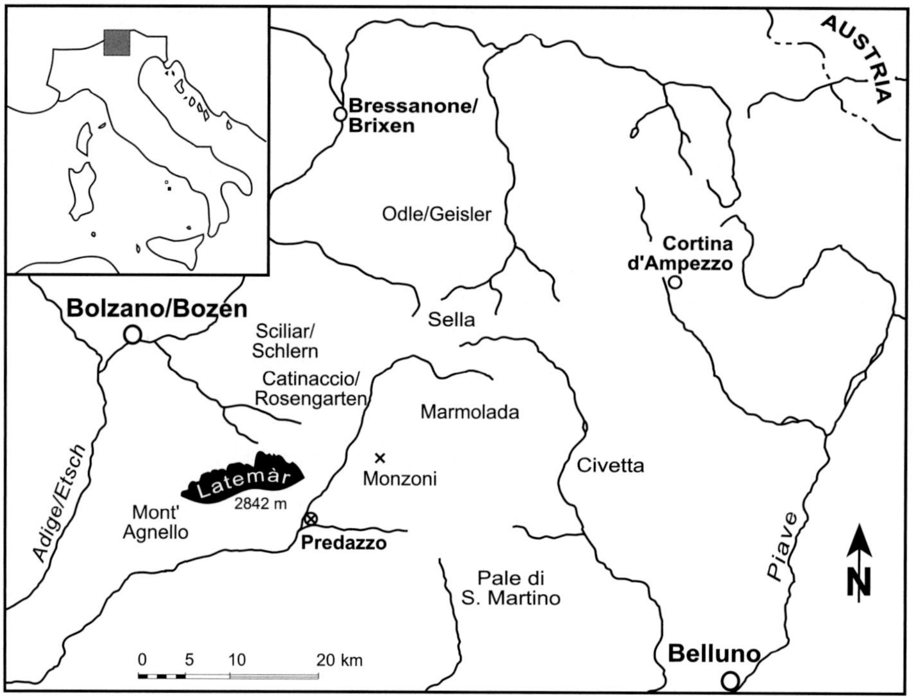

FIG. 1.—Location of the study area and other important Middle Triassic carbonate platforms in the Dolomites. Both German and Italian names are used in the South Tyrol area. Monzoni and Predazzo represent Ladinian magmatic centers (after Bosellini, 1991).

graphic high of the Latemàr platform circumscribing the peritidal lagoon (Egenhoff et al., 1999). This perimeter, 250–500 m wide, is characterized by a higher abundance of tepee structures, both vertical and lateral, than coeval deeper lagoonal areas, where tepees are comparably rare or absent. Interpretations of Latemàr interior facies are listed in Table 1.

In the early late Ladinian a magmato-tectonic event in the central Dolomites terminated a final progradation phase recorded by the youngest deposits of the Latemàr and the neighboring Rosengarten (Viel, 1979; Bosellini, 1984; Brack and Rieber, 1993, Maurer, 2000). Carbonate production ceased as volcanics and volcaniclastics covered many platform tops and partly filled the surrounding basins (Fig. 2; Bosellini and Rossi, 1974).

THE LATEMÀR CONTROVERSY

The Latemàr buildup is one of the most thoroughly studied carbonate platforms. Milankovitch-band eccentricity and superimposed precession rhythms were thought to be the controlling mechanism to account for the stratal pattern of the 720-m-thick lagoonal interior of the buildup (Hardie et al., 1986; Goldhammer et al., 1987, 1990; Goldhammer and Harris, 1989; Hardie et al., 1991; Hinnov and Goldhammer, 1991; Goldhammer et al., 1993). An allocyclic origin of the more than 600 "basic Latemàr precession cycles", which are bounded by marine flooding surfaces and are usually of meter scale, has been further advocated by more recent studies of Preto et al. (2001), Zühlke et al. (2003), and Zühlke (2004). These authors used time-series analysis and Fischer diagrams (Fischer, 1964; but see also Sadler et al., 1993) to support their allocyclic model for the Latemàr lagoonal succession. For these rocks, orbital forcing and cycle deposition in bundles of five, or "pentacycles", has been widely embraced by the sedimentological community, and the Latemàr has been persistently quoted in textbooks as the premier example (Tucker and Wright, 1990; Reading, 1996). However, this interpretation has been challenged or modifications have been suggested by various authors since it was first proposed: detailed biostratigraphic calibration of platform-to-basin evolution combined with radiometric age dating of zircons from volcaniclastic beds in both basinal and platform deposits reduce the average Latemàr cycle duration to less than 8 kyr (Brack et al., 1996; Mundil et al., 1996), less than 3 kyr (Mundil et al., 2003) or less than 2 kyr (Emmerich et al., 2005) instead of 20 kyr (Goldhammer et al., 1987, 1990; Goldhammer et al., 1993). This indicates partly sub-Milankovitch control,

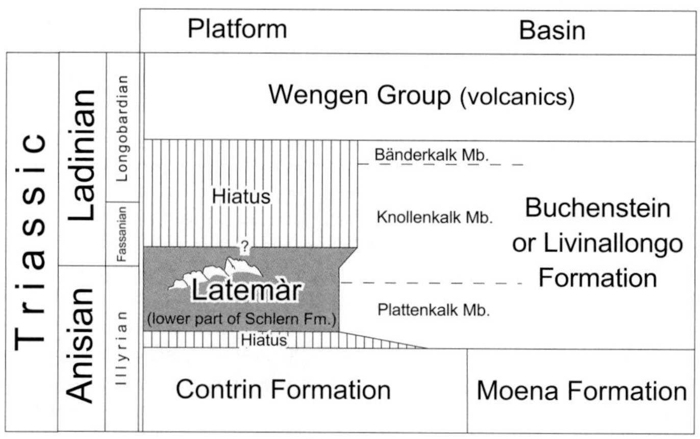

FIG. 2.—Stratigraphic relationship of the Latemàr succession to adjacent units in the central western Dolomites (stratigraphic correlations based on Maurer, 1999; Mundil et al., 2003).

TABLE 1.—Comparison of facies interpretations of the Latemàr interior presented by various authors.

	Goldhammer et al. (1987,1990, 1993)	**Preto et al. (2001)**	**Zühlke et al. (2003)**	**Blendinger (2004)**	**Egenhoff et al. (1999); this paper**
(1) Cycle type	Depositional cycle with diagenetic cap	Shallowing-upward facies cycles	Shallowing-upward accommodation cycles	Depositional cycle with top overprinted by hydrothermal fluids	Shallowing-upward cycles bounded by marine flooding surfaces
(2) Depositional environment	Shallow subtidal deposition	Deeper subtidal to subaerial	Shallow subtidal to supratidal	Subtidal deposition in photic water around storm wave base	Shallow subtidal to supratidal deposition.
(3) Vertical and lateral cycle completeness	**All** cycles with various thicknesses of solely subtidal facies and capped by thin dolomitic crust (or tepee); individual cycles are not studied for lateral variability	Vertical facies variations are detected through subfacies-ranking approach.	Most cycles are laterally continuous and can be physically traced across the platform "layer-cake" geometry).	Refers to Egenhoff et al. (1999)	Not every facies is (fully) developed in every cycle; lateral changes in cycle thickness and facies (partly pinching out)
(4) Microfacies arrangement	Basal cycle member: bioturbated shallow subtidal (within the photic zone) skeletal–lithoclast packstone or grainstone with coarsening-upward trend; occasionally storms produced graded "tempestites	Cycles show four sub-environments: (1) Deep subtidal packstones to grainstones with abundant and various fossils, overlain by (2) restricted subtidal fine-grained wackestones with scarce and/or oligotypic biota, overlain by (3) weakly laminated supratidal-flat deposits, apped by (4) caliche soils consisting of yellow dolostones with vadose pisoids and pendant/ meniscus cements.	Cycles consist of subtidal facies, overlain by intertidal- to supratidal floatstones, rudstones, bindstones and teepees, supratidal residual sediment orperitidal dolomitic caps.	Subtidal facies with coarsening-upward trend; refers to Gaetani et al. (1981) and Goldhammer et al. (1990)	Shallow subtidal dasycladalean-bearing pelois packstone to wackestone overlain by a peritidal fenestral packstone to wackestone with peloids and lumps, occasionally oncoids; coarse storm deposits in andaround teepees; lagoonal facies are organized in subconcentric belts following the outline ofthemargin
(5) Diagenetic overprinting	Upper cycle member: thin yellow dolomitic cap of subaerial origin with vadose cements, solution vugs, and caliche fabrics; tepees indicate continuous subaerial exposure and represent two or more cycles	Caliche and meniscus cements as well as dolomitization of cycle tops indicate early diagenetic overprinting during late stages of cycle evolution.	Intertidal and supratidal tepees, supratidal residual sediment and intertidal to supratidal dolomitic caps represent diagenetic overprint	Cycle top consists of thin dolomitic cap of subtidal origin with hydrothermally induced dripstone cements, subtidal tepees, and hydrothermal dissolution features resulting in reddening of carbonates.	Upper part of cycle **may be** diagenetically overprinted during peritidal (dolomitic cap and tepees) and subaerial conditions (calichification and karstification): peritidal conditions lead to dolomitization (Carballo et al., 1987), sheet cracks and cementation to tepee formation, eolian transport of ferruginous material and/or dissolution to red internal sediment (terra rossa), calichification to enlargement of components.

which according to Zühlke et al. (2003) may be related to rapid paleoceanographic and atmospheric changes. They attributed a bundle of four to five Latemàr cycles to precessional forcing (20 kyr) in contrast to Goldhammer et al.'s (1987, 1990) and Goldhammer et al.'s (1993) short-eccentricity interpretation (100 kyr) of these pentacycle sets. Kent et al. (2004), based on lithostratigraphic, biostratigraphic, and magnetostratigraphic correlations, proposed millennial-scale tidal-amplitude variations an order of magnitude faster than the Milankovitch band to explain the meter-scale basic Latemàr cycles. Also, they saw striking coincidence of subaerial exposure surfaces about every 10 m (Egenhoff et al., 1999) and the most prominent spectral peak of Preto et al.'s (2001) untuned time-frequency analysis centered at 10 m. This correlation would suggest precessional forcing for these recurring events (Kent et al., 2004) and, yet again, for a thicker depositional interval than in Zühlke et al. (2003). The two sequences making up the general facies units (LCF–MTF–UCF–UTF; Fig. 3; Egenhoff et al., 1999) then correspond to long-eccentricity forcing (400 kyr), which would imply a total duration of 800 kyr for the entire cyclic succession at Latemàr (Kent et al., 2004). Despite the heterogeneity of the recent interpretations with respect to cycle duration, there appears to be strong concurrent evidence against pure orbital forcing of the meter-scale basic Latemàr cycles.

Blendinger (2004) questioned sea-level changes and eustasy as the governing factors for cycle formation at Latemàr and offered an alternative interpretation involving intermittent hydrothermal influence alternating with normal marine deposition. On the basis of correlation between abundance of cycles and younger volcanic dikes crosscutting the buildup, he hypothesized that fluids with a composition similar to that of normal seawater were forced periodically by elevated heat flow from an underlying magma chamber to circulate to the seafloor. These fluids produced stratiform diagenetic features, including tepees, and favored growth of cyanobacterial mats and early dolomitization. However, the Latemàr facies and its distribution patterns argue against Blendinger's (2004) hydrothermal model, and stable-isotope data from Latemàr are equivocal (Preto et al., 2005; Peterhänsel and Egenhoff, 2005).

METHODOLOGY AND OBSERVATIONS

When "walking out" or tracing lagoonal flooding surfaces (e.g., Rogers, 1998; Rankey et al., 1999), Egenhoff et al. (1999) detected facies variations in individual platform-wide meter-scale cycles (now assigned to sub-Milankovitch sea-level oscillations; see above discussion), which reflect paleorelief on the depositional surface. A tepee belt near the platform margin was thought to represent the paleotopographic high circumscribing the peritidal lagoon. The tepee belt is characterized by intermittent tepee formation. In contrast to Egenhoff et al. (1999), who investigated lateral facies variations in intervals with laterally continuous meter-scale Latemàr cycles, this study focuses on strata that split laterally into several smaller-scale cycles. An approximately 60-m-thick section forming the base of the relatively tepee-poor Upper Cyclic Facies (UCF, Fig. 3; Egenhoff et al., 1999) is characterized by a lateral increase in cycle numbers. It was analyzed to characterize these cycles and to unravel the circumstances that resulted in their formation. This interval represents the lower 60 m of the section that was used for time-series analysis by Hinnov and Goldhammer (1991) and Preto et al. (2001). This section was said to "contain the longest uninterrupted sequence of cycle couplets [, subtidal base and subaerial exposure cap,] in the [Schlern] Formation that was measured at a single location" (Hinnov and Goldhammer, 1991).

The measured interval is stratigraphically well constrained: it is marked at its top by a distinct tuffite, which has been dated to 241.5 Ma (LAT 32 of Mundil et al., 2003), and it is internally controlled by various continuous marker beds, which are characterized either by prominent weathering or by subaerial emergence features. They are easily recognizable in the field and can be walked out within structural blocks of the buildup and correlated across faults or scree cover. A minimum of 10 carbonate beds above and below the markers were traced and correlated (e.g., marker bed N in Figures 3 to 5). The interval represents the base of the upper cyclic facies (Egenhoff et al., 1999) and includes strata that exemplify platform-wide continuity without splitting of cycles.

In the upper part of the 60 m interval, half a dozen zones, 2–3 m thick, display lateral splitting of meter-scale cycles into two to six centimeter- to decimeter-scale (small-scale) cycles, for example at 32–34 m, 39–41 m, and 43–46 m (Figs. 6 and 7). In the lower part of the measured section, tepees with discontinuous small-scale cycles are comparatively rare, for example at 17 m. Small-scale cycles are associated with abundant tepees belonging to the backreefal tepee belt, but obviously they are also restricted to certain episodes within the 60 m interval (Fig. 4), thus implying a spatial and temporal restriction of variation in numbers of lateral cycles.

Small-scale cycles are composed of shallowing-upward facies successions bounded by flooding surfaces. The bases of such cycles are usually composed of wackestones to packstones with varying amounts of dasycladalean algae, which grade upwards into peloid-lump wackestones and packstones, which are in turn overlain by oncoid packstones to grainstones (cf. Goldhammer et al., 1990; Egenhoff et al., 1999; Preto et al., 2001). A dolomitic cap forms the topmost part, if the cycle is fully developed. This facies succession, with a decrease in mud content towards the top, characterizes both the meter-scale cycles that are laterally traceable throughout the platform interior (Egenhoff et al., 1999) and the centimeter- to decimeter-thick cycles that are discontinuous and developed exclusively in association with tepees. Many of these very thin cycles can be followed throughout the tepee belt. Within the detailed measured interval of 60 m at Cimón Latemàr, 105 shallowing-upward cycles are recorded compared to 79 at the Cima del Forcellone, a decrease of some 25%.

DISCUSSION

Lateral Distribution and Origin of Discontinuous Small-Scale Cycles

There appears to be unanimous consensus on external forcing of the basic meter-scale Latemàr cycles, even if the specific driver has not yet been agreed upon (see controversy above). For the discontinuous centimeter- to decimeter-scale cycles, both, allocyclic and autocyclic forcing has to be considered:

Autocyclicity.—

Lateral variations of cycle stacking patterns in carbonate deposits have been attributed to lateral sediment transport: theoretical approaches of Ginsburg (1971) and Burgess et al. (2001) advocated shoreline and island progradation as possible cycle-producing mechanisms. Pratt and James (1986) and Satterley (1996) provided outcrop data supporting these interpretations. At Latemàr, distinct progradation geometries and lateral accretion in small-scale cycles, usually seen as an indicator of autocyclic influence, are absent, which argues against exclusive autocyclic control. Small-scale cycles are concentrated in the tepee belt

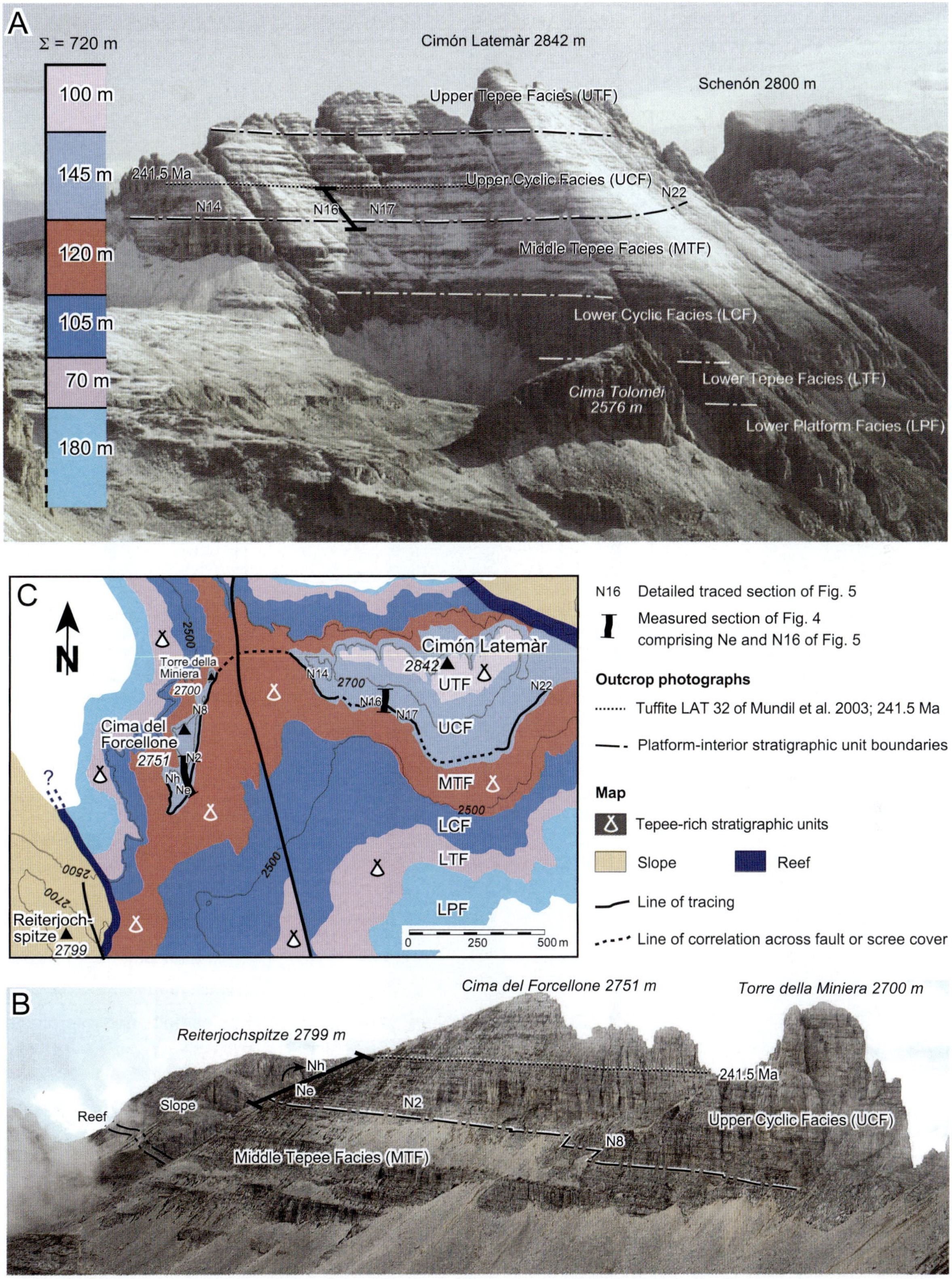

FIG. 3.—Location of traced and correlated interval displayed in Figure 4, linking Cimón Latemàr (A) and Cima del Forcellone (B). Stratigraphic units are in accordance with Egenhoff et al. (1999). The section is capped at the top by a distinct tuffite (LAT 32 of Mundil et al. 2003; 241.5 Ma). Locations of individual segments Nh, Ne, N2, N8, N14, N16, N17, and N22 composing the curved cross section in Figure 5 are indicated. The Forcellone and Latemàr blocks are separated by a fault with a displacement of 35 m (C).

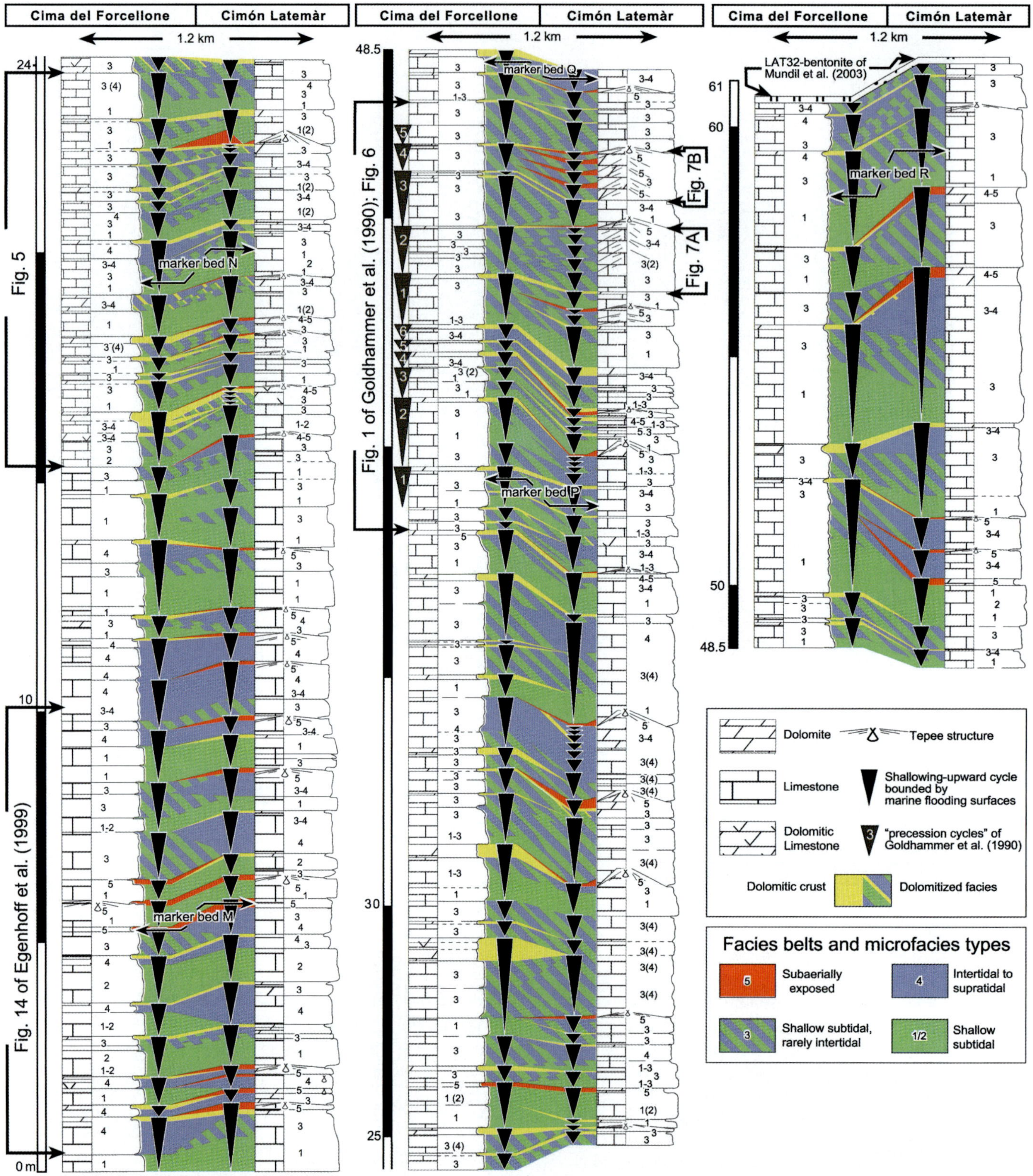

FIG. 4.—Vertical and lateral facies of a detailed traced and correlated interval illustrated with time-equivalent sections from Cima del Forcellone and Cimón Latemàr including detailed traced segments Ne and N16 of Figure 5 at 15–24 m. Locations of measured sections are shown in Figure 3. Cycle stacking pattern represented by black triangles is based on vertical facies development. Two sets of "pentacycles" measured at the same locality by Goldhammer et al. (1990) are indicated at 38–47 m and are portrayed in Figure 6. The interval comprising discontinuous small-scale cycles associated with tepees shown in Figure 7 is equivalent to part of the upper "pentacycle" set.

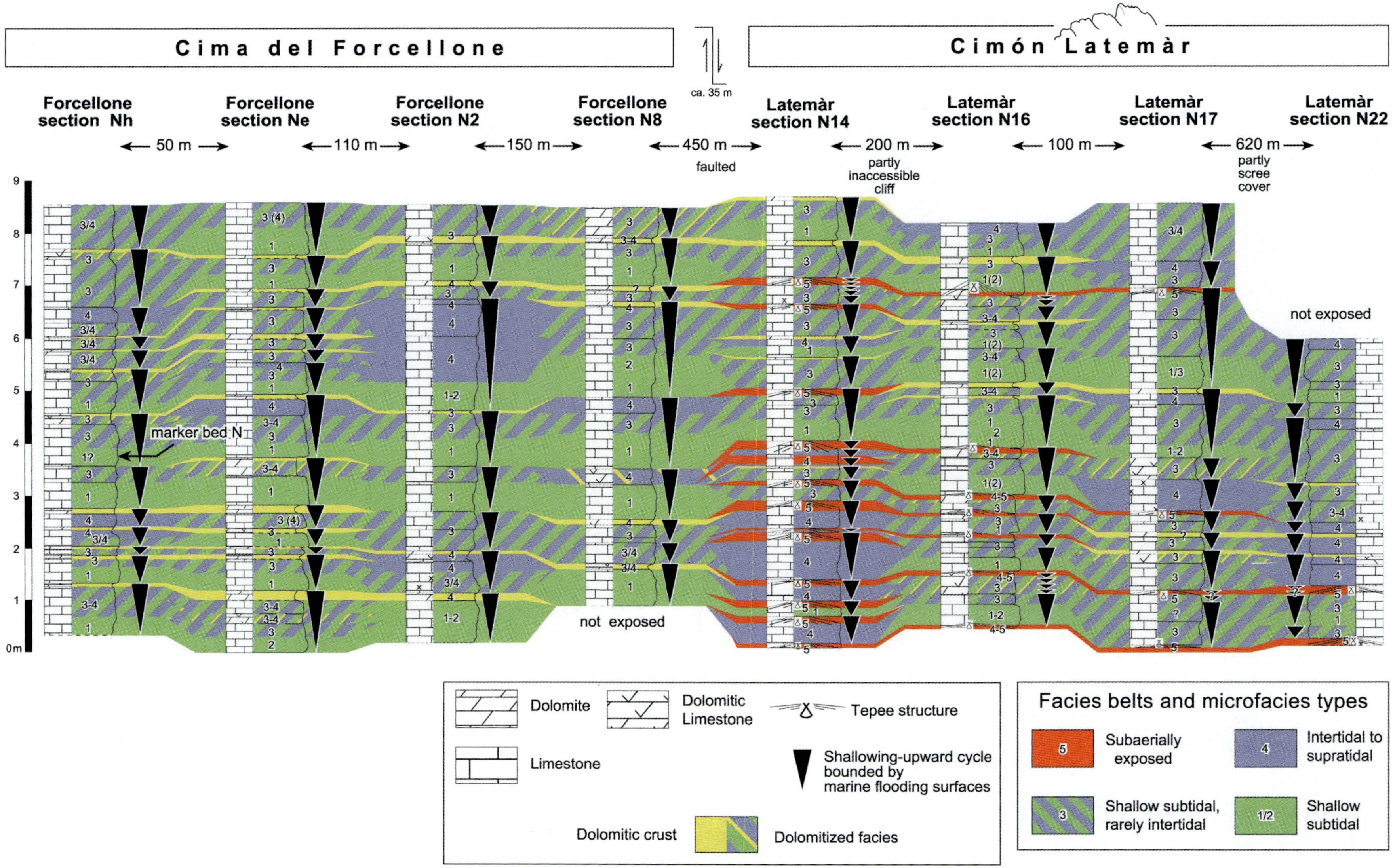

FIG. 5.—Lateral changes in facies and cycle stacking pattern within traced and correlated beds from the basal Upper Cyclic Facies (Fig. 3). Segments Ne and N16 form part of the sections portrayed in Figure 4, 15–24 m. The profile represents a curved cross section connecting the tectonic blocks of the Cima del Forcellone in the west and the Cimón Latemàr in the north, which are divided by a fault with 35 m offset. All beds were traced except where indicated (see also map Fig. 3). The four Forcellone sections (left) represent the outer lagoon whereas the four Latemàr sections (right) are situated in proximity to or within the tepee belt. Black triangles indicate shallowing-upward cycles bounded by marine flooding surfaces. Cycle numbers vary throughout the measured sections owing to amalgamation or splitting; lateral variation is greatest in sections with abundant tepees. Locations of individual segments Nh, Ne, N2, N8, N14, N16, N17, and N22 are in Figure 3.

FIG. 6.—Outcrop photograph of a part of the section measured at Cima del Forcellone (38.5–47.5 m of Figure 4) looking to the north with Rosengarten/Catinaccio massif in the background. The portrayed interval comprises the two "fourth-order cycles" or "pentacycles" of Figure 1 of Goldhammer et al. (1990). The numbers and dashed lines reflect their "fifth-order precession cycles". Cycles 2 and 4 in the upper part of the photograph correspond to tepee beds at Cimón Latemàr comprising several small-scale cycles (Figure 7).

rather than distributed more randomly across other areas of the Latemàr interior. This may be a further argument against pure internal forcing. However, redistribution of sediment towards the up to 500-m-wide tepee belt, for example during storms, may have contributed to shallowing of facies, particularly inasmuch as this elevated area may have functioned as a suitable sediment trap: laterally extensive tepee megapolygons, up to several meters in diameter and several decimeters in relief (Fig. 7), easily collected lagoon-derived sediment with only limited possibilities of it being removed again. Ammonite specimens, which are important for biostratigraphic analyses, were found in increased abundance within the megapolygons as well. Storm derivation of the small-scale cycles would imply that storm deposits or their preservation are causally related to tepee formation. Conversely, in the Latemàr succession tepees do exist that do not comprise small-scale cycles (Fig. 4), and storm layers have been identified in non-tepee strata forming part of a meter-scale Latemàr cycle (Egenhoff et al., 1999). This observation argues against a storm origin of the small-scale cycles. Furthermore, gradational upward coarsening of carbonate particles and decreasing portions of micrite within the small-scale cycles question deposition during a single storm event. Some small-scale cycles, however, do show sharp contacts between basal wackestones to packstones and overlying grainstones, reflecting local erosion during high-energy events.

Allocyclicity.—

Lateral variations in cycle stacking pattern may be related to high-frequency sea-level changes with comparably low amplitude, which are recorded only on topographic highs. This high zone is commonly represented by an entire platform or isolated buildup where surrounding basinal areas remain unaffected by sea-level changes. However, cycle recording across the interior of a buildup could also vary, if topographic relief exceeds the amplitude of the sea-level changes. Then, deeper lagoonal parts remain largely untouched by fluctuations in water depth. At Latemàr, superimposed high-frequency, low-amplitude sea-level fluctuations may have affected only the elevated tepee belt area, where reduced accommodation resulted in the formation of laterally discontinuous cycles (Figs. 8A, B3, C3, D3). Deeper parts of the lagoon remained largely unaffected and thus show no intermittent subaerial exposure and/or deposits of marine flooding caused by this allocyclicity (Goldhammer et al.'s subtidal missed beats).

With respect to the above discussion, small-scale cycles in the Latemàr tepee belt likely formed as a result of external forcing. Upward shallowing may have been further expedited by sediment redistribution towards the tepee belt.

In essence, the paleotopography of the buildup with the elevated position of the tepee belt and its saucer-shaped sedimentary structures can be seen as the foremost reason for the restriction of small-scale cycles to that area.

Vertical Distribution of Discontinuous Small-Scale Cycles

Small-scale cycles form clusters of discontinuity surfaces (Goldhammer et al., 1990; see also Hillgärtner, 1998) associated with tepees, and they occur intermittently throughout the middle and upper part of the Latemàr section measured in the immediate backreef (Fig. 4). Although tepee formation may have been influ-

FIG. 7.—Tepee structures from the Upper Cyclic Facies at Cimón Latemàr comprising several small-scale cycles. These strata grade laterally into one basic Latemàr shallowing-upward cycle. Figure 7A corresponds to 43.5–45 m of Forcellone section, and Figure 7B is equivalent to 47–47.5 m. The numbers represent "fifth-order precession cycles" of Goldhammer et al. (1990) portrayed in Figures 4 and 6. The vertical part of the measuring stick has a length of 1.6 m.

enced or advanced by the redistribution of lagoonal deposits (see discussion above), there is further evidence for predominant allocyclic forcing: in intervals with increased occurrence of tepee stacks, the meter-scale basic Latemàr cycles are often thinner than average, indicating a period of reduced overall accommodation (Goldhammer et al., 1990).

Long-term variations in accommodation may be responsible for vertical, that is, temporal, constraints for formation of small-scale cycles (see Lehrmann and Goldhammer, 1999). With changes in overall accommodation, various facies and cycle stacking patterns should form which are all represented in the Latemàr succession. Figure 8 demonstrates the influence of different levels of overall accommodation (B) on cycle stacking patterns (C) and facies distribution across the Latemàr platform (A and D). Vertical variability of formation of small-scale cycles and thus long-term variations in accommodation can be explained if a lower-order sea-level fluctuation with comparably larger amplitude is invoked (Vail et al., 1977; Haq et al., 1987). This interpretation largely follows the model for vertical cycle stacking of Goldhammer et al. (1987, 1990) and Goldhammer et al. (1993) but on a different scale and with important repercussions for tepee formation. Possible variations in the subsidence rate also have to be taken into account as a controlling factor for long-term accommodation (Fig. 8B). The nearby Predazzo and Monzoni magmatic

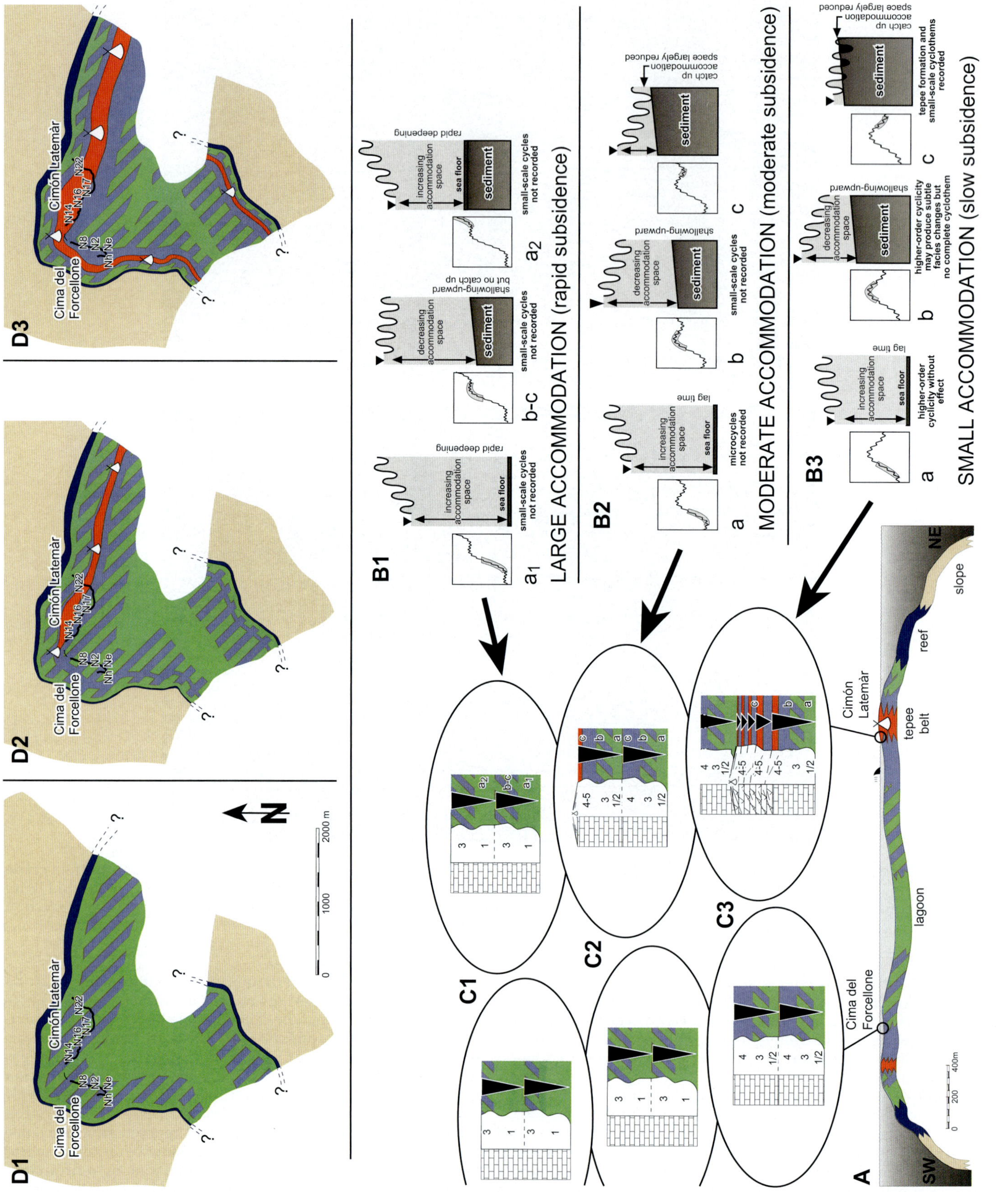
D1
D2
D3
Cima del Forcellone
Cimón Latemàr
N14
N16
N17
N22
N8
N2
Nh
Ne
N
0
1000
2000 m
B1
LARGE ACCOMMODATION (rapid subsidence)
a1
increasing accommodation space
sea floor
rapid deepening
small-scale cycles not recorded
b-c
decreasing accommodation space
sediment
shallowing-upward but no catch up
small-scale cycles not recorded
a2
increasing accommodation space
sea floor
sediment
rapid deepening
small-scale cycles not recorded
B2
MODERATE ACCOMMODATION (moderate subsidence)
a
increasing accommodation space
sea floor
lag time
microcycles not recorded
b
decreasing accommodation space
sediment
shallowing-upward
small-scale cycles not recorded
c
sediment
catch up
accommodation space largely reduced
B3
SMALL ACCOMMODATION (slow subsidence)
a
increasing accommodation space
sea floor
lag time
higher-order cyclicity without effect
b
decreasing accommodation space
sediment
shallowing-upward
higher-order cyclicity may produce subtle facies changes but no complete cyclothem
c
sediment
catch up
accommodation space largely reduced
tepee formation and small-scale cyclothems recorded
C1
C2
C3
A
SW
NE
lagoon
Cima del Forcellone
Cimón Latemàr
tepee belt
reef
slope
0
200
400m

FIG. 8 (facing page).—**A)** Paleotopography in the immediate backreef is attributed to higher amounts of carbonate fixation and enhanced growth in the tepee belt area likely related to small-scale cyclicity and sediment trapping in tepees and inter-tepee depressions and favored by microbial binding and cementation. Asymmetric paleorelief generated a broader tepee belt in the north and northeast of the buildup (Egenhoff et al., 1999). Influence of different levels of accommodation **(B)** related to two orders of superimposed sea-level oscillations and subsidence on cycle stacking patterns **(C)** and facies distribution across the Latemàr platform **(D)**. Long-term accommodation changes in the Latemàr lagoonal succession may be a composite of low-order sea-level oscillations and subsidence variations. B1–D1) High accommodation levels result in largely continuous deposition of meter-scale basic Latemàr cycles across the platform interior. Small-scale cycles are absent. B2–D2) With moderate accommodation levels, thin tepees may form on a narrow backreef belt in the north and northeast. B3–D3) Low levels of accommodation facilitate recording of high-frequency, low-amplitude cycles in the high areas of the platform during falling low-order sea level. The generated centimeter- to decimeter-scale cycles are associated with tepees and restricted to the tepee belt. Individual segments of traced section (Nh, Ne, N2, N8, N14, N16, N17, and N22) are in Figure 4.

centers (Fig. 1) were controlling regional subsidence and accommodation, producing deep faults and doming (Doglioni, 1988). Emmerich et al. (2005) have modeled subsidence variations for the Latemàr to fluctuate between 330 and 450 m Myr^{-1} between the central (MTF; Fig. 3) and upper (UTF) cyclic part of the lagoonal succession. Local changes in subsidence would imply that the assumed long-term accommodation changes are not only eustatic in origin. It is thus reasonable to assume that long-term accommodation changes in the Latemàr lagoonal succession are a composite of low-order sea-level oscillations and subsidence variations. With high levels of accommodation, high-frequency, low-amplitude sea-level oscillations are not recorded across the platform. Instead, meter-scale basic Latemàr cycles are deposited across the platform interior (Fig. 8—C1). It is noteworthy that rapid subsidence reduces the regressive time interval owing to convergence of the accommodation high and the onset of a new cycle (Fig. 8—B1), thus potentially shortening the period formation of small-scale cycles. Moderate accommodation levels result in the formation of thin tepees on a narrow backreef belt in the north and northeast (Fig. 8—B2, C2, D2). This distribution is related to the asymmetric paleorelief of the Latemàr buildup (Egenhoff et al., 1999). Centimeter- to decimeter-scale cycles are generated with comparably small accommodation levels when high-frequency, low-amplitude sea-level changes affect the elevated tepee belt (Fig. 8—A, B3, C3, D3). Further reduced overall accommodation space would shift the Goldilocks window (term of Goldhammer et al., 1990) of small-scale cycle and tepee formation to a deeper area on the buildup, thus widening the tepee belt to topographically lower areas, whereas higher areas would face somewhat longer subaerial exposure.

Tepees, Cycle Formation, and Missed Beats

Accommodation space varies in time and also laterally owing to paleorelief. At Latemàr, backreef strata characterized by tepees grade laterally lagoonward into layers without tepees, suggesting such paleorelief (Egenhoff et al., 1999; Fig. 8). Thus, Goldhammer et al.'s (1987, 1990) and Goldhammer et al.'s (1993) subaerial missed beats associated with tepee structures, although unrecorded on the elevated part of the platform, should be represented by carbonate deposits in deeper parts of the lagoon where enough accommodation space is provided. Cycle numbers in the deeper parts of the platform should therefore exceed the number of cycles recorded on elevated areas, and meter-scale basic Latemàr cycles should pinch out when they are traced from the paleotopographically deeper outer lagoon (Cima del Forcellone) into the elevated tepee belt. This is in contrast to our observations, in which cycle numbers increase laterally towards the elevated tepee belt by splitting of meter-scale cycles into several centimeter- to decimeter-scale cycles within the tepee belt (Figs. 5, 8). Thus, more sea-level changes, which means more marine flooding surfaces, are recorded in the tepee belt than elsewhere on the platform. Many of the small-scale cycles are of a higher order than the coeval meter-scale cycle of the deeper areas in the lagoon. It is therefore concluded that tepee stacks in general do not correspond to subaerial missed beats or represent more than one condensed basic Latemàr cycle as envisioned by Goldhammer et al. (1987, 1990) and Goldhammer et al. (1993) or Zühlke et al. (2003). An estimate of cycle numbers following their assumptions results in a significant overestimation of cycle numbers and corresponding time when measuring sections situated in the tepee belt. However, during subaerial exposure of the whole platform some cycles may have not been recorded, resulting in emergence horizons (Egenhoff et al., 1999).

Conversely, the recordability of centimeter- to decimeter-scale cycles in the tepee belt implies that tepees at Latemàr respond to relatively short pulses of sea-level fluctuations: brief phases of sedimentation likely enhanced by sediment redistribution from deeper lagoonal areas resulted in deposition of small-scale cycles separated by marine flooding surfaces. This involves a multistage process of deposition and syndepositional diagenesis including pervasive cementation. Once the tepee belt has been established, it forms a "symbiotic" relationship with small-scale cyclicity: centimeter- to decimeter-scale cycles are recorded only within the tepee belt and the related high-frequency sea-level fluctuations enhance alternations of diagenetic processes that favor tepee formation.

CONCLUSIONS

The cyclic Latemàr platform interior is characterized by meter-scale paleorelief, with a tepee belt in the immediate backreef representing its topographic high. Tepee formation is spatially limited and further temporally restricted to certain intervals. The number of cycles increases in the vicinity of these tepee zones, owing to a fanning out of cycles where discontinuous marine flooding surfaces appear. A laterally traced section from a relatively tepee-poor interval in comparison to the entire succession, that is, the Upper Cyclic Facies (UCF in Figure 3) reveals a 25% variation in cycle numbers. Low-amplitude sea-level fluctuations are likely responsible for the increase of cycle numbers in tepee zones. These oscillations influenced only the paleotopographically elevated tepee belt, whereas deeper parts of the lagoon remained largely unaffected owing to facies amalgamation without intermittent subaerial exposure and/or marine flooding. Sediment redistribution towards tepee areas, for example during storms, likely contributed to shallowing of facies as these saucer-shaped megapolygons acted as suitable sediment traps. The lateral increase in cycle numbers and the occurrence of tepee structures are indicators of subtle paleorelief on the Latemàr platform top.

Given the correlation between cycle-number increase and tepees, cycle-number variation in tepee-rich intervals of the succession, which are the Middle Tepee Facies (MTF) and Upper Tepee Facies (UTF), should exceed variations in cycle numbers in tepee-poor units at Latemàr, which are the lower and upper cyclic facies (Fig. 3) and thus amount to even larger variations than the reported 25%. The vertical variations of cycle stacking patterns are likely to be responding to long-term accommodation changes, whereas small-scale cycles associated with tepees form during overall regressive periods. Long-term accommodation was probably a composite of low-order eustatic sea level and regional subsidence.

Tepees do not comprise subaerial missed beats but instead form within a topographic range where low-amplitude sea-level fluctuations were recorded, in contrast to topographically lower lagoonal areas, which remained unaffected by them. It is postulated that such alternations of submergence and intermittent emergence favored tepee formation on the Latemàr platform.

Lateral variations in small-scale cycle stacking patterns are almost certainly a common feature of the interiors of carbonate buildups throughout the rock record: on these platforms, paleorelief delimited the recording of low-amplitude shallowing-upward cycles to topographically elevated areas. This cyclic setting may be further added to scenarios favoring tepee formation.

ACKNOWLEDGMENTS

This research has been supported by a Marie Curie individual fellowship from the European Community Programme "Improving Human Potential" to AP under contract number HPMF-CT-2002-01912. We are indebted to Thilo Bechstädt and Rainer Zühlke for introducing us to the Latemàr during our M.Sc. projects, and to Jürgen Grötsch and Volker Vahrenkamp for calling our attention to lateral issues in platform strata. We thank André Strasser, Brian Pratt, Guy Spence, Adrian Immenhauser, Wolfgang Blendinger, Dan Lehrmann, Tony Dickson, Isabel Montañez, David Osleger, and Ali Jaffri for discussion and valuable suggestions on an earlier draft. Dan Bosence and Mark Harris provided helpful reviews of the manuscript. Grappa-warm thanks go to our friend Tonio Gabrielli at the Rifugio "Torre di Pisa". This is Cambridge Earth Science Contribution 8867.

REFERENCES

ADAMS, R.W., AND GROTZINGER, J.P., 1996, Lateral continuity of facies and parasequences in Middle Cambrian platform carbonates, Carrara Formation, southeastern California, U.S.A.: Journal of Sedimentary Research, v. 66, p. 1079–1090.

BLENDINGER, W., 2004, Sea level changes versus hydrothermal diagenesis: origin of Triassic carbonate platform cycles in the Dolomites, Italy: Sedimentary Geology, v. 169, p. 21–28.

BORKHATARIA, R., AIGNER, T., AND PIPPING, K., 2006, An unusual, muddy, epeiric carbonate reservoir: The Lower Muschelkalk (Middle Triassic) of the Netherlands: American Association of Petroleum Geologists, Bulletin, v. 90, p. 61–89.

BOSELLINI, A., 1984, Progradation geometries of carbonate platforms: examples from the Triassic of the Dolomites, northern Italy: Sedimentology, v. 31, p. 1–24.

BOSELLINI, A., 1991, Geology of the Dolomites—an introduction: Dolomieu Conference on Carbonate Platforms and Dolomitization: Sankt Ulrich / Ortisei, Tourist Office Grödental / Val Gardena, 43 p.

BOSELLINI, A., AND ROSSI, D., 1974, Triassic carbonate buildups of the Dolomites, Northern Italy, *in* Laporte, L., ed., Reefs in Time and Space: Society of Economic Paleontologists and Mineralogists, Special Publication 18, p. 209–233.

BRACK, P., MUNDIL, R., OBERLI, F., MEIER, M., AND RIEBER, H., 1996, Biostratigraphic and radiometric age data question the Milankovitch characteristics of the Latemar cycles (Southern Alps, Italy): Geology, v. 24, p. 371–375.

BRACK, P., AND RIEBER, H., 1993, Towards a better definition of the Anisian / Ladinian boundary: New biostratigraphic data and correlations of boundary sections from the Southern Alps: Eclogae Geologicae Helvetiae, v. 86, p. 415–527.

BURGESS, P.M., WRIGHT, V.P., AND EMERY, D., 2001, Numerical forward modelling of peritidal carbonate parasequence development: implications for outcrop interpretation: Basin Research, v. 13, p. 1–16.

CASTELLARIN, A., LUCCHINI, F., ROSSI, P.L., SELLI, L., AND SIMBOLI, G., 1988, The Middle Triassic magmatic–tectonic arc development in the Southern Alps: Tectonophysics, v. 146, p. 79–89.

DELLA PORTA, G., KENTER, J.A.M., IMMENHAUSER, A., AND BAHAMONDE, J.R., 2002, Lithofacies character and architecture across a Pennsylvanian inner-platform transect (Sierra de Cuera, Asturias, Spain): Journal of Sedimentary Research, v. 72, p. 898–916.

DOGLIONI, C., 1988, Examples of strike-slip tectonics on platform-basin margins: Tectonophysics, v. 156, p. 293–302.

EGENHOFF, S., PETERHÄNSEL, A., BECHSTÄDT, T., ZÜHLKE, R., AND GRÖTSCH, J., 1999, Facies architecture of an isolated carbonate platform: tracing the cycles of the Latemàr (Middle Triassic, northern Italy): Sedimentology, v. 46, p. 893–912.

EMMERICH, A., GLASMACHER, U.A., BAUER, F., BECHSTÄDT, T., AND ZÜHLKE, R., 2005, Meso- / Cenozoic basin and carbonate platform development in the SW-Dolomites unraveled by basin modelling and apatite FT analysis: Rosengarten and Latemar (Northern Italy): Sedimentary Geology, v. 175, p. 415–438.

FISCHER, A.G., 1964, The Lofer cyclothems of the Alpine Triassic: Kansas Geological Survey, Bulletin 169, p. 107–149.

GAETANI, M., FOIS, E., JADOUL, F., AND NICORA, A., 1981, Nature and evolution of Middle Triassic carbonate buildups in the Dolomites (Italy): Marine Geology, v. 44, p. 25–57.

GINSBURG, R.N., 1971, Landward movement of carbonate mud: new model for regressive cycles in carbonates (abstract): American Association of Petroleum Geologists, Bulletin, v. 55, p. 340.

GOLDHAMMER, R.K., AND HARRIS, M.T., 1989, Eustatic controls on the stratigraphy and geometry of the Latemar buildup (Middle Triassic), the Dolomites of northern Italy, *in* Crevello, P.D., Wilson, J.L., Sarg, J.F., and Read, J.F., eds., Controls on Carbonate Platform and Basin Development:, SEPM, Special Publication 44, p. 323–338.

GOLDHAMMER, R.K., DUNN, P.A., AND HARDIE, L.A., 1987, High-frequency glacio-eustatic sea level oscillations with Milankovitch characteristics recorded in Middle Triassic platform carbonates in northern Italy: American Journal of Science, v. 287, p. 853–892.

GOLDHAMMER, R.K., DUNN, P.A., AND HARDIE, L.A., 1990, Depositional cycles, composite sea-level changes, cycle stacking patterns, and the hierarchy of stratigraphic forcing: examples from Alpine Triassic platform carbonates: Geological Society of America, Bulletin, v. 102, p. 535–562.

GOLDHAMMER, R.K., HARRIS, M.T., DUNN, P.A., AND HARDIE, L.A., 1993, Sequence stratigraphy and systems tract development of the Latemar Platform, Middle Triassic of the Dolomites (northern Italy): outcrop calibration keyed by cycle stacking patterns, *in* Loucks, R.G.,and Sarg, J.F., eds., Carbonate Sequence Stratigraphy: Recent Developments and Applications: American Association of Petroleum Geologists, Memoir 57, p. 353–387.

GROTZINGER, J.P., 1986, Cyclicity and paleoenvironmental dynamics, Rocknest platform, northwest Canada: Geological Society of America, Bulletin, v. 97, p. 1208–1231.

HAQ, B.U., HARDENBOL, J., AND VAIL, P.R., 1987, Chronology of fluctuating sea levels since the Triassic: Science, v. 235, p. 1156–1167.

HARDIE, L.A., BOSELLINI, A., AND GOLDHAMMER, R.K., 1986, Repeated subaerial exposure of subtidal carbonate platforms, Triassic, northern Italy: Evidence for high-frequency sea level oscillations on a 10^4 year scale: Paleoceanography, v. 1, p. 447–457.

HARDIE, L.A., WILSON, E.N., AND GOLDHAMMER, R.K., 1991, Cyclostratigraphy and dolomitization of the Middle Triassic Latemar buildup, the Dolomites, northern Italy: Conference on Carbonate Platforms and Dolomitization: St. Ulrich/Gröden, Guidebook Excursion F, 56 p.

HILLGÄRTNER, H., 1998, Discontinuity surfaces on a shallow-marine carbonate platform (Berriasian, Valanginian, France and Switzerland): Journal of Sedimentary Research, v. 68, p. 1093–1108.

HINNOV, L.A., AND GOLDHAMMER, R.K., 1991, Spectral analysis of the Middle Triassic Latemar Limestone: Journal of Sedimentary Petrology, v. 61, p. 1173–1193.

KENT, D.V., MUTTONI, G., AND BRACK, P., 2004, Magnetostratigraphic confirmation of a much faster tempo for sea-level change for the Middle Triassic Latemar platform carbonates: Earth and Planetary Science Letters, v. 228, p. 369–377.

LEHRMANN, D.J., AND GOLDHAMMER, R.K., 1999, Secular variation in parasequence and facies stacking patterns of platform carbonates: a guide to application of stacking-patterns analysis in strata of diverse ages and settings, *in* Harris, P.M., Saller A.H., and Simo, J.A., eds., Advances in Carbonate Sequence Stratigraphy: Application to Reservoirs, Outcrops and Models: SEPM, Special Publication 63, p. 187–225.

LUKASIK, J.J., JAMES, N.P., MCGOWRAN, B., AND BONE, Y., 2000, An epeiric ramp: low-energy, cool-water carbonate facies in a Tertiary inland sea, Murray Basin, South Australia: Sedimentology, v. 47, p. 851–881.

MASETTI, D., AND NERI, C., 1980, L'Anisico della Val di Fassa (Dolomiti occidentali): sedimentologia e paleogeografia: Universitá di Ferrara, Annali, Nuova Serie, Sezione IX. Scienze Geologiche e Mineralogiche, v. 7, p. 1–19.

MAURER, F., 1999, Wachstumsanalyse einer mitteltriadischen Karbonatplattform in den westlichen Dolomiten (Südalpen): Eclogae Geologicae Helvetiae, v. 92, p. 361–378.

MAURER, F., 2000, Growth mode of Middle Triassic carbonate platforms in the Western Dolomites (Southern Alps, Italy): Sedimentary Geology, v. 134, p. 275–286.

MAURER, F., AND SCHLAGER, W., 2003, Lateral variations in sediment composition and bedding in Middle Triassic interplatform basins (Buchenstein Formation, southern Alps, Italy): Sedimentology, v. 50, p. 1–22.

MUNDIL, R., BRACK, P., MEIER, M., RIEBER, H., AND OBERLI, F., 1996, High resolution U–Pb dating of Middle Triassic volcaniclastics: Time scale calibration and verification of tuning parameters for carbonate sedimentation: Earth and Planetary Science Letters, v. 141, p. 137–151.

MUNDIL, R., ZÜHLKE, R., BECHSTÄDT, T., PETERHÄNSEL, A., EGENHOFF, S.O., OBERLI, F., MEIER, M., BRACK, P., AND RIEBER, H., 2003, Cyclicities in Triassic platform carbonates: synchronizing radio-isotopic and orbital clocks: Terra Nova, v. 15, p. 81–87.

PETERHÄNSEL, A., AND EGENHOFF, S.O., 2005, Sea level changes versus hydrothermal diagenesis: origin of Triassic carbonate platform cycles in the Dolomites, Italy: Discussion: Sedimentary Geology, v. 178, p. 145–149.

PRATT, B.R., AND JAMES, N.P., 1986, The St George Group (Lower Ordovician) of western Newfoundland: tidal flat island model for carbonate sedimentation in shallow epeiric seas: Sedimentology, v. 33, p. 313–343.

PRETO, N., HINNOV, L.A., HARDIE, L.A., AND DE ZANCHE, V., 2001, Middle Triassic orbital signature recorded in the shallow marine Latemar carbonate buildup (Dolomites, Italy): Geology, v. 29, p. 1123–1126.

PRETO, N., HINNOV, L.A., HARDIE, L.A., AND HARRIS, M.T., 2005, Sea level changes versus hydrothermal diagenesis: origin of Triassic carbonate platform cycles in the Dolomites, Italy—Discussion: Sedimentary Geology, v. 178, p. 135–139.

RANKEY, E.C., BACHTEL, S.L., AND KAUFMAN, J., 1999, Controls on stratigraphic architecture of icehouse mixed carbonate–siliciclastic systems: a case study from the Holder Formation (Pennsylvanian, Virgilian), Sacramento Mountains, New Mexico, *in* Harris, P.M., Saller A.H., and Simo, J.A., eds., Advances in Carbonate Sequence Stratigraphy: Application to Reservoirs, Outcrops and Models: SEPM, Special Publication 63, p. 127–150.

RASPINI, A., 2001, Stacking pattern of cyclic carbonate platform strata: Lower Cretaceous of Southern Apennines, Italy: Geological Society of London, Journal, v. 158, p. 353–366.

READ, J.F., 1995, Overview of carbonate platform sequences, cycle stratigraphy and reservoirs in greenhouse and ice-house worlds, *in* Read, J.F., Kerans, C., Weber, L.J., Sarg, J.F., and Wright, F.M., eds., Milankovitch Sea Level Changes, Cycles and Reservoirs on Carbonate Platforms in Greenhouse and Icehouse Worlds: SEPM, Short Course Notes 35, p. 1–102.

READ, J.F., GROTZINGER, J.P., BOVA, J.A., AND KOERSCHNER, W.F., 1986, Models for generation of carbonate cycles: Geology, v. 14, p. 107–110.

READING, H.G., 1996, Sedimentary Environments, Third Edition: London, Blackwell Science, 704 p.

Rogers, R.R., 1998, Sequence analysis of the Upper Cretaceous Two Medicine and Judith River formations, Montana: nonmarine response to the Claggett and Bearpaw marine cycles: Journal of Sedimentary Research, v. 68, p. 615–631.

RÜFFER, T., AND ZÜHLKE, R., 1995, Sequence stratigraphy and sea-level changes in the Early to Middle Triassic of the Alps: a global comparison, *in* Haq, B.U., ed., Sequence Stratigraphy and Depositional Response to Eustatic, Tectonic and Climatic Forcing: Amsterdam, Kluwer, p. 161–207.

SADLER, P.M., OSLEGER, D.A., AND MONTAÑEZ, I.P., 1993, On the labeling, length, and objective basis of Fischer plots: Journal of Sedimentary Petrology, v. 63, p. 360–368.

SANDULLI, R., AND RASPINI, A., 2004, Regional to global correlation of lower Cretaceous (Hauterivian–Barremian) shallow-water carbonates of the southern Apennines (Italy) and Dinarides (Montenegro), southern Tethyan Margin: Sedimentary Geology, v. 165, p. 117–153.

SATTERLEY, A.K., 1996, The interpretation of cyclic successions of the Middle and Upper Triassic of the Northern and Southern Alps: Earth-Science Reviews, v. 40, p. 181–207.

SONNENFELD, M.D., AND CROSS, T.A., 1993, Volumetric partitioning and facies differentiation within the Permian Upper San Andres Formation of Last Chance Canyon, Guadalupe Mountains, New Mexico, *in* Loucks, R.G., and Sarg, J.F., eds., Carbonate Sequence Stratigraphy: Recent Developments and Applications: American Association of Petroleum Geologists, Memoir 57, p. 435–474.

STRASSER, A., AURELL, M., BADENAS, B., MELENDEZ, G., AND TOMAS, S., 2005, From platform to basin to swell: orbital control on sedimentary sequences in the Oxfordian, Spain: Terra Nova, v. 17, p. 407–413.

STRASSER, A., HILLGÄRTNER, H., HUG, W., AND PITTET, B., 2000, Third-order depositional sequences reflecting Milankovitch cyclicity: Terra Nova, v. 12, p. 303–311.

TUCKER, M.E., AND WRIGHT, V.P., 1990, Carbonate Sedimentology: London, Blackwell Science, 496 p.

VAIL, P.R., MITCHUM, R.M., JR., AND THOMSON, S., III, 1977, Seismic stratigraphy and global changes of sea level, Part 4: Global cycles of relative changes of sea level, *in* Payton, C.E., ed., Seismic Stratigraphy—Applications to Hydrocarbon Exploration: American Association of Petroleum Geologists, Memoir 26, p. 83–97.

VIEL, G., 1979, Lithostratigrafia ladinica: una revisione. Riconstruzione paleogeografica e paleostrutturale dell' area dolomitica–cadorina (Alpi Meridionali), I. e II. Parte: Rivista Italiana di Paleontologia, v. 85, p. 88–125 and p. 297–352.

WRIGHT, V.P., 1992, Speculations on the controls on cyclic peritidal carbonates: icehouse versus greenhouse eustatic controls: Sedimentary Geology, v. 76, p. 1–5.

ZÜHLKE, R., 2004, Integrated cyclostratigraphy of a model Mesozoic carbonate platform—the Latemar (Middle Triassic, Italy), *in* D'Argenio, B., Fischer, A.G., Premoli Silva, I., Weissert, H., and Ferreri, V., eds., Cyclostratigraphy: Approaches and Case Histories: SEPM, Special Publication 81, p. 183–212.

ZÜHLKE, R., BECHSTÄDT, T., AND MUNDIL, R., 2003, Sub-Milankovitch and Milankovitch forcing on a model Mesozoic carbonate platform—The Latemar (Middle Triassic, Italy): Terra Nova, v. 15, p. 69–80.

CARBONATE–BIOSILICEOUS SEDIMENTATION IN EARLY OLIGOCENE ESTUARIES DURING A TIME OF GLOBAL CHANGE, PORT WILLUNGA FORMATION, ST. VINCENT BASIN, SOUTHERN AUSTRALIA

NOEL P. JAMES
Department of Geological Sciences and Geological Engineering, Queen's University, Kingston Ontario, K7L 3N6, Canada
e-mail: james@geol.queensu.ca
AND
YVONNE BONE
School of Earth and Environmental Sciences, The University of Adelaide, South Australia 2005, Australia

ABSTRACT: The Port Willunga Formation is a cool-water, marine, quartzose, clay-rich, biosiliceous, and calcareous sedimentary succession of Early Oligocene age that accumulated in a series of proximal estuarine paleoenvironments along the eastern side of the St. Vincent Basin, South Australia. Coeval strata in two of the paleo-embayments are interpreted to record deposition during one ~ 3.5 My-long eustatic sea-level fluctuation. Transgressive facies above a ravinement surface comprise quartzose sands (subaqueous marine tidal dunes) that grade upward into fossiliferous floatstones and mudstones (shoreface to shallow basin-floor environments) that accumulated in a protected embayment. Highstand sediments are distinctly cyclic at the meter scale and consist of epifaunal bryozoan–pecten–echinoid clay-rich floatstones that become less fossiliferous but more spiculitic and chert-rich upward in each cycle. Whereas cyclic sediments in one embayment (Willunga) are interpreted to have accumulated on a current-swept, illuminated seafloor, those in the other (Noarlunga) are thought to have been deposited in a lower-energy, sub-photic setting. Cyclicity is interpreted to record the increasing influence of fluvial fresh water in the system during each sea-level fluctuation. Comparison with underlying strata reveals a striking similarity in depositional style and stratigraphic packaging between Late Eocene and Early Oligocene deposits; both are interpreted as paleoestuarine. Differences between the dark, organic-rich, biosiliceous, and low-diversity Eocene highstand deposits and the light, more calcareous, and more diverse Oligocene highstand deposits are interpreted to be due to local depositional controls. An important implication of local controls is that several postulated unconformities in the succession are not due to global eustatic changes but are ravinement surfaces related to estuarine sedimentation dynamics. Such controls, specifically terrestrial climate, hydrodynamic energy, and trophic resource levels were more important in determining sediment composition than eustasy and Southern Ocean cooling. Similar biosiliceous-carbonate sedimentary facies are a recurring feature of cool-water deposition throughout the Phanerozoic.

INTRODUCTION

The modern icehouse climate began during Late Eocene and Early Oligocene time (Zachos et al., 2001; Ivany et al., 2003) as Antarctica became progressively isolated and refrigerated, causing a profound change in global ocean circulation. This process seems to have started at about 37 Ma (Middle–Late Eocene boundary), accelerated at ~ 33.5 Ma, (Eocene–Oligocene boundary), and culminated with cooling and sea-level fall at ~ 30 Ma (early to late Oligocene boundary). The Early Oligocene is generally regarded as the period of most rapid and intense cooling, in response to buildup of continental glaciers on Antarctica (Berggren and Prothero, 1992).

Southern Australia lies along the northern margin of the Southern Ocean and was the site of neritic cool-water carbonate deposition during this important time period (McGowran et al., 1997; Gallagher et al., 1999; James and Bone, 2000). Middle to Late Eocene and Early Oligocene carbonates in the St. Vincent Basin, because they are well-exposed and near a large metropolitan area, have been actively studied from a paleontological and biostratigraphic perspective (Glaessner, 1951; Lindsay, 1967; Lindsay and McGowran, 1986; McGowran, 1987; McGowran and Beecroft, 1986a, 1986b) and are a critical site in recording neritic sedimentary response to this acute oceanographic change. Thus, they have been pivotal in understanding the evolution of the Southern Ocean and in turn global ocean circulation. While Eocene strata in southern Australia are now relatively well understood from a sedimentologic–paleoceanographic viewpoint (McGowran 1989, 1990; McGowran et al., 1992; McGowran et al., 1997; James and Bone, 2000) the Early Oligocene in the St. Vincent Basin has never been documented from a similar perspective. The purpose of this paper is to describe and interpret these strata and place them in the context of evolving Paleogene climate and oceanography in the Southern Ocean region.

SETTING

Southern Australia has been a site of cool-water carbonate sedimentation since the Middle Eocene (Lowry, 1970; McGowran et al., 1997), and continues to be so today (Wass et al., 1970; Boreen et al., 1993; James et al., 1992; James et al., 1994; James et al., 1997; James et al., 2001). The Paleogene and early Neogene part of this stratigraphic succession underlies much of the continental shelf and fills a series of onshore basins where, as a result of Miocene tectonic inversion, the rocks are now variably exposed along sea cliffs and in river valleys. Oligocene strata detailed in this study lie within the St. Vincent Basin, a fault-bounded depression in the central part of southern Australia that is now mostly covered by Gulf St. Vincent (Fig. 1). The basin was a large gulf facing the nascent Southern Ocean during the Paleogene and early Neogene (Lindsay and Alley, 1995). Whereas much of the succession lies in the subsurface, excellent sections are exposed in two northeast–southwest oriented half grabens along the eastern shore of Gulf St. Vincent, just south of Adelaide (Fig. 1). These small rift basins are called the Noarlunga Embayment and the Willunga Embayment (Reynolds, 1953; Cooper, 1979, 1985). The Willunga Embayment is

Controls on Carbonate Platform and Reef Development
SEPM Special Publication No. 89, Copyright © 2008
SEPM (Society for Sedimentary Geology), ISBN 978-1-56576-130-8, p. 231–253.

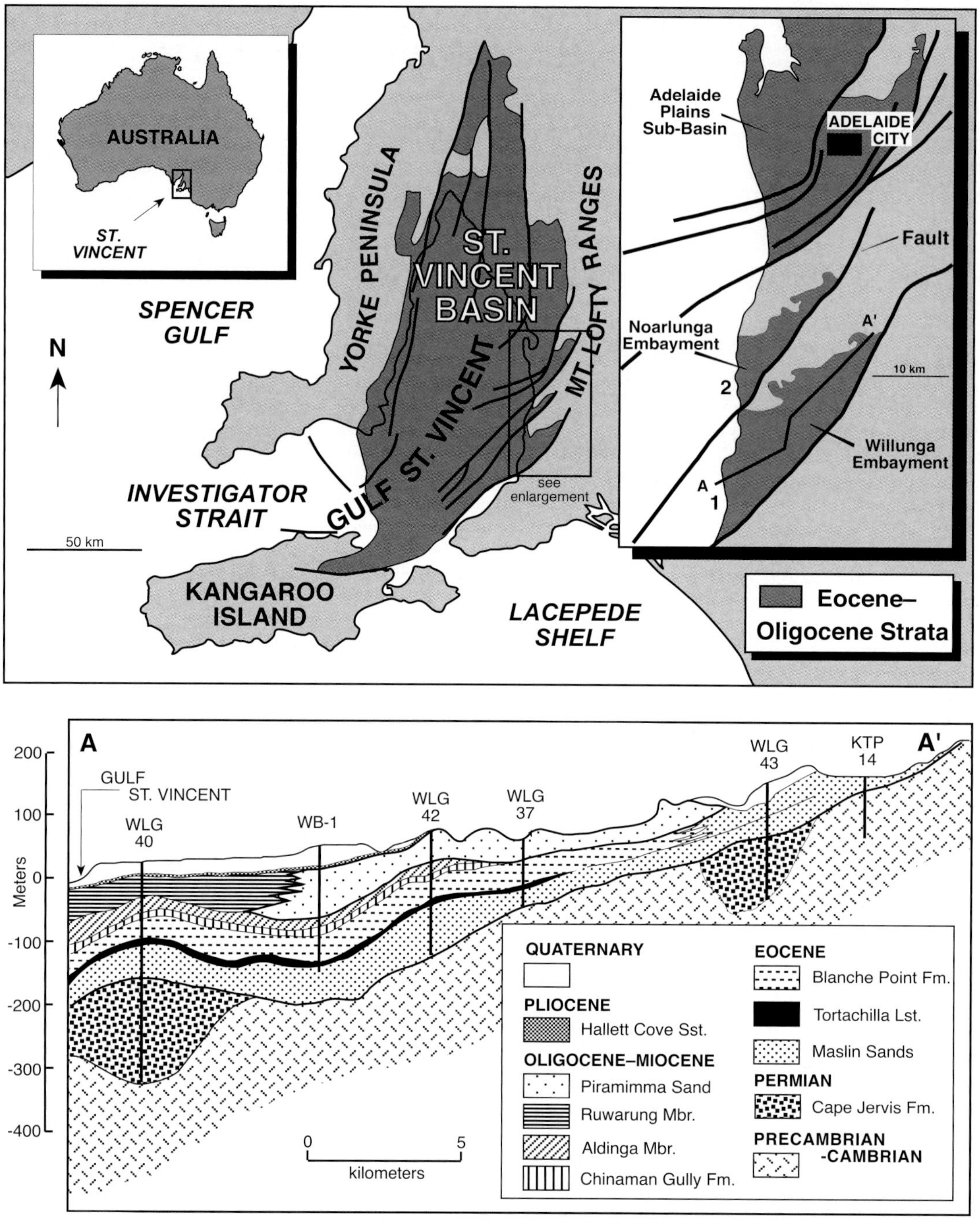

FIG. 1.—Maps illustrating the location of St. Vincent Basin and embayments along the eastern side of the basin. Numbers refer to specific localities; 1 = Port Willunga, 2 = Onkaparina Estuary and sea cliffs. A–A′ = the line of section illustrated in the cross section below (modified from Cooper, 1979). Core locations designated WB, WLG, and KTP are also used to construct Fig. 11. Solid lines = faults.

14 km wide at the present shoreline outcrop cliffs and extends inland in a NE direction some 26 km (Fig. 1). It is asymmetric, because of Neogene tectonics, such that the SW side is downthrown and fault bounded. The Noarlunga Embayment is 10 km wide at the entrance and extends inland 20 km in the same orientation. The basins to the north, the Adelaide Plains Basin and Golden Grove Embayment (Fig. 1), are larger, but all sediments are subsurface with the record mainly as soft sediment samples.

Exposures (Figs. 2, 3) form low sea cliffs at the western ends of the Noarlunga and Willunga embayments (Fig. 1). The deposits range from unlithified sediment to hard sedimentary rock. Although outcrops inland are scarce, there is enough subsurface borehole information to reveal a picture of the areal distribution of units (Cooper, 1979; Fairbain, 1998, 2000). The outcrops (Fig. 3A) are part of a thicker mid-Eocene to mid-Oligocene section at each of the two localities (Figs. 2, 3). Late Oligocene to Middle

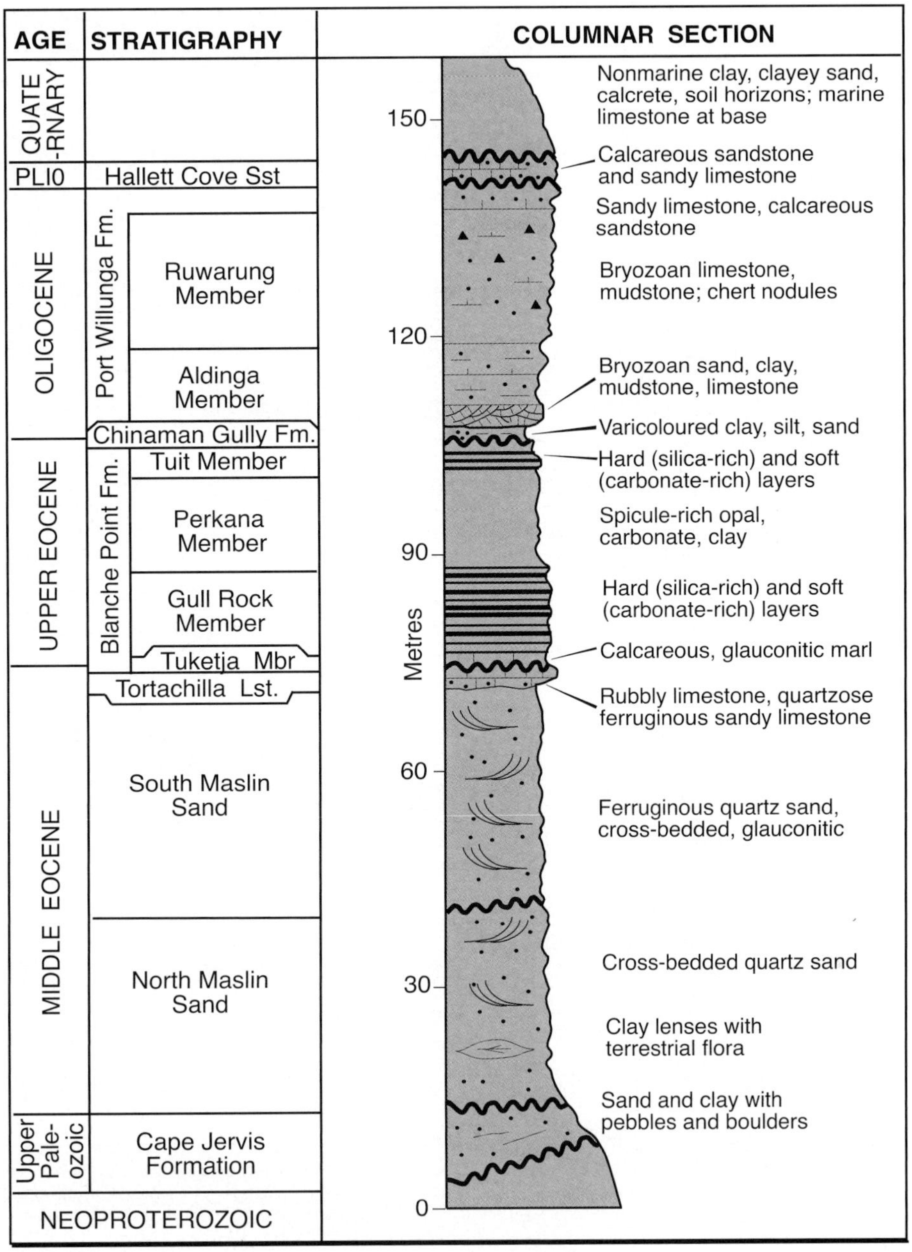

FIG. 2.—A generalized stratigraphic column of strata exposed along seacliffs in the Willunga and Noarlunga Embayments (after Cooper, 1979).

Miocene strata are, however, obscured by cover except at the southern end of the Willunga Embayment where Middle Miocene limestone is exposed. The largely Paleogene sediments are unconformably overlain by either Pliocene fossiliferous sandstone (*Hallett Cove Formation*) or Pleistocene sands and clays.

Initial Cenozoic deposits in these half-graben embayments are Middle Eocene sands and sandstones (*North Maslin Sand and South Maslin Sand*) (Dyson, 1998). Carbonate deposition began with a profusely fossiliferous Middle Eocene limestone (*Tortachilla Limestone*) that is punctuated by numerous omission surfaces and hardgounds (James and Bone, 2000). It is unconformably overlain by Upper Eocene fossiliferous and biosiliceous marls and mudstones (*Blanche Point Formation*) that are commonly silicified (Jones and Fitzgerald, 1986; James and Bone, 2000). The Eocene succession is truncated by an erosional unconformity that cuts down into the underlying Blanche Point. Basal Oligocene deposits that cover the unconformity and fill depressions on it (*Chinaman Gully Formation;* Fig. 3B) are paralic silty mudstones with quartzose sands and sandstones inland that contain wood fragments (Lindsay and Alley, 1995). The uppermost of these fine-grained deposits are marine and overlain by marine sand, sandstone, mudstone, marl, and limestone of the *Port Willunga Formation* (PWF) (Fig. 3A, B–F), the topic of this study. These strata are subdivided into a lower *Aldinga Member* (Fig. 3D–F), which is heterolithic sand, sandstone, mudstone, and marl, and an upper *Ruwarung Member*, which is a succession of mostly limestone and marl with conspicuous chert nodules and layers (Cooper, 1979).

FIG. 3.— **A)** Exposure of the Port Willunga Formation (PWF) at the mouth of the Onkaparinga Estuary in the Noarlunga Embayment (scale = person at left). **B)** Brightly colored, burrowed, sparsely fossiliferous silty mudstones at the top of the Chinaman Gully Formation (CGF), Willunga Embayment. **C)** Cross-bedded bryozoan-rich, quartzose sands with prominent inclined layering dipping to the left (southwest) and smaller-scale cross-laminae dipping to the right (east), Willunga Embayment. **D)** Sparsely fossiliferous brown and dark gray mudstones near the top of the Aldinga Member, Noarlunga Embayment, scale bar increments = 10 cm. **E)** Clay-rich floatstone with numerous bryozoans, Ruwarung Member, Noarlunga Embayment. **F)** Cycles in the middle of the Ruwarung Member (arrows), with conspicuous silicified upper parts; scale bar at center, increments = 10 cm.

Embayment geometry and the overall sedimentary succession point to deposition during flooding of incised valleys (cf. Zaitlin et al., 1994). Further, the facies gradation from eastern terrestrial and fluvial to western fossiliferous and marine implies deposition within an estuarine system (Dalrymple et al., 1992), an interpretation embraced by Dyson (1998). The two embayments discussed in this paper are, therefore, viewed as early Oligocene valleys that were filled with marine sediment. The term estuary is used here in the broad sense of Pritchard (1967), which includes salinity differences, but bearing in mind the more geological definition of Dalrymple et al. (1992).

Understanding the sedimentology and paleoenvironments of deposition within these embayments is important because sedimentary rocks and their contained fossils from these sections have been pivotal when unraveling the Cenozoic history of this part of the continent (Tate, 1878; McGowran et al., 1992; McGowran et al., 1997). In recent years they have been correlated with coeval strata in western Australia and Victoria (McGowran et al., 2004) to provide a more complete model for the oceanographic, biotic, tectonic, and climatic evolution of the northern part of the Southern Ocean.

Although the Eocene–Oligocene age of these sediments and sedimentary rocks has long been confirmed (Glaessner, 1951), the precise ages have subsequently proven to be a thorny problem, largely because the foraminifers, particularly the neritic forms, are endemic. This conundrum has been clearly outlined by McGowran et al. (1997). Simply put, the Eocene–Oligocene boundary within the Willunga Embayment was, for many years, placed at the Aldinga Member–Ruwarung Member contact in the Port Willunga Formation (Lindsay, 1967; Lindsay and McGowran, 1986). McGowran et al. (1992) realized that while worldwide this boundary coincided with extensive changes in all biotic realms, there was very little biotic change in southern Australia (McGowran and Beecroft, 1986b). Hence the boundary was shifted downward to the erosional base of the Chinaman Gully Formation, which did coincide with major changes in the microbiota (Moss and McGowran, 1993). Thus, the Chinaman Gully and Port Willunga formations as exposed in shoreline sections are currently regarded as Early Oligocene (Rupelian) (McGowran et al., 2004), except for the uppermost limestones in the northern Willunga Embayment, which seem to be Late Oligocene (Chattian, or locally Janjucian) in age (McGowran and Li, 1997).

METHODOLOGY

Stratigraphic sections were logged and photographed in the field. Sections were easily accessible and mostly continuous, but minor faulting, local dissolution, and unit lenticularity, particularly in the Willunga Embayment, necessitated detailed analysis. A total of 162 samples were collected. From these, 132 thin sections were prepared and stained (cf. Dickson, 1966), and 8 samples were examined under scanning electron microscope. On the basis of petrography, X-ray diffraction analyses were carried out on 16 samples using standard techniques on powder mounts with quartz as an internal standard (Tucker, 1988). Duplicate analyses were performed on whole samples and on those in which carbonate had been removed by HCl dissolution, specifically to confirm clay-mineral and Opal-CT identifications.

SEDIMENTARY FACIES

The bulk of the Early Oligocene succession (Fig. 3) is a series of meter-scale units bounded by burrowed omission surfaces and composed of bryozoan, pectenid bivalve, epifaunal echinoid floatstone with a spiculitic, clay-rich matrix. This basic motif is augmented by variable amounts of quartzose sand, glauconite, and chert. Basal lithologies are mostly sand and clay, middle lithologies are floatstone-dominated, and uppermost units are biofragmental limestone.

Sediment Composition

The finest material (Fig. 4) is calcareous silty clay, much of which is aggregated into medium to very coarse sand-size peloids. The 10–30% silt and very fine sand fraction is angular quartz (minor orthoclase). Clay (Fig. 4B, D, F), which forms 70–90% of the matrix, is mostly smectite (montmorillonite), clinoptilolite, and glauconite. There is also minor illite, kaolinite, and orthoclase. Glauconite occurs as peloids, inside skeletal pores, and fills bryozoan zooecia. The fine carbonate is mostly biofragmental material and coccoliths. Siliceous diatoms are present but not numerous.

The quartz sand is poorly to medium sorted, and medium to coarse grained. Carbonate and all coarser sediment is wholly biogenic. Macrofossils are mostly bryozoans (Fig. 3E), epifaunal echinoid pieces, and pectenid bivalves. Other biofragments such as turritellid gastropods (molds and aragonite shells), epifaunal and infaunal bivalves, brachiopods, serpulid worm tubes, and barnacles are abundant locally. Sand-size carbonate grains are mostly bryozoan remains, benthic foraminifer tests, echinoid pieces, and coralline algae (articulated) fragments.

Siliceous sponge spicules locally form as much as 50% of the sediment. Although generally molds (Fig. 4B, C), some are still opal-A, while others are replaced by opal-CT lepispheres (Fig. 4E) or microquartz.

Diagenesis

Since most sediment and sedimentary rocks have a clay matrix, cement is not obvious, even on SEM. This lack of cementation is manifest by the overall soft character of the materials. There is little obvious calcite cement implying minor meteoric diagenesis except that aragonitic turritellids, ascidian spicules, and infaunal bivalves are molds. The sediment has been only shallowly buried (< 200 m). Some obvious hard omission surfaces, however, are cemented with fibrous to peloidal calcite cement, implying synsedimentary lithification (cf. Nelson and James, 2000).

The most profound and conspicuous alteration is silicification. It is characteristically nodular, commonly beginning in *Thalassinoides* burrows and spreading outward. It ranges in composition from porcellanite to hard gray chert wherein the whole rock is silicified. Those units in which chert is common typically have numerous sponge-spicule molds, many of which are covered with opal-CT lepispheres. This opal-CT is also conspicuous in the silicified matrix.

STRATIGRAPHIC SEDIMENTOLOGY

Both embayments show the same basic stratigraphic succession (Fig. 5), but with important local differences. The lithological character is relatively simple at the large scale but complex at the small scale.

Basal Quartzose Sands and Sandstones

Terrestrial sediments of the Chinaman Gully Formation (Fig. 3B) are sharply overlain by a 2–3 m thick white to Fe-stained, cross-bedded, quartzose sand and sandstone (Figs. 3C, 6). Basal beds are granule to coarse-grained sand with abundant

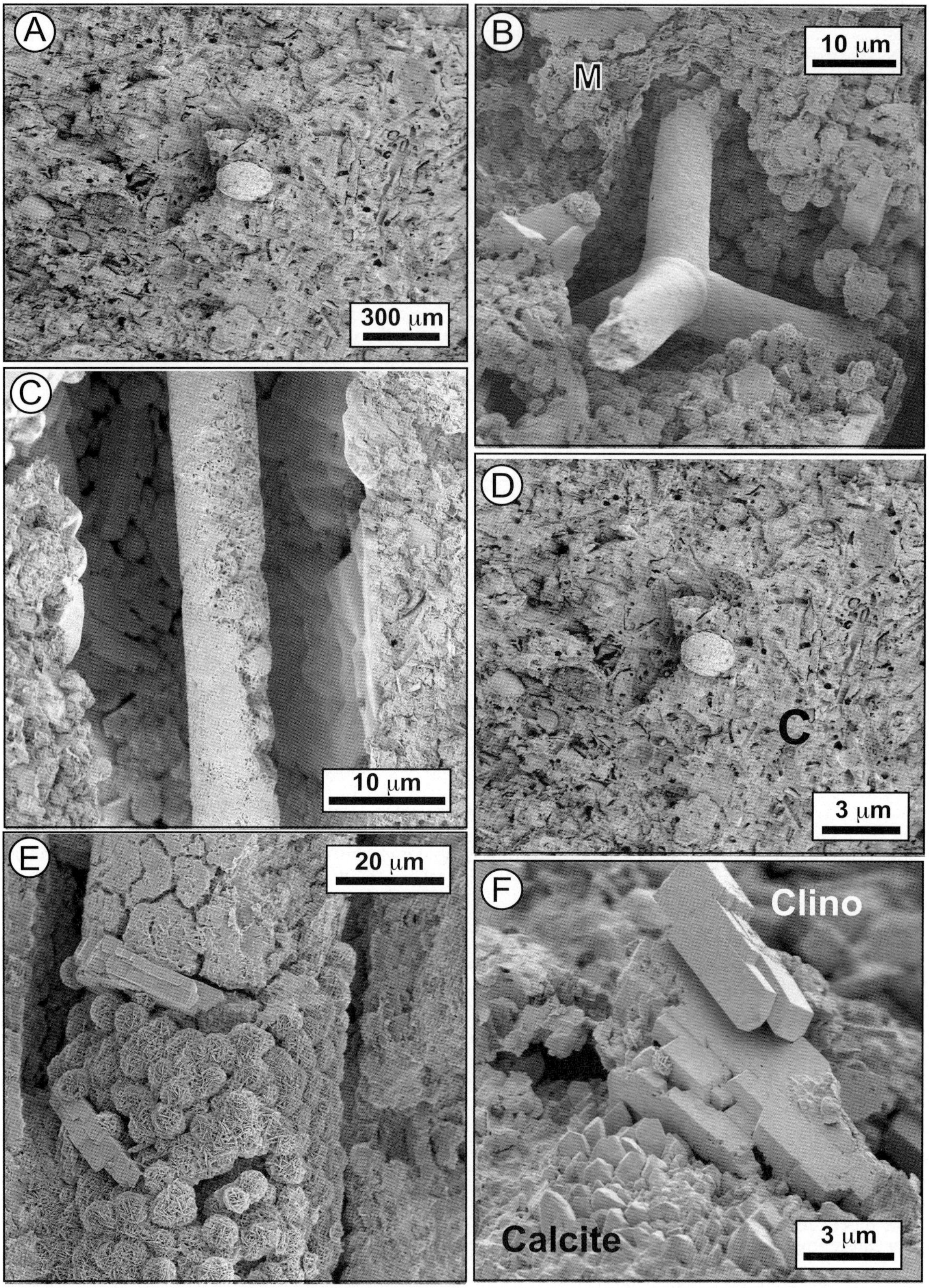

FIG. 4.—Scanning electron photomicrographs of sedimentary components and diagenesis. **A)** Fine-grained matrix of floatstone with numerous spicule molds, **B)** sediment-filled triaxon sponge-spicule mold surrounded by opal-CT lepispheres and minor detrital montmorillonite (M) at upper left, **C)** sponge-spicule mold in which the axial canal of the spicule is preserved and filled with opal-CT lepispheres, **D)** floatstone matrix composed of opal-CT lepispheres and clinoptilolite (C), **E)** spicule axial canal filled with an intergrowth of lepispheres and clinoptilolite, **F)** clinoptilolite (clino) and calcite cement in floatstone matrix.

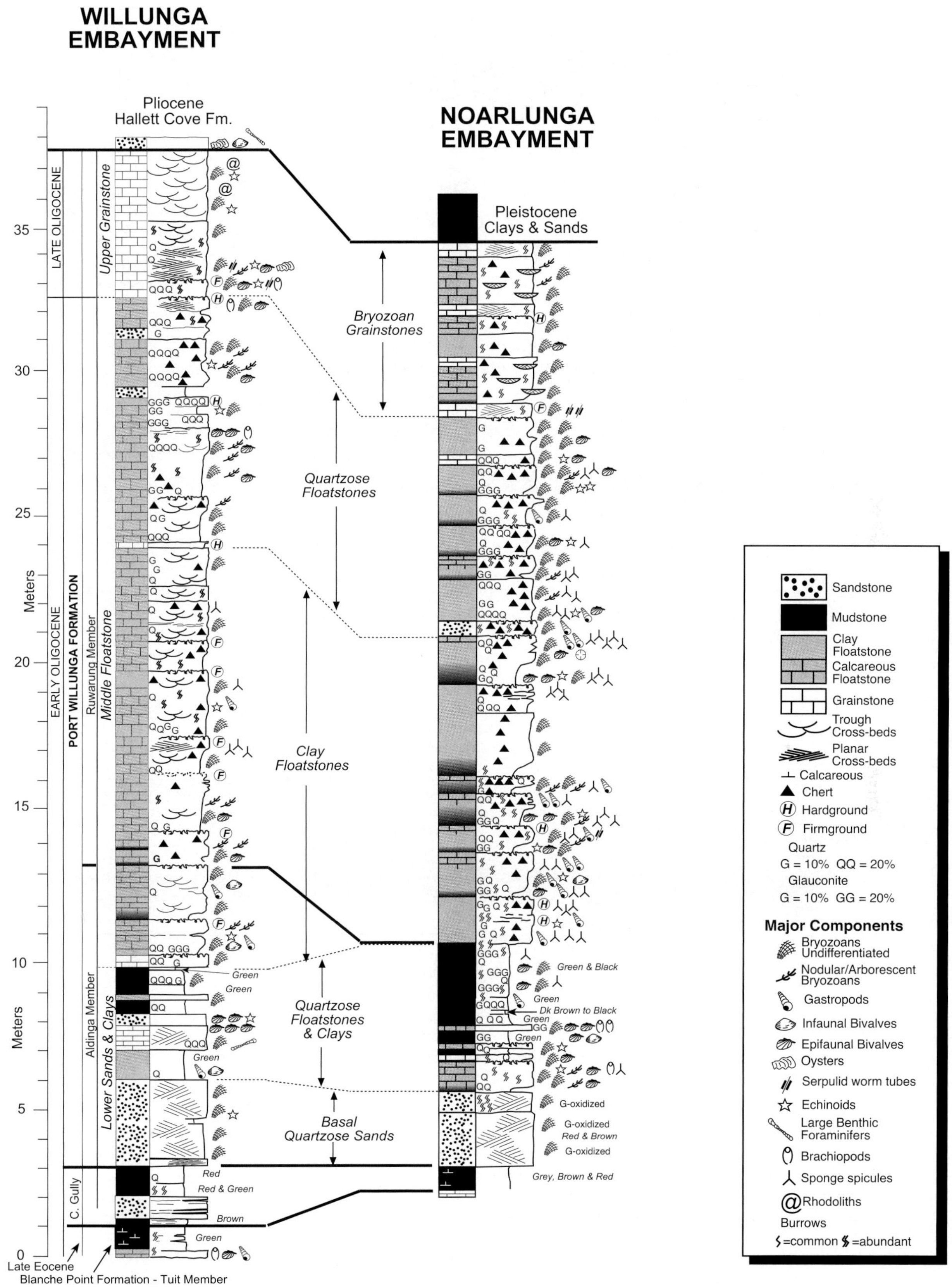

FIG. 5.—Stratigraphic sections of the Chinaman Gully Formation (CGF), and Port Willunga Formation (PWF) along coastal cliffs in the Willunga and Noarlunga Embayments.

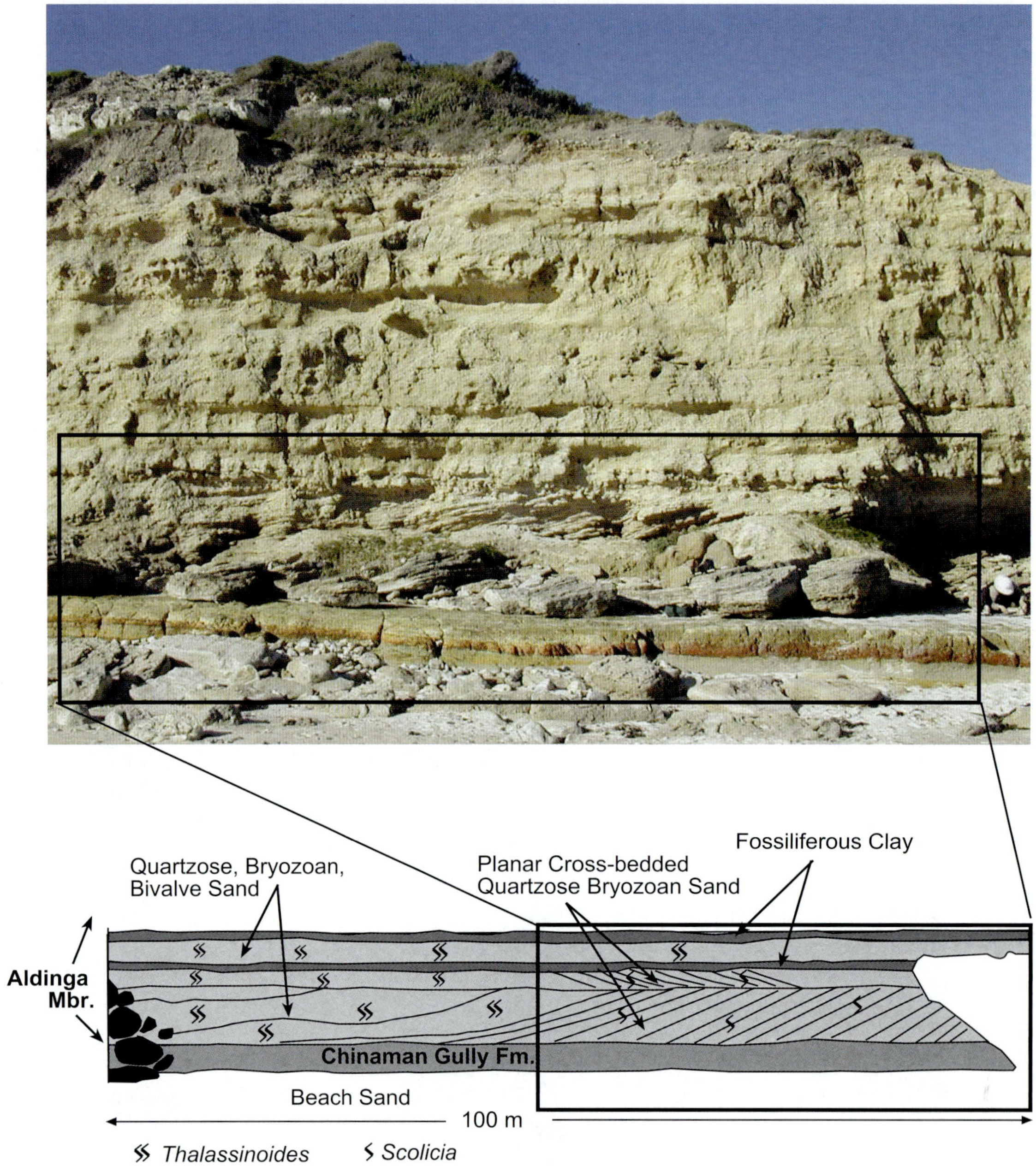

FIG. 6.—Coastal cliff section exposure of the CGF and overlying Aldinga Member of the PWF in the Willunga Embayment, (person for scale at right). Lower part of the PWF is conspicuously cross-bedded whereas most of the cliff is interbedded fossiliferous quartzose grainstone and floatstone or dark green variably fossiliferous clay (Fig. 5). Field sketch at base illustrates the bipolar nature of the cross-bedding and its relationship to burrowed fossiliferous sands.

Scolicia (Fig. 7A), conspicuous brachiopods, and numerous bryozoans (Fig. 7B). Most of the unit is planar cross-bedded up to angles of 20° that are regionally bidirectional; most indicate current directions from the N or SW. These are locally overlain by trough cross-beds, 2–4 cm thick, with flow directions from the N and NNE (Fig. 3C), mud drapes, and thin layers of wave ripples with mud drapes. Most bryozoans are articulated branching, delicate branching, and flat robust branching, with minor fenestrate types.

These sands in both embayments contain the large benthic foraminifers *Linderina, Halkyardia,* and *Masslinella* (Lindsay, 1985; this study) and coralline algal rods. The bryozoans also include *Densipora*, a form that today occurs only epiphytically on seagrass.

Willunga Embayment.—

At Aldinga Beach, where bedforms with opposing paleocurrent directions are exposed (Fig. 6), the sands are clearly subaqueous dunes that are draped by and locally grade laterally into *Thalassinoides*-burrowed sands rich in echinoid biofragments and pectens. The quartzose sands are also interbedded with highly fragmented skeletal grainstones.

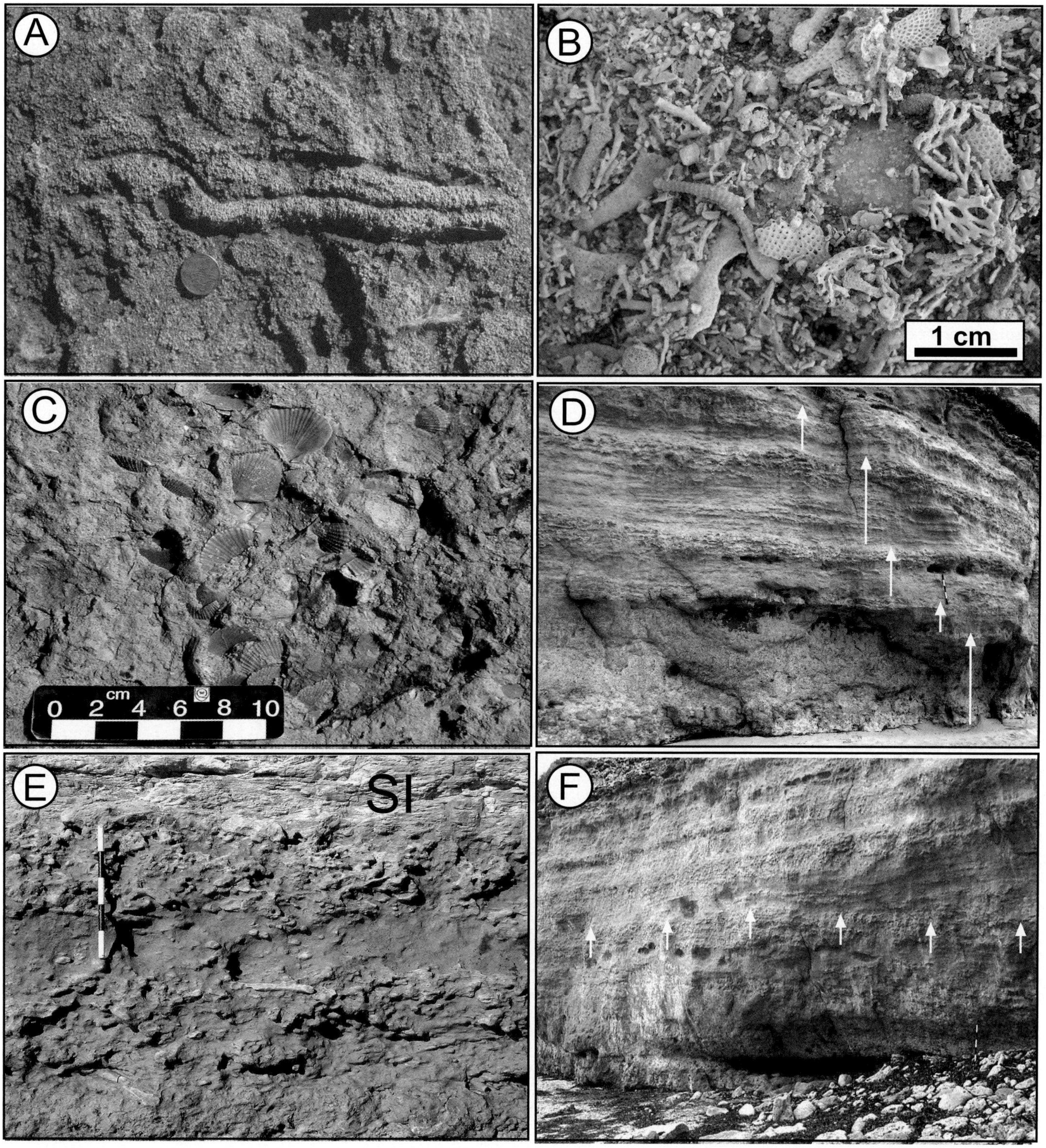

FIG. 7.—Port Willunga Formation: **A)** *Scolicia* burrow in basal cross-bedded sands, Port Willunga, coin 3 cm diameter, **B)** quartzose bryozoan grainstone, in basal cross-bedded sands, Port Willunga, **C)** pectenid bivalves littering the surface of a clay bed, Onkaparinga Estuary, **D)** cyclic bedding (arrows) in clay floatstone, Willunga Embayment, scale-bar increments at right = 10 cm, **E)** *Thalassinoides* burrows in clay floatstone; upper part is silicified (SI), scale-bar increments = 10 cm, Noarlunga Embayment, **F)** trough cross-bedded clay floatstones (arrows at base of trough), Willunga Embayment, scale-bar increments = 10 cm.

Noarlunga Embayment.—

The sands here are thinner and less fossiliferous than those in the Willunga Embayment. Together, these attributes of the sediments in both embayments indicate a relatively warm-water, marine, subtropical, (large benthic foraminifers; Hallock and Glenn, 1986), photic (corallines; Wray, 1977), high-energy, tidal (bedforms; Dalrymple, 1992) environment.

Quartzose Floatstones and Mudstones

The interpreted tidal sands are overlain in both embayments by a clay-rich unit (Fig. 3D) of floatstones and mudstones that are variably rich in quartzose sand with detrital orthoclase and illite, and diverse macrofossils.

The green to gray mudstones contain ~ 10% very fine-grained, angular quartzose sand (up to 40%), a mixture of illite, smectite, and authigenic clinoptilolite, coccoliths, and diatoms, with rare bryozoans. Pectens occur in layers that are only one valve thick but with no preferred orientation (Fig. 7C). The uppermost clay is virtually devoid of biota but contains ferruginous decalcified molds.

Floatstones are rich in epifaunal elements such as large and diverse bryozoans (Fig. 8A, B) (e.g., *Celleporaria* spp., various fenestrates, and encrusting forms), pectenid bivalves, echinoid spines, and locally large brachiopods. The large infaunal components are echinoid spines and rare bivalve molds. Grainy beds are burrowed by *Scolicia,* whereas muddy layers, particularly the tops of clay beds, are burrowed by *Thalassinoides* (Fig. 7E). The matrix is mostly clay (detrital illite), coccoliths, bryozoan biofragments, rare sponge spicules, and barnacle pieces. These sediments contain no large benthic foraminifers. Lower beds are quartz-rich at the base whereas upper beds are glauconite-rich at the base. Glauconite peloids are locally altered to limonite.

Willunga Embayment.—

The basal lithologies here are alternating decimeter-scale beds of biofragmental grainstone, quartzose sand, and green glauconitic clay locally rich in pectens with no apparent vertical organization. Grainy facies contain coralline algae. Quartzose sands include clay flasers.

Noarlunga Embayment.—

The lower beds are several meter-scale cycles comprising basal grainy, vaguely planar cross-bedded, fossil-rich floatstones that grade up into muddy, sparse floatstones. The lower parts contain a prolific bryozoan biota with particularly conspicuous foliose forms. The upper portions of the cycles contain a volumetrically smaller but still diverse bryozoan biota with numerous large *Celleporaria.*

The upper part of this the quartzose floatstone and mudstone facies comprises burrowed clay-dominated sediment containing abundant glauconite and prominent pectens. Cycles here comprise decimeter-thick grainstones or floatstones grading up to thicker burrowed clays.

The finer-grained nature of the sediments and dominance of burrowing over hydrodynamic bedforms imply relatively low energy. Lack of large benthic foraminifers suggests cooler waters (Hallock and Glenn, 1986), whereas the prolific infaunal and epifaunal macrobiota points to abundant trophic resources (Lukasik et al., 2000; Mutti and Hallock, 2003). The presence/absence of coralline algae indicate an illuminated seafloor in the Willunga embayment but not in the Noarlunga embayment.

Middle Floatstones

Fossiliferous floatstones (Figs. 5, 7, 8, 9) that form the bulk of the formation are generally similar in composition but dramatically different in depositional style in the two embayments. Both sections, however, are divisible into (1) lower clay floatstones and (2) upper quartzose floatstones (Fig. 5).

Willunga Embayment.—

Individual beds, 1–3 m thick, are typically cyclic (Figs. 7D, 10). Cream-colored floatstones range from richly fossilifeous, with almost a rudstone texture, to lean, wherein only a few macrofossils float in mud (Fig. 8A, B). The matrix is highly variable but can be up to 40% clay. Some floatstones have a fine fragmental grainstone to packstone matrix with only minor clay. Green clay lenses and seams 10–50 cm thick are, however, common within many beds, attesting to variable hydrodynamics. Such clays are locally burrowed into underlying sediment. Alternatively, there are also centimeter-thick layers of fine skeletal grainstone within the floatstone. Some beds are conspicuously glauconite-rich or quartz-rich at the base. In other instances, beds are punctuated or capped by layers rich in pectenid bivalves and concentrations of echinoid spines. Most burrows within the sediments are *Thalassinoides.*

Cyclic clay-rich floatstones form the lower part of the facies. They contain 10–20% angular to rounded, fine- to medium-grained quartz sand and are cyclic at the scale of 0.5–2.0 m (Fig. 7D). *Facies A* (Figs. 8A, 10), the lower trough cross-bedded part of the cycle, is grainstone to rich floatstone and typically constitutes two-thirds of the unit. Bedforms (Fig. 7F) are 1.5–2.0 m high and 8.0–10.0 m trough wide, with poorly constrained current directions to the SW. Deposits contain a volumetrically high proportion of pectens, epifaunal echinoid spines and tests, together with foliose, erect, articulated and fenestrate bryozoans. *Facies B* (Fig. 10), the upper part of the cycle, is intensely bioturbated (*Thalassinoides*) sparse floatstone wherein chert conspicuously replaces burrows. As the sediment becomes increasingly muddy upwards in this facies (Fig. 8B) the concentration of bryozoans decreases, foliose forms decline, but robust arborescent (*Celleporaria*) forms (Fig. 8C) increase. *Thalassinoides* burrows extend down as much as 30 cm from the overlying surface into facies B, and are filled with coarse biofragmental grainstone of the overlying cycle. The upper part of these clay-rich floatstones contains numerous chert nodules. The sponge-spicule content here ranges from 10 to 40%, but is usually ~ 10%.

Quartzose floatstones that constitute the upper part of the facies are grainy and contain > 30% fine- to medium-grained, subangular to subrounded quartz sand, with some being calcareous sandstone wherein the sand is locally medium- to coarse-grained; clay content is rarely > 20%. There are fewer obvious sponge spicules in the sediment (< 10%) than below, but chert nodules are abundant. The deposits are not as clearly cyclic; there is only minor trough cross-bedding, and most sediment is homogeneous or burrowed by *Thalassinoides.* Unlike clay-rich floatstones below, most bedding contacts are flat and not burrowed, but with local hardgrounds. The bryozoan biota is similar to that below except that *Celleporaria* is smaller and occurs together with foliose forms. At the top of this unit bryozoans (especially *Celleporaria* spp.) (Fig. 8D) and brachiopods are conspicuously more numerous, larger, and more diverse than in the lower clay-rich floatstones.

Ostracodes throughout the middle floatstone facies in the Willunga Embayment are similar to those in the underlying Eocene Tortachilla Limestone, are commonly fragmented, and

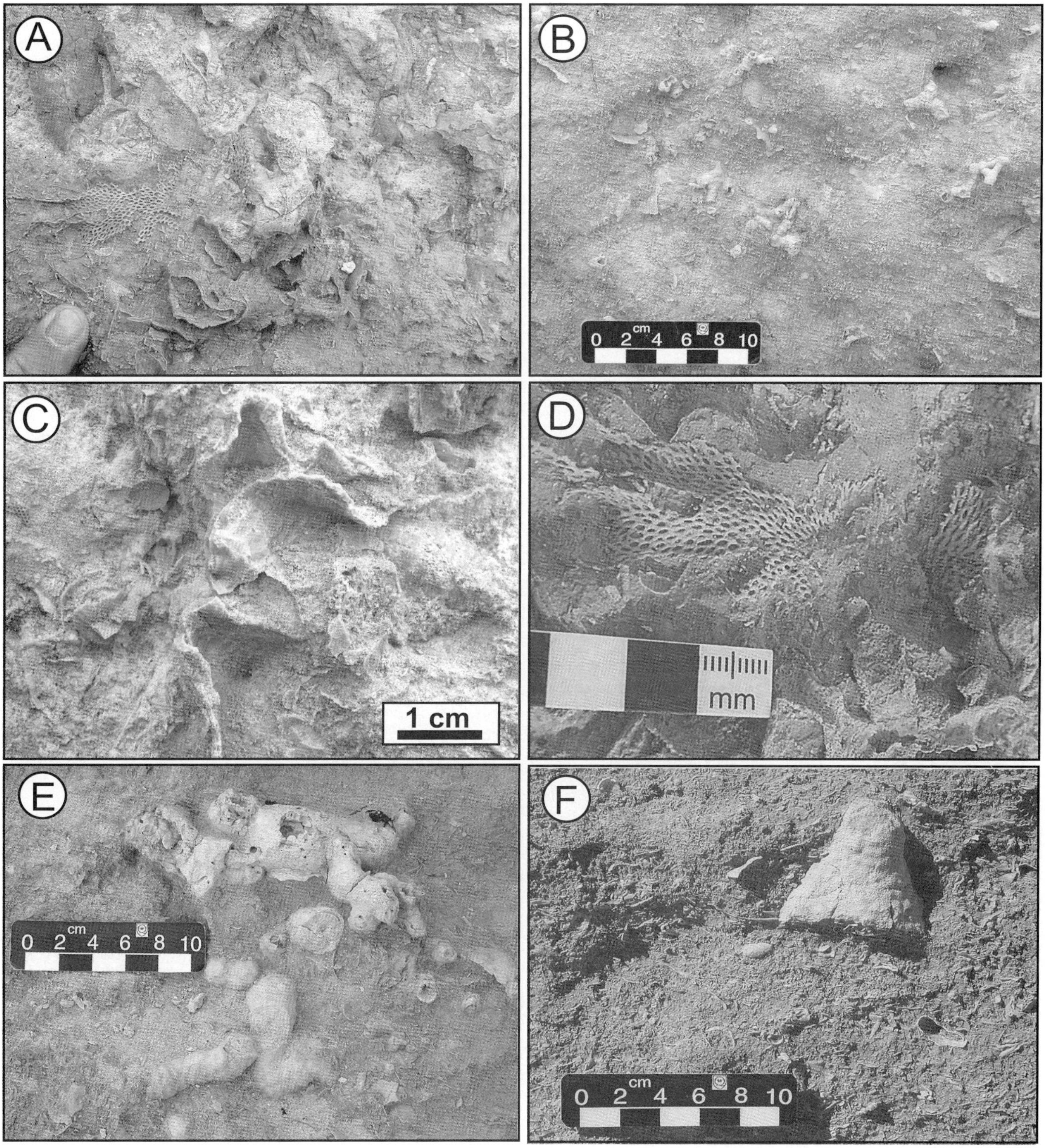

FIG. 8.—Middle Floatstones: **A)** Diverse bryozoans, especially fenestrate and foliose forms, in fossil-rich floatstone, Willunga Embayment, finger 1.5 cm wide, **B)** small *Celleporaria* in fossil-lean floatstone, Willunga Embayment, **C)** large foliose bryozoan in facies A, Noarlunga Embayment, **D)** complex to fenestrate and foliose bryozoans in facies A, Noarlunga Embayment, **E)** large arborescent *Celleporaria* colony, hollow digits likely formed around finger-like demosponges, Noarlunga Embayment, **F)** large domal *Celleporaria* in uppermost grainy floatstone carbonates, Noarlunga Embayment.

FIG. 9.—Middle Floatstones: **A)** Clay-rich floatstone with numerous bryozoans, pencil at left 2 cm long, **B)** sparse clay-rich floatstone with large chert nodule, hammer is 30 cm long, **C)** m-scale cycle (arrow) in lower floatstones, Noarlunga Embayment, **D)** upper part of the cycle illustrated in Part C wherein dark green glauconitic quartzose grainstones from above fill *Thalassinoides* burrows excavated into sparse clay floatstone, scale-bar increments = 10 cm, **E)** planar contact between Lower Oligocene floatstone below and Upper Oligocene low-angle, planar cross-laminated (beach) grainstone above, Willunga Embayment, scale-bar increments = 10 cm, **F)** view down onto low-tide platform at Aldinga Reef, Willunga Embayment, showing plan view of large-scale trough cross-beds and current directions (arrows) in Upper Oligocene grainstones.

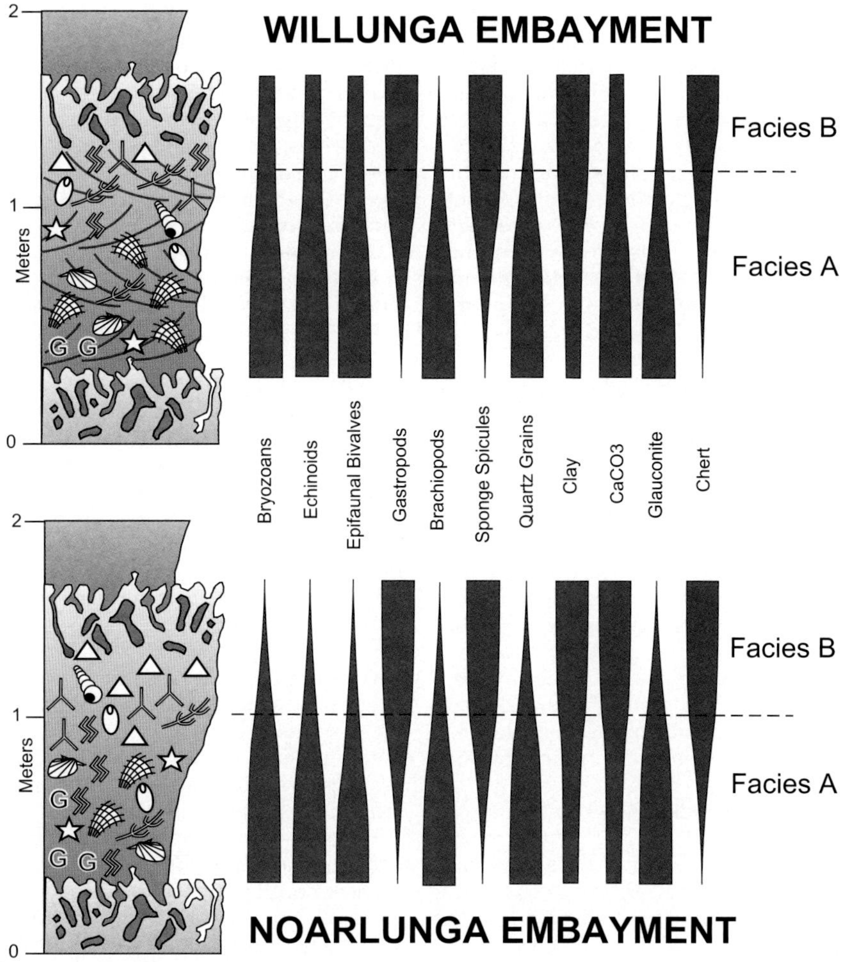

FIG. 10.—Synthesis of attributes in meter-scale cycles of the Port Willunga Formation. Facies A is richly fossiliferous floatstone, facies B is sparse clay-rich biosiliceous floatstone.

some of the taxa are found today growing on plants and algae. The presence of the planktic foraminifer *Gumbetrilina* in this facies indicates significant seawater cooling (McGowran and Beecroft, 1985) compared to waters in which the underlying sediments were deposited. The sand-size particles contain numerous small benthic foraminifers, dominated by high numbers of the epifaunal *Cibicides* spp. as well as articulated coralline algal rods.

Overall these clay-rich and quartzose floatstones are interpreted to have accumulated on an illuminated seafloor with an active burrowing infauna that locally obliterated sedimentary structures. The prolific epifauna of filter feeders confirms an oxygenated water column with high trophic resources capable of supporting abundant plankton. The trough cross-laminations indicates active bottom currents that, given the setting, were likely tide-induced. Scattered wave ripples indicate episodic wave reworking. Episodic clay layers attest to episodic influx of muds. Layers of macrofossils also suggest intermittent winnowing (Kidwell, 1991). Abundant fine- to medium-grained quartzose sand in the upper parts of this floatstone succession implies increasing fluvial influence.

Noarlunga Embayment.—

Sediments in the Noarlunga embayment are also cyclic (Fig. 9). Overall the beds are like those at Willunga, 0.5 to 3.0 m thick, bounded by burrowed omission surface contacts, and burrows with sediment from above extending 30–50 cm down into underlying units (Fig. 9D). The major differences are (1) they are rarely if ever cross-laminated, (2) they contain no coralline algae, (3) they are much more spiculitic (20–50%), and (4) there is more chert in the form of nodules and as completely silicified beds. Cycles comprise two discrete facies (Fig. 10). *Facies A* (Fig. 9A), at the base, is a calcareous, glauconitic, spiculitic, fossiliferous floatstone. It forms the lower two-thirds of the cycle, is buff to green in color, and contains 10–30% glauconite, especially in burrows. The biota is conspicuously epifaunal; bryozoans are particularly abundant (especially fenestrate and foliose forms (Fig. 8C, D) and *Celleporaria* as medium-sized arborescent branching colonies), as are epifaunal echinoid spines and large parts of tests (e.g., cidaroids) and pectens. Other macrofossils are brachiopods, as well as turritellid gastropod and azooxanthellate coral molds. Glauconite decreases in abundance upwards (and can locally be green clay). *Facies B*, forming the upper part of the cycle, is a light gray to buff spiculitic marl. It is generally poorly fossiliferous, with the only conspicuous megafossils being turittellid molds. This sediment is totally bioturbated, but *Thalassinoides* is excellent where preserved. Silicification (Figs. 2F, 9B) is more pervasive in this facies than below. Digitate sponge body fossils occur at the tops of some cycles. These two facies are end members; some cycles in facies B, for example, have occasional bryozoans, echinoids, and pectens.

Thicker cycles (> 1 m) tend to have the best symmetry, whereas thinner cycles are mostly facies A.

Clay-rich floatstones that form the lower part of the middle floatstone facies contain ~ 30% quartz silt. The clay matrix ranges from 40 to 60% but is locally vaguely interlaminated with carbonate mud that is conspicuously peloidal (medium to very coarse sand size). Whereas most are muddy, there are also scattered thin layers of bryozoan grainstone containing echinoid biofragments and miliolid foraminifers. Most bed contacts are burrowed, but some are planar and erosional, implying synsedimentary lithification. Spicules constitute from 20 to 60% of the sediment, and most are molds, surrounded by a mixture of opal-CT lepispheres and authigenic clinoptilolite, smectite clays, and detrital illite. Silicification ranges from nodules to partial (e.g., porcellanite) to complete silicification of beds. Fabric preservation is generally excellent.

Quartzose floatstones that form the upper part of the middle floatstone facies contain 20–60% very fine- to fine-grained angular quartz sand that is locally medium to coarse grained, ~ 20% glauconite peloids, 10–30% spicules, and 20–40% clay matrix. Silicification is similar to that below but tends to be concentrated at bed tops. Bryozoans are numerous but of low diversity and conspicuously large, with big *Celleporaria* spp. colonies up to 5 cm across (Fig. 8C). Brachiopods are likewise large. Body sponges are present but scarce. There is no obvious infauna except for scattered turritellid molds. The matrix contains conspicuous coccoliths and the same clays as in the underlying unit. The microbiotas in these deposits have not been studied in detail, but the sand-size carbonate particles are mostly bryozoans and benthic foraminifers, with no coralline algae.

These sediments and sedimentary rocks overall have many of the same attributes as those in the Willunga Embayment, except that their more clay-rich aspect and lack of physical sedimentary structures implies a quieter-water setting. Absence of physical disturbance may be in part why cycles are so distinct (for the origin of cycles, see below). Lack of corallines indicates a subphotic seafloor, reflecting either deeper water or a murky water column. Abundance of spicules suggests a richer sponge biota that reflects either water depth and environmental tranquility or abundant nutrients and source of silica.

Upper Bryozoan Grainstones

The top (Upper Oligocene) portions of both sections is an ~ 5-m-thick limestone which, although different at both localities, contains only minor quartzose sand and glauconite, has a higher bryozoan diversity than in underlying floatstones, and well-developed bedforms. The bryozoan biota is different than in the Lower Oligocene; there are numerous vagrants (whereas there were none below), prolific fenestrates, with abundant erect and foliose forms, all of which are much smaller.

Willunga Embayment.—

The top of the section is cross-bedded grainstone that is not obviously cyclic and has no chert. There are two units. The lower unit is planar cross-bedded at the base (Fig. 9E) with bedforms that are very low angle, and contain conspicuous reactivation surfaces and *Scolicia* burrows. Sediment is bryozoan rudstone with oysters, serpulid worm tubes, and numerous vagrant bryozoans. The upper part is trough cross-bedded bryozoan–echinoid grainstone containing numerous delicate branching bryozoan fragments or bryozoan rudstone with obvious *Celleporaria*.

The top unit, beautifully exposed on a wide rocky tidal flat (Aldinga Reef), is a spectacularly trough-cross-bedded grainstone–rudstone (Fig. 9F) with bed sets ~ 0.5 m high and ~ 20 m wide. Paleocurrent measurements indicate flow from the E-NE (040°–100°) with opposing directions from the WSW (230°–250°). *Scolicia* is abundant, with macrofossils dominated by small irregular, fragmented, and worn rhodoliths, numerous large echinoid spines, fragmented pectens, *Celleporaria* spp., *Nudicella*, and other fenestrate bryozoans. The coarse sand itself is an abraded bryozoan–echinoid grainstone with numerous small benthic foraminifers, abundant coralline algae, and local barnacle pieces.

Initial nearly flat-bedded sands are interpreted to record short-lived shoreface sedimentation above a cryptic unconformity. The large bidirectional trough cross-bedded sands likely reflect tide-dominated sedimentation.

Noarlunga Embayment.—

Uppermost sediments comprise grainy and muddy limestones with chert. They are in the form of several meter-scale cycles (2–3 m) comprising a lower, trough cross-bedded bryozoan grainstone and an upper burrowed bryozoan floatstone.

The basal grainstone in each cycle, ~ 0.5 m thick, overlies a burrowed or planed firmground, is burrowed throughout by *Scolicia*, and is composed almost entirely of bryozoans, with a few benthic foraminifers and echinoderm pieces. Some beds contain serpulid clusters and rolled bryozoans encrusted with serpulids.

The upper bryozoan floatstone part is heterogeneous, burrowed (*Thalassinoides*), with 10–20% green clay matrix containing ~ 10% sponge spicules. Within these floatstones there are lenses, 10–20 cm thick, of trough cross-bedded bryozoan rudstone. Green clay occurs as centimeter-scale laminae, as 10–20 cm layers interbedded with the floatstone, and in *Thalassinoides* burrows. The sediment is poorly swaly and hummocky cross-laminated (30 cm amplitude and 2 m wavelength). The matrix is variably silicified as chert nodules. Spectacular foliose bryozoans, colonies of fenestrates, delicate branching, articulated branching, as well as a range of nodular and arborescent celleporarids, vagrants including scalloped *Celleporaria* and "*Lichenopora*" types occur throughout. There are numerous pectens (all asymmetric *Chlamys* and *Pecten*), plentiful oysters and their spat, and limpets. Brachiopods are rare, and there are no gastropods. There are small marsupiate echinoids.

These facies and cycles are similar to those in the middle floatstone part of the Willunga at Port Willunga, only with a more obvious cyclicity and diverse biota. The grainy lithologies resemble those at the base of the section, but with much less quartz. Overall they are interpreted to represent shallowing, as in the Willunga Embayment, but still in somewhat deeper environments overall.

THE EMBAYMENTS AS FLOODED OLIGOCENE TIDE-DOMINATED ESTUARIES

The Model

The Chinaman Gully Formation (CGF) and the Port Willunga Formation (PWF) are generally regarded as passing laterally landward into the Pirramimma Sand member of the PWF (Cooper, 1979; also see Fairbain, 1998). The Pirramimma is a well-sorted, quartzose sand with indistinct cross-bedding and wood fragments.

The lower bounding surface is the unconformity beneath the Chinaman Gully Formation (Fig. 11) that signals a major sea-level fall near the Eocene–Oligocene boundary. In the Adelaide Plains sub-basin erosion here removed as much as 50 m of underlying strata (Lindsay, 1981). Terrestrial sedimentation is

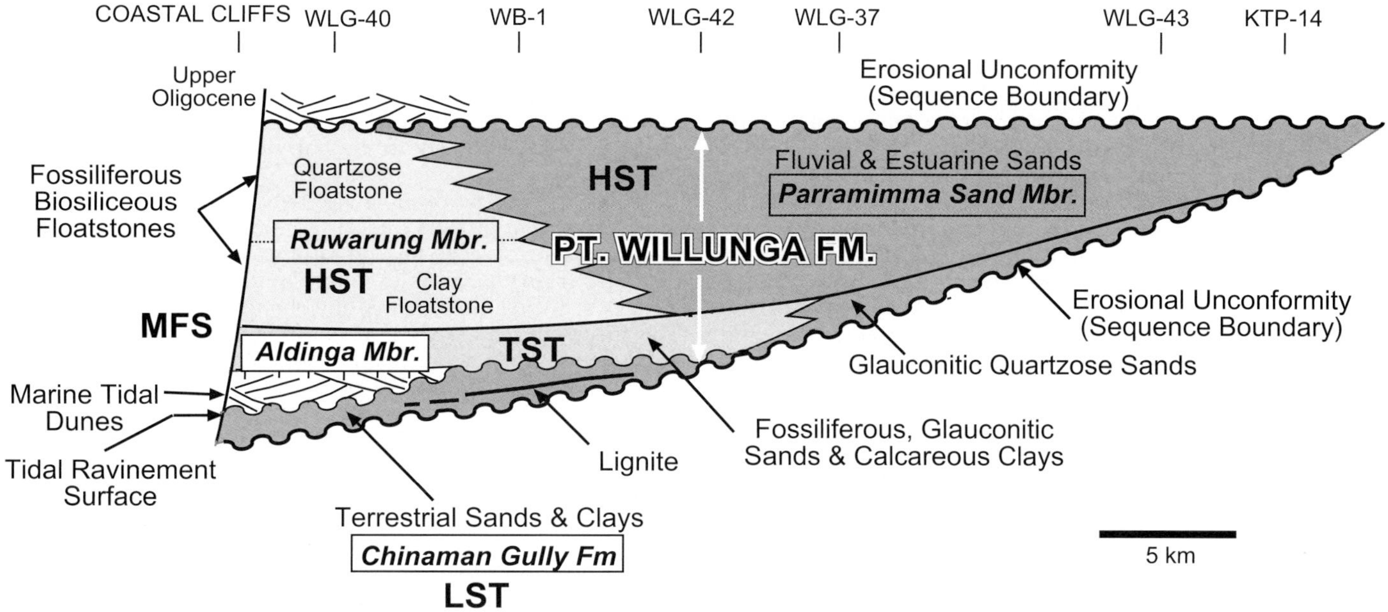

Fig. 11.—A diagrammatic cross section down the axes of the embayments, illustrating an interpretation of the sedimentary succession as an Early Oligocene embayment. HST = highstand systems tract, MFS = maximum flooding surface, TST = transgressive systems tract. Boreholes used to construct cross section = WLG, WB, and KTP.

represented by the Chinaman Gully Formation, which in both of the basins contains lignite, and carbonaceous silts and clays inland, implying terrestrial swamp and bog environments. This sedimentation was likely caused by rising baselevel, inasmuch as uppermost silts and clays are intensely burrowed and do contain a few marine fossils. This succession is interpreted as a series of freshwater swamps and coastal marshes that was transgressed by an estuarine lagoon (cf. Reinson, 1992; Dalrymple, 1992). The abrupt erosional surface beneath Aldinga Member sands is interpreted as a tidal ravinement surface that has eroded variable amounts of underlying Chinaman Gully middle- to inner-estuary facies. The Aldinga sands themselves have all the attributes of marine tidal shoals at the entrance to the estuary (Dalrymple and Rhodes, 1995), which became progressively more marine and less quartzose upward as sea level continued rising. They pass upward into clay-rich fossiliferous marine deposits. There is no clear wave ravinement surface between the two, implying, as do the clays themselves, a relatively quiet offshore marine setting. The comparatively abrupt transition into fossiliferous floatstones signals the beginnings of progradation with sedimentation largely matching sea-level change. The proportion of quartzose sand increases markedly towards the top of the succession, signaling increasing terrigenous influence in this still open marine environment.

Overall this succession has most of the attributes of a classic estuarine succession, especially the transgressive portion (Reinson, 1992; Dalrymple et al., 1992; Reading and Collinson, 1996). The stratigraphic sections would correspond to "segment 1" of Zaitlin et al. (1994), the outer portion of the flooded incised valley. In siliciclastic systems the regressive or progradational upper part of the succession is generally capped by open-marine highstand mudstones. This Paleogene succession departs from this model in that these highstand deposits are biosiliceous and carbonate sediment.

The top of the succession is marked by a planar erosion surface and limestones interpreted as beach deposits, which in turn are overlain by carbonate tidal sands (Upper Oligocene Janjucian units).

Sequence Stratigraphy

The overall PWF stratigraphy in the Willunga and Noarlunga sub-basins is the same, but with local differences. Given the motif, the PWF succession can be interpreted as a single decameter (~ 30 m) third-order sedimentary sequence (Fig. 11), within which there are variably developed meter-scale, higher-order cycles.

In this conceptualization the erosional unconformity at the base is interpreted as a sequence boundary (*sensu* Posamentier et al., 1988), which, in this case, based on current understanding, would include the Eocene–Oligocene boundary. The succeeding CGF and quartzose and heterolithic siliciclastics-rich interval of the PWF would represent the transgressive systems tract with an early freshwater, then warm-water, subtropical phase and the MFS lying within a cooler-water series of condensed, fossiliferous mudstones (cf. Myers and Milton, 1996). Floatstones that form most of the PWF are a highstand systems tract deposited in even cooler waters with shallowing towards the top reflected by an increasing content of detrital quartz sand. The upper sequence boundary is marked by the pronounced change in facies and higher-energy sediments of the Upper Oligocene upper carbonate unit.

The major difference in the sequence stratigraphy between sub-basins could be due to local tectonics, which allowed the Willunga sequence to remain in shallow-water, high-energy environments, while the Noarlunga sequence accumulated in somewhat quieter and presumably deeper-water settings. Alternatively, the Willunga sequence may represent deposition in a more distal or outer part of an estuary compared to the Noarlunga succession. This later interpretation is partly supported by the thinner basal subaqueous dune facies and more muddy MFS in the Noarlunga embayment.

Comparison with the Eocene

Documentation of the PWF in both embayments has allowed us to view the Eocene–Oligocene succession along the eastern margin of the St. Vincent Basin in a somewhat different way than previous interpretations. This is because although the PWF in the Noarlunga Embayment, dominated by clay-rich floatstones and chert, is different from the more calcareous and well-studied sediments in the Willunga Embayment; in many ways it is strikingly similar to the middle–late Eocene of both areas. This has not previously been appreciated, partly or mainly because the Noarlunga Embayment has received considerably less attention than the Willunga Embayment. Similarities between temporally different successions were not realized because of the carbonate-rich character of the PWF in the Willunga Embayment, wherein the differences instead of the similarities were emphasized (McGowran et al., 2004).

When placed against one another (Fig. 12) strong similarities emerge: (1) Each is bounded at the base by an unconformity; (2) each has basal terrigenous quartzose sands (North Maslin Sand = Chinaman Gully Formation); the estuarine character of the Maslins Sands in the Willunga Embayment has been clearly documented by Dyson (1998); (3) each has an overlying tidal marine biofragmental sand (South Maslin Sand and Tortachilla Limestone = Lower Aldinga Formation); (4) each has fossiliferous units near the base with warm-water biotic elements (Tortachilla Limestone = Lower Aldinga Member); (5) each has a glauconitic clay-rich unit above these lower shallow-water facies (Tuketja Member of Blanche Point Formation = Upper Aldinga Member clays); (6) each has an upper chert–carbonate–clay portion rich in sponge spicules (Blanche Point Formation [Gull Rock, Perkana, and Tuit members] = Ruwarung Member).

Against this template there are nevertheless differences in composition and depositional style. The thick Maslins Sands are likely the result of prolonged Paleogene weathering and subsequent fluvial transport of large amounts of stored sediment into the estuarine system. In contrast, the relatively thin Chinaman Gully–Lower Aldinga sands and clays represent a shorter weathering period and thus stored material, and possibly less accommodation space. The contact at the base of the South Maslin Sand is interpreted, in agreement with Dyson (1998), to be a tidal ravinement surface, like the base of the Aldinga member. The Tortachilla Limestone, representing shallow shoreface and subtidal accumulation, has, unlike genetically comparable upper Aldinga member sediments, numerous hardgrounds (James and Bone, 2000), probably the result of the numerous small-scale eustatic fluctuations in the middle Eocene (White, 2004) or small tectonoeustatic events (Li et al., 2003). Such eustatic fluctuations were not present in the early Oligocene. The lower contact of the Tortachilla is not considered a hiatus but a decalcification phenomenon (James and Bone, 2000). The base of the Tuketja glauconitic clays is complex, with the important surface likely lying in the top of the Tortachilla (James and Bone, 2000) and is likely a wave ravinement surface. Regardless, it is not an unconformity related to eustasy or tectonism, but merely represents marine erosion. This interpretation obviates the necessity of tying this event to the global sea-level model (Haq and Al-Qahtani, 2005), always a troubling fit (McGowran et al., 1997). In the revised interpretation herein the maximum flooding interval is likely within the glauconitic clays and not in the Perkana Member as postulated by McGowran et al. (1997) and James and Bone (2000). Finally, while biosiliceous muds and marls occupy the upper half of the succession, they are somewhat different, with the Eocene darker, more organic-rich and biotically less diverse sediments reflecting more dysoxic conditions than the Oligocene deposits. This may be part of the larger-scale oceanographic transition, as long postulated (Moss and McGowran, 1993; McGowran et al., 1997), from a stratified Australian–Antarctic Gulf with sluggish circulation in the Late Eocene to a well-ventilated Southern Ocean following opening of the Tasman Gateway in the Early Oligocene (Kennett and Exon, 2004).

Thus, the Eocene can likewise be viewed as a single third-order sequence, with lithological and compositional variations in the units being of secondary importance to the overall motif and resulting from the complex interplay of tectonics, short-term eustatic sea-level fluctuations, and changes in ocean circulation. This means that there are two, not four, sequences (McGowran and Li, 1997; McGowran et al., 2004) in the succession along the eastern side of St. Vincent Basin. The major sequence boundaries at the bases of the two sequences and at the tops are also recognized in deep-water ODP Leg 182 cores (Li et al., 2003). The boundary at the base of the Tuketja is poorly constrained, and the base of the Tuit does not appear to be present offshore. Furthermore, these similarities demand that there be an underlying long-lived tectonic–sedimentary–paleobiological–oceanic theme to the whole succession that lasted ~ 8 My from middle Eocene to Middle Oligocene. In this context McGowran et al. (2004) are correct: the early Oligocene as currently understood is the last part of a depositional style that existed throughout the later Eocene.

DISCUSSION

Introduction

Discovery that the Middle Eocene to Early Oligocene succession in two embayments along the eastern margin of the St. Vincent Basin comprise a single depositional phase with similar attributes throughout has important implications. First, it allows two coeval Oligocene estuarine systems to be compared. Second, it permits two temporally separate depositional systems to be analyzed, thus revealing critical information about the evolution of the southern margin of Australia during this time of global change.

The Early Oligocene Port Willunga Estuarine Embayments

The Depositional System.—

The two Early Oligocene embayments are part of a larger depositional system within the St. Vincent Basin. As currently understood (Lindsay and Alley, 1995; James and Bone, 2000) the basin mouth was partly obstructed by Kangaroo Island (Cooper and Lindsay, 1978; Lindsay, 1983) (Fig. 1) upon and around which were a series of carbonate sand shoals that served to somewhat isolate the Basin from the nascent Southern Ocean. The deposits in the basin center were open marine in character (Lindsay and Alley, 1995). The Early Oligocene terrestrial vegetation was largely *Nothofagus* (Southern Beech)-dominated cool, temperate rainforest with extensive marshes (Benbow et al., 1995). The northern apex of the basin was the site of coal deposition and associated fluvial sands, lacustrine silts, marine silty sands, and estuarine carbonaceous silts (Lindsay and Alley, 1995). In epicontinental basins to the east in South Australia and Victoria the PWF is coeval with neritic bryozoan-rich marls, calcisiltites, and calcarenites that are locally chert-rich (James and Bone, 1989; Li et al., 2000; Gallagher et al., 1999; Gallagher and Holdgate, 2000). To the west in the Eucla Basin of South Australia and Western Australia the Oligocene was likewise a

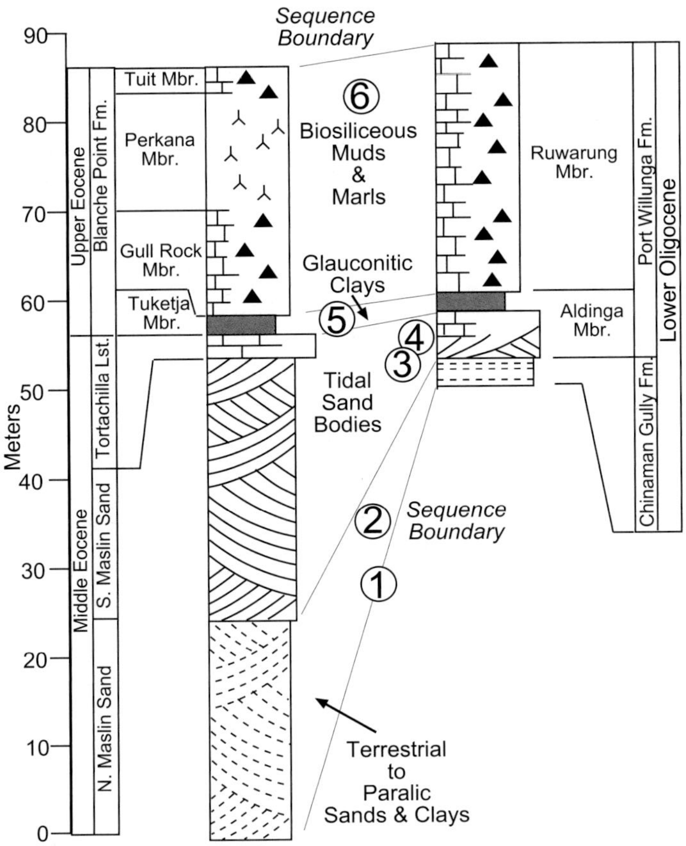

FIG. 12.—Comparison between the middle–late Eocene and Early Oligocene successions in embayments along the eastern side of the St. Vincent Basin and their interpretation as similar estuarine successions. Numbers designate specific segments and units that are similar in both the Eocene and Oligocene sequences.

time of extensive marine, bryozoan-dominated carbonate deposition, but with little biosiliceous sedimentation (James and Bone, 1991, 1994; Li et al., 1996).

The two embayments were flooded estuaries, in the sense that marine deposits filled the original rias. The clay-rich composition of the marine sediments speaks to active suspension transport into the embayment and into open marine environments. Abundant montmorillonite suggests active chemical weathering on land. The TST took place in warm marine waters. The environment during HST was cooler, somewhat unusual except when viewed in the light of overall progressive Early Oligocene Southern Ocean cooling. The sediments, however, are characterized by their cyclicity, and chert-rich character.

Origin and Implications of Meter-Scale Cyclicity.—

The meter-scale cycles in PWF sediments are wholly subtidal, with no evidence of subaerial exposure. The repetition of similar cycles at the meter scale strongly implies neritic sedimentation in response to glacio-eustasy (cf. Goldhammer et al., 1987, 1989). This interpretation is compatible with extensive Early Oligocene glaciation of east Antarctica (Zachos et al., 2001; Ivany et al., 2003). As such these are amongst the first neritic "icehouse" monotonic glacio-eustatic, meter-scale cycles in the Cenozoic geological record.

Although compositionally different in the two embayments, each cycle has a similar motif (Fig. 10). They are all (1) meter-scale, (2) bounded by *Thalassinoides*-burrowed firmgrounds or hardgrounds, (3) dominated by a rich epifaunal biota of bryozoans, pectenid bivalves, echinoids, and benthic foraminifers, (4) grainy to muddy upwards, (5) decreasingly fossiliferous upwards, (6) glauconite- and quartz-rich at the base, and (7) increasingly spicule- and chert-rich upwards.

The contrasting nature of the Willunga and Noarlunga meter-scale cycles (Fig. 10) is probably the result of different energy regimens wherein lower energy in the Noarlunga embayment did not mix facies of slightly different environments, allowing the accumulation to reflect subtleties of temporally changing sedimentation. The burrowed firmgrounds or hardgrounds as cycle boundaries imply reduced sedimentation possibly via wave sweeping (cf. James and Bone, 1994), which is typical of cool-water neritic carbonate settings (James, 1997).

Facies A indicates fluvial input of quartzose sand and clay, whereas abundant glauconite suggests relatively low accumulation rates (Odin, 1988). The prolific calcareous macrobiota, dominated by epifaunal suspension feeders, confirms high trophic resources and a thriving near-surface plankton population. Yet scarcity of planktic foraminifers (McGowran et al., 1992; McGowran et al., 1997; McGowran and Li, 1997) implies environmental stress, whereas numerous diatoms (see below) and coccoliths indicate abundant nutrients (Haq, 1998). The biota, particularly echinoderms, further confirms normal open marine conditions (Jones and Desrochers, 1992). Trough cross-bedding, with mud intercalations and numerous coralline algae, point to an illuminated seafloor swept by tidal currents in the Willunga Embayment. In contrast, the lack of corallines, more clay in the floatstones, and lack of bedforms indicates lower energy, and either a deeper seafloor or a more turbid and dark water column in the Noarlunga Embayment.

Facies B, on the other hand, indicates that the environment in which it accumulated was muddier and thus of lower energy, with proportionally more mud influx from land. It was also starved of quartzose sand and supported a less prolific benthic biota. Lack of bedforms in the Noarlunga embayment supports the notion of lower hydrodynamic energy.

Identification of diatoms in SEM mounts reveals two species, *Cocconeis* sp. and *Rhopalodia* sp.; most *Cocconeis* live as epiphytes in shallow brackish to marine environments, especially on seagrasses (Eddsbagge, 1968); most *Rhopalodia* also occur in marine to brackish environments along marine coasts (Krammer and Lange-Bertalot, 1986–1991).

The bryozoan biota within cycles reflects the decreasing energy upwards in each cycle, with basal highest-energy sediments composed of broken fragments of small, erect bryozoans (cellarids, fenestrates, foliose forms), a typical sand biota (Cook, 1966); the middle part of the cycle with large celleporarids, large foliose forms, and more erects, points to reduced energy; low-energy muddy sediments at the top contain the fewest bryozoans except for large celleporarids. Thus, the bryozoan biota, to a first approximation, appears to be reflecting energy and mud content. These relationships are compatible with observations from the modern and rock record (Nelson et al., 1988; Bone and James, 1993). Foliose forms do not live well in high-energy settings, but they can survive in an episodically high-energy environment, e.g., tidal. *Celleporaria*, particularly common in the upper parts of Willunga Embayment cycles, is compatible with its documented ecology living in low-energy settings on a mud to silt substrate, typically preserved in regressive facies (Lukasik et al., 2000). This relationship in turn implies medium to high mesotrophy (Hageman et al., 2003). Furthermore, the numerous digitate hollow growth forms that today grow around demosponges (Hageman et al., 2003) confirm a flourishing sponge biota.

In the Noarlunga embayment Facies B environments did not support a prolific carbonate epifauna. Instead the biota was restricted to semi-infaunal turritellid gastropods (much like the late Eocene), slightly greater carbonate mud production, and profuse glass sponges. Although having a relatively wide environmental tolerance, turritellid gastropods today live in normal to slightly below-normal marine salinities, usually between depths of 10 and 100 m, and appear to prefer temperatures between 15 and 20°C in areas with a high nutrient flux. Most species are sedentary and infaunal to semi-infaunal in soft, preferably siliciclastic substrates (Allmon, 1988). Allmon and Knight (1993) and Allmon et al. (1995) have argued that fossil turritellids also had similar environmental constraints. Furthermore, numbers of turritellids in some estuarine settings increase as salinity and total biomass decreases (Wu and Richards, 1981).

The change upward in each cycle (Fig. 10) from a prolific open marine to an impoverished biota implies that many organisms were subject to environmental stress (Hallock and Schlager, 1986; Brasier, 1995). Presence of coals onshore during the TST (Holdgate and Clarke, 2000) indicates humid conditions and thus fluvial runoff. The most likely stress in this situation would have been salinity fluctuations. The stress was felt more strongly in the Noarlunga Embayment, where fresh-water outflow seems to have contained a higher suspended load, as indicated by the greater proportion of clays, and by implication more fresh water overall. The Willunga Embayment appears, in this interpretation, to have had less fresh-water runoff, because it remained more marine throughout. This is further exemplified by the sponge biota, which is more prolific in the Noarlunga Embayment. While most demosponges are marine, they are euryhaline and extend into fresh-water environments (Rigby, 1987). Alternatively, the Noarlunga environment may have been farther up the estuary and so been more affected by fresh-water outflow.

Thus, the cycles are interpreted as "freshening-upward", driven by glacioeustasy but compositionally controlled by local factors, especially climate. During rising sea level the region was normal marine, but at sea-level highstand and fall estuarine circulation would have had a greater fresh-water influence, possibly becoming marginally brackish. While such conditions may have developed in the Noarlunga Embayment, the presence of numerous bryozoans, organisms that do not tolerate hyposaline conditions (cf. McKinney and Jackson, 1989), in the cycle tops in the Willunga Embayment indicate that freshening was not as pervasive there. Active tidal currents probably resulted in thorough water-column mixing.

Such an interpretation would fit with the notion that the older Eocene package was deposited in a dysoxic situation wherein fresh-water-induced stratification was the cause (McGowran and Beecroft, 1986a; James and Bone, 2000). High nutrient levels and abundant terrestrial organic material being exported to the seafloor would further reduce benthic oxygen levels. In contrast, during the Early Oligocene HST, lack of coal swamps (Holdgate and Clarke, 2000) would have resulted in less organic material being transported into the estuary while active circulation via tidal currents not only allowed a more marine biota to flourish but also promoted water mixing and nutrient dispersion.

Biosiliceous Sedimentation.—

Silica concentration is a first-order limiting nutrient for modern marine siliceous sponges (Maldonado et al., 1999). The proliferation of siliceous sponges and the abundance of silica in these sediments remain perplexing. As currently accepted (Jones and Fitzgerald, 1984, 1986, 1987) silicification and mineralogy of the BPF is thought to result from early alteration of volcanic ash. It is proposed that glass shards were extremely unstable in seawater and altered, either in the water column or just below the seafloor, to smectite, clinoptilolite, and opal-CT. Silica solubility was enhanced by chelating ligands that were produced during decomposition of the seafloor biota killed during episodic ashfalls. Lack of any glass shards or ash in the sediment today implies either that much breakdown occurred in the water column or that it took place prior to delivery to the environment of deposition. Silica thus released to seawater helped satisfy biotic demands of the prolific sponge biota, while Si released during alteration in the sediment led to early, episodic silicification. Numerous sponge-spicule molds imply that the dissolution of the spicules contributed to silicification, but it is not considered to have been significant.

This hypothesis is both compelling and defensible, and all of the macroscopic and microscopic attributes described for the BPF are equally applicable to the PWF. This period is one of distant small-scale volcanism northwest of Melbourne, Victoria (Price et al., 2003), and thus there was a potential source of ash. Nevertheless, there are puzzling contradictions: (1) there are macrofossils in layers with abundant chert, especially turritellid gastropods, where there should not be if ash killed off the benthic biota, (2) there is a clear correlation in the PWF between increasing sponge spicules and chert intensity, (3) the dysaerobic conditions deemed necessary at the seafloor were not present in the PWF in the Willunga embayment, and (4) there is no chert in the lower parts of the sequences, i.e., it is facies specific, whereas if volcanism was ongoing it should not be, unless there had been dramatic differences in sedimentation rate, with Facies A accumulating rapidly and Facies B accumulating slowly. Furthermore, if ashfall alone was the delivery mechanism, then most Eocene and early Oliogocene sediments outside the St. Vincent Basin should be chert-rich, and they are only locally so (McGowran et al., 2004). Finally, this biosiliceous and chert estuarine facies occurs, at least in middle–late Eocene sediments, westward some 1500 km across the Eucla Basin (Gammon and James, 2003; Clarke et al., 2003; Hou et al., 2006). This is so far distant from any known ashfall that a more local, terrestrial source of silica via weathering seems more plausible, possibly similar to that proposed by Gammon and James (2003).

In this hypothesis fresh-water-supplied silica (via runoff and groundwater seepage) supplied silica for biogenic opal, which reacted with kaolinite in the suboxic subsurface to form clinoptilolite and glauconite, and recycled silica back into seawater. In such a model clinoptilolite is accounted for, and the silica kept being recycled back into marine waters for sponge growth. This situation is also in many ways similar to middle Eocene estuarine glauconite-rich siliciclastic deposits in southern England that were rich in sponge spicules and feldspar but now contain quartz-spicule pseudomorphs, spicule molds surrounded by opal-CT lepispheres, and clinoptilolite (Huggett et al., 2005). In this situation the source as volcanic ash is clearly eliminated, and the relationship to coals and paleosols suggests dissolution of biogenic SiO_2, feldspars, and perhaps clays by terrestrial organic acids. Subsequent increased alkalinity in the shallow subsurface (< a few hundred meters deep) is thought to have triggered precipitation of clinoptilolite and opal-CT. These relations are presented as unusual, but given the widespread chert, opal-CT, and clinoptilolite sediments across southern Australia in the Eocene and Oligocene, again associated with estuaries and coals, this tripart facies may be more common than realized heretofore. Regardless, whatever the process, it was operative in these flooded estuaries for ~ 8 My throughout the Late Eocene and Early Oligocene.

Late Eocene and (Versus) Early Oligocene Sedimentation

The Cenozoic succession in southern Australia is made up of four depositional phases (Quilty, 1977; McGowran et al., 2004), and the rocks described herein would correspond to Phase II, Late Eocene to Early Oligocene. This study indicates that within the St. Vincent Basin margin succession, Phase II is composed of two unconformity-bounded sequences (*sensu* Posamentier et al., 1988). Each sequence records flooding of an incised-valley estuary, transforming it into a marine embayment, but one with a persistent albeit small fresh-water influence. In this interpretation the so-called Tuketja–Tuit and Aldinga transgressions (McGowran, 1989; McGowran et al., 1997), although related to sea-level rise, are instead ravinement surfaces and not related to global eustatic events. This conclusion is supported by the fact that they are not convincingly demonstrated in Leg 182 cores (Li et al., 2003).

Neritic Biogenic Sedimentation and the Eocene–Oligocene Boundary.—

The Paleogene succession in the eastern St. Vincent Basin has long been regarded as one of the best and most critical in the Southern Ocean area for understanding Cenozoic geohistory (McGowran and Li, 1997). This study has found that in spite of major oceanographic reorganization, basin-margin paleoenvironments across the Eocene–Oligocene boundary were not much affected. The biogenic theme of TST siliciclastic–carbonate and HST biosiliceous–carbonate is present both before and after the Terminal Eocene Event. The major differences are small and, as long postulated, related in part to a more invigorated Southern Ocean circulation in the Early Oligocene (Moss and McGowran, 1993). Thus, while there is biotic extinction and replacement, the main changes within the sediment-producing biota, as Prothero (1994) and Ivany et al. (2003) conclude elsewhere, had already taken place at the Middle–Late Eocene boundary.

Specifically (Table 1), the middle–late Eocene succession accumulated in relatively warm marine waters throughout. The climate was warm, and wet with coal swamps and bogs onshore. TST facies hosted a diverse calcareous biota that became sharply reduced with sea-level rise and deepening. HST facies are almost entirely biosiliceous with a very low-diversity biota. The environments are envisaged as dysoxic to anoxic beneath a strongly stratified water column with export of terrestrial organic material. The sediments show little evidence of cyclicity.

In contrast, the Early Oligocene succession accumulated in initially warm (TST) but later cool (HST) waters. The prevailing climate was wet enough for coal swamps to develop during the TST. TST facies hosted a diverse calcareous biota that remained largely the same in HST facies except in the Noarlunga Embayment, where they were of reduced diversity to absent in the tops of cycles, implying good circulation of normal marine waters. HST facies are enriched in biosiliceous components. The seafloor was oxygenated to locally dysoxic. The sediments are strongly cyclic.

These differences are not large and are ascribed principally to the interaction of changing climate and ocean cooling. Acting in concert these controls led to a strongly stratified Eocene estuary–embayment and a well-mixed Oligocene estuary–embayment. These differences affected the style but not the composition of the benthic biota and hence the sedimentation. The nutrient flux was high in both areas, as is reflected by the prolific heterozoan biota. The biota in turn was particularly sensitive to subtleties such as silica availability, substrate, freshwater influx, and tidal hydrodynamics. Although glaciation in Antarctica began in the middle Eocene (Birkenmajer et al. 2005) the dramatic increase in continental glaciation in the Early Oligocene is manifest in these neritic sediments by better-developed cycles.

IMPLICATIONS FOR THE COOL-WATER CARBONATE DEPOSITIONAL SYSTEM

There is a persistent chert–carbonate sedimentary association throughout the Phanerozoic (James and Bone, 2000) wherein the carbonates are both heterozoan and neritic while the chert is biosiliceous in origin (Pope and Steffen, 2003; Beauchamp and Baud, 2002). While many such sediments were clearly related to oceanic upwelling or polar environments, others seem remote from or decoupled from the global ocean. The Paleogene sediments of this study fall into this latter category.

Late Eocene–early Oligocene facies from the St. Vincent Basin westward for ~ 1500 km across the inner part of the Paleogene continental shelf comprise this cool-water carbonate biosiliceous couplet (Clarke et al., 1996; Gammon et al., 2000; Clarke et al., 2003; Hou et al., 2006). When placed in a larger context, the shallow-water cores of Paleogene sediments from ODP Leg 182 do not contain chert whereas those in deeper water do, suggesting active biosiliceous sedimentation in somewhat stratified waters in slope to basin settings (Feary et al., 2004). Thus, during this period, silica deposition was partitioned into deep-water facies and strandline–estuarine facies, with clean carbonate facies on the shelf between. This supports the notion of two sources of silica, terrestrial and deep ocean via upwelling, with both being mineralized as sponge spicules and diatoms.

In both paleoestuarine sequences studied here, however, it is the HST that is biosiliceous. The TST, containing large benthic foraminifers, is the warmest-water part of the sequence, which is to be expected if sea-level rise reflects climatic amelioration accompanying reduced Antarctic ice volumes. What is unexpected is the lack of such indicators in the HST, where the marine environment should also be comparatively warm but compositional proxies indicate a locally illuminated seafloor covered by waters of nearly normal salinity and containing abundant nutrients. The waters ranged from oxic to hypoxic.

An inevitable conclusion is that this HST environment with increased silica is a product of the marine–estuarine setting itself. In this case it was likely the result of prolonged terrestrial weathering and restriction of silica to the estuary because of the circular nature of estuarine water movement, which would have prevented Si and other nutrient elements (N and P) from escaping into the open ocean.

Carbonate Sediment and Paleoestuaries

The deposition of carbonate sediments in nearshore environments has traditionally been interpreted from the perspective of muddy tidal flats and beach–dune complexes (Scholle et al., 1983; James and Kendall, 1992; Tucker and Wright, 1990). While such deposits probably dominate the rock record, other depositional settings are possible. This study shows that, at least for this Paleogene segment of the rock record in southern Australia, cool-water carbonate deposition in marine embayments related to flooded estuaries was widespread. This occurrence emphasizes that such depositional environments are not restricted to siliciclastic sediments.

Estuarine sedimentation requires fluvial erosion of the valley during sea-level lowstand and deposition during the subsequent sea-level rise and highstand (Zaitlin et al., 1994). During global

TABLE 1.—Properties of Eocene versus Oligocene paleoestuarine facies.

	Middle to Late Eocene		Early Oligocene	
	TST	HST	TST	HST
Marine waters	Warm	Warm	Warm	Cool
Sea floor	Oxygenated	Dysoxic–anoxic?	Oxygenated	Oxygenated–dysoxic
Cyclicity	Condensed	Poor	Poor	Excellent
Climate	Warm and humid	Warm and Humid	Cool and humid	Cool and humid
Calcareous biota	Prolific	Restricted	Prolific	Prolific
Siliceous biota	Sparse	Prolific	Sparse	Moderate

greenhouse periods, when sea-level fluctuations were small and warm-water carbonate deposition was widespread, peritidal mudflats and strandline sands were abundant. In icehouse times, however, when large excursions of sea level led to profound fluvial incision and warm-water carbonates were more restricted to near-equatorial environments and cool-water carbonates were more widespread, carbonate deposition in flooded estuaries should be expected. The caveat of course is that fluvial siliciclastic outflow was not so great as to suppress carbonate production either through smothering by terrigenous sediment or killing by fresh waters. Such estuarine sedimentation may be a signature of icehouse cool-water carbonate sediments.

SUMMARY AND CONCLUSIONS

1. Early Oligocene strata along the eastern side of the St. Vincent Basin formed in a series of embayments interpreted herein as flooded paleoestuaries. Similarities between Middle–Late Eocene and Early Oligocene successions in two of these, the Willunga and Noarlunga embayments, indicate that these estuaries persisted as depocenters of similar style as the warm Australo-Antarctic Gulf evolved into the cold nascent Southern Ocean.

2. Sedimentary facies of the early Oligocene Port Willunga Formation are a series of quartzose, clay-rich sands, muds, and fossiliferous floatstones. Biotic components are dominantly epifaunal bryozoans, pectens, echinoids, and benthic foraminifers, together with siliceous sponges. The rich assemblage indicates overall normal marine waters throughout with abundant trophic resources and elevated silica levels.

3. The Chinaman Gully (CGF) and Aldinga and Ruwarung members of the Port Willunga Formation (PWF) constitute one unconformity-bounded sequence. Basal CGF terrestrial sands and muds become more marine upwards and are truncated by a tidal ravinement surface that is sharply overlain by marine tidal sand bodies, which in turn pass upward into open marine quartzose floatstones and clays. This TST is overlain by a thick HST composed of meter-scale, clay-rich floatstone cycles that are distinctly rich in quartzose sand at the top, reflecting marginally increased riverine sediment input.

4. The early Oligocene meter-scale cycles, reflecting the interaction between autogenic and allogenic controls, have the same general attributes; they (1) are conspicuously bryozoan-rich, (2) are glauconitic and quartzose at the base, (3) contain decreasing numbers of calcareous macrofossils upwards, and (4) contain an increasing proportion of spicules and chert upwards. Those in the Willunga Embayment are trough cross-bedded and contain coralline algae, implying active traction currents and an illuminated seafloor. Noarlunga cycles, in contrast, have few bedforms, are more burrowed, and have more clay and more chert, but no corallines, suggesting more terrestrial input and a tranquil, sub-photic seafloor.

5. Comparison of these two Eocene–Oligocene successions suggests that differences between them are due more to local climate and circulation than major ocean changes. Evolving Southern Ocean oceanography during this critical period in Earth history had minimal effect on facies composition, implying that neritic communities had already been strongly preconditioned during the earlier middle Eocene faunal turnover. Furthermore, some of the postulated unconformities within the sequences are not due to global eustatic changes but instead are part of the depositional mechanics of the system.

6. Recognition that this succession has equivalents across large parts of the Paleogene southern Australian margin has larger implications for the cool-water carbonate record. At certain times like this both the deep ocean and the strandline sedimentary rocks have a significant chert component, whereas intervening facies are heterozoan carbonate. The inboard biosiliceous facies of sponges and diatoms is here proposed to reflect extensive weathering under humid conditions and corresponding high flux of silica and nutrients into an estuarine but marine-dominated setting.

ACKNOWLEDGMENTS

This research is funded by the Natural Sciences and Engineering Research Council of Canada and by the Australian Research Council. We thank G. Braybrook, C. Kobernick, and I. Malcolm for technical assistance. A. Michels kindly identified the diatoms. The manuscript was greatly improved by reviews from R. Dalrymple, D. Lehrmann, P. Pufahl, and B. Saylor.

REFERENCES

ALLMON, W.D., 1988, Ecology of Recent turritelline gastropods (Prosobranchia, Turitellidae): current knowledge and paleontological implications: Palaios, v. 3, p. 259–284.

ALLMON, W.D., AND KNIGHT, J.L., 1993, Paleontological significance of a turitelline gastropod–dominated layer in the Cretaceous of South Carolina: Journal of Paleontology, v. 67, p. 355–360.

ALLMON, W.D., SPIZUCO, M.P., AND JONES, D.S., 1995, Taphonomy and paleoenvironment of two turitellid gastropod–rich beds, Pliocene of Florida: Lethaia, v. 28, p. 75–84.

BEAUCHAMP, B., AND BAUD, A., 2002, Growth and demise of Permian biogenic chert along northwest Pangea: evidence for end-Permian collapse of thermohaline circulation: Palaeogeography, Palaeoclimatology, Palaeoecology, v. 184, p. 37–63.

BENBOW, M.C., ALLEY, N.F., CALLEN, R.A., AND GREENWOOD, D.R., 1995, Geology and Palaeoclimate, *in* Drexel, J.F., and Preiss, W.V., eds., The Geology of South Australia, vol. 2, The Phanerozoic: Geological Survey of South Australia, Bulletin 54, p. 208–217.

BERGGREN, W.A., AND PROTHERO, D.R., 1992, Eocene–Oligocene climatic and biotic evolution, *in* Prothero, D.R., and Berggren, W.A., eds., Eocene–Oligocene Climatic and Biotic Evolution: Princeton, New Jersey, Princeton University Press, p. 1–28.

BIRKENMAJER, K., GAZDZICKI, A., KRAJEWSKI, K.P., PRZBYCIN, A., SOLECKI, A., TATUR, A., AND YOON, H.I., 2005, First glaciers in west Antarctica: Polish Polar Research, v. 26, p. 3–12.

BONE, Y., AND JAMES, N.P., 1993, Bryozoans as carbonate sediment producers on the cool-water Lacepede Shelf, southern Australia: Sedimentary Geology, v. 86, p. 247–271.

BOREEN, T., JAMES, N.P., HEGGIE, D., AND WILSON, C., 1993, Surficial cool-water carbonate sediments on the Otway continental margin, southeastern Australia: Marine Geology, v. 112, p. 35–56.

BRASIER, M.D., 1995, Fossil indicators of nutrient levels. 2. Evolution and extinction in relation to oligotrophy, *in* Bosence, D.W.J., and Allison, P.A., eds., Marine Paleoenvironmental Analysis from Fossils: Geological Society of London, Special Publication 83, p. 133–150.

CLARKE, J.D.A., BONE, Y., AND JAMES N.P., 1996, Cool-water carbonates in an Eocene tide-dominated paleoestuary, Norseman Formation, Western Australia: Sedimentary Geology, v. 10, p. 213–226.

CLARKE, J.D.A., GAMMON, P.R., HOU, B., AND GALLAGHER, S.J., 2003, Middle to Upper Eocene stratigraphic nomenclature and deposition in the Eucla Basin: Australian Journal of Earth Sciences, v. 50, p. 231–248.

COOK, P.L., 1966, Some 'sand fauna' Polyzoa (Bryozoa) from eastern Africa and the northern Indian Ocean: Cahiers de Biologie Marine, v. 7, p. 207–223.

COOPER, B.J., 1979, Eocene to Miocene stratigraphy of the Willunga Embayment: Geological Survey of South Australia, Reports of Investigation, no. 50, 101 p.

COOPER, B.J., 1985, The Cainozoic St. Vincent Basin—tectonics, structure, stratigraphy, *in* Lindsay, J.M., ed., Stratigraphy, Palaeontology, Malacology; Papers in Honour of Dr. Nell Ludbrook: South Australia Department of Mines and Energy, Special Publication 5, p. 35–51.

COOPER, B.J., AND LINDSAY, J.M., 1978, Marine entrance to the Cenozoic St. Vincent Basin: Geological Survey of South Australia, Quarterly Notes, v. 67, p. 4–6.

DALRYMPLE, R.W., 1992, Tidal Depositional Systems, *in* Walker, R.G., and James, N.P., eds., Facies Models: St. John's, Geological Association of Canada, p. 195–218.

DALRYMPLE, R.W., AND RHODES, R.N., 1995, Estuarine dunes and bars, *in* Perillo, G.M.E., ed., Geomorphology and Sedimentology of Estuaries: Amsterdam, Elsevier, Developments in Sedimentology 53, p. 359–422.

DALRYMPLE, R.W., ZAITLIN, B.A., AND BOYD, R., 1992, Estuarine facies models: conceptual basis and stratigraphic implications: Journal of Sedimentary Petrology, v. 62, p. 1130–1146.

DICKSON, J.A.D., 1966, Carbonate identification and genesis as revealed by staining: Journal of Sedimentary Petrology, v. 36, p. 491–505.

DYSON, I.A., 1998, Estuarine facies of the North Maslin Sand and South Maslin Sand, Maslin Beach: MESA Journal, v. 11, p. 42–46.

EDDSBAGGE, H., 1968, Some problems in the relationship between diatoms and seaweeds: Botanica Marina, v. 11, p. 64–67.

FAIRBAIN, B., 1998, The Willunga Embayment—a stratigraphic revision: MESA Journal, v. 11, p. 35–41.

FAIRBAIN, B., 2000, Cainozoic stratigraphy of the Noarlunga Embayment: MESA Journal, v. 18, p. 42–47.

FEARY, D.A., HINE, A.C., JAMES, N.P., AND MALONE, M.J., 2004, Leg 182 synthesis: exposed secrets of the Great Australian Bight, *in* Hine, A.C., Feary D.A., and Malone, M.J., eds., Proceedings of the Ocean Drilling Program, Scientific Results, v. 182, College Station, Texas, p. 1–30.

GALLAGHER, S.J., AND HOLDGATE, G., 2000, The paleogeographic and paleoenvironmental evolution of a Paleogene mixed carbonate–siliciclastic cool-water succession in the Otway Basin, southeast Australia: Paleogeography, Paleoclimatology, Paleoecology, v. 156, p. 19–50.

GALLAGHER, S.J., JONASSON, K., AND HOLDGATE, G.R., 1999, Foraminiferal biofacies and paleoenvironmental evolution of an Oligo-Miocene cool-water carbonate succession in the Otway basin, southeast Australia: Journal of Micropaleontology, v. 18, p. 143–168.

GAMMON, P.R., AND JAMES, N.P., 2003, Paleoenvironmental controls on Upper Eocene biosiliceous neritic sediments, southern Australia: Journal of Sedimentary Research, v. 73, p. 957–972.

GAMMON, P.R., JAMES, N.P., BONE, Y., AND CLARKE, J.D.A., 2000, Sedimentology and lithostratigraphy of a late Eocene sponge-dominated sequence, southern Western Australia: Australian Journal of Earth Sciences, v. 47, p. 1067–1105.

GLAESSNER, M.F., 1951, Three foraminiferal zones in the Tertiary of Australia: Geological Magazine, v. 88, p. 273–283.

GOLDHAMMER, R.K., DUNN, P.A., AND HARDIE, L.A., 1987, High frequency glacio-eustatic sea level oscillation with Milankovitch characteristics recorded in Middle Triassic platform carbonates in northern Italy: American Journal of Science, v. 287, p. 853–892.

GOLDHAMMER, R.K., DUNN, P.A., AND HARDIE, L.A., 1989, Depositional cycles, composite sea-level changes, cycle stacking patterns, and the hierarchy of stratigraphic forcing: examples from Alpine Triassic platform carbonates: Geological Society of America, Bulletin, v. 102, p. 535–562.

HAGEMAN, S.J., LUKASIK, J., MCGOWRAN, B., AND BONE, Y., 2003, Paleoenvironmental significance of (Bryozoa) from Modern and Tertiary cool-water carbonates of Southern Australia: Palaios, v. 18, p. 510–527.

HALLOCK, P., AND GLENN, E.C., 1986, Larger foraminifera: a tool for paleoenvironmental analysis of Cenozoic carbonate depositional facies: Palaios, v. 1, p. 55–64.

HALLOCK, P., AND SCHLAGER, W., 1986, Nutrient excess and the demise of coral reefs: Palaios, v. 1, p. 389–398.

HAQ, B.U., 1998, Calcareous nannoplankton, *in* Haq, B.U., and Boersma, A., eds, Introduction to Marine Micropaleontology: Singapore, Elsevier, p. 79–108.

HAQ, B.U., AND AL-QAHTANI, A.M., 2005, Phanerozoic cycles of sea-level change on the Arabian Platform: Geo Arabia, v. 10, p. 127–160.

HOLDGATE, G., AND CLARKE, J.D.A., 2000, A review of Tertiary brown coal deposits in Australia—their depositional factors and eustatic correlations: American Association of Petroleum Geologists, Bulletin, v. 84, p. 1129–1151.

HOU, B., ALLEY, N.F., FRAKES, L.A., STOIAN, L., AND COWLEY, W.M., 2006, Eocene stratigraphic succession in the Eucla Basin of South Australia and correlation to major regional sea-level events: Sedimentary Geology, v. 183, p. 297–319.

HUGGETT, J.M., GALE, A.S., AND WRAY, D.S., 2005, Diagenetic clinoptilolite and opal-CT from the middle Eocene Wittering Formation, Isle of White, U.K.: Journal of Sedimentary Research, v. 75, p. 585–595.

IVANY, L.C., NESBITT, E.A., AND PROTHERO, D.R., 2003, The marine Eocene–Oligocene transition: a synthesis, *in* Prothero, D.R., Ivany, L.C., and Nesbitt, E.A., eds., From Greenhouse to Icehouse; The Marine Eocene–Oligocene Transition: New York, Columbia University Press, p. 522–534.

JAMES, N.P., 1997, The cool-water carbonate depositional realm, *in* James N.P., and Clarke, J.A.D., eds., Cool-Water Carbonates: SEPM, Special. Publication 56, p. 1–22.

JAMES, N.P., AND BONE, Y., 1989, Petrogenesis of Cenozoic, temperate water calcarenites, South Australia: a model for meteoric/shallow burial diagenesis of shallow water calcite sediments: Journal of Sedimentary Petrology, v. 59, p. 191–203.

JAMES, N.P., AND BONE, Y., 1991, Origin of a cool-water, Oligo-Miocene deep shelf limestone, Eucla Platform, Southern Australia: Sedimentology, v. 38, p. 323–341.

JAMES, N.P., AND BONE, Y., 1994, Paleoecology of cool-water, subtidal cycles in mid-Cenozoic limestones, Eucla platform, Southern Australia: Palaios, v. 9, p. 457–476.

JAMES, N.P., AND BONE, Y., 2000, Eocene cool-water carbonate and biosiliceous sedimentation dynamics, St. Vincent Basin, South Australia: Sedimentology, v. 47, p. 761–786.

JAMES, N.P., AND KENDALL, A.C., 1992, Introduction to carbonate and evaporite facies models, *in* Walker, R.G., and James, N.P., eds., Facies Models: St. John's, Geological Association of Canada, p. 265–276.

JAMES, N.P., BONE, Y., VON DER BORCH, C.C., AND GOSTIN, V.A., 1992, Modern carbonate and terrigenous clastic sediments on a cool water, high energy, mid-latitude shelf: Lacepede, Southern Australia: Sedimentology, v. 39, p. 877–903.

JAMES, N.P., BOREEN, T.D., BONE, Y., AND FEARY, D.A., 1994, Holocene carbonate sedimentation on the West Eucla Shelf, Great Australian Bight: a shaved shelf: Sedimentary Geology, v. 90, p. 161-177.

JAMES, N.P., BONE, Y., HAGEMAN S., GOSTIN, V.A., AND FEARY, D.A., 1997, Cool-water carbonate sedimentation during the Terminal Quaternary, high-amplitude, sea-level cycle: Lincoln Shelf, Southern Australia, *in* James N.P., and Clarke, J.A.D., eds., Cool-Water Carbonates: SEPM, Special Publication 56, p. 53–76.

JAMES, N.P., BONE, Y., COLLINS, L.B., AND KYSER, T.K., 2001, Surficial sediments of the Great Australian Bight: facies dynamics and oceanography on a vast cool-water carbonate shelf: Journal of Sedimentary Research, v. 71, p. 549–568.

JONES, B., AND DESROCHERS, A., 1992, Shallow platform carbonates, *in* Walker, R.G., and James, N.P., eds., Facies Models: St. John's, Newfoundland, Geological Association of Canada, p. 277–302.

JONES, J.B., AND FITZGERALD, M.J., 1984, Extensive volcanism associated with the separation of Australia and Antarctica: Science, v. 226, p. 346–348.

JONES, J.B., AND FITZGERALD, M.J., 1986, Silica-rich layering at Blanche Point, South Australia: Australian Journal of earth Sciences, v. 33, p. 529–551.

JONES, J.B., AND FITZGERALD, M.J., 1987, An unusual and characteristic sedimentary mineral suite associated with the evolution of passive margins: Sedimentary Geology, v. 52, p. 45–63.

KENNETT, J.P., AND EXON, N.F., 2004, Paleoceanographic evolution of the Tasmanian Seaway and its climatic implications, *in* Exon, N., Kennett, J., and Malone, M., eds., The Cenozoic Southern Ocean; Tectonics, Sedimentation, and Climate Change between Australia and Antarctica: American Geophysical Union, Geophysical Monograph 151, p. 345–367.

KIDWELL, S.M., 1991, Chapter 5, The stratigraphy of shell concentrations, *in* Allison, P.A., and Briggs, D.E.G., eds., Taphonomy: New York, Plenum Press, p. 212–290.

KRAMMER, K., AND LANGE-BERTALOT, H., 1986–1991, Süßwasserflora von Mitteleuropa, Bacillariophyceae.—2/1: Naviculaceae, 1–876; 2/2: Bacillariaceae, Epithmiaceae, Surirellaceae, 1–596; 2/3: Centrales, Fragilariaceae, Eunotiaceae, 1–576; 2/4: Achnanthaceae: Stuttgart, Springer-Verlag, p. 1–437.

LI, G., JAMES, N.P., BONE, Y., AND MCGOWRAN, B., 1996, Foraminiferal biostratigraphy and depositional environments of the Mid-Cenozoic Abrakurrie Limestone, Eucla Basin, Southern Australia: Australian Journal of Earth Sciences, v. 43, p. 437–450.

LI, Q., MCGOWRAN, B.N., AND WHITE, M.R., 2000, Sequences and biofacies packages in the mid-Cenozoic Gambier Limestone, South Australia: reappraisal of foraminiferal evidence: Australian Journal of Earth Sciences, v. 47, p. 955–970.

LI, Q., JAMES, N.P., AND MCGOWRAN, B., 2003, Middle and Late Eocene Great Australian Bight lithobiostratigraphy and stepwise evolution of the southern Australian continental margin: Australian Journal of Earth Sciences, v. 50, p. 113–128.

LINDSAY, J.M., 1967, Foraminifera and stratigraphy of the type section of Port Willunga Beds, Aldinga Bay, South Australia: Royal Society of South Australia, Transactions, v. 91, p. 93–109.

LINDSAY, J.M., 1981, Tertiary Stratigraphy and Foraminifera of the Adelaide City Area, St. Vincent Basin, South Australia: Unpublished M.Sc. thesis, The University of Adelaide, 781 p.

LINDSAY, J.M., 1983, Late Eocene to Late Oligocene age of the Kingscote Limestone, Kangaroo Island, S.A, Royal Society of South Australia, Transactions, v. 107, p. 127–128.

LINDSAY, J.M., 1985, Aspects of South Australian Tertiary foraminiferal biostratigraphy, with emphasis on studies of *Massilina* and *Subbotina*, *in* Lindsay, J.M., ed., Stratigraphy, Palaeontology, Malacology; Papers in Honour of Dr. Nell Ludbrook: South Australia Department of Mines and Energy, Special Publication 5, p. 187–232.

LINDSAY, J.M., AND ALLEY, N.F., 1995, St. Vincent Basin, *in* Drexel, J.F., and Preiss, W.V., eds., The Geology of South Australia: Geological Survey of South Australia, Bulletin 54, p. 163–177.

LINDSAY, J.M., AND MCGOWRAN, B., 1986, Eocene/Oligocene boundary, Adelaide region, South Australia, *in* Pomeral, C., and Premoli-Silva, I., eds., Terminal Eocene Events: New York, Elsevier, p. 165–173.

LUKASIK, J.J., JAMES, N.P., MCGOWRAN, B., AND BONE, Y., 2000, An epeiric ramp: Low-energy, cool-water carbonate facies in a Tertiary inland sea, Murray Basin, South Australia: Sedimentology, v. 47, p. 851–882.

LOWRY, D.C., 1970, Geology of the Western Australian part of the Eucla Basin: Geological Survey of Western Australia, Bulletin 122, 199 p.

MALDONADO, M., CARMONA, M.C., URIZ, M., AND CRUZADO, A., 1999, Decline in Mesozoic reef-building sponges explained by silicon limitation: Nature, v. 401, p. 785–788.

MCGOWRAN, B., 1987, Late Eocene perturbations: foraminiferal biofacies and evolutionary overturn, southern Australia: Paleoceanography, v. 2, p. 715–727.

MCGOWRAN, B., 1989, The later Eocene transgressions in southern Australia: Alcheringa, v. 13, p. 45-68.

MCGOWRAN, B., 1990, Fifty million years ago: American Scientist, v. 78, p. 30–39.

MCGOWRAN, B., AND BEECROFT, A., 1985, *Guembelitria* in the Early Tertiary of southern Australia and its paleoceanographic significance, *in* Lindsay, J.M. ed., Stratigraphy, Palaeontology, Malacology; Papers in Honour of Dr. Nell Ludbrook: South Australia Department of Mines and Energy, Special Publication 5, p. 247–262.

MCGOWRAN, B., AND BEECROFT, A., 1986a, Foraminiferal biofacies in a silica-rich neritic sediment, Late Eocene, South Australia: Paleogeography, Paleoclimatology, Paleoecology, v. 52, p. 321–345.

MCGOWRAN, B., AND BEECROFT, A., 1986b, Neritic southern extratropical foraminifera and the terminal Eocene Event: Paleogeography, Paleoclimatology, Paleoecology, v. 55, p. 23–34.

MCGOWRAN, B., AND LI, Q., 1997, Stratigraphic Excursion to Maslin and Aldinga Bays: University of Adelaide, 41 p.

MCGOWRAN, B., MOSS, G., AND BEECROFT, A., 1992, Late Eocene and Early Oligocene in Southern Australia: Local neritic signals of global oceanic changes, *in* Prothero, D.R., and Berggren, W.A., eds., Eocene–Oligocene Climatic and Biotic Evolution: Princeton, New Jersey, Princeton University Press, p. 160–177.

MCGOWRAN, B., LI, Q., AND MOSS, G., 1997, The Cenozoic neritic record in southern Australia: the biogeohistorical framework, *in* James N.P., and Clarke, J.A.D., eds., Cool-Water Carbonates: SEPM, Special Publication 56, p. 185–204.

MCGOWRAN, B., HOLDGATE, G.R., LI, Q., AND GALLAGHER, S.J., 2004, Cenozoic stratigraphic succession in southeastern Australia: Australian Journal of Earth Sciences, v. 51, p. 459–496.

MCKINNEY, F.K., AND JACKSON, J.B.C., 1989, Bryozoan Evolution: Boston, Unwin Hyman, 238 p.

MOSS, G.D., AND MCGOWRAN, B., 1993, Foraminiferal turnover in neritic environments: the end-Eocene and mid-Oligocene events in southern Australia, *in* Jell, P.A., ed., Palaeontological Studies in Honour of Ken Campbell: Association of Australasian Paleontologists, Memoir 15, p. 407–416.

MUTTI, M., AND HALLOCK, P., 2003, Carbonate systems along nutrient and temperature gradients: some sedimentological and geochemical constraints: International Journal of Earth Sciences, v. 92, p. 465–475.

MYERS, K.J., AND MILTON, N.J., 1996, Concepts and principles of sequence stratigraphy, *in* Emery, D., and Myers, K.J., eds., Sequence Stratigraphy: Oxford, U.K., Blackwell Scientific Publications, p. 11–44.

NELSON, C.S., AND JAMES, N.P., 2000, Marine cements in mid-Tertiary, cool-water, shelf limestone of New Zealand and Australia: Sedimentology, v. 47, p. 609–630.

NELSON, C.S., HYDEN, F.M., KEANE, S.L., LEASK, W.L., AND GORDON, D.P., 1988, Application of bryozoan zooarial growth-form studies in facies analysis of non-tropical carbonate deposits in New Zealand: Sedimentary Geology, v. 60, p. 301–322.

ODIN, G.S., 1988, Green Marine Clays: Oolitic Ironstone Facies, Verdine Facies, Glaucony Facies and Celadonite-Bearing Facies—A Comparative Study: Amsterdam, Elsevier, Developments in Sedimentology, v. 45, 445 p.

POPE, M.C., AND STEFFEN, J.B., 2003, Widespread, prolonged late Middle to Late Ordovician upwelling in North America: A proxy record of glaciation?, Geology, v. 31, p. 63–66.

POSAMENTIER, H.W., JERVEY, M.T., AND VAIL, P.R., 1988, Eustatic controls on clastic deposition; I, conceptual framework, *in* Wilgus, C.K., Hastings, B.S., Kendall, C.G.St.C., Posamentier, H.W., Ross C.A., and Van Wagoner, J.C., eds., Sea Level Changes: An Integrated Approach: SEPM, Special Publication 42, p. 110–124.

PRICE, R.C., NICHOLLS, I.A., AND GRAY, C.M., 2003, Cenozoic igneous activity, *in* Birch, W.D., ed., Geology of Victoria: Geological Society of Australia, Special Publication 23, p. 361–375.

PRITCHARD, D.W., 1967, What is an estuary? Physical viewpoint, *in* Lauff, G.H., ed., Estuaries: American Association for the Advancement of Science, Publication 83, p. 3–5.

PROTHERO, D.R., 1994, The Eocene–Oligocene Transition; Paradise Lost: New York, Columbia University Press, 568 p.

QUILTY, P.G., 1977, Cenozoic sedimentation cycles in Western Australia, Geology, v. 5, p. 336–340.

READING, H.G., AND COLLINSON, J.D., 1996, Clastic coasts, *in* Reading, H.G., ed., Sedimentary Environments; Processes, Facies and Stratigraphy: Oxford, U.K., Blackwell Scientific, p. 154–231.

REINSON, G.E., 1992, Transgressive barrier island and estuarine systems, *in* Walker, R.G., and James, N.P., eds., Facies Models: St. John's, Geological Association of Canada, p. 179–194.

REYNOLDS, M.A., 1953, The Cainozoic succession of Maslins and Aldinga Bays, South Australia: Royal Society of South Australia, Transactions, v. 76, p. 114–140.

RIGBY, J.K., 1987, Phylum Porifera, *in* Boardman, R.S., Cheetham, A.H., and Rowell, A.J., eds., Fossil Invertebrates: Oxford, U.K., Blackwell Scientific Publications, p. 116–139.

SCHOLLE, P.A., BEBOUT, D.G., AND MOORE, C.H., 1983, Carbonate Depositional Environments: American Association of Petroleum Geologists, Memoir 33, 708 p.

TATE, R., 1878, Notes on the correlation of the coral-bearing strata of South Australia, with a list of the fossil corals occurring in the Colony: Philosophical Society of Adelaide, Transactions for 1877–78, p. 120–123.

TUCKER, M.E., 1988, Techniques in Sedimentology: Palo Alto, CA, Blackwell Scientific Publications, 394 p.

TUCKER, M.E., AND WRIGHT, V.P., 1990, Carbonate Sedimentology: Oxford, U.K., Blackwell Scientific Publications, 482 p.

WASS, R.E., CONNELLY, J.R., AND MACINTYRE, J., 1970, Bryozoan carbonate sand continuous along southern Australia, Marine Geology, v. 9, p. 6373.

WHITE, T.S., 2004, A chemostratigraphic and geochemical facies analysis of strata deposited in an Eocene Australo-Antarctic seaway: is a cyclicity evidence for glacioeustasy?, *in* Exon, N., Kennett, J., and Malone, M., eds., The Cenozoic Southern Ocean: Tectonics, Sedimentation, and Climate Change between Australia and Antarctica: American Geophysical Union, Geophysical Monograph 151, p. 153–172.

WRAY, J.L., 1977, Calcareous Algae: Amsterdam, Elsevier Publishing Company, 185 p.

WU, R.S.S., AND RICHARDS, J., 1981, Variations in benthic community structure in a sub-tropical estuary: Marine Biology, v. 64, p. 191–198.

ZACHOS, J.C., PAGANI, M., SLOAN, L., THOMAS, E., AND BILLUPS, K., 2001, Trends, rhythms, and aberrations in global climate 65 Ma to present, Science, v. 292, p. 686–693.

ZAITLIN, B.A., DALRYMPLE, R.W., AND BOYD, R., 1994, The stratigraphic organization of incised-valley systems associated with relative sea-level changes, *in* Dalrymple, R.W., Boyd, R., and Zaitlin, B.A., eds., Incised-Valley Systems: Origin and Sedimentary Sequences: SEPM, Special Publication 51, p. 45–62.

POST-TURONIAN RUDIST-BEARING LIMESTONES OF THE PERI-TETHYAN REGION: EVOLUTION OF THE SEDIMENTARY PATTERNS AND LITHOFACIES IN THE CONTEXT OF GLOBAL VERSUS REGIONAL CONTROLS

GABRIELE CARANNANTE
Dipartimento di Scienze della Terra, Università di Napoli "Federico II", Largo S. Marcellino 10, 80138 Napoli, Italy
e-mail: gabcaran@unina.it
ANTONIETTA CHERCHI
Dipartimento di Scienze della Terra, Università di Cagliari, Via Trentino 51, 09127 Cagliari, Italy
e-mail: acherchi@unica.it
ROBERTO GRAZIANO
Dipartimento di Scienze della Terra, Università di Napoli "Federico II", Largo S. Marcellino 10, 80138 Napoli, Italy.
e-mail: rgrazian@unina.it
DANIELA RUBERTI
Facoltà di Scienze Ambientali, II Università di Napoli, Via Vivaldi 43, 81100 Caserta, Italy
e-mail: daniela.ruberti@unina2.it
AND
LUCIA SIMONE
Dipartimento di Scienze della Terra, Università di Napoli "Federico II", Largo S. Marcellino 10, 80138 Napoli, Italy.
e-mail: lusimone@unina.it

Abstract: Post-Turonian (Late Cretaceous) rudist-bearing limestones of the Nurra region in northwestern Sardinia (northern Tethyan margin) and in the central-southern Apennines and Apulia (central Tethyan domain) have recorded relevant changes in the characteristics of the carbonate platforms following the "middle" Cretaceous crisis events which affected the peri-Tethyan region as well as other regions worldwide.

Rudist bivalves became the dominant lithogenetic taxon owing to their proliferation in shallow-water environments and strong dominance of Late Cretaceous carbonate factories. Their inception, evolution, and demise were seemingly controlled by a complex interplay of environmental processes that, acting on a global scale, profoundly modified the Early Cretaceous hydrosphere–atmosphere system and forced Tethyan depositional systems to change their organization, internal architecture, and facies patterns. As a result, wide, open shelves developed where the almost ubiquitous mode of carbonate fixation was that of foramol factories.

In this paper, evidence of the remarkable regional variability in the rudist-bearing carbonate platforms of the Mediterranean Tethys is presented. The analysis of the resulting shallow-water facies has demonstrated that, in spite of several stratigraphic similarities and common sedimentological features, some remarkable differences occurred between the northern Tethyan margin and the central Tethyan banks as regards the areal partitioning of the main paleoecologic controlling factors. This resulted in the deposition of rhodalgal successions in Sardinia (northern Tethyan margin) and rudist-rich foramol facies in the Apennine–Apulia (central Tethys) regions, respectively. Such Late Cretaceous carbonate systems can be viewed as geological products which have closely and coherently recorded the globally changing environmental conditions of the oceanic realm. In spite of this, the difference of the facies partitioning in different Tethyan regions according to a latitudinal gradient is interpreted as derived mainly from local variable paleoceanographic and paleoclimatic conditions.

INTRODUCTION

Dramatic environmental changes affected the Tethyan carbonate factories during the "middle" Cretaceous (e.g., Masse, 1989; Schlager and Philip, 1990; Simo et al., 1993), when shallow-water domains were forced to change sediment production modes, depositional patterns, and internal sedimentary architecture. These changes resulted from a complex evolution of the atmosphere–hydrosphere system which was triggered by geodynamic processes (e.g., Masse, 1989; Kerr, 1998; Larson and Erba, 1999; Skelton, 2003a). The "middle" Cretaceous (Aptian–Turonian interval) was a period of unusually active geodynamic activity that led to sustained volcanism, elevated seafloor spreading, and major paleogeographic and paleoceanographic changes at the beginning of the eoalpine cycle; these changes were coincident with relevant geological events: major turnovers of biotic communities (Masse, 1989; Harries and Little, 1999), oceanic anoxic events (OAEs) (Arthur et al., 1990), thermal maximum leading to greenhouse episodes, and major disruptions of geochemical cycles (Veizer et al., 1999; Hansen and Wallmann, 2003).

In particular, Turonian times are largely regarded as the interval of thermal maximum of Earth's climate since the beginning of the Cretaceous (e.g., Clarke and Jenkyns, 1999), as also suggested by the marked peak in atmospheric CO_2 content (Hansen and Wallmann, 2003). It has been postulated that the environmental disruption above terminated the rudist-rich, aragonite-dominated chlorozoan factories of the latest Albian–Cenomanian (mostly caprinids, ichthyosarcolitids, and corals), which changed to an early–middle Turonian microbial mode adapted to a hyper-greenhouse interval (Graziano et al., 2007). On the whole, the intense "middle" Cretaceous oceanic volcanic activity and the related hydrothermal fluids may have forced the trophic structure of the

Controls on Carbonate Platform and Reef Development
SEPM Special Publication No. 89, Copyright © 2008
SEPM (Society for Sedimentary Geology), ISBN 978-1-56576-130-8, p. 255–270.

oceans, fostering a relevant primary production and favoring heterotroph, *r*-mode opportunist filter-feeding organisms. A major Early–Middle Turonian event (Frakes, 1999) concludes the "middle" Cretaceous changes and represents a crucial interval for the ongoing development of shallow-water factories starting in the Late Turonian, when rudists played a major role as the relevant lithogenetic taxon. The inception of post-Turonian carbonate platforms in the Tethyan seaway was seemingly controlled by the complex interplay of different paleoceanographic and paleoclimatic processes (Graziano et al., 2007).

Data from Upper Turonian–Lower Campanian shallow-water limestones from the peri-Mediterranean area document important changes in the characterization of the related carbonate platforms in respect to the Early–middle Cretaceous platforms (e.g., Philip, 1982, 1993; Gusic and Jelaska, 1990; Gili, 1993; Gili et al., 1995a; Gili et al., 1995b; ; Gili et al., 2003; Skelton et al., 1995; Carannante et al., 1995; Carannante et al., 1997; Carannante et al., 1999; Moro, 1997; Steuber, 2002; Skelton, 2003a; Philip and Gari, 2005; among others). The post-Turonian carbonate systems evolved with a general transgressive trend undergoing local drowning events. By contrast, progradation of the depositional systems characterized the following Late Campanian–Maastrichtian *p.p.* time interval (Carannante et al., 1999, and references therein).

The main goals of this paper are to characterize the sedimentary and paleoecologic features of some peri-Mediterranean post-Turonian carbonate successions (Fig. 1), which reflect drastic environmental changes in the facing water masses, and to speculate on the possible global versus regional causes that have triggered main changes in the depositional systems. In previous papers, Carannante and Simone (1987), Carannante et al. (1997), and Carannante et al. (1999) described facies analogies and similarities in the evolutionary trends of the sedimentary sequences cropping out in the northwestern Sardinia (northern Tethyan domain) and central-southern Apennines–Apulia (central Tethyan domain) regions. Nonetheless the authors also found significant differences in the main features of the Late Cretaceous depositional systems that developed there.

In this paper, overall variabilities in the main features of the rudist carbonate platforms of the Mediterranean Tethys (Fig. 1) are investigated in a regional perspective in order to derive possible local controls within an evolving global paleoenvironmental frame.

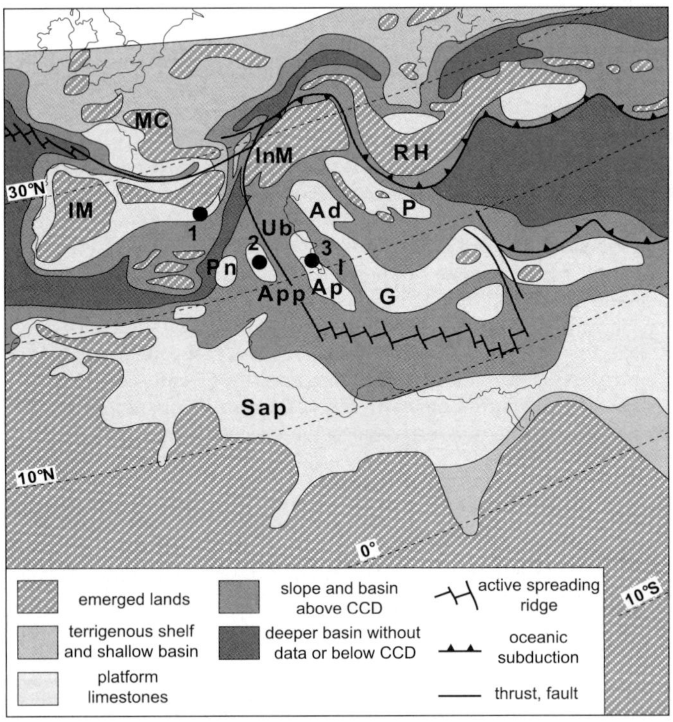

FIG. 1.—Late Cenomanian paleogeography of the Western Tethys (modified after Philip et al., 2000, and Skelton, 2003b) and location of the carbonate platform-to-basin transitions examined here: (1) Nurra carbonate platform (NW Sardinia); (2) Apennine carbonate platform; (3) Apulia carbonate platform. Abbreviations: Ad, Adriatic–Dinaric platform; Ap, Apulia platform; App, Apennine platform; G, Gavrovo platform; I, Ionian Basin; IM, Iberian Massif; InM, Insubrian Massif; MC, Massif Central; P, Pelagonian platform; Pn, Panormide platform; RH, Rhodope; Sap, Saharian platform; Ub, Umbria-Marchean Basin.

METHODS

Previously analyzed sections were reviewed and confronted with new sections in Sardinia, Apulia, and the central-southern Apennines (Fig. 1). Biostratigraphic–paleontologic, sedimentologic–taphonomic, and diagenetic analyses of the Upper Cretaceous shallow-water rudist-bearing carbonate lithofacies have been based on a combination of field observations and study of thin sections and polished slabs under the optical microscope. Similar investigations have been carried out on coeval deeper-water successions to attempt a genetic correlation with the shallow-water record. SEM analyses were performed in order to obtain data to interpret the nature of the pelagic and hemipelagic fines and their diagenetic history. The resulting data have also been used to trace the dispersal patterns of shallow-water sediments as well as the evolutionary path of the whole depositional system during relative-sea-level cycles.

POST-TURONIAN RUDIST-BEARING CARBONATE PLATFORMS OF THE PERI-TETHYAN REGION

Post-Turonian shallow-water limestones cropping out in the peri-Mediterranean region are largely characterized by rudists that spread out over the carbonate shelves and colonized various environments. Analyzed sequences from northwestern Sardinia, the central-southern Apennines, and Apulia in Italy (e.g., Carannante and Simone, 1987; Carannante et al., 1995; Carannante et al., 1997; Carannante et al., 1999; Stössel and Bernoulli, 2000; Simone et al., 2003), as well as literature data from other peri-Mediterranean countries (Spain, France, Croatia, Greece, Tunisia) (e.g., Accordi et al., 1986; Cestari and Sartorio, 1995; Negra et al., 1995; Skelton et al., 1995; Moro, 1997; Philip and Gari, 2005; Pomar et al., 2005; Pomoni-Papaioannou and Photiades, 2008; among others), document significant changes in biotic assemblages with respect to the previous sedimentary settings.

Sedimentary Features and Early Diagenesis

In the Late Turonian–Early Campanian peri-Tethyan carbonate platforms, rhodalgal and foramol assemblages replaced chlorozoan ones whereas nonskeletal grains tended to disappear. The different nature and organization of the sediment-producing (rhodalgal and foramol versus chlorozoan) assemblages as well as the abundance of the bioerosion-derived skeletal detritus resulted in sheets of loose bioclastic sediments that covered sea bottoms. Continuous and limiting bars were lacking (Fig. 2). Intense current and storm washing and win-

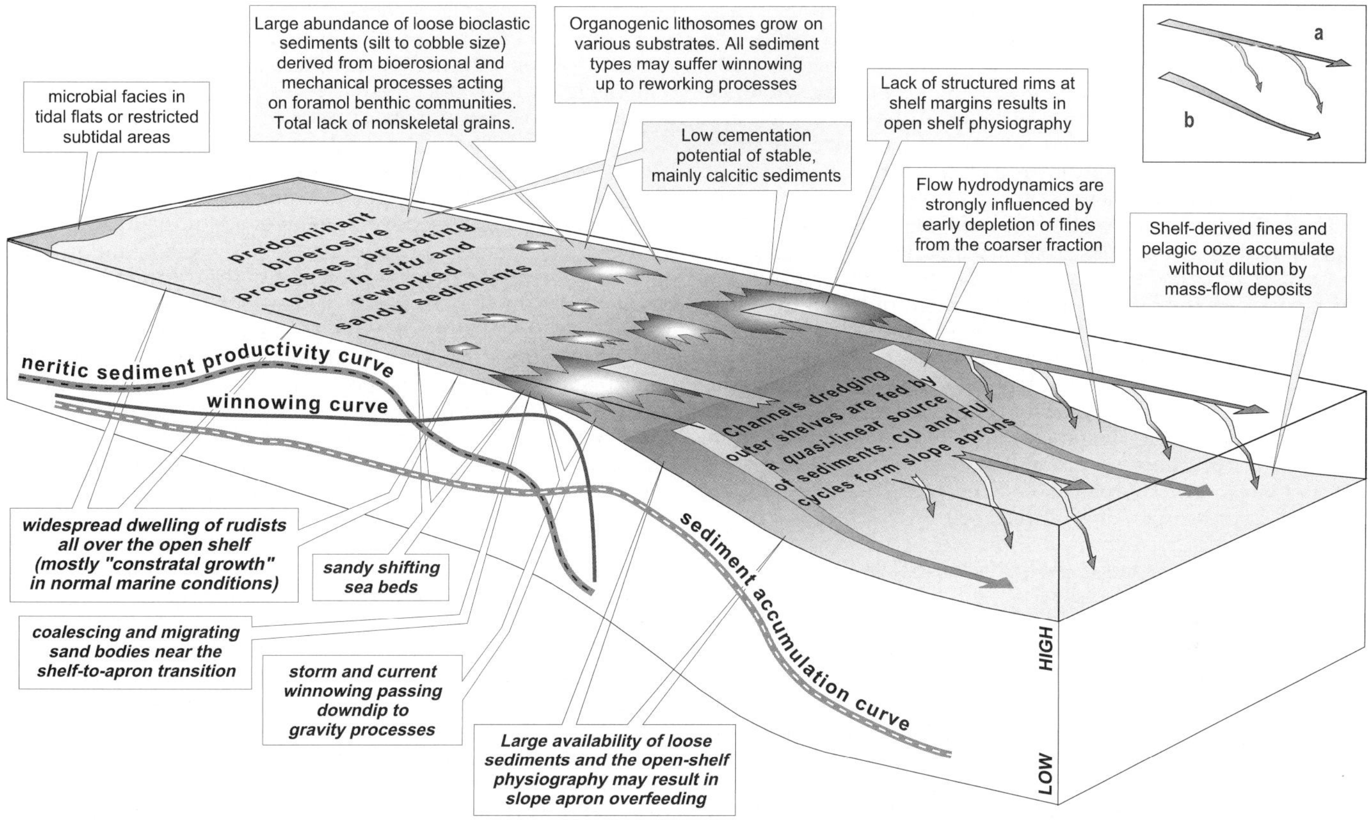

FIG 2.—Block diagram showing the main sedimentary features of the Late Turonian–Maastrichtian rudist-rich carbonate platform-to-basin transitions (modified after Carannante et al., 1997; Carannante et al., 1999). Inferred trends of the main sedimentary processes operating along the transition are also represented. Symbols in the inset: a) off-shelf sweeping and fallout of shelf-derived fines (mostly by density cascading and/or other hyperpycnal flows); b) off-shelf transport of coarse bioclastic sediments (sand to cobble sizes) by means of mass flows (mostly sand-supported debris flows). See text for discussion.

nowing resulted in shifting substrates mainly in the marginal regions of the shelves, where rudists were only rarely preserved in their original growth position. Storm events and currents flowing on the marginal sectors of these open shelves swept away the bioerosion-derived fine fraction, which accumulate mainly in hemipelagic deposits (Carannante et al., 1999; Simone et al., 2003). Fine sand (mainly foram tests and fine rudist debris) and silt, removed by storms and waves, also contributed to the shelf interior, where they accumulated to form thin episodic layers. More internal and/or leeward sectors, which are well represented in the southern Apennines and Apulia outcrops, are characterized by rudist-rich beds alternating with finer-grained foraminiferal limestones. Rudists are frequently found in growth position, although a large quantity of shells appears to be toppled and oriented by local hydrodynamics with minimal sediment reworking. Complex networks of channel-like depressions locally characterize the internal and transitional areas of the shelf (Carannante et al., 2003, Ruberti et al., 2006, 2007). Locally, the lateral bars in these current pathways were colonized, and possibly stabilized, by rudist congregations. Reworked rudist shells and rudist debris built up secondary shell concentrations, deposited under the influence of hydraulic processes within the channel axes (Carannante et al., 2003; Ruberti et al., 2006; Ruberti et al., 2007).

The decline in aragonite-shelled bioclast producers in the Late Cretaceous carbonate platforms (Van de Poel and Schlager, 1994; Carannante et al., 1995; Carannante et al., 1997) established important controls on the early diagenesis of these deposits, with significant implications for their redistribution phenomena. Early dissolution features are scarce in the preponderantly calcite shells; the intragranular cavities only locally contain early calcite cement fillings (Ross 1991; Ross and Skelton, 1993; Carannante et al., 1995). Owing to its stable mineralogical nature, the bioclastic fraction (fine to very coarse sands) was not or only scarcely involved in early cementation processes. As a consequence, a large quantity of loose bioclastic sediments resulted, thus strongly influencing the sedimentary budget and dispersal patterns and contributing to produce high dispersion rates (Carannante et al., 1999) (Fig. 2). Moreover, the almost total lack of aragonite donor biota resulted in the calcite-rich nature of the neighboring hemipelagic deposits, as documented by SEM observations of coeval hemipelagic limestones. These latter lack evidence of pristine aragonite needles. The finer sand grains as well as the coarse silt fraction are largely bioerosion-derived bioclasts, mainly rudist micro-fragments (Carannante et al., 1999).

Depositional Systems

The post-Turonian benthic assemblages dwelled in loose substrata in a patchy distribution and did not build real reefs at shelf margins. This allowed open water circulation on the shelf and exerted a driving control on the sedimentary arrangement of the shifting biogenic sediments, promoting the formation of large coalescing sheets of winnowed fine to coarse skeletal sands. The foramol bioclastic fraction, produced by intense bioerosive and mechanical processes acting on mainly diagenetically stable calcite-shelled organisms, blanketed most of the open shelf and provided the shelf with a huge amount of loose, largely uncemented, silt- to sand-size bioclastic debris, which was ready to be transported toward the deeper areas.

This interpreted depositional model (Fig. 2) is in striking contrast with that for chlorozoan, "Bahamian"-type carbonate platforms. Unlike on "tropical-type" platforms, the shelf interior, which supports the highest neritic sediment productivity, does not correspond to the site of maximum sediment accumulation. The latter, in fact, is located in deeper areas in response to the intense off-bank sediment shedding due to the open physiography of the factory. This peculiarity of the sediment budget is one of the major sedimentary and stratigraphic deviations from the chlorozoan "tropical-type" rimmed shelves. Nonetheless, it plays an extremely important role on the winnowing or redepositional processes in these open depositional settings.

Transgressive–Regressive Cycle

The post-Turonian carbonate platforms investigated here show a distinct bistadial evolution in a long-term, Late Turonian–Maastrichtian transgressive–regressive cycle. They are characterized by aggradation or backstepping of foramol-type carbonate platforms during the Late Turonian–Early Campanian followed by widening and progradation during the Late Campanian–Maastrichtian *p.p.* (Carannante et al., 1997; Carannante et al., 1999). According to Carannante et al. (1999) the Tethyan carbonate factories generally exported large amounts of skeletal debris, mainly during times of reduction in accommodation space (terminal highstand to lowstand intervals). By contrast, they tended to retreat and sometimes drown as a response to increased accommodation space (rapid transgressive phases). This contrasts with the tropical chlorozoan rimmed-shelf behavior and is consistent with that of the foramol carbonate system (Fig. 3; see also Brachert et al., 1996; Carannante et al., 1996). Foramol systems generally display a ramp-like morphology; shelf regions of net sediment production grade laterally into deeper areas of net sediment accumulation, particularly in the absence of a steep (tectonically or erosionally induced) margin (Fig. 2). These systems tend to prograde similarly to siliciclastic systems (see Carannante and Simone, 1988; Carannante et al., 1999), mainly in response to: (1) lower accommodation during the terminal highstand and lowstand of relative sea level; and (2) large availability of loose sediments that easily bypass the outer shelves toward the basin (Figs. 2, 3).

PERI-TETHYAN LATE CRETACEOUS CARBONATE SYSTEMS

Although the investigated depositional systems show, on a large peri-Mediterranean scale, compositional and geometrical analogies and similarities in their evolutionary trends, some differences have been observed among the outcropping successions. Such differences suggest major regional and / or local complexities in the paleoceanographic and paleoclimatic processes that acted on the Tethyan Realm during post-Turonian times.

The Northwestern Sardinia Cretaceous Sequence

Sardinia was part of the main European foreland, which, starting from the Oligocene, was affected by extensional tectonics leading to the opening of back-arc basins (Provençal, Sardinian, and Tyrrenian Basins). Shallow-water "Urgonian" Cretaceous sequences crop out widely in northwestern Sardinia (Nurra region, Fig. 1). They document non-restricted depositional environments whose sea bottoms supported a loose biodetritic cover (D'Argenio et al., 1985). During the mid-Cretaceous, tectonically driven regressive events (see Cherchi and Trémolières, 1984; Cherchi, 1985) resulted in regional exposure of Lower Cretaceous basement strata and the development of a fault-controlled karst network. The exposed lowland had shallow relief, and bauxite deposits, derived from the weathering of the exhumed Variscan basement (Mameli et al., 2007), accumulated over more deeply eroded regions. Patchily distributed Cenomanian, locally brecciated, black limestone rich in characean remains ("Munieria" limestones) became the first post-tectonic sediment resting over bauxite and substratum limestone. These continental and transitional deposits document transgressive phases, followed by repeated and complex erosive events. Subsequent fully marine deposition resulted in up to 140 m of post-Turonian shallow-water rudist-rich sediments. Partially reworked, rudist- and bryozoan-dominated deposits characterize the upper part of the successions. The sporadic presence of planktonic forams and green glauconite grains in these remobilized sediments suggests a deeper depositional environment. In some localities, more rapid deepening of the depositional environment occurred in Santonian time, resulting in localized drowning episodes. This is locally demonstrated by the evolution, upwards in the successions, to hemipelagic blackish marls and clay (see Olmedo and Punta Negra sections in Cherchi, 1985; Cherchi and Schroeder, 1987) and / or by the establishment of impressive gravity sedimentation processes (see Punta Negra and Mt. Miale Ispina sections in Cherchi, 1985). Based on different thicknesses and vertical evolution within a contiguous area, a complex paleomorphological–paleotectonic control has been hypothesized (Cherchi, 1985). Oligo-Miocene ignimbrites locally cap the Cretaceous calcareous sequences.

The Post-Turonian Nurra Carbonate System.—

In the post-Turonian sequence of the northern Tethys margin we observe a characteristic trend in the evolving depositional textures of the rudist limestone (see Carannante et al., 1995). In spite of some lateral variations, all sections are generally characterized by less than 10 m of bioclastic wackestone and packstone (Coniacian *p.p.* in age) with benthonic forams (including very large and thick miliolids), mollusc (rudist) fragments, chaetetids, large amounts of green algae (dasycladaceans), and a few corals in a pellet-rich, muddy matrix with a very limited contribution of nonskeletal grains (Table 1A, B). The overlying limestone is made

→

FIG 3 (facing page).—Comparison of "foramol / temperate-type" and "chlorozoan / tropical-type" carbonate depositional systems and their response to relative-sea-level oscillations. These systems have been chosen as two end members of a continuum of variability in order to stress their differences. Minor differences can be expected when comparing the foramol and chlorozoan systems in a "leeward" setting. The stratigraphic architecture and the sedimentary features (namely sediment types and their dispersal patterns) of the foramol systems support platform progradation and upbuilding of slope aprons during both lowstands and late highstands of sea level. On the contrary, the base-of-slope aprons in the chlorozoan systems could be fed only during highstands (see text for discussion).

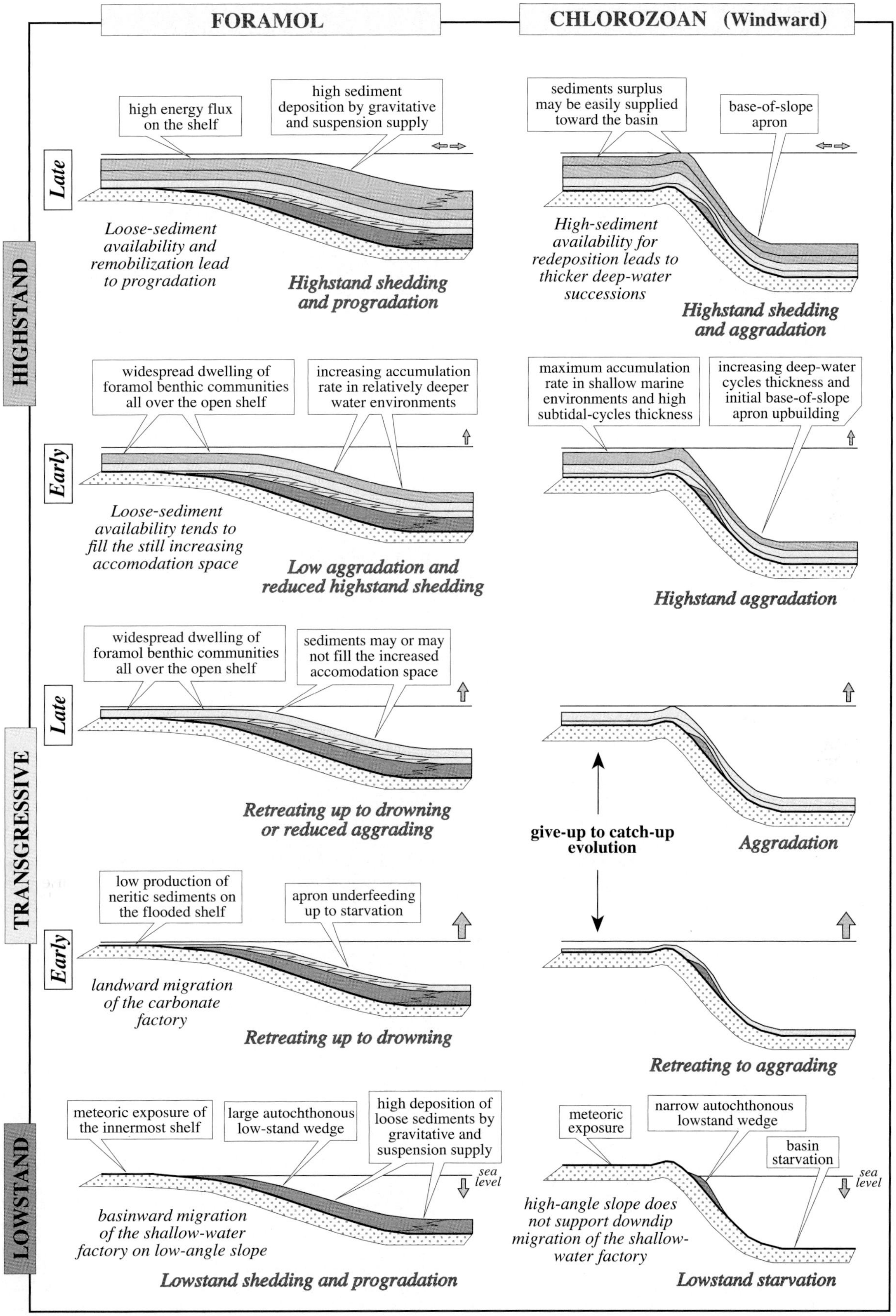
FOREMOL
CHLOROZOAN (Windward)
HIGHSTAND
Late
high energy flux on the shelf
high sediment deposition by gravitative and suspension supply
Loose-sediment availability and remobilization lead to progradation
Highstand shedding and progradation
sediments surplus may be easily supplied toward the basin
base-of-slope apron
High-sediment availability for redeposition leads to thicker deep-water successions
Highstand shedding and aggradation
Early
widespread dwelling of foramol benthic communities all over the open shelf
increasing accumulation rate in relatively deeper water environments
Loose-sediment availability tends to fill the still increasing accomodation space
Low aggradation and reduced highstand shedding
maximum accumulation rate in shallow marine environments and high subtidal-cycles thickness
increasing deep-water cycles thickness and initial base-of-slope apron upbuilding
Highstand aggradation
TRANSGRESSIVE
Late
widespread dwelling of foramol benthic communities all over the open shelf
sediments may or may not fill the increased accomodation space
Retreating up to drowning or reduced aggrading
give-up to catch-up evolution
Aggradation
Early
low production of neritic sediments on the flooded shelf
apron underfeeding up to starvation
landward migration of the carbonate factory
Retreating up to drowning
Retreating to aggrading
LOWSTAND
meteoric exposure of the innermost shelf
large autochthonous low-stand wedge
high deposition of loose sediments by gravitative and suspension supply
sea level
basinward migration of the shallow-water factory on low-angle slope
Lowstand shedding and progradation
meteoric exposure
narrow autochthonous lowstand wedge
basin starvation
sea level
high-angle slope does not support downdip migration of the shallow-water factory
Lowstand starvation

TABLE 1.—Main lithofacies of the contrasting chlorozoan–foramol depositional systems in the post-Turonian transgressive successions of northwestern Sardinia (northern Tethyan margin). **A, B)** Shallow-water, chlorozoan lithofacies in basal Coniacian rudist-bearing deposits. **C, D)** Shallow-water, foramol (rhodalgal-type) lithofacies in Santonian rudist-bearing deposits. **E, F)** deep-water, foramol lithofacies in Late Santonian–Early Campanian (?) drowning intervals. A) Rudist–coral floatstone in a matrix of bioclastic and peloidal wackestone–packstone. B) Bioclastic and peloidal wackestone with green algae (arrowed), large miliolids, *Dicyclina* sp., and gastropods. C, D) Coarse bioclastic grainstone with bryozoans, red algae, echinoids, and sparse rudist debris. E, F) Reworked grains as in Parts C and D. Note the large syntaxial cements around the echinoid fragments of the coarse grainstone (E) and the large sponge spicule in organic carbon–rich and burrowed packstone–grainstone (F). Scale bar is 1 mm.

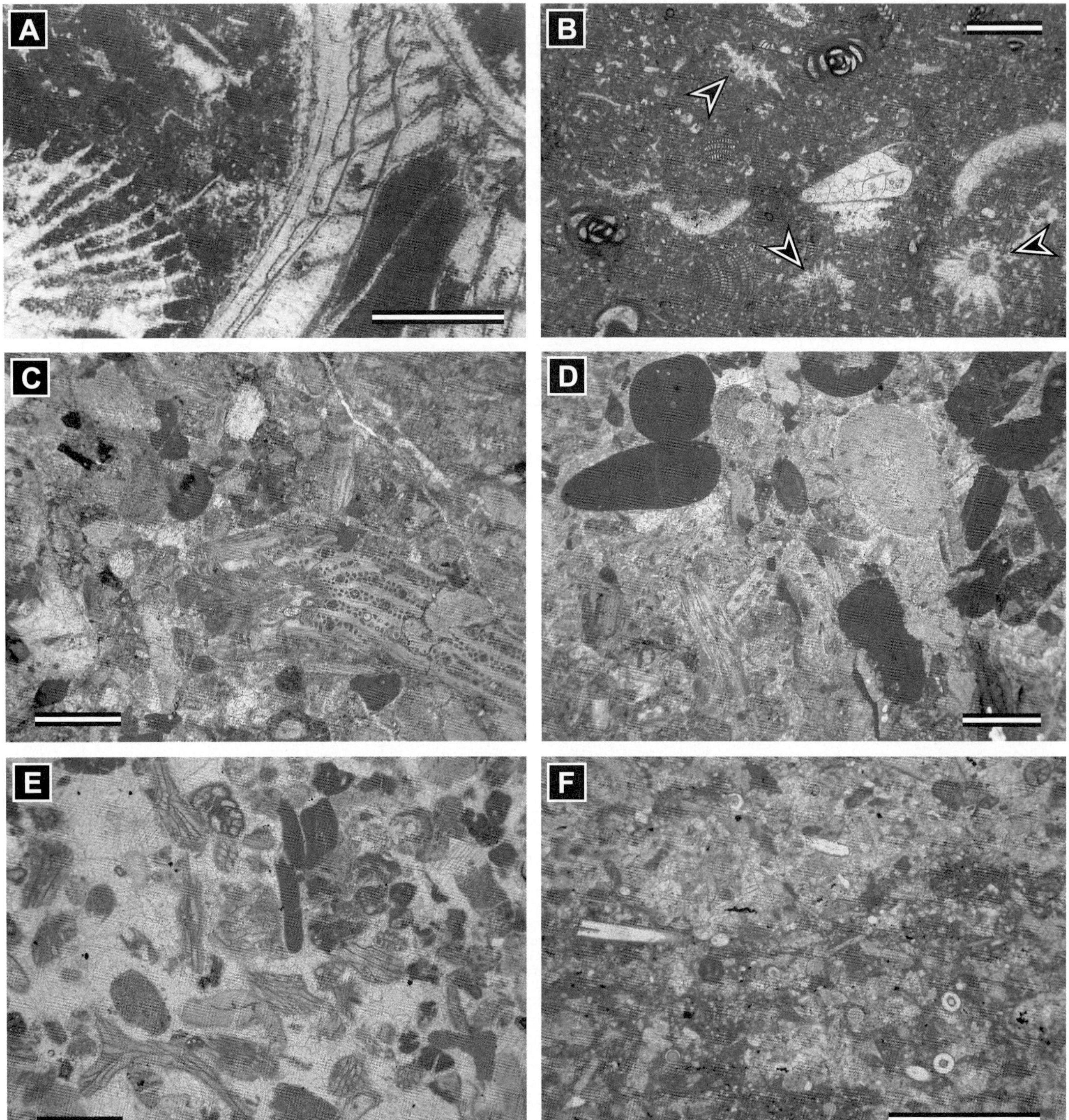

up of fine to coarse bioclastic grainstone and more rarely silty packstone, in which bryozoans and coralline red algae become more and more abundant upwards, accompanied by echinoids, mollusc (nearly all rudists) debris, and benthonic forams (Coniacian *p.p.*–Santonian in age) (Table 1C, D). A sharp reduction, up to disappearance, of green algae, corals, and chaetetids as well as of nonskeletal grains is observed. Rudists locally form complex primary shell concentrations (see Kidwell et al., 1986; Kidwell, 1991; Kidwell and Holland, 1991), but not-rigid frameworks characterized these rudist accumulations (Carannante et al., 1995), which suggests that the rudist inhabited loose mobile sediment with a constratal growth arrangement (*sensu* Gili et al., 1995a). Among the red coralline algae, *Sporolithon* is a common constituent of rhodolithic envelopes and a major component of the skeletal matrix. Rhodoliths are locally very abundant and occasionally form rhodolithic rudstone–floatstone which may contain abundant bryozoans, echinoids, benthic and planktonic forams, and green grains in the skeletal matrix. Early calcite cement has been recognized by Ross (1991), Ross and Skelton (1993), and Carannante et al. (1995).

In some of the studied outcrops (see Olmedo and Punta Negra sections in Carannante et al., 1995) as well as in cores drilled in neighboring areas, the above skeletal limestones evolve upwards (and locally laterally) into hemipelagic deposits. These latter deposits are Upper Santonian-Lower Campanian gray and blackish marls and clay, rich in sponge spicules (Table 1F), planktonic forams, among which are *Globotruncana concavata* (Brotzen), *Sigalia deflaensis* (Sigal), *Heterohelix reussi* Cushman, and *Hedbergella delrioensis* Carsey (Cherchi and Schroeder, 1987), and coccoliths (among which is *Lucianhorabdus cayeuxi* Deflandre). Fine and coarser skeletal debris (Table 1E) locally intercalate with plankton-rich deposits. Following the deposition of the hemipelagic deposits, coarse bioclastic grainstone was laid down. It is rich in rudist remains with bioeroded and poorly abraded grains. On the basis of foram associations these sediments are interpreted to be Late Campanian in age (Cherchi, 1985; Carannante et al., 1999).

Locally (see Capo Caccia section in Carannante et al., 1995), uppermost Santonian–Campanian rudist-bearing limestone documents persistent shallow-water conditions. At the top of this limestone a limited contribution of corals has been observed along with the presence of blackish clasts, micro-dissolution features, and crystal silt.

Environmental Meaning and Depositional Trend.—

The studied sections provide evidence of Early Coniacian low-energy, shallow, inner-shelf environments, inhabited by organisms living in well lighted areas (euphotic marine conditions) locally supporting rudist growth. They rapidly evolved upward to more open and deeper environments characterized by higher energy and inhabited by organisms adapted to low light intensity (oligophotic marine conditions). There, red algae and bryozoans become more and more significant in a sediment that continued to host rudist growth. The deepening trend resulted in the drowning of areas of the Santonian rudist-bearing carbonate factory under less productive conditions. The presence in the hemipelagic cover of *Lucianhorabdus cayeuxi* Deflandre indicates, according to Henricksson and Malmgren (1997), relatively shallow (not more than 300 m deep) and cold (9–11° C) conditions in outermost shelf areas or in distal ramps for the uppermost interval of the Early Campanian hemipelagites. A minor intermittent supply of bioclastic fraction contributed to the sedimentation in the marginal sectors of the shelf during the deposition of the Upper Santonian–Lower Campanian plankton-rich marls and clay. These latter appear to be related to the terrestrial runoff, which is largely documented by the frequency and abundance of terrestrial supply along the northern Tethyan margin over the Late Cretaceous (Philip and Gari, 2005). Only in the Late Campanian was shallow-water carbonate sediment deposited in the drowned areas, with clean skeletal sands devoid of pelagic intercalations. In outcrops where the limestone shows persistent shallow-water conditions up to the uppermost Santonian–? lower Campanian the compositional and diagenetic characteristics suggest a Campanian fresh- to brackish-water influx (see Capo Caccia section in Carannante et al., 1995).

On the whole, the studied Sardinia limestone documents the recovery of a carbonate platform after middle Cretaceous tectonics-related emersion phases (see Cherchi and Trémolières, 1984; Cherchi, 1985), which occurred during a time interval in which global crisis events developed (see Jenkyns, 1980; Graciansky et al., 1986; Vogt, 1989; Arthur et al., 1990; Carannante et al., 1995; D'Argenio and Mindszenty, 1995; Premoli Silva and Sliter, 1999; among others). In their sedimentary record, the response to changing water conditions is documented. The presence of very thick-shelled miliolids among the benthic forams, which characterize the basal muddy deposits but which are also present in the following more open and high-energy-related sediments, documents (according to Hallock, 1985) an increased nutrient supply in the waters. Mesotrophic to eutrophic water conditions as well as variations in the water temperature are also suggested by the change from the photophilous and oligotrophic-adapted chlorozoan assemblages to the sciophilous and mesotrophic rhodalgal ones (Carannante et al., 1988; James, 1997; Mutti and Hallock, 2003; Halfar et al., 2004). In addition, opportunistic forms as heterohelicids and *Hedbergella* characterize the hemipelagic deposits, to which the rudist-bearing shelves had vertical and, locally, lateral transition (Cherchi, 1985; Carannante et al., 1995). Similar forams are also present in the Santonian pelagic successions cropping out in the eastern sectors of Sardinia (Dieni and Massari 1985). According to Hallock et al. (1991) and Premoli Silva and Sliter (1999) these opportunistic forams also indicate mesotrophic to eutrophic water conditions in the pelagic realm (see Carannante et al., 1995; Carannante et al., 1997 for details). Mesotrophic up to oligotrophic conditions seem to characterize the source area of the following Upper Campanian prograding limestones (Carannante et al., 1999) as well as the uppermost Santonian?–Lower Campanian shallow-water limestone cropping out in the uppermost interval of the Capo Caccia section.

The Central-Southern Apennine and Apulia Cretaceous Sequences

Shallow-water Cretaceous sequences crop out widely in both the deformed Apennine chain and the undeformed Apulia region, which is part of the Adria–Africa foreland. Post-Turonian shallow-water carbonate successions unconformably overlie complex middle Cretaceous substrata showing contrasting evidence of episodic drowning and subaereal exposure, reflecting the action of early eoalpine tectonics (e.g., Carannante et al., 1994; Graziano, 1994; D'Argenio and Mindszenty, 1995; Vlahovic et al., 2005). Complex stratigraphic gaps of different characteristics and duration characterize many of the Cretaceous successions. In some localities of the central-southern Apennines, part of the Albian up to all of the Turonian sedimentary record is missing, whereas in other localities Cenomanian shallow-water deposits are found bounded by two unconformities, both characterized by paleokarst and/or bauxite. Moreover, at a present distance of a few tens of kilometers away, the same carbonate platform domain shows no evidence of long-lasting exposure or of bauxites. In the shallow-water sections of the Apulia region a unique long-lasting exposure event is recorded at the top of the gently tilted Upper Cenomanian substratum (e.g., Luperto Sinni and Reina, 1996). Deep karst relief, locally infilled by bauxite, developed during the subaerial exposure. Where present, in the southern Italian outcrops, the Turonian interval is characterized by a general and marked abundance of fine-grained limestones with very poorly diversified fossil associations made up of Miliolidae and Discorbidae (e.g., Chiocchini et al., 1994). In addition, microbial carbonates made up mostly of cyanoid-, stromatolite-, and thrombolite-rich fine-grained limestone became widespread. *Thaumatoporella* sp. and *Aeolisaccus*, recently reinterpreted as cyanobacteria (Cherchi and Schroeder, 2005; Golubic et al., 2006), are locally dominant, associated with rudists (hippuritids and radiolitids), which locally represent the only macrofaunal element along with nerineids and acteonids.

Starting in the Late Turonian, the Late Cretaceous sea encroached on the karstified substratum and the related bauxite cover as well as on Turonian restricted paralic deposits, resulting in the deposition of rudist-rich shallow marine sediments. The studied post-Turonian rudist-bearing limestones vary in thickness from a few meters to several hundred meters.

The Post-Turonian Central-Southern Apennine and Apulia Carbonate Systems.—

Since the Late Turonian, the first shallow-water sediments disconformably overlying bauxite-rich karstic substrata or paralic and/or restricted Turonian lithofacies are characterized by cryptalgal laminites and calcareous mud, either azoic or showing a limited biogenic content (Crescenti and Vighi, 1964; Luperto Sinni and Borgomano, 1989; Carannante et al., 1993). Ostracods, characeans, *Aeolisaccus kotori* Radoicic, and *Thaumathoporella* sp. are locally very abundant in these muddy and cryptalgal lithofacies (Table 2A), which tend (in the Santonian interval) to decrease upward in the successions. By contrast, rudists and related skeletal deposits became more and more widespread, reaching their acme during the Santonian *p.p.*–Maastrichtian (Cestari and Sirna, 1987; Luperto Sinni and Borgomano, 1989; Bosellini and Parente, 1994).

The rudist-rich shallow-water successions of the central-southern Apennines and Apulia areas have been described in great detail (e.g., Carbone and Sirna, 1981; Laviano, 1984; Luperto Sinni and Borgomano, 1989; Carannante et al., 1993; Carannante et al., 1997; Carannante et al., 2000; Carbone, 1993; Sanders, 1996; Ruberti, 1997; Simone et al., 2003; among others). Despite local small-scale differences, we observe that the main sedimentary and paleoecological features record a common stratigraphic trend and depositional pattern. Rudists (radiolitids and hippuritids) do represent the main, almost exclusive sediment producers (Fig. 4A; Table 2B, C, D) with a minor contribution from gastropods, mainly nerineids and acteonids (Fig. 4B; Table 2B). Rare corals are locally present. Green algae disappear, whereas a few coralline red algae appear together with some bryozoans and echinoid fragments. These taxa become significant in the coeval reworked bioclastic deposits described from margin to slope successions (Ruberti, 1993; Graziano, 1994). Presumably they characterized the biogenic components of the outermost sectors of the shelf. The wide spreading of the rudists is recorded all over the shelf since the Santonian. Rudists gave rise to a variety of lithosomes, the development and growth of which was controlled mainly by hydrodynamic conditions (high-energy versus low-energy settings; see Simone et al., 2003).

In those paleoenvironments characterized by very frequent to persistent conditions of high energy, storm- and wave-related events largely controlled the texture and distribution of rudist-related sedimentary bodies. The resulting sedimentary bodies are characterized by prevailing coarse grainstone to rudstone lithologies (Table 2C, D), in which only a few rudists retained their original life position. Associated sediments contain almost exclusively bioerosion-derived rudist fragments (hippuritids and radiolitids), rare benthic foraminifers, and very rare corals.

In the more interior and/or low-energy areas, the sedimentary record contains rudist-rich beds rhythmically alternating with finer-grained foraminiferal limestones (cf. Carannante et al., 1993; Carannante et al., 1998, 2000; Cestari et al., 1992; Simone et al., 2003). Peloidal laminites or peloidal wackestone with fenestrae and/or dissolution cavities, a few benthonic foraminifers, *Aeolisaccus* spp., thin-shelled ostracods, and thaumatoporellaceans

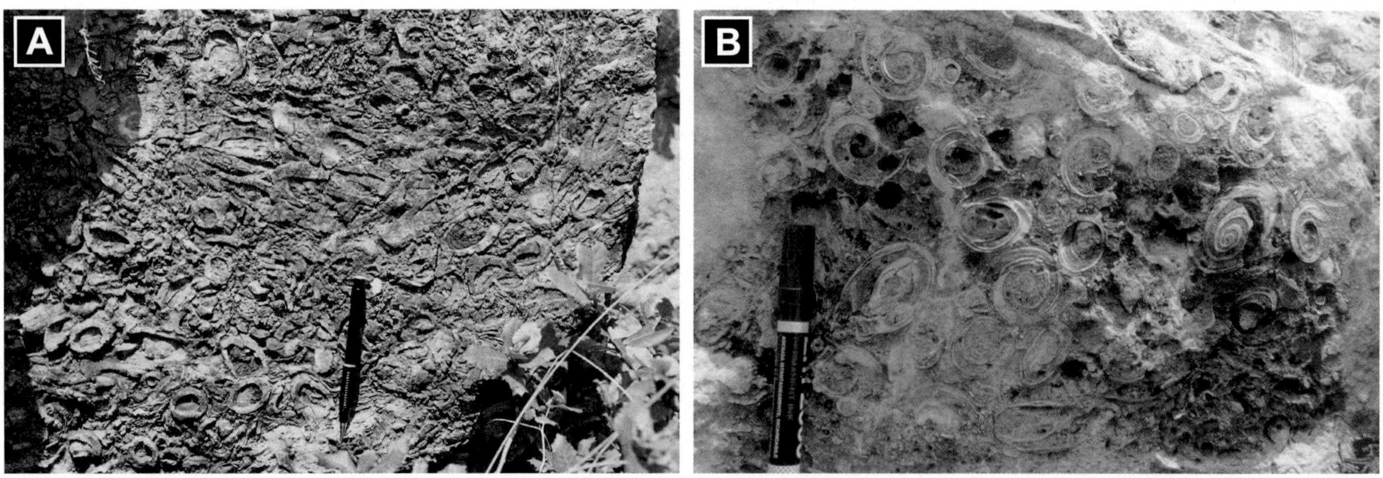

FIG 4.—Representative facies of the post-Turonian foramol benthic associations in central Tethyan domains. **A)** Hydraulic shell concentration of mostly unbroken rudists (radiolitids). Such a lithofacies typically surrounds the core of rudist biogenic beds (Santonian of Cilento Mts., southern Apennines). **B)** Gastropod (acteonids), deep-water floatstone in a matrix of rudist and gastropod grainstone. The reworked shells suggest transport in the rigid plug of a sand-supported debris flow (*sensu* Shanmugan, 2000) (Late Santonian–Early Campanian of the Gargano Promontory, Apulia). These organisms stress the role of grazing in the trophism of benthic communities inhabiting particular areas of the foramol open shelves.

TABLE 2.—Main lithofacies of the rudist-rich foramol depositional systems in the post-Turonian transgressive successions of the central-southern Apennines and Apulia (central Tethyan domains). **A, B)** Shallow-water lithofacies of inner-shelf settings in Coniacian rudist-rich deposits. **C, D)** Shallow-water lithofacies of outer-shelf settings in Coniacian–Santonian rudist-rich deposits. **E, F)** Deep-water lithofacies of upper-slope (E) and mid-slope (F) Coniacian–Santonian apron settings. A) *Thaumatoporella* sp. and *Aeolisaccus* sp. packstone. B) Rudist and gastropod floatstone in a matrix of bioclastic and peloidal wackestone. Note the intense bioerosion of the shells. C) Poorly sorted bioclastic grainstone made up of rounded rudist debris. D) Winnowed, well-sorted and rounded grainstone made up of rudist debris (several grains are micritized). E) Fine-grained bioclastic and silty packstone showing well-rounded rudist debris (note the lump denoted by the arrows made up of micritized rudist sands). F) Coarse bioclastic packstone of well-rounded rudist fragments in a silty matrix (note the high moldic porosity). Scale bar is 1 mm.

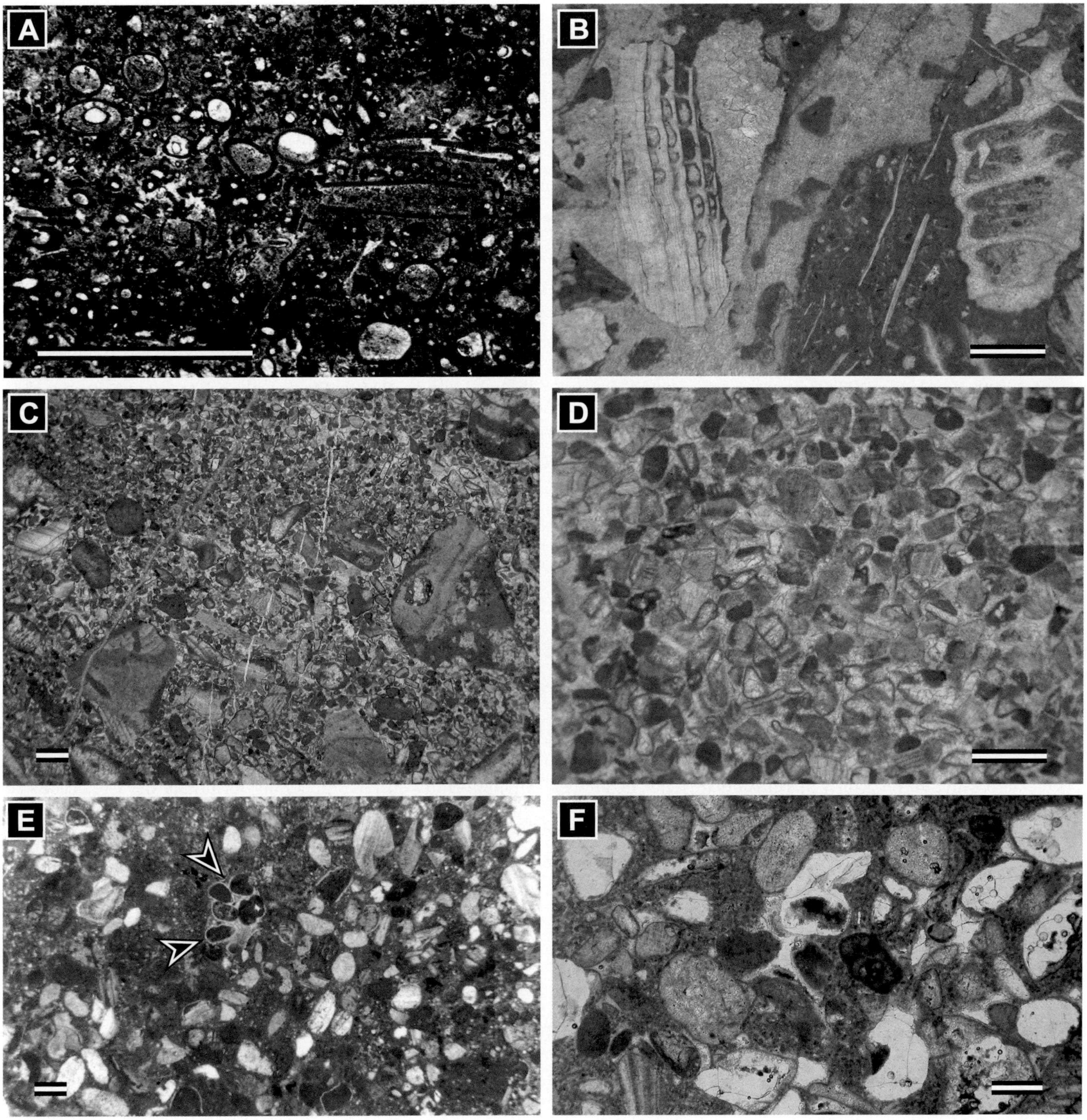

are observed in the Upper Turonian–Lower Santonian strata. Filaments of *Aeolisaccus* may contribute to form laminae. In these sequences dominated by silt / mud-rich lithotypes, small elevator radiolitids with oligospecific diversity are dominant. In the related depositional settings, the preservation of the original rudist congregations, which usually form laterally continuous biostromal shell beds, was associated with the *in situ* maintenance of the bioerosion-derived finer skeletal debris. Locally, nests or lenses rich in gastropod shells (nerineids and acteonids) occur.

These Upper Turonian–Lower Campanian successions record a general transgressive trend documented by the increase in thickness of the depositional cycles (see Ruberti, 1997). In addition the gradual change from peritidal and shallow-subtidal cycles to dominantly subtidal cycles document an increase in the amount of accommodation space which resulted from a long-term rise in relative sea level (Ruberti, 1997). The resulting meter-scale depositional cycles do not show any evidence of emersion. Submarine exposure surfaces are identified by firmground-related features.

Although little evidence of well documented drowning episodes has been found in the Apennine–Apulia shallow-water successions, one of the main features of the Upper Turonian–Lower Campanian margin to slope successions is the occurrence of pelagic intercalations rich in planktonic forams in an overall transgressive trend (Graziano, 1994; Carannante et al., 1999). R-strategic planktonic forams (e.g., *Hedbergella* and heterohelicids) are occasionally present in some Apennines outcrops. Slope facies are composed mainly of an intermittent supply of bioclastic material, which tends to diminish, and finally stops, in the Apulia carbonate successions, with the deposition of a pure pelagic interval in the (early?) middle Campanian. (Luperto Sinni and Borgomano, 1989, 1994; Graziano, 1994).

The Late Campanian–Maastrichtian shallow-water sedimentary record is sparse in the central-southern Apennines, but some sections have been documented (e.g., Cestari et al., 1992). By contrast, this interval is more widely documented in the Apulia carbonate successions (Cestari and Sirna, 1987; Reina and Luperto Sinni, 1993; Parente, 1994). On the whole this interval is typically represented by a few meters of restricted, muddy limestones. Aragonite-shelled rudists (e.g., *Sabinia* sp.), a few corals, and other originally aragonitic biotic components (gastropods) occur. Moldic cavities filled geopetally by crystal silt indicate fresh-water diagenesis apparently related to a later shifting of the water table in the sediment. Huge amounts of rudist-rich skeletal debris were displaced to the coeval slope settings (e.g., Accordi et al., 1982; Bosellini et al., 1993; Ruberti, 1993; Luperto Sinni and Borgomano, 1994; Graziano, 1994; Borgomano, 2000; Table 2E, F), thus suggesting that healthy carbonate factories flourished in the shallow-water domains. These skeletal sediments were transported across an inferred sandy margin and are made up almost exclusively of rudist debris with minor amounts of echinoids, gastropods, benthic foraminifers, and coral fragments.

Environmental Meaning and Depositional Trend.—

On the whole, the studied central-southern Apennine and Apulia post-Turonian sequences document the recovery of a complex system of carbonate platforms that developed after the middle Cretaceous tectonics-driven substrata modeling (see Carannante et al., 1994; D'Argenio and Mindszenty, 1995). The analyzed central Tethyan carbonate systems developed following an Early–Middle Turonian hyper-greenhouse interval, during which cyanobacterial consortia, with variable contributions from rudists, largely prevailed in the shallow-water domains (Graziano, 2007; Graziano et al., 2007). Due to the paucity of significant biostratigraphic markers, the Late Turonian through the Santonian epochs are difficult to discern in the carbonate sequences of the central Tethyan areas. On the whole, the studied sequences show persistent evidence of microbial-related carbonate deposition in the basal intervals (presumably Latest Turonian–Early Coniacian in age). *Thaumathoporellales* and *Aeolisaccus* sp. are locally found to form *in situ* within intertidal–shallow subtidal laminites or as dispersed particles in largely prevailing muddy subtidal facies with thrombolitic textures. Some *Thaumathoporellales*, previously considered as green algae on the basis of their reproductive features (De Castro, 1990), have recently been reinterpreted as colonial calcified cyanobacteria in a preliminary paper by Cherchi and Schroeder (2005), and Golubic et al. (2006) document the cyanobacterial nature of *Aeolisaccus kotori* Radoicic (renamed as *Decastronema*) due to its correspondence with recent intertidal–shallow subtidal colonies of shitonemaceans. In addition, the presence of grazer gastropods (mainly nerineids and acteonids) in the associated subtidal facies and in small channel-like rills supports the hypothesis that large cyanobacterial covers had occurred. Upward in the succession, subtidal rudist-rich lithofacies tend to prevail.

The post-Turonian recovery of the central Tethyan carbonate factories shows very limited contributions of chlorozoan assemblage components. Very peculiar foramol depositional systems developed, inasmuch as biotic assemblages were largely dominated by rudists along with benthonic forams. Rudists occupied practically all of the shallow-water domains and produced up to 90% of the coarse skeletal debris, as well as, presumably, most of the fines. The lack of modern analogues makes understanding the paleoecology of the rudist-dominated assemblages difficult. Although abundantly fossiliferous, the related limestones are characterized by only moderately diverse biotic components. After the initial richness in cyanobacterial consortia, which seems to suggest persistently high pCO_2, alkalinity, and temperature (see Graziano, 2007; Graziano et al., 2007), the uppermost Turonian–Santonian limestone becomes dominated by rudists which usually grew in monospecific or oligospecific gatherings. The flora–fauna assemblage was paucispecific, although benthic foraminifers document normal marine waters. Based on the above, it is inferred that stressed conditions were still present in the ameliorated environmental contexts developing during the Late Cretaceous. Presumably, rudists were more tolerant than other shallow-water taxa in the latest Turonian–Santonian carbonate factories (see discussion below).

A general deepening trend characterizes the whole Upper Turonian–Lower Campanian successions. It is documented by the typology of the depositional facies (Carannante et al., 1999) as well as by the change from peritidal–shallow-subtidal cycles to dominantly subtidal cycles mentioned above (Ruberti, 1997). Unconformity surfaces generally terminate the Cretaceous sequences, but where present, the discontinuous outcrops of Upper Campanian–Maastrichtian limestone suggest a very shallow, paralic depositional setting (e.g., Reina and Luperto Sinni, 1993). In addition, early diagenesis features document an influx of fresh–schizohaline water. This regressive trend is also documented in the slope areas by the large amount of Late Campanian–Maastrichtian rudist-rich skeletal debrites, testifying to significant persistent off-bank transport in a generally prograding system (Graziano, 1994; Carannante et al., 1999).

DISCUSSION

The recovery of the analyzed post-Turonian Tethyan carbonate platforms (see Graziano et al., 2007) followed the Early–Middle Turonian time, which was characterized by hyper-greenhouse conditions (e.g., Clarke and Jenkyns, 1999). During this time the carbonate depositional systems underwent major modifications with deposition of microbial-related muddy deposits (with variable contributions of hippuritids and radiolitids) largely prevailing in the shallow-water domains. Environmental disruption and microbial bloom following OAE2 were seemingly related to high pCO_2 and elevated alkalinity of oceanic waters (Graziano, 2007), as similarly proposed by Immenhauser et al. (2005), Graziano (2007), and Graziano et al. (2006) in the case of the Early Toarcian and late Early Aptian (*Selli*) OAEs. The genetic link between elevated pCO_2 and alkalinity, and the diffusion of cyanobacteria in marine waters, is well established by recent microbiological studies (e.g., Paerl, 1999; Arp et al., 2003) and sedimentary analysis of the fossil record (Kempe, 1990).

Such biotic behaviors seem to match the pattern of the Late Cretaceous thermal and trophic evolution of the oceans, suggesting a cause-and-effect relationship. The demise of the rudist-rich (mostly caprinids), chlorozoan carbonate factory close to the Cenomanian–Turonian boundary (OAE2) is not a causal result (Philip and Airaud-Crumiere, 1991; Graziano et al., 2007).

The Late Turonian–Coniacian *p.p.* recovery of the Sardinia carbonate factories by chlorozoan assemblages suggests less extreme environmental conditions in a context of a moderate greenhouse. A subsequent shift toward foramol depositional systems occurred with rhodalgal assemblages flourishing in the newly established shallow-water domains during the post-Turonian. Relatively cool and mesotrophic up to eutrophic conditions became dominant in the water mass impinging on the marginal areas of the shelf or distal ramp of the Nurra carbonate system (see Carannante et al., 1995; Carannante et al., 1997, for details) in the Coniacian *p.p.*–Santonian times. This was a time interval in which only minor oceanic anoxic events occurred in the Tethyan Realm (OAE3 of Arthur et al., 1990), being restricted to small areas in epeiric and marginal peri-Tethyan domains (Bosellini et al., 1978; Jenkyns, 1980). This long-lived, ca. 2.7 My OAE was in fact typically recorded along the moderately shallow continental margins of the opening Atlantic Ocean (see Wagner et al., 2004) rather than in the narrowing Tethyan domain. Deteriorated water conditions presumably drove the demise of large areas of the Nurra carbonate factory, which underwent local drowning episodes (see Cherchi and Trémolières, 1984; Cherchi, 1985). Since the middle Campanian, ameliorated water conditions presumably allowed the reestablishment of carbonate production with significant supply of skeletal debris to the deeper neighboring areas.

As noted above, no reliable information exists regarding the limiting factors affecting rudist growth, because of the lack of a suitable modern analog. Although an obvious link should exist between these mollusc assemblages and the rapidly changing Cretaceous environmental conditions, it is difficult to compare in great detail the fluctuating proliferation of rudists as relevant lithogenetic taxa and the short-term oscillations of the environmental parameters. Graziano et al. (2007) observe that the increasing proliferation of rudists as relevant, up to almost exclusive, sediment-producing biota correlates to the long-term decline of pCO_2 and temperature during Late Cretaceous times after the peak reached in Early–Middle Turonian times. In addition, a cause-and-effect relationship can been postulated with a low Mg/Ca ratio, which was close to 1.0 during most of the Late Cretaceous. Presumably, rudists were more tolerant than other shallow-water taxa of the high pCO_2 associated with the early–middle Turonian hyper-greenhouse, because they became present in the poorly diversified, high pCO_2 and temperature–adapted biota assemblages dominated by microbial consortia (Fig. 5). In those times, stagnant ocean waters coupled with elevated pCO_2, temperature (26–30°C), and salinity (> 35‰) characterized the southern Mediterranean Tethys (Poulsen et al., 2001). The environmental conditions above possibly limited most of the biological requirements for both chlorozoan and foramol associations (Carannante et al., 1988; Reynaud et al., 2003) and presumably supported thriving microbial associations (e.g., Kazmierczak et al., 1996; Hallock, 2005). Green algae were only limited contributors in the Turonian shallow-water domains, whose salinity may have locally favored impoverished chloralgal assemblages (see Lees and Buller, 1972; Lees, 1975). In the following latest Turonian–Santonian times, the recovery and ensuing development in the central Tethyan Realm of rudist-dominated foramol carbonate factories could have been a result of the slight decline in pCO_2 and temperature in a general context which still presents anomalous values of pCO_2 and temperature and, presumably, still elevated salinity. As stated above, a clear deepening trend characterized the related successions although rare well-documented episodes of drowning have been recorded in the platform domains as in the case of the Late Santonian (CC17 zone) of the Apulia platform (Spalluto and Pieri, 2008). In the studied central Tethyan rudist-rich foramol carbonate systems, the relatively high production rates of skeletal particles through the bioerosion and breakdown of the ubiquitous rudist shells could partially compensate the increasing accommodation space during the Late Turonian–Early Campanian transgressive interval. The sustained bioclast production overcame the typical drowning tendency of the foramol open shelves; in spite of this, low aggradation and/or retrogradation occurred and local starvation episodes characterized the neighboring slopes and basins.

Unconformity surfaces generally terminate the Cretaceous sequences. However, discontinuous outcrops of Upper Campanian–Maastrichtian shallow-water limestone suggest a very shallow, paralic depositional setting. In addition, early-diagenesis features document influx of fresh–schizohaline water (Carannante et al., 1999; Simone et al., 2003).

Late Campanian–Maastrichtian rudist-rich skeletal debrites exhibit a clear regressive trend which was caused by significant off-bank transport episodes in an overall progradational system (Graziano, 1994; Carannante et al., 1997).

SUMMARY AND CONCLUSIONS

Cretaceous carbonate platforms, as reported by Simo et al. (1993), differ widely from recent platforms as well as from those of many other stratigraphic intervals. The tendency of rudist bivalves to become dominant, proliferating into all the shallow-water environments, strongly characterized the Late Cretaceous carbonate factories. The inception, evolution, and demise of these latter were seemingly controlled by a complex interplay of processes that, acting on a global scale, profoundly changed the previous conditions of the hydrosphere–atmosphere system. Such changes were quite effective in controlling the development of new depositional systems, which were forced to modify their organization, internal architecture, and facies patterns. Looking at the major global-scale variations in the chemical and physical parameters of the Cretaceous hydrosphere–atmosphere system, it is easy to infer that middle and early Late Cretaceous carbonate factories faced oceanic waters whose environmental properties exerted a strong impact on the biological assemblages. In particular, the Early–Middle Turonian seems to represent a time interval in which hyper-greenhouse conditions prevailed on a global scale, forcing anomalous environmental conditions (see discussion in Graziano, 2007, and Graziano et al., 2007). As a consequence, the carbonate depositional systems underwent a major shift to microbial-related

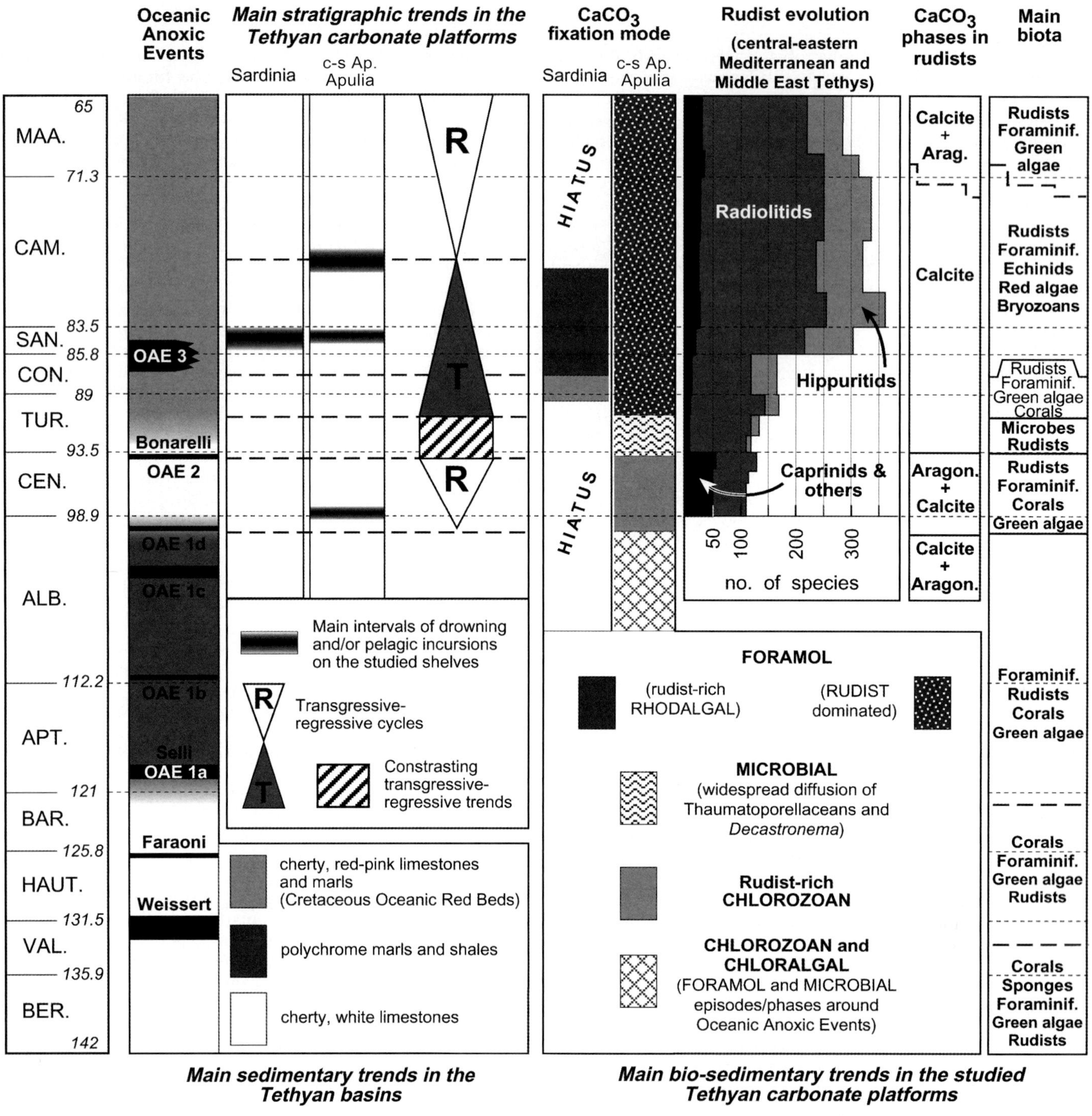

FIG 5.—Synthesis of the evolving middle–Late Cretaceous shallow- and deep-water carbonate depositional systems in the Tethyan realm (modified after Graziano et al., 2007). From left to right: chronostratigraphy of the oceanic anoxic events (after Arthur and Sageman, 1994; Coccioni et al., 1998; Erba, 2004; and Wagner et al., 2004) and evolving sedimentary trend in the Tethyan basins; main stratigraphic trends in Sardinia (northern Tethys) and central-southern Apennines–Apulia (central Tethys) carbonate factories; carbonate production systems (modified after Carannante et al., 1995; Carannante et al., 1997; Carannante et al., 1999); species richness of Late Cretaceous rudists in part of the Tethyan realm (modified after Steuber, 2000); mineralogy of the rudist shells and composition of the main shallow-water biotic assemblages.

muddy deposits (and variable contributions of rudists) in the shallow-water domains. According to this interpretation, the recovery of the post-Turonian carbonate platforms in the Tethyan seaway closely reflects profoundly changing hydrosphere–atmosphere conditions. Since the Late Turonian a cooling phase coupled with ameliorated environmental conditions (decreasing alkalinity, temperature, and, locally, salinity) allowed the rudists to recover rapidly, becoming predominant among the carbonate-secreting biota in the carbonate factories. Net latitudinal thermal gradients and current paths in the Tethys as well as the effects of cooler and/or nutrient-rich waters related to terrestrial runoff allowed rudists to share their lithogenetic role with bryozoans and red algae in the foramol (rhodalgal) factories of the northern Tethyan margin starting in the Late Coniacian. During the Late Turonian–Late Coniacian transgressive event, these northern Tethyan carbonate factories shifted from chlorozoan aragonite-rich lithofacies to rhodalgal (foramol *sensu lato*), calcite-dominated lithofacies. Tolerant mesotrophic communities prevailed and opportunistic planktonic forams thriving in the upper water column characterized the coeval hemipelagic deposits. Rudists, even if abundant, did not reach the total dominance as in the central Tethyan domains, nor is there evidence of microbial laminites or grazer molluscs. According to the foramol–rhodalgal depositional model, this resulted in a reduced aggrading potential and, as a consequence, a tendency to drown (see Simone and Carannante, 1988), as in the case of the Santonian peripheral areas of the Nurra carbonate platform (Carannante et al., 1995). By contrast, coralline red algae and bryozoans were subordinate sediment contributors only in the outermost sectors of the Apennine–Apulia open shelves (central Tethyan domain). In these carbonate platforms very peculiar foramol assemblages may be related to the ubiquitous dominance of the rudists. In spite of the ameliorated environmental conditions, water temperature and, presumably, salinity were still sustained in post-Turonian times. Significantly, microbial consortia and the related grazer communities still occur in the Coniacian–Santonian inner-shelf successions. Rudists furnished nearly all of the skeletal grains as well as the silt-size fraction via extensive bioerosion. The high production rates of sediments provided by these molluscs resulted in a more abundant availability of skeletal sands, which presumably reduced the tendency of the Late Cretaceous foramol open shelves to drown. Only limited, tectonically controlled peripheral areas underwent a drowning tendency during episodes of slope underfeeding.

In the Late Campanian–Maastrichtian times a widening and prograding trend occurred in the studied Tethyan factories with flourishing of rudist-rich assemblages. A new, though limited, contribution by hermatypic corals and green algae as well as aragonitic rudists has been detected. Large amounts of skeletal debris were exported from the resulting open shelves to deeper areas. According to the physiographic features of the inherited depositional settings along with the nature of their controlling mechanism (i.e., synsedimentary tectonics versus sea-level changes), prograding wedges on distal ramp areas or deep fans could develop.

In conclusion, the post-Turonian rudist-rich carbonate platforms can be viewed as geological products which have closely recorded the global changes of environmental conditions within the oceanic realm. This is consistent with the fluctuations in the physical and chemical state of the oceanic waters at global scale during middle–Late Cretaceous times. Nevertheless a clear partition of the carbonate fixation modes can be detected from northern to central Tethyan domains (rhodalgal versus rudist-rich foramol, respectively), presumably reflecting regional-scale variations of the overall oceanographic and climatic conditions.

ACKNOWLEDGMENTS

Constructive comments by reviewers Jeff Pietras and Peter Drzewiecki are gratefully acknowledged. We also thank John Southard and Bob Clarke for their careful editorial suggestions. This research was supported by a PRIN grant (2002) to Professor Lucia Simone.

REFERENCES

ACCORDI, G., CARBONE, F., AND SIRNA, G., 1982, Relationship among tectonic setting, substratum and benthonic communities in the Upper Cretaceous of north-eastern Matese (Molise, Italy): Geologica Romana, v. 21, p. 755–793.

ACCORDI, G., CARBONE, F., AND SIRNA, G., 1986, Some affinities between the Ionian Islands and the Apulian Upper Cretaceous rudist facies: Società Geologica Italiana, Memorie, v. 32, p. 5–48.

ARP, G., REIMER, A., AND REITNER, J., 2003, Microbialite formation in seawater of increased alkalinity, Satonda Crater Lake, Indonesia: Journal of Sedimentary Research, v. 73, p. 105–127.

ARTHUR, M.A., AND SAGEMAN, B.B., 1994, Marine black shales: depositional mechanisms and environments of ancient deposits: Annual Review of Earth and Planetary Sciences, v. 22, p. 499–551.

ARTHUR, M.A., JENKYNS, H.C., BRUMSACK, H.-J., AND SCHLANGER, S.O., 1990, Stratigraphy, geochemistry and paleoceanography of organic carbon-rich Cretaceous sequences, *in* Ginsburg, R.N., and Beaudoin, B., eds., Cretaceous Resources, Events and Rhythms: Dordrecht, The Netherlands, Kluwer, p. 75–119.

BORGOMANO, J., 2000, The Upper Cretaceous carbonates of the Gargano–Murge region, southern Italy: a model of platform-to-basin transition: American Association of Petroleum Geologists, Bulletin, v. 84, p. 1561–1588.

BOSELLINI, A., AND PARENTE, M., 1994, The Apulia platform margin in the Salento Peninsula (southern Italy): Giornale di Geologia, v. 56, p. 167–177.

BOSELLINI, A., BROGLIO LORIGA, C., AND BUSETTO, C., 1978, I bacini cretacei del Trentino: Rivista Italiana di Paleontologia, v. 84, p. 897–946.

BOSELLINI, A., NERI, C., AND LUCIANI, V., 1993, Guida ai carbonati cretaceo–eocenici di scarpata e di bacino del Gargano (Italia meridionale): Università di Ferrara, Annali (Nuova Serie), Sezione Scienze della Terra, v. 4, p. 1–81.

BRACHERT, T.C., BETZLER, C., BRAGA, J.C., AND MARTIN, J.M., 1996, Record of climatic change in neritic carbonates: turnover in biogenic associations and depositional modes (Late Miocene, southern Spain): Geologische Rundschau, v. 85, p. 327–337

CARANNANTE, G., AND SIMONE, L., 1987, "Temperate" versus "tropical" Cretaceous carbonate platforms in Italy: Società Geologica Italiana, Rendiconti, v. 9, p. 153–156.

CARANNANTE, G., AND SIMONE, L., 1988, Foramol carbonate shelves as depositional site and source area: recent and ancient examples from the Mediterranean: American Association of Petroleum Geologists, Bulletin, v. 72, p. 993–994.

CARANNANTE, G., ESTEBAN, M., MILLIMAN, J.D., AND SIMONE, L., 1988, Carbonate lithofacies as paleolatitude indicators: problems and limitations: Sedimentary Geology, v. 60, p. 333–346.

CARANNANTE, G., RUBERTI, D., AND SIMONE, L., 1993, Rudists and related sediments in Late Cretaceous open shelf settings. A case history from Matese Area (Central-Southern Apennines, Italy): Giornale di Geologia, v. 55, p. 21–36.

CARANNANTE, G., D'ARGENIO, B., MINDSZENTY, A., RUBERTI, D., AND SIMONE, L., 1994, Cretaceous–Miocene shallow water carbonate sequences: Regional unconformities and facies patterns, *in* Carannante, G., and Tonielli, R., eds., International Association of Sedimentologists, 15th Regional Meeting, Ischia, Italy, Pre-Meeting Field Trip Guidebook, p. 27–60.

CARANNANTE, G., CHERCHI, A., AND SIMONE, L., 1995, Chlorozoan versus foramol lithofacies in Upper Cretaceous rudist limestones: Palaeogeography, Palaeoclimatology, Palaeoecology, v. 119, p. 137–154.

CARANNANTE, G., SEVERI, C., AND SIMONE, L., 1996, Off-shelf carbonate transport along foramol (temperate type) open shelf margins: an example from the Miocene of the central-southern Apennines (Italy), *in* Bourrouilh–Le Jan, G., ed., Carbonates Intertropicaux: Société Géologique de France, Mèmoires, v. 169, p. 277–288.

CARANNANTE, G., GRAZIANO, R., RUBERTI, D., AND SIMONE, L., 1997, Upper Cretaceous temperate-type open shelves from northern (Sardinia) and southern (Apennines–Apulia) Mesozoic Tethyan Margins, *in* James, N.P., and Clarke, J.D.A., eds., Cool-Water Carbonates: SEPM, Special Publication 56, p. 309–325.

CARANNANTE, G., RUBERTI, D., AND SIRNA, G., 1998, Senonian rudist limestones in the Sorrento Peninsula sequences (Southern Italy): Géobios, Memoire Speciale, v. 22, p. 47–68.

CARANNANTE, G., GRAZIANO, R., PAPPONE, G., RUBERTI, D., AND SIMONE, L., 1999, Depositional system and response to sea-level oscillations of the Senonian rudist-bearing carbonate shelves. Examples from Central Mediterranean areas: Facies, v. 40, p. 1–24.

CARANNANTE, G., RUBERTI D., AND SIRNA, M., 2000, Upper Cretaceous ramp limestones from the Sorrento Peninsula (southern Apennines, Italy): micro- and macrofossil associations and their significance in the depositional sequences: Sedimentary Geology, v. 132, p. 89–123.

CARANNANTE, G., RUBERTI, D., AND SIMONE, L., 2003, Sedimentological and taphonomic characterization of low energy rudist-dominated Senonian carbonate shelves (Southern Apennines, Italy), *in* Gili, E., Negra, M.H., and Skelton, P.W., eds., North African Cretaceous Carbonate Platform Systems: NATO Sciences Series, Earth and Environmental Sciences, Kluwer Academic Press, v. 28, p. 189–201.

CARBONE, F., 1993, Cretaceous depositional systems of the evolving Mesozoic carbonate platform of central Apennine thrust belt, Italy: Geologica Romana, v. 29, p. 31–53.

CARBONE, F., AND SIRNA, G., 1981, Upper Cretaceous reef models from Rocca di Cave and adjacent areas in Latium, Central Italy, *in* Toomey, D.F., ed., European Fossil Reef Models: SEPM, Special Publication 30, p. 427–445.

CESTARI, R., AND SIRNA, G., 1987, Rudist fauna in the Maastrichtian deposits of southern Salento (Southern Italy): Società Geologica Italiana, Memorie, v. 40, p. 133–147.

CESTARI, R., AND SARTORIO, D., 1995, Rudists and facies of the Periadriatic Domain: Milano, AGIP Spa, 207 p.

CESTARI, R., REALI, S., AND SIRNA, M., 1992, Biostratigraphical characteristics of the Turonian–? Maastrichtian pp. (Upper Cretaceous) deposits in the Simbruini–Ernici Mts. (Central Apennines, Italy): Geologica Romana, v. 28, p. 359–372.

CHERCHI, A., ed., 1985, 19th European Micropaleontological Colloquium, Sardinia, October 1–10, 1985, 338 p.

CHERCHI, A., AND SCHROEDER, R., 1987, Biostratigraphie du Crétacé de la Nurra: Groupe Française du Crétacé, Excursion en Sardaigne, 24–29 May 1987, p. 26–60.

CHERCHI, A., AND SCHROEDER, R., 2005, Calcimicrobial oncoid coatings from the Pliensbachian Massone Member (Calcari Grigi Formation, Trento Platform, Italy), *in* Fugagnoli, A., and Bassi, D., eds., Giornata di Studi Paleontologici "Prof. C. Loriga Broglio". Ferrara, June 18th, 2004: Università di Ferrara, Annali, Sezione Museologia Scientifica e Naturalistica, Special Volume, p. 45–49.

CHERCHI, A., AND TRÉMOLIÈRES, P., 1984, Nouvelle données sur l'évolution structurale au Mésozoique et au Cénozoique de la Sardaigne et leur implications géodynamiques dans le cadre méditerranéen: Académie des Sciences, (Paris), Comptes Rendus, v. 298, p. 889–894.

CHIOCCHINI, M., FARINACCI, A., MANCINELLI, A., MOLINARI, V., AND POTETTI, M., 1994. Biostratigrafia a Foraminiferi, Dasicladali e Calpionelle delle successioni carbonatiche mesozoiche dell'Appennino centrale (Italia): Studi Geologici Camerti, Volume Speciale, "Biostratigrafia dell'Italia Centrale", p. 9–129.

CLARKE, L.J., AND JENKYNS, H.C., 1999, New oxygen isotope evidence for long-term Cretaceous climatic change in the southern hemisphere: Geology, v. 27, p. 699–702.

COCCIONI, R., BAUDIN, F., CECCA, F., CHIARI, M., GALEOTTI, S., GARDIN, S., AND SALVINI, G., 1998, Integrated stratigraphic, palaentological and geochemical analysis of the uppermost Hauterivian Faraoni Level in the Fiume Bosso section, Umbria–Marche Apennines, Italy: Cretaceous Research, v. 19, p. 1–23.

CRESCENTI, U., AND VIGHI, L., 1964, Caratteristiche, genesi e stratigrafia dei depositi bauxitici cretacici del Gargano e delle Murge: cenni sulle argille con pisoliti bauxitiche del Salento (Puglie): Società Geologia Italiana, Bollettino, v. 83, p. 285–338.

D'ARGENIO, B., AND MINDSZENTY, A., 1995, Bauxites and related paleokarst: tectonic and climatic event markers at regional unconformities: Eclogae Geologicae Helvetiae, v. 88, p. 453–499.

D'ARGENIO, B., SIMONE L., AND CARANNANTE, G., 1985, Sedimentological evolution of Jurassic and Cretaceous carbonates, *in* Cherchi, A., ed., 19th European Micropaleontological Colloquium, Sardinia, October 1–10, 1985, p. 57–65.

DE CASTRO, P., 1990, Thaumatoporelle: conoscenze attuali e approccio all'interpretazione: della Società Paleontologica Italiana, Bollettino, v. 29, p. 179–206.

DE GRACIANSKY, P.C., DEROO, G., HERBIN, J.P., JAQUIN, T., MAGNIEZ, F., MONTADERT, L., PONSOT, C., SCHAAF, A., AND SIGAL, J., 1986, Ocean-wide stagnation episodes in the Late Cretaceous: Geologische Rundschau, v. 75, p. 17–41.

DIENI, I., AND MASSARI, F., 1985, Mesozoic of eastern Sardinia, *in* Cherchi, A., ed., 19th European Micropaleontological Colloquium, Sardinia, October 1–10, 1985, p. 66–78.

ERBA, E., 2004, Calcareous nannofossils and Mesozoic oceanic anoxic events: Marine Micropaleontolgy, v. 52, p. 85–106.

FRAKES, L.A., 1999, Estimating the global thermal state from Cretaceous sea surface and continental temperature data, *in* Barrera, E., and Johnson, C.C., eds., Evolution of the Cretaceous Ocean–Climate system: Geological Society of America, Special Paper 332, p. 49–57.

GILI, E., 1993, Facies and geometry of Les Collades de Basturs carbonate platform, Upper Cretaceous, South-central Pyrenèes, *in* Simo, J.A., Scott, R.W., and Masse, J.P., eds., Cretaceous Carbonate Platforms: American Association of Petroleum Geologists, Memoir 56, p. 343–352.

GILI, E., MASSE, J.P., AND SKELTON, P.W., 1995a, Rudists as gregarious sediment-dwellers, not reef-builders, on Cretaceous carbonate platforms: Palaeogeography, Palaeoclimatology, Palaeoecology, v. 118, p. 245–267.

GILI, E., SKELTON, P.W., VICENS, E., AND OBRADOR, A., 1995b, Coral to rudists—an environmentally induced assemblage succession: Palaeogeography, Palaeoclimatology, Palaeoecology, v. 119, p. 127–136.

GILI, E., EL HÉDI NEGRA, M., AND SKELTON, P., eds., 2003, North African Cretaceous Carbonate Platforms Systems: Earth and Environmental Sciences, NATO Science Series 28, Kluwer, 252 p.

GOLUBIC, S., RADOICIC, R., AND LEE, S.-J., 2006, *Decastronema kotori* gen. nov., comb. nov.: a mat-forming cyanobacterium on Cretaceous carbonate platforms and its modern counterpart: Carnets de Géologie / Notebooks on Geology, Brest, Article 2006 / 2, (CG2006_A02).

GRAZIANO, R., 1994, Evoluzione cretacica del sistema "Plataforma carbonatica Apula / Bacino Est-Garganico" nel Promontorio del Gargano. Sedimentologia e stratigrafia sequenziale: Unpublished Ph.D. Thesis, University of Napoli "Federico II", 253 p.

GRAZIANO, R., 2007, Cyanobacteria blooms and drowning unconformities of carbonate platforms: signs of Earth's endogenic, global control on the productivity of carbonate depositional systems; Examples from the Jurassic–Cretaceous of the mediterranean Tethys, *in* Herrle, J., Erba, E., and Weissert, H., eds., Understanding the l
Linkages of Geosphere and Biosphere Evolution During Cenozoic and Mesozoic Times: European Geosciences Union General Assembly, Vienna, Austria, 15–20 April 2007, Geophysical Research Abstract, v. 9, A09465.

GRAZIANO, R., BUONO, G., AND TADDEI RUGGIERO, E., 2006, Lower Toarcian (Jurassic) brachiopod-rich carbonate facies of the Gran Sasso range (central Apennines, Italy): Società Paleontologica Italiana, Bollettino, v. 45, p. 61–74.

Graziano, R., Carannante, G., and Simone, L., 2007, The inception and evolution of the Late Cretaceous rudist bearing carbonate platforms in the Mediterranean Tethys: mirror of geodynamically induced biosphere–geosphere interactions, *in* Mutti, M., and Samankassou, E., eds., Paleo-Environmental Indicators in Carbonate Systems: European Geosciences Union General Assembly, Vienna, Austria, 15–20 April 2007: Geophysical Research Abstracts, v. 9, A08010.

Gusic, I., and Jelaska, V., 1990, Upper Cretaceous stratigraphy of the island of Brac within the geodynamic evolution of the Adriatic carbonate platform: Jugoslavenska Akademija Znanosti i Umjetnosti, Institut za Geoloska Istrazivanja, OOUR za Geologiju, 160 p.

Halfar, H., Godinez-Orta, L., Mutti, M., Valdez-Holguin, J.E., and Borges, J.M., 2004, Nutrients and temperature controls on modern carbonate production: an example from the Gulf of California, Mexico: Geology, v. 32, p. 213–216.

Hallock, P., 1985, Why are larger foraminifera large?: Paleobiology, v. 11, p. 195–208.

Hallock, P., 2005, Global change and modern coral reefs: New opportunities to understand shallow-water carbonate depositional processes: Sedimentary Geology, v. 175, p. 19–33.

Hallock, P., Premoli Silva, I., and Boersma, A., 1991, Similarities between planktonic and larger foraminiferal evolutionary trends through Paleogene paleoceanographic changes: Palaeogeography, Palaeoclimatology, Palaeoecology, v. 83, p. 49–64.

Hansen, K.W., and Wallmann, K., 2003, Cretaceous and Cenozoic evolution of seawater composition, atmospheric O_2 and CO_2: a model perspective: American Journal of Science, v. 303, p. 94–148.

Harries, P.J., and Little, C.T.S., 1999, The early Toarcian (Early Jurassic) and the Cenomanian–Turonian (Late Cretaceous) mass extinctions: similarities and contrasts: Palaeogeography, Palaeoclimatology, Palaeoecology, v. 154, p. 39–66.

Henriksson, A.S., and Malmgren, B.A., 1997, Biogeographic and ecologic patterns in calcareous nannoplankton in the Atlantic and Pacific Oceans during the terminal Cretaceous: Salamanca, Studia Geologica Salmanticensia, v. 33, p. 17–40.

Immenhauser, A., Hillgartner, H., and Van Bentum, E., 2005, Microbial–foraminiferal episodes in the Early Aptian of the southern Tethyan margin: ecological significance and possible relation to oceanic anoxic event 1a: Sedimentology, v. 52, p. 77–99.

James, N.P., 1997, The cool-water carbonate depositional realm, *in* James, N.P., and Clarke, J.A.D., eds., Cool-Water Carbonates: SEPM, Special Publication 56, p. 1–20.

Jenkyns, H.C., 1980, Cretaceous anoxic events: from continents to oceans: Geological Society of London, Journal, v. 137, p. 171–188.

Kazmierczak, J., Coleman, M.L., Gruszczynski, M., and Kempe, S., 1996, Cyanobacterial key to the genesis of micritic and peloidal limestones in ancient seas: Acta Paleontologica Polonica, v. 41, p. 319–338.

Kempe, S., 1990, Alkalinity: the link between anaerobic basins and shallow water carbonates?: Naturwissenschaften, v. 77, p. 426–427.

Kerr, A.C., 1998, Oceanic plateau formation: a cause of mass extinction and black shale deposition around the Cenomanian–Turonian boundary?: Geological Society of London, Journal, v. 155, p. 619–626.

Kidwell, S.M., 1991, The stratigraphy of shell concentration, *in* Allison, P.A., and Briggs, D.E.G., eds., Taphonomy: Releasing the Data Locked in the Fossil Record: New York, Plenum Press, Topics in Geobiology, 9, p. 211–290.

Kidwell, S.M., and Holland, S.M., 1991, Field description of coarse bioclastic fabrics: Palaios, v. 6, p. 426–434.

Kidwell, S.M., Fürsich, F.T., and Aigner, T., 1986, Conceptual framework for the analysis and classification of fossil concentrations: Palaios, v. 1, p. 228–238.

Larson, R.L., and Erba, E., 1999, Onset of the mid-Cretaceous greenhouse in the Barremian–Aptian: igneous events and the biological, sedimentary, and geochemical responses: Paleoceanography, v. 14, p. 663–678.

Laviano, A., 1984, Preliminary observations on the Upper Cretaceous coral–rudist facies of Ostuni (south-eastern Murge, Apulia): Rivista Italiana di Paleontologia e Stratigrafia, v. 90, p. 177–204.

Lees, A., 1975, Possible influence of salinity and temperature on modern shelf carbonate sedimentation: Marine Geology, v. 19, p. 159–198.

Lees, A., and Buller, A.T., 1972, Modern temperate-water and warm-water shelf carbonate sediments contrasted: Marine Geology, v. 13, p. 67–73.

Luperto Sinni, E., and Borgomano, J., 1989, Le Crétacé supérieur des Murges sud-orientales (Italie Méridionale): stratigraphie et evolution des paléoenvironnements: Rivista Italiana di Paleontologia e Stratigrafia, v. 95, p. 95–136.

Luperto Sinni, E., and Borgomano, J., 1994, Stratigrafia del Cretaceo superiore in facies di scarpata di Monte S. Angelo (Promontorio del Gargano, Italia meridionale): Società Geologica Italiana, Bollettino, v. 113, p. 355–382.

Luperto Sinni, E., and Reina, A., 1996, Nuovi dati stratigrafici sulla discontinuità mesocretacea delle Murge (Puglia, Italia meridonale): Società Geologica Italiana, Memorie, v. 51, p. 1179–1188.

Mameli, P., Mongelli, G., Oggiano, G., and Dinelli, E., 2007, Geological, geochemical and mineralogical features of some bauxite deposits from Nurra (Western Sardinia, Italy): insights on conditions of formation and parental affinity: International Journal of Earth Sciences, v. 96, p. 887–902.

Masse, J.P., 1989, Relations entre modifications biologiques et phénoménes géologiques sur les plates-formes carbonatées du domaine périméditerranéen au passage Bédulien–Gargasien: Géobios, Mémoire Spécial 11, p. 279–294.

Moro, A., 1997, Stratigraphy and paleoenvironments of rudists biostromes in the Upper Cretaceous (Turonian–upper Santonian) limestones of southern Istria, Croatia: Palaeogeography, Palaeoclimatology, Palaeoecology, v. 131, p. 113–131.

Mutti, M., and Hallock, P., 2003, Carbonate systems along nutrient and temperature gradients: a review of sedimentological and geochemical constraints: International Journal of Earth Sciences, v. 92, p. 465–475.

Negra, M.H., Purser, B.H., and M'Rabet, A., 1995, Sedimentation, diagenesis and syntectonic erosion of Upper Cretaceous rudists mounds in central Tunisia, *in* Monty, C.L.V., Bosence, D.W.J., Bridges, P.H., and Pratt, B.R., eds., Carbonate Mud-Mounds; Their Origin and Evolution: International Association of Sedimentologists, Special Publication 23, p. 401-419.

Paerl, H.W., 1999, Physical–chemical constraints on cyanobacterial growth in the Oceans: Institute Océanographique, Monaco, Bulletin, Numero Spécial 19, p. 319–349.

Parente, M., 1994, A revised stratigraphy of the Upper Cretaceous to Oligocene units from southeast Salento (Apulia, southern Italy): Società Geologica Italiana, Bollettino, v. 33, p. 155–170.

Philip, J., 1982, Paleobiogeographie des Rudistes et geodinamique des marges mesogèennes au Crètacè supèrieur: Sociètè Geologique de France, Bulletin, v. 24, p. 995–1006.

Philip, J., 1993, Late Cretaceous carbonate–siliciclastic platforms of Provence, Southeastern France, *in* Simo, T., Scott, R.W., and Masse, J.P., eds., Cretaceous Carbonate Platforms: American Association of Petroleum Geologists, Memoir 56, p. 375–385.

Philip, J., and Airaud-Crumiere, C., 1991, The demise of rudist-bearing carbonate platforms at the Cenomanian–Turonian boundary: a global control: Coral Reefs, v. 10, p. 115–125.

Philip, J., and Gari, J., 2005, Late Cretaceous heterozoan carbonates: palaeoenvironmental setting, relationships with rudist carbonates (Provence, south-east France): Sedimentary Geology, v. 175, p. 315–337.

Philip, J., Floquet, M., Platel, J.P., Bergerat, F., Sandulescu, M., Baraboshkin, E., Amon, E.O., Guiraud, R., Vaslet, D., Le Nindre, Y., Ziegler, M., Poisson, A., and Bouaziz, S., 2000, Map 14—Late Cenomanian (94.7 to 93.5 Ma), *in* Dercourt, J., Gaetani, M., Vrielynck, B., Barrier, E., Biju-

Duval, B., Brunet, M.F., Cadet, J.P., Crasquin, S., and Sandulescu, M., eds., Atlas Peri-Tethys: Palaeogeographical Maps, CCGM/CGMW, Paris, p. 153–178.

Pomar, L., Gili, E., Obrador, A., and Ward, W., 2005, Facies architecture and high-resolution sequence stratigraphy of an Upper Cretaceous platform margin succession, southern central Pyrenees, Spain: Sedimentary Geology, v. 175, p. 339–365.

Pomoni-Papaioannou, F., and Photiades, A., 2008, Carbonate sedimentary diversification in a Cretaceous Pelagonian margin: Rhodiani area (west Macedonia, Greece): Società Geologica Italiana, Bollettino, in press.

Poulsen, C.J., Barron, E.J., Arthur, M.A., and Peterson, W.H., 2001, Response of the mid-Cretaceous global oceanic circulation to tectonic and CO_2 forcings: Palaeoceanography, v. 16, p. 576–592.

Premoli Silva, I., and Sliter, W.V., 1999, Cretaceous paleoceanography: evidence from planktonic foraminiferal evolution, *in* Barrera, E. and Johnson, C.C., eds., Evolution of the Cretaceous Ocean–Climate System: Geological Society of America, Special Paper 332, p. 301–328.

Reina, A., and Luperto Sinni, E., 1993, Depositi maastrichtiani di piattaforma carbonatica interna affioranti nell'area delle murge baresi (Puglia, Italia meridionale): Società Geologica Italiana, Bollettino, v. 112, p. 837–844.

Reynaud, S., Leclerq, N., Romaine-Lioud, S., Ferrier-Pages, C., Jaubert, J., and Gattuso, J.-P., 2003, Interacting effects of CO_2 partial pressure and temperature on photosynthesis and calcification in a scleractinian coral: Global Change Biology, v. 9, p. 1660–1668.

Ross, D.J., 1991, Botryoidal high-magnesium calcite marine cements from the Upper Cretaceous of the Mediterranean region: Journal of Sedimentary Petrology, v. 61, p. 349–353.

Ross, D.J., and Skelton, P.W., 1993, Rudist formations of the Cretaceous: a palaeoecological, sedimentological and stratigraphical review, *in* Wright, V.P., ed., Sedimentology Review, v. 1, p. 73-91, Oxford, U.K., Blackwell Scientific Publications.

Ruberti, D., 1993, Late Cretaceous carbonate shelf-to-slope facies in the central-western Matese (central Apennines, Italy): Giornale di Geologia, serie 3, v. 55, p. 117–129.

Ruberti, D., 1997, Facies analysis of an Upper Cretaceous high-energy rudist-dominated carbonate ramp (Matese Mountains, central-southern Italy): subtidal and peritidal cycles: Sedimentary Geology, v. 113, p. 81–110.

Ruberti, D., Carannante, G., Simone, L., and Toscano, F., 2006, Rudist lithosomes related to current pathways in Upper Cretaceous, temperate-type, inner shelves: a case study from the Cilento area, southern Italy, *in* Pedley, H.M., and Carannante, G., eds., Cool-Water Carbonates: Depositional Systems and Palaeoenvironmental Controls: Geological Society of London, Special Publication 255, p. 181–197.

Ruberti, D., Carannante, G., Simone, L., Sirna, G., and Sirna, M., 2007, Sedimentary processes and biofacies of Late Cretaceous carbonate low energy ramp systems (Southern Italy), *in* Scott, R.W., ed., Cretaceous Rudists and Carbonate Platforms: Environmental Feedback: SEPM, Special Publication 87, p. 57–70.

Sanders, D.G.K., 1996, Rudist biostromes on the margin of an isolated carbonate platform: the Upper Cretaceous of Montagna della Maiella, Italy: Eclogae Geologicae Helvetiae, v. 89, p. 845–871.

Schlager, W., and Philip, J., 1990, Cretaceous carbonate platforms, *in* Ginsburg, R.N., and Beaudoin, B., eds., Cretaceous Resources, Events and Rhythms: Dordrecht, The Netherlands, Kluwer, p. 173–195.

Shanmugan, G., 2000, 50 years of the turbidite paradigm (1950s–1990s): deep-water processes and facies models—a critical perspective: Marine and Petroleum Geology, v. 17, p. 285–342

Simo, J.A., Scott, R.W., and Masse, J.P., 1993, Cretaceous carbonate platform: an overview, *in* Simo, J.A., Scott R.W., and Masse, J.P., eds., Cretaceous Carbonate Platforms: American Association of Petroleum Geologists, Memoir 56, p. 1–14.

Simone, L., and Carannante, G., 1988, The fate of foramol ("temperate-type") carbonate platforms: Sedimentary Geology, v. 60, p. 347–354.

Simone, L., Carannante, G., Ruberti, D., Sirna, M., Sirna, G., Laviano, A., and Tropeano, M., 2003, Development of rudist lithosomes in the Coniacian–Lower Campanian carbonate shelves of central-southern Italy: high energy vs low-energy settings: Palaeogeography, Palaeoclimatology, Palaeoecology, v. 200, p. 5–29.

Skelton, P., 2003a, Rudist evolution and extinction: a North African perspective, *in* Gili, E., Negra, E.H.M., and Skelton, P., eds., North African Cretaceous Carbonate Platform Systems: Earth and Environmental Sciences, NATO Science Series, v. 28, p. 215–227, Kluwer.

Skelton, P., ed., 2003b, The Cretaceous World: The Open University, Milton Keynes, and Cambridge University Press, Cambridge, U.K., 360 p.

Skelton, P., Gili, E., Vicens, E., and Obrador, A., 1995, The growth fabric of gregarious rudist elevator (hippuritids) in a Santonian carbonate platform in the southern Central Pyrenees: Palaeogeography, Palaeoclimatology, Palaeoecology, v. 119, p. 107–126.

Spalluto, L. and Pieri, P., in press, Rilevamento geologico delle unità carbonatiche mesozoiche e cenozoiche affioranti nel Gargano sud-occidentale: nuovo quadro stratigrafico e nuovi vincoli stratigrafici per l'evoluzione tettonica dell'area: Memorie Descrittive della Carta Geologica d'Italia, Roma.

Steuber, T., 2000, Skeletal growth rates of Upper Cretaceous rudist bivalves: implications for carbonate production and organism–environment feedbacks, *in* Insalaco, E., Skelton, P.W., and Palmer, T.J., eds., Carbonate Platform Systems; Components and Interactions: Geological Society of London, Special Publication 178, p. 21–31.

Steuber, T., 2002, Plate tectonic control on the evolution of Cretaceous platform carbonate production: Geology, v. 30, p. 259–262.

Stössel, I., and Bernoulli, D., 2000, Rudist lithosome development on the Maiella Carbonate Platform margin, *in* Insalaco, E., Skelton, P.W., and Palmer, T.J., eds., Carbonate Platform Systems; Components and Interactions: Geological Society of London, Special Publication 178, p. 177–190.

Van de Poel, H.M., and Schlager, W., 1994: Variations in Mesozoic–Cenozoic skeletal carbonate mineralogy: Geologie en Mijnbouw, v. 73, p. 31–51.

Veizer, J., Ala, D., Azmy, K., Bruckschen, P., Buhl, D., Bruhn, F., Carden, G.A.F., Diener, A., Ebneth, S., Godderis, Y., Jasper, T., Korte, C., Pawellek, F., Podlaha, O.G., and Strauss, H., 1999, $^{87}Sr/^{86}Sr$, $\delta^{13}C$, $\delta^{18}O$ evolution of Phanerozoic seawater: Chemical Geology, v. 161, p. 59–88.

Vlahovic, I., Tisljar, J., Velic, I., and Maticec, D., 2005, Evolution of the Adriatic carbonate platform: palaeogeography, main events and depositional dynamics: Palaeogeography, Palaeoclimatology, Palaeoecology, v. 220, p. 333–360.

Vogt, P.R., 1989, Volcanogenic upwelling of anoxic, nutrient rich water: possible factor in carbonate-bank/reef demise and benthic faunal extinction?: Geological Society of America, Bulletin, v. 101, p. 1225–1245.

Wagner, T., Sinninghe Damsté, J.S., Hofmann, P., and Beckmann, B., 2004, Euxinia and primary production in Late Cretaceous eastern equatorial Atlantic surface waters fostered orbitally driven formation of marine black shales: Paleoceanography, v. 19, PA3009, doi:10.1029/2003PA000898.

NUTRIENT-GRADIENT CONTROLS ON DEVONIAN REEFS: INSIGHT FROM THE RAMP-SITUATED ALEXANDRA REEF SYSTEM (FRASNIAN), NORTHWEST TERRITORIES, CANADA

ALEX J. MacNEIL
Department of Earth and Atmospheric Sciences, University of Alberta, Edmonton, Alberta T6G 2E3, Canada
Present address: Imperial Oil Resources, Ltd., 237 4th Ave. SW, Calgary, Alberta T2P 3M9, Canada
e-mail: alex.j.macneil@esso.ca

AND

BRIAN JONES
Department of Earth and Atmospheric Sciences, University of Alberta, Edmonton, Alberta T6G 2E3, Canada

ABSTRACT: Devonian reef systems are thought to represent the greatest phase of global reef development in the Phanerozoic. Despite this, ecological and environmental controls on the sedimentary nature of these vast systems have scarcely been investigated and remain enigmatic. The Late Devonian (Frasnian) Alexandra Reef System, exposed in the Northwest Territories of Canada, developed on a ramp that was situated on the western margin of Laurussia. The system consists of two reef complexes. The second reef complex developed basinwards of the first after sea level fell ~ 17 m. In contrast to stromatoporoid (± coral)–dominated reef facies in the first reef complex and the upper part of the second reef complex, reef facies in the lower part of the second reef complex are dominated by stromatoporoid–microbe associations. These include significant renalcid boundstone and stromatolite accumulations that are not found elsewhere in the reef system. It is concluded that the occurrence of the stromatoporoid–microbe reef facies indicates that a shift in the reef environment from oligotrophic to mesotrophic conditions took place. The mechanisms of nutrification were linked to the platform geometry, sea-level position, and oceanographic system, indicating that on carbonate ramps, systems tracts of falling sea level (forced regression) and sea-level lowstand may be particularly susceptible to nutrification. A nutrient-gradient model developed to explain different types of reef facies in the Alexandra Reef System indicates that trophic resources were an important control on the composition of Devonian reef-building communities, and that Devonian reefs and carbonate platforms were not highly susceptible to nutrient-invoked drowning.

INTRODUCTION

Devonian reef systems and associated carbonate platforms, especially of the Givetian–Frasnian, represent the most significant phase of global reef development in the Phanerozoic (Copper, 1989, 2002). Although numerous studies have catalogued the gross fossil composition of Devonian reefs (e.g., stromatoporoids, corals, calcimicrobes), the relative lack of detailed studies has limited our understanding of Devonian reef paleoecology (Moore, 1988; Wood, 1993; Kiessling et al., 1999). Because ecological dynamics control the composition of any carbonate-producing community (Wood, 1999), and thereby the types of carbonate facies that are deposited, this lack of information fundamentally limits our understanding of the controls on Devonian carbonate systems. Nutrients (e.g., nitrogen, phosphate, iron), in particular, are an important control on carbonate systems (Hallock, 1987; Mutti and Hallock, 2003), and yet their role in the development of Devonian reef systems has been investigated in only a few studies (e.g., Wood, 1993; Eliuk, 1998; Whalen et al., 2002), and they are not addressed in existing facies models for these systems (Machel and Hunter, 1994).

The Upper Devonian (Frasnian) Alexandra Reef System is located in the Western Canada Sedimentary Basin. With the perspective that carbonate depositional systems and platform evolution are controlled by the interaction of several dynamic variables, this study examines the relationships between platform geometry, sea-level change, and nutrification on its development. A fall in relative sea level that caused a basinward jump of reef development in the Alexandra Reef System (MacNeil and Jones, 2006a) is interpreted to have facilitated a combination of nutrient-enrichment (= nutrification) mechanisms that caused a change in its reef facies from being metazoan dominated to metazoan–microbe dominated. The mechanisms of nutrient enrichment probably included terrestrial runoff, groundwater seepage, and storm disturbance of offshore nutrient reservoirs. It is possible that seasonal climate changes also triggered periodic nutrification. Rising sea level diminished the synergy of the nutrification mechanisms, resulting in nutrient depletion and return of the reef facies to being metazoan-dominated.

The sequence stratigraphic context of nutrification in the Alexandra Reef System demonstrates some important mechanisms of nutrient enrichment that apply to forced regressive and lowstand systems on carbonate ramps, and that nutrient enrichment is not necessarily detrimental to reef-building communities and platform evolution. The study also demonstrates that the common assumption (cf. Wood, 1998a) that microbial carbonates are ubiquitous in Devonian reefs is false. A nutrient-gradient model, developed to explain the contrasting facies types in the Alexandra Reef System, is proposed as a new approach for understanding variations in Devonian reef facies with significant implications for how these systems are evaluated. In particular, the nutrient-gradient model implies that Devonian reef systems were not limited to oligotrophic (nutrient-deficient) environments and their platforms were not highly susceptible to nutrient-invoked drowning. This, and the ability of stromatoporoids to grow on different types of substrates (Wood, 1995; Kershaw, 1998), may partly explain the global expansiveness of these systems and the preponderance of these reefs to form broad, gently sloping platforms (cf. Wilson, 1975), quite different from

Controls on Carbonate Platform and Reef Development
SEPM Special Publication No. 89, Copyright © 2008
SEPM (Society for Sedimentary Geology), ISBN 978-1-56576-130-8, p. 271–289.

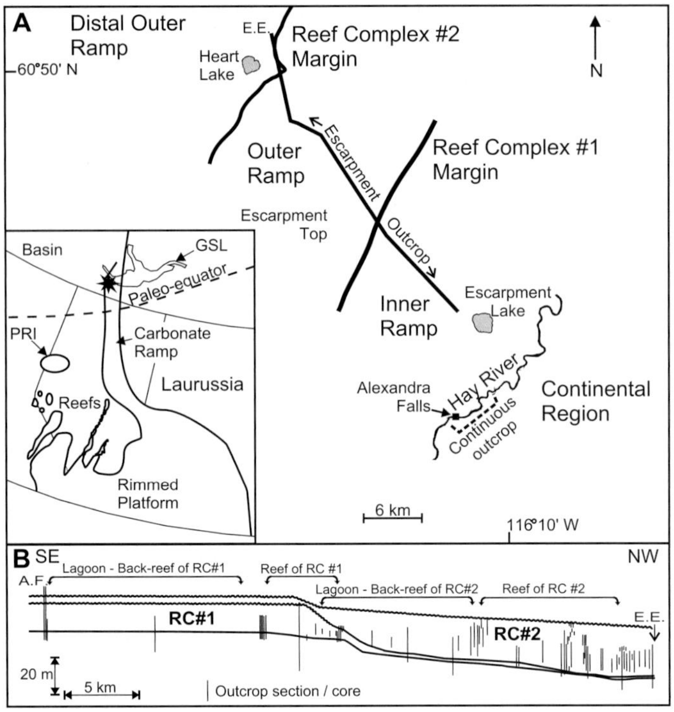

FIG. 1.—Regional setting. **A)** General map of study area in southern part of Northwest Territories with paleogeography of the Alexandra Reef System (modified from MacNeil and Jones, 2006a). Inset shows study area (star) relative to Late Devonian paleogeography of western Canada, overlaid on geopolitical boundaries. Paleoequator position is from Scotese (2004). GSL = Great Slave Lake; PRI = Peace River Island. **B)** Cross section through Alexandra Reef System with distribution of 62 measured outcrop sections and 11 cores plotted. For clarity, 20 closely spaced outcrop sections not figured. RC#1 = Reef Complex #1; RC#2 = Reef Complex #2; A.F. = Alexandra Falls; E.E. = End of escarpment northeast of Heart Lake.

the geometries of modern carbonate platforms constructed by coralgal reefs in oligotrophic realms.

METHODOLOGY

Strata that belong to the Alexandra Reef System (= Alexandra Formation, ~ 35 m thick), which is located in the southern part of the Northwest Territories of Canada (Fig. 1A), are exposed along a southwest–northeast gorge system cut by Hay River and along a southeast–northwest trending escarpment. Strata along Hay River show little lateral variation because the exposure trend is nearly parallel to depositional strike. In contrast, the escarpment trend is oblique to depositional strike, providing a nearly continuous section across different parts of the reef system. Eighty-two stratigraphic sections (Fig. 1B), measured from 47 localities along the escarpment edge, 5 road-cuts on the escarpment top, a quarry, and 3 localities along Hay River, combined with 11 continuous cores, provided the database for this study. Section localities, along with detailed aspects of the sequence stratigraphy, are given in MacNeil and Jones (2006a). Most (> 90%) of the oriented samples ($n = 827$) collected from the measured sections, at ~ 1 m intervals, were slabbed and etched in dilute HCl for study of macroscopic features. Large (5 cm x 7.5 cm) thin sections were made from outcrop (n = 134) and core (n = 160) samples for detailed petrography.

GEOLOGICAL SETTING OF THE ALEXANDRA REEF SYSTEM

The Alexandra Reef System developed during the mid-Frasnian (Klapper and Lane, 1989; McLean and Klapper, 1998) on an epicontinental ramp that was situated near the equator on the western edge of Laurussia (Fig. 1A inset; cf. Kiessling et al., 1999; Scotese, 2004). The continental margin was located to the north and west, and continental hinterland lay to the east. To the south, inundation of western Canada through a series of sea-level rises (Johnson et al., 1985) had formed a long and narrow intracratonic basin in which extensive, warm-water carbonate platforms existed (Moore, 1988; Whalen, 1995).

The Alexandra Reef System overlies regressive, mixed siliciclastic–carbonate ramp deposits that belong to Member E of the Escarpment Formation (Fig. 2A; Hadley and Jones, 1990). Member E includes green shales that were deposited in deeper depositional environments, marls and sparsely fossiliferous quartzose limestones from shallower environments, and quartz siltstones deposited in the shallowest environments (Hadley and Jones, 1990). MacNeil and Jones (2006a) determined that the ramp formed by these deposits in the area of the Alexandra Reef System had a very gentle slope (< 0.1°) and could be divided into an *inner ramp* and an *outer ramp* on the basis of a step in the ramp profile, termed the *ramp step*. Whereas the inner ramp and ramp step entirely underlie the Alexandra Reef System, the outer ramp extends beyond the Alexandra Reef System, probably several tens to hundreds of kilometers. Most of the Alexandra Reef System therefore developed above the shallow marine quartz siltstones and sparsely fossiliferous quartzose limestones that had been deposited in the inner to proximal outer-ramp regions, with marls, green shales, and deeper basinal shales deposited beyond the study area.

The Alexandra Reef System is formed of two stromatoporoid-dominated reef complexes (Reef Complex #1 and Reef Complex #2), separated by a sequence boundary and its correlative conformity (Fig. 2B–E). Conodont zonations and the magnitude of sea-level change recorded in these deposits indicate that these reefs were deposited during two relatively high-frequency fourth-order sea-level cycles (MacNeil and Jones, 2006a) in the upper part of T–R cycle IIc of Johnson et al. (1985). The second reef complex developed on the outer ramp, basinwards of Reef Complex #1, after a sea-level fall of ~ 17 m terminated development of Reef Complex #1 (Fig. 2C; MacNeil and Jones, 2006a). This sea-level fall appears to correlate with a sea-level fall identified by Whalen et al. (2000) and Potma et al. (2001) in other parts of the Western Canada Sedimentary Basin (MacNeil and Jones, 2006a), indicating a basin-scale fluctuation in sea-level position. MacNeil and Jones (2006a) suggested that punctuated global cooling in the Frasnian, the existence of which has been inferred from biodiversity trends and fossil assemblages (Stanley, 1988; Copper, 2002), and indicated by stable isotopes (Joachimski et al., 2004), could account for the magnitude of this sea-level fall and its subsequent rise. As sea level fell, the gentle slope of the ramp was a primary control on development of Reef Complex #2 because it provided a broad region of shallow marine environments, in front of Reef Complex #1, that were conducive to new colonization by carbonate-producing communities (MacNeil and Jones, 2006a). The stratigraphy of Reef Complex #2 is complex relative to that of Reef Complex #1 because it subsequently developed through an entire cycle of sea-level rise (Fig. 2D), and a second sea-level fall

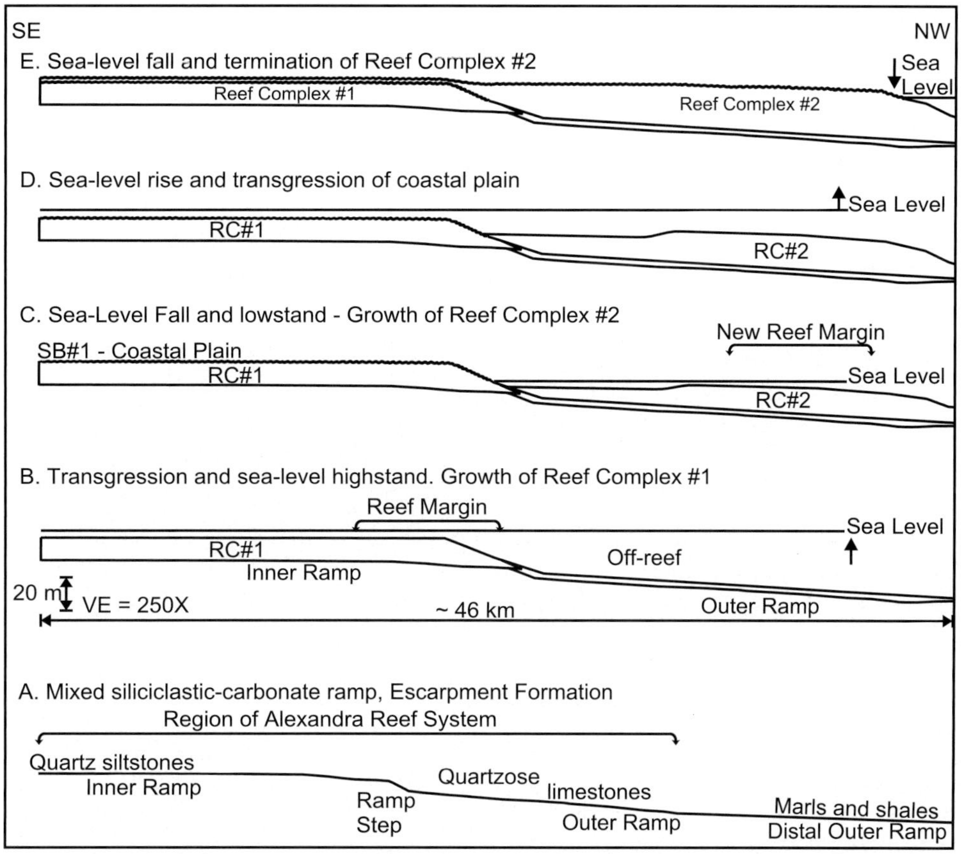

FIG. 2.—Simplified stratigraphic evolution of Alexandra Reef System. **A)** General ramp profile and distribution of mixed carbonate–siliciclastic facies at top of Escarpment Formation. **B–E)** Evolution of Alexandra Reef System. Scale at base of Part B. Note Reef Complex #2 (RC#2) developed basinward of Reef Complex #1 (RC#1), after sea-level fall, while sequence boundary #1 (SB#1) developed at top of exposed Reef Complex #1 (part C). Final succession (Part E) consists of two depositional sequences separated by sequence boundary #1. A second sequence boundary caps top of formation.

(Fig. 2E), before being terminated by subaerial exposure. Bioclastic limestones of the Twin Falls Formation unconformably overlie the Alexandra Reef System.

SEQUENCE STRATIGRAPHIC FRAMEWORK OF REEF FACIES IN THE ALEXANDRA REEF SYSTEM

Reef facies in the Alexandra Reef System can be divided into stromatoporoid-dominated facies, stromatoporoid–coral-dominated facies, and stromatoporoid–coral–microbe-dominated facies. The spatial–temporal distribution of these reef facies is delineated by the sequence stratigraphic framework of the Alexandra Reef System (cf. MacNeil and Jones, 2006a).

Reef Complex #1

Reef Complex #1 developed on the inner ramp in a transgressive and highstand systems tract (Fig. 3A), above Member E of the Escarpment Formation. The transgressive systems tract (= basal onlapping succession of MacNeil and Jones, 2006a) comprises muddy carbonate deposits with scattered stromatoporoids and corals on the outer ramp (off-reef), and a stromatoporoid–coral biostrome (~ 3.5 m thick) on the inner ramp, where shallow marine conditions conducive to reef development existed. The biostrome is dominated by laminar, tabular, and domal stromatoporoids, and corals (Fig. 3B, C) that include solitary and colonial rugosans (*Macgeea, Phillipsastrea, Thamnophyllum, Disphyllum, Smithiphyllum*) and tabulates (*Alveolites, Thamnopora*).

The main stage of reef development (~ 21 m thick) began during the transition from the transgressive to highstand systems tract. Reef development took place at the outer edge of the inner ramp, at the top of the ramp step, with the *in situ* accumulation of a broad, stromatoporoid-dominated buildup (Fig. 3A). Parts of this buildup are dominated by stromatoporoid patch reefs laden with *Stachyodes*-dominated floatstones and rudstones, whereas other parts are characterized by a more massive framework of densely packed domal and tabular stromatoporoids (Fig. 3D–H). Tabulate and rugose corals (e.g., *Phillipsastrea, Alveolites*) are present locally in the reef-front to fore-reef area, but the reef core, with the exception of *Thamnopora*, lacks colonial corals (MacNeil and Jones, 2006a). Back-reef deposits include small stromatoporoid-dominated patch reefs and thin- to medium-bedded *Amphipora* wackestones and floatstones that contain calcispheres and small bulbous stromatoporoids.

In a basinal direction, the reef front of Reef Complex #1 passes into thin-bedded, gently dipping, rubbly recessive-weathering fore-reef wackestones and floatstones that contain numerous peloids, brachiopods, and crinoids. In deeper water on the outer ramp, solitary and colonial corals and stromatoporoids grew in

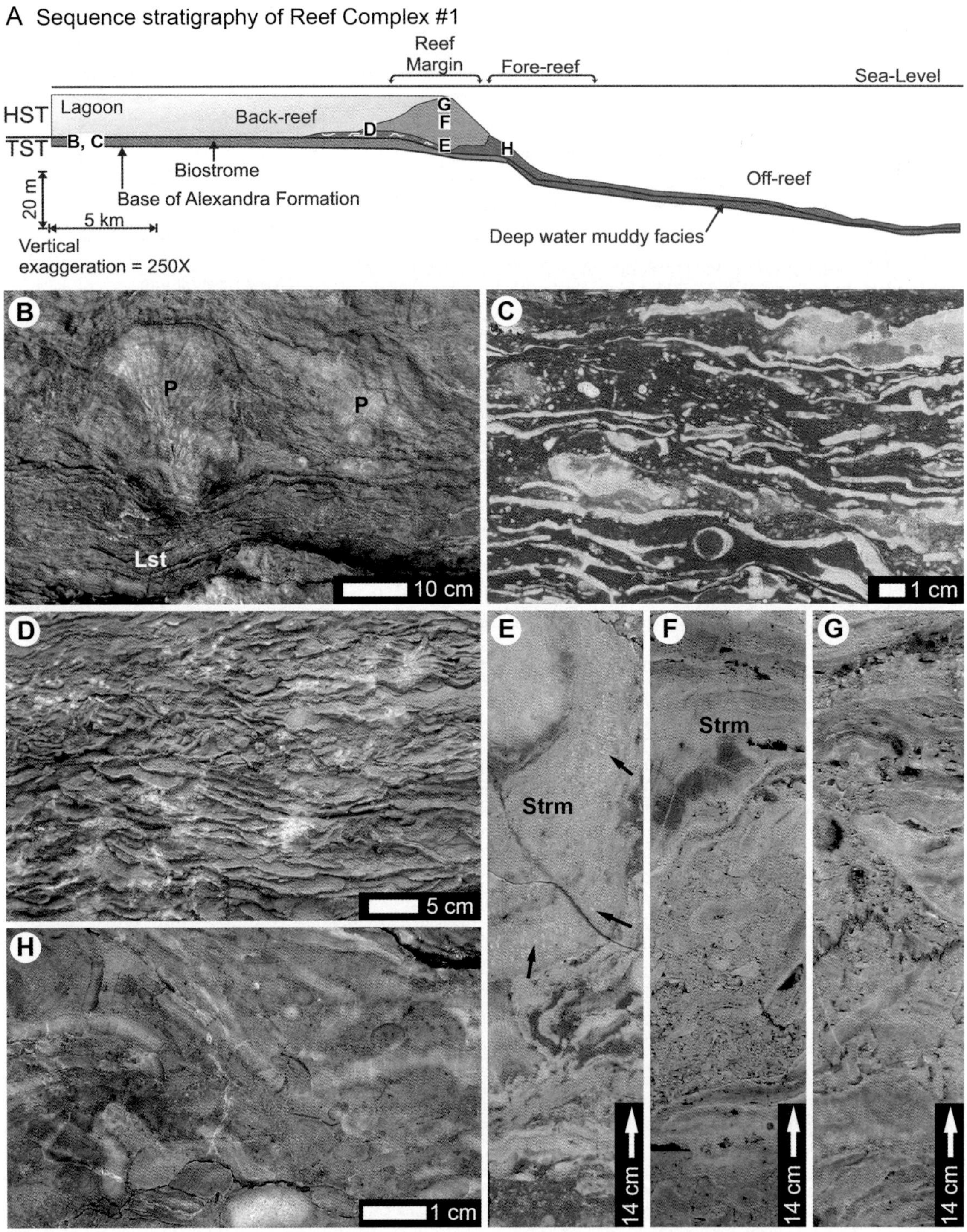

FIG. 3.—Metazoan-dominated reef facies of Reef Complex #1. B, C = transgressive systems tract. D–H = highstand systems tract. **A)** Reef Complex #1 sequence stratigraphy and environment zonations (modified from MacNeil and Jones, 2006a). TST = transgressive systems tract; HST = highstand systems tract. Positions of photos B–H are indicated. **B)** Outcrop photograph of laminar stromatoporoid–coral bindstone facies in biostrome on inner ramp. P = *Phillipsastrea*; Lst = laminar stromatoporoids. **C)** Cut and polished laminar stromatoporoid bindstone with dark *Amphipora* wackestone bound between plates. **D)** Outcrop photograph of laminar stromatoporoid bindstone. Cryptic spaces are filled with mudstone–wackestone. E–G) Core photographs; interval lengths in bottom right corners. **E)** Stromatoporoid-dominated reef facies with toppled domal stromatoporoid (Strm). Note caunopores (arrows) in stromatoporoid. **F)** *Stachyodes*-dominated floatstone–rudstone capped by tabular stromatoporoid (Strm). **G)** Stromatoporoid rubble mixed with *Stachyodes*-dominated floatstone and capped by laminar stromatoporoids. **H)** Cut and polished laminar stromatoporoid bindstone–floatstone with wackestone matrix from reef front.

scattered, isolated positions. Locally, low-profile bioherms developed. Most of the outer-ramp deposits consist of brown, muddy (carbonate mud) wackestones and floatstones.

Apart from rare specimens (e.g., a few scattered pieces in a few thin sections) of *Renalcis* in the reef deposits, and minor clots of *Girvanella* on stromatoporoids in the reef and on corals and stromatoporoids situated off-reef, in the outer-ramp deposits, no microbial deposits have been found in the reef facies of Reef Complex #1. The calcimicrobes are so rare that they do not form visible clots in cut and polished hand samples, and they did not contribute to development of the reef framework. Rather, the reef developed through stromatoporoid–coral (transgressive systems tract) and stromatoporoid-dominated (highstand systems tract) associations, and the trapping of significant amounts of *Stachyodes*-dominated sediment beneath tabular and irregular stromatoporoids (Fig. 3F). As such, the reef-building communities in Reef Complex #1 were metazoan-dominated (for brevity, stromatoporoids and corals are collectively referred to as metazoans). Falling sea level terminated Reef Complex #1 (MacNeil and Jones, 2006a).

Reef Complex #2

Reef Complex #2 was initiated as sea level fell, and subsequently developed through an entire cycle of sea-level rise and fall. Its development involved (1) initiation and differentiation stages, which took place in a falling-stage systems tract and lowstand systems tract, respectively, (2) a flooding stage, which consists of a transgressive systems tract and highstand systems tract, and (3) shallowing and termination stages, which took place in a second falling-stage systems tract, as relative sea level started to drop from its highest position (MacNeil and Jones, 2006a).

Initiation and Differentiation Stages.—

Sea-level fall and termination of Reef Complex #1 coincided with shallowing of the marine conditions over the outer ramp. In this falling-stage systems tract (Fig. 4A), an expansive stromatoporoid–microbe biostrome (maximum of ~ 4 m thick) developed, which forms the basal part of Reef Complex #2. Stromatolites capped the top surfaces of stromatoporoids (Fig. 4B, C), whereas renalcid boundstone accumulated in cavities under and between framebuilders (Fig. 4D). Encrusting of stromatoporoid and coral fragments by *Girvanella* and *Rothpletzella* was common but less significant. Whereas stromatolites were present across the extent of biostrome development, renalcid boundstone was most common in the more basinward areas of biostrome development.

In the middle region of the biostrome, broad bioherms 300–400 m across with 5–7 m of synoptic relief subsequently developed on top of the biostrome (Fig. 4A). Laminar stromatoporoids, commonly only 2–3 mm thick, capped by laminar and domal stromatolites, generally ~ 2 cm thick, are a conspicuous part of these bioherms. Growth of an inner buildup above these bioherms, and an outer buildup in a slightly basinwards position, defined a margin to the evolving reef complex and marked the start of its environmental differentiation (Fig. 5A). This took place when relative sea level reached its lowest position and the falling-stage systems tract passed into the lowstand systems tract (MacNeil and Jones, 2006a). Concurrent with growth of the inner and outer buildups, a broad reef front developed in front of the outer buildup, and a back-reef region and shallow lagoon developed landwards of the inner buildup (Fig. 5A). Between the inner and outer buildups was a low area, termed the intra-buildup low, where bioherms developed. Microbial reef deposits are a volumetrically significant part, and in some cases dominate, the intra-buildup low, outer buildup, and reef-front facies, but are less significant in the inner-buildup facies.

The inner buildup (~ 8 m thick) is characterized by stromatoporoid–coral deposits (Fig. 5B). Stromatolites on top of stromatoporoids, and renalcid boundstone, are common in some areas. Bioherms in the intra-buildup low were dominated by laminar stromatoporoids and laminar stromatoporoids that amalgamated into outward- and upward-growing platy complexes. Thick (up to 4 cm) masses of renalcid boundstone accumulated on the undersides of these plates, in crypts between vertically stacked plates, and between laterally adjacent plates (Fig. 5C).

The outer buildup (maximum of ~ 14 m thick) is formed of stromatoporoid rubble and intraclastic wackestones–packstones. Large (up to ~ 1 m) fragmented pieces of *Actinostroma*, in all possible orientations relative to their original growth position, are common, but *in situ* stromatoporoids and corals are minor (Fig. 5D). As *in situ* actinostromids are commonly found in the reef front, their occurrence as fragmented pieces in the outer buildup, with the intraclastic-rubble-dominated sediment, is interpreted to indicate that the outer buildup accumulated through intense storm activity that included pruning of skeletal framebuilders in the reef front. This mechanism of reef development, whereby storms devastate thriving reef-front environments and build up a rubble-dominated ridge in a landwards location, is well known from numerous modern reef systems (Longman, 1981; Blanchon et al., 1997; Blanchon and Perry, 2004).

In the outer buildup, stromatolites cap stromatoporoid fragments and renalcid boundstone is ubiquitous (Fig. 5E–G). In some cases, stromatolites developed upwards from renalcid boundstone, which locally forms >75% of the framework (Fig. 5E, G). Syndepositional fractures in these deposits (Fig. 5G), filled with intraclastic sediment, support the interpretation that renalcid boundstone formed hard, lithified masses. Marine cements are common throughout the outer buildup.

The reef front is dominated by massive (up to 2.2 m across) *in situ* stromatoporoids that grew in tabular, concave-up tabular, expanding-conical, laminar, and anastamosing-laminar growth forms (Fig. 6A, B). Colonial corals, including *Alveolites*, *Phillipsastrea*, and *Thamnopora*, are also present. The stromatoporoids did not coalesce into a rigid framework, but rather, existed as densely packed individual entities. Floatstones and rudstones of intensely fragmented laminar stromatoporoids that contain large, overturned coral heads (Fig. 6C) attest to the periodic storm activity that led to the formation of the outer buildup. Thin (≤ 2 cm thick) beds of green siliciclastic mud that periodically cover the tops of the reef-front stromatoporoids (Figs. 6D, 7) were probably deposited from suspension as the storms waned. Renalcid boundstone and stromatolites are common in the lower part of the reef-front succession, especially in crypts and between stromatoporoids and corals (Fig. 6E, F). Upwards through the reef-front facies, the renalcid component decreases significantly. Only near the top of the succession, however, does the abundance of stromatolites start to decrease.

Due to limited data, less is known about the deeper reef-front and fore-reef environments. A "valley" that divides the continuity of the reef front and opens into the fore-reef (Fig. 5A), however, offers some insight. It appears that transition of the reef front into deeper water of the valley was marked by the development of stromatoporoid–coral bioherms, with stromatolites and lesser renalcid boundstone.

Flooding Stage.—

An increase in the rate of rise of sea level flooded Reef Complex #2, marking its passage into a transgressive systems

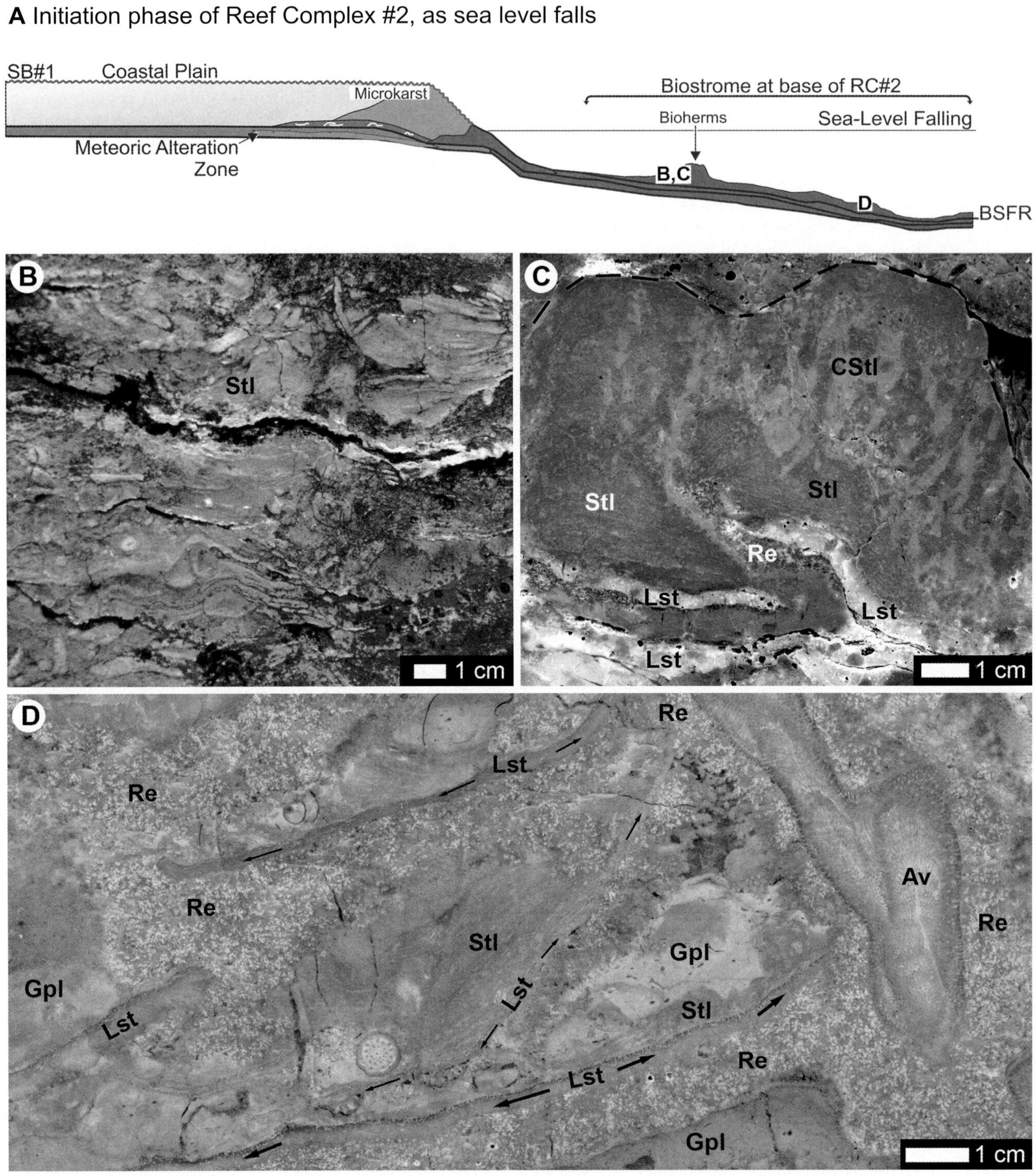

FIG. 4.—Metazoan–microbial reef facies in initiation phase of Reef Complex #2. **A)** Sequence stratigraphy of Alexandra Reef System at time of Reef Complex #2 (RC#2) initiation phase (modified from MacNeil and Jones, 2006a). Scale as in Fig. 3A. BSFR = basal surface of forced regression at base of falling-stage systems tract. Positions of photos B–D are indicated. **B)** Outcrop photograph of laminar stromatoporoid bindstone with *Thamnopora* fragments in basal biostrome. Note stromatolite (Stl) capping laminar stromatoporoid. **C)** Cut and polished sample of laminar stromatoporoids (Lst) with minor clots of *Renalcis* (Re), capped by domal stromatolites (Stl) that grade up into columnar stromatolites (CStl). Top of stromatolites is outlined (black dashed line). **D)** Cut and polished laminar stromatoporoid–coral–*Renalcis* boundstone. Branch of *Alveolites* (Av) and thin, delicate laminar stromatoporoids (Lst) are bound into rigid framework by thick renalcid (Re) masses. Renalcids encrust under stromatoporoid plates, entire skeleton of *Alveolites*, and bridge between these elements, dominating framework. Stromatolites (Stl) cap some tops of laminar stromatoporoids. Geopetal mudstone (Gpl) and calcite cement fill remaining cavity spaces.

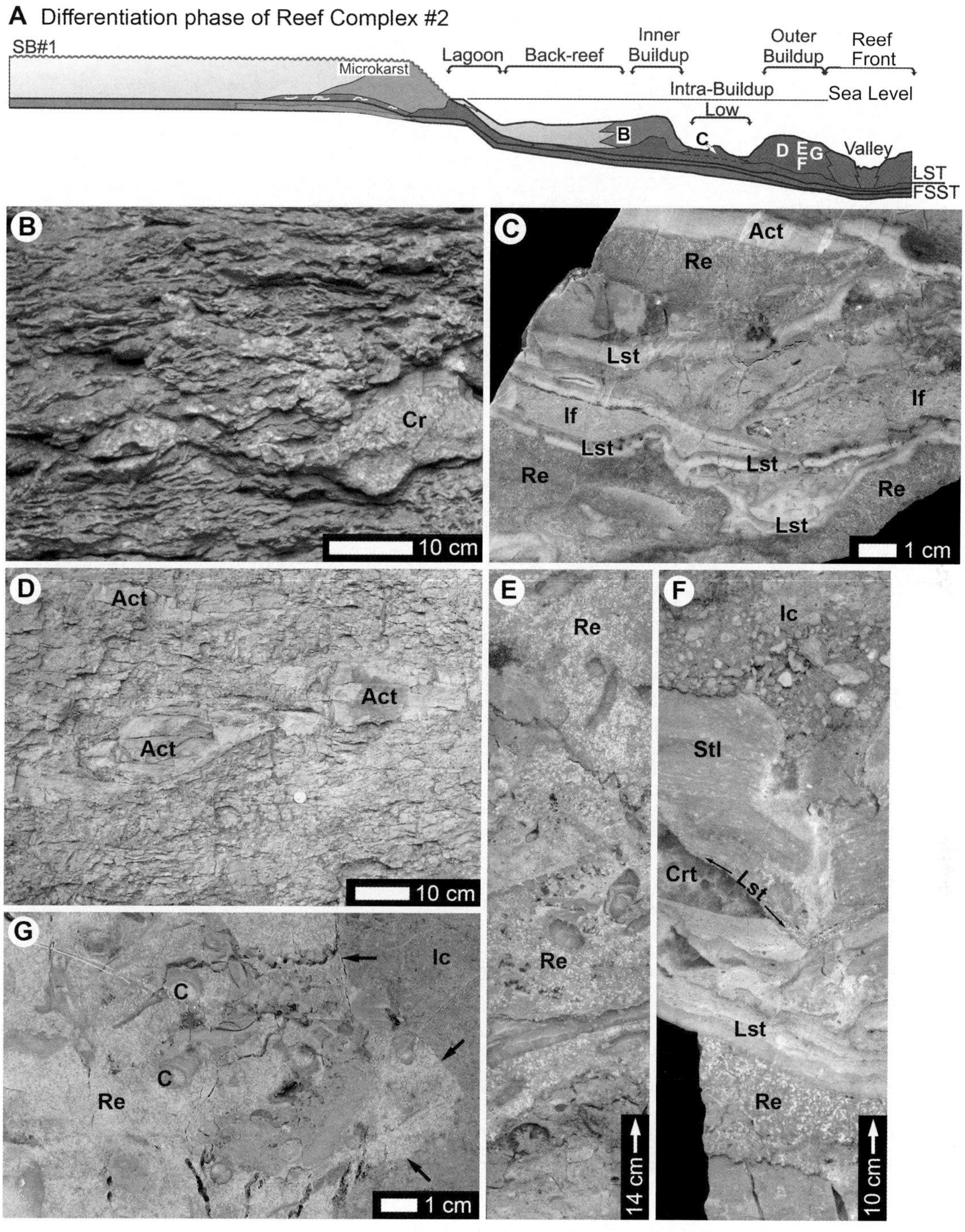

FIG. 5.—Metazoan–microbe reef facies in differentiation phase of Reef Complex #2. **A)** Sequence stratigraphy of Alexandra Reef System at time of Reef Complex #2 (RC#2) differentiation (lowstand systems tract; modified from MacNeil and Jones, 2006a). Scale as in Fig. 3A. LST = lowstand systems tract; FSST = falling-stage systems tract. Positions of photos B–G are indicated. **B)** Outcrop photograph of laminar stromatoporoid bindstone characteristic of inner buildup. Partly encrusted coral (Cr) is present. **C)** Cut and polished framework of amalgamated laminar stromatoporoid plates (Lst) with thick renalcid (Re) encrustations. Thin tabular *Actinostroma* (Act) is present. Remaining cavity spaces in platy complex are infilled (If) by muds and wackestone. **D)** Outcrop photograph of rubble deposit of outer buildup. Large, fractured actinostromids (Act) are hosted in renalcid-bound intraclastic packstones–rudstones with fragmented pieces of corals and stromatoporoids. **E)** Core photograph of renalcid (Re)–bound laminar-stromatoporoid floatstone of outer buildup. Drill location is near outcrop shown in Part D. **F)** Core photograph of laminar stromatoporoid–stromatolite–renalcid boundstone of outer buildup, capped by intraclastic (Ic) wackestone–packstone. Crypt (Crt) beneath laminar stromatoporoid is filled by geopetal mud and cement. **G)** Cut and polished renalcid (Re) boundstone from outer buildup with horn corals (C) capped by small domal stromatolites. Syndepositional fracture (edge marked by arrows) is filled with intraclastic (Ic) packstone, indicating that renalcid accumulation formed lithified mass in depositional realm.

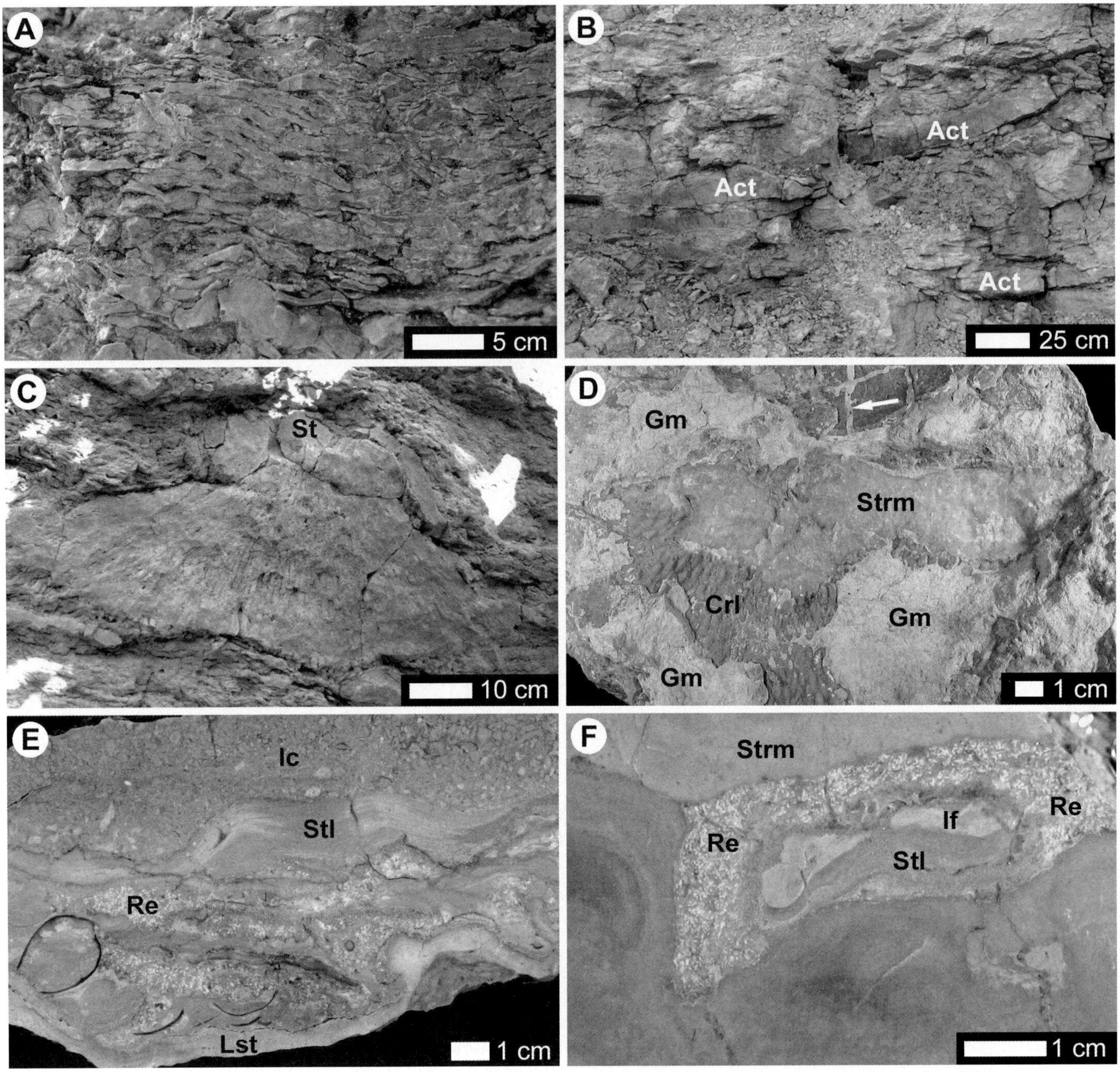

Fig. 6.—Metazoan–microbe reef-front facies in differentiation phase of Reef Complex #2. **A)** Outcrop photograph of densely packed tabular stromatoporoids. Note that each stromatoporoid is an individual entity; they do not coalesce into framestone. **B)** Outcrop photograph of massive tabular and concave-up *Actinostroma* (Act) and thinner tabular stromatoporoids. Each stromatoporoid is an individual entity. **C)** Outcrop photograph of laminar stromatoporoid rudstone with large toppled head of fasciculate rugosan. Top of toppled coral is encrusted by stromatoporoid (St). **D)** Bedding-plane view of laminar stromatoporoid (Strm)–coral (Crl) facies. Thin deposit of green mud (Gm), partly weathered away, covered top surface of skeletons and filled physical compaction cracks (arrow at top center). **E)** Cut and polished laminar stromatoporoid (Lst)–*Renalcis* (Re)–stromatolite (Stl) reef facies with > 50% of framework of microbial origin, buried by intraclastic (Ic) packstone. **F)** Cut and polished stromatoporoid (Strm). Growth cavity partly lined by renalcid boundstone (Re). A thin stromatolite (Stl) developed on cavity floor before being buried by infilling (If) mud. Figure parts A, B, and D are modified from MacNeil and Jones (2006a).

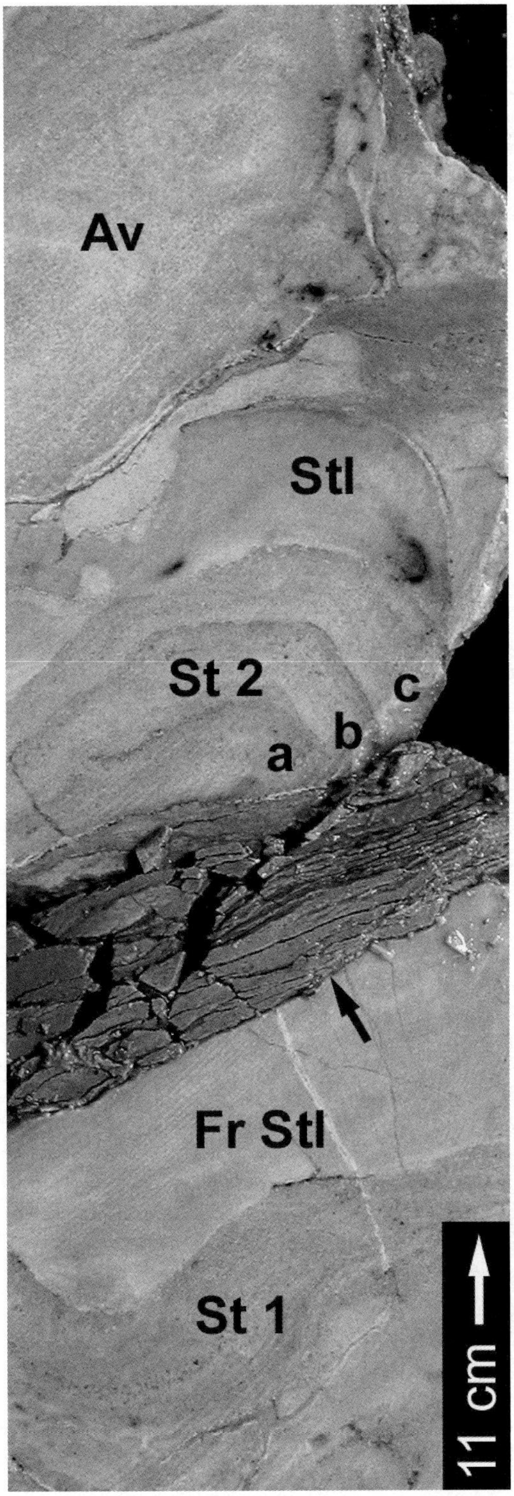

FIG. 7.—Core photograph of storm deposit in reef front of Reef Complex #2. Deposit of green mud ~ 2 cm thick covers fractured, upside-down stromatoporoid (St 1)–stromatolite (Fr Stl) association. Note sharp edge of fracture (arrow) at top of stromatolite. Three latilaminae (a, b, c) of new stromatoporoid growth (St 2) subsequently overgrew green mud. Stromatolite (Stl) caps second stromatoporiod, in partly sheltered cavity formed by *Alveolites* (Av).

tract (Fig. 8A; MacNeil and Jones, 2006a). Some areas of the reef complex were able to maintain growth whereas other areas drowned. Associated with passage into the transgressive systems tract was disappearance of the stromatolites in the reef framework and near-disappearance of all renalcid components. Only in some areas of the outer buildup did minor clots of *Renalcis* and related calcimicrobes form. These were not, however, significant, and did not contribute to framework development. Likewise, *Girvanella* and *Rothpletzella* encrusted some stromatoporoids and corals, but only as minor elements. Most of the deeper reef front (Fig. 8B, C) was dominated by decimeter- to meter-scale thickets of fasciculate rugosans (*Smithiphyllum*) whereas the outer buildup was dominated by laminar stromatoporoids. Over the outer part of the outer buildup, stromatoporoids coalesced to form dense framestone locally divided by large (e.g., 1–2 m across, 1 m high) thickets of fasciculate rugosans; no microbial components have been found in this framestone or bafflestone. In drowned and nearly drowned areas of the reef front, intra-buildup low, and back-reef, dark, thin-bedded carbonate mudstones and wackestones with cricoconarids and other deep-water biota were deposited (Fig. 8C; MacNeil and Jones 2006a). As such, passage from the lowstand systems tract to the transgressive systems tract corresponded with loss of calcimicrobial contribution to framework development, complete loss of peri-reefal stromatolites, and domination of framework development, in those parts that did not drown, by stromatoporoids and corals. As sea-level position peaked in the highstand systems tract, there were no significant facies changes and highstand shedding was not significant (MacNeil and Jones, 2006a). As such, most of Reef Complex #2 remained "underproductive".

Regional Shallowing and Termination Stages.—

As sea level began falling, which marks the beginning of the second falling-stage systems tract (Fig. 8D), extensive carbonate sedimentation over Reef Complex #2 resumed (MacNeil and Jones, 2006a). The falling-stage systems tract was characterized by strongly progradational "clean" carbonate deposits of *Amphipora* and *Stachyodes* floatstones and rudstones (Fig. 8E) and growth of stromatoporoid-dominated patch reefs (Fig. 8F) and bioherms, like the reef facies that characterized Reef Complex #1. Corals are minor in these deposits, except in the outer fore-reef location, where stromatoporoid–coral deposits accumulated. Microbial components were not significant in these deposits; framework in the bioherms and patch reefs was not built through any microbial contributions, and unlike the outer buildup in the lowstand systems tract, rubbly deposits were not bound by renalcids (Fig. 8E). Termination of Reef Complex #2 is attributed to subaerial exposure (MacNeil and Jones, 2006a).

NUTRIENTS AS A CONTROL ON THE DIFFERENT TYPES OF REEF FACIES IN THE ALEXANDRA REEF SYSTEM

Contrasting types of reef framework are conspicuous in the Alexandra Reef System (Fig. 9A). Given that seawater chemistry in the Late Devonian was markedly conducive to the calcification of microbial communities (Riding, 1992; Riding and Liang, 2005), these facies contrasts must be linked to ecological and environmental controls on the composition of the reef-building communities. Indeed, traces of *Renalcis*, *Girvanella*, and *Rothpletzella* in Reef Complex #1, and the upper parts of Reef Complex #2, indicate that physicochemical conditions were conducive to the calcification of these microbes throughout the evolution of the Alexandra Reef System, if they were present.

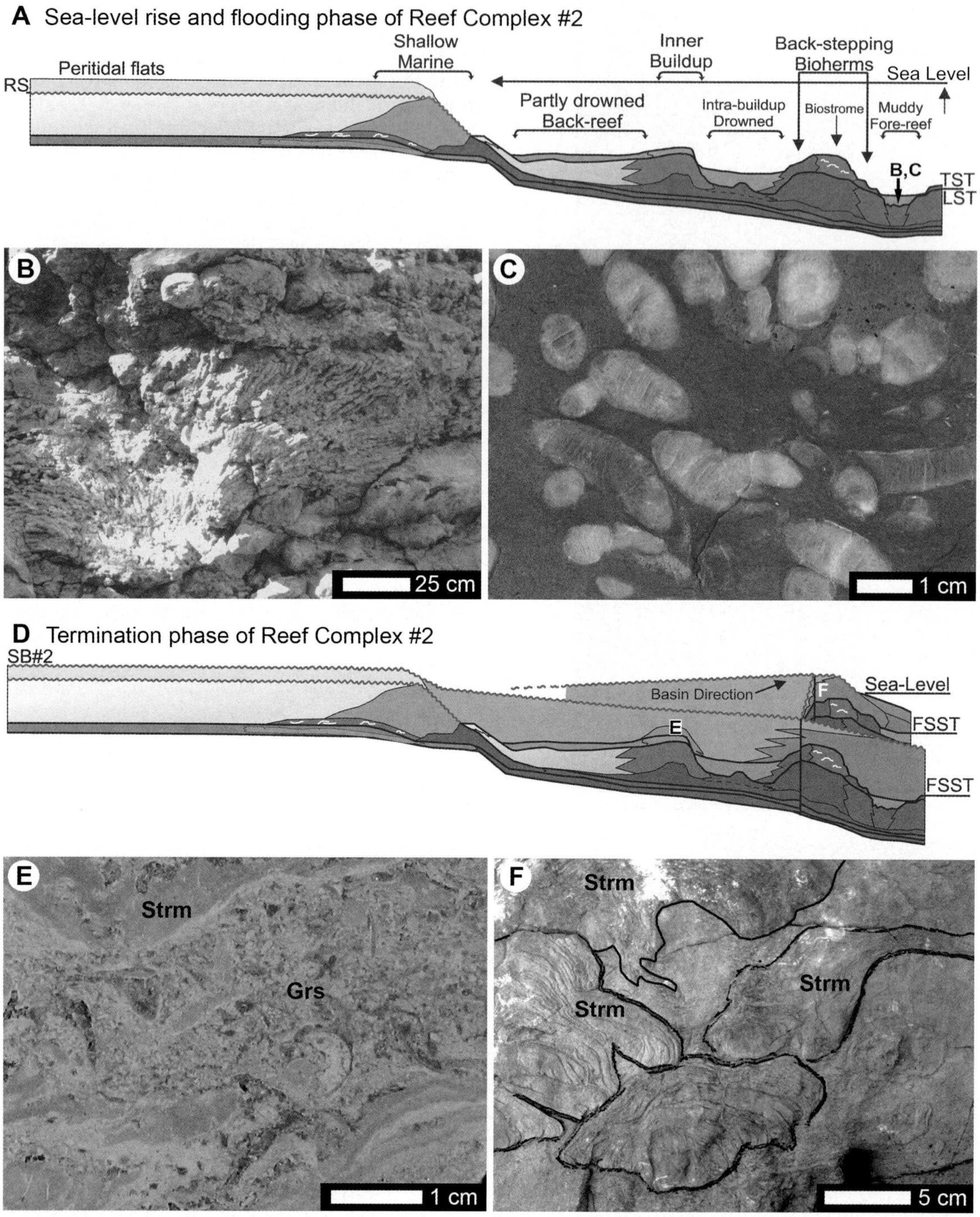

FIG. 8.—Metazoan-dominated reef facies in flooding phase and termination phase of Reef Complex #2. **A)** Sequence stratigraphy of Reef Complex #2 flooding (transgressive systems tract), with significant facies changes across reef complex and flooding of coastal plain on top of Reef Complex #1 (modified from MacNeil and Jones, 2006a). Microbial reef facies are not present. Scale as in Fig. 3A. LST = lowstand systems tract; TST = transgressive systems tract; RS = ravinement surface. Positions of photographs B and C are indicated. **B)** Outcrop photograph of *Smithiphyllum* thicket in reef front. **C)** Photograph of cut and polished sample of *Smithiphyllum*. Dark mudstone, without microbial constituents, fills area between coral stalks. **D)** Sequence stratigraphy of Reef Complex #2 termination phase (modified from MacNeil and Jones, 2006a). Scale as in Fig. 3A. FSST = Falling-stage systems tract; SB#2 = sequence boundary #2. Positions of photographs E and F are indicated. **E)** Cut and polished intraclastic–peloidal grainstone (Grs) bound by irregular laminar stromatoporoids in patch reef facies. No microbial constituents are present. **F)** Outcrop photograph of domal–irregular stromatoporoid patch reef, akin to reef-facies in Reef Complex #1, with no microbial constituents. Outlines of individual stromatoporoids are drawn with black marker. Wackestone fills cavities between stromatoporoids.

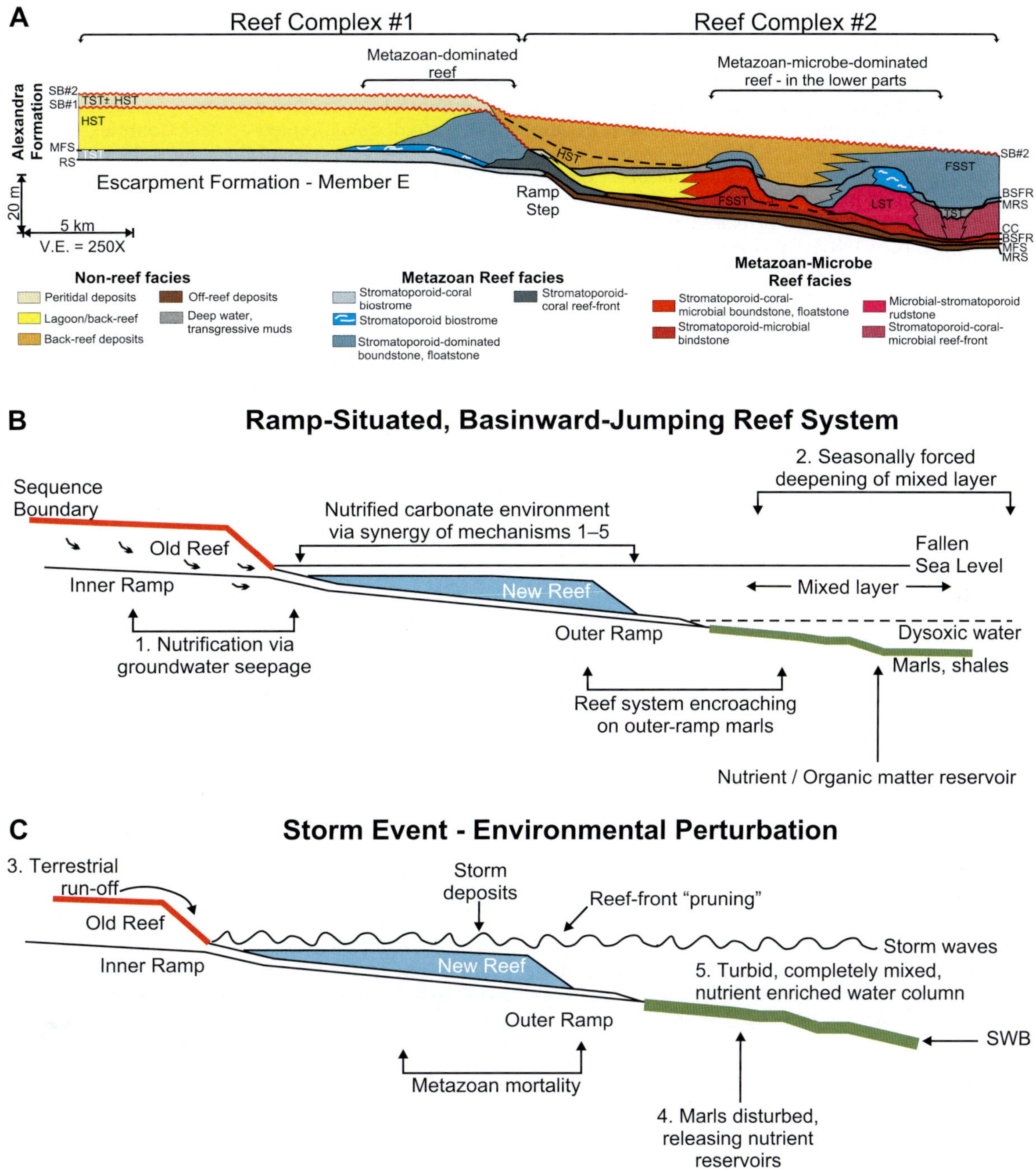

FIG. 9.—Summary of reef facies in Alexandra Reef System and models for nutrification on carbonate ramps. **A)** Sequence stratigraphic distribution of metazoan-dominated and metazoan–microbe-dominated reef facies in Alexandra Reef System. Metazoan–microbe reef facies are found only in lower parts of Reef Complex #2. **B, C)** Nutrification mechanisms on a carbonate ramp as sea level falls and in lowstand systems tracts. Note deposits of new carbonate complex, in this case a reef, separated from old reef by sequence boundary and its correlative conformity. Synergy of potential nutrification mechanisms 1–5 bathe new carbonate complex in nutrient-enriched seawater, with potential implications for types of carbonates produced. No scale is provided, but affected area, because of low depositional gradient of ramp, may cover several thousand square kilometers. **B)** Basinward-jumping reef system encroaches on distal outer-ramp deposits as sea level falls. Potential nutrification mechanisms include (1) nutrification via groundwater seepage through porous, now-exposed old reef deposits, and (2) seasonally forced deepening of mixed layer. **C)** Storm event perturbs reef complex and metazoans experience significant mortality. Concurrent with metazoan mortality, environment may be nutrified by mechanisms (continued from Part B) including (3) terrestrial runoff, (4) release of nutrient reservoirs hosted in marls and shales on outer ramp, and (5) mixing of surface layer with nutrient-rich parts of the water column at depth. Nutrified water that bathes post-storm carbonate complex, as metazoans begin to recover, may have long residence time. SWB = storm wave base.

The shift to the metazoan–microbe-dominated reef framework of Reef Complex #2, from the metazoan-dominated framework of Reef Complex #1, is herein attributed to elevated nutrients in the ambient environment of the reef system, for a number of reasons. First, physiological requirements limit most types of algae and microbes to environments with available nitrogen, phosphate, and/or iron (Lobban et al., 1985). If required nutrients are not available, these organisms will not be present. Second, community-structure models for modern reefs predict algal–microbial blooms to be positively correlated with increasing nutrients (Littler and Littler, 1984). Third, numerous studies have documented that nutrification results in coral–algal phase shifts and algal–microbial blooms, on the scale of days, months, and years (e.g., Smith et al., 1981; D'Elia and Wiebe, 1990; Hallock et al., 1993; Genin et al., 1995; Lapointe, 1997; Miller et al., 1999; Costa et al., 2000; Lapointe et al., 2004; Hallock, 2005; Paul et al., 2005). Relating the shift from metazoan-dominated to metazoan–microbe dominated reef facies to increased nutrient levels is therefore consistent with the physiological needs of algae and microbes, models of reef-community structure, and the conclusions of numerous studies of modern reefs. Increased trophic resources were also invoked to explain shifts from metazoan-dominated to metazoan–microbe-dominated reef facies in Jurassic reef systems (Dupraz and Strasser, 2002), and similar conclusions have been made for these types of shifts in Quaternary (Camoin et al., 1999) and modern reef facies (Sprachta et al., 2001).

Alternative ecological and/or environmental controls on the microbial components in Reef Complex #2, such as the availability of preferred habitat types (e.g., cryptic spaces for the renalcids, hard substrates for the stromatolites), differences in bathymetry, changes in salinity levels or temperature ranges, or the degree of wave energy, can all be precluded. Reasons include:

- All systems tracts of both reef complexes contained habitats of the same composition (e.g., laminar stromatoporoids forming hard substrates with sheltered cavities underneath) in comparable spatial arrangements, and yet these habitats were not exploited by microbial constituents.

- A range of water depths and depositional conditions, comparable to the early stages of development of Reef Complex #2, also existed in Reef Complex #1 and the later stages of Reef Complex #2. Microbial constituents, however, did not contribute to framework development in these other stages of the reef system.

- Environmental conditions did not preclude the growth of stromatoporoids, corals, and brachiopods in any systems tract, indicating that values of temperature and salinity fell within ideal, normal-marine ranges.

- There is no evidence in any part of the Alexandra Reef System for mass terrigenous influx, which is often accompanied by a reduction in salinity that is detrimental to reef-building metazoans (Hallock and Schlager, 1986).

Thus, a state of elevated nutrient levels is the only reasonable explanation for the microbial framework constituents in the lower parts of Reef Complex #2. The lack of algal laminites in the reef deposits (as in soft, noncalcified organic matter from plankton blooms), which would be expected with nutrient enrichment, can be attributed to scavengers (e.g., gastropods), grazing and scavenging fish, and the generally poor preservation potential of nonmineralized organic matter in shallow marine, well-oxygenated environments.

NUTRIFICATION ON CARBONATE RAMPS

Carbonate depositional systems are controlled by the interaction of dynamic variables such as climate, platform geometry, water depth, temperature, and sea-level change. In the Alexandra Reef System, development of Reef Complex #2 was controlled largely by the geometry of the ramp-type platform, which provided a broad region of shallow marine environments suitable for carbonate-producing communities as and after sea level fell (MacNeil and Jones, 2006a). The magnitude of sea-level fall determined the location of new reef development, and the new reef complex was distinct from the older Reef Complex #1. This type of basinward-jumping reef system, however, would be precluded from developing on a high-relief, steep-sided carbonate shelf, or isolated platform, because of the different platform geometry (Sarg, 1988; Hunt and Tucker, 1993).

Just as several variables interact to control the depositional nature of carbonate systems and platform evolution, it is to be expected that these interactions and changes will also carry implications for potential nutrification mechanisms. The result of falling sea level on carbonate ramps is that (1) their depositional systems encroach upon potential sediment-hosted nutrient reservoirs situated in distal ramp positions, (2) new carbonate-producing communities are increasingly susceptible to nutrification via terrestrial runoff, eolian transport of sediment with nutrients, and groundwater seepage from older, exposed deposits, and (3) decreasing water volume in the basin (assuming that sea-level fall is either basin-scale or eustatic), as the system encroaches basinward, increases the likelihood that storm and/or seasonal deepening of the mixed layer will impinge upon nutrient reservoirs in deeper waters, and these nutrient-enriched waters may be advected onto the region of carbonate production (Fig. 9B, C). If the basin becomes significantly restricted, eutrophication and a shift to evaporite-dominated facies (James and Kendall, 1992) may take place. In contrast, rising sea level and transgression diminishes many of these nutrification mechanisms because (1) increasing water volume in the basin dilutes and expands the water column and nutrient reservoirs may be sequestered to greater depths, (2) terrestrial erosion decreases as base level rises, and (3) weathering profiles and soils, which are sources of nutrients, may not be mature and their transgressive erosion in the coastal zone is likely to be highly diachronous (Catuneanu, 2002; MacNeil and Jones, 2006a), which limits their deliverability of nutrients at any specific time. Thus, transgressive shifts on carbonate ramps may not be balanced by the same degree of nutrification.

Relating Reef Complex #2 to a model for basinward-jumping reef systems (Fig. 9B, C) reveals that nutrient enrichment of its initiation and differentiation stages was most probably caused by a combination of (1) terrestrial runoff from the coastal plain on top of Reef Complex #1, (2) groundwater seepage through recently exposed deposits of Reef Complex #1, (3) storm disturbance of the offshore oceanographic system, and possibly (4) seasonally forced deepening of the mixed layer. Not all of these mechanisms were active at the same time, and the punctuated events (storms and terrestrial runoff) probably supplemented longer-term events (groundwater seepage) and seasonal-climate events. Thus, it was likely that the synergy of mechanisms, rather than one specific mechanism, facilitated the microbial contribution to development of the reef framework. Different mechanisms also likely supplied different types and concentrations of nutrients, and nutrient variability, in itself, can be an important control on the composition of algal–microbial blooms (e.g., Kuffner and Paul, 2001). Passage into the transgressive systems tract probably diminished and turned off nutrification as the top of Reef Complex #1 flooded (turning off direct terrestrial runoff and reducing groundwater seepage through Reef Complex #1), the climate warmed (assuming that a short-term cooling event was associated with the sea-level fall; cf. MacNeil and Jones, 2006a), and storm events either became less frequent or, with rising sea level, no longer impinged upon nutrient reservoirs at depth. Erosion of soils as the coastal plain flooded probably contributed nutrients, but not in the supply needed to facilitate microbial blooms in the reef framework.

Terrestrial Runoff

Terrestrial runoff from the coastal plain that developed on top of Reef Complex #1 (Fig. 9C), as sea level fell, may have delivered nutrients and organic matter to the outer ramp where Reef Complex #2 was developing. An indication that runoff took place is the presence of blackened lithoclasts in storm beds in the basal parts of Reef Complex #2, which are also found in the coastal-plain deposits above Reef Complex #1 (MacNeil and Jones, 2006a; MacNeil and Jones, 2006b). Consideration, however, that the climate appears to have been semiarid (Golonka et al., 1994; MacNeil and Jones, 2006b) suggests that nutrification due to terrestrial runoff was probably sporadic and short-lived. This mechanism also would have been present for the growth of Reef Complex #1, which lacks peri-reefal microbial components, suggesting that it was probably an insignificant mechanism on its own.

Groundwater Seepage

The geographic position of Reef Complex #2 made it particularly susceptible to nutrification via groundwater seepage through the recently exposed deposits of Reef Complex #1 (Fig. 9B). As sea level fell below the margin of Reef Complex #1, this was the first time that these deposits, which were probably a rich source of decaying organic matter and trapped nutrients, were subject to groundwater percolation and a shifting water table. Groundwater seepage is a well-documented mechanism of nutrification on modern carbonate platforms (Marsh, 1977; Johannes, 1980; D'Elia et al., 1981; Lewis, 1987; Cuet et al., 1988; Lapointe, 1997), and in some cases it is of significantly greater magnitude than terrestrial runoff (Johannes, 1980; Lapointe, 1997).

A distinct horizon of vuggy porosity in the lower parts of Reef Complex #1 (Fig. 4A), resulting from extensive dissolution of stromatoporoid and coral skeletons, at the approximate level of sea-level position after its fall, was interpreted by MacNeil and Jones (2006a) to be the result of mixing-zone diagenesis in an aquifer established during or after the fall in sea level. The development of such aquifers is predictable with falls in sea level (Sarg, 1988), and these may effectively discharge nutrient-enriched groundwater into the environs of attached platforms. Even though the climate was semiarid, this mechanism of nutrification would not have been limited by precipitation levels in the same manner that terrestrial runoff was limited, because sea level (and thereby the water table) did not fall below the base of Reef Complex #1.

Groundwater seepage from Reef Complex #1 probably supplied significant amounts of nutrients in the form of ammonia and nitrate, derived from the decay of organic matter trapped in the deposits of Reef Complex #1, and in ephemeral marshes at its top (cf. MacNeil and Jones, 2006b). Nitrogen-fixing microbes in these deposits would also have enriched the nitrogen content in percolating water, and consistent with this, MacNeil and Jones (2006b) documented skeletal stromatolites of *Rivularia haematites*, an efficient nitrogen-fixing cyanobacterium, in the marsh deposits at the top of Reef Complex #1.

Storm-Disturbance of the Offshore Oceanographic System

Storms are a viable, potentially high-impact mechanism of nutrient enrichment (Fig. 9C). Storm-induced vertical mixing of the oceanic surface layer, such that normally nutrient-depleted surface water is mixed with deeper, nutrient-rich seawater, can lead to a short-term (e.g., 1–3 weeks) but significant enrichment in the seawater and rapid blooms of algae and microbes (Holloway et al., 1985; Ostrander et al., 2000; Shiah et al., 2000; Subrahmanyam et al., 2002; Lin et al., 2003; Babin et al., 2004). The impact of storm-induced enrichment is magnified for seasonal storm activity, when repeated storm events compound the extent of nutrient enrichment (e.g., Holloway et al., 1985; Babin et al., 2004). In addition to deepening the mixed surface layer, storms may also disturb sediment-hosted nutrient reservoirs on the platform and in off-reef locations, especially if they lie above storm wave base (Fig. 9C). In addition to dissolved and particulate organic matter and nitrogen, clay-rich sediments may contain large volumes of iron, in a variety of states (e.g., reduced, colloidal) that can be utilized by microbes (Entsch et al., 1983), facilitating significant microbial blooms.

In the lowstand systems tract of Reef Complex #2, the rubble deposits of fragmented stromatoporoids, toppled and fragmented corals, and intraclastic sediment that form the outer buildup attest to intense storm (hurricanes?) activity. Stromatoporoid and coral rubble in the inner buildup and reef front support the interpretation that storms affected development of the lowstand systems tract, and the green clays that cover stromatoporoids in the reef front were probably deposited from suspension as storm activity waned (cf. MacNeil and Jones, 2006a). The most likely source for these muds are green marls and shales situated on the ramp basinwards of Reef Complex #2, in deeper water, that were put into suspension by storm wave energy (Fig. 9C). Such deposits characterize deeper-water deposits of Member E in the underlying Escarpment Formation, which may have been exposed or only thinly covered as the Alexandra Reef System developed. Fluvial input of the mud is discounted because it is not found in landward portions of the lowstand systems tract, including the protected area of the intra-buildup low, or in the coastal-plain deposits at the top of Reef Complex #1 (cf. MacNeil and Jones, 2006b).

Storm disturbance of the offshore oceanic system therefore may have caused significant nutrient enrichment of the seawater circulating over the lowstand systems tract by (1) mixing of the surface layer with stratified layers of nutrient-enriched seawater at depth, and advection of these waters onto the reef complex, and (2) release of nutrient reservoirs (e.g., iron) hosted in the marls and green shales situated further out on the ramp (Fig. 9C). With respect to the latter, the gentle slope ($< 0.1°$) of the ramp was particularly important because the sea-level fall of ~ 17 m may have brought several thousand square kilometers of deeper, outer-ramp marls and shales above storm wave base, especially if the lowstand systems tract was characterized by exceptionally intense storms with deeper-than-average wave bases.

The synergy of storms with other mechanisms was also probably important. Heavy rains associated with the storms would have increased the volume of terrestrial runoff and groundwater seepage, thereby increasing nutrient delivery from these mechanisms. In addition, the havoc that was wreaked upon the reef-building communities in the lowstand systems tract must have caused significant stromatoporoid and coral mortality. It would have taken several weeks to months for new stromatoporoids and corals to start colonizing the decimated reef complex and to start developing new framework. It is in this post-storm window of significant metazoan mortality, positively correlated with nutrient enrichment from the different sources, that microbial blooms would have flourished in the reefal environs (e.g., renalcid boundstone lithifying new rubble deposits at the outer buildup). From modern reefs, this type of post-storm shift to algae and microbes (for a period of weeks to several months), because of coral mortality and the slow speed of new coral growth, relative to the fast growth rates of algae and microbes, is well documented (e.g., Stoddart, 1969; Woodley et al., 1981; Ostrander et al., 2000; McManus and Polsenberg, 2004).

Seasonally Forced Deepening of the Mixed Layer

Advection of seawater that has been enriched in nutrients through seasonally forced deepening of the mixed layer is a mechanism of nutrient supply that carbonate ramps located in intracratonic basins and marginal seas, with weakly stratified water columns, may be particularly susceptible to. This is especially true for periods of sea-level lowstand, when water depth is relatively low and the basins may be increasingly restricted. A modern example is the arid, partly restricted Gulf of Aqaba, where vertical stratification of the water column, because of inflow of warm water from the Red Sea and minimal seasonal variations in salinity, is weak (Reiss and Hottinger, 1984; Genin et al., 1995). As a result, vertical variations in water density are determined primarily by temperature, and the mixed layer deepens in the winter when the sea-surface temperature cools by ~ 5° C and surface waters sink. The sinking, oxygenated waters mix with relatively nutrient-rich waters below the euphotic zone, and nutrients are transported upwards into the photic zone (Genin et al., 1995). This process in the Gulf of Aqaba deepens the mixed layer every winter to depths of 500–600 m and causes significant nutrient enrichment in the spring. The potential impact of this mechanism was demonstrated during an unusually cool winter in 1991–1992 with a deepening of the mixed layer to ~ 850 m (Genin et al., 1995). The level of nutrification that resulted, when the nutrient-rich seawater was advected onto reefs along the northern margin of the gulf in March of 1992, caused significant algal blooms that blanketed the reefs with algal mats that exceeded 15 cm in thickness and caused notable coral death (Genin et al., 1995).

Relating this type of mechanism to the region that contained the Alexandra Reef System is speculative, but not unreasonable, given that a number of parallels can be drawn between the inland sea on the western margin of Laurussia with the Gulf of Aqaba. First, Whalen (1995) argued that, like the partly restricted Gulf of Aqaba, tectonic barriers along the western margin of Laurussia partly restricted its inland sea from open-ocean circulation, such that contrary to its eastern-ocean position, vast carbonate platforms were able to develop. An implication of this interpretation is that the oceanic water flowing into the basin, compensating for evaporative drawdown, would have been limited to less-dense waters closest to the ocean's surface (because of barriers at depth). Second, similarly to the arid climate of the Gulf of Aqaba, this region of Laurussia, within 0–10° of the equator, was hot and semiarid (Golonka et al., 1994). Third, akin to the Gulf of Aqaba, the basin configuration was long, narrow, and deep. Whalen (1995) argued that such configurations do not favor intrabasin gyral circulation, which may have limited the strength of its stratification. Weak stratification, such that cooling of the sea surface by only a few degrees Celsius results in significant vertical mixing, is integral to this mechanism of nutrient enrichment.

The punctuated global cooling in the Frasnian (cf. Stanley, 1988; Copper, 2002; Joachimski et al., 2004), which MacNeil and Jones (2006a) inferred to have caused the sea-level fall that terminated Reef Complex #1 and exposed platforms throughout the basin, may have perturbed the climate such that a seasonally forced mechanism of nutrification, similar to the one that operates in the Gulf of Aqaba, was activated. This may have been particularly important in the falling-stage systems tract, when storm events appear to have been less important. This mechanism does not require the surface layer to mix with waters at hundreds of meters of depth. The nutricline in inland and marginal seas can be < 100 m deep (e.g., Tyson and Pearson, 1991; Wignall and Newton, 2001) and Chow et al. (1995) interpreted dysoxic–anoxic, presumably nutrient-rich water masses in the Late Devonian of western Canada at depths as shallow as 39 m (Fig. 9B). Such water masses are easily within the range of seasonally forced mixing (and storm events), especially when sea-level position is lowered.

DISCUSSION

Carbonate platforms and reef systems from the Devonian have been studied extensively, but as Kiessling et al. (1999) demonstrated, consistent with the opinions of Moore (1988) and Wood (1993), there are few Devonian reef systems with "good data". even though some specific reef systems (e.g., the Lennard Shelf in the Canning Basin, Australia) are well studied, the paleoecology of Devonian reef-building communities remains scarcely investigated and largely enigmatic (Moore, 1988; Wood, 1993; Kershaw, 1998). In this regard, our understanding of Devonian reefs and carbonate platforms, despite their global significance, is based largely on assumptions (cf. Mountjoy, 1980).

A common, prevalent assumption (Wood, 1998a), exemplified by existing facies models (e.g., Machel and Hunter, 1994), is that stromatoporoid–microbe reef facies were the norm in pre-Famennian Devonian reefs (e.g., Kershaw, 1998; Wood, 1999). This may be partly the result of a bias towards the well-studied and famous Canning Basin reef systems in Australia, from which numerous types and fabrics of microbial carbonates have been documented (e.g., Wray, 1967; Playford et al., 1976; Playford, 1980; George, 1999; Wood, 1998b, 2000, 2004; Chow and George, 2004). With the advent of sequence stratigraphy, attempts have also been made to correlate specific microbial facies, assumed to be the norm, with specific systems tracts (cf. George, 1999). Whalen et al. (2002), for example, concluded that Late Devonian highstand systems tracts in Devonian reefs of western Canada are characterized by stromatoporoid–*Renalcis* associations. Although this may apply to highstand systems tracts in the Ancient Wall and Miette reef systems of Alberta, it is incorrect to apply this conclusion across the basin to all platforms, as indicated by the lack of *Renalcis* in the highstand systems tract of Reef Complex #1 in the Alexandra Reef System. Indeed, the general assumption that stromatoporoid–microbe associations characterize Devonian reef facies ignores early observations that the prevalence of microbes in Devonian reefs is highly variable between geographic regions (Wray, 1967; Krebs, 1974; Burchette, 1981). Krebs (1974), for example, noted that *Renalcis* and related microbes (referred to as algae in that paper) are rarely found in Devonian reef complexes in Germany. Similarly, various studies of Devonian reefs in the Western Canada Sedimentary Basin (e.g., Fischbuch, 1968; Leavitt, 1968; Noble, 1970; Dolphin and Klovan, 1970; Vopni and Lerbekmo, 1972) have not identified *Renalcis*, and related forms, as being present.

The lack of microbial constituents in some Devonian reefs is exemplified by the sharp facies contrasts in the Alexandra Reef System (Fig. 9A). The significant accumulations of renalcid boundstone and peri-reefal stromatolites found in the lower parts of Reef Complex #2 disappear upwards and are not found

Nutrient-Gradient Model

Increasing nutrient levels →			
	Oligotrophic	Mesotrophic	Eutrophic
Biotic composition of framework facies	Stromatoporoid-dominated reef facies Stromatoporoid-tabulate coral reef facies Stromatoporoid-tabulate-rugose coral reef facies	Stromatoporoid-microbe[1] reef facies Stromatoporoid-coral[2]-microbe[1] reef facies	Megalodont-microbe[3] facies (replace reef facies)

[1] *Renalcis*, *Shuguria*, *Izhella*, stromatolites
[2] Tabulate and/or rugose corals
[3] Calcimicrobes, stromatolites, thrombolites

FIG. 10.—Nutrient-gradient model for Devonian reef systems. Devonian reef communities are capable of inhabiting oligotrophic and mesotrophic environments. Only in eutrophic environments does platform demise take place, marked by replacement of reef facies with megalodont–microbe associations (cf. Eliuk, 1998).

in any part of Reef Complex #1. Consistent with the complex ecology of modern reefs, such variations are probably not the result of random coincidences. Rather, they may offer valuable insight into how different ecological and environmental variables controlled these carbonate systems. The interpretation that the metazoan–microbe reef facies in Reef Complex #2 represent mesotrophic conditions therefore forms the premise for a new, nutrient-gradient model that explains variations in the biotic compositions of Devonian reef facies (Fig. 10). The use of changes in reef-facies composition to identify fluctuating nutrient levels in these systems, if integrated with potential mechanisms for such fluctuations, should be a powerful tool for interpreting the significance of lateral and vertical facies changes in these systems and understanding aspects of the local climate, sea-level change, and the surrounding oceanographic system (e.g., seasonality, importance of storm events). As part of this, other variables that may accompany mechanisms of nutrification, such as changes in salinity and temperature, can also be considered. The nutrient-gradient model supports the first-order conclusion of Kiessling et al. (1999) that Givetian–Frasnian reefs lived in environments with high nutrient concentrations. This was based on their continent-reconstruction maps, which placed most studied reefs in regions with > 70% probability of upwelling.

The breadth of oligotrophic to eutrophic environments in the nutrient-gradient model (see Hallock, 1987, for a detailed review of the trophic-resource continuum) is based on consideration of the nutritional mode (e.g., phototrophic, heterotrophic) of the different reef-building constituents and the resources that are required to fulfill their nutritional needs. Reef-building metazoans in the Alexandra Reef System, like many other Devonian reefs, include stromatoporoids, tabulate corals, and rugose corals. The nutritional mode of these fossil metazoans is debatable, but the general consensus is that stromatoporoids and tabulate corals were probably mixotrophs (Coates and Jackson, 1987; Wood, 1993, 1995), and stromatoporoid–tabulate coral associations are therefore considered indicative of oligotrophic environments (Hallock and Schlager, 1986; Hallock, 1987; Wood, 1993, 1995; Kershaw, 1998; Whalen et al., 2002). Rugose corals, in contrast, are generally interpreted to have been heterotrophic (Coates and Jackson, 1987; Wood, 1993, 1995), although it should be noted that this interpretation is biased towards solitary and pseudocolonial morphologies (Coates and Jackson, 1987). As heterotrophs, rugose corals are generally treated as indicative of mesotrophic environments (Wood 1993, 1995).

Rugose corals, however, are well documented from the soft-substrate reef-front and fore-reef environments of a number of Devonian reef complexes that consisted of stromatoporoid (± tabulate coral)–dominated reef facies in coeval, shallower environments (e.g., Klovan, 1964; Embry and Klovan, 1972; Burchette, 1981; MacNeil and Jones, 2006a). As such, the occurrence of rugose corals cannot be treated as an unequivocal indicator of mesotrophic conditions in a systems tract (e.g., as done in Mallamo and Geldsetzer, 1991) and they may have been specially adapted to soft-substrate oligotrophic environments. Furthermore, mixed stromatoporoid–tabulate–rugose-coral asso-

TABLE 1.—Trophic ranges of biota found in Devonian reefs.

	Oligotrophic	Mesotrophic	Eutrophic
Stromatoporoids	X	X	
Tabulate Corals	X	X	
Rugose Corals	X	X	
Renalcis		X	X?
Shuguria		X	X?
Izhella		X	X?
Girvanella	X	X	X
Rothpletzella	X?	X	X?
Stromatolites		X	X
Megalodonts	X?	X	X

ciations are common in Devonian reefs, indicating that the ecological niches of these organisms overlapped. Therefore, in contrast to general arguments that rugose corals indicate mesotrophic environments, and stromatoporoids and tabulate corals indicate oligotrophic environments, all three of these metazoans are treated by the nutrient-gradient model as having been adapted to living in mesotrophic *and* oligotrophic environments (Table 1). The presence or absence of microbial components determines which condition existed. Stromatoporoid–tabulate-coral associations, for example, are treated by the nutrient-gradient model as indicative of oligotrophic environments (cf. Hallock and Schlager, 1986, Hallock, 1987, and Wood, 1993, 1995). Likewise, stromatoporoid–tabulate–rugose-coral associations, without microbial components, are treated in the model as indicative of oligotrophic conditions (Fig. 10). In oligotrophic environments, the nutrient-gradient model does not preclude lateral variations between different types of metazoans (e.g., as may be expected between different reef zones) or community succession. In Reef Complex #1, for example, stromatoporoid–coral-dominated frameworks are vertically succeeded by stromatoporoid-dominated frameworks.

Microbial components and deposits that contributed to reef-framework development in Reef Complex #2, and that indicate mesotrophic conditions in the nutrient-gradient model (Fig. 10, Table 1), include *Renalcis, Shuguria, Izhella*, a clotted form of *Renalcis*, and stromatolites with a micro-peloidal fabric. *Epiphyton* is not found in the Alexandra Reef System (with the possible exception of one specimen), and *Rothpletzella* and *Girvanella* are minor, with a broad distribution. The broad distribution of *Rothpletzella* and *Girvanella* may suggest that these microbes are poor indicators of fluctuating nutrient levels. Certainly, *Girvanella* is well known throughout most of the Paleozoic (Pratt 2001), and *Girvanella*–oncolites, characteristic of low-light, deeper-water environments with low sedimentation rates, are common deposits in the Paleozoic (Peryt, 1981; Riding, 1983).

The nutrient-gradient model indicates that stromatoporoids, tabulate corals, and rugose corals were capable of living in a broader range of nutrient levels than traditionally considered and broader ranges than modern reef-building corals, which are generally restricted to oligotrophic environments (Hallock and Schlager, 1986; D'Elia and Wiebe, 1990; Mutti and Hallock, 2003). In addition to the ability of stromatoporoids to live in both oligotrophic and mesotrophic environments, many stromatoporoids also appear to have been capable of colonizing and developing over soft substrates (e.g., Fig. 7) as individual entities (Wood, 1995; Kershaw, 1998). This was the case in the initiation and differentiation stages of Reef Complex #2, in which the stromatoporoids did not coalesce to form framestone (MacNeil and Jones, 2006a). Rather, each skeleton existed as an individual entity, commonly bound together by microbial elements. This ability of stromatoporoids to colonize a wider range of substrates than modern scleractinian corals, which are restricted to areas where hard substrates are available (Wood, 1993), may partly explain why many Devonian reefs, akin to the Alexandra Reef System, have low-profile, ramp-like geometries with reef development spread over a broad region (Wilson, 1975; Mountjoy, 1980; Walls and Burrowes, 1985; Eliuk, 1998; Kershaw, 1998; da Silva and Boulvain, 2004). This type of platform geometry is very different from that of modern coralgal-reef platforms with narrow reef zones and steep marginal slopes (Wilson, 1975; James and Ginsburg, 1979; Enos and Moore, 1983).

A broad implication of the nutrient-gradient model is that Devonian reefs and their platforms were not highly susceptible to nutrient-invoked drowning (cf. Kiessling et al., 1999). As metazoan-dominated oligotrophic reef communities could be replaced by metazoan–microbe-communities as nutrients became more available (Fig. 10), elevated nutrient levels should not have caused reef demise or platform drowning, except in cases of extreme nutrification. This supports the idea of Eliuk (1998) that replacement of Devonian reef facies by megalodont–microbe associations (a non-reef-building phase) reflects significant eutrophication. Accordingly, replacement of reef facies by megalodont–microbe deposits is treated as an end member in the nutrient-gradient model, indicative of nutrient-invoked platform drowning. The broad range of trophic resources that Devonian reef-building communities could utilize may also partly account for the global expansiveness of these systems, including their ability to develop in regions of upwelling (Kiessling et al., 1999), which normally preclude reefs.

The prediction that Devonian reefs were expansive and resilient to nutrification because of the wide range of trophic resources that they could exploit finds support in the Canning Basin reef systems of northwestern Australia. These reef systems, with siltstones to pebble conglomerates deposited periodically across the span of back-reef to fore-reef environments (Read, 1973; Playford et al., 1989; George, 1999; Wood, 2000), were almost certainly nutrified (also suggested by Wood, 2000) because significant terrestrial input is a well-established cause of nutrification (Garrels and MacKenzie, 1971; Hallock and Schlager, 1986; Hallock, 1987; Mutti and Hallock, 2003; McManus and Polsenberg, 2004). Given the tectonically active extensional setting of the Canning Basin in the Devonian, it is also possible that many of the neptunian dikes in these reefs were connected with deeper-seated fractures and faults (Dörling et al., 1996), spewing nutrients as subsurface fluids vented (cf. Playford and Wallace, 2001). Thus, it is unlikely that these reefs developed in oligotrophic conditions, which is consistent with the volume of siliciclastic–stromatoporoid–microbial carbonate facies (e.g., Wood, 2000) that characterize these reef systems. Kiessling et al. (1999) also concluded that these reef systems developed in mesotrophic conditions, given their close proximity to a region that was susceptible to upwelling.

CONCLUSIONS

The Alexandra Reef System demonstrates that the interaction of platform geometry, sea-level change, and nutrient levels was an important control on Devonian reefs. The nutrient-gradient model indicates that Devonian reefs and associated platforms were formed by dynamic carbonate-producing communities, and carries significant implications for interpreting Devonian reef facies. Conclusions from this study include:

(1) Ramp-situated carbonate systems—of any geological age—may be increasingly susceptible to nutrification as sea level falls and in subsequent lowstand systems tracts. In contrast, transgressive systems on ramps may not be as susceptible to nutrification. Because nutrients are a fundamental control on the types of carbonates deposited, different systems tracts on carbonate ramps may be characterized by different types of deposits (e.g., photozoan, heterozoan assemblages).

(2) Stromatoporoids, tabulate corals, and rugose corals lived in oligotrophic and mesotrophic environments. The occurrence of stromatolites and calcimicrobes in Devonian reef framework is not characteristic of all Devonian reefs, and is important because these types of deposits indicate that nutrients were not limited in the environment. The mixed carbonate–siliciclastic Canning Basin reef systems were almost certainly nutrified (cf. Kiessling et al., 1999; Wood, 2000).

(3) Devonian platforms were not highly susceptible to nutrient-invoked drowning because changes in community composition facilitated continued carbonate production and accumulation as nutrient levels increased. The wide range of environments and substrates that could be colonized by Devonian reef builders may have been an important factor in the global expansiveness of these systems, and it probably explains why these platforms were generally characterized by broad areas of reef-framework development with stratigraphic architectures that are markedly different from modern reef-rimmed carbonate platforms.

(4) Use of the nutrient-gradient model to recognize fluctuating nutrient levels, and interpretation of the mechanisms causing such fluctuations, is a critical step towards understanding variations in Devonian reef facies and local controls on the initiation, development, and demise of Devonian carbonate platforms.

ACKNOWLEDGMENTS

The authors are grateful for the financial support of this project by the Canadian Circumpolar Institute of the University of Alberta, Northern Science Training Program of the Department of Indian Affairs and Northern Development, the American Association of Petroleum Producers Grant-in-Aid Foundation (to MacNeil), Natural Sciences and Engineering Research Council of Canada (PGS B to MacNeil and grant A6090 to Jones), the CR Stelck Chair of Petroleum Geology (held by Jones), Anadarko Canada Corporation, and Canadian Natural Resources Limited. All fieldwork was conducted with permits granted by the Aurora Research Institute of the Northwest Territories. N. Darlow is thanked for many insightful discussions on current concepts in ecology. Comments on an earlier draft of this manuscript by W. Morgan and B. Witzke were greatly appreciated.

REFERENCES

Babin, S.M., Carton, J.A., Dickey, T.D., and Wiggert, J.D., 2004, Satellite evidence of hurricane-induced phytoplankton blooms in an oceanic desert: Journal of Geophysical Research–Oceans, v. 109, C03043, doi:10.1029/2003JC001938.

Blanchon, P., and Perry, C.T., 2004, Taphonomic differentiation of *Acropora palmata* facies in cores from Campeche Bank Reefs, Gulf of Mexico: Sedimentology, v. 51, p. 53–76.

Blanchon, P., Jones, B., and Kalbfleisch, W., 1997, Anatomy of a fringing reef around Grand Cayman: storm rubble, not coral framework: Journal of Sedimentary Research, v. 67, p. 1–16.

Burchette, T.P., 1981, European Devonian reefs: a review of current concepts and models, *in* Toomey, D.F., ed., European Reef Models: SEPM, Special Publication 30, p. 85–142.

Camoin, G.F., Geutret, P., Montagionni, L.F., and Cabioch, G., 1999, Nature and environmental significance of microbialites in Quaternary reefs: the Tahiti paradox: Sedimentary Geology, v. 126, p. 271–304.

Catuneanu, O., 2002, Sequence stratigraphy of clastic systems: concepts, merits, and pitfalls: Journal of African Earth Sciences, v. 35, p. 1–43.

Chow, N., and George, A.D., 2004, Tepee-shaped agglutinated microbialites: An example from a Famennian carbonate platform on the Lennard Shelf, northern Canning Basin, Western Australia: Sedimentology, v. 51, p. 253–265.

Chow, N., Wendte, J., and Stasiuk, L.D., 1995, Productivity versus preservation controls on two organic-rich carbonate facies in the Devonian of Alberta—sedimentological and organic petrological evidence: Bulletin of Canadian Petroleum Geology, v. 43, p. 433-460.

Coates, A.G., and Jackson, J.B.C., 1987, Clonal growth, algal symbiosis, and reef formation by corals: Paleobiology, v. 13, p. 363–378.

Copper, P., 1989, Enigmas in Phanerozoic reef development, *in* Jell, P.A., and Pickett, J.W., eds., Fossil Cnidaria 5: Association of Australasian Palaeontologists, Memoir, p. 371–385.

Copper, P., 2002, Reef development at the Frasnian/Famennian mass extinction boundary: Palaeogeography, Palaeoclimatology, Palaeoecology, v. 181, p. 27–65.

Costa, O.S., Leao, Z., Nimmo, M., and Attrill, M.J., 2000, Nutrification impacts on coral reefs from northern Bahia, Brazil: Hydrobiologia, v. 440, p. 307–315.

Cuet, P., Naim, O., Faure, G., and Conan, J.Y., 1988, Nutrient-rich groundwater impact on benthic communities of La Saline fringing reef (Reunion Island, Indian Ocean): preliminary results, *in* Choat, J.H., ed., 6th International Coral Reef Symposium, Australia, Proceedings, p. 207-212.

da Silva, A.C., and Boulvain, F., 2004, From palaeosols to carbonate mounds: facies and environments of the middle Frasnian platform in Belgium: Geological Quarterly, v. 48, p. 253–265.

D'elia, C.F., and Wiebe, W.J., 1990, Biogeochemical nutrient cycles in coral-reef ecosystems, *in* Dubinsky, Z., ed., Coral Reefs: Ecosystems of the World 25: Amsterdam, Elsevier, p. 49-74.

D'elia, C.F., Webb, K.L., and Porter, J.W., 1981, Nitrate-rich groundwater inputs to Discovery Bay, Jamaica: A significant source of N to local coral reefs?: Bulletin of Marine Science, v. 31, p. 903–910.

Dolphin, D.R., and Klovan, J.E., 1970, Stratigraphy and paleoecology of an Upper Devonian Carbonate Bank, Saskatchewan River Crossing, Alberta: Bulletin of Canadian Petroleum Geology, v. 18, p. 289–331.

Dörling, S.L., Dentith, M.C., Groves, D.I., Playford, P.E., Vearncombe, J.R., Muhling, P., and Windrim, D., 1996, Heterogeneous brittle deformation in the Devonian carbonate rocks of the Pillara Range, Canning Basin: implications for the structural evolution of the Lennard Shelf: Australian Journal of Earth Sciences, v. 43, p. 15–29.

Dupraz, C., and Strasser, A., 2002, Nutritional modes in coral–microbialite reefs (Jurassic, Oxfordian, Switzerland): Evolution of trophic structure as a response to environmental change: Palaios, v. 17, p. 449–471.

Eliuk, L.S., 1998, Big bivalves, algae, and the nutrient poisoning of reefs: a tabulation with examples from the Devonian and Jurassic of Canada, *in* Johnston, P.A., and Haggart, J.W., eds., Bivalves: An Eon of Evolution—Paleobiological Studies Honoring Norman D. Newell: Calgary, University of Calgary Press, p. 157–184.

Embry, A.F., and Klovan, J.E., 1972, Absolute water depth limits of Late Devonian paleoecological zones: Geologische Rundschau, v. 61, p. 672-686.

Enos, P., and Moore, C.H., 1983, Fore-reef slope, *in* Scholle, P.A., Bebout, D.G., and Moore, C.H., eds., Carbonate Depositional Environments: American Association of Petroleum Geologists, Memoir 33, p. 507–537.

Entsch, B., Sim, R.G., and Hatcher, B.G., 1983, Indications from photosynthetic components that iron is a limiting nutrient in primary producers on coral reefs: Marine Biology, v. 73, p. 17–30.

Fischbuch, N.R., 1968, Stratigraphy, Devonian Swan Hills Reef Complexes of central Alberta: Bulletin of Canadian Petroleum Geology, v. 16, p. 444–556.

Garrels, R.M., and Mackenzie, F.T., 1971, Evolution of Sedimentary Rocks: New York, Norton and Company, 397 p.

Genin, A., Lazar, B., and Brenner, S., 1995, Vertical mixing and coral death in the Red-Sea following the eruption of Mount-Pinatubo: Nature, v. 377, p. 507–510.

George, A.D., 1999, Deep-water stromatolites, Canning Basin, Northwestern Australia: Palaios, v. 14, p. 493–505.

Golonka, J., Ross, M.I., and Scotese, C.R., 1994, Phanerozoic paleogeographic and paleoclimatic modeling maps, *in* Embry, A.F., Beauchamp, B., and Glass, D.J., eds., Pangea: Global Environments and Resources: Canadian Society of Petroleum Geologists, Memoir 17, p. 1–47.

HADLEY, M.G., AND JONES, B., 1990, Lithostratigraphy and nomenclature of Devonian strata in the Hay River area, Northwest Territories: Bulletin of Canadian Petroleum Geology, v. 38, p. 332–356.

HALLOCK, P., 1987, Fluctuations in the trophic resource continuum: A factor in global diversity cycles?: Paleooceanography, v. 2, p. 457–471.

HALLOCK, P., 2005, Global change and modern coral reefs: New opportunities to understand shallow-water carbonate depositional processes: Sedimentary Geology, v. 175, p. 19–33.

HALLOCK, P., AND SCHLAGER, W., 1986, Nutrient excess and the demise of coral reefs and carbonate platforms: Palaios, v. 1, p. 389–398.

HALLOCK, P., MULLERKARGER, F.E., AND HALAS, J.C., 1993, Coral-reef decline: Research & Exploration, v. 9, p. 358–378.

HOLLOWAY, P.E., HUMPHRIES, S.E., ATKINSON, M., AND IMBERGER, J., 1985, Mechanisms for nitrogen supply to the Australian north west shelf: Australian Journal of Marine and Freshwater Research, v. 36, p. 753–764.

HUNT, D., AND TUCKER, M.E., 1993, Sequence stratigraphy of carbonate shelves with an example from the mid-Cretaceous (Urgonian) of southeast France, *in* Posamentier, H.W., Summerhayes, C.P., Haq, B.U., and Allen, G.P., eds., Sequence Stratigraphy and Facies Associations: International Association of Sedimentologists, Special Publication 18, p. 307–341.

JAMES, N.P., AND GINSBURG, R.N., 1979, The Seaward Margin of the Belize Barrier and Atoll Reefs: International Association of Sedimentologists, Special Publication 3, 191 p.

JAMES, N.P., AND KENDALL, A.C., 1992, Introduction to carbonate and evaporite facies models, *in* Walker, R.G., and James, N.P., eds., Facies Models—Response to Sea Level Change: St. John's, Geological Association of Canada, p. 265–275.

JOACHIMSKI, M.M., VAN GELDERN, R., BREISIG, S., BUGGISCH, W., AND DAY, J., 2004, Oxygen isotope evolution of biogenic calcite and apatite during the Middle and Late Devonian: International Journal of Earth Sciences, v. 93, p. 542–553.

JOHANNES, R.E., 1980, The ecological significance of the submarine discharge of groundwater: Marine Ecology Progress Series, v. 3, p. 365–373.

JOHNSON, J.G., KLAPPER, G., AND SANDBERG, C.A., 1985, Devonian eustatic fluctuations in Euramerica: Geological Society of America, Bulletin, v. 96, p. 567–587.

KERSHAW, S., 1998, The applications of stromatoporoid palaeobiology in palaeoenvironmental analysis: Palaeontology, v. 41, p. 509–544.

KIESSLING, W., FLÜGEL, E., AND GOLONKA, J., 1999, Paleoreef maps: Evaluation of a comprehensive database on Phanerozoic reefs: American Association of Petroleum Geologists, Bulletin, v. 83, p. 1552–1587.

KLAPPER, G., AND LANE, H.R., 1989, Frasnian (Upper Devonian) conodont sequence at Luscar Mountain and Mount Haultain, Alberta Rocky Mountains, *in* McMillan, N.J., Embry, A.F., and Glass, D.J., eds., Devonian of the World: Canadian Society of Petroleum Geologists, Memoir 14, p. 469–478.

KLOVAN, J.E., 1964, Facies analysis of the Redwater Reef complex, Alberta, Canada: Bulletin of Canadian Petroleum Geology, v. 12, p. 1–100.

KREBS, W., 1974, Devonian carbonate complexes of central Europe, *in* Laporte, L.F., ed., Reefs in Time and Space: SEPM, Special Publication 18, p. 155–208.

KUFFNER, I.B., AND PAUL, V.J., 2001, Effects of nitrate, phosphate and iron on the growth of macroalgae and benthic cyanobacteria from Cocos Lagoon, Guam: Marine Ecology Progress Series, v. 222, p. 63–72.

LAPOINTE, B.E., 1997, Nutrient thresholds for bottom-up control of macroalgal blooms on coral reefs in Jamaica and southeast Florida: Limnology and Oceanography, v. 42, p. 1119–1131.

LAPOINTE, B.E., BARILE, P.J., YENTSCH, C.S., LITTLER, M.M., LITTLER, D.S., AND KAKUK, B., 2004, The relative importance of nutrient enrichment and herbivory on macroalgal communities near Norman's Pond Cay, Exumas Cays, Bahamas: a "natural" enrichment experiment: Journal of Experimental Marine Biology and Ecology, v. 298, p. 275–301.

LEAVITT, E.M., 1968, Petrology, paleontology, Carson Creek North Reef Complex: Bulletin of Canadian Petroleum Geology, v. 16, p. 298–413.

LEWIS, J.B., 1987, Measurements of groundwater seepage flux onto a coral reef: spatial and temporal variations: Limnology and Oceanography, v. 32, p. 1165–1169.

LIN, I., LIU, W.T., WU, C.C., WONG, G.T.F., HU, C.M., CHEN, Z.Q., LIANG, W.D., YANG, Y., AND LIU, K.K., 2003, New evidence for enhanced ocean primary production triggered by tropical cyclone: Geophysical Research Letters, v. 30(13), 1718, doi:10.1029/2003GL017141, 2003.

LITTLER, M.M., AND LITTLER, D.S., 1984, Models of tropical reef biogenesis: the contribution of algae: Progress in Phycological Research, v. 3, p. 323–364.

LOBBAN, C.S., HARRISON, P.J., AND DUNCAN, M.J., 1985, The Physiological Ecology of Seaweeds: Cambridge, U.K., Cambridge University Press, 237 p.

LONGMAN, M.W., 1981, A process approach to recognizing facies of reef complexes, *in* Toomey, D.F., ed., European Fossil Reef Models: SEPM, Special Publication 30, p. 9–40.

MACHEL, H.G., AND HUNTER, I.G., 1994, Facies models for Middle to Late Devonian shallow-marine carbonates, with comparisons to modern reefs: a guide for facies analysis: Facies, v. 30, p. 155–176.

MACNEIL, A.J., AND JONES, B., 2006a, Sequence stratigraphy of a Late Devonian ramp-situated reef system in the Western Canada Sedimentary Basin: dynamic responses to sea-level change and regressive reef development: Sedimentology, v. 53, p. 321–359.

MACNEIL, A.J., AND JONES, B., 2006b, Palustrine deposits on a Late Devonian coastal plain—sedimentary attributes and implications for concepts of carbonate sequence stratigraphy: Journal of Sedimentary Research, v. 76, p. 292–309.

MALLAMO, M.P., AND GELDSETZER, H.H.J., 1991, The western margin of the Upper Devonian Fairholme reef complex, Banff–Kananaskis area, southwestern Alberta: Geological Survey of Canada, Current Research, Part B, Paper 91-1B, p. 59–69.

MARSH, J.A., 1977, Terrestrial inputs of nitrogen and phosphorus on fringing reefs of Guam, *in* Taylor, D.L., ed., 3rd International Coral Reef Symposium, Miami, Proceedings, p. 331–336.

MCLEAN, R.A., AND KLAPPER, G., 1998, Biostratigraphy of Frasnian (Upper Devonian) strata in western Canada, based on conodonts and rugose corals: Bulletin of Canadian Petroleum Geology, v. 46, p. 515–563.

MCMANUS, J.W., AND POLSENBERG, J.F., 2004, Coral–algal phase shifts on coral reefs: ecological and environmental aspects: Progress in Oceanography, v. 60, p. 263–279.

MILLER, M.W., HAY, M.E., MILLER, S.L., MALONE, D., SOTKA, E.E., AND SZMANT, A.M., 1999, Effects of nutrients versus herbivores on reef algae: A new method for manipulating nutrients on coral reefs: Limnology and Oceanography, v. 44, p. 1847–1861.

MOORE, P.F., 1988, Devonian reefs in Canada and some adjacent areas, *in* Geldsetzer, H.H.J., James, N.P., and Tebbutt, G.E., eds., Reefs—Canada and Adjacent Areas: Canadian Society of Petroleum Geologists, Memoir 13, p. 367–390.

MOUNTJOY, E.W., 1980, Some questions about the development of Upper Devonian buildups (reefs), western Canada: Bulletin of Canadian Petroleum Geology, v. 28, p. 315–344.

MUTTI, M., AND HALLOCK, P., 2003, Carbonate systems along nutrient and temperature gradients: some sedimentological and geochemical constraints: International Journal of Earth Sciences (Geologische Rundschau), v. 92, p. 465–475.

NOBLE, J.P.A., 1970, Biofacies analysis, Cairn Formation of Miette Reef Complex (Upper Devonian) Jasper National Park, Alberta: Bulletin of Canadian Petroleum Geology, v. 18, p. 493–543.

OSTRANDER, G.K., MEYER AMSTRONG, K., KNOBBE, E.T., GERACE, D., AND SCULLY, E.P., 2000, Rapid transition in the structure of a coral reef community: The effects of coral bleaching and physical disturbance:

National Academy of Sciences (U.S.A.), Proceedings, v. 97, p. 5297–5302.

PAUL, V.J., THACKER, R.W., BANKS, K., AND GOLUBIC, S., 2005, Benthic cyanobacterial bloom impacts the reefs of South Florida (Broward County, U.S.A.): Coral Reefs, v. 24, p. 693–697.

PERYT, T.M., 1981, Phanerozoic oncoids—an overview: Facies, v. 4, p. 197–214.

PLAYFORD, P.E., 1980, Devonian Great Barrier-Reef of Canning Basin, Western-Australia: American Association of Petroleum Geologists, Bulletin, v. 64, p. 814–840.

PLAYFORD, P.E., AND WALLACE, M.W., 2001, Exhalative mineralization in Devonian reef complexes of the Canning Basin, Western Australia: Economic Geology, v. 96, p. 1595–1610.

PLAYFORD, P.E., COCKBAIN, A.E., DRUCE, E.C., AND WRAY, J.L., 1976, Devonian stromatolites from the Canning Basin, Western Australia, *in* Walter, M.R., ed., Stromatolites: Amsterdam, Elsevier, p. 543–563.

PLAYFORD, P.E., HURLEY, N.F., KERANS, C., AND MIDDLETON, M.F., 1989, Reefal platform development, Devonian of the Canning Basin, Western Australia, *in* Crevello, P.D., Wilson, J.L., Sarg, J.F., and Read, J.F., eds., Controls on Carbonate Platform and Basin Development: SEPM, Special Publication 44, p. 187–202.

POTMA, K., WEISSENBERGER, J.A.W., WONG, P.K., AND GILHOOLY, M.G., 2001, Toward a sequence stratigraphic framework for the Frasnian of the Western Canada Basin: Bulletin of Canadian Petroleum Geology, v. 50, p. 341–349.

PRATT, B.R., 2001, Calcification of cyanobacterial filaments: *Girvanella* and the origin of lower Paleozoic lime mud: Geology, v. 29, p. 763–766.

READ, J.F., 1973, Paleo-environments and paleogeography, Pillara Formation (Devonian), Western Australia: Bulletin of Canadian Petroleum Geology, v. 21, p. 344–394.

REISS, Z., AND HOTTINGER, L., 1984, The Gulf of Aqaba: Ecological Studies, v. 50: Berlin, Springer-Verlag, 354 p.

RIDING, R., 1992, Temporal variation in calcification in marine cyanobacteria: Geological Society of London, Journal, v. 149, p. 979–989.

RIDING, R., 1983, Cyanoliths (cyanoids): oncoids formed by calcified cyanophytes, *in* Peryt, T.M., ed., Coated Grains: Berlin, Springer-Verlag, p. 276–283.

RIDING, R., AND LIANG, L.Y., 2005, Geobiology of microbial carbonates: metazoan and seawater saturation state influences on secular trends during the Phanerozoic: Palaeogeography, Palaeoclimatology, Palaeoecology, v. 219, p. 101–115.

SARG, J.F., 1988, Carbonate sequence stratigraphy, *in* Wilgus, C.K., Hastings, B.S., Kendall, C.G.St.C., Posamentier, H.W., Ross, C.A., and Van Wagoner, J.C., eds., Sea-Level Changes: An Integrated Approach: SEPM, Special Publication 42, p. 155–182.

SCOTESE, C.R., 2004, A continental drift flipbook: Journal of Geology, v. 112, p. 729–741.

SHIAH, F.K., CHUNG, S.W., KAO, S.J., GONG, G.C., AND LIU, K.K., 2000, Biological and hydrographical responses to tropical cyclones (typhoons) in the continental shelf of the Taiwan Strait: Continental Shelf Research, v. 20, p. 2029–2044.

SMITH, V.S., KIMMERER, W.J., LAWS, E.A., BROCK, R.E., AND WALSH, T.W., 1981, Kaneohe Bay sewage diversion experiment: perspectives on ecosystem responses to nutritional perturbation: Pacific Science, v. 35, p. 279–402.

SPRACHTA, S., CAMOIN, G., GOLUBIC, S., AND LE CAMPION, T., 2001, Microbialites in a modern lagoonal environment: nature and distribution, Tikehau atoll (French Polynesia): Palaeogeography, Palaeoclimatology, Palaeoecology, v. 175, p. 103–124.

STANLEY, S.M., 1988, Climatic cooling and mass extinction of Paleozoic reef communities: Palaios, v. 3, p. 228–232.

STODDART, D.R., 1969, Ecology and morphology of recent coral reefs: Cambridge Philosophical Society, Biological Reviews, v. 44, p. 433–498.

SUBRAHMANYAM, B., RAO, K.H., RAO, N.S., MURTY, V.S.N., AND SHARP, R.J., 2002, Influence of a tropical cyclone on Chlorophyll-a concentration in the Arabian Sea: Geophysical Research Letters, v. 29(22), 2065, doi:10.1029/2002GL015892, 2002.

TYSON, R.V., AND PEARSON, T.H., 1991, Modern and ancient continental shelf anoxia: an overview, *in* Tyson, R.V., and Pearson, T.H., eds., Modern and Ancient Continental Shelf Anoxia: Geological Society of London, Special Publication 58, p. 1–24.

VOPNI, L.K., AND LERBEKMO, J.F., 1972, The Horn Plateau Formation: A Middle Devonian Coral Reef, Northwest Territories, Canada: Bulletin of Canadian Petroleum Geology, v. 20, p. 498–548.

WALLS, R.A., AND BURROWES, G., 1985, The role of cementation in the diagenetic history of Devonian reefs, western Canada, *in* Schneidermann, N., and Harris, P.M., eds., Carbonate Cements: SEPM, Special Publication 36, p. 185–220.

WHALEN, M.T., 1995, Barred basins—A model for eastern ocean-basin carbonate platforms: Geology, v. 23, p. 625–628.

WHALEN, M.T., EBERLI, G.P., VAN BUCHEM, F.S.P., MOUNTJOY, E.W., AND HOMEWOOD, P.W., 2000, Bypass margins, basin-restricted wedges, and platform-to-basin correlation, Upper Devonian, Canadian Rocky Mountains: Implications for sequence stratigraphy of carbonate platform systems: Journal of Sedimentary Research, v. 70, p. 913–936.

WHALEN, M.T., DAY, J., EBERLI, G.P., AND HOMEWOOD, P.W., 2002, Microbial carbonates as indicators of environmental change and biotic crises in carbonate systems: examples from the Late Devonian, Alberta basin, Canada: Palaeogeography, Palaeoclimatology, Palaeoecology, v. 181, p. 127–151.

WIGNALL, P.B., AND NEWTON, R., 2001, Black shales on the basin margin: a model based on examples from the Upper Jurassic of the Boulonnais, northern France: Sedimentary Geology, v. 144, p. 335–356.

WILSON, J.L., 1975, Carbonate Facies in Geologic History: New York, Springer Verlag, 471 p.

WOOD, R., 1993, Nutrients, predation and the history of reef-building: Palaios, v. 8, p. 526–543.

WOOD, R., 1995, The changing biology of reef-building: Palaios, v. 10, p. 517–529.

WOOD, R., 1998a, The ecological evolution of reefs: Annual Review of Ecology and Systematics, v. 29, p. 179–206.

WOOD, R., 1998b, Novel reef fabrics from the Devonian Canning Basin, Western Australia: Sedimentary Geology, v. 121, p. 149–156.

WOOD, R., 1999, Reef Evolution: Oxford, U.K., Oxford University Press, 414 p.

WOOD, R., 2000, Palaeoecology of a Late Devonian back reef: Canning Basin, Western Australia: Palaeontology, v. 43, p. 671–703.

WOOD, R., 2004, Palaeoecology of a post-extinction reef: Famennian (Late Devonian) of the Canning Basin, north-western Australia: Palaeontology, v. 47, p. 415–445.

WOODLEY, J.D., CHORNESKY, E.A., CLIFFORD, P.A., JACKSON, J.B.C., KAUFMAN, L.S., KNOWLTON, N., LANG, J.C., PEARSON, M.P., PORTER, J.W., ROONEY, M.C., RYLAARSDAM, K.W., TUNNICLIFFE, V.J., WAHLE, C.M., WULFF, J.L., CURTIS, A.S.G., DALLMEYER, M.D., JUPP, B.P., KOEHL, M.A.R., NEIGEL, J., AND SIDES, E.M., 1981, Hurricane Allen's impact on Jamaican coral reefs: Science, v. 214, p. 749–755.

WRAY, J.L., 1967, Upper Devonian calcareous algae from the Canning Basin, western Australia: Golden, Colorado, Colorado School of Mines, Professional Contributions 3, 76 p.

MAGNETIC SUSCEPTIBILITY, BIOSTRATIGRAPHY, AND SEQUENCE STRATIGRAPHY: INSIGHTS INTO DEVONIAN CARBONATE PLATFORM DEVELOPMENT AND BASIN INFILLING, WESTERN ALBERTA, CANADA

MICHAEL T. WHALEN
Dept. of Geology and Geophysics, University of Alaska, Fairbanks, Alaska 99775–5780, U.S.A.
e-mail: mtwhalen@gi.alaska.edu
AND
JAMES E. (JED) DAY
Dept. of Geography–Geology, Illinois State University, Normal, Illinois 61790–4400, U.S.A.
e-mail: jeday@ilstu.edu

ABSTRACT: This study applies high-resolution sequence stratigraphy, biostratigraphy, and magnetic susceptibility (MS) stratigraphy to better constrain correlation of upper Middle and Upper Devonian strata and geologic events in western Alberta, Canada. We also explore the potential of MS stratigraphy as a long-range correlation tool and paleoclimatic or oxygen isotope proxy. High-resolution MS data from slope and basin deposits near the isolated Miette and Ancient Wall platforms provide insight into patterns of carbonate-platform development and infilling of the Devonian Alberta basin. Our MS data, combined with conodont and brachiopod biostratigraphic data and sequence stratigraphy, provides additional control on the relative timing of five major and fifteen higher-frequency MS excursions and nine depositional sequences. Sea-level events that initiated deposition of seven of nine late Givetian–early Famennian third order depositional sequences in western Alberta coincide with Devonian transgressive–regressive (T–R) cycles IIa-2 to IIe. Eight of these form the main sequence stratigraphic architectural units of the isolated Miette and Ancient Wall platforms. Sea-level events were identified based on significant sequence stratigraphic horizons, including exposure and marine flooding surfaces, and were biochronologically calibrated using combined conodont and brachiopod biostratigraphy. Identification of sequence boundaries and differentiation of highstand and lowstand slope and basinal deposits was based on the geometry, mineralogy, and clast content of redeposited carbonate units. The magnetic-susceptibility signature of slope and basin facies is also shown to vary systematically within the sequence stratigraphic framework. Spikes in the MS record coincide with events associated with lowstand or initial transgression. The MS stratigraphy displays a consistent pattern across the Alberta basin, with generally higher MS values toward the east. The MS signature is generally low in the late Givetian and early Frasnian (through MN Zone 9) but displays a major bimodal MS increase in the middle to late Frasnian (MN zones 10–11). MS values return to generally lower levels during the late Frasnian (MN zones 12–13) and early Famennian. This general pattern of increasing followed by decreasing MS is interpreted to indicate variations in delivery of magnetically susceptible terrigenous material. The highest MS values correlate directly to the lithologic change associated with an influx of fine-grained siliciclastics in the Mount Hawk Formation. The generally consistent pattern of MS change across the Alberta basin points toward the utility of MS stratigraphy as a regional correlation tool.

Several other positive MS excursions documented here are also associated with increased detrital input and are coeval with decreasing or low oxygen isotope values (increasing or high paleotemperatures) reported from both Laurasia and Gondwana. This relationship implies a paleoclimatic linkage with increasing temperatures and weathering rates resulting in higher detrital input and higher MS values. Published oxygen isotope data are too coarse to conduct high-resolution comparison with our MS data, but the parallel trends noted here suggest that further research on the use of MS as an oxygen isotope or paleoclimate proxy is warranted. The MS signature of coeval Devonian rocks from highly condensed sections in Morocco displays a shape structure similar to our data and reinforces arguments that MS stratigraphy has potential as a long-range correlation tool.

INTRODUCTION

Carbonate-platform systems are particularly sensitive to tectonics, influx of siliciclastic sediments, sea-level change, and biotic or climatic events. Frasnian (early Late Devonian) reef ecosystems represent the maxima in Paleozoic reef development that ended with the Frasnian–Famennian (F–F) mass extinction crises (Lower and Upper Kellwasser events) , which fundamentally altered the composition of Paleozoic reef and carbonate-platform systems (McLaren, 1982; Stearn, 1987; McGhee, 1996; Copper, 2002). One of the most areally extensive Frasnian reef systems in the world was located along the western continental margin of Laurussia in western Canada (Fig. 1).

We report here results of ongoing high-resolution analyses of isolated carbonate reef platforms and associated basin sediments in Devonian outcrops of the Canadian Rocky Mountains, Alberta, and attempt to improve stratigraphic correlation of records of regional and global events during the Middle–Late Devonian. To better document the timing and effects of local and eustatic sea level and of climatic and biotic events, we applied an integrated stratigraphic approach employing several independent methods to improve correlations between Upper Devonian stratigraphic successions in western North America. Our research entails integration of biostratigraphy, high-resolution cyclostratigraphy and sequence stratigraphy, and magnetic-susceptibility (MS) stratigraphy to reach a better understanding

Controls on Carbonate Platform and Reef Development
SEPM Special Publication No. 89, Copyright © 2008
SEPM (Society for Sedimentary Geology), ISBN 978-1-56576-130-8, p. 291–314.

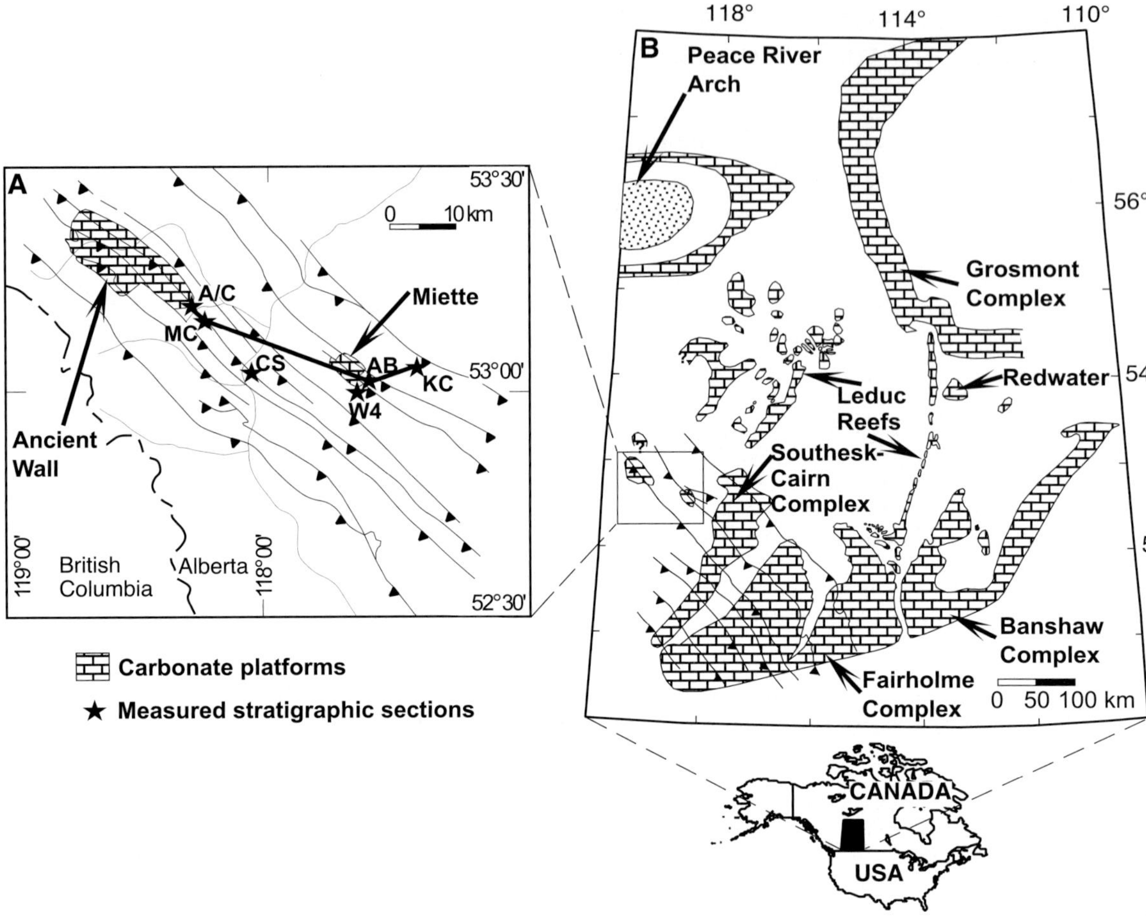

FIG. 1.—**A)** Location map of the overthrust belt in western Alberta, showing the locations of the Miette and Ancient Wall platforms (after Mountjoy, 1965). Lettered stars indicate the location of measured stratigraphic sections along the southeast margin of the Ancient Wall and Miette platforms (see Figs. 3, 4). Gray line indicates line of cross section illustrated in Figure 9. **B)** Palinspastically restored middle Frasnian paleogeographic map, illustrating the locations of Upper Devonian isolated and attached carbonate platforms in the Alberta basin. Buildups located west of the barbed line, indicating the eastern limit of Laramide thrusting, are exposed in the Canadian Rocky Mountains (after Mountjoy, 1980; Geldsetzer, 1989; Switzer et al., 1994). Key to stratigraphic sections: MC = Marmot Crack, A / C = Thornton Cr., CS = Cold Sulphur Springs, AB and W4 = Poachers Cr., KC = Klapper Cr. See Fig. 3 and 4 for the location of stratigraphic sections in relation to platform-margin cross sections.

of one of the most profound periods of change and restructuring of Paleozoic carbonate-platform systems in Earth history. The high-resolution database from this project is used to evaluate the timing and interrelationship of tectonic, sea-level, climatic, and biotic events during the Late Devonian.

F–F successions on several cratons record a second-order transgressive–regressive (T–R) cycle implying a significant episode of tectono-eustatic sea-level change (Johnson et al., 1985; Johnson et al., 1996; House and Ziegler, 1997, and papers therein). Initiation of latest Devonian glaciation in Gondwana (Isaacson and Diaz-Martinez, 1995; Isaacson et al., 1999) marks the beginning of the end of an extended paleoclimatic greenhouse and is probably related to the observed trend in progressive Famennian global sea-level fall. The Johnson et al. (1985) and Johnson et al. (1996) Devonian sea-level curve stands out as one of the best-constrained eustatic curves for the Paleozoic. Subsequent research has supported the eustatic interpretation of most third-order sea-level events designated as Devonian T–R cycles, but others require further investigation. This paper builds upon prior high-resolution sequence stratigraphic investigations (van Buchem et al., 1996; Whalen et al., 2000a; Whalen et al., 2000b; Uyeno and Wendte, 2005a; Wendte and Uyeno, 2005) in western Canada and expands the database of records of Late Devonian sea-level events.

GEOLOGIC SETTING OF WESTERN CANADA

The rocks discussed here are from two isolated reefal platforms and nearby basinal sequences that developed on the western margin of Late Devonian Laurussia in the Alberta basin (Figs. 1, 2). Western Canada was located at near-equatorial latitudes, within the southern hemisphere trade-wind belt, along a meridionally oriented western continental margin (Witzke and Heckel, 1988; Scotese and McKerrow, 1990). A regionally extensive carbonate ramp first developed in the Alberta basin during the Middle (late Givetian) and Late (early Frasnian) Devonian transgressions across a widespread subaerial unconformity (Figs. 3, 4). A system of attached and isolated carbonate platforms developed atop the ramp during the Frasnian Age of the Late Devonian (Figs. 1–4; Andrichuk, 1961; Mountjoy, 1980; Moore, 1989). To the east and south, broad areas characterized by oceanographic restriction were the sites of coeval evaporite deposition

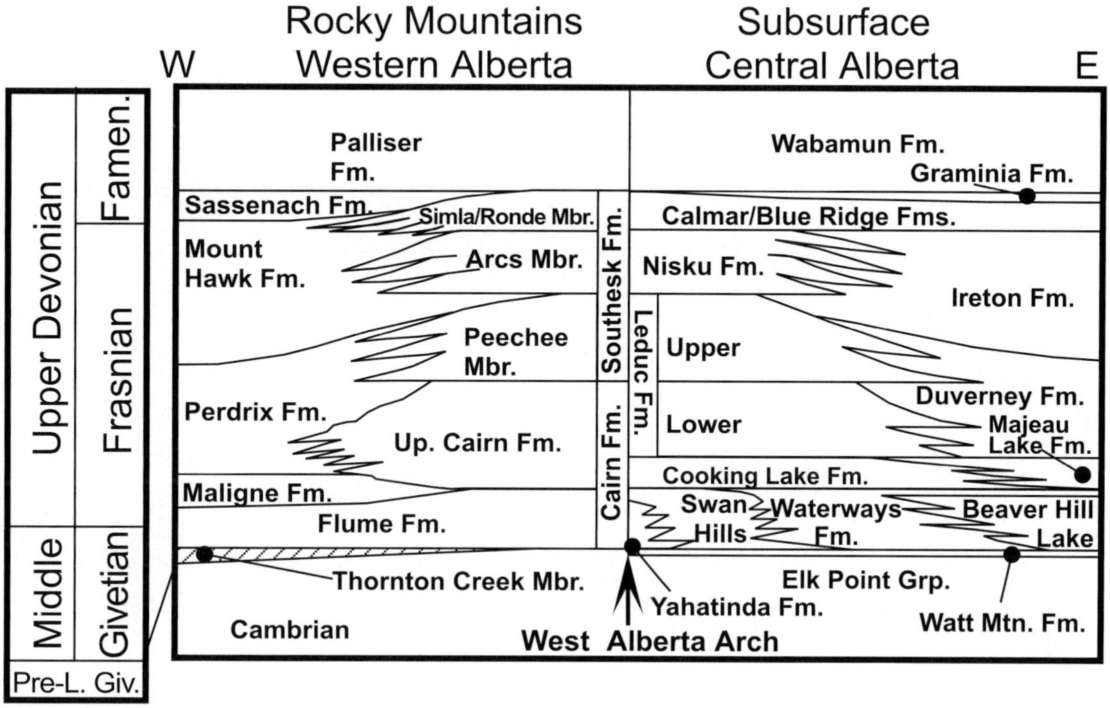

FIG 2.—Upper Devonian stratigraphic nomenclature for the Rocky Mountains, western Alberta and the central Alberta subsurface (after Switzer et al., 1994). Note differing stratigraphic terminologies between platform and basin units and between Alberta Rocky Mountain outcrops and the central Alberta subsurface. Units of interest in this study are mainly basinal units exposed in Rocky Mountain outcrops. Fm. = Formation, Mbr. = Member, Famen. = Famennian.

(Fig. 1; Switzer et al., 1994). Fine-grained siliciclastic sediment (Woodbend shales), interpreted to have been shed from the Ellesmerian Fold Belt to the northeast in the Canadian Arctic Archipelago, mixed with platform-derived carbonates around the platforms in the Alberta basin (Fig. 1; Oliver and Cowper, 1963; Stoakes, 1980; Switzer et al., 1994). Carbonate-platform growth kept pace with a prolonged second-order and shorter-duration third-order sea-level rises and outpaced basinal sedimentation during most of the Frasnian (Johnson et al., 1985; Johnson et al., 1996; Day et al., 1996; Day, 1998; Whalen et al. 2000a; Whalen et al., 2000b). The F–F extinction event brought about the demise of most of the Frasnian framework-building fauna and marks the end of extensive Devonian reef development in western Canada (McLaren, 1982; Stearn, 1987; Copper, 2002). However, stromatoporoid patch reefs in western Canada (Eliuk, 1984; Stearn et al., 1987; Copper, 2002) and Germany (Herbig and Weber, 1996), and microbial reefs in the Canning basin, Australia (George et al., 1997; George, 1999; Wood, 2000a, 2004) persisted into the Famennian. Early Famennian mixed carbonate–siliciclastic sediments of the Sassenach Formation filled the Alberta basin prior to widespread middle and late Famennian (Palliser platform) through Carboniferous carbonate-ramp development (Richards, 1989; Stoakes, 1992; Mountjoy and Savoy, 1995; Caplan and Bustin, 1998; Mountjoy and Becker, 2000).

Stratigraphy

Devonian rocks exposed in western Alberta and eastern British Columbia range from middle Givetian to Famennian in age, whereas those in the subsurface, to the east, range in age from late Eifelian to Famennian (Fig. 2). Different formation names are given to basinal and platform lithologies, and an independent system of stratigraphic nomenclature was applied to coeval rocks in the subsurface (Fig. 2). Major platform carbonate units in the Rocky Mountains are the Flume, Cairn, Southesk, and Palliser formations whereas basinal units include the Maligne, Perdrix, Mount Hawk, and Sassenach formations (Fig. 2)

The Miette and Ancient Wall platforms, located in the western part of the Alberta basin, are 400–500 m thick (Figs. 1, 3, 4). Miette had an areal extent of about 165 km^2, and Ancient Wall was approximately 1200 km^2 (Geldsetzer, 1989; Mountjoy, 1989). The Redwater platform is located in the eastern Alberta basin (Fig. 1), is approximately 350 m thick (including the Cooking Lake platform), and had an areal extent of about 500 km^2 (Klovan, 1964; Wendte, 1994; Chow et al., 1995). The dominant organisms in these platforms were tabular, bulbous, and domal stromatoporoids, *Amphipora,* and tabulate and rugose corals (Klovan, 1964; Kobluk, 1975; Wendte, 1994). Platform margins are locally characterized by a rim of stromatoporoid biostromes surrounding a typical back-reef interior with subtidal and peritidal meter-scale shoaling-upward cycles (Klovan, 1964; Mountjoy, 1965; Mountjoy and Mackenzie, 1974; Kobluk, 1975; 1994; Whalen et al., 2000a).

METHODS

In this project we employ the sequence stratigraphic methods in conjunction with conodont and brachiopod biostratigraphy and magnetic-susceptibility (MS) stratigraphy (Crick et al., 1997; Crick et al., 2001a; Ellwood et al., 1999; Ellwood et al., 2000; Ellwood et al., 2001) to better constrain the timing of sea-level, biotic, and other geologic events. Conodont and brachiopod biostratigraphic samples were collected where appropriate, and

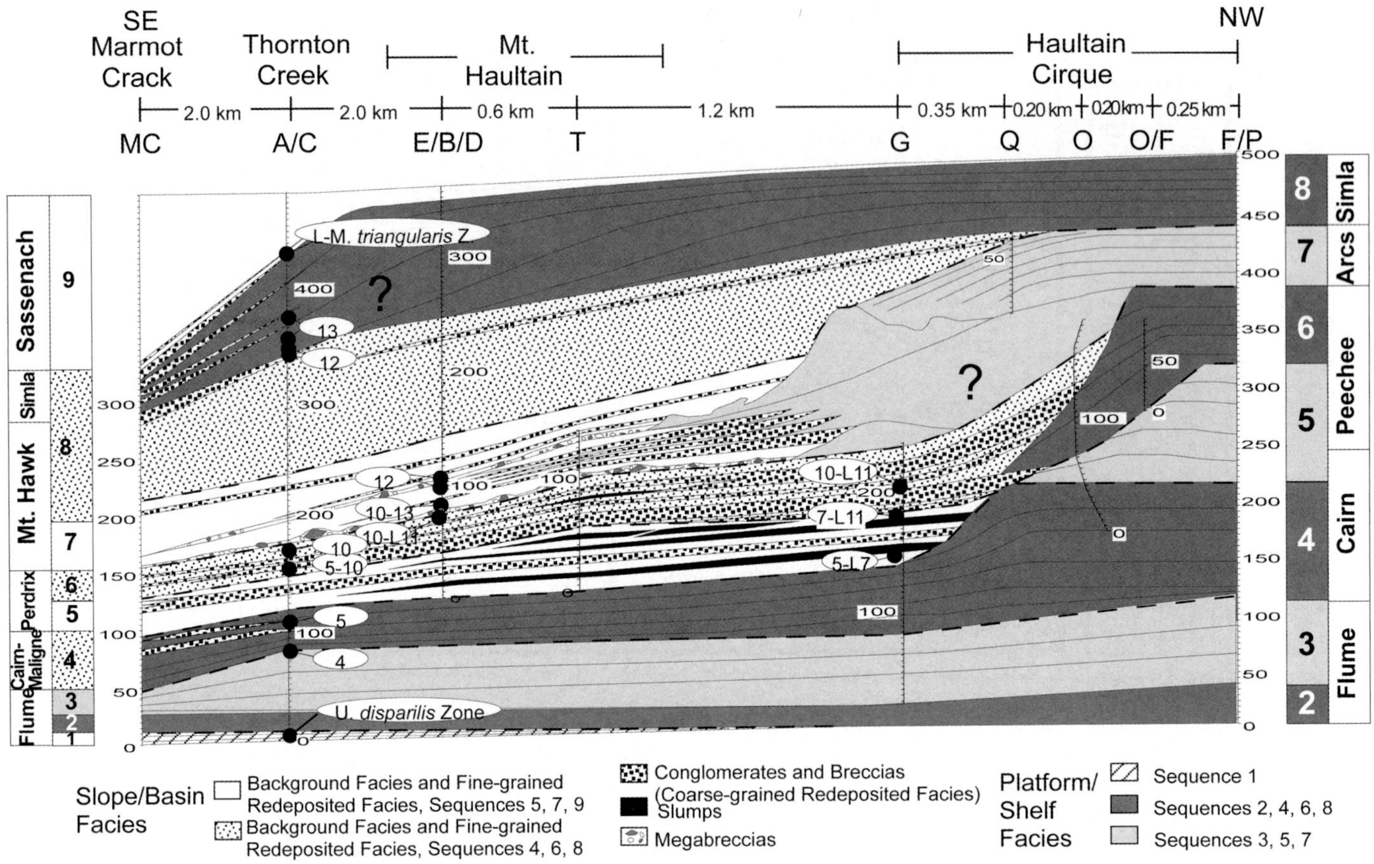

FIG 3.—Cross section of the southeast margin of the Ancient Wall platform, illustrating lithostratigraphy, the sequence-stratigraphic subdivision, and general lithofacies. Locations of measured stratigraphic sections are indicated by letters and graduated lines. Conodont sample locations and zones are indicated in ovals next to measured sections (see Tables 1, 2). Platform sequences are indicated by alternating dark and light shades of gray. Slope and basin sequences are indicated by alternating white and stippled patterns. Thick, dashed black lines indicate sequence boundaries. Thin black lines in platform sequences indicate parasequence boundaries (after Whalen et al., 2000a; Whalen et al., 2000b).

MS samples were collected every 0.5 to 1.0 m. MS samples were weighed to within 0.001 g and measured on a KLY-3 Kappa bridge magnetic-susceptibility meter at Brooks Ellwood's laboratory (Louisiana State University). Three measurements were taken on each sample, and the average of those measurements is reported here. MS values are reported as mass-normalized bulk magnetic susceptibility with units of m^3/kg. New stratigraphic data and MS samples were collected from several exposures for this study (Fig. 1), including a proximal locality to the east of Miette, a basinal section southwest of Ancient Wall. MS samples for this study were also collected from stratigraphic sections examined previously along the southeast margins of Ancient Wall and Miette (Figs. 1, 3, 4).

Sequence Stratigraphy, Cyclostratigraphy, and Biostratigraphy

A second-order eustatic T–R cycle served as a major control on the development of Late Devonian carbonate platforms in western Canada (Johnson et al., 1985; Johnson et al., 1996; McLean and Mountjoy, 1994; van Buchem et al., 1996; Whalen et al., 2000a; Whalen et al., 2000b; Wendte et al., 1995a). Third-order sequences appear to have been controlled by basin-wide relative sea-level changes, and comparisons with other coeval successions around the world permits evaluation of their status as eustatic variations. Basinal sequences became physically detached during platform aggradation, and onlapping basinal wedges display variations in the carbonate content of fine-grained hemipelagic facies and in clast content of coarse-grained, redeposited carbonates (Whalen et al., 2000b). These lithologic and mineralogic trends permitted sequence stratigraphic correlation of different basinal successions and between platform and basin sections (Whalen et al., 2000a; Whalen et al., 2000b).

The third-order platform and basin sequences are composed of higher-frequency cycles (McLean and Mountjoy, 1994; van Buchem et al., 1996; Whalen et al. 2000a; Whalen et al., 2000b). Deep-water cycles far outnumber platform shoaling-upward cycles and appear to be controlled by the interplay of relative sea-level change and higher-frequency events like climate change, variations in continental weathering, and variations in styles of platform development (Whalen et al. 2000a; Whalen et al., 2000b). Identification of deep-water cycles, manifested in the magnetic-susceptibility events described below, thus provides a refined degree of stratigraphic resolution that is exploited in this study to refine regional correlations.

Determination of the timing of events recorded in the sedimentary record hinges on accurate chronostratigraphy and/or biostratigraphy. Chronostratigraphic data are not available for

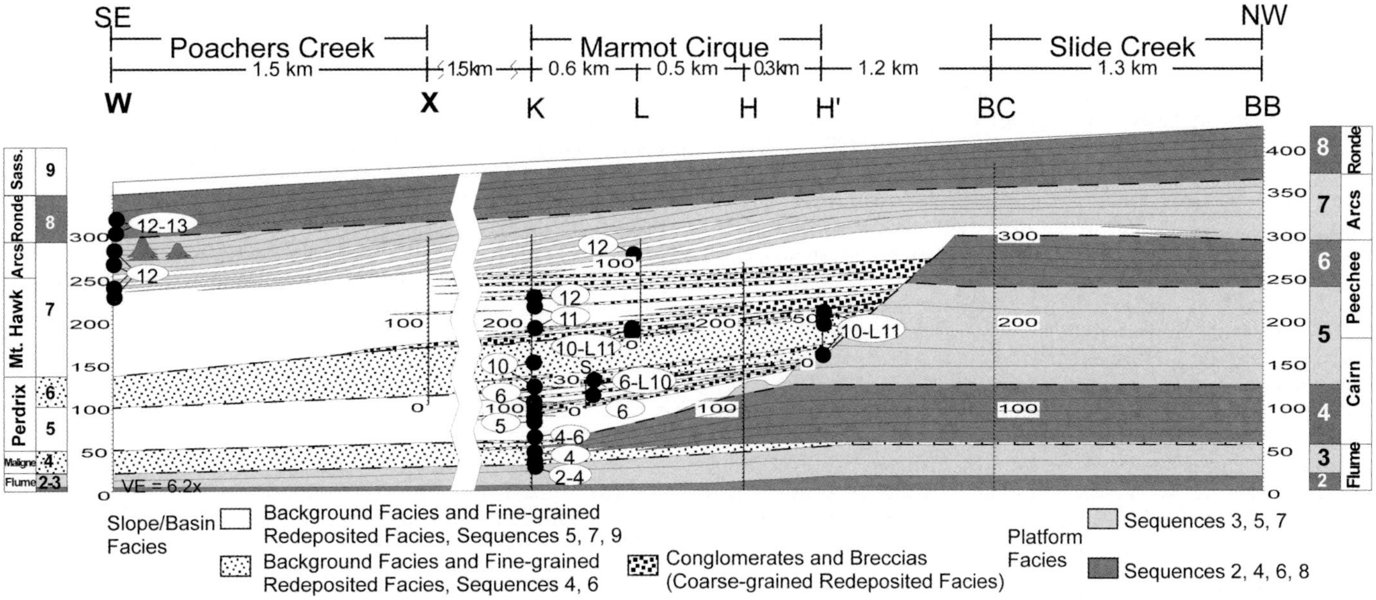

FIG 4.—Cross section of the southeast margin of the Miette platform, illustrating lithostratigraphy, the sequence-stratigraphic subdivision, and general lithofacies. Locations of measured stratigraphic sections are indicated by letters and graduated lines. Conodont sample locations and zones are indicated in ovals next to measured sections (see Tables 3, 4). Platform sequences are indicated by alternating dark and light shades of gray. Slope and basin sequences are indicated by alternating white and stippled patterns. Thick, dashed black lines indicate sequence boundaries. Thin black lines in platform sequences indicate parasequence boundaries. The abbreviation C/M in the lithostratigraphy column on the left stands for the Cairn and Maligne formations (after Whalen et al., 2000a; Whalen et al., 2000b).

most of the Devonian succession in the Alberta basin, due to a lack of radiometrically datable material. However, significant refinement in biostratigraphic resolution has resulted from the application of graphic correlation and the conodont-based Montagne Noire (MN) zonation (Klapper, 1989; Klapper and Lane, 1989; Klapper et al., 1995; McLean and Klapper, 1998; Whalen et al., 2000a).

Magnetic-Susceptibility Stratigraphy

Ellwood and colleagues have pioneered the magnetic-susceptibility event and cyclostratigraphy (MSEC) method (Crick et al., 1997; Crick et al., 2001a; 2002; Ellwood et al. 1999; Ellwood et al., 2000). Their work demonstrates that the MS signature of sediments varies with the proportion of detrital-dominated paramagnetic and ferrimagnetic minerals present. MS differs from magnetic-polarity data in that it is less sensitive to low-temperature thermal events that would remagnetize polarity (Crick et al., 1997). Early studies of magnetic susceptibility in sedimentary rocks concentrated on Quaternary loess deposits, where climatic changes were invoked to explain variations in susceptibility (Heller and Liu, 1984; Kukla et al., 1988; Beget et al., 1990; Verosub et al., 1993). More recently, magnetic-susceptibility studies have focused on MS as a climatic signal in Paleozoic loessites (Soreghan et al., 1997; Soreghan et al., 2002; Tramp et al., 2004), for interregional or global correlation of Paleozoic deep marine sediments (Crick et al., 1997; Crick et al., 2001a; Ellwood et al., 1999; Ellwood et al., 2000), as a proxy for oxygen isotope variations associated with Pleistocene ice volume and climate variations (Shackleton, 1999), and as a paleoclimatic indicator in Holocene archeological sites (Ellwood et al., 1997).

Rates of siliciclastic input into marine basins are controlled by climate, tectonic uplift, weathering of continental rocks, and relative sea-level change (Davies et al., 1977; Worsley and Davies, 1979; Raymo et al., 1988). During episodes of pronounced tectonism, heightened weathering rates may result in a regional or even global increase in the delivery of fine-grained siliciclastic sediments to marine basins (Mackenzie and Pigott, 1981; Wold and Hay, 1990). Links between rates of sea-floor spreading and eustatic sea-level change and between major orogenic events, continental weathering, and climate change have been proposed (Mackenzie and Pigott, 1981; Fischer, 1983; Raymo et al., 1988; Edmond, 1992; Raymo and Ruddiman, 1992; Richter et al., 1992). Variations in the argillaceous content of deep marine facies likely results from linked tectonic, sea-level, and climatic changes. These events should be accompanied by changes in magnetic susceptibility of deep marine sediments because the flux of magnetic minerals varies with changes in weathering patterns (Crick et al., 1997; Crick et al., 2001a; Ellwood et al. 1999).

Controls on the MS signature of marine rocks thus include eustatic sea-level change, climate (Ellwood et al., 1999) and diagenetic events that can remagnetize the rocks or alter the MS signature (Burton et al., 1993; Katz et al., 1998; Schneider et al., 2004). Eustatic sea-level changes result in base-level fluctuations that cause variations in erosion and detrital influx to the world ocean (Worsley and Davies, 1979). A eustatic fall in sea level results in high MS magnitudes associated with increased detrital supply when more continental land mass is exposed to erosion, whereas eustatic sea-level rise results in low MS values due to decreased detrital supply (Ellwood et al., 1999; Ellwood et al., 2000). Changes in global climate result in variations in rainfall or glaciation accompanied by changes in erosion rates, which also influence detrital flux to ocean basins. The recognition of pronounced interregional anomalies and cyclic patterns in MS signatures of deep marine rocks led to the development of the MSEC method (Crick et al., 1997; Crick et al., 2001a; Ellwood et al., 1999;

Ellwood et al., 2000). Use of MSEC as an interregional or global correlation tool therefore hinges on the assumption that if base-level changes or long-term climatic effects are global, then MS variation in deep marine facies should be similar even if the absolute magnitude of the variations are influenced by local tectonics, climate, or sedimentation (Crick et al., 1997).

Documentation of the history of siliciclastic input into the Alberta basin is thus crucial to the interpretation of MS data. Late Givetian siliciclastic sediments in western Alberta include silt and fine sand, derived partially from the exposed west Alberta arch, that mixed with carbonate-ramp sediments (Williams and Krause, 2000; Day and Whalen, 2005). Frasnian siliciclastic sediments consist of clay and silt derived from the distant Ellesmerian fold belt (Stoakes, 1980; Stevenson et al., 2000) with a likely contribution from eolian transport across Laurussia. These sediments mixed with fine-grained platform-derived carbonate material and form low-angle, east-to-west prograding clinoforms that progressively infilled the Alberta basin (Stoakes, 1980). During the Famennian coarser-grained, west-derived siliciclastic sediments were shed during uplift of the Antler orogenic belt (Mountjoy and Savoy, 1995; Stevenson et al., 2000).

FACIES ASSOCIATIONS AND DEPOSITIONAL ENVIRONMENTS

Detailed descriptions and interpretations of facies and depositional environments are beyond the scope of this paper; the reader is referred to McLean and Mountjoy (1994), Whalen et al. (2000a), Whalen et al. (2000b), Wendte (1994), and Chow et al. (1995) for more information. Upper Devonian facies (Fig. 5) of the Miette and Ancient Wall platforms can be differentiated into five facies associations: (1) peritidal platform, (2) subtidal platform, (3) platform margin, (4) forereef and slope, and (5) basin (Whalen et al. 2000a; Whalen et al., 2000b). These subdivisions are somewhat gradational, and facies from adjoining zones are commonly interbedded.

Peritidal Platform Facies Association

Peritidal facies include: laminated mudstone and wackestone, silty or sandy peloidal wackestone, *Amphipora* wackestone and packstone, peloidal packstone and grainstone, and oncoid packstone and rudstone. Locally these facies contain stromatolites, fenestral pores (Fig. 5H), and dripstone cements, which supports their interpretation as peritidal facies. The paucity of open marine faunal elements in these facies implies a restricted peritidal origin.

Subtidal Platform Facies Association

Shallow to deep subtidal platform facies consist of bioclastic mudstone and wackestone, bioclastic, peloidal wackestone, packstone, and grainstone, bioturbated wackestone and packstone, nodular wackestone and packstone, and *Amphipora* floatstone and rudstone. These facies contain a diverse fauna with stromatoporoids, brachiopods, crinoids, solitary and colonial rugose corals (Fig. 5F), tabulate corals, bivalves, and calcispheres. Locally grains are partially to thoroughly micritized. The increase in abundance and diversity of skeletal fauna and the incidence of burrowing and bioturbation in these facies implies open marine conditions and distinguishes them from peritidal facies.

Platform-Margin Facies Association

Platform-margin facies comprise stromatoporoid framestones, *Stachyodes* floatstones and rudstones, bioclastic, intraclastic, peloidal floatstones and rudstones, and bioclastic, intraclastic, peloidal packstones and grainstones. These facies commonly consist of tabular or bulbous stromatoporoids (Fig. 5G), *Stachyodes*, *Amphipora*, tabulate and rugose corals, crinoids, calcispheres, foraminifera, peloids, and intraclasts. The massive reef framestones (Fig. 5G) and associated facies, which are commonly coarse grained and moderately to poorly sorted, indicate moderate to strong agitation and strongly turbulent water above fair-weather wave base.

Forereef and Slope Facies Association

Facies on the forereef and slope were deposited either from suspension (background sediments, described below) with *in situ* faunal components or as platform- or slope-derived material redeposited by various types of gravity-flow mechanisms (redeposited carbonates). Redeposited facies include megabreccias, lithoclastic breccias, oncolitic, bioclastic, and lithoclastic floatstone and rudstone, coarse- and fine-grained bioclastic–lithoclastic turbidites, and slumps (Fig. 5A, B, E; Whalen et al., 2000b; Whalen et al., 2002). Redeposited carbonates exhibit a wide range of grain sizes and sedimentary structures indicative of deposition from turbidity currents, debris flows, and slumping (Whalen et al., 2000b). Fauna and clasts in redeposited carbonates that are indicative of platform environments include bulbous, branching, and thick tabular stromatoporoids, *Amphipora*, tabulate corals, colonial rugose corals, calcareous algae, small oncoids, calcimicrobes like *Renalcis* and *Girvanella*, light gray, subangular to

→

FIG 5 (opposite page).—Facies photographs. **A)** Slope facies, Mount Hawk Formation, section E, Ancient Wall platform margin, LST sequence 7. Note the rhythmic bedding and alternation of coarse-grained (resistant) and fine-grained (non-resistant) lithologies of the slope facies association. **B)** Closeup of coarsening-upward packages of turbidites and debris flows illustrated in Part A. Non-resistant intervals are dominated by background facies, including argillaceous wackestone and crinoid wackestone–packstone deposited from suspension and bio-lithoclastic wackestones–packstones interpreted as fine-grained turbidites. Resistant units are coarse-grained bio-lithoclastic floatstones and rudstones interpreted as turbidites and debris flows. **C)** Burrow-mottled bioclastic wackestone–packstone of the subtidal-platform facies association, Mount Hawk Formation, section MC, Ancient Wall platform margin, HST sequence 7. **D)** Background lithofacies of the slope and basin facies associations, including laminated calcareous shale, argillaceous mudstone–wackestone, and crinoid wackestone–packstone, lower Perdrix Formation, section B, Ancient Wall platform margin, LST sequence 6. **E)** Lithoclastic breccia with dark gray, slope-derived lithoclasts and a few bioclasts in the slope facies association, middle Perdrix Formation, section K, Miette platform margin, HST sequence 6. **F)** Coral wackestone–packstone, subtidal-platform facies association, Mount Hawk Formation, section KC-2, HST sequence 7. **G)** Domal stromatoporoid framestone, platform-margin facies association, section MC, HST sequence 3. **H)** Stromatolite with low-amplitude, laterally linked hemispheroids, overlain by peloid–bioclast packstone with fenestral texture, peritidal facies association, section BB, Cairn Formation, Miette platform, HST sequence 5.

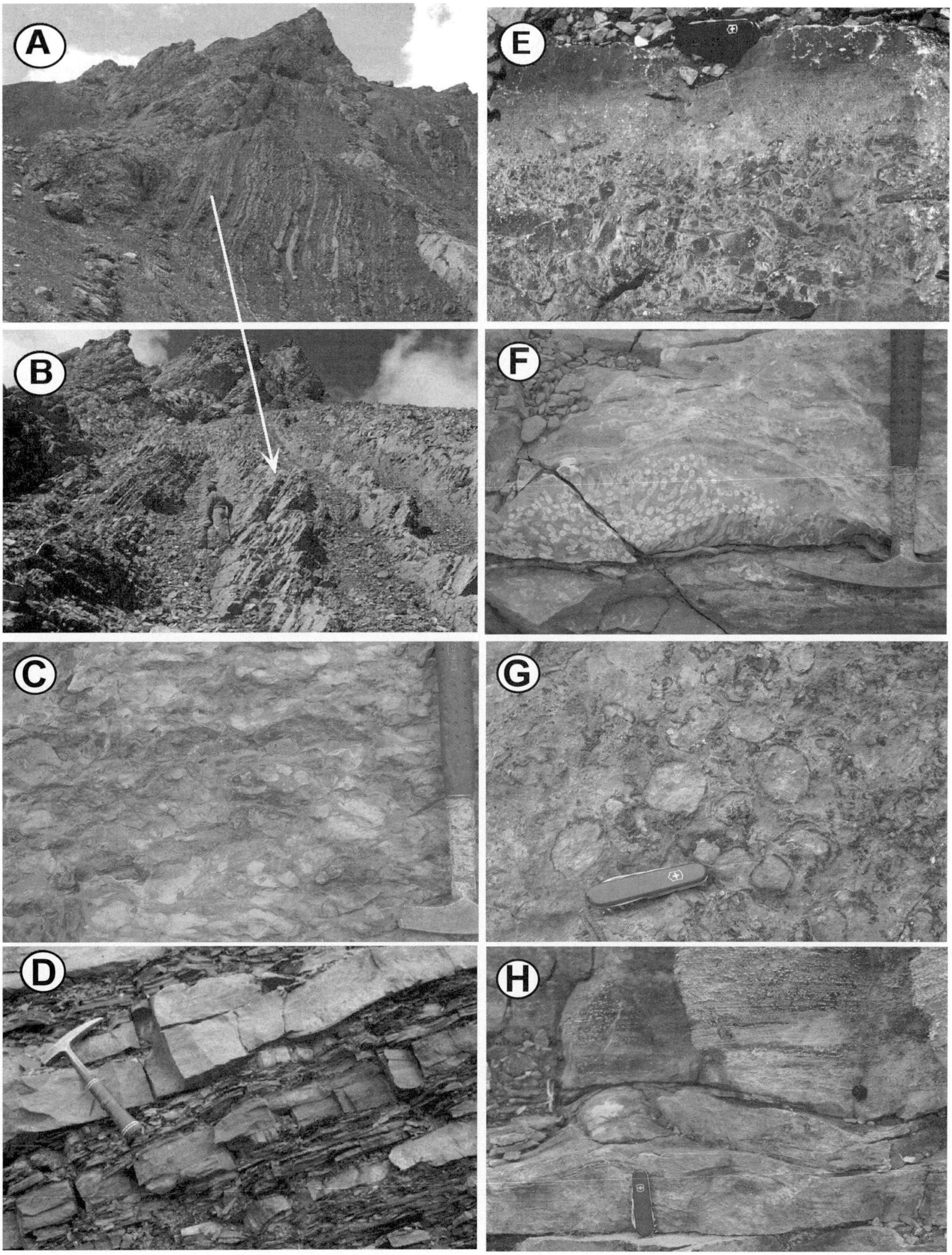
A
B
C
D
E
F
G
H

subrounded limestones of variable textures, reddened or blackened limestone, and micritized bioclasts (Klovan, 1964; Playford, 1980; Machel and Hunter, 1994; Whalen et al., 2000a; Whalen et al., 2000b; Whalen et al., 2002; Wood, 2000b). Slope-derived fauna and grains include crinoids, solitary rugose corals, some tabulate corals, thin laminar stromatoporoids, many brachiopod species, and dark gray, tabular to subangular limestone clasts (Klovan, 1964; Playford, 1980; Machel and Hunter, 1994; Whalen et al., 2000a; Whalen et al., 2000b).

Background facies include bioturbated, nodular, bioclastic, and silty bioclastic wackestone and packstone (Fig. 5C), silty marl, and calcareous siltstone (Whalen et al., 2000a; Whalen et al., 2000b). These facies contain no evidence of traction transport or reworking by waves and currents and are dominantly hemipelagic. Bioturbation and the shelly fauna in many of these facies indicate that they were deposited in oxygenated environments above the local oxygen-minimum zone (OMZ). Background sediments locally exhibit soft-sediment deformation features on a variety of scales (Whalen et al., 2000b).

Basin Facies Association

Basin facies include argillaceous mudstone and wackestone, calcareous shale (marl), and organic-rich mudstone and shale (Fig. 5D; Whalen et al., 2000a). These facies are characterized by their fine grain size, thin internally laminated beds, and a dominantly hemipelagic fauna. Argillaceous mudstone, and wackestone and marl contain autochthonous fauna (mainly crinoids and rhynchonelloid, atrypoid, and spiriferoid brachiopods) as well as pelagic biota (criccoconarids and calcispheres) that settled from suspension. The generally fine-grained texture and lack of sedimentary structures indicative of wave or current energy suggest that these facies were deposited in a low-energy hemipelagic environment.

SEQUENCE STRATIGRAPHY AND PLATFORM DEVELOPMENT

High-resolution sequence stratigraphic analysis of the Miette and Ancient Wall platforms originally indicated that they were deposited as six depositional sequences that record four geomorphic phases of development (van Buchem et al., 1996; Whalen et al., 2000a; Whalen et al., 2000b). Lowstand facies of a seventh sequence were deposited subsequent to the demise of the Frasnian platforms (Mountjoy and Becker, 2000; Whalen et al. 2000a; Whalen et al., 2000b). Subsequent fieldwork has identified two additional sequences. One is Givetian in age (Day and Whalen, 2005), and the other is early Frasnian in age and was lumped with the first sequence in earlier publications (van Buchem et al., 1996; Whalen et al., 2000a; Whalen et al., 2000b). This brings the total number of late Givetian to early Famennian third-order sequences to nine (Figs. 3, 4, 6).

Sequences at Miette and Ancient Wall were defined on the basis of their geometries, stratal stacking patterns, sediment composition, and identification of important platform and slope stratal horizons (van Buchem et al., 1996; Whalen et al., 2000a; Whalen et al., 2000b). Stratal horizons interpreted as platform sequence boundaries are either subaerial exposure surfaces or significant transgressive, marine-flooding surfaces that overlie cryptic exposure horizons or thin lowstand carbonates that are difficult to differentiate from shallow-water highstand facies (van Buchem et al., 1996; Whalen et al. 2000a; Whalen et al., 2000b).

Mountjoy (1965) and Mountjoy and Mackenzie (1974) were the first to note the successive stages of development and the long-term transgressive–regressive trend of the Miette and Ancient Wall buildups. Chow et al. (1995) documented a similar overall transgressive–regressive pattern in the Redwater platform. The Miette and Ancient Wall platforms are characterized by a lower biostromal phase (sequences 2–4) and an upper sandy, peritidal phase (sequences 5–8) (Mountjoy, 1965; Mountjoy and Mackenzie, 1974; Whalen et al., 2000a). The first sequence (Fig. 3) records the initial onlap of the west Alberta Arch in western Alberta Rocky Mountains and includes mixed carbonate–siliciclastic facies and a restricted to normal marine fauna (Day and Whalen, 2005). The next three sequences (Figs. 3, 4) record construction of a rimmed platform and infill of a subtidal to peritidal lagoon, whereas the last four sequences indicate deposition on a relatively flat-topped peritidal platform (Whalen et al., 2000a).

The four phases of Miette and Ancient Wall platform development vary both lithologically and in the geometry of platform and basin units (Figs. 3, 4; van Buchem et al., 1996; Whalen et al., 2000a; Whalen et al., 2000b). They include: a regionally extensive, aggrading carbonate ramp (sequences 1–3), an isolated prograding platform (sequence 4), a backstepping and aggrading isolated platform (sequences 5 and 6), and a second ramp-like phase (sequences 7 and 8) that prograded across basin fill (van Buchem et al., 1996; Whalen et al., 2000a; Whalen et al., 2000b). The entire succession was deposited during a second-order T–R cycle (Fig. 6), designated as Devonian Depophase II by Johnson et al. (1985, 1996), that ultimately controlled the patterns of platform-margin development (van Buchem et al., 1996; Whalen et al., 2000a; Whalen et al., 2000b). Individual sequences are interpreted to represent third-order packages (Figs. 3, 4, 6), the geometry of which is influenced by second-order eustatic sea-level change and differential subsidence depending on paleogeographic position relative to the continental-margin hingeline (Whalen et al., 2000b). Maximum regression and the F–F extinction horizon (Upper Kellwasser horizon; Becker, 1993) coincide with a second-order, terminal Frasnian sea-level lowstand and early Famennian transgression (Fig. 6; initiation of Devonian T–R cycle IIe; Johnson et al., 1985; Johnson et al., 1996) , which eventually reestablished carbonate deposition atop the former Frasnian platforms (Mountjoy and Becker, 2000).

MAGNETIC-SUSCEPTIBILITY EVENTS IN THE DEVONIAN OF WESTERN CANADA

MS data from four relatively continuous, upper Middle to Upper Devonian sections, three of which span the F–F boundary, are presented here (Figs. 1, 3, 4, 7–9). Section A/C is from a basin-to-slope section along the southeast margin of the Ancient Wall platform (Figs. 1, 3), and section AB/W4 is from a similarly positioned section along the southeast Miette platform margin (Figs. 1, 4). Section MC (Figs. 1, 3) is located farther basinward (southeast) of section A/C. Section KC is located in the Nikanassin Range, east of the Miette platform, at a position more proximal to terrestrial source areas (Fig. 1). F–F boundary exposures are present in sections A/C, W4, and KC (Figs. 1, 3, 4).

Several key events designated A to E are recorded in the MS data from slope and basin successions in the Canadian Rockies, within which two or more smaller-scale events are observed (Figs. 7–9, designated by number: A1, A2, and so on). Magnetic-susceptibility events are represented either by high-frequency peaks and troughs in MS values or longer-term stable or somewhat variable MS signatures with trends toward higher or lower MS values (Figs. 7–9). Trends and spikes in MS data are directly related to trends in detrital input (Fig. 8). The magnitude of the MS signal also increases along the distal-to-proximal transect (Figs. 7, 9), with the peak MS values registering in the

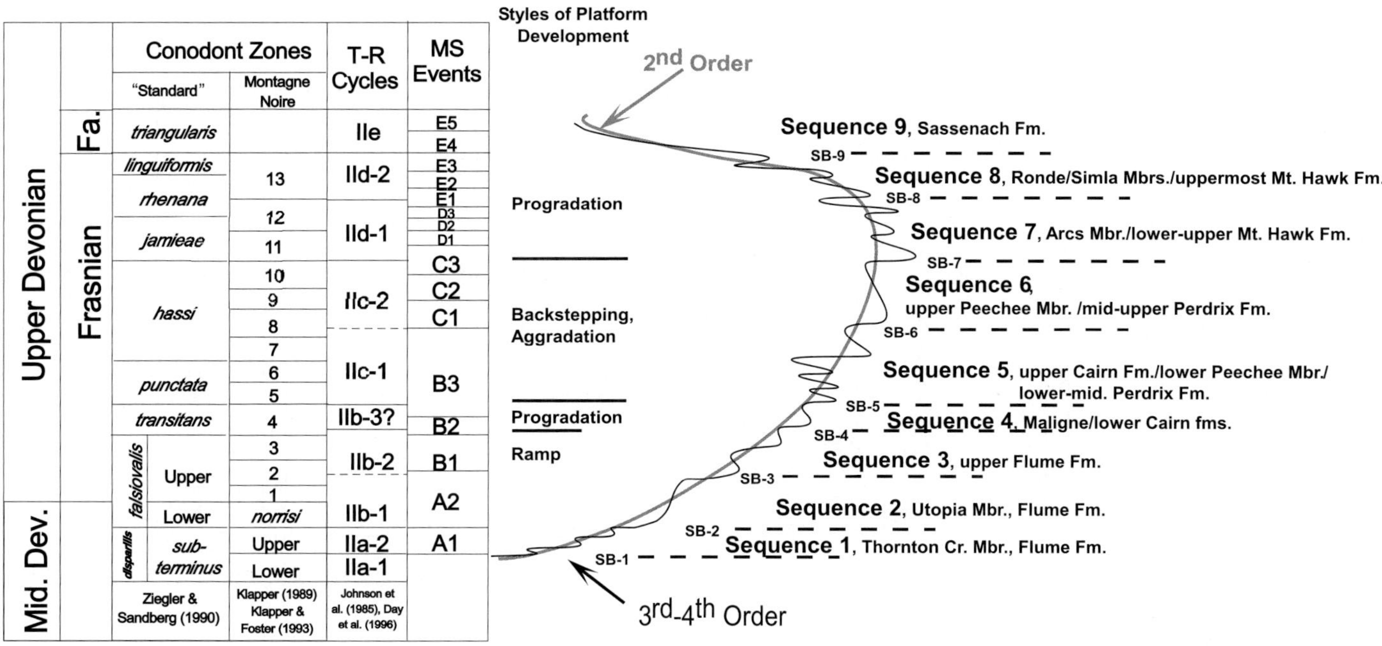

FIG 6.— Late Devonian conodont biostratigraphy, T–R cycles, and MS events plotted against second- and third- and fourth-order sea-level variations, sequence stratigraphy, and lithostratigraphy for western Canada. Different styles of platform development are also noted. Montagne Noir conodont biostratigraphic zonation is after Klapper (1989) and Klapper and Foster (1993), and "standard" zonation is after Ziegler and Sandberg (1990). Devonian T–R Cycles are after Johnson et al. (1985) and Johnson et al. (1996). North American Devonian cratonic T–R cycles IIa-2, IIb-1, IIb-2 are after Day et al. (1996), and divisions of T–R cycle IId are after Day (1998). Proposed subdivision of T–R cycle IIb (IIb1-3) is after Day and Whalen (2003) and Day (2004). Fm. = Formation, Mbr. = Member, Fa. = Famennian.

10^{-8} m^3/kg range in the west and over 2 x 10^{-7} m^3/kg in the east (Figs. 7, 9).

In total we have identified 14 individual MS events that we can correlate between well-exposed sections (Figs. 7, 9). Signatures of individual events display different magnitudes but similar trends that permit their identification and correlation from section to section. Based on a Frasnian stage duration of approximately 6 My (Tucker et al., 1998), these events appear to be third- or fourth-order cycles similar to the magnetic-susceptibility zones defined by Crick et al. (2002).

Significant MS events appear to coincide with important sequence stratigraphic horizons or systems tracts. In particular, LSTs and TSTs tend to record a peak in MS probably related to lowstand delivery and subsequent transgressive reworking of more susceptible minerals (Fig. 8). Platform aggradation and progradation during HSTs results in high proportions of redeposited coarse-grained carbonate and fine-grained carbonate settling from suspension in slope and basin settings, resulting in reduced MS values or signatures (Fig. 8). In some cases MS values may increase slightly during late HST, probably associated with maximum shoreline progradation.

The initial Givetian onlap of the Thornton Creek Member of the Flume Formation records higher susceptibility than the rest of the overlying Flume (MS event A1, Figs. 3, 7, 9; Day and Whalen, 2005). Sequences 2 and 3 display dominantly low but fluctuating MS values that may be controlled by higher-frequency climatic events (Figs. 7–9). The last phases of dominantly aggradational Flume carbonate-ramp deposition displays reduced MS values likely related to increased carbonate input (Figs. 7–9, MS event A2). Initiation of the flooding event that began sequence 4 (Maligne Fm.) coincides with a significant MS spike in distal sections that is recorded with lower amplitude at sections proximal to carbonate-platform margins (Figs. 7–9, MS event B1). Platform progradation later in sequence 4 results in decreasing MS values (lowermost Perdrix, Figs. 7–9). Sequence 5 (lower Perdrix) and 6 (upper Perdrix to lowermost Mount Hawk) display patterns similar to the Maligne to lower Perdrix succession below. The TST is characterized by elevated MS values with a general drop in MS within the HST (Figs. 7–9, MS events B2 and B3). The top of sequence 5 in the platform facies is a subaerial exposure horizon with microkarst and a well developed red paleosol (Whalen et al., 2000b). The base of sequence 7 (lower Mount Hawk, Figs. 7–9, MS event C3–C4) records an abrupt increase in MS that appears to be related to an influx in fine-grained siliciclastics. The MS increase coincides with a change in rock color from black or gray nodular mudstones and calcareous shales of the upper Perdrix to brownish gray calcareous shales and mudrocks of the Mount Hawk Formation. As the Arcs Member (Southesk Fm.) ramp prograded during the HST of sequence 7, MS values decreased, likely related to the increased carbonate input. The top of the Arcs Member (and subsurface equivalent Nisku) is a well-documented exposure horizon (Fejer and Narbonne, 1992; Shields and Geldsetzer, 1992; Whalen et al., 2000b; Potma et al., 2001) and records an increased MS signature that appears to be related to this lowstand of relative sea level (Figs. 7–9). This pattern is repeated in sequence 8, where MS values are slightly elevated near the base and top of the Ronde and Simla members (Southesk Fm.) with generally lower values during most of the HST. Exposure at the tops of these units is well documented (Whalen et al., 2000a) and likely explains the MS signature. MS values increase in sequence 9, coinciding with a major influx of silici-

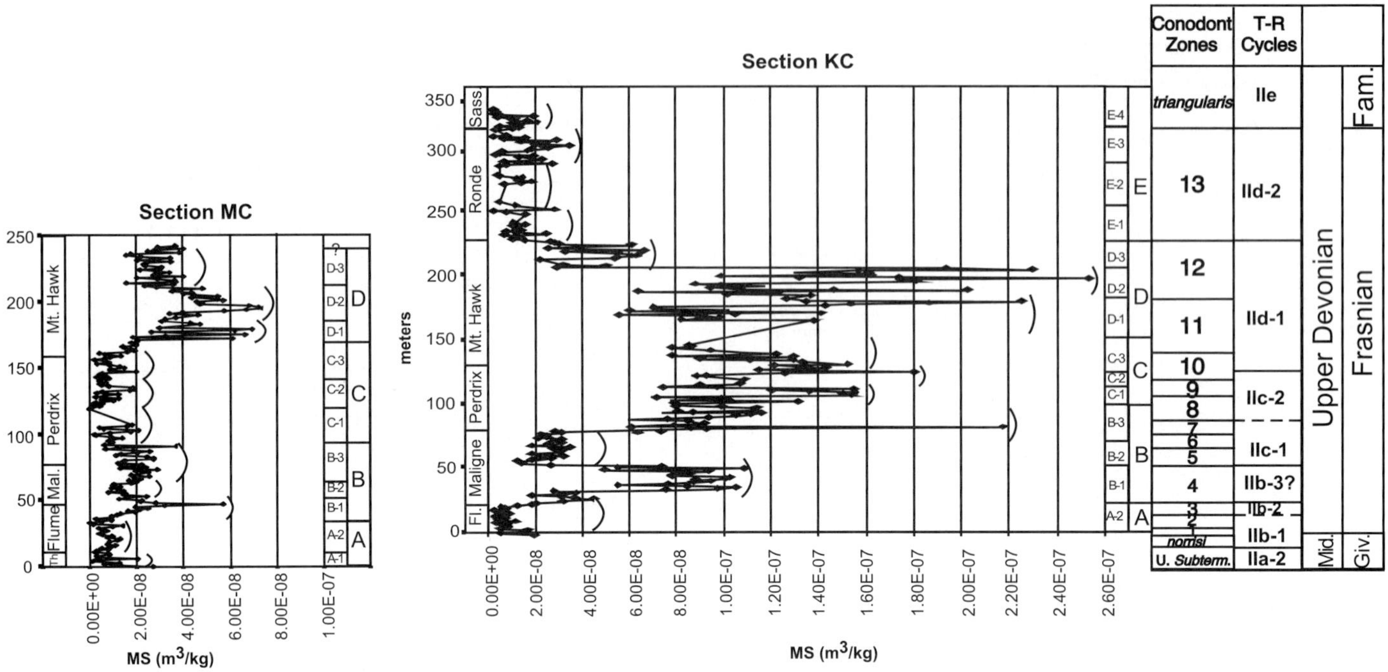

FIG 7.—Lithostratigraphy, MS events, conodont biostratigraphy, and T–R cycles (Johnson et al., 1985; Johnson et al., 1996; Day et al., 1996) in sections MC and KC, western Alberta. MS events are defined on the basis of positive, negative, or relatively constant MS signature. Fifteen high-frequency and 5 lower-frequency MS events are defined. All MS samples were collected within our high-resolution biostratigraphic framework, which provides biochronologic control for correlation of MS events. U. *subterm.* = Upper *subterminus* zone, Mid. = Middle Devonian, Giv. = Givetian, Fam. = Famennian.

clastics during deposition of LST and TST facies of the Sassenach Formation (Figs. 7, 9, MS events E4 and E5).

Timing of Middle and Late Devonian Sea-Level and Magnetic-Susceptibility Events

Existing (Uyeno, 1987; Klapper and Lane, 1989; Mclean and Klapper, 1998; Johnston and Chatterton, 2001) and new conodont data reported here (Tables 1–4) establish a framework for correlation of the late Givetian to early Famennian platform and basinal facies at Miette and Ancient Wall and constrain the timing of most major sea-level changes that initiated deposition of Alberta Rocky Mountain Devonian depositional sequences 1–9, and magnetic-susceptibility events A–E (Figs. 7, 9). Some of these new data (Tables 2, 3) provide the first conodont-based correlations of the early–late Frasnian basinal and prograding-ramp successions from the Miette Platform.

The Flume Formation, above the basal Thornton Creek Member (Day and Whalen, 2005), did not yield offshore conodont faunas and do not permit detailed correlations of two of the three late Givetian–early Frasnian Flume carbonate-ramp sequences at Ancient Wall and Miette. Provisional correlations of the Flume are based on correlations of the basal Thornton Creek Member (Day and Whalen, 2005), the conodont sequence in the Flume and Maligne formations at the Maligne type section between Miette and Ancient Wall at Cold Sulphur Springs (Fig. 1; Uyeno, 1987), and new data from Ancient Wall (Fig. 3, sequences 2 and 3; Table 1). Faunas recovered from the early Frasnian to Famennian succession above the Flume Formation (Figs. 3, 6, sequences 4–9; data from Johnston and Chatterton, 2001) indicate correlations with the upper part of Frasnian Montagne Noire Zone 4 through the Lower *crepida* Zone (Fig. 6; Tables 1–4).

Sequence 1.—

Deposits of sequence 1 are included in the Thornton Creek Member of the Flume Formation, and are thus far known from the vicinity of the Ancient Wall Platform (Figs. 2, 3, sections A/C and MC; Day and Whalen, 2005). At the type section of the Thornton Creek Member (Fig. 3, section A/C) sequence 1 yielded the conodont *Icriodus subterminus* (Table 1, sample 3F) in association with brachiopods of the *Eleutherokomma–Schizophoria* Fauna (Day and Whalen, 2005). The occurrence of *I. subterminus* below the base of the Flume carbonate-ramp succession with conodonts of the *Pandorinellina insita* Fauna suggests a pre-*norrisi* Zone correlation of the Thornton Creek Member and MS Event A-1(Table 1, sample AC-103). Uyeno (1987) recovered *P. insita* from the lower two meters of the Flume carbonate-ramp facies at a locality approximately 25 kilometers southeast of our sections at Ancient Wall (GSC samples 12NBd and 12NBe). The occurrence of *I. subterminus* in sequence 1, below the lowest occurrence of *P. insita*, indicates a position in the upper part of the *I. subterminus* Fauna (Fig. 6). The *I. subterminus* Fauna is a shallow-water conodont fauna considered to be equivalent to all or part of the *disparilis* Zone of the Devonian conodont zonation (Witzke et al., 1985; Klapper and Johnson, *in* Johnson, 1990; Rogers, 1998; Day and Whalen, 2005).

The late Givetian deepening event that initiated sequence 1 coincides with North American cratonic Devonian T–R cycle IIa-2 (Day et al., 1996; Uyeno, 1998), and the initial late Givetian marine transgression in central Alberta subsurface (basal Beaverhill Lake Group T-R cycle Lower A, Wendte et al., 1995b, 1997; Uyeno and Wendte, 2005a; Wendte and Uyeno, 2005). This same event initiated deposition of the Amco Member of the Slave Point Formation in northeast Alberta, and along the south shore

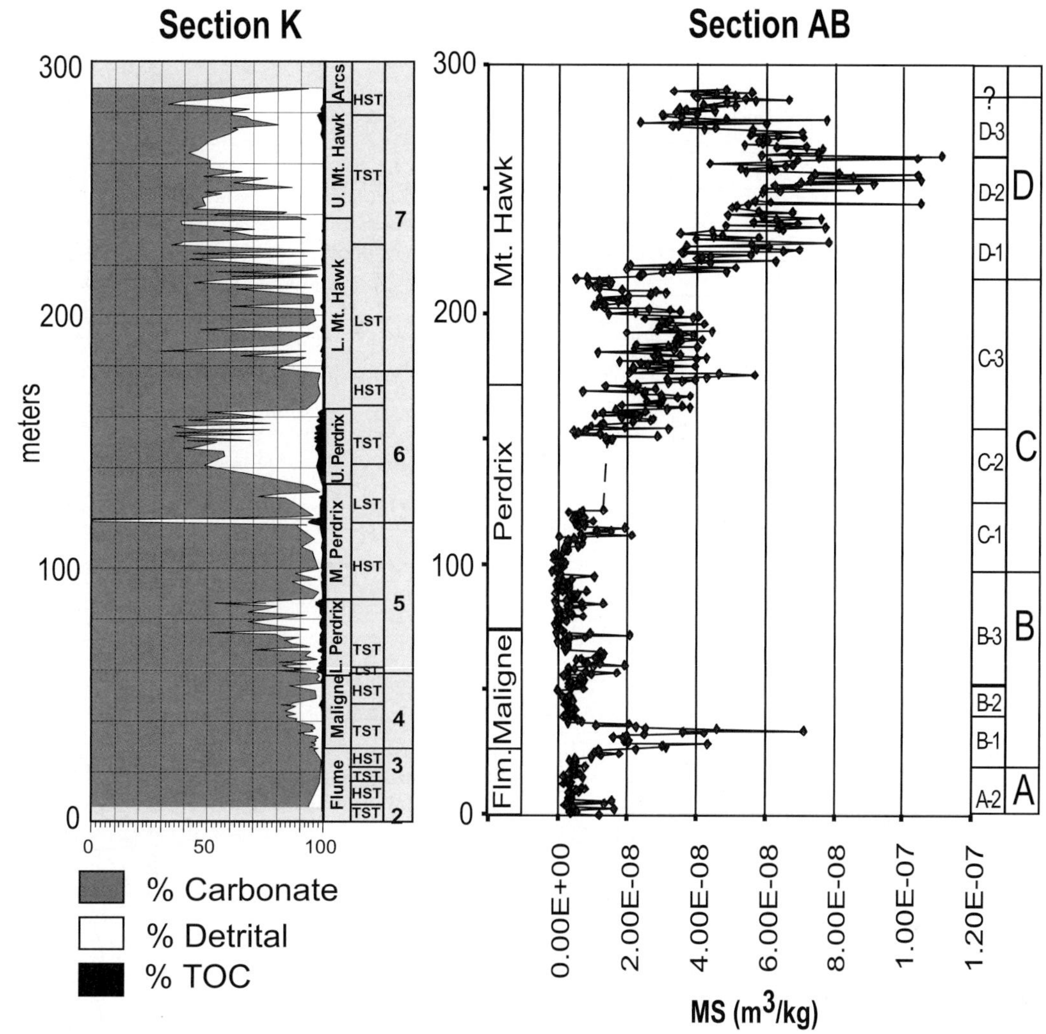

FIG 8.—Relationship between lithology, mineralogy, and the MS signature in two correlative sections, Miette platform. Sections K and AB (Fig. 3) are in adjacent thrust sheets along the southeast margin of the Miette platform and record identical lithostratigraphy. The column on the left (after Whalen et al., 2000b) illustrates the lithostratigraphy and sequence stratigraphy, percentage of total carbonate (gray shaded area), detrital material (white area), and total organic carbon (TOC; solid black area) through stratigraphic section K. Data are derived from gasometric carbonate analyzer (calcite and detrital) and LECO carbon analyzer (TOC) data. On the right are MS data for section AB. Note the close relationship between MS response and detrital-mineral content, especially in the Maligne, Perdrix, and Mount Hawk formations.

of the Great Slave Lake region of southern Northwest Territories (Uyeno and Norris, *in* Day et al., 1996; Uyeno, 1998). Sequence 1 strata at Ancient Wall (Figs. 3, 7) appear to correlate with the lower part of Beaverhill Lake Group (BHL) sequence 1 of Potma et al. (2001).

Sequences 2–3.—

Above the Thornton Creek Member at the Ancient Wall platform the base of sequence 2 coincides with the base of the lowest ramp carbonates of the Flume Formation (Figs. 3, 6). Sequence 2 includes as many as three shallow-ramp fourth-order T–R cycles (parasequences) capped by a prominent exposure surface on the top of the Utopia Member (Figs. 3, 4). The base of the carbonate-ramp facies of the Flume in the Alberta Rocky Mountains were portrayed by McLean and Klapper (1998) as being diachronous, spanning the interval of the *disparilis* to *norrisi* zones (no evidence is cited in that study to confirm their interpretation). In our field area the oldest Flume carbonates appear to be no older than the *norrisi* Zone (Klapper, in Johnson, 1990, = Lowermost *asymmetricus* Zone of the older conodont zonation). We recovered *Pandorinellina insita* at 39 meters above the base of the Flume Formation in the upper part of sequence 2 at Ancient Wall (Table 1, sample AC-103). Uyeno (1987) also recovered *P. insita* in the lower five meters of the Flume Formation at the Maligne Formation type section (GSC samples 12NBd and 12Nbe). In that study Uyeno provisionally correlated the entire Flume with the "Lowermost? *asymmetricus* Zone" of the old Frasnian Standard Zonation. The lowest occurrence of *P. insita* is within or near the base of the *norrisi* Zone. Consequently the major marine deepening event that initiated Alberta Rocky Mountain Devonian depositional sequence 2 most likely coincides with the initial marine flooding event of Devonian T–R Cycle IIb (Fig. 6; Johnson et al., 1985; Johnson et al., 1996), North American cratonic Devonian T–R Cycle IIb-1 of Day et al.

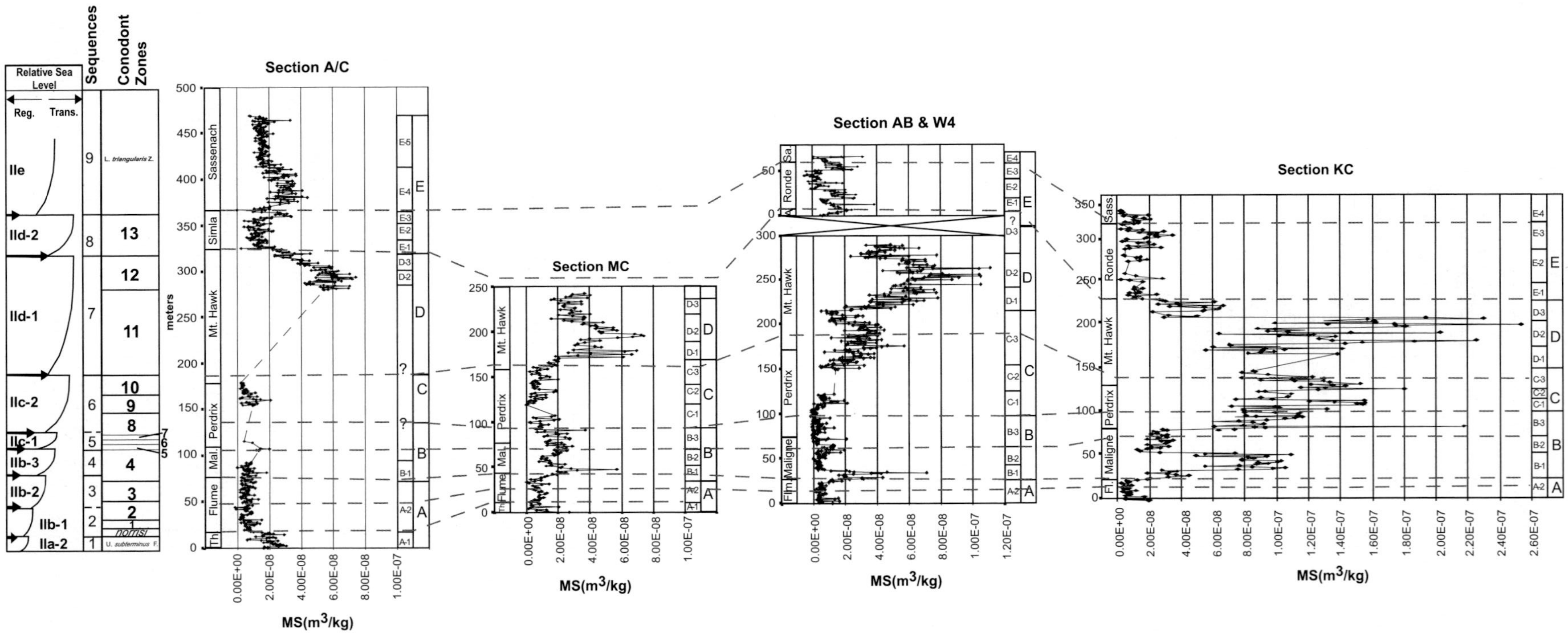

FIG 9.—Cross section illustrating magnetic-susceptibility records from across the Alberta basin, Rocky Mountains, western Alberta. Sequence-stratigraphic subdivision is illustrated on the left, and dashed lines indicate sequence boundaries. Individual MS sections illustrate the thickness in meters, stratigraphic units, MS data in m^3/kg, and both short-term and long-term MS events. Note the major MS excursions in the Maligne, Perdrix, and Mount Hawk formations. Note also that most significant positive excursions are in LST, TST, or late HST deposits. Devonian T–R Cycles are after Johnson et al. (1985) and Johnson et al. (1996). North American Devonian cratonic T–R cycles IIa-2, IIb-1, IIb-2 are after Day et al. (1996), and divisions of T–R cycle IId are after Day (1998). Proposed subdivision of T–R cycle IIb (IIb1-3) is after Day and Whalen (2003) and Day (2004). Sa. and Sass. = Sassenach Formation.

TABLE 1.—Late Givetian to middle Frasnian (Late Devonian) conodont faunas from the Flume and Perdrix formations in Sections A1 and A2 at Ancient Wall, Thornton Pass, Jasper National Park, Alberta.

Ancient Wall, Thornton Creek, Section A1 - A2 Composite									
Sample	A1-3F	AC-102	AC-103	AC-105	A7C	A6C	A2-106	A2-154.2	175.5
Interval above base (m)	5.3	34.5	39	59.4	82.6	83.3	106	154.2	175.5
Conodont Zone/Fauna	*subterm.*	?	*insita* F.	?	?	Z. 4	Z. 5	Z. 5-10	Z. 10
Icriodus subterminus	X						X	X	
indet. ramiform elements		X		X					
indet. coniform elements			X						
Pandorinellina insita			X						
Ozarkodina sp.					X				
Palmatolepis transitans						X			
Mesotaxis asymmetrica						X			
Meso. johnsoni						X	X		
Polygnathus dubius						X			
P. sp.						X	X		X
Pa. sp.						X		X	
P. timanicus							X		
Pa. punctata								X	
Icriodus symmetricus								X	
Ozarkodina postera								X	
Pa plana									X
Pa. proversa									X
Ancyrodella curvata early form									X
Ancyrognathus coeni									X
Stratigraphic Unit			**Flume Formation**			**Maligne**	**Lower**	**Perdrix**	**M-U P.F.**
Depostional Sequence	**1**	**Seq.**	**2**	**Seq.**	**3**	**Seq. 4**	**Seq.**	**5**	**Seq. 6**

Abbreviations for conodont zones in Tables 1–4: *T.* = Lower *triangularis* Zone (early Famennian) of Ziegler and Sandberg (1984); zones 12 and 13 are the Frasnian Montagne Noire (MN) zones 12 and 13 of Klapper (1989). Abbreviations of conodont taxa in Tables 1–4: *A.* = *Ancyrodella*, *An.* = *Anycrognathus*, *I.* = *Icriodus*, *Meso.* = *Mesotaxis*, *O.* = *Ozarkodina*, *Pa.* = *Palmatolepis*, *P.* = *Polygnathus*, *Pelek.* = *Pelekygnathus*.

(1996), central Alberta subsurface third-order T–R cycle Upper A (Uyeno and Wendte, 2005a; Wendte and Uyeno, 2005), and the base of BHL2 (Potma et al., 2001). This event initiated Waterways Formation deposition (Firebag and Peace Point members) in northeastern Alberta (Norris and Uyeno 1981, 1982; Day et al., 1996; Uyeno and Wendte, 2005a).

The timing of the marine transgression that initiated sequence 3 deposition cannot be constrained with confidence at this time, and is tentatively aligned with a position within MN Zone 2 (Fig. 6). It is likely that the subaerial exposure surface on the top of the Utopia Member in the Alberta outcrop succession coincides with the widespread subsurface R4 unconformity within Beaverhill Lake T–R cycle C (i.e., upper C) in the Swan Hills and Beaverhill Lake platform successions. Wendte and Uyeno (2005) correlated the R4 unconformity in the upper part of their T–R cycle C within MN Zone 2. This tentative correlation suggests widespread carbonate platform emergence in western Canada at that time (Fig. 6). Conodonts in the lower part of Maligne Formation at section K (Figs. 4, 6, Table 2) constrain the age of the top of Flume Formation (upper sequence 3 boundary) within MN Zone 4. Consequently, magnetic-susceptibility event A2 appears to span the interval of the *norrisi* Zone through the lower part of MN Zone 4 (Figs. 6, 9). BHL3 of Potma et al. (2001) appears to correlate with our sequence 3; however, several of the sequence boundaries that they identify above this interval do not appear to coincide with boundaries that we have identified, and the biostratigraphic data that they present are insufficient to firmly establish correlations of their remaining sequences with our sequence stratigraphic framework.

Sequence 4.—

Sequence 4 includes the Maligne and lower Cairn formations (Figs. 3, 4, 6). At Ancient Wall the fauna at 0.3 meters above the base of the Maligne Formation includes *Palmatolepis transitans*, *Mesotaxis asymmterica*, *M. Johnsoni*, and *Polygnathus dubius*, permitting a direct correlation within MN Zone 4 (Table 1, sample A6C). The lowest occurrence of *Pa. transitans* defines the base of MN Zone 4 (Klapper, 1989). Thus far all conodont samples from the Maligne Formation at Miette indicate that sequence 4 is within MN Zone 4, as indicated by the occurrence of "*Scaphignathus*" homeomorph, *Pa. transitans* (Table 2, samples 7, 10), and *Ancryodella africana* (Table 2, sample 20). These occur below the lowest MN Zone 5 samples, with *M. johnsoni* and *Pa. punctata*, in the overlying lower Perdrix Formation (Table 1, samples 23–29).

Uyeno (1987) recovered *Pandorinellina insita* in the lower 12.5 meters of the Maligne type section (GSC samples 12NBaa to 12NBgg), with *Ancyrodella africana* occurring in the uppermost Maligne (GSC sample 12NBjjj). The latter species first occurs within the upper part of MN Zone 3 and ranges into MN Zone 6 (Klapper, 1989, samples 15, 19). Klapper and Lane (1989) reported a fauna with *Palmatolepis transitans*, *Mesotaxis asymmetrica*, and *A. africana* from the "uppermost Flume Formation" 0.5 meters be-

TABLE 2.— Frasnian (Late Devonian) conodont sequence in the Maligne, Perdrix, and Mount Hawk formations in Section K, at Marmot Cirque (Miette Platform—off-reef basinal section), Jasper Park, Alberta. Frasnian conodont zones are after Klapper (1989) as in Table 1. Abbreviations of conodont taxa are as in Table 1.

Miette, Marmot Cirque, Section K																									
Sample Number	K7	K10	K11	K20	K23	K24	K26	K29	K32	K32AC	K33	K36	K38	K40	K44	K45	K46	K47	K48	K50	K51	K52	K53	K57	K58
Sample Position (m)	34.5	44.4	46.8	60.2	66.5	72.8	82.7	89	97.8	103.2	110.2	120.8	132.8	148.6	178.8	182.7	187.4	195.1	200.8	209.5	214.9	220.7	223.7	243.6	250
Conodont Zone	4	4	?	?	5	?	5	up 5	6	6	?	10	10	10	?	?	11	11	?	?	11	?	12	?	?
Pandorinellina insita	x																								
"*Scaphignathodus* " homeomorph	x																								
Mehlina sp.	x														x										
Palmatolepis transitans		x							x																
Polygnathus dubius		x																							
Icriodus sp.		x		x											x										x
Mesotaxis asymmetrica			x																						
P. webbi			x																						
I. subterminus			x										x					x							
Ancryodella africana				x																					
P. sp.				x	x	x	x	x	x	x	x	x	x	x	x				x						
P. uchtensis					x																				
Mesotaxis johnsoni					x		x																		
A. sp.					x		x																		
I. symmetricus						x				x	x		x												
Pelekygnathus sp.						x																			
P. efimovae?							x																		
Pa. punctata								x	x																
Ancyrodella gigas form 1								x																	
A. curvata early form								x			x														
Pa. spinata									x																
Ancyrognathus primus									x																
An. Ancryognathoideus										x															
Pa. sp.											x	x	x				x								
Pa. plana												x	x	x											
Ozarkodina postera												x													
Palmatolepis orbicularis													x												
Pa. amplifacata													x												
Pa. mucronata													x												
Pa. aff. *Pa. proversa*													x												
Ancyrognathus barba													x												
Ancyrodella curvata late form													x												
Bellodella sp.															x										
P. sp. indet.																x				x		x			x
Pa. semichatovae																	x	x							
Pa. ljaschenokae																	x								
A. buckeyensis																	x								
Pa. aff. *P. winchelli*																					x				
P. samueli																							x		
P. aspelundi																							x		
P. unicornis?																							x		
Pa. winchelli?																								x	
Stratigraphic Unit	Maligne Fm.				Lower Perdrix Fm.							M.-U. Perdrix			Mt. Hawk Fm. - Lower Arcs Mbr.										
Depositional Sequence	Sequence 4				Sequence 5							Sequence 6			Sequence 7										

low the base of the Perdrix at Luscar Mountain correlated with a position high in MN Zone 4. That sample is actually from the uppermost Maligne Formation. Elevated rates of organic-matter burial and/or preservation, indicated by high total-organic-carbon (TOC) values, are seen in the transgressive systems tract of sequence 4 at Miette (van Buchem et al., 1996; Whalen et al., 2002). This likely resulted from expansion of eutrophic bottom waters of the upper part of the OMZ during the sequence 4 transgression, immediately below the fore-reef slope oncoid horizon in the lower Maligne (Whalen et al., 2002). The sequence 4 TOC peak coincides approximately with peak values of MS event B1 (Fig. 8).

The sequence 4 deepening event in MN Zone 4 (base of MS Event B1) coincides with the deepening that initiated Beaverhill Lake second-order T–R cycle G (Uyeno and Wendte, 2005a; Wendte and Uyeno, 2005), and was informally designated as North American Devonian T–R cycle IIb-3 (Fig. 6; Day and Whalen, 2003; Day, 2004). This sea-level event coincides with the deepening event in the *transitans* Zone (= intra-MN Zone 4), designated as T–R cycle IIb/c in the Late Devonian bank-to-reef complex in eastern Laurussia (Holy Cross Mountains, Poland; Racki, 1997: Racki and Baliński, 1998; Racki et al., 2004). This event can also be recognized in the early Frasnian epeiric carbonate-platform succession in the Iowa Basin, where the same deepening event in MN Zone 4 coincides with the base of the Buffalo Heights Member of the Lithograph City Formation, eastern Iowa, and upper part of the New Bloomfield Member of the Snyder Creek Shale, central Missouri (Day, 1996, 1997, 1998, 2004).

Sequences 5 and 6.—

Conodont faunas from the lower Perdrix Formation with *Polygnathus uchtensis, Mesotaxis johnsoni*, and *Palmatolepis punctata* at Miette (Figs. 3, 6; Table 2, samples K23, K26), and Ancient Wall (Table 1, sample A2-106) indicate a correlation of the lower part of the Perdrix with MN Zone 5. Overlying facies at Miette yield conodonts of MN Zone 6 (Table 2, samples K32 and K32AC). Klapper and Lane (1989) documented the middle Frasnian conodont sequence in the Perdrix at section E/B/D (Fig. 4) and near section KC (Fig. 1), demonstrating that it spans the interval of MN Zones 5–10, confirmed by graphic correlation (Klapper, 1997). Klapper and Lane (1989, p. 472) placed the base of their informal Alberta Frasnian zone 4 in the middle Perdrix, defined on the first occurrence of *Palmatolepis proversa,* whose lowest occurrence also defines the base of MN Zone 9 (Klapper, 1989). Consequently, the upper part of the lower Perdrix (basinal facies of sequence 5) is likely no younger than zone 9 and no older than Zone 8 (Fig. 6).

TABLE 3.— Late Devonian (late Frasnian) conodont sequence through the upper part of depositional sequences 7 and 8 sampled from the upper part of the Arcs and Ronde members of the Southesk Formation in sections W2, W3, and W4 (Section W composite) at Poachers Creek, Jasper Park, Alberta (prograding ramp sections south of the Miette detached carbonate platform). Conodont zones are after Klapper (1989) as in Table 1. Abbreviations of conodont taxa are as in Table 1.

Miette, Poachers Creek, Section W Composite																	
Sample Number	W2-1	W2-2	W2-3	W2-4	W2-5	W2-6	W2-9	W2-11	W3-1	W3-2	W3-3	W3-4	W3-6F	W4-1	W4-4	W4-6	W4-7
Conodont Zone	?	12-13	?	12	12	?	?	?	12	?	12	12	?	12-13	?	12-13	?
Palmatolepis sp.	X				X	X			X								
Polygnathus sp.	X	X		X		X	X	X									
Pa. rhenana		X	X														
Pa. winchelli		X	X						X								
P. politus		X															
Ozarkodina postera		X							X								
P. imparilis			X			X			X		X	X					
P. unicornis			X														
P. s p.			X		X				X	X	X	X	X	X	X	X	
P. samueli				X					X		X	X					
O. dissimilis				X		X			X			X		X	X		
Ancyrodella ioides					X												
Mehlina sp.					X										X		
Ancyrognathus amana						X											
Pa. rhenana									X								
Pa. bogaardi?, or Pa. orlovi?									X								
indet. ramiform elements										X							X
An. buckeyensis transitional with An. iodes											X						
Stratigraphic Unit	Arcs Member, Southesk Formation														Ronde Mb.		
Depositional Sequence	Sequence 7														Sequence 8		

Black shales and organic-rich hemipelaic carbonates of the lower Perdrix signify a significant middle Frasnian deepening at or near the base of MN Zone 5 (Fig. 6) coincident with the base of Devonian T–R cycle IIc (Johnson et al., 1985; Johnson et al., 1996). The lower Perdrix of sequence 5 is correlated with the middle and upper Cooking Lake Formation in the Redwater platform, yielding conodonts of MN Zone 5, above depositional cycle H (= upper Beaverhill Lake and lower Cooking Lake fms.; Uyeno and Wendte, 2005a). The limited data available at present do not permit us to constrain with certainty the upper age range of sequence 5 and magnetic-susceptibility event B3. As mentioned above, the base of MN Zone 9 falls within the middle Perdrix (sequence 6), and all of our samples from the middle–upper Perdrix thus far fall within MN Zone 10 (Table 1, sample A2-175; Table 2, samples K36, K38, K40). Given the available data, the major marine deepening of sequence 6 (middle–upper Perdrix Fm., Figs. 3, 4, 6–9) is no younger than MN Zone 9 and no older than Zone 8 (Klapper and Lane, 1989; Klapper 1997; this study). This tentatively aligns magnetic-susceptibility events C1 to lower part of C3 within the upper part of the Middle Frasnian (Fig. 6, 7).

Sequence 7.—

Late Frasnian deposits of sequence 7 include the Mount Hawk Formation, the Arcs Member, and the lower half of the Simla Member of the Southesk Formation in the study area (Figs. 1–4). The major marine deepening event that initiated sequence 7 deposition occurred at our near the base of MN Zone 11 (Fig. 6), referred to the *semichatovae* transgression on the basis of the widespread first occurrence of *Palmatolepis semichatovae* in the lower part of that zone, coinciding with the initial transgression of Devonian T–R cycle IId (Johnson et al., 1985; Johnson et al., 1996), and North American cratonic T–R cycle IId-1 (Day, 1998). Deposits of sequence 7 yield conodonts (Tables 3, 4) of MN zones 11–12. Peritidal deposits capping the upper part of sequence 7 (Arcs and lower Simla mbrs.) record ramp emergence following the rapid basin infilling by Mount Hawk Formation shales and prograding-ramp carbonates of the Arcs. Magnetic-susceptibility events upper C3 and D1 to D3 fall within sequence 7, spanning MN zones 11–12 (Figs. 6–9).

Sequence 8.—

Deposits of sequence 8 record a very late Frasnian sea-level deepening event that resulted in progradation of ramp facies of the Ronde and upper part of the Simla members of the Southesk Formation. This deepening occurred at or near the base of MN Zone 13, as documented by conodont faunas of that zone in the upper Simla at Ancient Wall (Fig. 6; Table 4, sample 34). At Miette late Frasnian faunas in the upper part of composite section W are developed in the *Polygnathus* biofacies and thus far do not allow recognition of MN Zone 13 (Table 3). Magnetic-susceptibility events E1–E3 are entirely within the interval of MN Zone 13. The deepening event recorded by sequence 8 coincides with a major global sea-level rise designated as North American cratonic Devonian T–R cycle IId-2 (Day, 1998) and is also coincident with the Lower Kellwasser extinction bioevent.

Sequence 9.—

Deposits of sequence 9 include the Sassenach Formation (Figs. 3, 4, 6). The tops of the extinct Frasnian platforms in western Alberta were emergent during a latest Frasnian–earliest Famennian lowstand, although subtidal conditions persisted in off-reef basinal settings. The early Famennian conodont fauna from an oncoid debris-flow bed (Whalen et al., 2002), 1.8 meters above the base of the Sassenach Formation type section at section A/C (Table 4, sample 48.8) indicates that initial Famennian transgression and resumption of carbonate deposition on the flanks and tops of the older extinct Frasnian platforms occurred during the Lower to Middle *Palmatolepis triangularis* Zone. The initial records of magnetic-susceptibility

TABLE 4.— Late Devonian (late Frasnian and early Famennian) conodont sequence in Section C at Thornton Pass (Ancient Wall off-reef section), Jasper Park, Alberta. This is the type section of the Sassenach Formation. Frasnian conodont zones after Klapper (1989) as in Table 1, Famennian conodont zones are after Ziegler and Sandberg (1984). Abbreviations of conodont taxa are as in Table 1. Abbreviations of stratigraphic units: Mt. H = Mount Hawk Formation.

Ancient Wall, Thornton Creek, Section C

Sample Interval (m)	**0.5**	**13**	**16**	**21.5**	**23**	**25.5**	**34**	**35**	**45.5**	**47**	**48.8**	**68.4**
Conodont Zone	**Zone 12**				**Zone 13**						***L-M triang.***	
Palmatolepis winchelli	X	X			X		X	X	X			
Polygnathus politus	X	X	X		X		X					
Mehlina sp.	X			X	X							
P. imparilis		X	X		X	X	X		X	X		
P. samueli		X	X									
Pa. sp.		X	X				X					X
Ancyrognathus aff. *An. altus*			X									
Pa. rhenana			X		X		X					
Pa. boogaardi				X	X		X		X			
P. lodiensis				X	X			X				
Pelekygnathus planus					X							
Ancyrodella ioides [homeomorph]					X		X					
A. nodossa					X							
P. brevicarina					X		X		X			
P. sp						X						X
P. unicornis							X					
Pa. sp. - juveniles								X				
A. buckeyensis								X				
P. n. sp. R of Klapper and Lane								X				
P. sp. cf. *P. unicornis*									X			
An. cf. *An. asymmetricus*									X			
Icriodus alternatus									X	X		
I. lowaensis									X			
Pa. triangularis											X	
P. brevilaminus											X	X
P. praecursor?											X	
I. alternatus											X	
I. iowaensis											X	
Pelek. sp.												X
Stratigraphic Unit	**Mt.H**	**Simla Member, Southesk Formation**									**Sassenach**	
Depostional Sequence	**Seq. 7**				**Seq. 8**						**Seq. 9**	
Upper Devonian Stage	**Frasnian Stage**										**Famennian**	

events E4 and E5 were developed from the samples collected at the Sassenach type section (Fig. 3, section A/C). There the Sassenach spans the interval of the Lower–Middle *triangularis* Zone (Table 4) through the Lower *Palmatolepis crepida* Zone, on the basis of brachiopod-based correlations by Raasch (1989; Famennian zones DFM1 to DFM3), and conodont faunas documented by Johnston and Chatterton (2001) that place the base of the overlying Palliser Formation within the Lower *P. crepida* Zone. Consequently magnetic-susceptibility event E4 is entirely within the *triangularis* Zone (undifferentiated), and the lower part of E5 is in the upper part of the *triangularis* Zone, with the upper part of event E5 in part of the Lower *crepida* Zone (Fig. 6).

DISCUSSION

Sequence Boundaries, Flooding Surfaces, and the Transgressive Systems Tract

All sequences in the late Givetian to early Famennian succession shallow upward, and some provide evidence of subaerial exposure (Whalen et al., 2000b; Day and Whalen, 2005). Only a few of these surfaces appear to indicate actual base-level fall. Evidence supporting sea-level fall below the carbonate-platform top includes karst and/or paleosol development at the tops of sequences 2, 5, 7, and 8 (Whalen et al., 2000a; Whalen et al., 2000 b). In most cases lowstand deposits are not preserved on the platform top and in many cases consist of wedges of platform-onlapping sediment confined to the basin (Whalen et al. 2000a; Whalen et al., 2000b). The absence of lowstand deposits on the platform top means that the sequence boundary and transgressive marine flooding surface are often coincident. These surfaces commonly juxtapose nodular marls with slope fauna over platform-margin or lagoonal facies (Whalen et al. 2000a; Whalen et al., 2000b). Transgressive surfaces are commonly erosional and truncate underlying platform facies, and overlying facies locally contain clasts eroded from the underlying platform.

The transgressive systems tract on top of the platforms is usually relatively thin, and the platform commonly catches up and platform-margin facies were reestablished within several meters of sediment accumulation. Whalen et al. (2002) documented the role of microbial components in the transgressive system. They demonstrated the preponderance of mesotrophic taxa during transgressive events and implied that nutrification was an important transgressive process. There are several likely sources of nutrients associated with sequence boundaries and transgressive surfaces. Lowstand progradation of the shoreline could have delivered more terrigenous material to the basin and increased nutrient levels. Recent work in southwest Florida documented transgressive carbonate muds with up to 30% organic matter from modern microbial activity and erosion of mangrove and freshwater peat (Wanless et al., 2005). Upwelling is another potential nutrient source, although other typical upwelling-related lithofacies are not present (Parrish, 1982). Eolian transport is another significant nutrient-delivery system, the importance of which we are only beginning to understand (Ridgwell and Watson, 2002; Soreghan and Soreghan, 2002; Irino and Tada, 2003; Werner et al., 2003; Ikehara et al., 2004). Lowstand erosion is thought to be the source for much eolian silt and clay and is a likely driver of the magnetic-susceptibility signal in deep marine sediments (Ellwood et al., 1999; Ellwood et al., 2000). This part of western Canada was likely in the southern tropical to subtropical tradewind belt, and coeval evaporites indicate arid conditions to the east and south (Fig. 1; Witzke and Heckel, 1988; Scotese and McKerrow, 1990; Switzer et al., 1994). Eolian transport across the "Old Red Continent" was likely significant. The role of eolian nutrients in the marine system has only recently gained widespread attention, and their contribution during lowstand and reworking during early transgression may explain both the nutrification and the increased MS signature that we see in transgressive systems.

Our preliminary MS data display remarkable consistency across the Alberta basin (Fig. 9). Specific MS events can be tied directly to the sequence stratigraphic framework, with lowstand, early transgressive, and late highstand intervals preferentially recording elevated MS signatures and late transgressive and early highstand intervals with a lower or constant MS signal (Fig. 8; Day and Whalen, 2005).

Controls on the MS Signal and Potential for Long-Range Correlation

Correlation and paleoclimatic interpretation of marine, lacustrine, and loess successions using MS is readily accepted in Quaternary and Tertiary sediments (Kukla et al., 1988; Beget et al., 1990; Verosub et al., 1993; Nowaczyk et al., 2002; McFadden et al., 2005) and has even been used to astronomically calibrate Tertiary sedimentary successions (Shackleton et al., 1999), but the technique has been criticized for use in ancient rocks because of potential alteration of the MS signal during diagenesis. Several studies have demonstrated that the MS signature has been altered due to diagenetic clay mineralization (Katz et al., 1998) or the migration of petroliferous formation fluids (Schneider et al., 2004). The MS data presented here are quite robust across at least five thrust sheets and over 30 km. Once palinspastically restored, this would have represented a much greater distance across the Late Devonian continental margin. The similarity in the patterns across a proximal-to-distal transect and the general increase in MS signature toward proximal localities in the east supports our contention that we are seeing a signature dominated by original detrital input (Fig. 8). It seems unlikely that such consistent patterns would result from diagenesis. The fact that the Alberta basin has experienced several episodes of diagenetic fluid flow (Machel and Mountjoy, 1987; Machel et al., 1996; Machel and Cavell, 1999) resulting in dolomitization and hydrocarbon migration indicates that the primary MS signature is not always destroyed by such diagenetic events.

At present there are limited data for comparison to test the utility of MS for long-range correlation. Several studies of the MS signature across the F–F boundary have been published (Crick et al., 2001b; Crick et al., 2002; Racki et al., 2002), but very few studies present data that span the entire Frasnian and F–F interval. In an MS study of Upper Devonian platform carbonates in Belgium da Silva and Boulvain (2002) identified four complete and one partial MS event during the hassi conodont zone (MN zones 7-10). This compares well with the four MS events we have identified during the same interval (Figs. 7–9). Data collected at 5 cm intervals from thin condensed sections in Morocco provide another point of comparison (Fig. 10; Ellwood et al., 1999). The spline-smoothed MS signal from these Upper Devonian rocks displays a shape structure quite similar to that of the MS signal from western Alberta. Our considerably thicker outcrop sections were sampled at 0.5–1.0 m intervals, but, despite the difference in lithofacies, sampling interval, and extreme distance, major MS events constrained by biostratigraphy can be reasonably correlated (Fig. 10).

If the MS signature is controlled by sea level and climate change there is the possibility that it may be a useful paleoclimate proxy. If, as Shackleton (1999) suggested, the MS signature of deep marine sediments is a proxy for oxygen isotopes then we should see covariation of ^{18}O and the MS signal. Recently published ^{18}O data from Devonian conodont apatite and brachiopod calcite (Joachimski et al., 2004; van Geldern et al., 2005) can be compared to the MS signature of strata from western Canada (Fig. 11). The shape structure of the MS curve reveals several peaks in common between the data sets (Fig. 11). Joachimski et al. (2004) argue convincingly that the ^{18}O data can largely be interpreted in terms of temperature, since the fauna present in the rocks do not indicate significant changes in salinity (Day, 1996). The oldest oxygen isotope excursions are represented only by facies deposited at Ancient Wall or even farther to the west, because sea level did not onlap areas to the east prior to the uppermost Givetian *norrisi* zone (Figs. 3, 9, 11). Latest Givetian ^{18}O data point to paleotemperatures of about 25°C with a general warming of about 8°C through the Frasnian and into the lowermost Famennian (Joachimski et al., 2004). This general trend is punctuated by several relatively short-term positive ^{18}O excursions interpreted as rapid low-latitude cooling events of up to 7°C (Joachimski et al., 2004). The most precipitous of these cooling events coincide with the Lower and Upper Kellwasser extinction horizons (Fig. 11; Joachimski et al., 2004; van Geldern et al., 2006).

A minor paleotemperature increase (negative ^{18}O excursion) is recorded near the boundary of MN zones 2 and 3 that appears to correlate to MS event A-2 in the Flume Formation. The next significant positive MS event (B-1) is represented by a minor blip in paleotemperature data (Joachimski et al., 2004) but does not register in the $^{18}O_{calcite}$ data of van Geldern et al. (2006) because of insufficient sampling (Fig. 11). A significant paleotemperature increase (negative ^{18}O excursion) is recorded in MN zone 7 that correlates with MS event C-1 in the Perdrix Formation (Fig. 11). A series of relatively rapid positive and negative ^{18}O excursions, interpreted as paleotemperature swings (Joachimski et al., 2004), occur between upper MN zone 11 and the Famennian Lower *triangularis* zone and correlate well with MS events (Fig. 11). Due to insufficient sample density, not all of these events are identified in $\delta^{18}O_{calcite}$ data of van Geldern et al. (2006). These episodes of warming with an intervening cooling event straddle the MN zone 11–12 boundary (Fig. 11). The two warming events correlate with Mount Hawk Formation MS events D-1 and D-2, and the cooling event corresponds to the trough that separates these two MS events (Fig. 11). A similar pattern is seen in MN zone 13, with two paleotemperature spikes that appear to correlate with MS

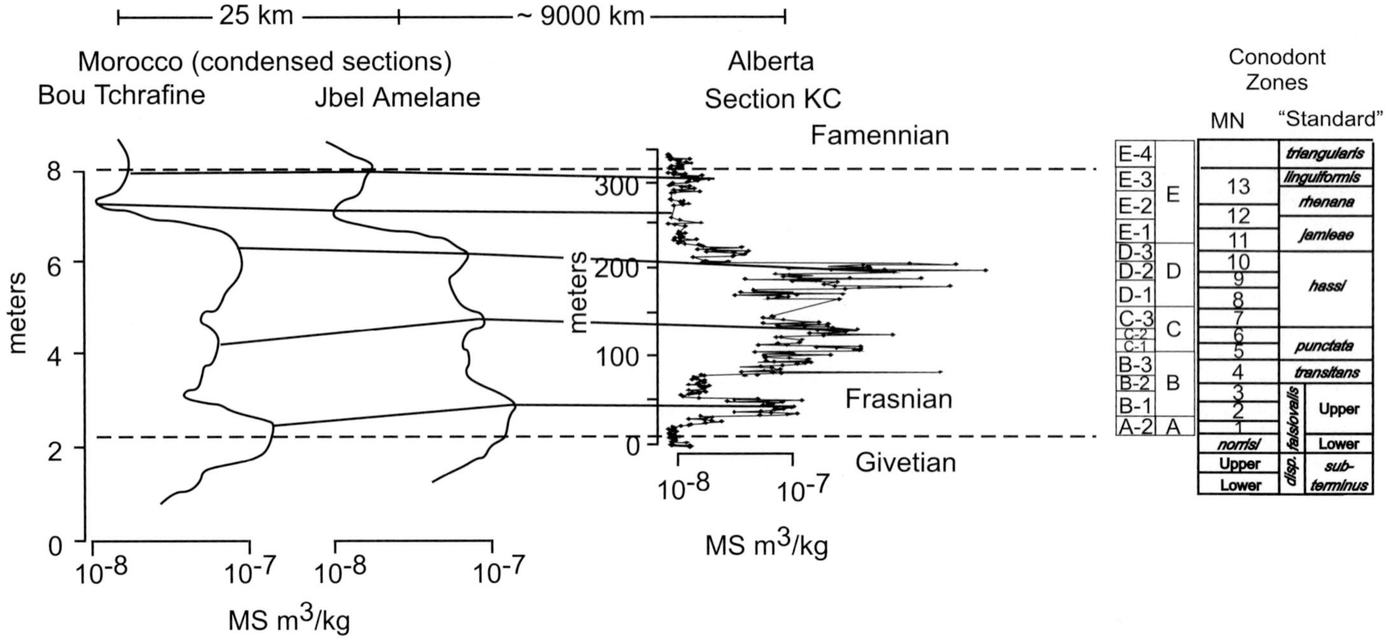

FIG 10.—Spline-smoothed MS data from condensed sections in Morocco (after Ellwood et al., 1999) and MS data from section KC, Alberta (this study). Note that the magnitude of individual MS events varies between localities but the general pattern and timing of major MS excursions displays remarkable correspondence between localities separated by approximately 9000 km. MN (Montagne Noir) conodont zonation is after Klapper (1989), and Klapper and Foster (1993); "standard" conodont zonation is after Ziegler and Sandberg (1990).

events E-2 and E-3 in the Ronde Member (Southesk Fm.) (Fig. 11). The intervening cooling event (positive $\delta^{18}O$ excursion) coincides with the Lower Kellewasser extinction horizon (Joachimski et al., 2004) and appears to correlate with the MS trough separating events E-1 and E-2 (Fig. 11). A decrease in paleotemperature is also associated with the F–F boundary and the Upper Kellewasser extinction event (Joachimski et al., 2004; van Geldern et al., 2006) and appears to correlate with the MS low in event E-4 (Fig. 11).

The $\delta^{18}O$ data reported to date are relatively coarse, and not all MS excursions correspond to documented oxygen isotope excursions. In addition, while the general shape structure of the MS and oxygen isotope data sets are similar, the magnitude of the variations in $\delta^{18}O$ are not always of the same magnitude as the shifts in MS values. This hints at a more complicated relationship between $\delta^{18}O$ and MS data, both of which could be influenced by local inputs (i.e., variations in salinity or freshwater input on $\delta^{18}O$ data or local changes in weathering on MS data). Higher-resolution sampling for oxygen isotopes will be necessary to test whether MS data can serve as an oxygen isotope proxy. The utility of MS data as an oxygen isotopic proxy thus remains equivocal. However, MS measurements are relatively quick and inexpensive, and demonstration of MS as a reliable paleoclimate proxy would provide a new and relatively simple tool for analysis of paleoclimate change.

CONCLUSIONS

Upper Givetian through Lower Famennian rocks in western Alberta were deposited as nine sequences separated by subaerial exposure surfaces and/or marine flooding surfaces. Each sequence records a relative rise and fall of sea level, and seven of nine sequences correspond to parts of Johnson et al. (1985) T–R cycles IIa to IIe, and subdivisions thereof by Day et al. (1996) and Day (1998). We report here high-resolution magnetic-susceptibility and conodont biostratigraphic data that provide further constraints on the timing and pattern of sea-level and biotic events.

Initial Givetian transgression of the West Alberta Arch is recorded by the Thornton Creek Member (sequence 1: T–R cycle IIa-2) of the Flume Formation at Ancient Wall. This unit displays a somewhat higher MS signature (event A-1) than the rest of the overlying Flume Formation This unit consists of three mixed carbonate–siliciclastic parasequences that record restricted to normal marine environments (Day and Whalen, 2005). Restricted environments of parasequence 1 contain a low-diversity *Athyris* fauna, while more open marine faces in parasequences 2 and 3 include a diverse brachiopod fauna dominated by *Eleutherokomma*, *Athyris*, and *Schizophoria* (Day and Whalen, 2005).

A second sea-level rise (T–R Cycle IIb-1) at the base of the *norrisi* Zone resulted in carbonate-ramp deposition of sequence 2, represented by the Flume Formation. Evidence of exposure of part of the Flume ramp is observed at the top of the Utopia Member (top sequence 2), followed by a deepening recorded by the rest of the upper Flume (sequence 3: T–R cycle IIb-2?). The Flume ramp was drowned by a significant early Frasnian deepening within MN Zone 4 that marked the initiation of a prograding rimmed platform recorded by the Maligne–Lower Cairn formations (sequence 4—upper subdivision of T–R cycle IIb = IIb-3?). Reworked lowstand and early transgressive facies of the Maligne commonly display an increase in MS signature (event B-1) compared to the underlying Flume Formation, but this progressively decreases as platform progradation progressed and carbonate input into the basin increased (Figs. 7–9). The profound middle Frasnian deepening event at the base of MN 5/*punctata* Zone resulted in deposition of the lower Perdrix–Upper Cairn formations (sequence 5) coincident with the initial flooding of T–R cycle IIc (base of MS event B-2). This sequence

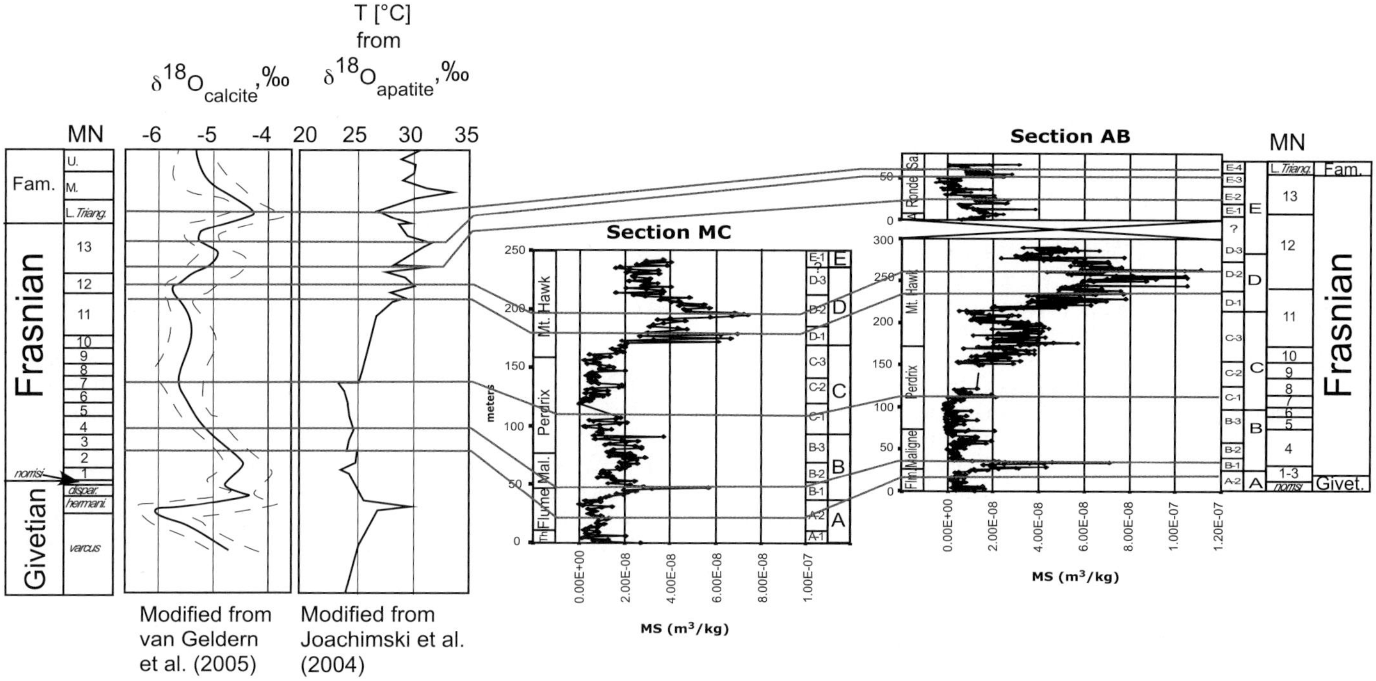

FIG 11. —Comparison of MS events from sections MC and KC, Alberta, $\delta^{18}O$ data from brachiopod calcite (van Geldern et al., 2006), and interpreted low-latitude sea-surface temperature derived from $\delta^{18}O$ of conodont apatite (Joachimski et al., 2004). Positive MS excursions commonly coincide with temperature increase (negative $\delta^{18}O$ excursions) while negative MS signatures are associated with decreasing paleotemperatures (positive $\delta^{18}O$ excursions). Our high-resolution MS sampling has identified several excursions that do not appear in the relatively coarsely sampled oxygen isotope data. Note that while general trends in the data are parallel the magnitude of the shifts in MS and temperature are not always coincident. See text for further discussion. MN (Montagne Noir) conodont zonation is after Klapper (1989) and Klapper and Foster (1993).

displays a highly variable MS signature that displays a general decrease followed by increase in susceptibility through the lower Perdrix Formation (MS events B-2 and B-3). Another middle Frasnian intra-IIc deepening event initiated deposition of the Middle Perdrix Formation–Peechee Member of the Southesk Formation (sequence 6) during the interval of MN zones 7?-10. The MS signature of this sequence (C-1 through lower C-3) is partially obscured by poor exposure in some sections (Fig. 9) but appears to display a general increase in susceptibility up section. Major deepening at or near the base of MN 11 initiated T–R cycle IId (IId-1) deposition as recorded by the Upper Perdrix–Mount Hawk formations / Arcs Member of the Southesk Formation (sequence 7). This basinal sequence exhibits a general increase in MS signature through the Mount Hawk Formation, which then decreases in the carbonate-dominated Arcs Member of the Southesk Formation, where it prograded over slope and basin facies (MS events upper C-3 and D1–3). Most of the typical late Frasnian Mount Hawk brachiopod fauna became extinct just prior to or during the major deepening coinciding with the Lower Kellwasser event in the very late Frasnian. This event coincides with a significant drop in sea-surface paleotemperature (Joachimski et al., 2004; van Geldern et al., 2006) and the MS trough separating events E-1 and E-2. At Ancient Wall the upper Mount Hawk Formation–Simla Member (sequence 8) was deposited during the interval of MN 13 (T–R cycle IId-2). Extinction of latest Frasnian shelly faunas occurs very high in MN 13, as indicated by a crisis fauna in the uppermost Simla. The F–F boundary and Upper Kellwasser event is also associated with a significant low-latitude cooling event (Joachimski et al., 2004; van Geldern et al., 2006) and the MS trough near the base of event E-4. The overlying early Famennian Sassenach Formation (sequence 9) records the initial post-extinction flooding of T–R cycle IIe, yielding survivor and post-extinction-recovery shelly faunas. The Sassenach records an influx of siliciclastics into the basin and thus records slightly higher MS signatures (MS events E-4 and E-5) than underlying Ronde–Simla facies.

This study demonstrates the utility of MS stratigraphy for correlation and to delineate relative changes in sea level, some of which may be eustatic in origin. Additional data from Devonian localities spanning the globe will be necessary to test this hypothesis fully. The general correlation between the MS signature in Alberta and condensed facies in Morocco (Fig. 10) hints that this method has the potential for long-range correlation and to provide stratigraphic resolution similar to or better than that provided by biostratigraphy. The potential of MS as an oxygen isotope proxy is also explored through the comparison of our MS data with available oxygen isotope data from brachiopod calcite and conodont apatite (Joachimski et al., 2004; van Geldern et al., 2006). Trends in the MS data appear to mirror and temporally coincide with similar trends in ^{18}O data and interpreted paleotemperatures; however, the magnitude of the shifts in the two data sets are not always comparable. The viability of MS to serve as an oxygen isotope proxy remains open to question, but the signal does appear to be modulated by paleoclimate, and the speed and relative simplicity of performing MS measurements make the technique an attractive alternative or supporting method if this relationship proves to be robust.

ACKNOWLEDGMENTS

This work benefited from thought-provoking discussions with J. Wendte, T. Uyeno (Geological Survey of Canada), and J. Over (SUNY Geneseo). D. Stone (U. of Alaska) provided insight into paleomagnetics and magnetic susceptibility. R. Missler and Z. Pearson (U. of Alaska) and A. Norton (SUNY Geneseo) performed MS measurements in B. Ellwood's magnetics laboratory at Louisiana State University. This research would not have been possible without Dr. Ellwood's assistance. We would like to acknowledge A. Fuhrman, M. Kuhn, J. Morris (Illinois State U.), J. Over, A. Norton (SUNY Geneseo), P. Mayer (U. of Wisconsin Madison), and A. Krumhardt (U. of Alaska) for assistance in the field. Thanks to Mitch Harris and Robert Erlich for thoughtful reviews and excellent suggestions for improving the manuscript. Acknowledgment is also made to the donors of The Petroleum Research Fund, administered by the American Chemical Society (Whalen), and to the National Geographic Society—Foundation for Exploration and Research (Day) for partial support of this research. The Department of Geology and Geophysics, University of Alaska Fairbanks, assisted with the cost of manuscript preparation. This research would not have been possible but for the cooperation of Parcs Canada, who provided access and permission to sample at field localities in Jasper National Park.

REFERENCES

ANDRICHUK, J.M., 1961, Stratigraphic evidence for tectonic and current control of Upper Devonian reef sedimentation, Dunhamel area, Alberta, Canada: American Association of Petroleum Geologists, Bulletin, v. 45, p. 612–632.

BECKER, T., 1993, Anoxia, eustatic changes, and Upper Devonian to lowermost Carboniferous global ammonoid diversity: Systematics Association, Special Volume 47, p. 115–163.

BEGET, J.E., STONE, D.B., AND HAWKINS, D.B., 1990, Paleoclimatic forcing of magnetic susceptibility variations in Alaskan loess during the late Quaternary: Geology, v. 18, p. 40–43.

BURTON, E.A., MACHEL, H.G., AND QI, J., 1993, Thermodynamic constraints on anomalous magnetization in shallow and deep hydrocarbon seepage environments, *in* Aissaoui, D.M., McNeill, D.F., and Hurley, N.F., eds., Applications of Paleomagnetism to Sedimentary Geology: SEPM, Special Publication 49, p. 193–207.

CAPLAN, M.L., AND BUSTIN, R.M., 1998, Sedimentology and sequence stratigraphy of Devonian–Carboniferous strata, southern Alberta: Bulletin of Canadian Petroleum Geology, v. 46, p. 487–514.

CHOW, N., WENDTE, J., AND STASIUK, L.D., 1995, Productivity versus preservation controls on two organic-rich carbonate facies in the Devonian of Alberta: sedimentological and organic petrological evidence: Bulletin of Canadian Petroleum Geology, v. 43, p. 433–460.

COPPER, P., 2002, Reef development at the Frasnian Famennian mass extinction boundary: Palaeogeography, Palaeoclimatology, Palaeoecology, v. 181, p. 27–65.

CRICK, R.E., ELLWOOD, B.B., HASSANI, A.E., FEIST, R., AND HLADIL, J., 1997, Magnetosusceptibility event and cyclostratigraphy of the Eifelian–Givetian GSSP and associated boundary sequences in north Africa and Europe: Episodes, v. 20, p. 267–175.

CRICK, R.E., ELMWOOD, B.B., HASSANI, A.E., HLADIL, J., HROUDA, F., AND CHLUPAC, I., 2001a, Magnetosusceptibility event and cyclostratigraphy (MSEC) of the Pridoli–Lochkovian GSSP (Klonk, Czech Republic) and coeval sequences in the Anti-Atlas Morocco: Palaeogeography, Palaeoclimatology, Palaeoecology, v. 167, p. 73–100.

CRICK, R.E., ELLWOOD, B.B., OVER, D.J., FEIST, R., AND GIRARD, C., 2001b, Magnetostratigraphic susceptibility of the Frasnian–Famennian boundary (Upper Devonian) in Southern Oklahoma and its relationship to the type area in Southern France: Oklahoma Geological Survey, Circular 105, p. 71–82.

CRICK, R.E., ELLWOOD, B.B., FEIST, R., HASSANI, A.E., SCHINDLER, E., DREESEN, R., OVER, D.J., AND GIRARD, C., 2002, Magnetostratigraphy susceptibility of the Frasnian/Famennian boundary: Paleogeography, Paleoclimatology, Palaeoecology v. 181, p. 67–90.

DA SILVA, A.C., AND BOULVAIN, F., 2002, Sedimentology, magnetic susceptibility and isotopes of a Middle Frasnian Carbonate platform: Tailfer section, Belgium: Facies, v. 46, p. 89–102.

DAVIES, T.A., HAY, W.W., SOUTHAM, J.R., AND WORSLEY, T.R., 1977, Estimates of Cenozoic oceanic sedimentation rates: Science, v. 197, p. 53–55.

DAY, J., 1996, Faunal signatures of Middle–Upper Devonian depositional sequences and sea level fluctuations in the Iowa Basin: US mid-continent, *in* Witzke, B.J., Ludvigson, G.A., and Day, J., eds., Paleozoic Sequence Stratigraphy: Views from the North American Craton: Geological Society of America, Special Paper 306, p. 277–300.

DAY, J., 1997, Phylogeny and biogeography of *Tecnocyrtina* (Brachiopoda—Spiriferinida) in the Devonian (Givetian–Frasnian) of North America, *in* Klapper, G., Talent, J., and Murphy, M.A., eds., Paleozoic Sequence Stratigraphy, Biostratigraphy and Biogeography: Studies in Honor of J. Granville (Jess) Johnson: Geological Society of America, Special Paper 321, p. 245–261.

DAY, J., 1998, Distribution of latest Givetian–Frasnian Atrypida (Brachiopoda) in central and western North America: Acta Palaeontologica Polonica, v. 43, p. 205–240.

DAY, J., 2004, Timing of late Givetian–early Frasnian (Middle and Late Devonian) sea level events and subtropical cratonic carbonate platform development, central North America (abstract): Geological Society of America, Abstracts with Programs, v. 36, no. 3, p. 39.

DAY, J., AND WHALEN, M.T., 2003, Western Laurussian record of regional and global Middle–Late Devonian sea level changes and bioevents: Alberta Rocky Mountains (abstract): Geological Society of America, Abstracts with Programs, v. 34, p. 208.

DAY, J., AND WHALEN, M.T., 2005, Thornton Creek Member (new) of the Flume Formation and the initial Middle Devonian Onlap of the West Alberta Arch: Canadian Rocky Mountains: Bulletins of American Paleontology 369, p. 123–149.

DAY, J., UYENO, T. C., NORRIS, A.W., WITZKE, B.J., AND BUNKER, B.J., 1996, Middle–Upper Devonian relative sea-level histories of North American cratonic interior basins, *in* Witzke, B.J., Ludvigson, G. A., and Day, J., eds., Paleozoic Sequence Stratigraphy; Views from the North American Craton: Geological Society of America, Special Paper 306, p. 259–276.

EDMOND, J.M., 1992, Himalayan tectonics, weathering processes, and the strontium isotope record in marine limestones: Science, v. 258, p. 1594–1597.

ELIUK, L.S., 1984, A hypothesis for the origin of hydrogen sulphide in Devonian Crossfield Member dolomite, Wabamun Formation, Alberta, Canada, *in* Eliuk, L.A. ed., Carbonates in Subsurface and Outcrop: Canadian Society of Petroleum Geologists, 1984 Core Conference, p. 245–289.

ELLWOOD, B.B., CRICK, R.E., AND HASSANI, A.E., 1999, The magnetosusceptibility event and cyclostratigraphy (MSEC) method used in Geological correlation of Devonian rocks from Anti-Atlas Morocco: American Association of Petroleum Geologists, Bulletin, v. 83, p. 1119–1134.

ELLWOOD, B.B., CRICK, R.E., HASSANI, A.E., BENOIST, S.L., AND YOUNG, R.H., 2000, Magnetosusceptibility event and cyclostratigraphy method applied to marine rocks: Detrital input versus carbonate productivity: Geology, v. 28, p. 1135–1138.

ELLWOOD, B.B, CRICK, R.E, GARCIA-ALCALDE, J.L., SOTO, F.M., TRUYOIS-MASSONI, M., HASSANI, A.E., AND KOVAS, E.J., 2001, Global correlation using magnetic susceptibility data from Lower Devonian rocks: Geology, v. 29, p. 583–586.

FEJER, P.E., AND NARBONNE, G.M., 1992, Controls on Upper Devonian, meter-scale carbonate cyclicity, Icefall Brook, southeast British Columbia: Bulletin of Canadian Petroleum Geology, v. 40, p. 363–380.

FISCHER, A.G., 1983, The two Phanerozoic subcycles, *in* Berggren, W., and Couvering, J., eds., Catastrophes in Earth History; The New Uniformitarianism: Princeton, New Jersey, Princeton University Press, p. 138–149.

GELDSETZER, H.H.J., 1989, Ancient Wall reef complex, Frasnian age, Alberta, *in* Geldsetzer, H.H.J., James, N.P., and Tebbutt, G.E., eds., Reefs, Canada and Adjacent Areas: Canadian Society of Petroleum Geologists, Memoir 13, p. 431–439.

GEORGE, A.D., 1999, Deep-water stromatolites, Canning Basin, northwestern Australia: Palaios, v. 14, p. 493-505.

GEORGE, A.D., PLAYFORD, P.E., POWELL, C.M., AND TORNATORA, P.M., 1997, Lithofacies and sequence development of an Upper Devonian mixed carbonate–siliciclastic fore-reef slope, Canning Basin, western Australia: Sedimentology, v. 44, p. 843–887.

HELLER, F., AND LIU, T.S., 1984, Magnetism of Chinese loess deposits: Royal Astronomical Society, Geophysical Journal, v. 77, p. 125–141.

HERBIG, H.G., AND WEBER, H.M., 1996, Facies and stromatoporoid biostromes in the Strunian (latest Devonian) of the Aachen region, Germany: Göttinger Arbeiten zur Geologie und Paläontologie, Sonderband, v. SB2, p. 359–364.

HOUSE, M.R., AND ZIEGLER, W., 1997, On sea-level fluctuations in the Devonian: Courier Forschungsinstitut Senckenberg, v. 199, 146 p.

IKEHARA, M., KAWAMURA, K., OHKOUCHI, N., MURAYAMA, M., NAKAMURA, T., AND TAIRA, A., 2004, Variations of terrestrial input and marine productivity in the Southern Ocean (48°S) during the last two deglaciations: Paleoceanography, v. 15, p. 170–180.

IRINO, T., AND TADA, R., 2003, High-resolution reconstruction of variation in aeolian dust (Kosa) deposition at ODP Site 797, the Japan Sea, during the last 200 ka: Global and Planetary Change, v. 35, p. 143–156.

ISAACSON, P.E., AND DÍAZ-MARTÍNEZ, E., 1995, Evidence for a middle and Late Paleozoic foreland basin and significant paleolatitudinal shift, central Andes, *in* Tankard, A.J., Suarez, R., and Welsink, H.J., eds., Petroleum Basins of South America: American Association of Petroleum Geologists, Memoir 62, p. 231–149.

ISAACSON, P.E., HLADIL, J., SHEN, J., KALDOVA, J., AND GRADER, G., 1999, Late Devonian (Famennian) glaciations in South America and marine offlap on other continent: Geologisches Bundesanstalt, Abhandlungen, v. 54, p. 239–257.

JOACHIMSKI, M.M., VAN GELDERN, R., BREISIG, S., BUGGISCH, W., AND DAY, J., 2004, Oxygen isotope evolution of biogenic calcite and apatite during the Middle and Late Devonian: International Journal of Earth Science, v. 93, p. 542–553.

JOHNSTON, D.I., AND CHATTERTON, B.D.E., 2001, Upper Devonian (Famennian) conodonts of the Palliser Formation and Wabamun Group, Alberta and British Columbia, Canada: Palaeontographica Canadiana, no. 19, 154 p.

JOHNSON, J.G., 1990, Lower and Middle Devonian brachiopod-dominated communities of Nevada, and their position in a biofacies-province-realm model, with a section on revision of Middle Devonian conodont zones, by G. Klapper and J.G. Johnson: Journal of Paleontology, v. 64, p. 902–941.

JOHNSON, J.G., KLAPPER, G., AND SANDBERG, C.A., 1985, Devonian eustatic fluctuations in Euramerica: Geological Society of America, Bulletin, v. 96, p. 567–587.

JOHNSON, J.G., SANDBERG, C.A., AND POOLE, F.G., 1991, Devonian lithofacies of western United States, *in* Cooper, J.D., and Stevens, C.H., eds., Paleozoic Paleogeography of the Western United States—II: SEPM, Pacific Section, Los Angeles, p. 83–106.

JOHNSON, J.G., KLAPPER, G., AND ELRICK, M., 1996, Devonian transgressive–regressive cycles and biostratigraphy, northern Antelope Range, Nevada, establishment of reference horizons for global cycles: Palaios, v. 11, p. 3–14.

KATZ, B., ELMORE, R.D., COGOINI, M., AND FERRY, S., 1998, Widespread chemical remagnetization: orogenic fluids or burial diagenesis of clays?: Geology, v. 26, p. 603–606.

KLAPPER, G., 1989, The Montagne Noire Frasnian (Upper Devonian) Conodont succession, *in* McMillan, N.J., Embry, A.F., and Glass, D.J., eds., Devonian of the World: Canadian Society of Petroleum Geologists, Memoir 14, v. 3, p. 449–468.

KLAPPER, G., 1997, Graphic correlation of Frasnian (Upper Devonian) sequences in Montagne Noire, France and western Canada, *in* Klapper, G., Murphy, M.A., and Talent, J.A., eds., Paleozoic Sequence Stratigraphy, Biostratigraphy, and Biogeography: Studies in Honor of J. Granville ("Jess") Johnson: Geological Society of America, Special Paper 321, p. 113–130.

KLAPPER, G., AND FOSTER, C.T., JR., 1993, Shape analysis of Frasnian species of the Late Devonian conodont genus *Palmatolepis*, Paleontological Society, Memoir 32 (Journal of Paleontology, v. 67, supplement to no. 4), 35 p.

KLAPPER, G., AND LANE, H.R., 1989, Frasnian (Upper Devonian) conodont sequence at Luscar Mountain and Mount Haultain, Alberta Rockies, *in* McMillan, N.J., Embry, A.F., and Glass, D.J., eds., Devonian of the World: Canadian Society of Petroleum Geologists, Memoir 14, v. 3, p. 469–478.

KLAPPER, G., KIRCHGASSER, W.T., AND BAESEMANN, J.F., 1995, Graphic correlation of a Frasnian (Upper Devonian) composite standard, *in* Mann, K.O., Lane, H.R., and Scholle, P.A., eds., Graphic Correlation: SEPM, Special Publication 53, p. 177–184.

KLOVAN, J.E., 1964, Facies analysis of the Redwater reef complex, Alberta, Canada: Bulletin of Canadian Petroleum Geology, v. 12, p. 1–100.

KOBLUK, D.R., 1975, Stromatoporoid paleoecology of the southeast margin of the Miette carbonate complex, Jasper Park, Alberta: Bulletin of Canadian Petroleum Geology, v. 23, p. 224–277.

KUKLA, G., HELLER, F., LIU, X.M., XU, T.C., LIU, T.S., AND AN, Z.S., 1988, Pleistocene climates in China dated by magnetic susceptibility: Geology, v. 16, p. 811–814.

MACHEL, H.G., AND CAVELL, P.A., 1999, Low-flux, tectonically-induced squeegee fluid flow ("hot flash") into the Rocky Mountain foreland basin: Bulletin of Canadian Petroleum Geology, v. 47, p. 510–533.

MACHEL, H.G., AND HUNTER, I.G., 1994, Facies models for Middle to Late Devonian shallow-marine carbonates, with comparisons to modern reefs: a guide for facies analysis: Facies, v. 30, p. 155–176.

MACHEL, H.G., AND MOUNTJOY, E.W., 1987, General constraints on extensive pervasive dolomitization—and their application to the Devonian carbonates of western Canada: Bulletin of Canadian Petroleum Geology, v. 35, p. 143–158.

MACHEL, H.G., CAVELL, P.A., AND PATEY, K.S., 1996, Isotopic evidence for carbonate cementation and recrystallization, and for tectonic expulsion of fluids into the Western Canada Sedimentary Basin: Geological Society of America, Bulletin, v. 108, p. 1108–1119.

MACKENZIE, F.T., AND PIGOTT, J.P., 1981, Tectonic controls of Phanerozoic sedimentary rock cycling: Geological Society of London, Journal, v. 138, p. 183–191.

MCFADDEN, M.A., PATTERSON, W.P., MULLINS, H.T., AND ANDERSON, W.T., 2005, Multi-proxy approach to long- and short-term Holocene climate-change: evidence from eastern Lake Ontario: Journal of Paleolimnology, v. 33, p. 371–391.

MCGHEE, G.H., JR., 1996, The Late Devonian Mass Extinction: the Frasnian / Famennian Crisis: New York, Columbia University Press, 303 p.

MCLAREN, D.J., 1982, Frasnian–Famennian extinction, *in* Silver, L.T., and Schultz, P.H., eds., Geological Implications of Large Asteroids and Comets on Earth: Geological Society of America, Special Paper 190, p. 477–484.

MCLEAN, R.A., AND KLAPPER, G., 1998, Biostratigraphy of Frasnian (Upper Devonian) strata in western Canada, based on conodonts and rugose corals: Bulletin of Canadian Petroleum Geology, v. 46, p. 515–563.

McLean, D.J., and Mountjoy, E.W., 1994, Allocyclic control on Late Devonian buildup development, southern Canadian Rocky Mountains: Journal of Sedimentary Research, v. B64, p. 326–341.

Moore, P.F., 1989, Devonian reefs in Canada and some adjacent areas, *in* Geldsetzer, H.H.J., James, N.P., and Tebbutt, G.E., eds., Reefs, Canada and Adjacent Areas: Canadian Society of Petroleum Geologists, Memoir 13, p. 367–389.

Mountjoy, E.W., 1965, Stratigraphy of the Devonian Miette reef complex and associated strata, eastern Jasper National Park, Alberta: Geological Survey of Canada, Bulletin 110, 132 p.

Mountjoy, E.W., 1980, Some questions about the development of Upper Devonian carbonate buildups (reefs) western Canada: Bulletin of Canadian Petroleum Geology, v. 28, p. 315–344.

Mountjoy, E.W., 1989, Miette Reef Complex (Frasnian), Jasper National Park, Alberta, *in* Geldsetzer, H.H.J., James, N.P., and Tebbutt, G.E., eds., Reefs, Canada and Adjacent Areas: Canadian Society of Petroleum Geologists, Memoir 13, p. 497–505.

Mountjoy, E.W., and Becker, S., 2000, Frasnian to Fammenian sea level changes and the Sassenach Formation, Jasper Basin, Alberta Rocky Mountains, *in* Homewood, P.W., and Eberli, G.P., eds., Genetic Stratigraphy on the Exploration and Production Scales—Case Studies from the Pennsylvanian of the Paradox Basin and the Upper Devonian of Alberta: Bulletin, Centre Recherche Elf Exploration-Production, Mémoire 24, p. 181–201.

Mountjoy, E.W., and Mackenzie, W.S., 1974, Stratigraphy of the southern part of the Devonian Ancient Wall complex, Jasper National Park, Alberta: Geological Survey of Canada, Paper 72-20, 121 p.

Mountjoy, E.W., and Savoy, L.E., 1995, Cratonic-margin and Antler-age foreland basin strata (Middle Devonian to Lower Carboniferous) of the southern Canadian Rocky Mountains and adjacent plains, *in* Dorobek, S.L., and Ross, G.M., eds., Stratigraphic Evolution of Forland Basins: SEPM, Special Publication 52, p. 211–226.

Norris, A.W., and Uyeno, T.T., 1981, Stratigraphy and paleontology of the lowermost Upper Devonian Slave Point Formation on Lake Claire and the lower Upper Devonian Waterways Formation on Birch River, northeastern Alberta: Geological Survey of Canada, Bulletin 334, 53 p.

Norris, A.W., and Uyeno, T.T., 1982, Biostratigraphy and paleontology of the Middle Upper Devonian boundary beds, Gypsum Cliffs area, northeastern Alberta: Geological Survey of Canada, Bulletin 313, 65 p.

Nowaczyk, N.R., Minyuk, P., Melles, M., Brigham-Grette, J., Glushkova, O., Nolan, M., Lozhkin, A.V., Stetsenko, T.V., Andersen, P.M., and Forman, S.L., 2002, Magnetostratigraphic results from impact crater Lake El'gygytgyn, northeastern Siberia; a 300 kyr long high-resolution terrestrial palaeoclimatic record from the Arctic: Geophysical Journal International, v. 150, p. 109–126.

Oliver, T.A., and Cowper, N.W., 1963, Depositional environments of the Ireton Formation, central Alberta: Bulletin of Canadian Petroleum Geology, v. 11, p. 183–202.

Parrish, J.T., 1982, Upwelling and petroleum source beds, with reference to the Paleozoic: American Association of Petroleum Geologists, Bulletin, v. 66, p. 750–774.

Playford, P.E., 1980, Devonian Great Barrier Reef of Canning Basin, western Australia: American Association of Petroleum Geologists, Bulletin, v. 64, p. 814–840.

Potma, K., Weissenberger, J.A.W., Wong, P.K., and Gilhooly, M.G., 2001, Toward a sequence stratigraphic framework for the Frasnian of the western Canada basin: Bulletin of Canadian Petroleum Geology, v. 49, p. 37–85.

Raasch, G.O., 1989, Famennian faunal zones in western Canada, *in* McMillan, N.J., Embry, A.F., and Glass, D.J, eds., Devonian of the World: Canadian Society of Petroleum Geologists, Memoir 14, v. 3, p. 619–631.

Racki, G., 1997, Devonian eustatic fluctuations in Poland: Courier Forschungsinstitut Senckenberg, 199, p. 1–12.

Racki, G., and Baliński, A., 1998, Late Frasnian Atrypida (Brachiopoda) from Poland and the Frasnian–Famennian biotic crisis: Acta Palaeontologica Polonica, v. 43, no. 2, p. 272–204.

Racki, G., Racka, M., Matyja, H., and Devleeschouwer, X., 2002, The Frasnian / Famennian boundary interval in the south Polish–Moravian shelf basins: integrated event-stratigraphical approach: Palaeogeography, Palaeoclimatology, Palaeoecology, v. 181, p. 251–298.

Racki, G., Piechota, A., Bond, D., and Wignall, P.B., 2004, Geochemical and ecological aspects of the lower Frasnian pyrite–ammonoid level of Kostomloty (Holy Cross Mountains, Poland): Geological Quarterly, v. 48, no. 3, p. 267–282.

Raymo, M.E., and Ruddiman, W.R., 1992, Tectonic forcing of late Cenozoic climate: Science, v. 359, p. 117–122.

Raymo, M.E., Ruddiman, W.R., and Froelich, P.N., 1988, Influence of late Cenozoic mountain building on ocean geochemical cycles: Geology, v. 16, p. 649–653.

Richards, B.C., 1989, Upper Kaskaskia Sequence: uppermost Devonian and Lower Carboniferous, Chapter 9, *in* Ricketts, B.D., ed., Western Canada Sedimentary Basin, a Case History: Canadian Society of Petroleum Geologists, p. 165–201.

Richter, F.M., Rowley, D.B., and DePaolo, D.J., 1992, Sr isotope evolution of seawater: the role of tectonics: Earth and Planetary Science Letters, v. 109, p. 11–23.

Ridgwell, A.J., and Watson, A.J., 2002, Feedback between aeolian dust, climate, and atmospheric CO_2 in glacial time: Paleoceanography, v. 17, p. 1059–1070.

Rogers, F. S., 1998, Conodont biostratigraphy of the Little Cedar and lower Coralville formations of the Cedar Valley Group (Middle Devonian) of Iowa and significance of a new species of *Polygnathus*: Journal of Paleontology, v. 72, p. 726–737.

Schneider, J., Bechstadt, T., and Machel, H.G., 2004, Covariance of C- and O-isotopes with magnetic susceptibility as a result of burial diagenesis of sandstones and carbonates: an example from the Lower Devonian La Vid Group, Cantabrian Zone, NW Spain: International Journal of Earth Science, v. 93, p. 990–1007.

Scotese, C.R., and McKerrow, W.S., 1990, Revised world maps and introduction, *in* McKerrow, W.S., and Scotese, C.R., eds., Palaeozoic Palaeogeography and Biogeography: Geological Society of London, Memoir 12, p. 1–21.

Shackleton, N.J., 1999, Will oxygen isotope stratigraphy survive the next millennium: Eos, v. 80, p. F505.

Shackleton, N.J., Crowhurst, S.J., Weedon, G.P., and Laskar, J., 1999, Astronomical calibration of Oligocene–Miocene time: Royal Society of London, Philosophical Transactions, A, v. 357, p. 1907–1929.

Shields, M.J., and Geldsetzer, H.H.J., 1992, The MacKenzie margin, Southesk–Cairn carbonate complex: Depositional history, stratal geometry and comparison with other Late Devonian platform-margins: Bulletin of Canadian Petroleum Geology, v. 40, p. 274–293.

Soreghan, G.S., Elmore, R.D., Katz, B., Cogoini, M., and Banerjee, S., 1997, Pedogenically enhanced magnetic susceptibility variations preserved in Paleozoic loessite: Geology, v. 25, p. 1003–1006.

Soreghan, G.S., Elmore, R.D., and Lewchuk, M.T., 2002, Sedimentologic–magnetic record of western Pangean climate in upper Paleozoic loessite (lower Cutler beds, Utah): Geological Society of America, Bulletin, v. 114, p. 1019–1035.

Soreghan, G.S., and Soreghan, M.J., 2002, Atmospheric dust and algal dominance in the late Paleozoic: a hypothesis: Journal of Sedimentary Research, v. 72, p. 457–461.

Stearn, C. W., 1987, Effect of the Frasnian–Famennian extinction event on the stromatoporoids: Geology, v. 15, p. 677–679.

Stearn, C.W., Halim-Dihardja, M. K., and Nishida, D.K., 1987, An oil-producing stromatoporoid patch reef in the Famennian (Devonian) Wabamun Formation, Normandville Field, Alberta: Palaios, v. 2, p. 560 –570

STEVENSON, R.K., WHITTAKER, S., AND MOUNTJOY, E.M., 2000, Geochemical and Nd isotopic evidence for sedimentary-source changes in the Devonian miogeocline of the southern Canadian Cordillera: Geological Society of America, Bulletin, v. 112, p. 531–539.

STOAKES, F.A., 1980, Nature and control of shale basin fill and its effect on reef growth and termination: Upper Devonian Duvernay and Ireton Formations of Alberta, Canada: Bulletin of Canadian Petroleum Geology, v. 28, p. 345–410.

STOAKES, F.A., 1992, Early Mississippian megasequences, *in* Wendte, J.C., Stoakes, F.A., and Campbell, C.V., eds., Devonian–Early Mississippian Carbonates of the Western Canada Sedimentary Basin: A Sequence Stratigraphic Framework: SEPM, Short Course 28, p. 241–249.

SWITZER, S.B., HOLLAND, W.G., CHRISTIE, G.S., GRAF, G.C., HEDINGER, A.S., MCAULEY, R.J., WIERZBICKI, R.A., AND PACKARD, J.J., 1994, Devonian Woodbend–Winterburn strata of the western Canada sedimentary basin, *in* Mossop, G., and Shetsen, I., eds., Geological Atlas of the Western Canada Sedimentary Basin: Canadian Society of Petroleum Geologists and Alberta Research Council, p. 165–202.

TRAMP, K.L., SOREGHAN, G.S., AND ELMORE, D.R., 2004, Paleoclimatic inferences from paleopedology and magnetism of the Permian Maroon Formation loessite, Colorado, USA: Geological Society of America, Bulletin, v. 117, p. 671–686.

TUCKER, R.D., BRADLEY, D.C., VER STRAETEN, C.A., HARRIS, A.G., EBERT, J.R., AND MCCUTCHEON, S.R., 1998, New U-Pb zircon ages and the duration and division of Devonian time: Earth and Planetary Science Letters, v. 158, p. 175–186.

UYENO, T.T., 1987, Report on conodonts from Cold Sulphur Spring (Stop 5), *in* Norris, A.W., and Pedder, A.E.H., eds., Devonian of Alberta Rocky Mountains between Banff and Jasper: International Union of Geological Sciences, Subcommission on Devonian Stratigraphy, Geological Survey of Canada, p. 62–63.

UYENO, T.T., 1998, Conodont faunas: Geological Survey of Canada, Bulletin 522, Part II, p. 149–191.

UYENO, T.T., AND WENDTE, J.C., 2005a, Conodont biostratigraphy and physical stratigraphy in two wells of the Beaverhill Lake Group, Upper Middle to Lower Upper Devonian, south-central Alberta, Canada: Bulletins of American Paleontology, 369, p. 123–149.

UYENO, T.T., AND WENDTE, J.C., 2005b, Conodont biostratigraphy of the Beaverhill Lake strata, upper Middle to lower Upper Devonian, south-central Alberta (abstract): American Association of Petroleum Geologists, Annual Meeting, v. 14, p. A144.

VAN BUCHEM, F.S.P., EBERLI, G.P., WHALEN, M.T., MOUNTJOY, E.W., AND HOMEWOOD, P., 1996, The basinal geochemical signature and platform margin geometries in the Upper Devonian mixed carbonate–siliciclastic system of western Canada: Société Géologique de France, Bulletin, v. 167, p. 685–699.

VAN GELDERN, R., JOACHIMSKI, M.M., DAY, J., JANSEN, U., ALVAREZ, F., YOLKIN, E.A., AND MA, X.-P., 2006, Carbon, oxygen, and strontium isotope records of Devonian brachiopod shell calcite: Palaeogeography, Palaeoclimatology, Palaeoecology, v. 240, p. 47–67.

VEROSUB, K.L., FINE, P., SINGER, M.J., AND TENPAS, J., 1993, Pedogenesis and paleoclimate: interpretation of the magnetic susceptibility record of Chinese loess–paleosol sequences: Geology, v. 21, p. 1011–1014.

WANLESS, H.R., VLASWINKEL, B., AND JACKSON, K.L., 2005, Transgressive recycling produces organic-rich carbonate muds (abstract): American Association of Petroleum Geologists, Abstracts with Programs, v. 14, p. A149.

WENDTE, J.C., 1992, Platform evolution and its control on reef inception and localization, *in* Wendte, J.C., Stoakes, F.A., and Campbell, C.V., eds., Devonian–Early Mississippian Carbonates of the Western Canada Sedimentary Basin: A Sequence Stratigraphic Framework: SEPM, Short Course 28, p. 41–88.

WENDTE, J.C., 1994, Cooking Lake platform evolution and its control on Late Devonian Leduc reef inception and localization, Redwater, Alberta: Bulletin of Canadian Petroleum Geology, v. 42, p. 499-528.

WENDTE, J.C., AND UYENO, T.T., 2005, Sequence stratigraphy and evolution of Middle to Upper Devonian Beaverhill Lake strata, south-central Alberta: Bulletin of Canadian Petroleum Geology, v. 53, p. 250–354.

WENDTE, J.C., STOAKES, F.A., AND CAMPBELL, C.V., 1992, Devonian–Early Mississippian Carbonates of the Western Canada Sedimentary Basin: A Sequence Stratigraphic Framework: SEPM, Short Course 28, 255 p.

WENDTE, J.C., STOAKES, F.A., BOSMAN, M., AND BERNSTEIN, L., 1995a, Genetic and stratigraphic significance of the Upper Devonian Frasnian Z Marker, west-central Alberta: Bulletin of Canadian Petroleum Geology, v. 43, p. 339–406.

WENDTE, J.C., UYENO, T.T., AND ABRAHAMSON, B., 1995b, Physical stratigraphy and conodont biostratigraphy of the Beaverhill Lake Group, Late Middle to Early Late Devonian, east-central Alberta, *in* Bell, J.S., Bird, T.D., Heller, T.L., and Greener, P.L., eds., Proceedings of the Oil and Gas Forum '95, Energy from Sediments: Geological Survey of Canada, Open File Report 3058, p. 141–144.

WENDTE, J., UYENO, T.T., AND ABRAHAMSON, B., 1997, Basin scale three dimensional sequence stratigraphy of the Beaverhill Lake Group, Late Middle to Early Late Devonian, central Alberta (abstract): Joint Convention of the Canadian Society of Petroleum Geologists and SEPM, Calgary, 1997, Program with Abstracts, p. 291.

WERNER, M., TEGEN, I., AND HARRISON, S. P., 2003, Dust–biosphere–atmosphere coupling on glacial–interglacial time scales: XVI INQUA congress; Shaping the Earth; a Quaternary Perspective, v. 16, p. 188.

WHALEN, M.T., AND DAY, J.E., 2000, Initial Devonian onlap of the West Alberta Arch, Canadian Rocky Mountains: Implications for paleogeography and sea-level history of western Alberta (abstract): American Association of Petroleum Geologists, Annual Meeting, Abstracts, v. 9, p. A-156.

WHALEN, M.T., EBERLI, G.P., VAN BUCHEM, F.S.P., AND MOUNTJOY, E.W., 2000a, Facies architecture of Upper Devonian carbonate platforms, Rocky Mountains, Canada, *in* Homewood, P.W., and Eberli, G.P., eds., Genetic Stratigraphy on the Exploration and Production Scales—Case Studies from the Pennsylvanian of the Paradox Basin and the Upper Devonian of Alberta: Bulletin, Centre Recherche Elf Exploration–Production, Mémoire 24, p. 139-178.

WHALEN, M.T., EBERLI, G.P., VAN BUCHEM, F.S.P., MOUNTJOY, E.W., AND HOMEWOOD, P.W., 2000b, Bypass margins, basin-restricted wedges and platform-to-basin correlation, Upper Devonian, Canadian Rocky Mountains: implications for sequence stratigraphy of carbonate platform systems: Journal of Sedimentary Research, v. 70, p. 913–936.

WHALEN, M.T., DAY, J., EBERLI, G.P., AND HOMEWOOD, P.W., 2002, Microbial carbonates as indicators of environmental change and biotic crises in carbonate systems: examples from the Upper Devonian, Alberta Basin, Canada: Palaeogeography, Palaeoclimatology, Palaeoecology, v. 181, p. 127–151.

WILLIAMS, C.A., AND KRAUSE, F.F., 2000, Paleosol chronosequences and peritidal deposits of the Middle Devonian (Givetian) Yahatinda Formation, Wasootch Creek, Alberta, Canada: Bulletin of Canadian Petroleum Geology, v. 48, p. 1–18.

WITZKE, B.J., AND HECKEL, P.H., 1988, Paleoclimatic indicators and inferred Devonian Paleolatitudes of Euramerica, *in* McMillan, N.J., Embry, A.F., and Glass, D.J., eds., Devonian of the World: Second International Symposium on the Devonian System: Canadian Society of Petroleum Geologists, Memoir 14, v. 1, p. 49–68.

WITZKE, B.J., BUNKER, B.J., AND KLAPPER, G., 1985, Devonian stratigraphy in the Quad Cities area, eastern Iowa–northwestern Illinois, *in* Hammer, W.R.. Anderson, R.C., and Schroeder, D.A., eds., Devonian and Pennsylvanian Stratigraphy of the Quad-Cities Region Iowa–Illinois: SEPM, Great Lakes Section, 15th Annual Field Conference, Guidebook, p. 19–64.

WOLD, C.N., AND HAY, W.W., 1990, Estimating ancient sediment fluxes: American Journal of Science, v. 290, p. 1069–1089.

WOOD, R., 2000a, Novel paleoecology of a postextinction reef: Famennian (Late Devonian) of the Canning basin, northwestern Australia: Geology, v. 28, p. 987–990.

WOOD, R., 2000b, Paleoecology of a Late Devonian back reef: Canning Basin, western Australia: Palaeontology, v. 43, p. 671–703.

WOOD, R., 2004, Paleoecology of a post-extinction reef: Famennian (Late Devonian) of the Canning basin, north-western Australia: Palaeontology, v. 47, p. 415–445.

WORSLEY, T.R., AND DAVIES, T.A., 1979, Sea-level fluctuations and deep-sea sedimentation rates: Science, v. 203, p. 455–456.

ZIEGLER, W., AND SANDBERG, C.A., 1984, *Palmatolepis*-based revision of the upper part of the standard Late Devonian conodont zonation, *in* Clark, D.L., ed., Conodont Biofacies and Provincialism: Geological Society of America, Special Paper 196, p. 179–194.

ZIEGLER, W., AND SANDBERG, C.A., 1990, The Late Devonian standard conodont zonation: Courier Forschungs-Institut Senckenberg, v. 121, p. 1–115.

Improved Characterization of Carbonate Platforms and Reefs

REMOTE SENSING AND COMPARATIVE GEOMORPHOLOGY OF HOLOCENE CARBONATE DEPOSITIONAL SYSTEMS

EUGENE C. RANKEY

Comparative Sedimentology Laboratory, Center for Southeastern Tropical Advanced Remote Sensing, Rosenstiel School of Marine and Atmospheric Sciences, University of Miami, Miami, Florida 33149, U.S.A.
e-mail: grankey@rsmas.miami.edu

AND

PAUL M. (MITCH) HARRIS

Chevron Energy Technology Company, 6001 Bollinger Canyon Road, San Ramon, California 94583, U.S.A.
e-mail: MitchHarris@chevron.com

ABSTRACT: Shallow-water carbonate systems encompass a spectrum of environments, from reefs to shoals to tidal flats. Although the sedimentologic characteristics of these systems have been studied for many years, new remote-sensing data provide unique, unparalleled perspectives on spatial heterogeneity in Holocene carbonate systems. The purpose of this paper is to describe the structure, content, and utility of a database of remote-sensing images that are included on a companion CD.

The twenty-five focus areas include a range of depositional systems from across the globe. The image database for most areas includes three scales of remote-sensing data: one that captures the regional context (Moderate Resolution Imaging Spectroradiometer [MODIS] data, with 250 m^2 pixels), another that illustrates the local setting (Landsat data, with ~ 30 m^2 pixels), and a third that includes high-resolution details of the area (IKONOS or QuickBird data, with 4 m^2 or 2.5 m^2 pixels). These are presented in an interactive, zoomable format, and are supplemented by supporting information on the setting and published work in each area.

The overall goal of the database is to provide the suite of new images to a broad audience, rather than to provide a detailed interpretation of any or all areas, or an overview of controls on carbonate depositional systems. These images can be used in several ways. For example, specific examples in the database have been applied as a learning tool in a classroom, and quantitative analysis illustrates enhanced student understanding of spatial complexity in these systems. Additionally, beyond the classroom, these images provide information on depositional models and the scales of depositional heterogeneity in carbonate systems for geologists, reservoir modelers, and geophysicists.

INTRODUCTION AND PURPOSE: SPACE—THE FINAL FRONTIER

Facies patterns in carbonates are distinctly heterogeneous, across a range of scales from that of textural changes within an individual facies to morphological variability among platforms through geologic time. Although stratigraphers have collected ample data on vertical facies patterns of carbonate successions (e.g., from core or outcrop; cf. summary in Lehrmann and Goldhammer, 1999), studies assessing spatial patterns are much less abundant, largely due to difficulties in observing depositional surfaces in ancient successions. In many regards, space represents the final frontier in the study of carbonate systems.

In light of this challenge, Holocene carbonate systems can provide a unique opportunity to examine facies patterns as they exist along a time line, i.e., the modern depositional surface. This paper provides an introduction to an image database containing a wide suite of examples of varied geomorphic character from several Holocene carbonate systems. Many examples contained in the image database build upon the more limited collection of satellite images published in hardcopy by Harris and Kowalik (1994). The aim of this paper is to describe the structure, content, and applications of the companion CD. The collective purposes of the paper and CD include: (1) to present an overview of geomorphic patterns present in modern coastal and shallow marine carbonate systems; (2) to offer a dynamic and interactive environment in which learners can explore different aspects of modern carbonate systems; (3) to provide examples of types of heterogeneity that might be present in ancient geologic analogs and the scales at which they occur; and (4) to show the value of some potential uses of these images and derivative data.

Limitations

The image database provides unparalleled perspectives on Holocene carbonate systems, most of which have never been published before. Nonetheless, it includes several limitations. First, images simply represent a "snapshot" of facies patterns at one point in time. There is no guarantee that the facies that enter the geologic record will be analogous to those captured in these images. For example, extensive coring of the ooid shoal complex at Joulters Cays, Bahamas, showed that the present configuration of the shoals with a vast stabilized sand flat to be the product of reworking of earlier-formed bar and channel systems (Harris, 1979). Second, the database includes a collection of images that were captured as "windows of opportunity," and should not be considered all-inclusive. There are many areas and types of systems for which complete data are simply not available. Nonetheless, many "classic" areas (e.g., Joulters Cays ooid shoals, Three Creeks tidal flats, Glovers atoll, Lily Bank ooid shoals) are included. Third, the database includes only the images of geomorphology of these systems, and does not explicitly deal with generative processes (e.g., Rankey et al., 2006a). For most areas, detailed process information is lacking. Fourth, some interpretations rely heavily on published literature, some of which is pre-GPS or pre-Landsat (1973). In these cases, some interpretive differences are likely simply due to positional errors in the older field studies. Finally, these images represent

Controls on Carbonate Platform and Reef Development
SEPM Special Publication No. 89, Copyright © 2008
SEPM (Society for Sedimentary Geology), ISBN 978-1-56576-130-8, p. 317–322.

Holocene accumulations, and there can always be questions of how representative the Holocene is of the geologic past. Because of its unique geologic and biologic aspects, the Holocene may not be the key to understanding every other carbonate system – so *caveat emptor*!

DATABASE—ORGANIZATION

The web-based system on the CD is organized hierarchically. From an introductory front page, users can explore the image database from several perspectives: by geomorphology, by geography, and by visual inspection of thumbnail images (Fig. 1). Each route contains links to specific locations in the image database.

For each location, the database includes interactive remote-sensing images ("Remote Sensing Overviews," left column of Figure 2) and information on key sedimentologic or geologic characteristics (right column of Figure 2). In most cases, the remote-sensing images include three scales of data: one that captures the regional context (Moderate Resolution Imaging Spectroradiometer [MODIS] data, with 250 m^2 pixels), another

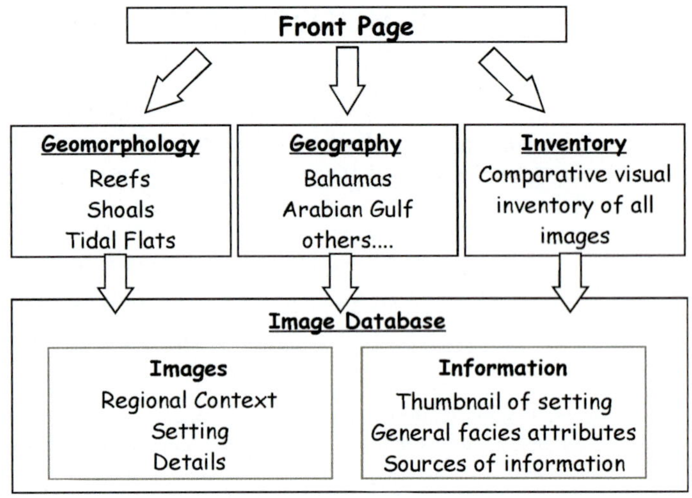

FIG. 1.—General organizational structure of the database.

FIG. 2.—An example page for one location (the tidal flats of western Andros Island, Bahamas). The left column contains links to remote-sensing images and interpretations. The right column provides information on the area, including general setting, facies attributes, and literature sources.

that illustrates the setting (Landsat data, with ~ 30 m^2 pixels), and a third that includes high-resolution details of the area (IKONOS or QuickBird data, with ~ 4 m^2 or 2.5 m^2 pixels). These images are accompanied by sedimentologic or geomorphic interpretations that provide a high-level overview.

In addition, each specific location includes a short summary of information on the area, which includes an overview of the setting (e.g., climate, tides, protection), generalized facies attributes, and important sources of information. Some well-studied areas have been documented in numerous publications, whereas other have little or no information in the published literature. The goal of this part of each web page is not to provide an all-inclusive literature review, only to highlight some of the most relevant or historically important papers on the area.

DATABASE—REMOTE-SENSING IMAGES

The database includes images from numerous Holocene carbonate systems (Fig. 3), and a classic ancient shelf margin (the Guadalupe Mountains, New Mexico and Texas). Twenty-one of the Holocene areas include MODIS, Landsat, and high-resolution images; four areas have only Landsat data. The images are presented in a pre-processed, zoomable format, but they do not include the original georeferenced multispectral data (those data are proprietary and copyright DigitalGlobe or GeoEye). A link from the front page provides a brief summary of each type of data and offers external links for those interested in learning more.

EXAMPLES DEMONSTRATING VALUE OF IMAGE MODULES

As discussed above, the image database provides new and improved perspectives on many classic (and not-so-classic) carbonate geomorphic systems. In this light, it provides information that could be used in many ways. This section briefly describes two examples that illustrate not only *that* these information can be used (which might be self-evident), but illustrate *the value of their use*.

Teaching

One obvious application of this database is use as an educational tool, to complement standard textbooks on carbonate sedimentology and stratigraphy. Students are commonly drawn

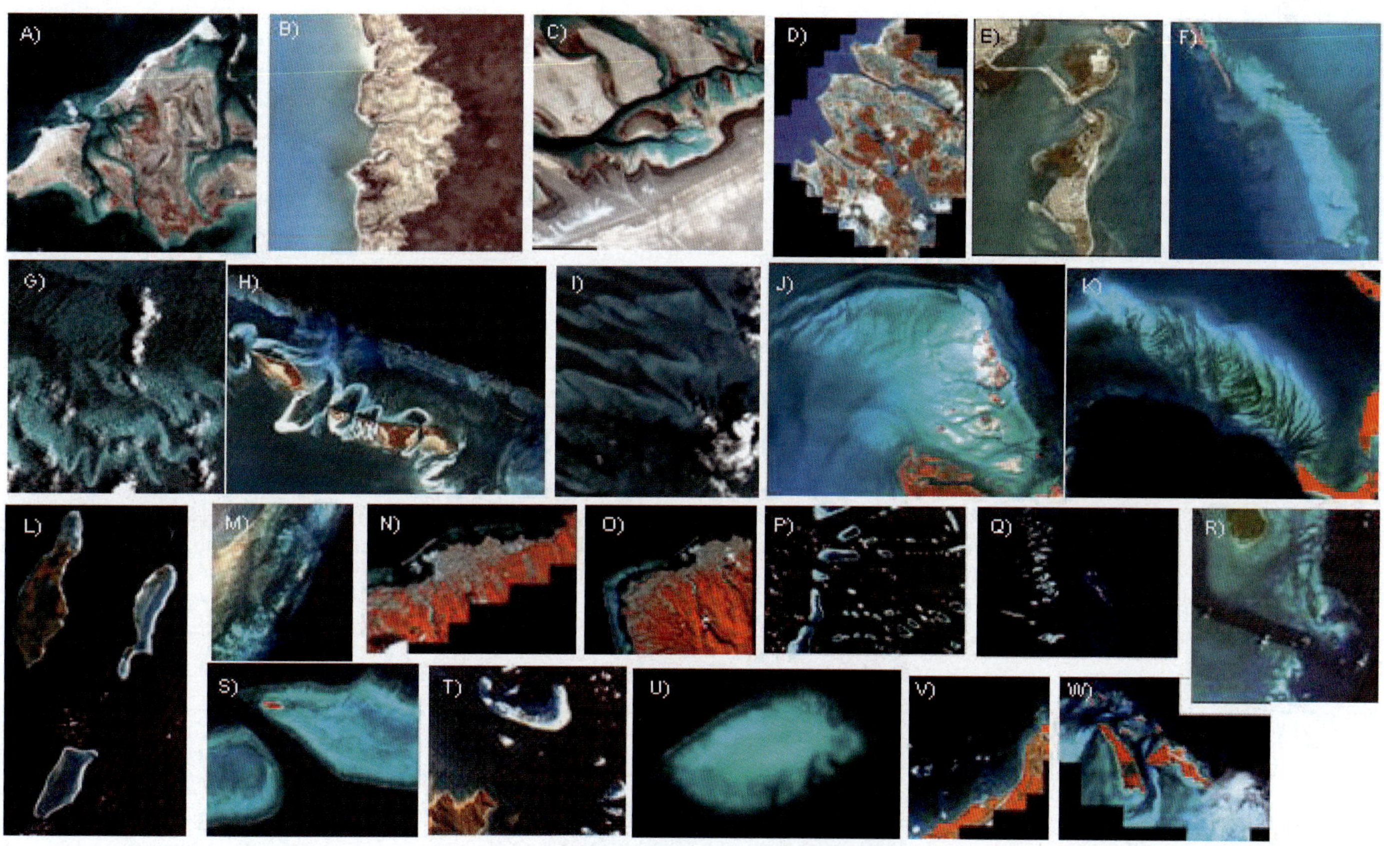

FIG. 3.—Images from each focus area. Long (x) axis on images = 11 km unless otherwise noted. **A)** Al Sadiyat, United Arab Emirates; **B)** western Andros Island, Bahamas; **C)** Khor al Bazam, United Arab Emirates; **D)** southwest Andros Island, Bahamas; **E)** Bear Cut, Miami, Florida, USA (x axis = 7 km); **F)** Cat Cay sand belt, Bahamas (x axis = 13 km); **G)** Lily Bank sand belt, Bahamas; **H)** northwestern Abaco Islands, Bahamas (x axis = 20 km); **I)** Tongue of the Ocean, Bahamas; **J)** Joulters ooid shoal, Bahamas (long axis = 30 km); **K)** Schooner Cays ooid shoal, Bahamas (long axis = 80 km); **L)** Turneffe, Lighthouse, and Glovers atolls, Belize (long axis = 55 km); **M)** Carysfort reef, Florida, USA (x axis = 7 km); **N, O)** Reefs, Tahiti, French Polynesia; **P)** Maldives (x axis = 30 km); **Q)** Palua Seribu, Indonesia (x axis = 50 km); **R)** English Cay channel, Belize (x axis = 7 km); **S)** Heron Island, Great Barrier Reef, Australia (x axis = 8 km); **T)** Great Barrier Reef, Australia (xa axis = 50 km); **U)** Sabah reef, Malaysia; **V)** Sabah, Malaysia; **W)** Lee Stocking Island, Bahamas. Images A, B, C, D, E, G, I, M, N, O, R, S, U, V, and W are copyright GeoEye.com.

to interactive, web-based learning tools, and our experiences suggest that these remote sensing image-based modules are no exception.

But do these tools work? To address this question, we assessed the effectiveness of using exploration of remote sensing data in an upper-level undergraduate sedimentary geology course. To start, the class split into two groups. Half the class completed a lab module that used Landsat and Ikonos remote-sensing data from the tidal flats of Andros Island in the Bahamas (a link to a similar enhanced module is provided on the CD). This module aided students' exploration by asking questions about the data, geomorphic characteristics, and different types of vegetation and surface cover for two different areas. They were then asked to compare and contrast the patterns in two areas of the tidal flat and to hypothesize on the geologic evolution of the system on the basis of their observations. They were actively encouraged to pose and discuss possible interpretations of their observations.

Next, the entire class, including those without the remote sensing module, had a "traditional" lecture on tidal flats. All students were then given a series of questions to evaluate their understanding.

To evaluate the utility of the module, anecdotal comments and quantitative data were captured in a student assessment. Student comments on use of remote-sensing data included "these data clarified my comprehension … [and] made it easier to visualize the geologic processes" and "we were able to … actually use our own brainpower to figure out what geomorphic processes were occurring or had occurred. I would love for every lab to be this educational!" Students who completed the remote sensing module performed significantly (30%) better on the question-and-answer part of the assessment than those that did not ($p = 0.90$; $n = 15$). Similarly, they had greater cognizance of geomorphic and sedimentologic patterns (Figure 4). These results clearly demonstrate how active participation in the discovery process enhanced student learning and enthusiasm.

Analogs for Reservoir Description, Characterization, and Modeling

Geological models are best constrained when they include information on facies attributes, including spatial distribution patterns. Surprisingly, this type of information commonly is not gathered quantitatively during studies of modern or outcrop carbonate systems. In hydrocarbon reservoir studies, the most common sources of data are vertical cores and well logs, from which lateral heterogeneity must be inferred. A common lack of observational data on lateral facies distributions can hinder understanding of reservoir complexity.

The study of modern analogs like those illustrated on the CD offers one means to help refine understanding of the possible reservoir heterogeneity and to provide quantitative data on the size, shape, and patterns of facies. This database provides geologists and engineers an overview of the spectrum of geomorphic patterns present in modern carbonate systems. Rankey (2002), Rankey et al. (2006b), and a companion study (Harris and Vlaswinkel, this volume) illustrate approaches to collecting such quantitative data from Holocene systems, data that can be used to better understand reservoir analogs.

But can these analogs be used to build more realistic reservoir models? One example of how these may be used is in Multiple-Point Statistics (MPS)/Facies Distribution Modeling (FDM) of carbonate reservoirs (e.g., Levy et al., 2007a). To explore use of data from the Holocene, they modeled an isolated platform using training images, FDM cubes and MPS simulations for varied grainstone and reef reservoir types.

MPS training images are constructed for individual facies and for combined facies associations drawing upon a dimensional database for input parameter ranges. The training image is a 3D conceptual model of the reservoir, containing information about facies dimensions and relationships among facies geobodies (Fig. 5A). This conceptual model can be constrained by qualitative trends and quantitative data from analog information, such as that contained in the image database or in Harris and Vlaswinkel (this volume). Facies depocenter maps are generated for the various facies, and the stratigraphy of the reservoir is modeled by digitizing a vertical proportion curve reflecting the variations of facies proportions with depth. The map data and the vertical data are combined to generate the facies probability cube (Fig. 5B). The facies probability cube allows controlling the spatial distribution of the facies in the MPS model when combined with the training image (Fig. 5C). These facies-based MPS earth models can be used to test which input parameters have the greatest impact on flow behavior for uncertainty management (e.g., Levy et al., 2007b).

SUMMARY

"The 3-D geologic model is first built in the mind…"
(Kerans and Tinker, 1997)

A fundamental challenge of many stratigraphic, hydrologic, and petroleum geology questions is developing an appropriate range of models of facies heterogeneity. The series of remote-

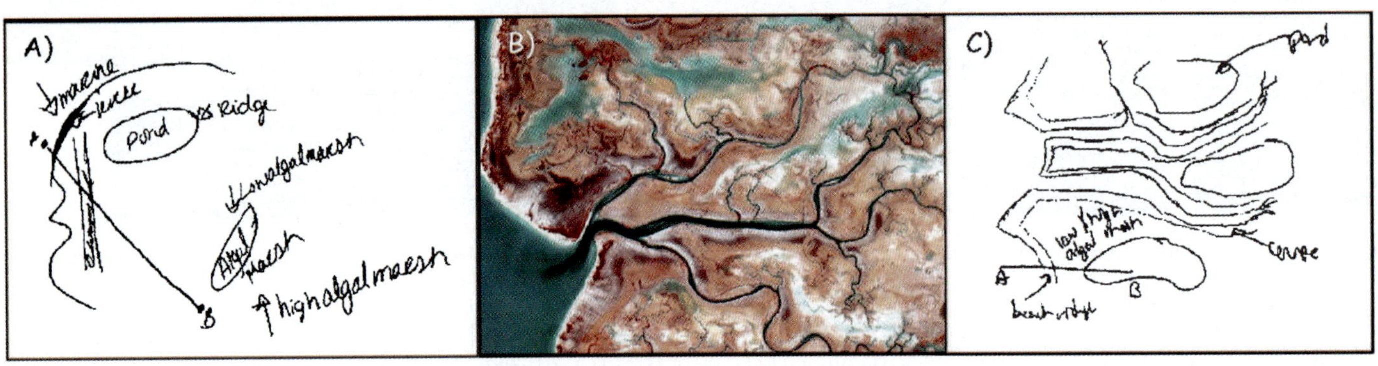

FIG. 4.—The impact of remote-sensing-based learning module on student learning. These results are: **A)** from a student who had a traditional lecture on tidal flats and **C)** one who worked the interactive remote-sensing data. They illustrate that the module enhanced students' understanding of relationships among different facies types (high algal marsh, low algal marsh, levee, pond; terminology of Hardie, 1977) and **B)** their real-world distribution patterns.

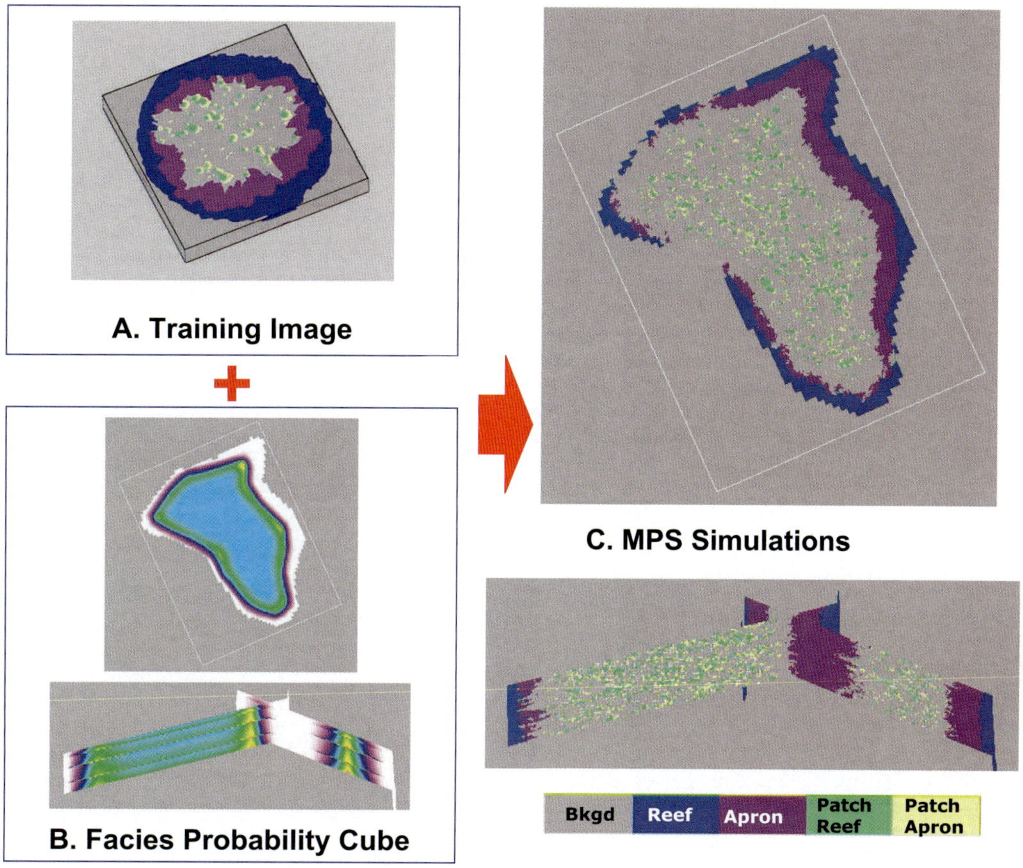

FIG. 5.—A conceptual example of a carbonate reservoir model where the reservoir is an isolated platform with a reef rim, a sand apron immediately behind the reef, and a platform interior containing numerous small patch reefs. The training image (A) is based on Glovers atoll from Belize (see Fig. 3L) The facies probability cube (B), when combined with the training image, allows controlling the spatial distribution of the facies in the MPS simulation. (C) shows the modeled facies distribution in map and cross-section views.

sensing images on the accompanying CD offers perspectives on the types of heterogeneity that are present in Holocene systems, and that could be present in ancient analogs.

ACKNOWLEDGMENTS

This effort has resulted from the collective sponsorship of numerous groups. Thanks mostly to Chevron Energy Technology Company, for their initial and continuing support and for permission to publish these results. ConocoPhillips, Shell International Exploration and Production, and Pioneer Natural Resources provided support during the early stages of this project. Data for this project were acquired largely through grants to Rankey from the National Aeronautics and Space Administration (through the Scientific Data Purchase program), the Universities Space Research Administration, National Science Foundation (Grant # EAR-0418815) and the Army Research Office (Grant # W911NF0510005). Some of the Landsat data illustrated herein are available free from the Global Land Cover Facility (http://glcf.umiacs.umd.edu/index.shtml); some of the high-resolution data were graciously provided by DigitalGlobe. The Center for Southeastern Tropical Advanced Remote Sensing (CSTARS) and the Comparative Sedimentology Laboratory (CSL) at University of Miami provided additional support. Alun Williams, Toni Simo, Jeff Lukasik, and an anonymous reviewer made numerous comments that improved the clarity and presentation of this contribution. This paper represents CSTARS Contribution Number 11.

REFERENCES

HARDIE, L.A., ed., 1977, Sedimentation on the modern carbonate tidal flats of northwest Andros Island, Bahamas: Johns Hopkins University, Studies in Geology 22, 202 p.

HARRIS, P.M., 1979, Facies Anatomy and Diagenesis of a Bahamian Ooid Shoal: Miami Beach, Florida, University of Miami, Comparative Sedimentology Laboratory, Sedimenta 7, 163 p.

HARRIS, P.M., AND KOWALIK, W.S., 1994, Satellite Images of Carbonate Depositional Settings: Examples of Reservoir- and Exploration-Scale Geologic Facies Variation: American Association of Petroleum Geologists, Methods in Exploration Series, no. 11, 147 p.

KERANS, C., AND TINKER, S.W., 1997, Sequence Stratigraphy and Characterization of Carbonate Reservoirs: SEPM, Short Course Notes no. 40, 165 p.

LEHRMANN, D.J., AND GOLDHAMMER, R.K., 1999, Secular variation in parasequence and facies stacking patterns of platform carbonates: a guide to application of stacking-patterns analysis in strata of diverse ages and settings, *in* Harris, P.M., Saller, A.H., and Simo, J.A., eds., Advances in Carbonate Sequence Stratigraphy; Application to Reservoirs, Outcrops, and Models: SEPM, Special Publication 63, p. 187–225.

LEVY, M., HARRIS, P.M., STREBELLE, S., AND RANKEY E.C., 2007a, Geomorphology of carbonate systems and reservoir modeling: carbonate training images, FDM cubes, and MPS simulations (abstract): American Association of Petroleum Geologists, Annual Convention, Abstracts Volume, p. 83.

LEVY, M., MILLIKEN, W., HARRIS, P.M., STREBELLE, S., AND RANKEY, E.C., 2007b, Importance of facies-based earth models for understanding flow behavior in carbonate reservoirs (abstract): American Association of Petroleum Geologists, Annual Convention, Abstracts Volume, p. 83.

RANKEY, E.C., 2002, Spatial patterns of sediment accumulation on a Holocene carbonate tidal flat, northwest Andros Island, Bahamas: Journal of Sedimentary Research, v. 72, p. 591–601.

RANKEY, E.C., RIEGL, B., AND STEFFEN, K., 2006a, Form, function, and feedbacks in a tidally dominated ooid shoal, Bahamas: Sedimentology, v. 53, p. 1191–1210. doi: 10.1111/j.1365-3091.2006.00807.x

RANKEY, E.C., REEDER, S.L., WATNEY, W.L., AND BYRNES, A., 2006b, Spatial trend metrics of ooid shoal complexes, Bahamas: Implications for reservoir characterization and prediction: 26th Annual Gulf Coast Section, SEPM Foundation, Bob F. Perkins Research Conference. CD-ROM.

MODERN ISOLATED CARBONATE PLATFORMS: TEMPLATES FOR QUANTIFYING FACIES ATTRIBUTES OF HYDROCARBON RESERVOIRS

PAUL M. (MITCH) HARRIS
Chevron Energy Technology Company, 6001 Bollinger Canyon Road, San Ramon, California 94583, U.S.A.
e-mail: MitchHarris@chevron.com

AND

BRIGITTE VLASWINKEL
University of Miami, Comparative Sedimentology Laboratory, Miami, Florida 33149, U.S.A.
Present Address: Shell International E&P, Rijswijk, 2288 GS, The Netherlands
e-mail: bvlaswinkel@shell.com

ABSTRACT: A geologic model is essential for characterization and modeling of hydrocarbon reservoirs, and depositional facies are a main parameter controlling heterogeneity in porosity and permeability. Quantification of facies attributes (size, shape, orientation, and distribution) can decrease uncertainty in a geologic model and therefore enhance the model's utility. This study uses Landsat images of modern isolated carbonate platforms to evaluate platform characteristics, facies distribution, and facies attributes, and assess quantitative relationships among these parameters. Facies include fully aggraded reef, partially aggraded reef, reef apron, shallow platform interior, shoals, intermediate platform interior, deep platform interior, forereef and outer platform, and land. Statistical analyses demonstrate the existence of relationships between the configuration and composition of facies tracts on and among carbonate platforms, which become useful predictive tools. In addition, several aspects of reef-belt facies have been quantified, including the abundance by platform size, width and variability, length and variability, and aspect ratio. The results offer insights that could be used for better prediction of facies distribution in ancient platforms and reservoirs.

INTRODUCTION

Modern carbonate platforms have long been studied by surface facies mapping, coring, and seismic profiling. Studies of modern environments show lateral facies relations along a single timeline and can provide a direct comparison between facies distribution and measured hydrodynamic processes like wind, waves, and currents. Facies maps derived from these studies also serve as templates for the ancient by illustrating facies-related heterogeneity. We build upon the value of modern environments as templates for understanding facies attributes of ancient platform systems and reservoirs by looking at facies maps derived from satellite imagery.

An important consideration for understanding spatial facies patterns well enough to be predictive about their characteristics is whether relationships can be established between platform size (Wang, 1998; Purdy and Winterer, 2001) and shape (Stoddard, 1965; Weins, 1962) with facies distribution, abundance, dimension, orientation, and shape. If such relationships do exist, are they then platform-scale dependent or independent? Quantification and prediction of these relationships would be helpful in building more realistic geologic models. This study attempts to gather this much-needed data on facies characteristics in relation to platform size and shape. The specific objectives are to provide an overview of the spectrum of facies patterns present in modern isolated carbonate platforms with an emphasis on reefs and their subenvironments, to obtain data on facies characteristics grouped by size and shape of carbonate platform, and to explore correlations and trends on the landscape (regional) and facies scale.

The key results of this study are the quantification of facies attributes, several general landscape relationships, and several trends for reef facies attributes. The results are applicable to research on modern environments as shown by the detailed mapping of reef facies using high-resolution satellite imagery by Andréfouët et al. (2003). We follow the work of Stoddard (1965), Purdy and Winterer (2001), and Bachtel (2005) and emphasize facies characteristics that are particularly valuable to subsurface geologists who are involved in integrating well and seismic data to formulate geologic models of isolated platform reservoirs.

Data

Landsat 5 and 7 satellite images of 19 isolated carbonate platforms are the basis of the analysis in this study. Landsat data are collected in seven spectral bands and have a pixel size of 28.5 meters. For this study, only bands 1 (blue), 2 (green), 3 (red), and 4 (near-infrared) are used. The study areas (Fig. 1) and references that were consulted for facies interpretation are: Heron Island and a platform informally named as Northwest platform in the Great Barrier Reef of northeast Australia (Maxwell et al., 1964; Maiklem, 1970; Harris and Kowalik, 1994); Bohol and Kalimantan in the Philippines; Sabah 1 (no name), Pulau Bodgaya and Pulau Danawan off the coast of Sabah and close to the northern border to Indonesia (Kirk, 1962; Wood et al., 1987; Wood, 2001); Bu Tini in the United Arab Emirates (Purser, 1973; Harris and Kowalik, 1994); Glovers and Lighthouse platforms in Belize (Wantland and Pusey, 1975; Harris and Kowalik, 1994; Gischler and Lomando, 1999; Gischler, 2003; Purdy and Gischler, 2005); Alacran reef (Folk and Robles, 1964) and Chinchorro Bank (Jordan and Martin, 1987; Harris and Kowalik, 1994; Gischler and Lomando, 1999) in Mexico; Raraka, Kauehi, and Fakarava Atolls in the South Pacific (Ranson, 1958; Chevalier, 1972; Montaggioni, 1985; Andréfouët and others, 2001); Pearl and Hermes Atoll in Hawai'i (Purdy and Gischler, 2005); Arno atoll (Hiatt, 1950; Wells, 1952; Maragos, 1994) and Bikini atoll (Tracey et al., 1948; Ladd et al., 1950; Emery et al., 1954; Wells, 1954, 1957; Maragos, 1994) in the Marshall Islands of Micronesia; and Caicos platform in the British West Indies (Wanless and Dravis, 1989; Harris and Kowalik, 1994).

Controls on Carbonate Platform and Reef Development
SEPM Special Publication No. 89, Copyright © 2008
SEPM (Society for Sedimentary Geology), ISBN 978-1-56576-130-8, p. 323–341.

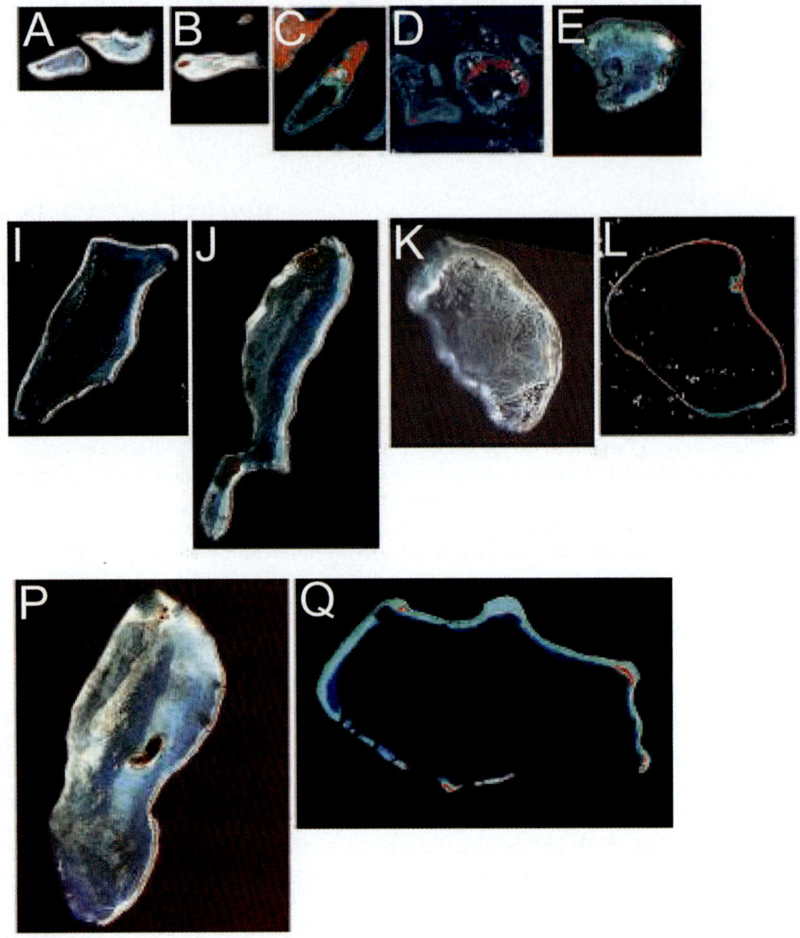

FIG. 1.—Landsat 5 and 7 images from the 19 isolated carbonate platforms analyzed in this study, organized by size. Small platforms are: **A)** Heron Island and an adjacent small platform from the Great Barrier Reef, **B)** Northwest platform also from the Great Barrier Reef, **C)** Bohol from the Philippines, **D)** Pulau Bodgaya from Indonesia, **E)** Sabah 1 (no name) from Indonesia, **F)** Bu Tini from the United Arab Emirates, **G)** Kalimantan from the Philippines, and **H)** Pulau Danawan from Indonesia. Intermediate-size platforms are: **I)** Glovers platform in Belize, **J)** Lighthouse platform in Belize, **K)** Alacran reef in Mexico, **L)** Atoll Kauehi from the South Pacific, **M)** Atoll Raraka from the South Pacific, **N)** Pearl and Hermes Atoll from Hawai'i, and **O)** Arno atoll from the Marshall Islands. Large platforms are: **P)** Chinchorro Bank in Mexico, **Q)** Bikini atoll from the Marshall Islands, **R)** Atoll Fakarava from the South Pacific, and **S)** Caicos platform in the British West Indies.

WORKFLOW

The grouping of facies for mapping purposes, the objective reproducible criteria for facies identification, and image-processing techniques are discussed as background to analyzing the facies of the platforms in this study.

Facies Grouping

Facies are grouped in nine different classes: fully aggraded reef, partially aggraded reef, reef apron, shallow platform with or without reefs, intermediate platform with or without reefs, forereef and outer platform, and land (Fig. 2), shoals (Fig. 3), and deep platform interior with or without reefs (Fig. 4).

Objective Reproducible Criteria

Color, texture, shape, and context relative to other facies are the objective reproducible criteria that are used to identify, map, and hand-digitize the nine facies classes on the satellite images using an image-analysis software program ERMapper. The facies mapping is subjective, so care was used to maintain consistency in defining facies boundaries and to test for repeatability. Mapping techniques were first tested on those areas where bathymetry maps and facies maps were available. Subfacies were grouped in some cases for mapping purposes, e.g., reef crest, reef flat, and immediate back reef, into fully aggraded reef, to further minimize the uncertainty in picking facies boundaries.

Fully aggraded reef, predominantly boundstone, includes the reef crest, reef flat, and immediate back reef (Fig. 5). The fully aggraded reef appears as a homogeneous, honey-brown feature when Landsat bands 1, 2, and 3 are combined. The orientation of the long axis of the reef is parallel to and at or near the platform break. Breaking waves are commonly observed on the images but are not a necessary criterion. "Striations" that are perpendicular to the platform edge can be observed within the reef.

Partially aggraded reef, boundstone and skeletal grainstone, also includes the reef crest, reef flat, and immediate back reef (Fig. 6). In comparison with the fully aggraded reef, when Landsat bands 1, 2, and 3 are combined the partially aggraded reef shows as a heterogeneous mixture of honey-brown features with bright patches. The orientation of the long axis is parallel to the platform break, but the location of the reef is not restricted to the platform break; partially aggraded reef also occurs in the platform interior.

Reef apron is bare-reef-derived sand or skeletal grainstone (Fig. 7). When combining Landsat bands 1, 2, and 3, reef apron shows as a highly reflective, homogeneous white color. Reef apron is always platformward of fully or partially aggraded reef, and its long axis is parallel to the reef. Grainstone shoals are reworked skeletal or oolitic sands that may be active or locally stabilized (Fig. 8). Shoals appear as a highly reflective white color when viewing a combination of Landsat bands 1, 2, and 3, and bedforms are commonly visible.

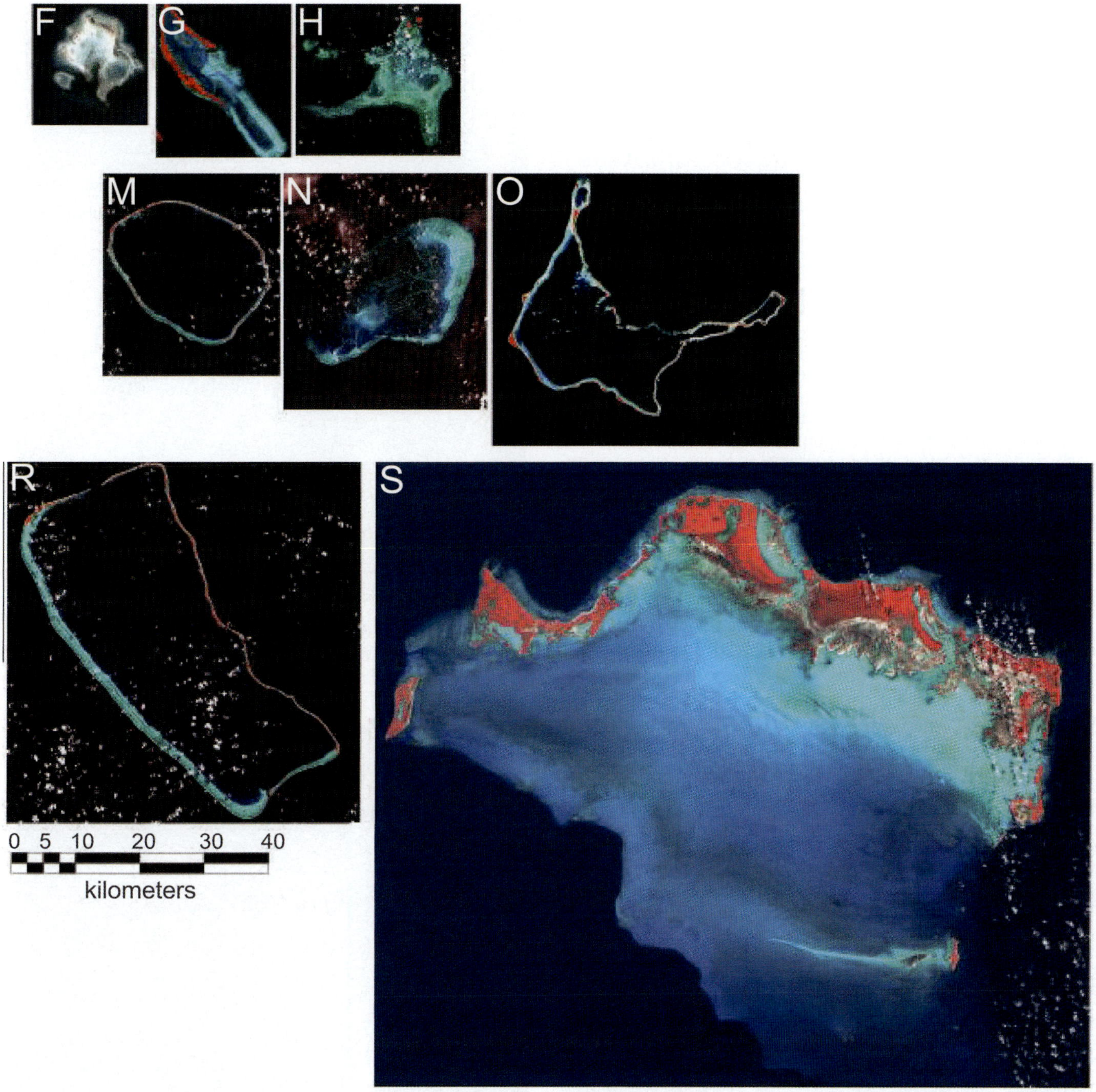

Shallow platform interior can occur with or without reefs. Shallow platform interior varies from a sandy to muddy bottom and can be sea-grass covered. Shallow platform interior without reefs, which is usually wackestone to packstone, has a homogeneous texturally and highly reflective turquoise color when Landsat bands 1, 2, and 3 are combined (Fig. 9). There is an absence of bedforms. Where reefs are present, Landsat bands 1, 2, and 3 show a heterogeneous, highly reflective mixture of turquoise with honey-brown areas (Fig. 10), and wackestone to boundstone is expected. The honey-brown features have a patchy texture, and they can be circular, elongate, or anastomosing. The shallow platform interior coincides with a water depth generally less than 3 meters, based on a comparison with examples where bathymetric data were available.

Intermediate platform interior also occurs with or without reefs and varies from a muddy to sandy bottom. The intermediate platform interior without reefs, commonly wackestone to packstone, appears on Landsat bands 1, 2, and 3 as a homogeneous deeper blue color than the shallow platform interior (Fig. 11). A heterogeneous mixture of darker blue and honey-brown features occurs where reefs are present and wackestone to boundstone is likely (Fig. 12). As within the shallow platform interior, the honey-brown features have a patchy texture and can be circular, elongate, or anastomosing. The intermediate platform interior occurs in water depths ranging from 3 to 18 meters, based on a comparison with examples where bathymetric data were available.

Deep platform interior, or lagoon, can also occur with or without reefs and usually has a muddy bottom. The deep platform interior (Fig. 13A) appears on Landsat bands 1, 2, and 3 as a homogeneous dark blue, and wackestone is expected. Green speckles mark the sites of isolated reefs (Fig. 13B), and wackestone and boundstone are likely. The deep platform interior generally occurs in water depths > 18 meters, based on a comparison with examples where bathymetric data were available.

The forereef or outer platform, composed of boundstone and skeletal grainstone/packstone, consists of spurs and grooves, local patch reefs, skeletal sands, or rocky bottom (Fig. 14A). Combining Landsat bands 1, 2, and 3, the forereef or outer platform is a deeper blue color commonly with a patchy texture. The facies always occurs seaward of either fully aggraded or partially aggraded reef and platformward of the shelf break.

Land or exposed terrain with diagenetic overprint such as cementation or dissolution is commonly Holocene or Pleistocene beach ridges and dunes, but in some cases Pleistocene or older reefs and aprons are also exposed. When combining Landsat bands 1 and 2 with band 4, healthy vegetation on land appears red (Fig. 14B)

Image Processing

All image-enhancement analyses and digitization of facies are made with the image-processing software ERMapper. Landsat bands 1, 2, 3 (and 4 for land) are imported as raster files and jointly

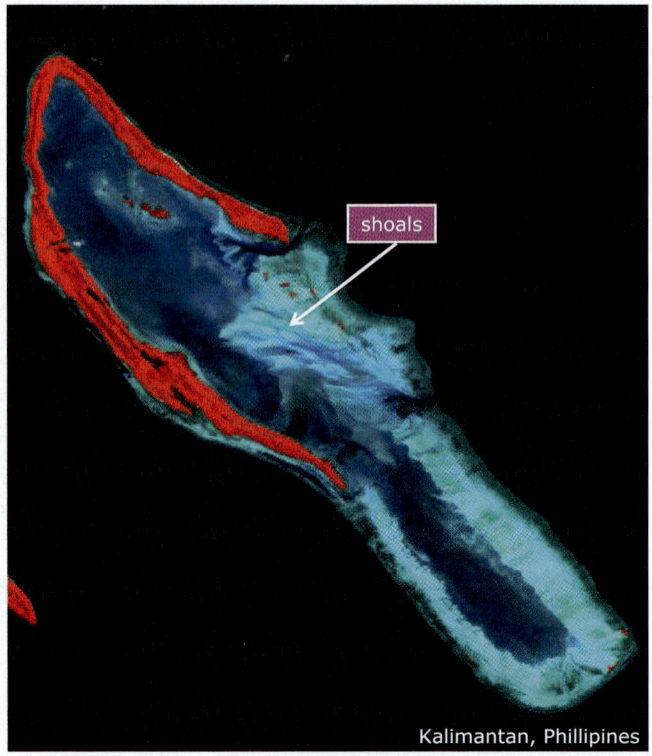

FIG. 3.—Landsat 7 image of platform from Kalimantan in the Philippines, showing an example of the shoal facies interpreted in this study.

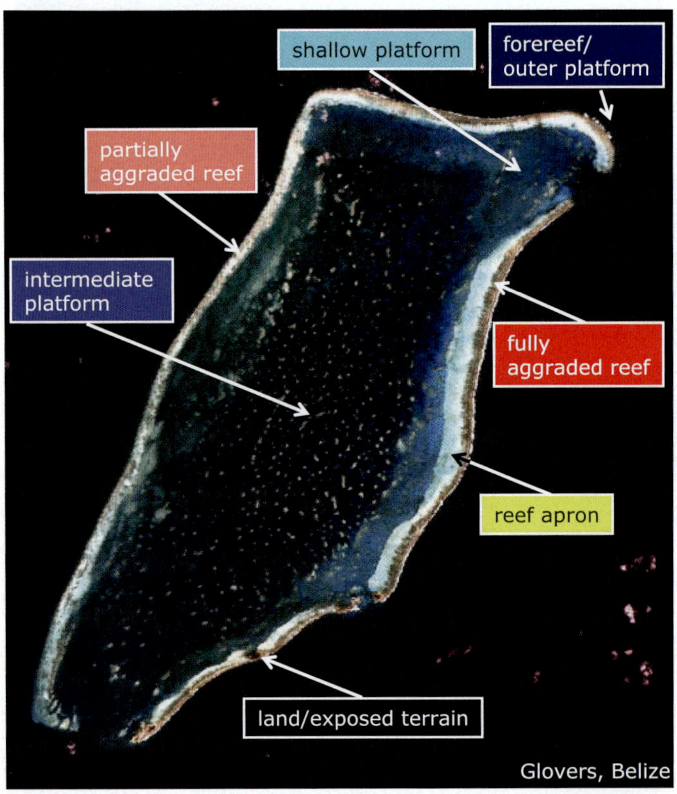

FIG. 2.—Landsat 7 image of Glovers platform in Belize, highlighting examples of most of the facies interpreted in this study.

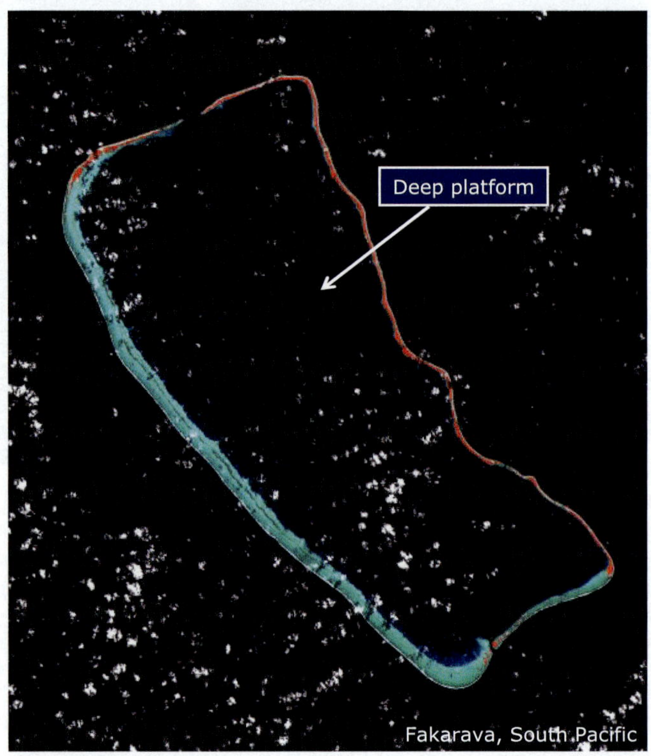

FIG. 4.—Landsat 7 image of Atoll Fakarava from the South Pacific, showing an example of the deep platform facies interpreted in this study.

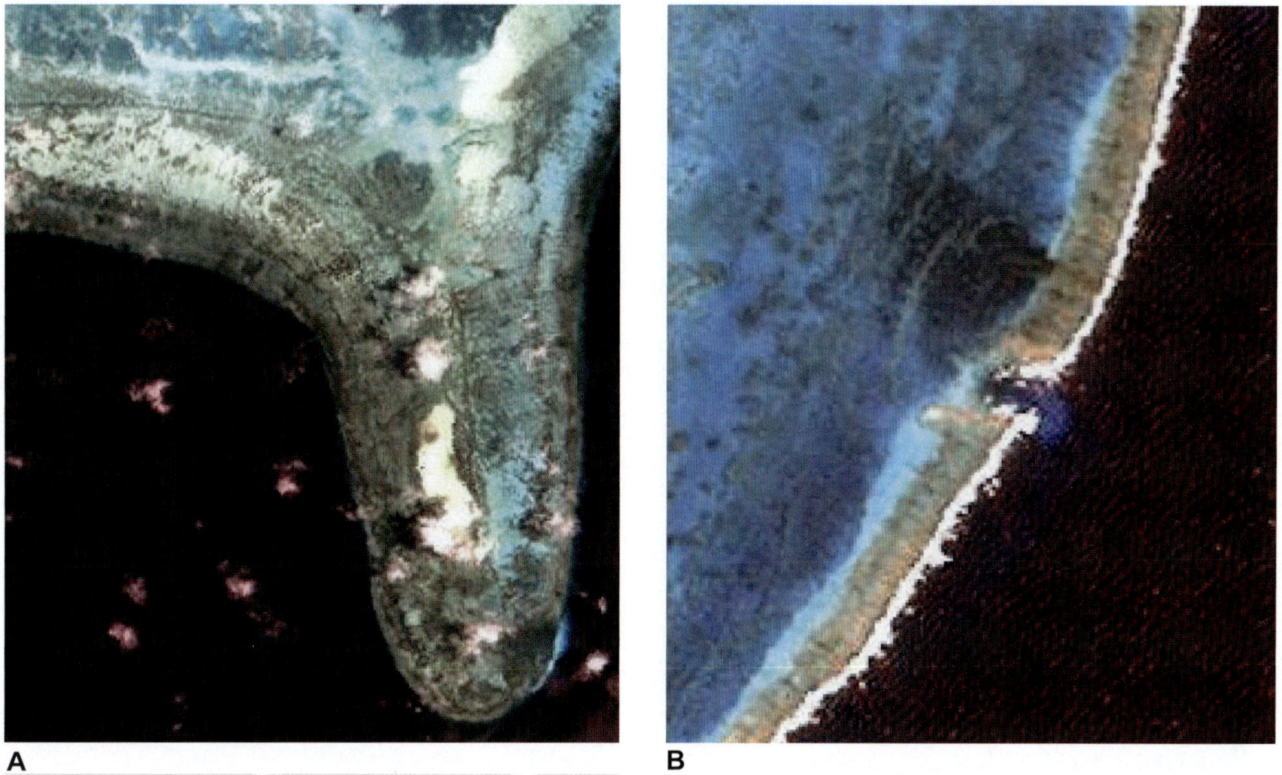

FIG. 5.—Subscenes of Landsat images showing the fully aggraded reef facies on **A)** Pulau Danawan from Indonesia and **B)** Chinchorro Bank in Mexico.

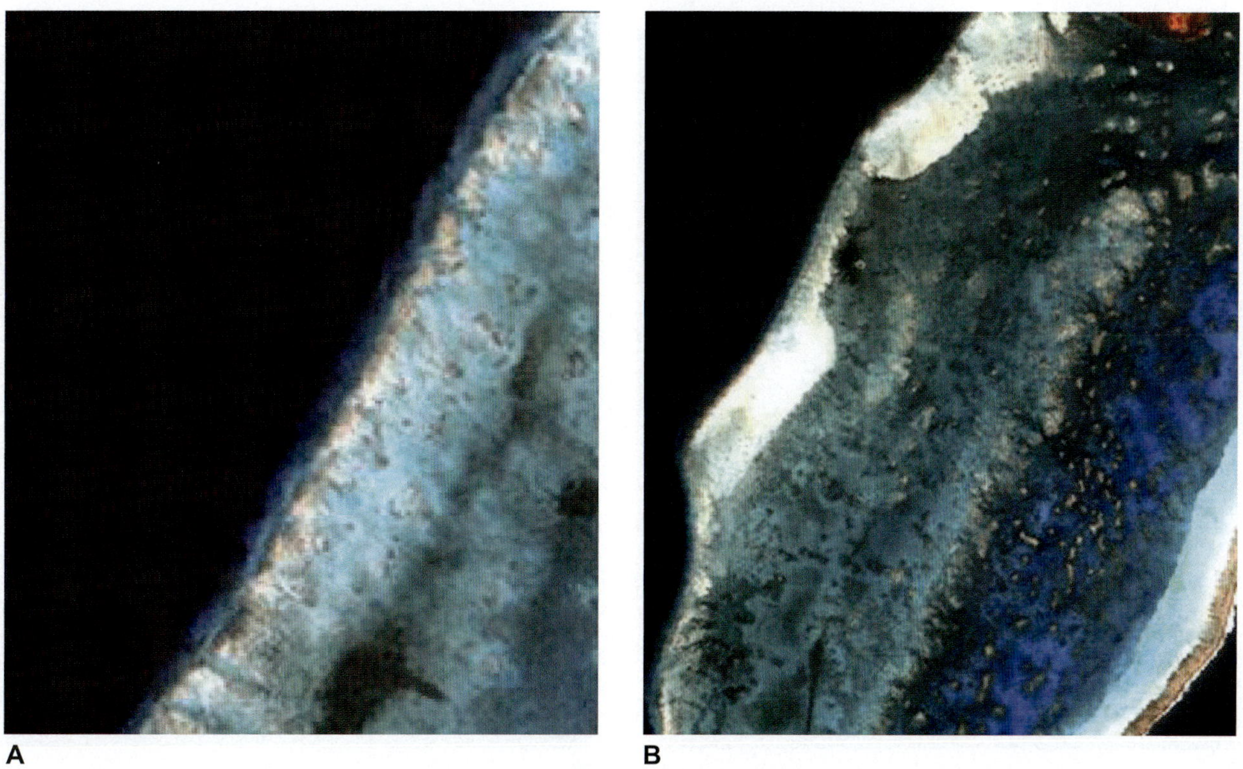

FIG. 6.—Landsat image subscenes showing the partially aggraded reef facies on **A)** Chinchorro bank from Mexico and **B)** Lighthouse platform in Belize.

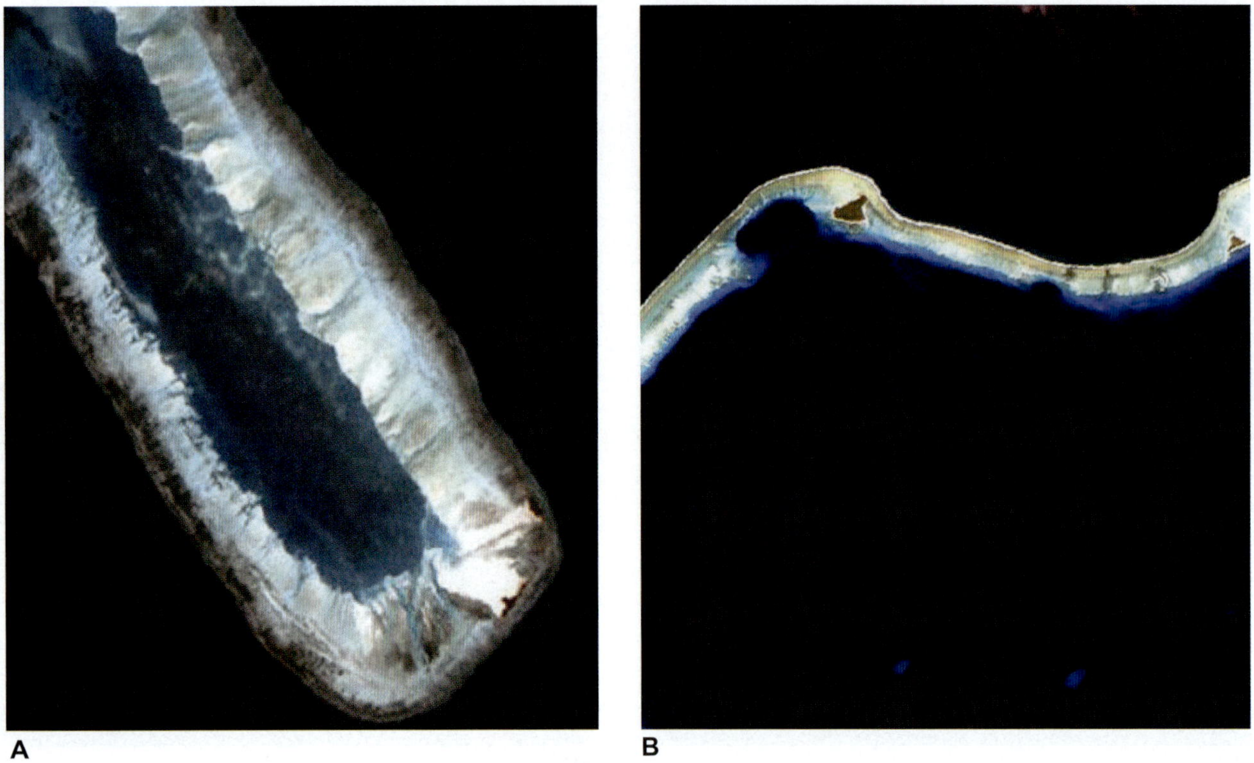

FIG. 7.—Landsat image subscenes showing the reef apron facies on **A)** Kalimantan from the Philippines and **B)** Bikini atoll from the Marshall Islands.

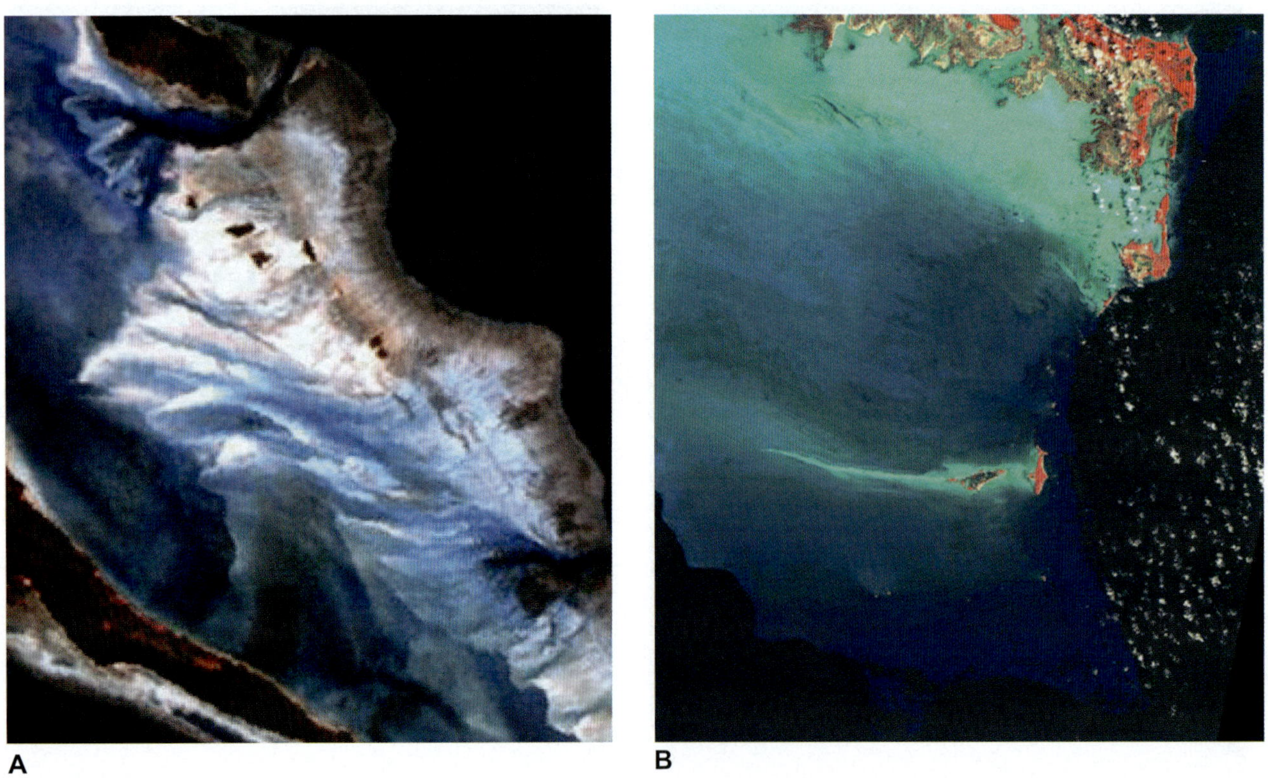

FIG. 8.—Subscenes of Landsat images showing the shoal facies on **A)** Kalimantan from the Philippines and **B)** Caicos platform (Ambergris shoal).

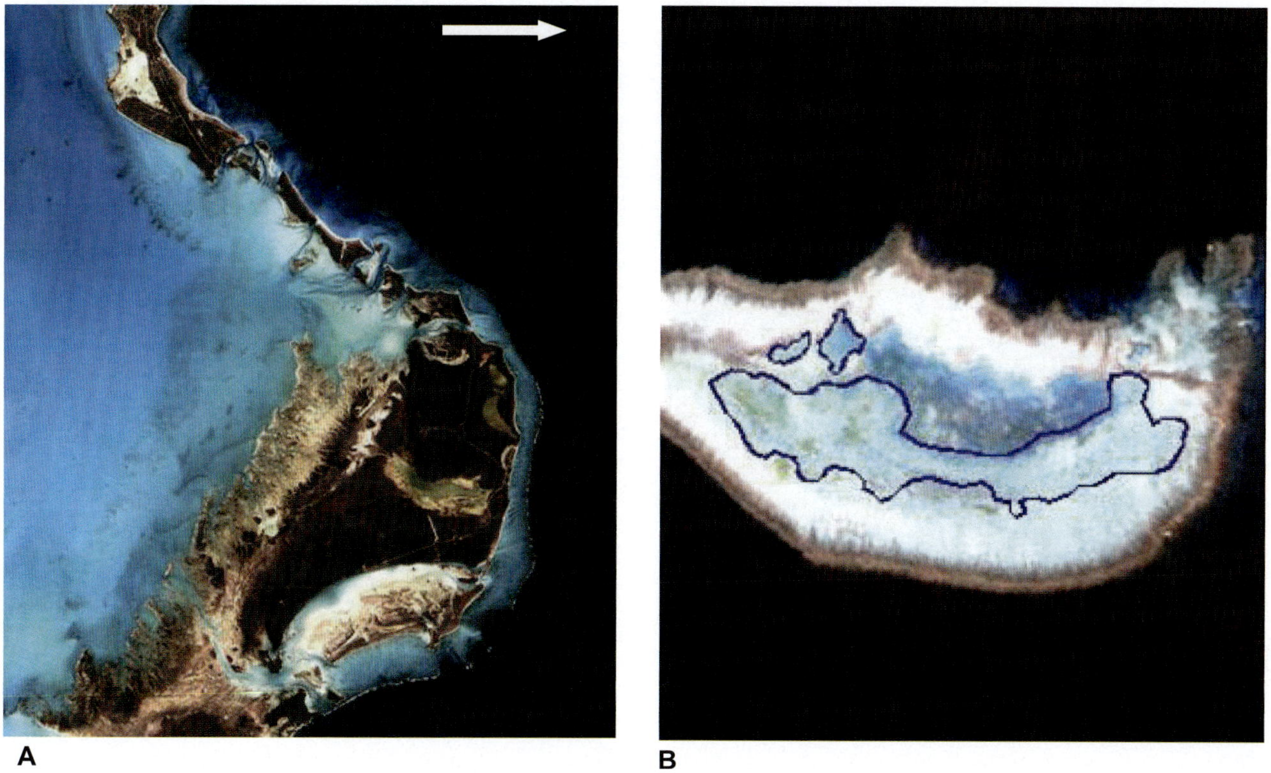

FIG. 9.—Subscenes of Landsat images showing the shallow platform interior facies without reefs on **A)** Caicos platform (north is to right) and **B)** Heron Island from northeast Australia, where the facies is outlined.

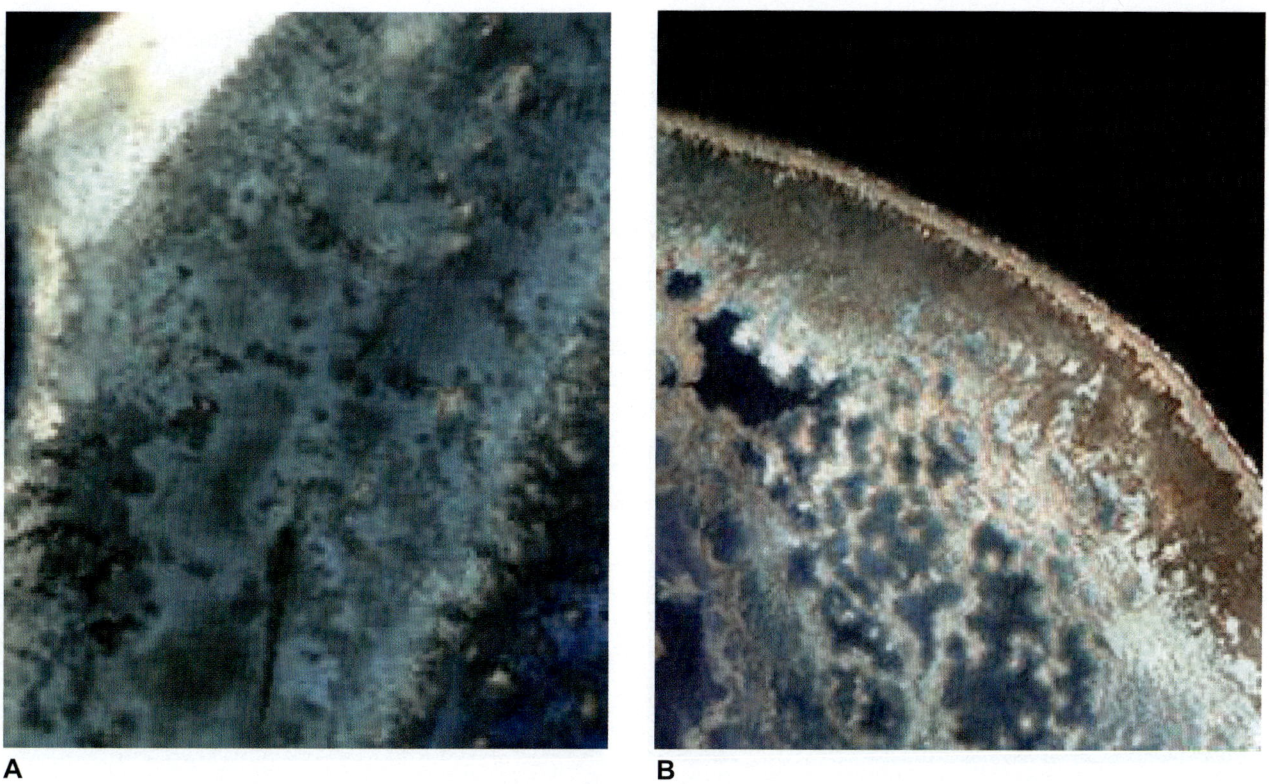

FIG. 10.—Landsat image subscenes showing the shallow platform interior facies with reefs on **A)** Lighthouse platform in Belize and **B)** Alacran reef from Mexico.

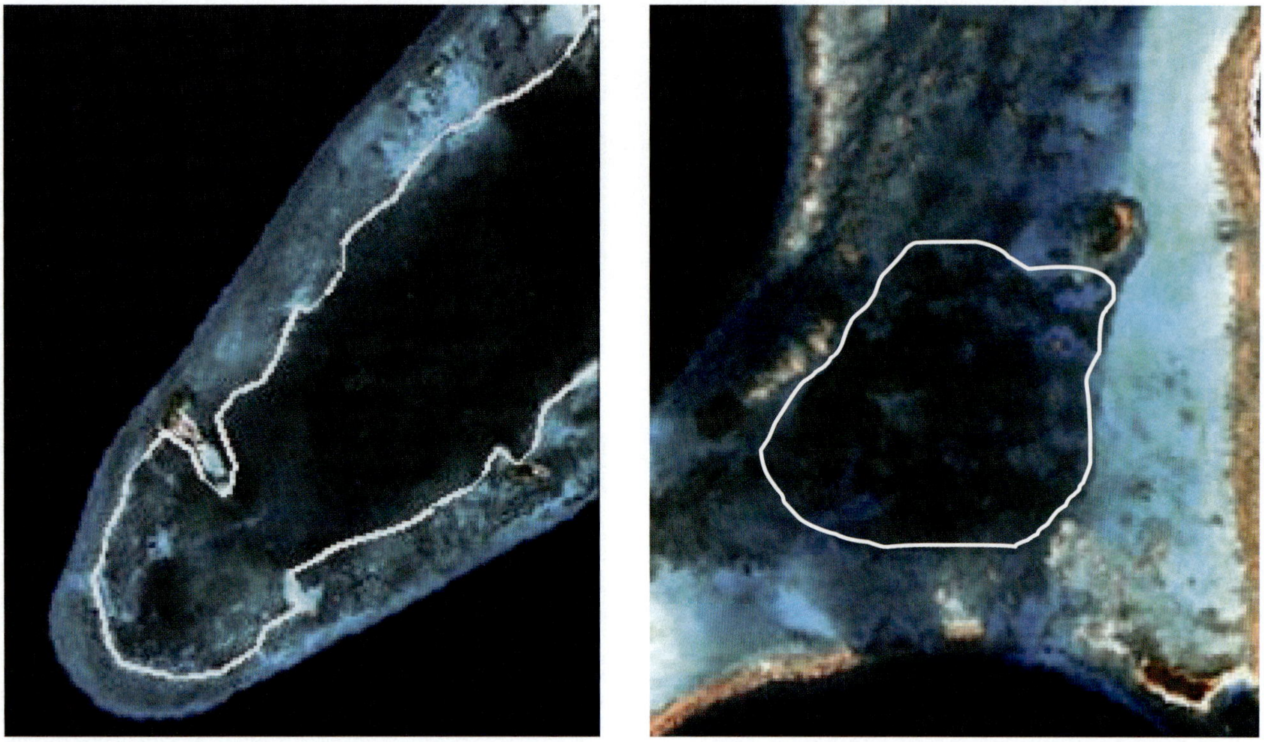

A B

FIG. 11.—Landsat image subscenes showing with outlines the intermediate platform interior facies without reefs on **A)** Bohol from the Philippines and **B)** Lighthouse platform in Belize.

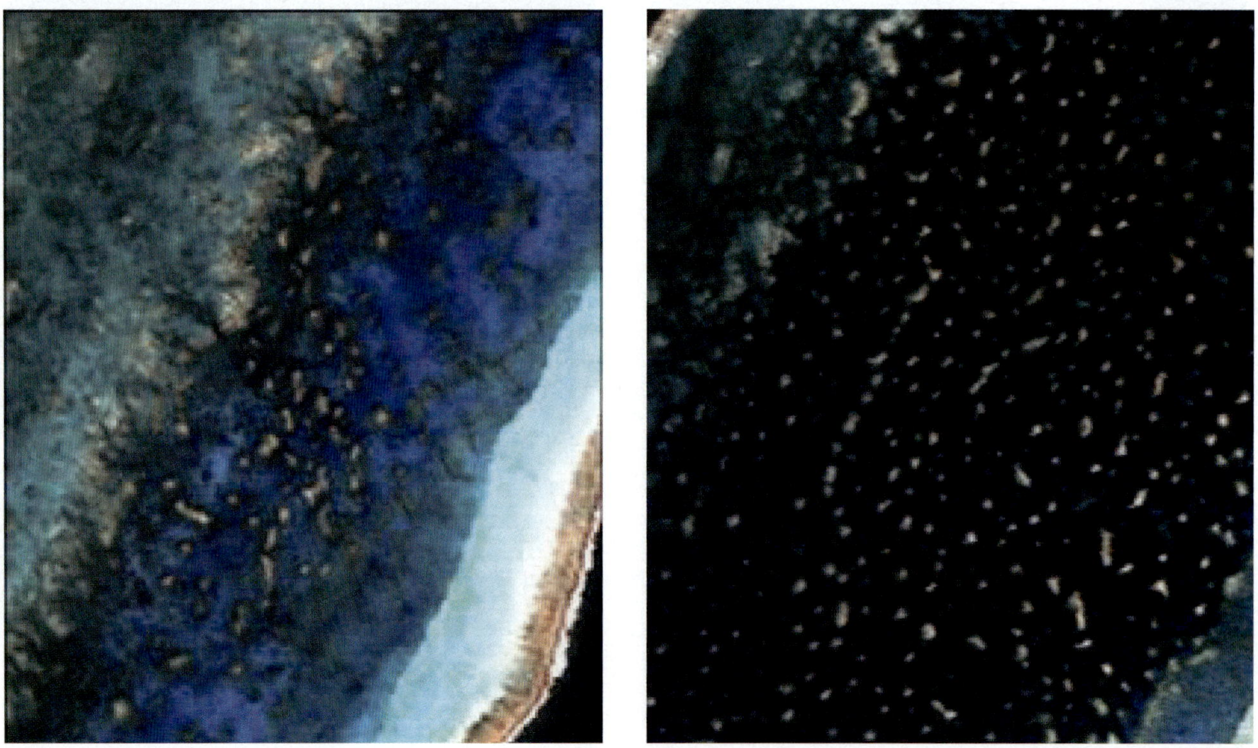

A B

FIG. 12.—Subscenes of Landsat images from Glovers platform in Belize, showing the intermediate platform interior facies with reefs.

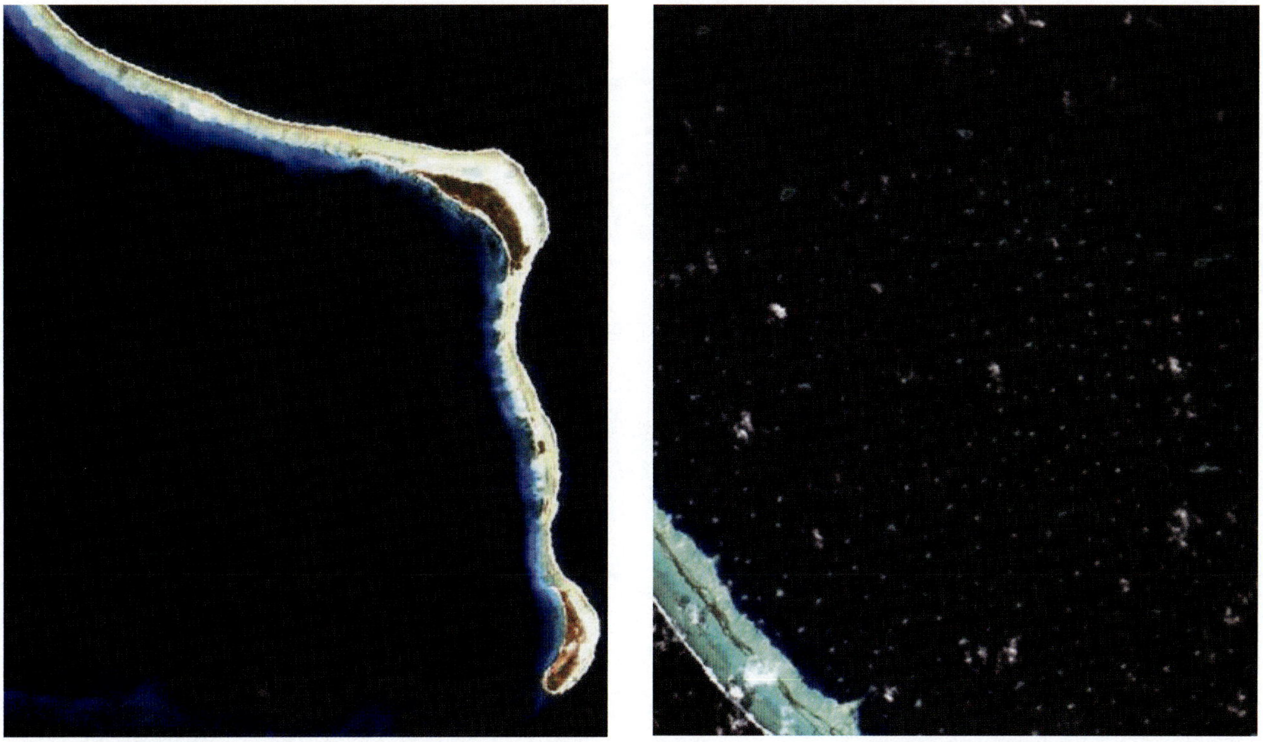

A B

FIG. 13.—Subscenes of Landsat images from **A)** Bikini atoll in the Marshall Islands, showing the deep platform interior facies without reefs and **B)** Atoll Fakarava from the South Pacific, showing the deep platform interior facies with reefs.

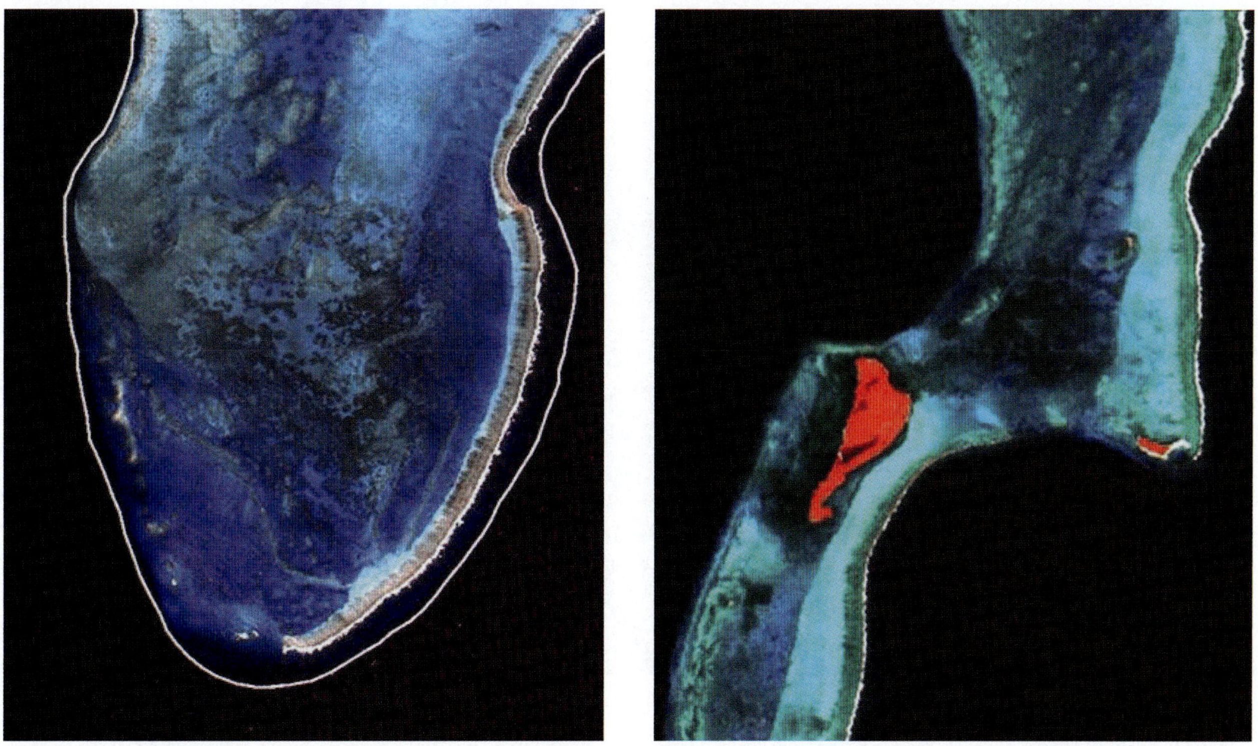

A B

FIG. 14.—Landsat image subscenes showing **A)** the forereef facies outlined along the windward margin of Chinchorro bank in Mexico and **B)** Lighthouse platform in Belize, showing the land or exposed terrain facies in red.

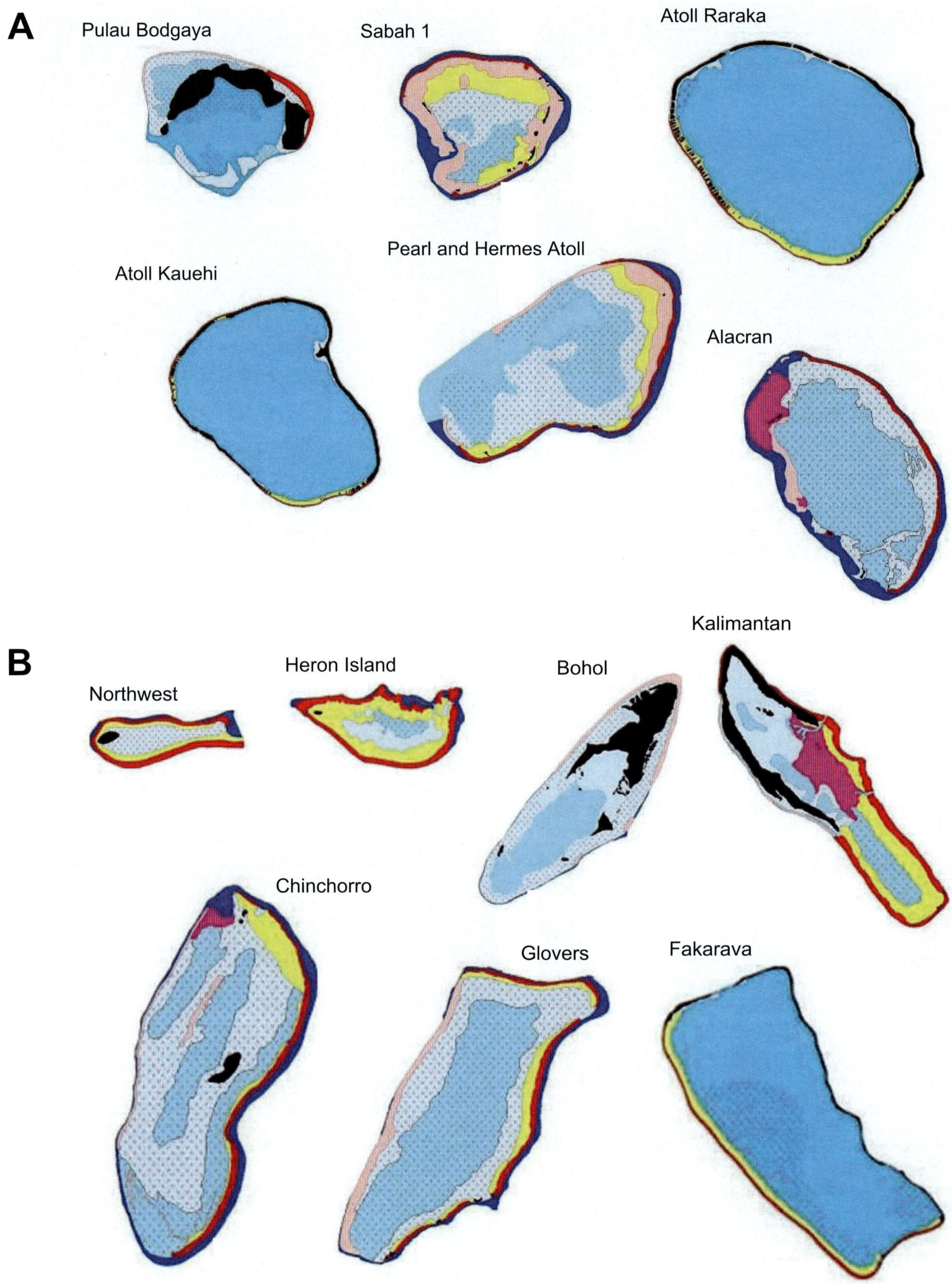

FIG. 15.— **A, B, C)** Facies maps interpreted for the various isolated platforms of Figure 1 following the workflow discussed in the text. Maps are not shown to scale for purposes of legibility; see Figure 1 for a scaled comparison of the platforms.

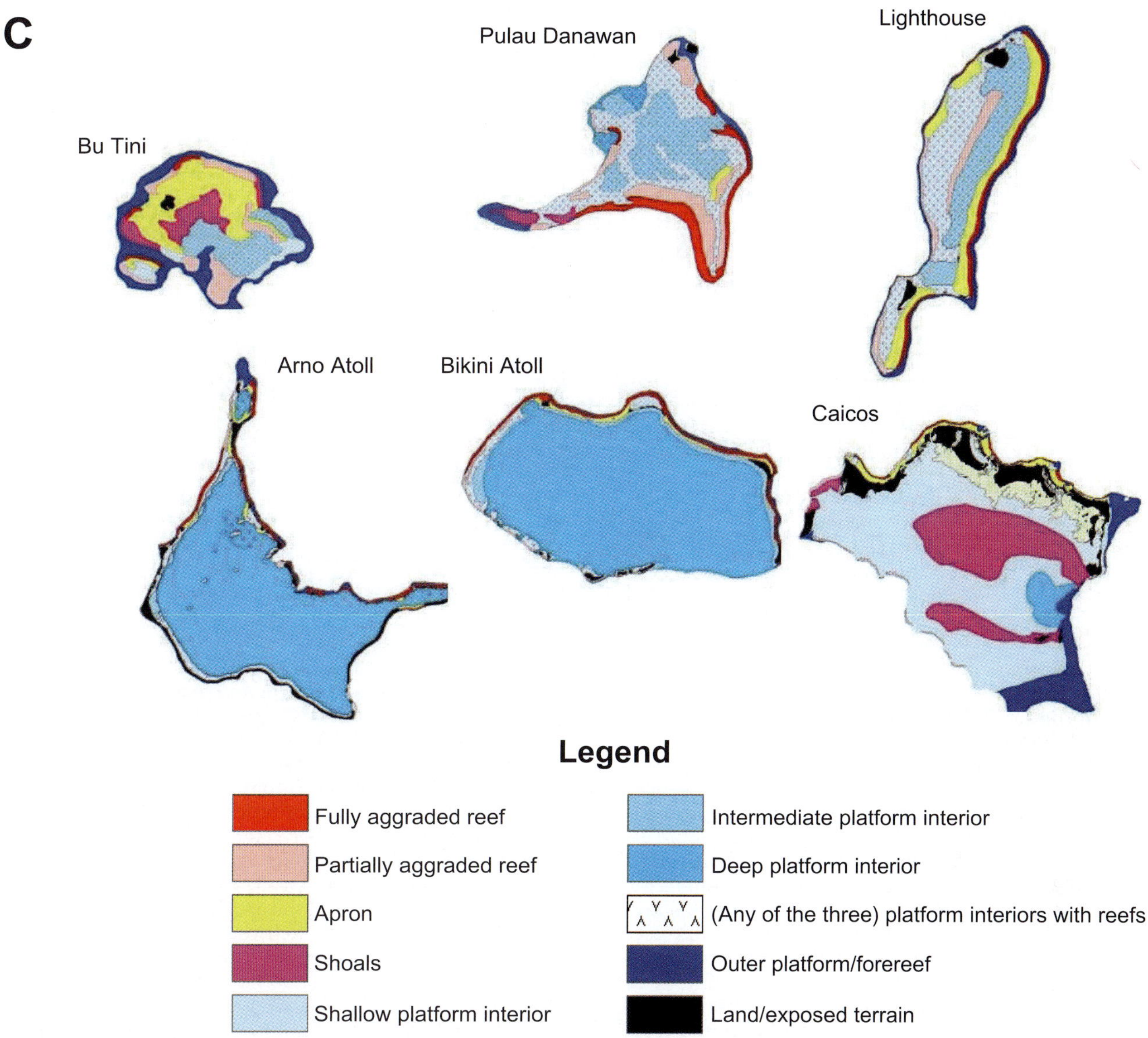

displayed. An algorithm, containing in particular some filtering and editing of transform limits (which simply controls contrast and color), is used with the sole purpose of highlighting the different facies. Each facies on each image required a slightly different image enhancement, and these individual algorithms are not documented. After a certain algorithm has enhanced the particular feature optimally, the feature outline is then identified by the interpreter and manually digitized. The digitized polygons are saved as vector files.

Polygons from each study area are imported as shapefiles into ArcMap, an ArcGIS 9 module. Where needed, the overlap between polygons is removed and polygons are merged and corrected. The facies map is created and attributes such as perimeter and area are directly measured. The shapefiles are transformed into raster files to calculate a centerline for each polygon. Length, azimuth, width, and variability of width (minimum, maximum, standard deviation) are calculated every 100 meters along the centerline determined for each polygon. All attribute data are exported to Excel.

PLATFORM AND FACIES ANALYSIS

Following the workflow outlined above, facies interpretations are made for each of the isolated carbonate platforms (Fig. 15A–C). The facies maps are not shown to scale in Figure 15 since there is such a wide range of platform sizes (Fig. 1). The maps were scrutinized based on existing literature and personal knowledge of the study area, and based on this "ground-truthing" were deemed acceptable for analysis of the facies within each platform. A comparison between the facies maps for Glovers platform, Kalimantan, and Fakarava atoll (Fig. 15B) with their respective Landsat images (Figures 1I, 1G, and 1R, respectively) gives a visual comparison of the quality of the facies mapping.

The platforms are grouped by size and shape and then analyzed statistically, but the groupings have somewhat arbitrary breaks, and the statistics should be considered with caution. The platforms are grouped into three size classes: small, intermediate, and large (Fig. 1, Table 1). The platforms are also grouped into circular, elongate, and irregular shape classes (Fig. 16). Because

TABLE 1.—Platform size in km^2 and number of facies identified within each platform.

Small Platforms	Area (km^2)	# of facies
Heron Island	32	6
Northwest	34	5
Bohol	58	6
Bodgaya	58	8
Sabah 1	102	7
Bu Tini	157	9
Kalimantan	174	8
Danawan	177	9
Intermediate Platforms		
Glovers	254	7
Lighthouse	291	8
Alacran	321	7
Kauehi	343	6
Raraka	401	7
Pearl and Hermes	445	8
Arno	447	9
Large Platforms		
Chinchorro	604	8
Bikini	726	9
Fakarava	1259	7
Caicos	6167	10

the objective of this study was analysis of facies attributes in the context of platforms of different size and shape, and not an analysis of the platform morphology itself, platform sizes and shapes are not compared with those examined by Wang (1998), Purdy and Winterer (2001), or other workers. The shape classification is done qualitatively, inasmuch as shape is difficult to categorize (see discussion by Weins, 1962; Stoddard, 1965). Figure 16 illustrates the shape groupings and shows that some overlap occurs when platform shape is categorized using the form factor shape parameter of Russ (1999). The circular platforms show uniquely high values of form factor, with a value of 1 being a perfect sphere, the elongate platforms have a fairly narrow range of moderate values, and the irregular platforms have generally lower form factor values. A plot of platform size and Russ' form factor (Fig. 17) shows an overlap between elongate and irregular platforms but a distinct cluster of circular platforms. Figure 17 also

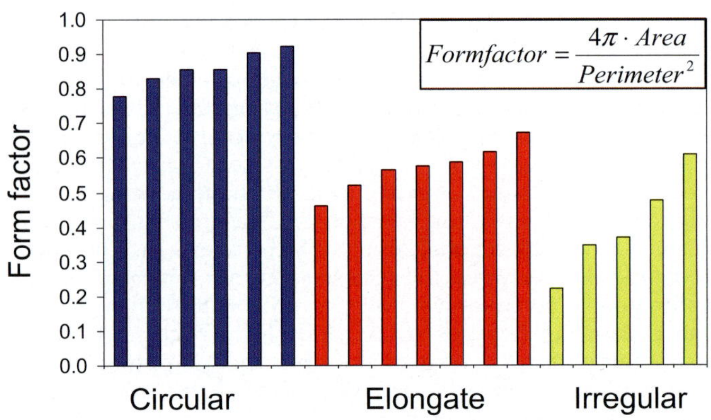

FIG. 16.—The isolated platforms from Figure 1 are grouped into circular, elongate, and irregular classes. Analysis of platform shapes by form factor (Russ, 1999), which is a shape parameter, highlights the groupings.

shows some shortcomings of the dataset, inasmuch as not every likely size–shape combination is represented.

Several facies characteristics are determined for each individual platform by analyzing the maps of Figure 15: perimeter and area for each facies polygon, number of occurrences for each facies, and relative abundance of facies in area percentage. Furthermore, width, length, sinuosity, and aspect ratio are calculated for the reef-associated facies, which includes the fully aggraded reef, partially aggraded reef, apron, and shoals. The results of platform and facies analysis of this study are presented as landscape relationships and facies metrics.

Landscape Relationships

Landscape relationships are of platform scale and are shown by the combined data of all platforms. This scale of relationship describes variability at the size of a hydrocarbon reservoir, and as such can be important to a geologist doing regional exploration assessments, applying multiple scenarios for geologic models, and calculating reservoir volumes.

The number of facies present on a platform is independent of platform size, with the average number of facies being 6–8. Table 1 shows that small platforms (e.g., Pulau Bodgaya is ~ 58 km^2) have a number of facies similar to that of platforms an order of magnitude larger (e.g., Chinchorro is ~ 604 km^2).

A fundamental characteristic of facies is size (Rankey, 2002), but plots of exceedance probability versus area of each facies, e.g., Figure 18, suggests there is no characteristic size for any of the facies mapped in this study. Each facies has a different slope on this log–log plot, indicating different size distributions.

Platforms commonly have asymmetric facies distribution over the platform top, as is shown by several of the platforms (Figs. 1, 15). This relationship has long been recognized and attributed to windward and leeward margins related to a dominant wind (or current) direction. Reefs are commonly best developed and aggraded to mean sea level on the windward margin (the fully aggraded reef of this study), whereas leeward reefs are

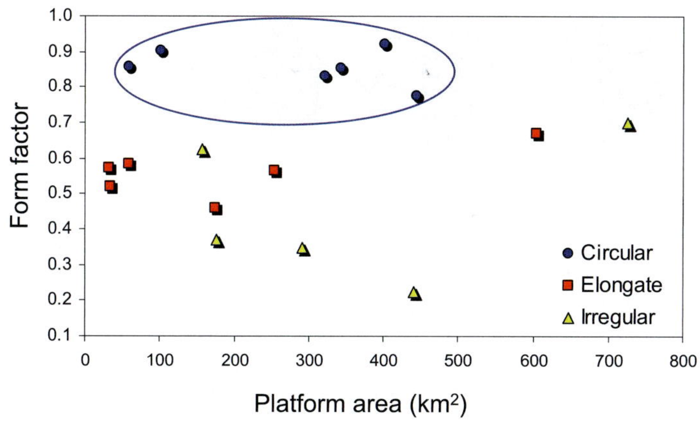

FIG. 17.—Platform size (area) and shape (form factor) shows a cluster of data from circular platforms, whereas data from elongate and irregular platforms is intermixed. Form factor = $4\pi \cdot$ area/perimeter^2).

less well developed (partially aggraded reef of this study). Data in this study quantify the correlation of reef-associated facies along a windward margin, showing a relation between the width of fully aggraded reef and its orientation with respect to wind direction. Figure 19 shows that fully aggraded reef (reef crest, reef flat, and immediate back reef) is wider where it is perpendicular to the dominant wind direction.

Quantification of the shapes of the facies within a platform shows them to be variable and to reflect the overall shape of the platform. Shape is a complex attribute that can be measured in many different ways (Mandelbrot, 1983; O'Neill et al., 1988); fractal dimension (D), which is one measure of shape complexity, is defined by the slope ($D/2$) of the power-law regression line between perimeter and area (O'Neill et al., 1988; McGarigal and Marks, 1995). In effect, the method determines the relative "edginess" of a polygon; a simple shape such as a square (low perimeter-to-area ratio) has $D = 1$, whereas a more complex polygon (high perimeter-to-area ratio) has D approaching 2. All nine facies when analyzed from all platforms display power-law relations between perimeter and area ($R^2 > 0.86$). Each facies has a slightly different slope of the regression line, however, and this is a measure of the shape complexity for that individual facies and thus a unique "mean" fractal dimension. Figure 20 shows the shape analysis of all of the facies considered during this study, and in general the two reef facies have the most complicated shapes ($D > 1.3$). Furthermore, a comparison between the shape of the platform and the combined fractal dimension of the facies within that platform in Figure 21 shows that a more regular-shaped platform (lower left black platform outline in Figure 21) contains more regular-shaped facies.

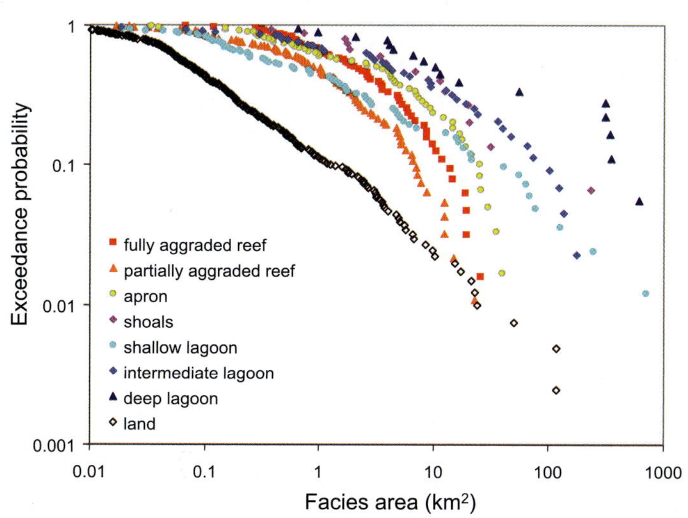

FIG. 18.—Different slopes of the exceedance probability versus area for each facies identified on the isolated platforms indicate different size distributions. Exceedance probability equals, for any particular size, the probability that the facies polygon in question will exceed it.

Small platforms contain proportionally more shallow platform facies (potential reservoir) than large platforms (Fig. 22). Approximately 50% of the area of "small" platforms consists of shallow platform facies, as opposed to 30% on "large" platforms. Shallow platform facies considered to be potential reservoir are fully aggraded reef, partially aggraded reef, reef apron, shoals, and shallow platform interior (with or without reefs); deep platform facies which are considered to be non-reservoir include all other facies. Postdepositional development of reservoir quality during diagenesis is excluded for this reservoir analogy, and the "small" platforms must, of course, still be large enough reservoirs to be economic.

Facies Metrics

Facies metrics summarize and quantify dimensions of key facies attributes, such as length, width, and aspect ratio. Reef-associated facies are emphasized due to their importance in many isolated platform-reservoirs. Tables 2 and 3 summarize the key facies metrics data for the size and shape groupings of the 19 platforms analyzed in this study. This scale of relationship describes variability at the facies scale, and as such can be important to the geologist applying multiple scenarios to geologic models and requiring quantitative input for reservoir models. These results also provide data that can be compared to characteristics of other modern reefs.

Platform size and reef-belt abundance are related, wherein reef (fully and partially aggraded) and reef-apron facies occupy a smaller percentage of the platform area with increasing platform size (Fig. 23). Taking into account biological first principles, it is logical that larger isolated platforms contain fewer reef facies than smaller systems, because the reach of favorable conditions for reef framework builders (nutrients, normal salinity, and low turbidity) is generally limited to a narrow belt (< 1000 m) along the outer perimeter of the platform.

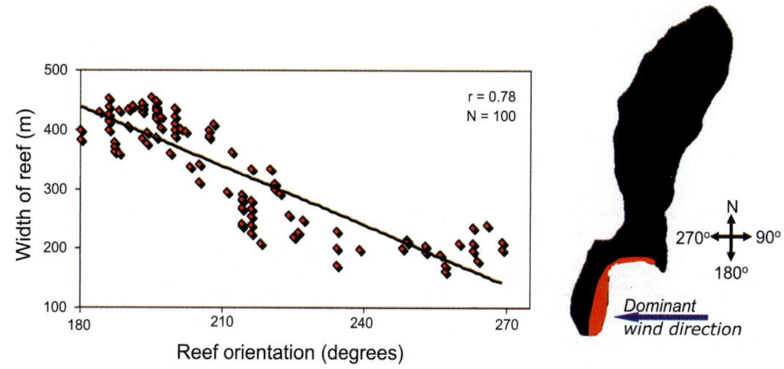

FIG. 19.—Reef width is related to reef orientation with respect to the wind direction. Data are from the windward margin of Lighthouse platform in Belize; map shows the relation between reef (red) orientation and the dominant wind direction.

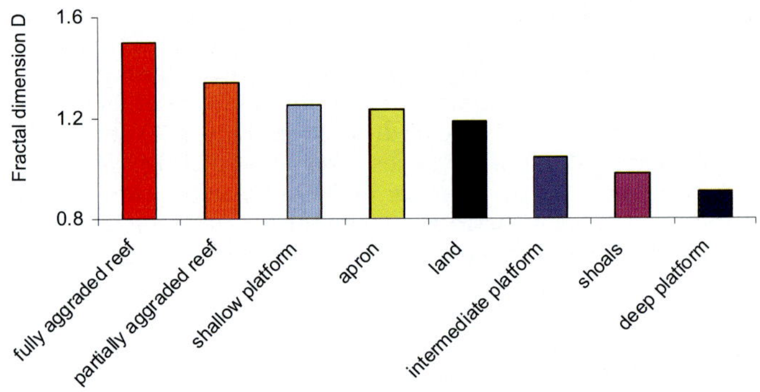

FIG. 20.—A shape analysis (fractal dimension *D*) for all of the facies mapped in this study, showing that reef-associated facies tend to have more complex shapes. *D* is a measure of shape complexity of a polygon; a high fractal dimension (1.4) illustrates a more complex facies shape than *D* of 1.

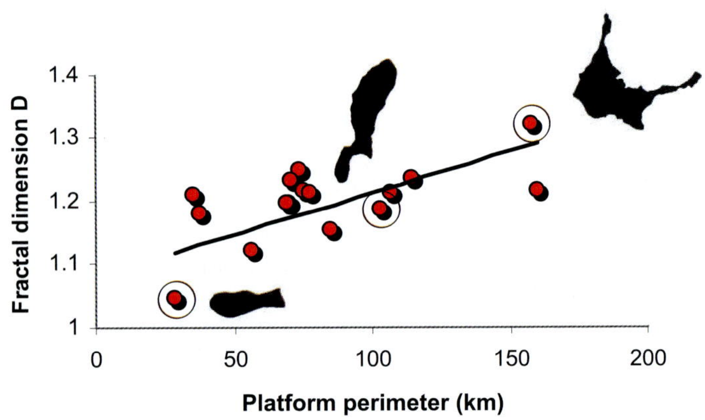

FIG. 21.—The shape of the facies (fractal dimension *D*) becomes more complex, as does the shape of the platform (perimeter). A higher *D*, which is a measure of shape complexity of a polygon, illustrates a more complex facies shape. The three platform profiles in black belong to the three points encircled.

Fully aggraded reefs become somewhat wider as length of their long axis increases (Fig. 24), although the positive correlation is poor. Partially aggraded reefs are discontinuous along their long axis, so the relation to reef width is more subtle. Irregular and elongate platforms display a moderately positive relationship between reef width and apron width (Fig. 25). On average, the apron is twice the width of the reef on irregular

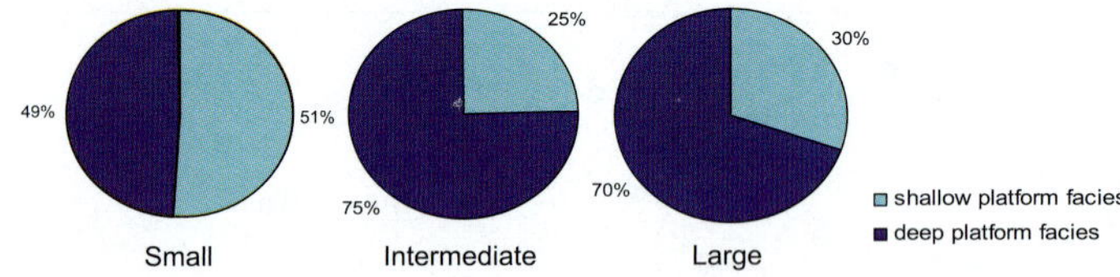

FIG. 22.—Small platforms in the data set of this study contain proportionally more shallow platform facies (potential reservoir) than large platforms. Shallow platform facies equal reservoir and deep platform facies equal non-reservoir; see text for discussion.

platforms, and elongate platforms tend to have very narrow aprons regardless of reef width. Circular platforms show no obvious relation between reef width and apron width.

Reef width varies with orientation relative to the dominant wind direction, where on Lighthouse reef in Belize, for example (Fig. 19), reef width doubles from ~ 200 to ~ 400 m as orientation changes from wind-parallel (east–west) to wind-perpendicular (north–south). This correlation is not observed on the leeward side of the platform. Generally, reef facies along the windward margins demonstrate larger variability in width (a larger standard deviation) than do reef facies on leeward margins.

Reef width, apron width, and reef length vary by shape of platform (Fig. 26). Circular platforms have the narrowest reefs (~ 230 m versus 400 m for elongate and irregular-shaped platforms), elongate platforms have the narrowest aprons (~ 480 m versus 800 m for circular and irregular-shaped platforms), and irregular platforms have the least continuous reefs (~ 10 km versus 14–16 km for circular and elongate platforms). The total length/average width aspect ratio of reef facies (fully aggraded reef, partially aggraded reef, and reef apron) for each shape of platform shows that irregular platforms have a lower aspect ratio for their reef facies and lower variability (standard deviation) than circular and elongate platforms (Fig. 27). This suggests that reefs on irregularly shaped platforms are consistently shorter and/or wider than those on the other platform types.

Reef width, apron width, and reef length vary by size of platform, and probabilities can be set for their expected dimensions (Fig. 28). In general one could expect to find the following size attributes for:

1) reef width (reef is fully or partially aggraded); 10% probability that reef width is > 410 m, 50% probability that reef width is > 270 m, and 90% probability that reef width is > 145 m.

2) apron width; 10% probability that apron width is > 945 m, 50% probability that apron width is > 395 m, and 90% probability that apron width is > 90 m.

TABLE 2.—Attribute data from reef associated facies obtained from platforms grouped by size as Small (S), Intermediate (I), and Large (L).

		% Relative Abundance			Avg. width (m) and Standard dev. (m)			Width range (m)			Average length (km)			Aspect ratio			Sinuosity		
FACIES	**Rock type**	**S**	**I**	**L**	**S**	**I**	**L**	**S**	**I**	**L**	**S**	**I**	**L**	**S**	**I**	**L**	**S**	**I**	**L**
Fully aggraded reef	Boundstone Rudstone	7	2	1	369 97	282 69	549 144	140– 585	115– 430	240– 890	9.8	9.4	23.9	28	38	49	0.25	0.09	0.11
Partially aggraded reef	Boundstone Rudstone Grainstone	6	2	1	479 158	262 115	294 63	150– 900	110– 570	140– 435	6.1	9.5	8	22	43	42	0.16	0.08	0.06
Apron	Rudstone Grainstone	1 2	3	2	657 254	460 168	829 303	175– 1265	150– 825	180– 1465	9.6	9.1	13.6	17	15	21	0.40	0.11	0.08

TABLE 3.—Attribute data from reef associated facies obtained from platforms grouped by shape as Circular (C), Elongate (E), and Irregular (I).

		% Relative Abundance			Avg. width (m) and Standard dev. (m)			Width range (m)			Average length (km)			Aspect ratio			Sinuosity		
FACIES	Rock type	C	E	I	C	E	I	C	E	I	C	E	I	C	E	I	C	E	I
Fully aggraded reef	Boundstone Rudstone	2	6	3	241 69	387 114	389 90	80–425	150–625	170–590	13.7	16.5	9.7	66	41	24	0.15	0.21	0.11
Partially aggraded reef	Boundstone Rudstone Grainstone	4	1	3	361 203	196 42	377 105	110–1115	115–300	150–600	13.0	11.3	4.8	77	71	15	0.22	0.09	0.05
Apron	Rudstone Grainstone	4	7	4	726 234	565 207	686 239	175–1135	130–965	185–1200	10.8	14.8	7.6	17	27	13	0.11	0.20	0.13

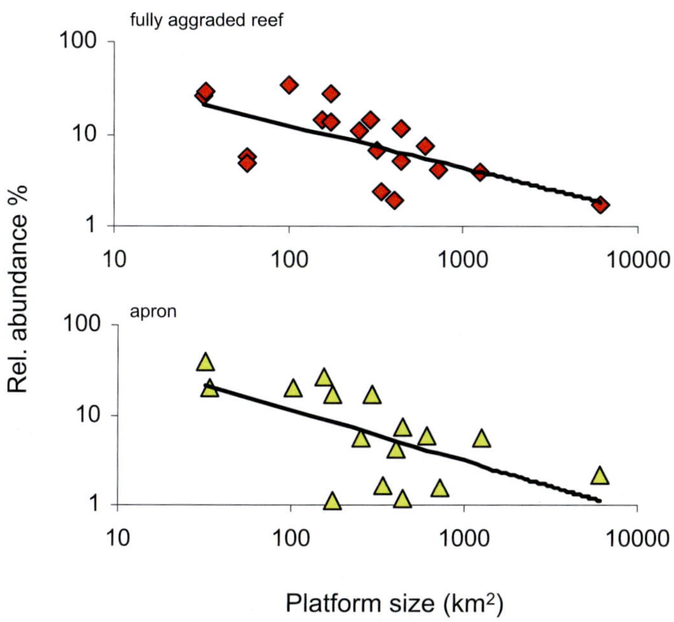

FIG. 23.—Fully aggraded reef and reef apron facies occupy a smaller percentage of the platform area with increasing platform size (km^2).

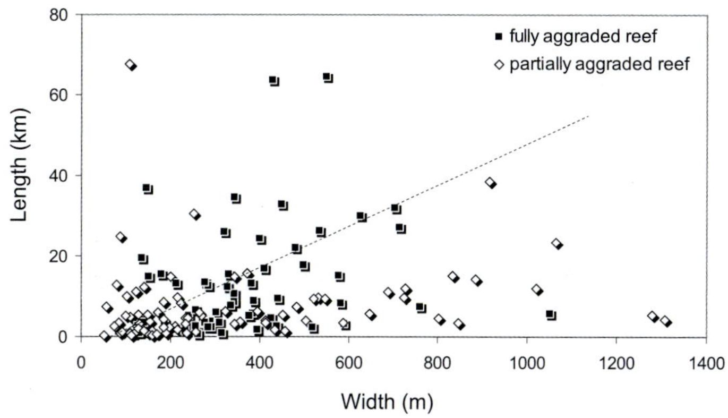

FIG. 24.—Fully aggraded reefs generally become wider as their long axis increases, whereas the relation between length and width is more subtle for the more discontinuous partially aggraded reefs.

3) reef length (fully aggraded); 10% probability that reef length is > 19.5 km, 50% probability that reef length is > 4.5 km, and 90% probability that reef length is > 1 km.

DISCUSSION AND SUMMARY

The key results of this study are the quantification of facies characteristics, general quantitative landscape relationships, and several trends for reef facies attributes. The most significant general landscape relationships are that the number of facies is independent of platform size, there is no characteristic size for any facies, platforms have asymmetric facies distribution, and facies shape reflects platform shape. These relationships imply that each platform studied, whether large or small, has the same complexity of landscape composition (facies diversity) and configuration (asymmetric). Knowing the shape of a facies body becomes particularly important when populating geologic models, because shapes of facies with similar reservoir rock characteristics (e.g., apron versus shoals) can be different. Secondly, estimation of probabilities of neighboring facies is more straightforward with a "simple" facies shape than with a complex one. Trends for reef-associated facies are documented for platform size and relative reef abundance, reef length and width, reef width and apron width by shape of platform, and reef width and orientation. Information on the expected range of sizes for varies facies is necessary input for building facies-based reservoir models.

Both the empirical and the more quantitative results of this study should be considered with caution, for several reasons. The dataset of 19 platforms is certainly too small to derive universal results; it covers only isolated platforms and not other platform types; and not all types of carbonate facies are fully covered. In addition the facies maps that form the basis for the quantitative analysis are derived from satellite imagery, which represents merely a snapshot in time; the facies interpretation is based on minimal ground-truthing; the facies mapping was done by hand and is thus somewhat interpretive; and the platform grouping by shape and size are subjective. Nevertheless, we feel that this study represents an important step in the collection of quantitative data on sizes and shapes of carbonate facies on extant carbonate platforms.

The primary focus in this study has been to analyze modern facies in a manner that is particularly valuable to subsurface workers who integrate well and seismic data to build a geologic model for isolated-platform reservoirs or who are involved in building quantitative reservoir models. Modern analogs can play an important role as conceptual facies models for characterization of a reservoir, and also in providing information on facies attributes to be used as input in building reservoir models. For example, the data of Tables 1 and 2 can be used to populate a reservoir model with facies having more realistic shapes and

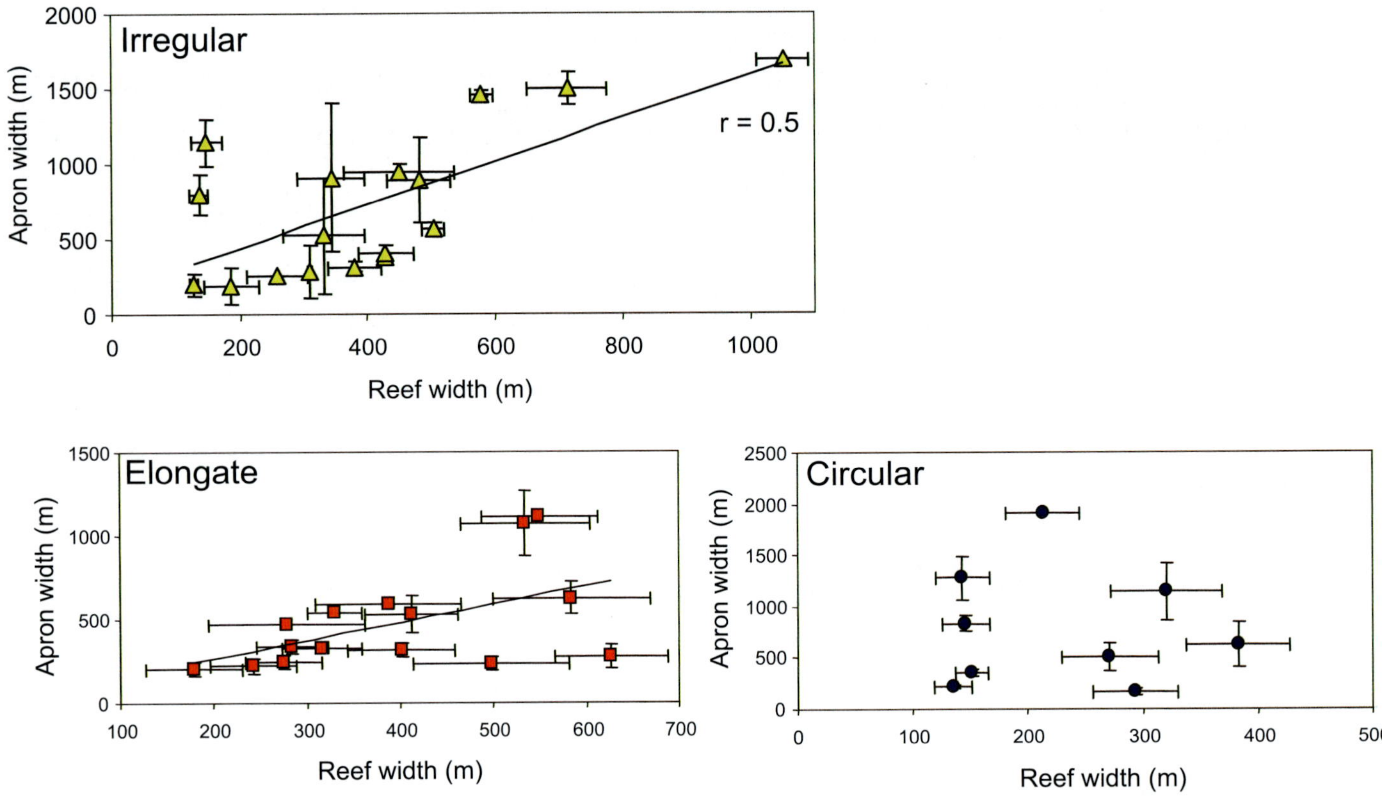

FIG. 25.—A comparison between the widths of the reef and apron for different-shaped platforms showsthat the apron is twice the width of the reef on irregular platforms, and elongate platforms tend to have very narrow aprons regardless of reef width.

sizes. To briefly illustrate this point, Figure 29 shows a map of the Malampaya reservoir (Grötsch and Mercadier, 1999; Fournier et al., 2005), which is an isolated carbonate platform with a partial reef rim. Well data are sparse in the reservoir, and most facies characteristics cannot be discerned from the seismic data. The empirical and quantitative information generated from modern facies studies improves the ability to conceptualize the position and size of subenvironments within a platform like Malampaya. The separate facies maps in Figure 29 postulate the distribution of a fully aggraded reef margin (red) and reef apron (yellow) along

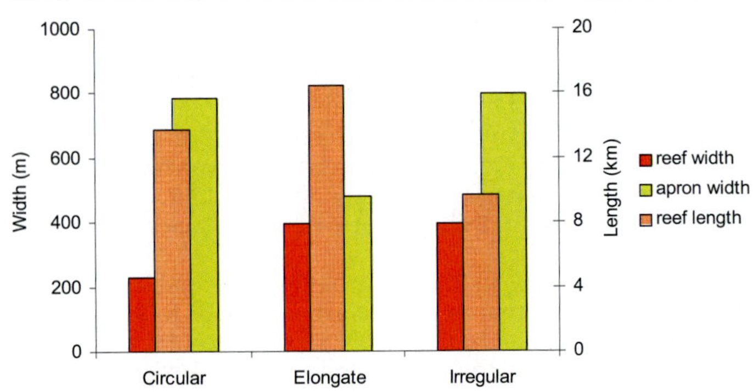

FIG. 26.—Comparing reef width, apron width, and reef length for the variously shaped platforms shows that circular platforms have the narrowest reef, elongate platforms have the narrowest apron, and irregular platforms have the least continuous reef.

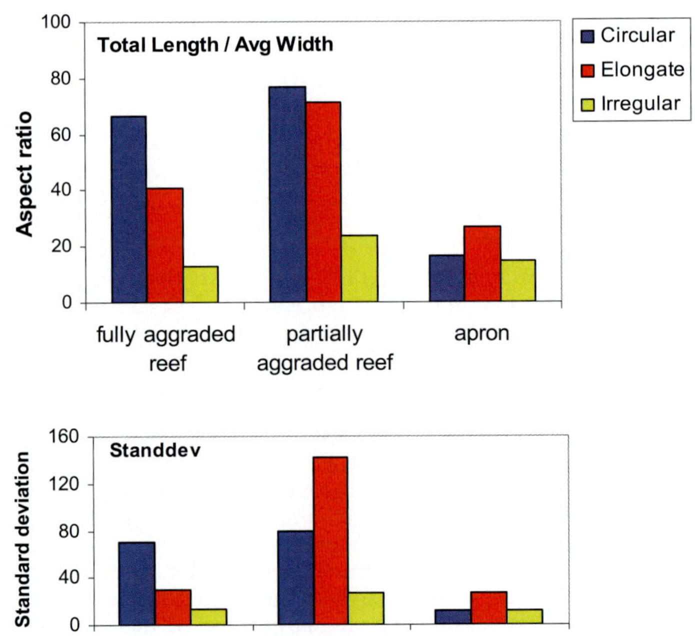

FIG. 27.—Aspect ratio (total length/average width) is calculated for fully aggraded reef, partially aggraded reef, and apron and shown for variously shaped platforms. Irregular platforms have a significantly lower aspect ratio for their reef facies (fully aggraded reef, partially aggraded reef, and reef apron) and lower variability (standard deviation) than those of their counterparts on circular and elongate platforms.

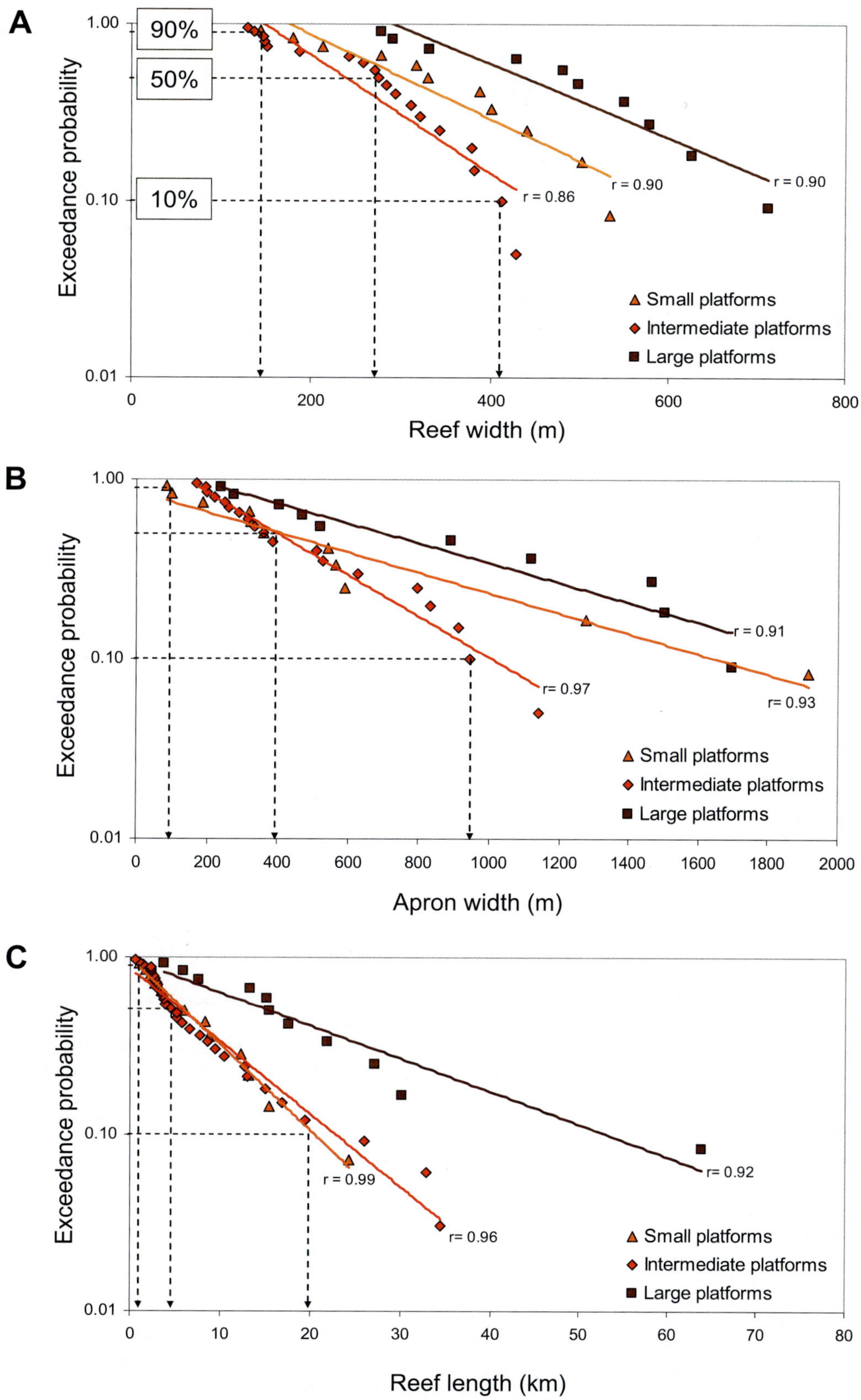

FIG. 28.—Exceedance probability for: **A)** reef width by size of platform showing:10% probability that reef width is > 410 m, 50% probability > 270 m, and 90% probability > 145 m; **B)** apron width by size of platform showing 10% probability that apron width is > 945 m, 50% probability > 395 m, and 90% probability > 90 m; and **C)** reef length by size of platform showing 10% probability that reef length is > 19.5 km, 50% probability > 4.5 km, and 90% probability > 1 km.

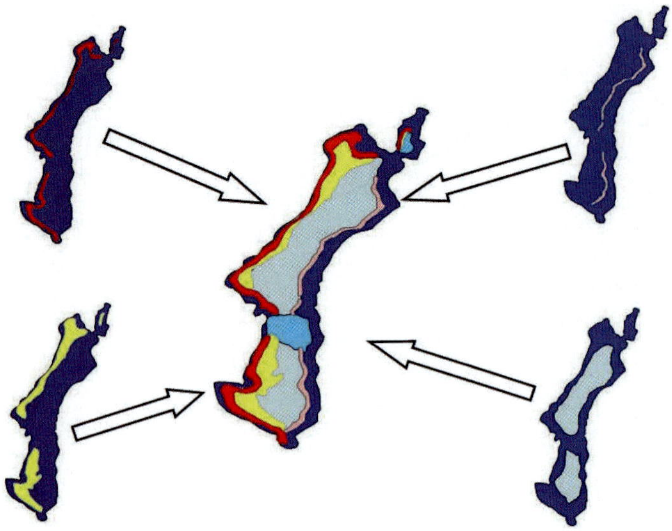

FIG. 29.—Simplified map of Malampaya, an isolated platform hydrocarbon reservoir with a partial reef rim, described by Grötsch and Mercadier (1999). Quantitative data on reef facies like that generated during this study can serve as input during reservoir modeling to better capture the key facies within the reservoir, e.g., fully aggraded reef margin (red) and reef apron (yellow) along one side of the platform, a partially aggraded reef trend (gray) along the other side, and a shallow platform interior (light blue).

one margin of the platform, a partially aggraded reef trend (gray) along the other margin, and a shallow platform interior (light blue). These important components of the platform relate directly to characteristic rock types and to expected ranges of primary depositional porosity and should be incorporated in any facies-based reservoir model.

ACKNOWLEDGMENTS

Gene Rankey (University of Miami) played a key role in all phases of this study by assisting in selecting the platforms included, finalizing the workflows, and providing insight into the parameters that are quantified. Dave Stodola (Chevron) developed a macro for converting shape files into raster files, which was essential during our analysis of facies polygons. SEPM reviewers Steve Bachtel and Amy Ruf provided extremely thoughtful reviews and numerous suggestions that greatly improved the quality of the manuscript. We thank Chevron Energy Technology Company for their support and permission to publish.

REFERENCES

ANDRÉFOUËT, S., CLAEREBOUDT, M., MATSAKIS, P., PAGÈT, J., AND DUFOUR, P., 2001, Typology of atoll rims in Tuamotu Archipelago (French Polynesia) at landscape scale using SPOT HRV images: International Journal of Remote Sensing, v. 22, p. 987–1004.

ANDRÉFOUËT, S., KRAMER, P., TORRES-PULLIZA, D., JOYCE, K.E., HOCHBERG, E.J., GARZA-PEREZ, R., MUMBY, P.J., RIEGL, B., YAMANO, H., WHITE, W.H., ZUBIA, M., BROCK, J.C., PHINN, S.R., NASEER, A., HATCHER, B.G., AND MULLER-KARGER, F.E., 2003, Multi-site evaluation of IKONOS data for classification of tropical coral reef environments: Remote Sensing of Environment, v. 88, p. 128–143.

BACHTEL, S.L., 2005, Platform-scale facies distributions using Landsat data from isolated carbonate platforms: methods to constrain lateral facies continuity for geologic modeling (abstract): American Association of Petroleum Geologists, Annual Convention, June 19–22, Calgary, Alberta, Canada.

CHEVALIER, J.P., 1972, Observations sur les chenaux incompletes appeles Hoa dans les atolls des Tuamotu: Symposium on Corals and Coral Reefs, India, Proceedings, p. 477–488.

EMERY, K.O., TRACEY, J.I., JR., AND LADD, H.S., 1954, Geology of Bikini and nearby atolls: U.S. Geological Survey, Professional Paper 260-A, 265 p.

FOLK, R.L., AND ROBLES, R.R., 1964, Carbonate sands of Isla Perez, Alacran Reef complex, Yucatan: Journal of Geology, v. 72, p. 255–292.

FOURNIER, F., BORGOMANO, J., AND MONTAGGIONI, L.F., 2005, Development patterns and controlling factors of Tertiary carbonate buildups: Insights from high-resolution 3D seismic and well data in the Malampaya gas field (Offshore Palawan, Philippines): Sedimentary Geology, v. 175, p. 189–215.

GISCHLER, E., 2003, Holocene lagoonal development in the isolated carbonate platforms off Belize: Sedimentary Geology, v. 159, p. 113–132.

GISCHLER, E., AND LOMANDO, A.J., 1999, Recent sedimentary facies of isolated carbonate platforms, Belize–Yucatan system, Central America: Journal of Sedimentary Research, v. 69, p. 747–763.

GRÖTSCH, J., AND MERCADIER, C., 1999, Integrated 3-D reservoir modeling based on 3-D seismic: the Tertiary Malampaya and Camago buildups, offshore Palawan, Philippines: American Association of Petroleum Geologists, Bulletin, v. 83, p. 1703–1728.

HARRIS, P.M., AND KOWALIK, W.S., eds., 1994, Satellite images of carbonate depositional settings: examples of reservoir- and exploration scale geologic facies variation: American Association of Petroleum Geologists, Methods in Exploration Series 11, 147 p.

HIATT, R.W., 1950, Marine and zoology study of Arno Atoll, Marshall Islands: Atoll Research Bulletin, no. 4, p 1–13.

JORDAN, E., AND MARTIN, E., 1987, Chinchorro: morphology and composition of a Caribbean atoll: National Museum of Natural History, Atoll Research Bulletin 310, 20 p.

KIRK, H.J.C., 1962, The Geology and Mineral Resources of the Semporna Peninsula, North Borneo: Geological Survey Department, British Territories in Borneo, Government Printing Office, Kuching, Memoir 14.

LADD, H.S., TRACEY, J.I., WELLS, J.W., AND EMERY, K.O., 1950, Organic growth and sedimentation on an atoll: Journal of Geology, v. 58, p. 410–425.

MAIKLEM, W.R., 1970, Carbonate sediments in the Capricorn reef complex, Great Barrier Reef, Australia: Journal of Sedimentary Petrology, v. 40, p. 55--80.

MANDELBROT, B., 1983, The Fractal Geometry of Nature: New York, W.H. Freeman, 468 p.

MARAGOS, J.E., 1994, Description of the reefs and corals for the 1988 protected area survey of the northern Marshall Islands: Atoll Research Bulletin, no. 419, p. 1–88.

MAXWELL, W.G.H., JELL, J.S., AND MCKELLAR, R.G., 1964, Differentiation of carbonate sediments in the Heron island reef: Journal of Sedimentary Petrology, v. 34, p. 294–308.

McGarigal, K., and Marks, B.J., 1995, FRAGSTATS: spatial pattern analysis program for quantifying landscape structure: U.S. Department of Agriculture, Foreign Service General Technical Report PNW-351.

MONTAGGIONI, L.F., 1985, Makatea Island, Tuamotu Archipelago: 5th International Coral Reef Congress, Tahiti, Proceedings, v. 1, p. 103–158.

O'NEILL, R.V., KRUMMEL, J.R., GARDNER, R.H., SUGIHARA, G., JACKSON, B., DEANGELIS, D.L., MILNE, B.T., TURNER, M.G., ZYGMUNT, B., CHRISTENSON, S.W., DALE, V.H., AND GRAHAM, R.L., 1988, Indices of landscape pattern: Landscape Ecology, v. 1, p. 153–162.

PURDY, E.G., AND WINTERER, E., 2001, Origin of atoll lagoons: Geological Society of America, Bulletin, v. 113, p. 837–854.

PURDY, E.G., AND GISCHLER, E., 2005, The transient nature of the empty bucket model of reef sedimentation: Sedimentary Geology, v. 175, p. 35–47.

Purser, B.H., ed., 1973, The Persian Gulf; Holocene Carbonate Sedimentation and Diagenesis in a Shallow Epicontinental Sea: New York, Springer-Verlag, 471 p.

Rankey, E.C., 2002, Spatial patterns of sedimentation on a Holocene carbonate tidal flat, Northwest Andros Island, Bahamas: Journal of Sedimentary Research, v. 72, p. 591–601.

Ranson, G., 1958, Coreaux et récifs coralliens; observations sur les îles coralliennes de l'archipel des Tuomoto (Océanie française): Cahier du Pacifique, v. 1, p. 15–36.

Russ, J.C., 1999, The Image Processing Handbook, Third Edition: Boca Raton, Florida, CRC Press, 717 p.

Stoddard, D.R., 1965, The shape of atolls: Marine Geology, v. 3, p. 369–383.

Tracey, J.E., Jr., Ladd, H.S., and Hoffmeister, J.E., 1948, Reefs of Bikini, Marshall Islands: Geologic Society of America, Bulletin, v. 59, p. 861–878.

Wang, G., 1998, Tectonic and monsoonal controls on coral atolls in the South China Sea, *in* Camion, G.F., and Davies, P.J., eds., Reefs and Carbonate Platforms in the Pacific and Indian Oceans: International Association of Sedimentologists, Special Publication 25, p. 237–248.

Wanless, H.R., and Dravis, J.J., 1989, Carbonate environments and sequences of Caicos Platform: 28th International Geological Congress, Field Trip Guidebook T374, 75 p.

Wantland, K.F., and Pusey, W.C., 1975, Belize shelf—carbonate sediments, clastic sediments, and ecology: American Association of Petroleum Geologists, Studies in Geology 2, 599 p.

Wells, J.W., 1952, The coral reefs of Arno Atoll, Marshall Islands: Atoll Research Bulletin, no. 9, 14 p.

Wells, J.W., 1954, Recent corals of the Marshall Islands, Bikini and nearby atolls: U.S. Geological Survey, Professional Paper 260, Part 2, p. 385–486.

Wells, J.W., 1957, Coral reefs: Geological Society of America, Memoir 67, v. 1, p. 609–631.

Wiens, H.J., 1962, Atoll Environment and Ecology: New Haven, Yale University Press, 532 p.

Wood, E.M., 2001, Corals of the Semporna Islands Park, Sabah, Malaysia: A report for the Semporna Islands project, Marine Conservation Society.

Wood, E., George, J.D., George, J., and Wood, C., 1987, The coral reefs of the Danawan Islands (Sabah: Malaysia) and Pulau Sipadan: Malayan Nature Journal, v. 40, p. 311–324.

ANALYZING SEQUENCE ARCHITECTURE AND RESERVOIR QUALITY OF ISOLATED CARBONATE PLATFORMS WITH FORWARD STRATIGRAPHIC MODELING

PHIL BASSANT AND PAUL M. (MITCH) HARRIS

Chevron Energy Technology Company, 6001 Bollinger Canyon Road, San Ramon, California 94583, U.S.A.
e-mail: phil.bassant@chevron.com; MitchHarris@chevron.com

ABSTRACT: Dionisos is a forward stratigraphic modeling package that can be used to analyze the facies distributions in an isolated carbonate platform. Two grainstone distributions are considered and used as "inspiration" for an unconditioned case study: grainstones around the platform margin and grainstones in the platform center. In one experiment a linear rate of sea-level rise is used and the rate is varied to generate a series of models. In a second experiment a cyclic sea-level curve is added, and the amplitude of the cycles is varied to generate a second series of models.

Platform-margin grainstone facies patterns develop during times of very shallow water depth on the platform top, and slow changes in bathymetry. Platform-center grainstone facies patterns develop during times of slightly deeper water depth and increasing bathymetry (flooding). Different facies patterns can occur at the same bathymetry, depending on whether water depth is increasing or decreasing. The difference is attributed to the persistence of local depocenters that depended on previous environments (stratigraphic inertia) and the roundness of the platform. In these experiments a platform will be flatter during deepening than during shallowing because of its previous history. Neither of these two factors is reflected in the average bathymetry value but is captured in the rate and sense of flooding.

The timing of interpreted sequence boundaries and maximum flooding surfaces is investigated. In some models these surfaces are clearly diachronous, and their timing relative to the accommodation cycle varies depending on the amplitude of the accommodation change. During "greenhouse" times the subtle environmental shifts created by changes in accommodation are more significant than the actual resulting bathymetry change, and the timing of sequences is poorly related to the highs and lows on the accommodation cycle. Conversely, during "icehouse" times bathymetry change is directly driven by the accommodation cycle and the resulting sedimentary cycle is in phase with accommodation.

INTRODUCTION

Forward Stratigraphic Modeling Alternatives

Forward numerical modeling of stratigraphy is the computer simulation of the sedimentological processes that produce stratigraphy. Numerous modeling packages are available, and each has distinct specificity to time and spatial scale and to different geological settings. For instance, CARBONATE3D (Warrlich et al., 2002) has been used effectively to model carbonate platforms and mixed carbonate–siliciclastic shelves at a fourth–third order time and spatial scale. Nordlund's (1996, 1999) fuzzy logic program (more recently known as FUZZIM) uses "common sense" rules to incorporate complex geological knowledge and ideas into the modeling of carbonate systems. The SEDSIM package (Tetzlaff, 1986) models grain-by-grain sediment transport to simulate facies distributions in clastic systems at a variety of scales, and CARB3D+ (Smart et al., 2005) is designed to model isolated carbonate platforms, with the significant addition that it models the impact of early diagenetic processes on porosity, permeability, and mineralogy of the platform. The Dionisos program of Granjeon (1997), which is currently being developed as a research consortium run by the Institut Français du Pétrole, is used in this study.

Dionisos is an acronym for Diffusive Oriented Normal and Inverse Simulation of Sedimentation. It is a diffusion-based, multi-lithology, three-dimensional simulator that has been designed for use at the basin to sub-basin scale (tens to hundreds of kilometers) and over time spans of millions of years. Its advantages are that it provides a very open structure in which many different kinds of systems can be modeled (e.g., clastics, carbonates, slope processes, deep water, and growth faulting), and in which interactions between lithologies can be incorporated (for example, carbonate growth being sensitive to clays in suspension). One of the key limitations is that it will not model individual "objects" which have some kind of statistical distribution, such as patch reefs or channels; rather, it tends to produce smooth facies belts or trends. Details of the Dionisos program can be found in Granjeon (1997) and Granjeon and Joseph (1999).

Applications of Forward Stratigraphic Modeling

Although stratigraphic concepts developed by geologists are typically qualitative and based on very incomplete observations, they are routinely used in the subsurface to predict reservoir, seal, and source-rock distributions. A tool like stratigraphic modeling that allows one to quantitatively generate volumes populated with geologically reasonable stratigraphic properties such as facies, lithology, porosity, and even permeability is potentially valuable in quantifying subsurface predictions and reducing uncertainty. Applications would include:

1. generation of alternative geological models for hydrocarbon exploration and reservoir prediction;

2. validation and improvement of existing geological models;

3. education of earth scientists about geological processes and stratigraphy;

4. quantification of reservoir-body dimensions, geometries, and distribution for input into geocellular models; and

5. fully conditioned simulation of subsurface dataset to make a 3D earth model for reservoir prediction, hydrocarbon volumes, and flow simulation purposes.

Controls on Carbonate Platform and Reef Development
SEPM Special Publication No. 89, Copyright © 2008
SEPM (Society for Sedimentary Geology), ISBN 978-1-56576-130-8, p. 343–359.

Warrlich et al. (2002) generated alternative geological models for a range of Tertiary-style isolated platform architectures. Their models predicted that different drowning mechanisms (drowning by fast sea-level rise, by decreases in production, and by high-amplitude short-term sea-level cycles) would produce different platform geometries. Barnett et al. (2002) validated existing geological models by investigating two competing geological hypotheses explaining large-scale stacking patterns observed in Late Visean platform carbonates from the British Isles. An example of education of earth scientists is the University of South Carolina, which has published a number of stratigraphic animations on the internet that explain stratigraphic evolution in different settings (http://strata.geol.sc.edu, and Kendall, 2003).

These first three applications typically use simulations which are unconditioned or only partially conditioned to data, making them simple to run, even with long run times, because large numbers of simulations are generally not required. Quantification for geocellular modeling involves extracting information from a forward stratigraphic modeling simulation for use as input into some kind of statistical geocellular modeling process. This requires the generation of models that capture both the concepts and correct ranges of scalar information. Databases of dimensions and shapes of reservoir bodies have long been a subject of great interest to the petroleum industry but have typically been very time consuming and costly to construct, because of sparse availability of truly 3D quantitative data in the public literature from outcrop and subsurface. Forward stratigraphic modeling may be an alternative and (relatively) quick way of generating such data, although it too would need to be calibrated to field data.

Conditioned simulation to create a 3D geocellular model for reservoir management planning and flow simulation is still not routine, simply due to the time required for multiple runs to optimize multi-dimensional parameters. Lessenger and Lerche (1999) discussed uses and challenges to inverse modeling, and Cross and Lessenger (1999) documented an inversion workflow and showed an example of where it was applied in clastics of the Mesa Verde Group of the San Juan Basin, United States). Bornholdt et al. (1999) applied an alternative inversion workflow to a carbonate example from the Upper Miocene of Mallorca, Spain. Run time is a key, and as computing power increases and the modeling packages become more efficient through better coding, stratigraphic inversion may eventually become a routinely used tool.

Background for This Study

In this study we are concerned primarily with the first two applications described above: generating alternative geological models, and validating existing models. The Carboniferous Tengiz isolated carbonate platform of southwest Kazakhstan is used as "inspiration" for an unconditioned case study. The stratigraphic architecture of Tengiz is described by Weber et al. (2003), and the modeling presented here is based on stratigraphic and facies information from that paper. The Tengiz isolated platform is approximately 20 km across and has vertical relief of over 1.5 km (Weber et al., 2003). The interval of interest to this study is the Late Visean to Bashkirian (Carboniferous), which is located at the top of the platform, and is the main reservoir interval (Fig. 1). This main platform reservoir is approximately 500 m thick and spans

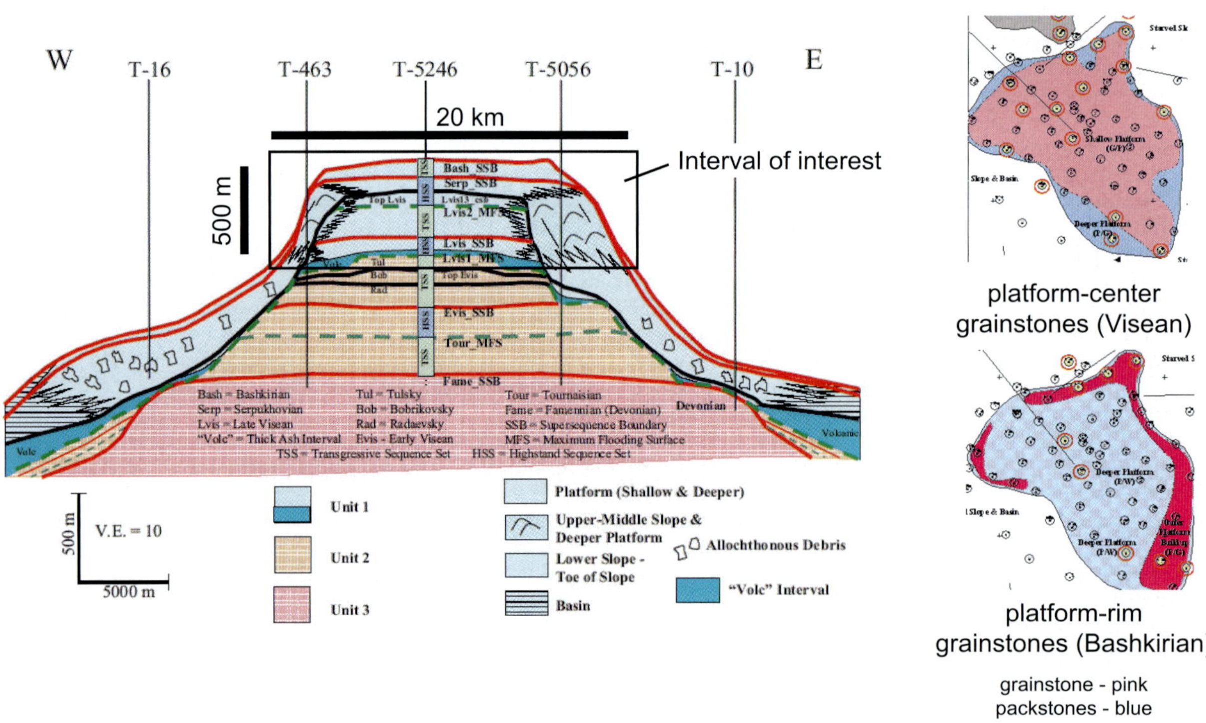

FIG. 1.—Tengiz cross section and maps of grainstone distribution modified from Weber et al. (2003). The cross section of Tengiz on the left highlights the interval of interest for this study, which is the Late Visean to Bashkirian (Carboniferous) top 500 m of the platform. Note that the facies distribution is very different from that of a Tertiary-type platform: no reefs develop at the margins, and the slopes are dominated by microbial boundstone deposits. The platform top is either packstone or grainstones. The right-hand maps show two typical facies distributions: grainstones (pink) in the platform center with deeper-water packstones around the margins at the top, and a grainstone rim (pink) and packstones across the platform top (blue) at the bottom.

roughly 20–30 My of time. The facies in Tengiz are distinct from Tertiary-style platforms in that a bioconstructed margin is absent. Microbial bioconstruction occurs, but dominantly on the slopes, and not on the shallow platform or margin. The Tengiz slope and other similar microbial boundstone slopes are discussed in Kenter et al. (2005). The shallow platform top is dominated by packstone and grainstone shoals of crinoids, brachiopods, peloids, and benthic foraminifera. Two distinct reservoir geometries are observed in the interval of interest. The first is a high-energy grainstone rim with lower-energy packstones in the platform center. The second is high-energy grainstones in the platform center, with more packstones towards the rim. The grainstones have generally higher porosity and permeability than the packstones, and for our modeling purposes are considered to be the best reservoir on the platform top. What exactly causes this change in reservoir distribution is uncertain, and is one of the subjects to be explored here.

The forward stratigraphic models are not intended to be exact simulations of the supergiant oil field. Rather, the models have similar facies, scale, thickness, bathymetry of deposition, slope angles, and time duration. Our experiments are driven by several questions. Can we model the changes in reservoir distribution that have been observed in Tengiz, and can we gain insight into why this happens? How do the facies distributions change within a sedimentary cycle, and how is that linked to sea level or other factors? Can one use conceptual models to see what other types of facies distribution might be expected? Our Dionisos models successfully reproduce the facies distributions observed at different times in the evolution of Tengiz, and they show how these facies variations through time are linked to accommodation cycles. Analysis of the models shows a relationship between gross platform morphology and net reservoir volume which may be of use in hydrocarbon exploration .

METHODOLOGY

Two distinct forward stratigraphic modeling experiments are formulated to generate a range of models and address the questions posed above. The first experiment (experiment 1) investigates the impact of rate of monotonous accommodation increase on stratigraphic architecture, whereas the second experiment (experiment 2) is designed to investigate the impact of changing the amplitude of accommodation cycles superimposed on top of a linear flooding rate. Accommodation means the amount of space that is created for sedimentation to potentially fill, whether it is filled or not. Eustatic and tectonic effects are combined into a single accommodation curve, and no differential subsidence is applied to the platform. The two experiments run with the same model size (20 km x 20 km), initial topography (see Figure 2), wave direction (from the south), time step (0.5 My), cell size (250 m), sediment production, and sediment transport rules. The only thing changing between experiments is the form of the accommodation curve. The initial topography of a flat-topped area approximately rounded in plan view with 100 m of relief off its flanks acts as the seed for the isolated platform. Accommodation cycles had a 4 My time period, which is an arbitrary choice corresponding approximately to the duration of the composite cycles defined by Weber et al. (2003) for the Tengiz platform. The total run time for each experiment is 20 My.

Three sediment types are produced during the modeling experiments: grainstones, packstones, and boundstones. There are two controls on sediment production rates in the models: bathymetry and wave energy. The production rates of the sediment types with depth are shown in Figure 3. Grainstones produce at their highest rate between 1 and 10 m water depth, with

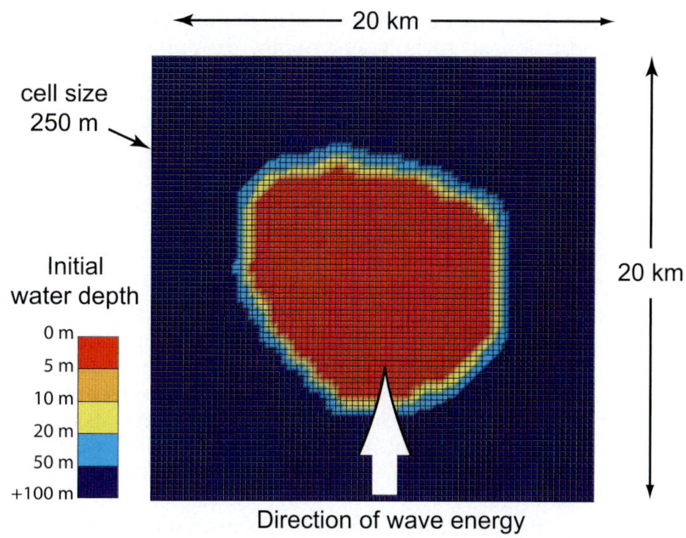

FIG. 2.—Parameters for the forward stratigraphic models. The model area covers 20 km by 20 km, with a rounded elevated platform of approximately 15 km diameter in the center. Initial topography was around 80 m, with the platform at sea level. Cell size for the model was 250 m.

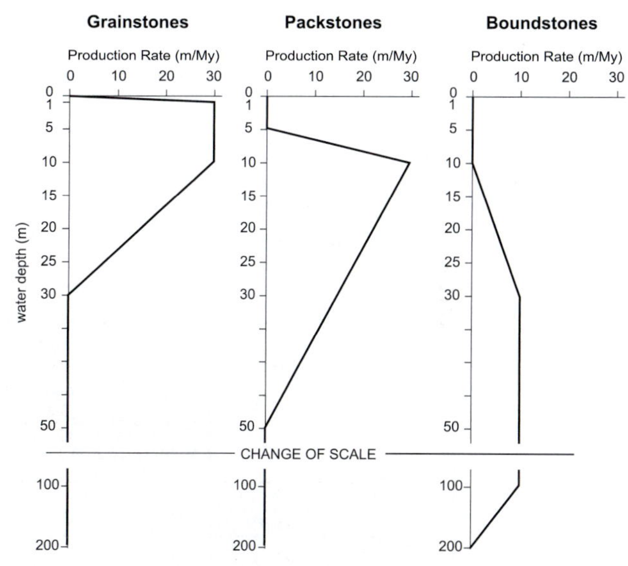

FIG. 3.—Accumulation rates for facies by depth. Grainstones produce at a maximum rate of 30 m / My from 1–10 m, decreasing linearly to 0 m / My at sea level, and decreasing downwards to 0 m / My at 30 m depth. Packstones do not accumulate in shallow waters from 0 to 5 m, increase up to a maximum accumulation rate of 30 m / My at 10 m water depth, and then decrease again to nothing at 50 m depth. Boundstones accumulate with slower rates in deeper water than either the grainstones or packstones, and do not produce in shallow water (1–10 m). Maximum boundstone accumulation rate is reached at 30 m depth, and stays at the low rate of 10 m / My down to 100 m depth. These depth-dependent accumulation rates will also be multiplied (by a factor of 0–1) by the wave-energy component. Grainstones will produce only in high energy, while packstones will produce in low energy. Boundstone production is energy independent, although little wave energy reaches these depths (30–100+ m).

production dropping off to zero at 30 m depth. Packstones have peak production at 10 m water depth, dropping off to zero production at 50 m and 5 m depth. These production profiles are chosen to approximately match the profiles for grainstones and packstones across Great Bahama Bank. The production profile for boundstones is based on the descriptions and interpretations of Kenter et al. (2005). In addition to a bathymetric control there was also an energy control over grainstone and packstone distribution; grainstones produce only in areas of high wave energy (50–100% of maximum wave energy), whereas packstones form in areas of low–medium energy (0–50% of maximum wave energy). Wave energy is distributed from a single direction (Fig. 2) using modeled refraction across the shallow platform top. The maximum sediment production rate for the grainstones and packstones is set at 30 m / My, while the boundstones are set at 10 m / My. 30 m / My was chosen for the platform top sediments to fill a maximum platform thickness of 600 m in 20 My, which corresponds approximately to the thickness and duration of the interval of interest (Fig. 1).

Dionisos uses a diffusion-based algorithm (Granjeon, 1997; Granjeon et al., 1999) to simulate transport of sediment. This means that the transportation rates are proportional to the topographic slope. A small amount of downslope sediment transport is modeled for both grainstones and packstones. No transport was applied to the boundstone because slope geometries were considered adequately simulated and the focus of the study is on the platform top.

As mentioned, the accommodation history was the only factor that was changed between experiments. In experiment 1 (runs 1.2 to 1.6 in Figure 4) the rate of accommodation increase is linear (no cyclicity) and varies from 10 m / My to 30 m / My (33–100% of maximum sediment production rate). In experiment 2 (runs 2.1 to 2.8 from Figure 4) the linear rate of accommodation increase is held constant at 20 m / My and an additional accommodation cyclicity of variable amplitude is added. In runs 2.1 to 2.8 the amplitude of the additional cyclicity starts off low at only 2 m and gradually increases to 40 m amplitude. The form is sinusoidal (the simplest case), and this produces additional rates of accommodation rise of up to ± 40 m / My (± 133% of maximum sediment production rates) in addition to the linear 20 m / My (67% of maximum sediment production rate).

OBSERVATIONS FROM EXPERIMENT 1 (LINEAR ACCOMMODATION INCREASE)

Simulations from experiment 1 (models 1.1–1.6) are shown in Figure 5 as cross sections through the 3D models with colors showing the sediment types. The high-energy wave-dominated margin is on the right side of each model. Packstones are shown in light blue, grainstone in purple, and boundstone slope in green. The models show a progressively faster accommodation rise from 1.1 to 1.6: note how the gross platform architecture changes. Low rates (1.1–1.2) have overall prograding geometries with increasing amounts of aggradation. Medium rates (models 1.3 to 1.5) show general aggradation and a thicker total accumulation. High rates (model 1.6) show rapid drowning of the platform to form a mounded geometry.

The internal distribution of grainstone, which we equate with better reservoir than packstone, shows surprisingly little change through most of the models except the last one (1.6). In each of the models 1.1 to 1.5 a grainstone-dominated margin occurs on the windward side, whereas the platform top is packstone-dominated with a patchy distribution of grainstone bodies (up to about a quarter of the rock volume). One has the impression that the proportion of grainstone on the platform increases slightly with flooding rate. To verify this notion the models were imported into Gocad in an S-grid format and then analyzed using volumetric and multivariate statistical analysis tools. Net-to-gross in percentage and the total volume of grainstone in km^3 are shown on Figure 6 for the various models of the first experiment. Flooding rate increases along the horizontal axis to the right. The net-to-gross value is the volumetric percentage of grainstone on the

CYCLIC ACCOMMODATION RATES

LINEAR ACCOMMODATION RATE

		0%	7%	13%	20%	33%	50%	66%	133%	additional accommodation rate due to cyclicity (% of max prod rate)
		0 m	2 m	4 m	6 m	10 m	15 m	20 m	40 m	amplitude of cyclic accommodation change
33%	10 m/My	1.2								
50%	15 m/My	1.3								
67%	20 m/My	1.4	2.1	2.2	2.3	2.5	2.6	2.7	2.8	Experiment 2
83%	25 m/My	1.5								
100%	30 m/My	1.6								
rate of rise (% of max production rate)	rate of rise (m/My)	Experiment 1								

FIG. 4.—Accommodation changes for linear and cyclic models. The vertical axis is changes of linear rate, expressed as both m / My and as a percentage of the maximum platform-top sedimentation rate (which is 30 m / My). These are models 1.2–1.6. The horizontal axis shows the amplitude of the cyclic accommodation changes (in m / My and as a percentage of maximum platform-top sedimentation rate). Models 2.1–2.8 have varying accommodation cycle amplitude in addition to the linear rate of 20 m / My.

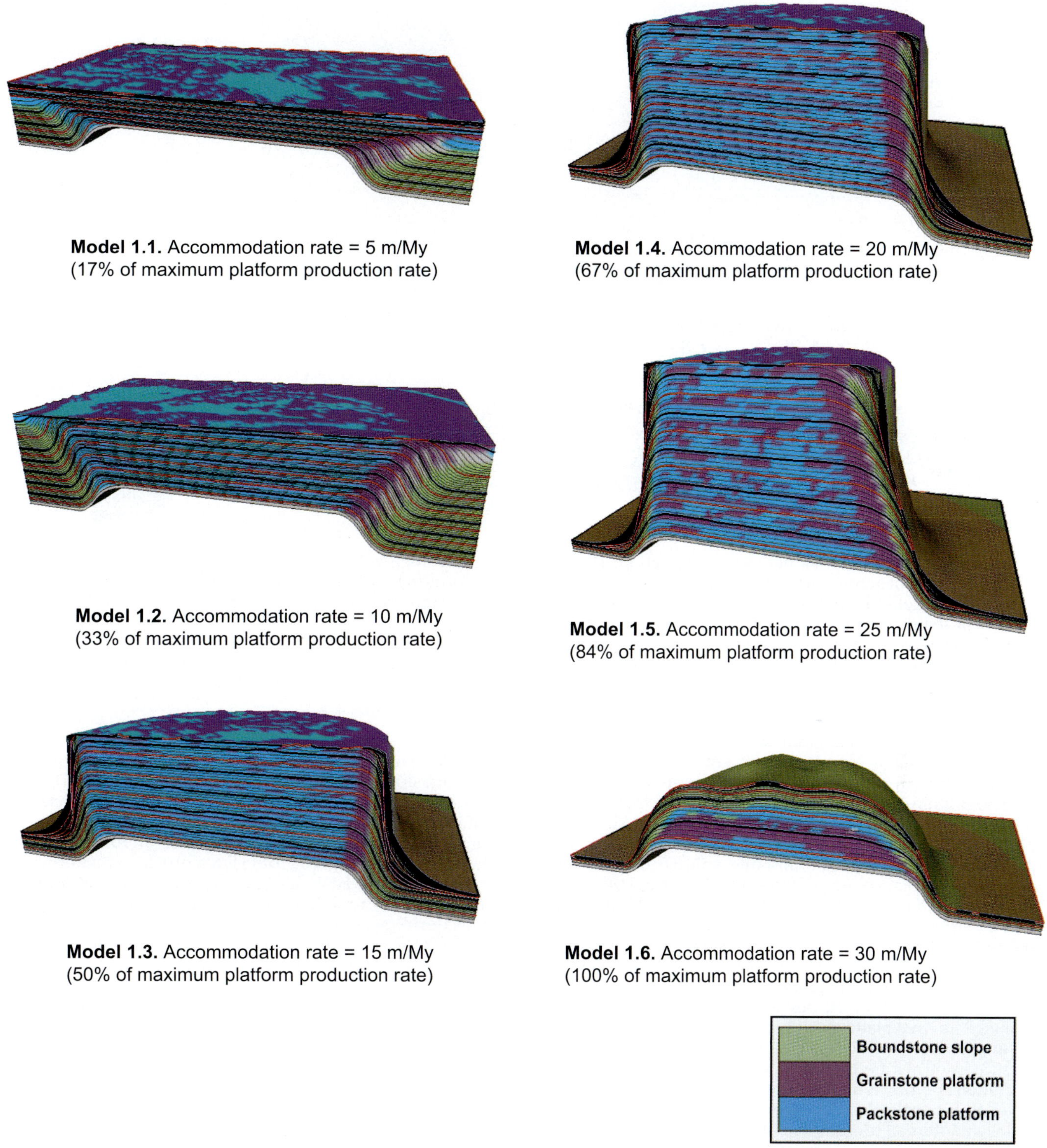

Fig. 5.—Linear-accommodation reservoir simulations. Models 1.1 to 1.6 cover a range of linear accommodation rates which are expressed in meters per million years and as a percentage of the maximum sediment production rate (which is 30 m / My). The dominant wind direction is from the right. Note how the gross platform geometry changes as the rate increases from prograding (models 1.1–1.2) to aggrading (models 1.3–1.5) to rapidly retrograding and drowning (model 1.6). The facies distributions for most of the models (1.1–1.6) are very similar, with a grainstone rim on the high-energy waveward margin, and a dominantly packstone platform interior. However, the proportion of grainstone in the platform interior increases visibly with accommodation rate. The one exception to this distribution is model 1.6, which shows a series of varying facies distributions vertically through the platform.

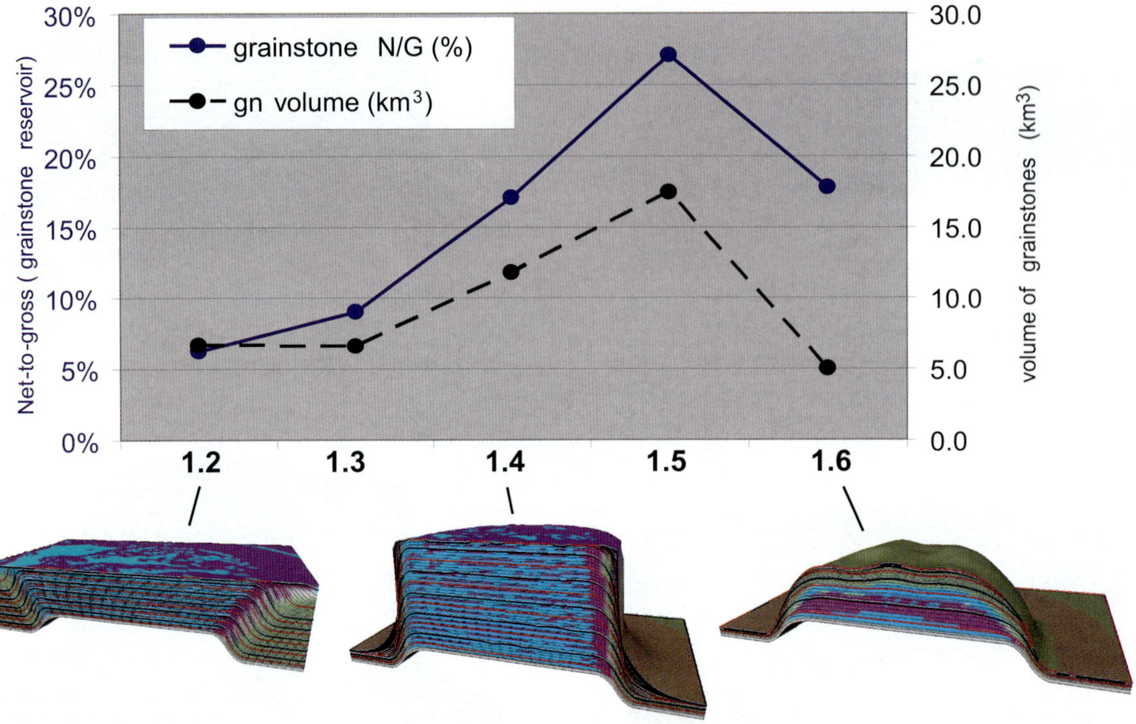

FIG. 6.—Linear-accommodation net-to-gross variation. The changes in grainstone volume and proportions are shown relative to increasing rate of accommodation rise. Volumetric percentage of platform-top sediments is shown on the left axis (blue curve), while volume of sediment is shown on the right axis (black curve). Grainstone volumes increase from around 7% to 27% of total platform sediment volume with increasing accommodation rate, up to a threshold (at model 1.5), beyond which the platform rapidly drowns and grainstone volume drops off again. The increase in total volume could be attributed only to the increasing amount of space created, but the increase in percentage volume indicates a fundamental change in the facies proportions on the platform top.

platform top, assuming that grainstones are reservoir and packstones are not. This is sometimes, but not always, the case in carbonate platform reservoirs; a notable exception is Tengiz, where both grainstones and packstones are principal reservoir facies in the platform top (Weber et al., 2003). Figure 6 shows that the net-to-gross and the proportion of grainstone increase with increasing flooding rate. The absolute volume of grainstone increases as the total volume of the model increases with higher flooding rates. The increase in net-to-gross is not limitless; the total platform volume decreases rapidly, as do the volumetric proportion and the volume of grainstones once the flooding rate reaches a certain threshold and the platform starts to drown.

While most of the simulations (1.1 to 1.5 in Figure 5) have the same spatial distribution of facies (grainstone margin, packstone-dominated internal platform with patchy grainstone bodies) the drowning platform of model 1.6 is an exception. During drowning a succession of different depositional profiles or spatial facies distributions occurs at a given time. Five distinct depositional profiles are recognized under different conditions (Fig. 7). There are likely additional profiles that either are variations on those identified or are produced under conditions not yet simulated. Profile A shows a grainstone margin with a raised rim, slightly deeper packstone-dominated platform top, and an abrupt break in slope at the margin. Profile B shows a similar facies distribution, with a grainstone margin and packstone platform, but in this case there is no elevated relief at the margin. The break in slope is abrupt but slightly more rounded than in depositional profile A. In profile C the facies distribution changes dramatically as platform-center grainstones develop. The platform top now has a rounded appearance, sloping off more gently to a packstone-dominated margin with a rounded slope break and transition to boundstone slope. Depositional profile D is similar to C in that the facies are platform-centered; however, the platform facies are packstones, and grainstones are rare to absent. The platform shape in D is highly rounded, and some undulation on the platform top may develop. Profile E is a drowned platform with only deepwater boundstones growing across the platform top. The form is highly rounded and very undulated, with bumps or mounds developing locally.

The suite of depositional profiles (Fig. 7) can be compared with Bachtel's (2005) classification of isolated platforms into three categories: filled (fully aggraded), underfilled (partially aggraded), and "drowned" platforms. Profiles A–B may be the filled (fully aggraded) platform, with a characteristic flat top and shallow bathymetries; profiles C–D may be the underfilled platform, with low-medium bathymetries and slightly rounded top, and the "drowned" platform equates with profile E, developing in deeper water and highly rounded and irregular in shape. The link between depositional profile and bathymetry can at times be complex and is discussed in more detail later.

OBSERVATIONS FROM EXPERIMENT 2 (CYCLIC ACCOMMODATION INCREASE)

The results of experiment 2 (increasing the amplitude of cyclic accommodation) are shown in Figure 8. Models 2.1 to 2.8 represent progressively increasing accommodation-cycle amplitudes superimposed on a background linear flooding rate of 20 m/My.

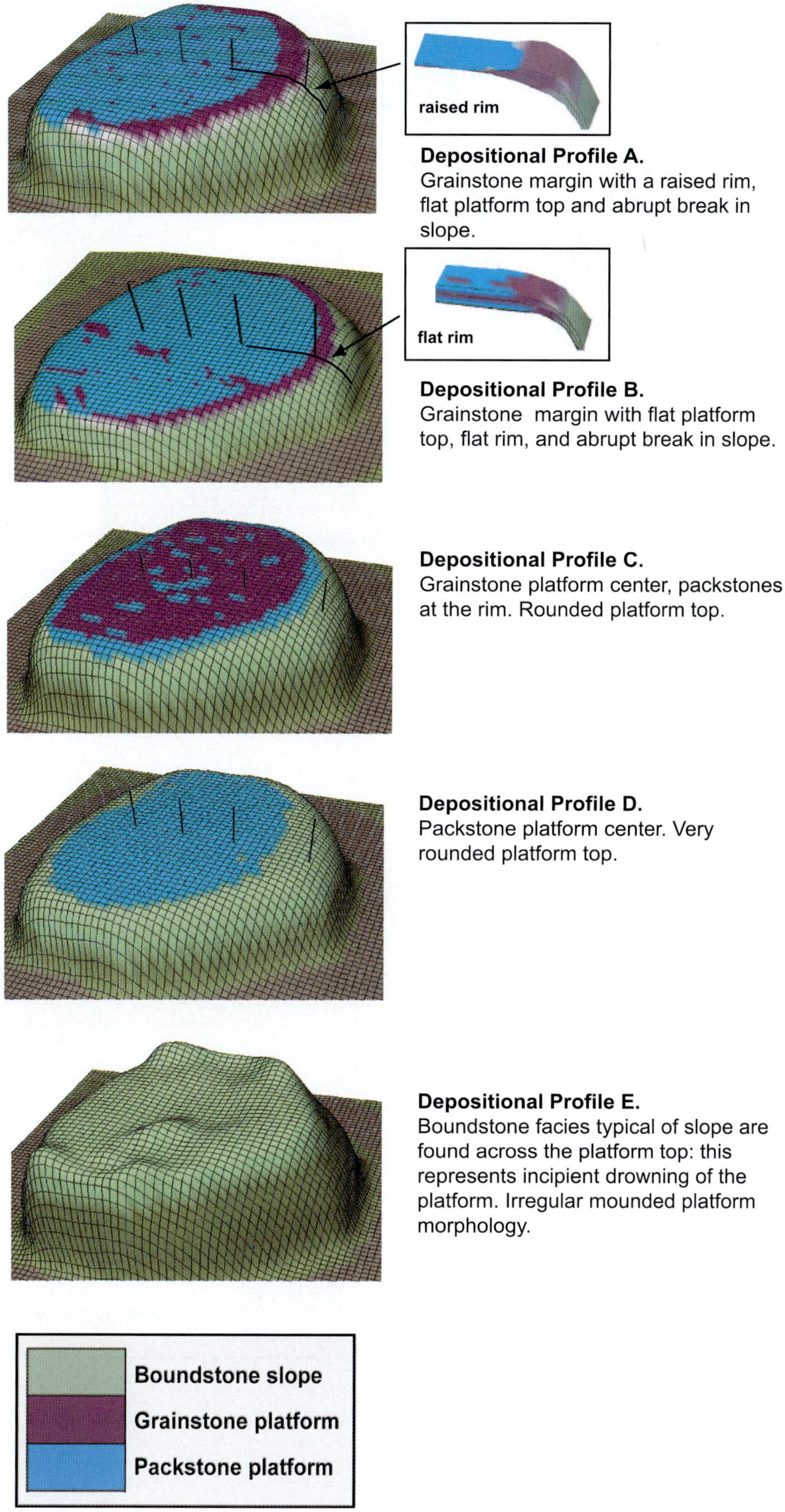

FIG. 7.—Examples of depositional profiles from the linear-accommodation models. Profile A has a raised (2–5 m water depth) grainstone rim on the waveward side of the platform, with a flat platform interior (4–8 m water depth) dominated by packstones. Platform morphology is flat-topped with a raised rim. Profile B also has a grainstone rim but with no topographic relief, and the platform interior (5–10 m water depth) is dominated by packstones, with varying amounts of grainstones. The platform is flat-topped. Profile C has dominantly grainstones at the platform center (with variable amounts of packstones) with some packstones at the rim in slightly deeper water. Water depth across the platform top is typically 5–12 m, and the platform top is slightly rounded. Profile D has dominantly packstones across the platform top (10–22 m water depth), and the platform has a very rounded morphology. Profile E has deeper-water microbial boundstone facies across the platform top (> 25 m), and a very rounded and irregular morphology.

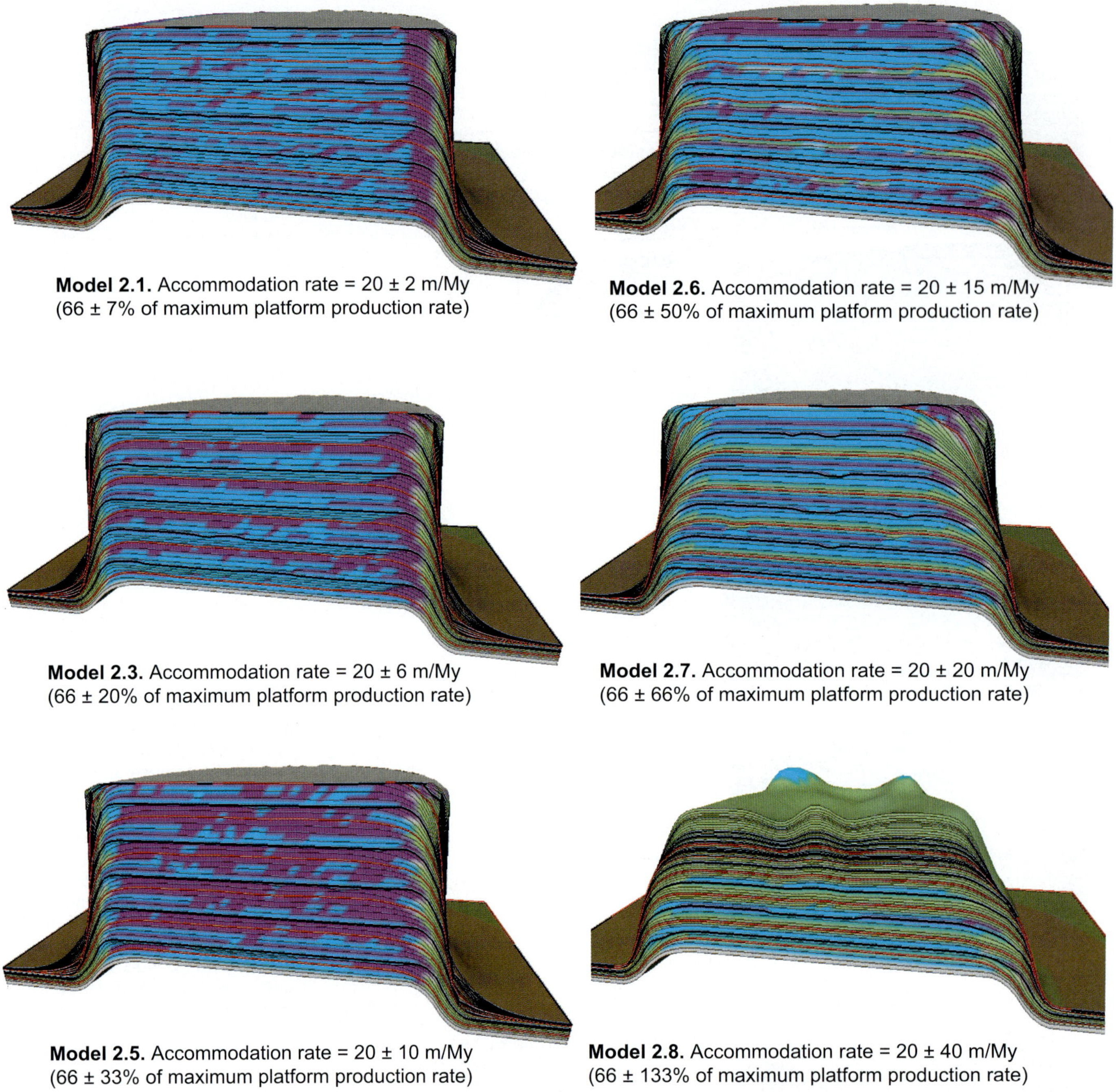

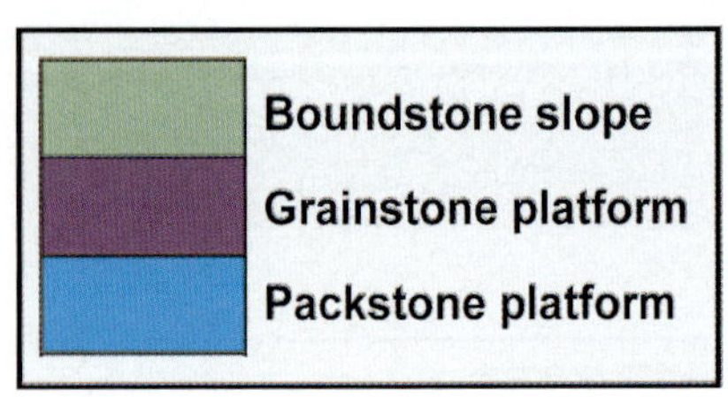

FIG. 8.—Cyclic-accommodation reservoir simulations. Models 2.1 to 2.8 cover a range of accommodation amplitudes superimposed on the linear accommodation rate of 20 m/My. The rates are expressed in meters per million years, and as a percentage of the maximum sediment production rate (which is 30 m/My). Green is boundstone slope, light blue is packstone, and purple is grainstone. The dominant wind direction is from the right. The gross platform geometry changes subtly as the amplitude of accommodation increases from aggrading (models 2.1–2.6) to retrograding (models 2.7–2.8). Compared to the linear accommodation increase models of Figure 5, facies distributions within the platform change rapidly with increasing amplitude. Also, the proportion of grainstone in the platform interior increases visibly with cycle amplitude from models 2.1 to 2.5, decreasing thereafter. Model 2.8 evolves rapidly towards drowning.

The linear accommodation rate used in experiment 2 corresponds to model 1.4 from the first experiment (Fig. 4): this was considered a reasonable mean-case value that had the potential to give a wide range of results without rapidly drowning or aggrading.

Note how the gross platform architecture changes in Figure 8. With increasing cycle amplitude there is a progressive retrogradation of the platform shape. In model 2.7 this is quite notable; an increasing proportion of deepwater microbial boundstone impinges on the platform top. In model 2.8 there is a rapid progression to drowning over a number of cycles represented by a microbial boundstone platform top (depositional profile E). Each progressive cycle gets successively deeper although shallow platform production of packstones persists for a number of cycles. This deepening might be interpreted as an accelerating flooding rate. In this case, however, the long-term flooding rate is constant and the short-term flooding rate is cyclic (compare with Warrlich et al., 2002).

A major difference between the results of the first and second experiments (Figs. 5, 8) is the distribution of facies within the platform. While the linear accommodation experiment showed little diversity of depositional profiles, the cyclic experiment starts to show increasing variability with cycle amplitude. From the platform overviews of Figure 8 we can see how the facies proportions shift dramatically, with increasingly continuous grainstone bodies appearing within the platform interior as cycle amplitude increases. This abruptly changes at model 2.5: subsequent models show a rapid decrease in grainstones and the addition of microbial boundstones over the platform top. As more cyclicity is added to the models, the depositional styles diverge further from the single facies distribution (depositional profile B) seen in the linear-accommodation experiment.

The reservoir net-to-gross ratio increases with increasing cycle amplitude up to a certain amplitude threshold (Fig. 9). Beyond this limit, the volume and proportion of grainstone decrease rapidly, as incipient drowning or exposure occurs across the platform top because of the high accommodation-cycle amplitude. With this incipient drowning the platform facies are dominated by either microbial boundstone or packstones.

The patterns of facies distributions change within the platform with increasing accommodation cycle amplitude (Fig. 10). With a low (6 m) accommodation-cycle amplitude (model 2.3 of Fig. 10), depositional profile B dominates, as in the linear accommodation models of experiment 1. However, other depositional styles occur: the plot on the left of Figure 10 shows how depositional style (horizontal axis) changes by layer throughout an accommodation cycle. The thick black line represents the lowest point on the accommodation curve, and the thick red line represents the highest point. Depositional profile C (platform-centered grainstones) is attained in model 2.3 following the time of fastest flooding. During times of slow flooding, depositional profile A (raised grainstone rim) develops.

Medium-amplitude (10 m) cyclicity (model 2.5 of Figure 10) produces results similar to model 2.3 in that depositional profile B dominates, but with deviations to depositional profiles A and C. However, some asymmetry starts to develop as more of profile C (generally representing deeper environments) develops than profile A. With high-amplitude (20 m) cyclicity (model 2.7 of Figure 10), depositional profiles D and E (packstones and micro-

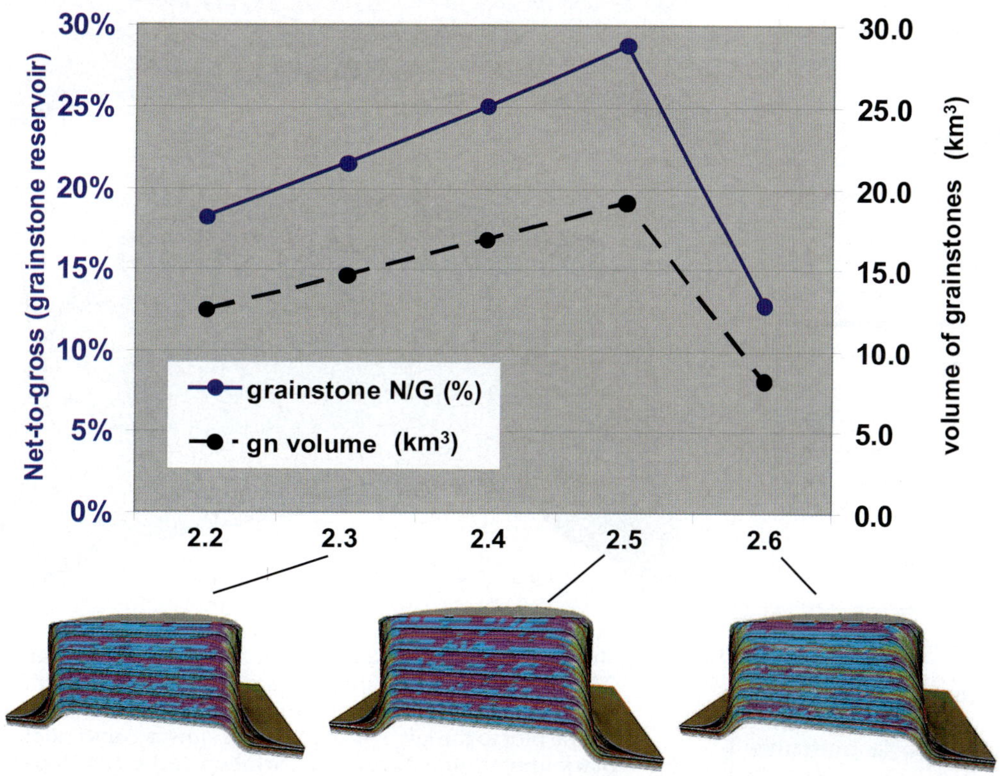

FIG. 9.—Cyclic-accommodation net-to-gross variation. The change in grainstone volume and proportions are shown in relation to increasing amplitude of accommodation cycles. Volumetric percentage of platform-top sediments is shown on the left axis (blue curve), and volume of sediment is shown on the right axis (black curve). Grainstone volumes increase from approximately 17% to 29% of total platform sediment volume with increasing accommodation rate, up to a threshold (at model 2.5), beyond which grainstone volume drops off again as platform-top bathymetry increases.

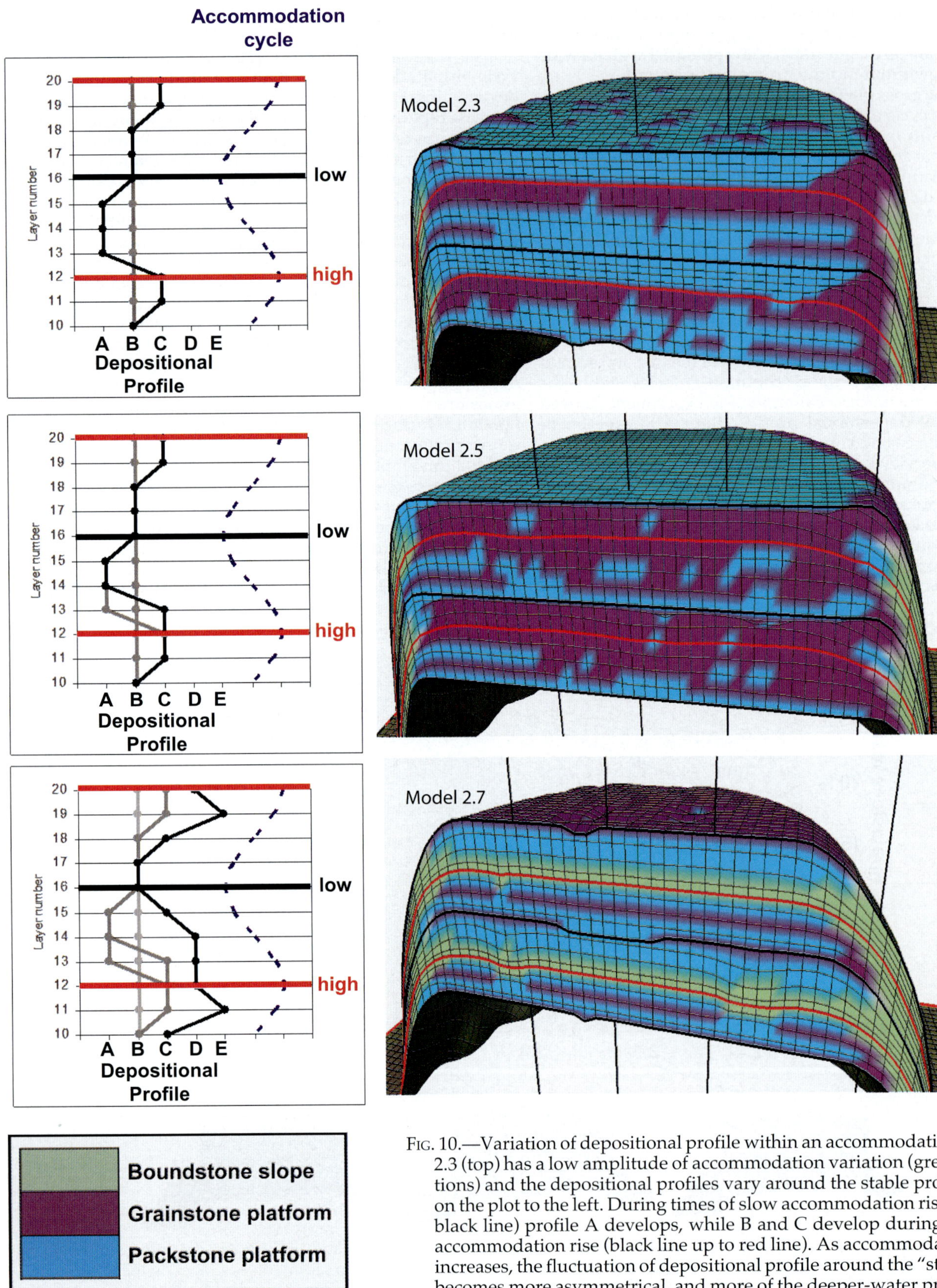

FIG. 10.—Variation of depositional profile within an accommodation cycle. Model 2.3 (top) has a low amplitude of accommodation variation (greenhouse conditions) and the depositional profiles vary around the stable profile B as shown on the plot to the left. During times of slow accommodation rise (red line up to black line) profile A develops, while B and C develop during times of faster accommodation rise (black line up to red line). As accommodation amplitude increases, the fluctuation of depositional profile around the "stable profile", B, becomes more asymmetrical, and more of the deeper-water profiles (D and E) develop. In model 2.7 (bottom) the stable profile B is reached only briefly just after the low point in the accommodation cycle.

bial boundstones across the platform top respectively) dominate. The platform top has a hard time producing enough sediment as accommodation rates peak at +40 m/My, which is well over the maximum sediment production rate of 30 m/My of the grainstones and packstones. The depositional profile representing the shallowest conditions is B (the stable state from experiment 1), and depositional style A (raised grainstone rim) is never produced.

INTERPRETATION AND DISCUSSION

Relationship between Depositional Profile and Bathymetry

Bathymetry is a controlling factor in defining which depositional profile develops in a simulation (Figure 11). Model 1.6 is the simulation that rapidly drowns during fast linear accommodation rise and is the only simulation from that experiment 1 to show depositional profiles other than the stable profile B. The depositional profiles B to E are deposited in model 1.6 in successively deeper bathymetries (Fig. 11, left). However, looking at depth by depositional profile for model 2.3 shows this depth relationship to be more poorly developed (Fig. 11, right). There is only a subtle deepening of less than a meter between depositional styles, suggesting that bathymetry provides a partial control on which depositional profiles develop, but other factors may also be important.

The impact of rate of change of water depth on the depositional profile is analyzed by cross-plotting values of bathymetry versus rate of bathymetry change from cells in one cycle (8 layers) for model 2.6 (Fig. 12). In this type of plotting the layers in one cycle form a loop, with depositional profile A (rimmed grainstone margin) developing at the top of the loop in the shallowest bathymetries during times of decreasing water depth. The overall platform morphology has a flat top with a slightly elevated rim, which is a direct result of the very shallow water depth and the production profiles chosen for this experiment (see Figure 3). At the shallowest depths grainstones develop in the high-energy setting along the waveward margin, absorbing all the wave energy and creating a sheltered platform top on which only packstones can be produced. Packstones are not produced at less than 5 m water depth (because of the rules for sediment production by depth; see Figure 3), though the grainstones will fill up to sea level, creating the elevated rim. Making the depth control on packstone production the same as for the grainstones would effectively change style A to style B. This minimum 5 m water depth for packstone production was chosen by direct analogy with the open-platform packstones across Great Bahama Bank, which are typically deposited in 5 m of water depth or greater, so profile A seems to be realistic.

As the rate of bathymetry change slows to around zero in shallow waters (Fig. 12) we see a transition to depositional profile B (grainstone margin with no elevated rim). Sediment production rates of both grainstones and packstones reach equilibrium with the imposed rate of accommodation change. The overall platform morphology is flat-topped with a sharp break in slope at the margin. As the accommodation rate increases, sedimentation rates can no longer keep up and water depth slowly increases. With increasing water depth, the wave energy over the platform top increases, less of the wave energy is absorbed at the margin, grainstones develop across the platform, and depositional profile C (grainstone-centered platform) develops. High production rates occur across the platform, and are not constrained by accommodation as the general trend is deepening. Bathymetry increases gently towards the platform

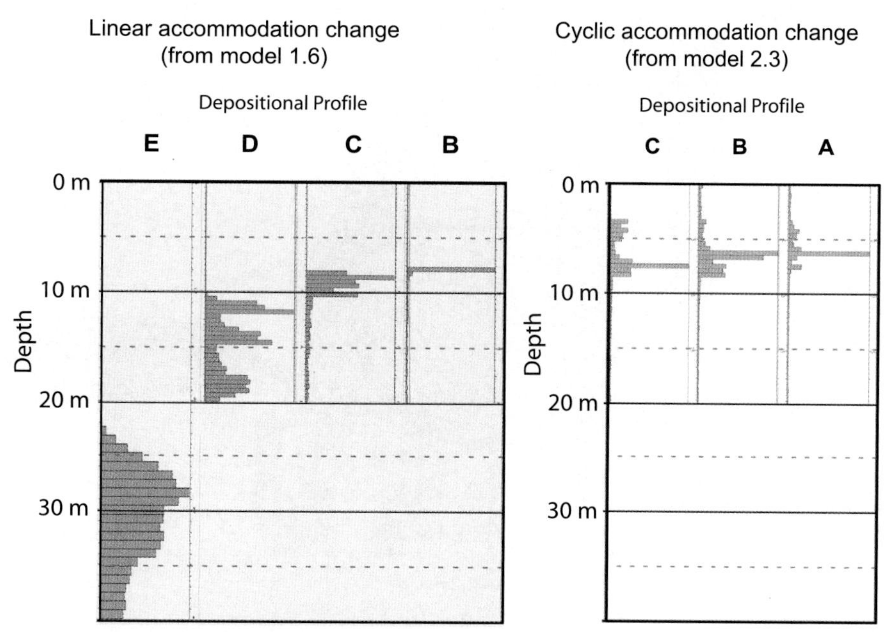

FIG. 11.—Depositional profile by depth. The left-hand set of histograms show how depth varies with depositional profile (B–E) for model 1.6, which is the model in which rapid drowning of the platform occurs during fast linear accommodation rise. The histograms clearly demonstrate a depth control on the formation of the deposition profiles. Additionally, the spread on the histograms with increasing depth is a measure of the roundness of the platform top, with more spread (profiles D and E) indicating more rounding. The right-hand set of histograms shows how depth varies with profiles (A–C) for model 2.3. These histograms show a poorer link between depositional profile and water depth, and suggest that other factors may also be important in determining which profile forms.

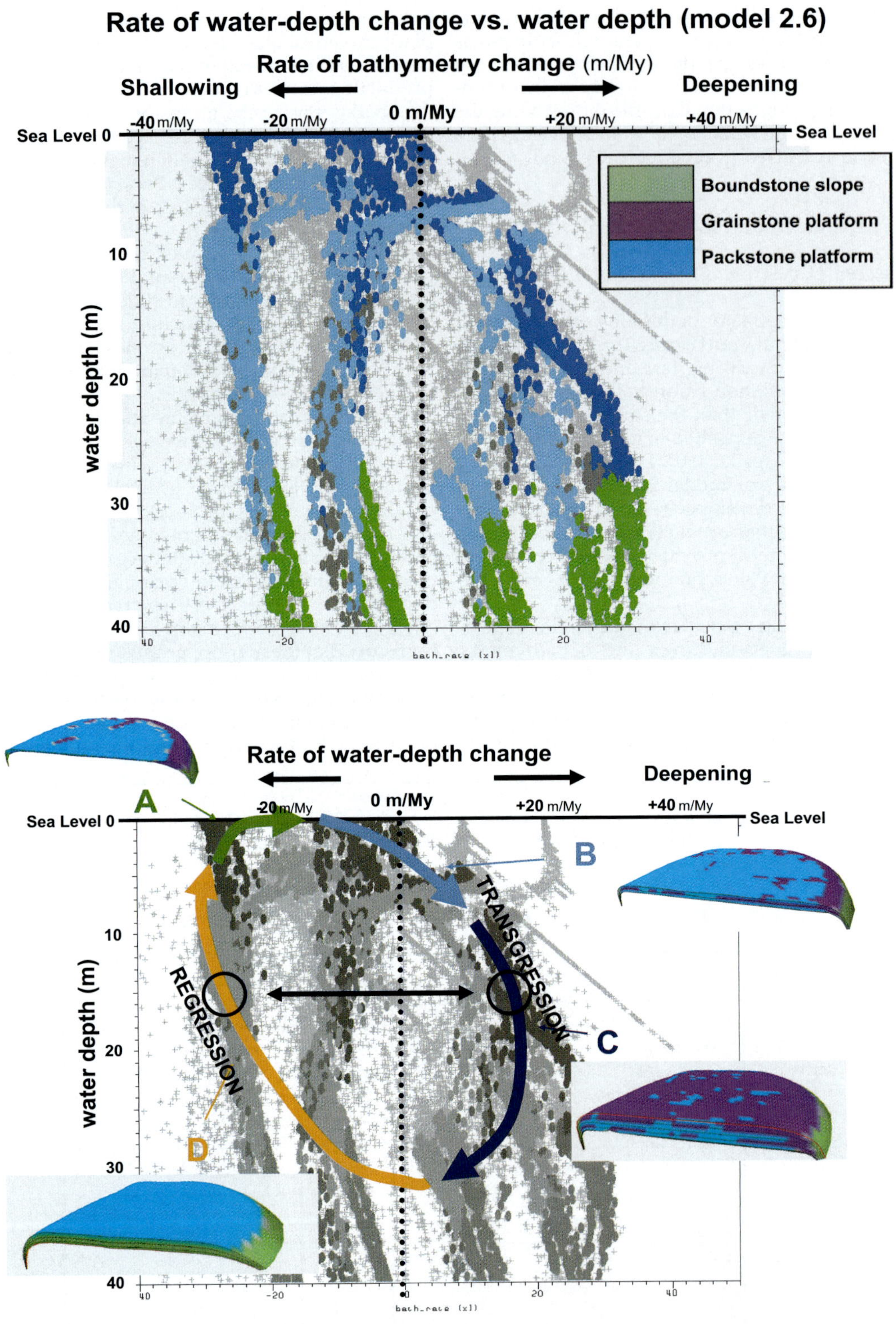

FIG. 12.—Crossplot of water-depth change versus water depth. An uninterpreted (top) and interpreted (bottom) cross plot of all cells from one cycle in model 2.6 is displayed. The horizontal axis is rate of change of water depth, and the vertical axis is water depth. The data points in the top example are colored in facies. Because each surface has cells from the platform top and the slope, a set of values from a surface is smeared along the vertical axis. It is the group of points at the shallow extreme from the platform top that are of interest. The consecutive layers form a loop in this space, and the loop is skewed and offset from the vertical axis. The depositional profile varies with position on the cycle, and these are indicated in the interpretation (bottom). Note that because of the loop a single water depth can have two different depositional profiles, depending on whether the positive or negative rate of water-depth change is picked. The progression of profiles is described in the text.

margins as platform-derived sediment is transported into the slope and basin, and the platform develops a slightly rounded morphology. This roundness further focuses wave energy at the platform center, reinforcing the platform-centered facies belts during this time.

With increasing depth the proportion of grainstones decreases (Fig. 12) due to both decreasing energy with depth and the bathymetric control on sedimentation rate shown in Figure 2, and platform-top deposition is dominated by packstones (depositional profile D). This has a highly rounded form with some local undulations developing. Soon the deepest point on the accommodation cycle is reached, and the platform starts to shallow. However, depositional profile D continues to dominate, even as the platform moves through depths at which profile C occurred during the flooding leg of the accommodation cycle.

This asymmetry of depositional profiles around the accommodation cycle is common to simulations 2.3, 2.5, and 2.6 in Figure 8, or to 2.3, 2.5, and 2.7 in Figure 10, although with different depositional profiles depending on the model. For instance, model 2.3 in Figure 10 has depositional profiles B and C during rising accommodation but style A during times of decreasing accommodation. As already noted, water depth in model 2.3 varies little (Fig. 11), but it is sufficient to change the distribution of wave energy across the platform and thus change the facies distribution. In model 2.3 the same water depths can have three alternative depositional profiles depending on the rate of change of accommodation. Similarly in model 2.6, similar water depths of 10–30 m develop profile C during times of increasing water depth and profile D during decreasing times.

Putting aside possible effects of the program coding, there may be some stratigraphic reasons to explain these observations. The shape of the platform may be a major factor in controlling facies distributions and the resulting depositional profile. Form clearly controls the transition from profile B to profile C (platform margin to center grainstones), with a very flat top (completely filling accommodation) limiting high-energy grainstones to the very edge of the platform. Some rounding of the platform but still with an abrupt drop-off, or perhaps just deepening, allows wave energy to reach the platform top without being dispersed by a long gentle incline and resulting in grainstones across the platform (profile C). Profile D has a significantly more rounded shape than profile C, possibly increasing dispersion of wave energy due to a gradually rising slope and keeping energy on the sea floor below the threshold required for grainstone formation. Thus the shallow platform is effectively protected from wave action.

Sediment transport may also be a cause of the difference in depositional style at similar water depths. Grainstone highs may locally produce large amounts of sediment during increasing accommodation, exporting it downslope to form sheets of allochthonous grainstone. When the grainstone production is finally drowned these sites are no longer available, and grainstones appear again only once water depth decreases sufficiently.

Implications for Sequence Stratigraphic Interpretation

The link between accommodation cycles and the interpretation of sequence stratigraphic surfaces has been a subject of much discussion and remains unclear (Haq et al., 1988), partly because of the confusion between "cycles" in the driving parameters (relative sea level rising and falling) and the resulting "cycles" observed in the rocks (sedimentary cycles, sequences, etc.). Forward stratigraphic models are a conceptual tool to investigate this link. The timing of sequence stratigraphic surfaces for two end-member models is shown in Figure 13. The upper example has low-amplitude cycles (model 2.3; "greenhouse-type" accommodation) whereas the lower example has high-amplitude cycles (model 2.7; "icehouse-type" accommodation). The left-hand panel for each shows a virtual well cross section through the platform, with the high-wave-energy margin on the right. The facies and bathymetry of deposition are shown at each well, and the timing of the accommodation-cycle high point (solid red line) and the low point (solid black line) are shown as surfaces in the wells. Additionally, a sequence stratigraphic interpretation has been added: the sequence boundary (SB) is shown as a thick dotted black line and the maximum flooding surface (MFS) is shown as a dotted blue line. The SB was picked in the shallowest facies (grainstones) and at the top of the grainstone package whereas the MFS was picked at or close to the deepest facies and preferably in the packstones, following what we consider to be reasonable interpretations that many geologists might pick.

The right-hand panels of Figure 13 show the positions of the interpreted SB and MFS relative to the accommodation cycle. The timing of SB and MFS formation in the high-amplitude example (lower Figure 13) are exactly in phase with the accommodation minima and maxima, respectively. The accommodation rates are high compared to sedimentation, varying from 0 to 133% of maximum sedimentation rate. Because the accommodation rates are so high they directly drive bathymetric variations on the platform: maximum water depth (MFS) occurs around the high point on the accommodation cycle, and minimum water depth (SB) occurs at the low point on the accommodation cycle. Accommodation, bathymetry, and facies are in phase, due to the dominance of accommodation.

However, a number of significant differences are observed when the low-amplitude accommodation example (the "greenhouse" setting, upper Figure 13) is considered. First, the SB interpretation is diachronous; shallow environments occur across the platform interior as grainstones are deposited and fill up to sea level. As the platform becomes more restricted, energy levels in the platform interior decrease and only quiet-water packstones are deposited, though grainstones continue to accumulate at the margin. The top of the grainstones at the margin is the obvious candidate for the SB, as is the top of the grainstones in the platform interior. This shows how changing water depth and facies across the platform may change the age of a SB pick made on these parameters.

A second difference is the timing of the SB and MFS relative to the accommodation cycles of the greenhouse example (upper Figure 13). Note how the MFS is situated halfway between the accommodation low and high on the rising limb of the accommodation cycle. This is an intuitive place for the MFS to be, inasmuch as it is the time when the rate of accommodation increase is greatest. In a system where sedimentation rates are generally larger than accommodation rates we might expect to see the deepest water depth occur at this point.

The position of the SB is more difficult to explain, because it is situated at the highest point of the accommodation cycle. Given the logic described above of balancing sedimentation and accommodation rates, the SB would be expected to occur halfway between the high and low points on the accommodation cycle: its position seems to be anomalous. To understand this, platform energy levels and base level must be considered for the two kinds of shallow-platform sediment (grainstone and packstone). The grainstones have a shallower base level than the packstones (0 m for grainstones compared to 5 m for the packstones). This is simply an interpretation (see Figure 3), which is based on observed distributions of sediment across the Great Bahama Platform and on facies interpretations of sediments in Tengiz, where packstones are usually limited to the deeper environments (We-

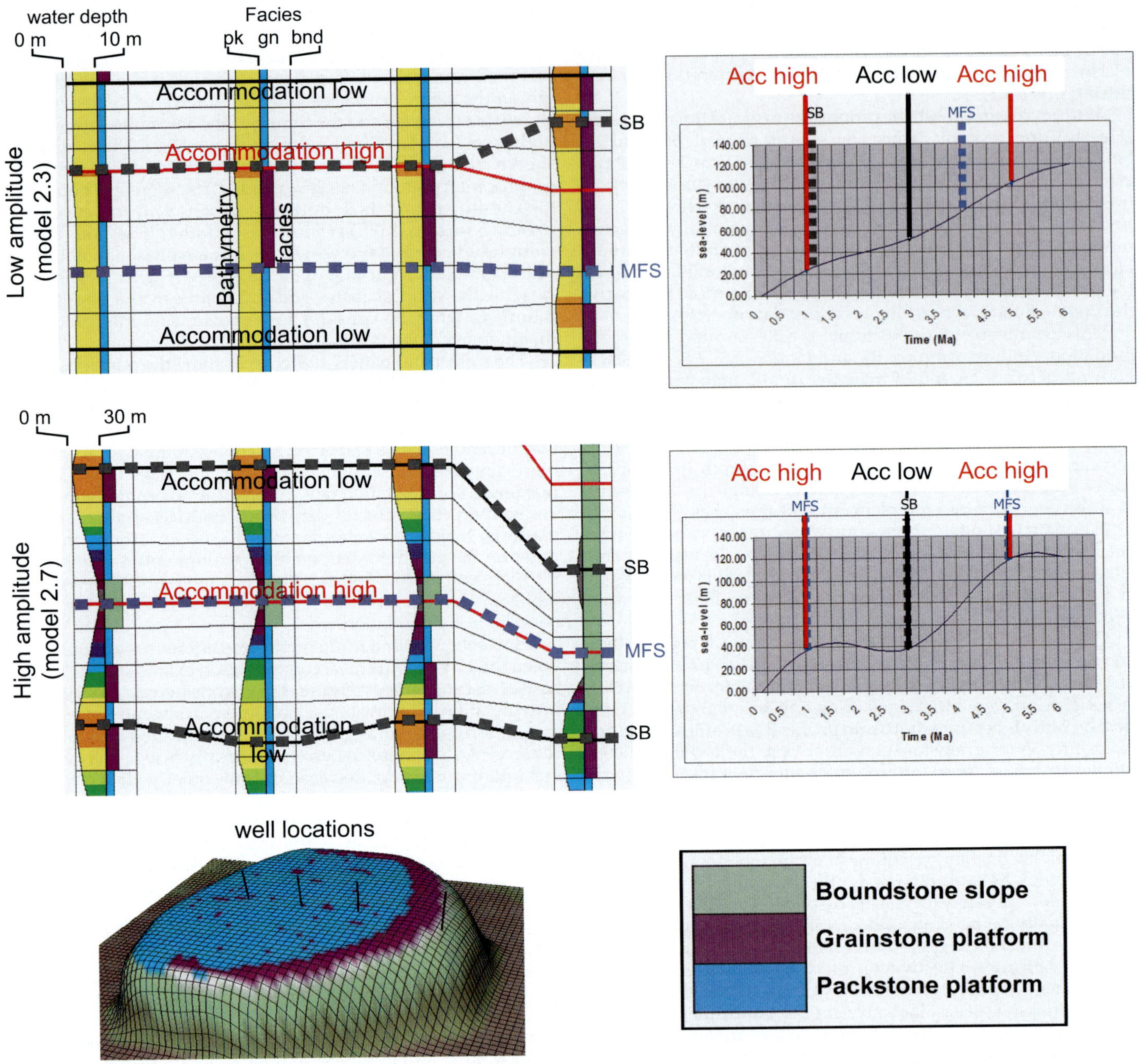

FIG. 13.—Sequence stratigraphic interpretation of sedimentary cycles. The left-hand figures show a four-well cross section through the platform; the location of the wells is shown in the figure at the base. Facies (right-hand column) and water depth (left-hand column) are indicated for each well. The high point (solid red line) and low point (solid black line) on the accommodation cycles are added as correlation surfaces, as are sequence stratigraphic interpretations of sequence boundary (SB; thick dotted black) and maximum flooding surface (MFS; thick dotted blue). Criteria for these interpretations are discussed in the text.

ber et al., 2003). Grainstones are thus deposited across the platform top only when accommodation rates are higher, the platform top is flooded, and energy levels increase. The grainstones across the platform then fill up to base level, which is effectively sea level, and form the SB. The subtle changes in environment brought about by accommodation change are more significant in this case than the actual change in water depth due to accommodation, since the maximum sediment rates can readily keep up with accommodation (which varies from 47 to 87% of maximum sedimentation rate), and the type of sediment (dependent on environment) is more significant.

The two cases of Figure 13 demonstrate that the timing of sequence stratigraphic surfaces relative to the accommodation cycle varies with the amplitude of the accommodation cycle. The position of SB and MFS changes with gradually varying accommodation amplitude for the set of models made during this study. The SB varies 180 degrees on the accommodation cycle from the high point to the low point. It changes from being completely out of phase to being in phase with accommodation. The MFS changes 90 degrees from the time of fastest rise to the point of highest accommodation. This change in relative position can be understood conceptually by considering the following: accommodation drives development of sequence stratigraphic surfaces through one of two mechanisms in this model. It can either directly increase or decrease water depth, or alternatively it can subtly change the depositional environment, affecting facies type and distribution. Where accommodation changes are low the environmental impact may be most significant, while where accommodation changes are high and largely surpass sedimentation rates, the direct bathymetric control is most important. Each of these two control mechanisms generate SB and MF surfaces in slightly different locations relative to accommodation, as we have shown. This concept is summarized on the right-hand side of Figure 14.

Explanation of Grainstone Distribution in Reservoir Example

In this study the Dionisos program successfully simulated the two grainstone distribution styles observed in Tengiz by Weber et al. (2003) (Fig. 1), that of platform-centered (depositional profile C) and platform-rimmed (depositional profiles A and B) grainstone belts. This is a significant result, inasmuch as the facies distributions were not explicitly pre-determined. The water depth of deposition of the different facies is predefined by the curves of production versus water depth (Fig. 3), and the energy conditions of deposition are defined by the user (grainstones in high energy, and packstones in low energy), so for a predefined topographic profile the distribution of facies could be calculated. However, the user does not control the topography of the sea floor: this is generated in the model by net deposition and erosion of sediment through time. Thus the link between topography and sedimentation rates is nonlinear, and the facies distributions are not preprogrammed but are a solution of the simulation.

The reasons for the transition from platform-rim to platform-center grainstones have been discussed. Grainstone platform rims occur during times of shallow water depth and low rates of increase of water depth, whereas platform-centered grainstones occur when water depth is low to moderate but increasing at a faster rate. One obvious question is whether this learning from the simulation can be applied toward better understanding of a reservoir like Tengiz. Is the Late Visean, which is typified by platform-centered grainstones in Tengiz, a period of higher accommodation-rate rise than the Bashkirian, when platform-rim grainstones occur? The Late Visean spans a time of approximately 14 My (Gradstein et al., 2004), and the thickness of the Late Visean interval at Tengiz is ~ 220 m (Weber et al., 2003). This gives a minimum long-term accommodation rate increase of 15 m/My. By comparison, the Bashkirian spans a time of approximately 10 My (Gradstein et al., 2004) and the thickness at Tengiz is typically 80–100 m across the platform. This gives a minimum long-term accommodation rate increase of 8–10 m/My for the Bashkirian, which is significantly lower than that calculated for the Visean. Though these accommodation-rate calculations are very approximate and by no means a proof, they are consistent with an explanation of the different facies distributions being related to rate of accommodation change.

Stacking Patterns and Platform Drowning

The stacking-pattern interpretation of the forward stratigraphic model simulations merits some examination. Models 1.2–1.5 show very similar stacking patterns within the platform. However, model 1.2 has a prograding motif and model 1.5 is retrograding. Model 1.5 would continue to retrograde until the surface area of the

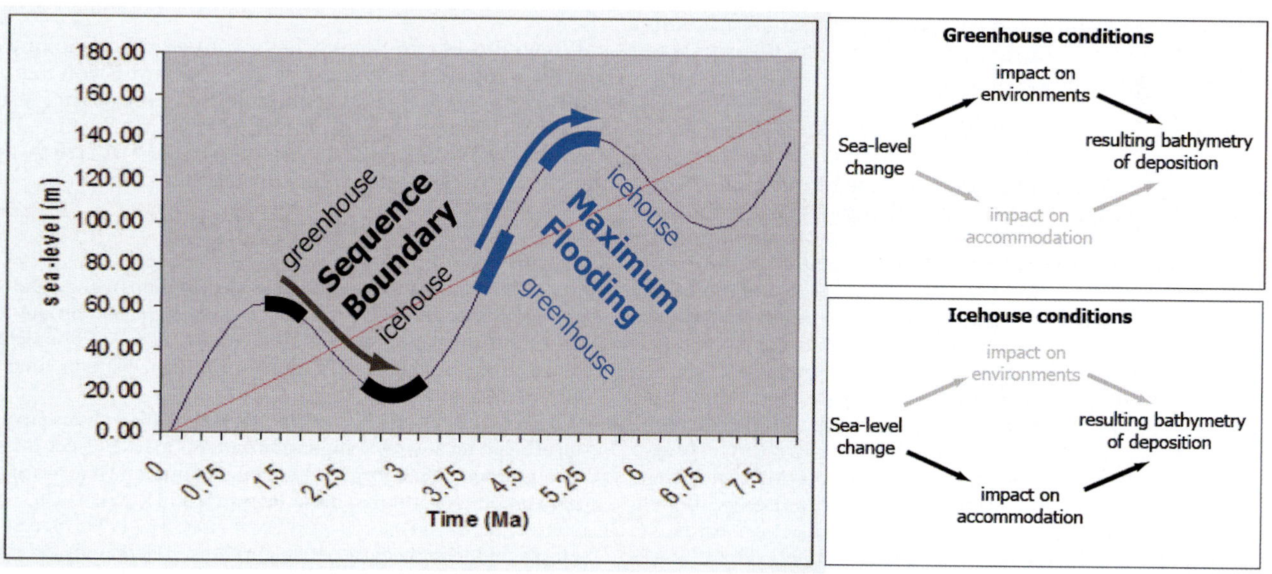

FIG. 14.—Link between sequence stratigraphic interpretation and cycle amplitude. Transition from greenhouse to icehouse conditions in the models presented here forces a shift in the position of the interpreted SB and MFS relative to the accommodation cycle. During greenhouse times the SB occurs when accommodation is at its highest on the platform top, while during icehouse times SB formation occurs at the accommodation low point. This is due to two different ways in which accommodation affects the stratigraphy. During greenhouse times when the accommodation change is small, the resulting changes in platform-top environment (here energy conditions) are most significant. During icehouse times the actual water-depth change created by a change in accommodation (and not filled by sedimentation) is the most important factor, and accommodation change directly drives the timing of SB and MFS formation.

platform is no longer large enough to accumulate shallow platform sediments, and then a drowning sequence similar to that observed in model 1.6 would occur. Observing the vertical facies evolution only from the platform center would give the impression that platform production was "healthy" and able to trace sea level until abrupt drowning occurred, and one might be tempted to attribute this to a catastrophic event. However, the modeling has shown this simply to be the inability of the platform to aggrade and keep up with sea level. Schlager (1981) identified this mechanism for isolated platform drowning when he pointed out that "upward growth of carbonate platforms with constant slope requires deposition of ever larger volumes of sediment on the flanks."

Our modeling suggests that there are effectively two important production-rate thresholds to be considered. The maximum sedimentation rate determines whether the platform top can temporarily keep up with the accommodation rate, but a higher (and increasingly higher) sediment production is required if the isolated platform is to continue to keep up over the long term. Abrupt and seemingly catastrophic "drowning" can appear to occur in a well where no signs of environmental stress appear in the underlying shallow platform sediments, though the "catastrophic" event may be no more than an ongoing sediment production problem culminating in drowning. Conversely, a vertical well through the platform center in model 2.8 (Fig. 8) would show progressively deeper facies on a cycle-by-cycle basis. Stacking-pattern analysis of this might suggest a progressively accelerating long-term transgression. However, long-term accommodation rates in this case are linear and at a constant rate (only the short-term rate is cyclic), and the progressive deepening is due to slow "drowning" of the platform because of low sediment production rates relative to the accommodation rate and the rate of the short-term cycles. So, stacking-pattern analysis in these two examples might lead to erroneous conclusions about accommodation change. This is not to say that stacking-pattern analysis is redundant: it is clearly of use in identifying cycles, trends, and composite packing of sedimentary cycles. However, its meaning in terms of genetic processes needs to be thought through carefully.

Timescale and Scale-Dependence of the Simulation

The timescale, rates, and geometric scale of the simulations were chosen to roughly match those identified for the Tengiz platform by Weber et al (2003). However, the simulation is in some respects scale-independent: as long as relative parameters of duration and thickness have the same proportions the rates will be the same, whatever the numerical values attributed. Conversely, if we want to keep the geometric dimensions constant, changing the timescale changes the rates. For instance, if the time for the simulation is divided by 100 the total runtime changes from 15 My to 150 ky. To conserve spatial dimensions (the same sediment thickness accumulated but in a shorter time), this would change the maximum sedimentation rate from 30 m/My to 3 m/ky, which is comparable to slow glacial rises (Miller et al., 2005), and comparable to carbonate production rates in shallow tropical seas (Bosscher and Schlager, 1992). The linear-accommodation-rise experiments would be 2 m/ky, with accommodation cycles of up to 40 m amplitude (unchanged) but with rates of up to ± 4 m/ky, once again similar to both tropical carbonate production rates and slow glacio-eustasy.

CONCLUSIONS

The depositional profiles simulated by Dionisos in this study resemble facies distributions observed in the Tengiz example and in other platform reservoirs. Platform-rim grainstone deposits (profiles A and B)—the "filled platforms" of Bachtel (2005)—develop during times of shallow water depth and slow water depth change, and are analogous to Bashkirian facies distribution in Tengiz as described by Weber et al. (2003). Platform-center grainstone and packstone deposits (profiles C and D)—the "underfilled platforms" of Bachtel (2005)—develop in shallow to medium water depths during times of increasing bathymetry and are analogous to the Late Visean of Tengiz.

The simulations show that similar water depths may have different depositional styles, so facies distribution is dependent not only on bathymetry: the rate of change of water depth is also a key factor, with times of increasing bathymetry developing facies distributions different from those in times of decreasing bathymetry. Stratigraphic "inertia" or memory occurs through the process of both bathymetry development and bathymetric profile. For example, a rounded platform has an energy distribution very different from that of a flat-topped platform with the same average water depth, and will produce different facies distributions.

The timing of a sequence stratigraphic interpretation relative to the accommodation cycle is not fixed, and may vary depending on the amplitude of the accommodation cycle. When amplitudes are low ("greenhouse" times), small changes in accommodation may greatly affect the platform environments, in this case energy distribution, leading to large changes in facies distributions. Cycles are preserved which show relatively little change in water depth, and are not necessarily in phase with the highs and lows of the accommodation cycle. During times of high amplitude ("icehouse" times) large changes in accommodation generate large changes in water depth as sedimentation does not keep up, and a strong water-depth signature which is in phase with accommodation cycles is preserved in the stratigraphy.

ACKNOWLEDGMENTS

We would like to thank Chevron Energy Technology Company for support of our research and for permission to publish. Additionally many thanks to the Institut Français du Pétrole for much constructive and rapid support in using the Dionisos program. We thank SEPM reviewers Georg Warrlich and an anonymous reviewer for their constructive comments.

REFERENCES

BACHTEL, S.L., 2005, Platform-scale facies distributions using LandSat data from isolated carbonate platforms: methods to constrain lateral facies continuity for geologic modeling (abstract): American Association of Petroleum Geologists, Annual Convention abstract, June 19–22, Calgary, Alberta, Canada.

BARNETT. A.J., BURGESS, P.M., AND WRIGHT, V.P., 2002, Icehouse world sea-level behavior and resulting stratal patterns in Late Visean (Mississippian) carbonate platforms: integration of numerical forward modeling and outcrop studies: Basin Research, v. 14, p. 417–438.

BORNHOLDT, S., NORDLAND, U., AND WESTPHAL, H., 1999, Inverse stratigraphic modeling using genetic algorithms, *in* Harbaugh, J.W., Watney, L., Rankey, E.C., Slingerland, R., Goldstein, R.H., and Franseen, E., eds., Numerical Experiments in Stratigraphy: Recent Advances in Stratigraphic and Sedimentologic Computer Simulations: SEPM, Special Publication 62, p. 85–90.

BOSSCHER, H., AND SCHLAGER, W., 1992, Accumulation rates of carbonate platforms: Journal of Geology, v. 101, p. 345–355.

CROSS, T.A., AND LESSENGER, M.A., 1999, Construction and application of a stratigraphic inverse model, *in* Harbaugh, J.W., Watney, L., Rankey, E.C., Slingerland, R., Goldstein, R.H., and Franseen, E., eds., Numerical Experiments in Stratigraphy: Recent Advances in Stratigraphic and Sedimentologic Computer Simulations: SEPM, Special Publication 62, p. 69–83.

Gradstein F.M., Ogg, J.G., and Smith, A.G., 2004, A geologic time scale: International Commission on Stratigraphy (ICS): under www.stratigraphy.org; 2004.

Granjeon, D., and Joseph, P., 1999, Concepts and applications of a 3-D multiple lithology, diffusive model in stratigraphic modeling, *in* Harbaugh, J.W., Watney, L., Rankey, E.C., Slingerland, R., Goldstein, R.H., and Franseen, E., Numerical Experiments in Stratigraphy: Recent Advances in Stratigraphic and Sedimentologic Computer Simulations, SEPM, Special Publication 62, p. 197–210.

Granjeon, D., 1997, Modélisation stratigraphique déterministe, conception et applications d'un modèle diffusif 3D multilithologique: Mémoires de Géosciences Rennes, v. 78, 189 p.

Haq, B.U., Hardenbol, J., and Vail, P.R., 1988, Mesozoic and Cenozoic chronostratigraphy and cycles of sea-level change, *in* Wilgus, C.K., Hastings, B.S., Kendall, C.G.St.C., Posamentier, H.W., Ross, C.A., and Van Wagoner, J.C., eds., Sea Level Changes: an Integrated Approach, SEPM, Special Publication 42, p. 71–108.

Kendall, C.G., 2003, Stratigraphic animations published on the internet [*sic*] at: http://strata.geol.sc.edu.

Kenter, J.A.M., Harris P.M., and Della Porta, G., 2005, Steep microbial boundstone–dominated platform margins—Examples and Implications: Sedimentary Geology, v. 178, p. 5–30.

Lessenger, M.A., and Lerche, I. 1999, Inverse modeling, *in* Harbaugh, J.W., Watney, L., Rankey, E.C., Slingerland, R., Goldstein, R.H., and Franseen, E., Numerical Experiments in Stratigraphy: Recent Advances in Stratigraphic and Sedimentologic Computer Simulations, SEPM Special Publication 62, p. 29–34.

Miller, K.G., Kominz, M.A., Browning, J.V., Wright, J.D., Mountain, S.W., Katz, M.E., Sugarman, P.J., Cramer, B.S., Christie-Blick, N., and Pekar, S.F., 2005, The Phanerozoic record of global sea-level change: Science, v. 310, p. 1293–1298.

Nordlund, U., 1996, Formalizing geological knowledge—with an example of modelling stratigraphy using fuzzy logic: Journal of Sedimentary Research, v. 66, p. 689–698.

Nordlund, U., 1999, Stratigraphic modeling using common-sense rules, *in* Harbaugh, J.W., Watney, L., Rankey, E.C., Slingerland, R., Goldstein, R.H., and Franseen, E., Numerical Experiments in Stratigraphy: Recent Advances in Stratigraphic and Sedimentologic Computer Simulations: SEPM, Special Publication 62, p. 245–251.

Schlager, W., 1981, The paradox of drowned reefs and carbonate platforms: Geological Society of America, Bulletin, v. 92, p. 197–211.

Smart, P.L., Waltham, D., Felce, D., and Whitaker, F.F., 2005, CARB3D+: a new forward simulation model for sedimentary architecture and near-surface diagenesis in isolated carbonate platforms (abstract): American Association of Petroleum Geologists, International Meeting, Paris, abstracts.

Tetzlaff, D.M., 1986, Clastic simulation model of clastic sedimentary processes: American Association of Petroleum Geologists, Bulletin, v. 70, p. 655.

Warrlich, G.M.D., Waltham, D.A., and Bosence, D.W.J., 2002, Quantifying the sequence stratigraphy and drowning mechanism of atolls using a new 3-D forward stratigraphic modeling program (CARBONATE 3D): Basin Research, v. 13, p. 379–400.

Weber, L.J., Francis, B.P., Harris, P.M., and Clark, M., 2003, Stratigraphy, facies, and reservoir characterization—Tengiz Field, Kazakhstan, *in* Ahr, W.M., Harris, P.M., Morgan, W.A., and Somerville, I., eds., Permo-Carboniferous Carbonate Platforms and Reefs: Joint SEPM Special Publication 78—American Association of Petroleum Geologists, Memoir 83, p. 351–394.

Index

A

B

C

D

Controls on Carbonate Platform and Reef Development
SEPM Special Publication No. 89, Copyright © 2008
SEPM (Society for Sedimentary Geology), ISBN 978-1-56576-130-8, p. 361–364.

S

T

V

W